Feng Yeh

Sep 11th 1988

Mechanical Dep.
U. of Houston

The Modern
Theory of Solids

The Modern Theory of Solids

FREDERICK SEITZ
The Rockefeller University, New York

DOVER PUBLICATIONS, INC., NEW YORK

This Dover edition is dedicated to two special individuals:

Elizabeth Marshall Seitz

Eugene Paul Wigner

Published in Canada by General Publishing Company, Ltd., 30 Lesmill Road, Don Mills, Toronto, Ontario.
Published in the United Kingdom by Constable and Company, Ltd., 10 Orange Street, London WC2H 7EG.

This Dover edition, first published in 1987, is an unabridged and unaltered republication of the work first published by the McGraw-Hill Book Company, New York, 1940, in its International Series in Pure and Applied Physics. The author has written a new Preface specially for the Dover edition.

Manufactured in the United States of America
Dover Publications, Inc., 31 East 2nd Street, Mineola, N.Y. 11501

Library of Congress Cataloging-in-Publication Data

Seitz, Frederick, 1911–
    The modern theory of solids.

    Reprint. Originally published: New York ; London : McGraw-Hill, 1940.
    Includes index.
    1. Solid state physics.   2. Crystallography.
I. Title.
QC176.S45   1987   530.4′1   87-9152
ISBN 0-486-65482-6 (pbk.)

PREFACE TO THE DOVER EDITION

Solid state physics, along with its generalized relative, condensed matter physics, has become a cornerstone field of modern physics. Among other things, it has fed into the technological aspects of our society in major ways. All the evidence suggests that it has an enduring future: it still attracts imaginatively inspired young scientists and engineers—some from the academic world and others from either government or industrial laboratories.

When this book was written, it had the limited but idealistic goal of attempting to pull together many diverse strands of basic and applied solid state science that could be regarded in a coherent way as a result of understanding derived from quantum theory. While my own roots at that time were in what might be called fundamental areas of science, I had begun to appreciate the challenges of applied research, which often accelerates scientific accomplishment to a far greater degree and in more diverse ways than are normally possible in an academic environment—if such achievement is possible there at all. What I did not appreciate at that time were the enormous advances that would soon result from greatly increased public and private support on a truly international basis.

Although this volume is antiquated in one sense because so much has occurred since it was written, it does appear that it still conveys the essence of the underlying logic of the field. Almost all areas to which various chapters are devoted are still alive and well and have blossomed in many interesting and productive ways. Presumably the starting student can still gain valuable perspective by becoming familiar with it. I trust that Dover Publications will find this reprint rewarding.

Among the various areas of solid state physics in which there has been enormous evolution or innovation, the following should be mentioned. First, the instrumentation available for physical measurements has expanded remarkably in both convenience and capability so that the experimenter now has much more powerful tools at his or her service. Second, there is far more intimate cooperation between experimenters and theoreticians than was common at the time this book was written. Third, the development of techniques of electron and nuclear magnetic resonance has made it possible to obtain deeper insight into a host of problems which were once beyond reach. Fourth, the evolution of methods of handling many-body problems has not only strengthened the ability of the theoretical physicists to handle problems of the interactions of electrons and phonons but has finally provided an understanding of the phenomenon of superconductivity, which was still a complete mystery at the time this volume was written. Fifth, the development of the laser has not only enhanced the utilization of crystal physics but has provided enormously valuable tools for the experimenters. Finally, the evolution of the electronic computer, which was in the very earliest stages of development at the University of Pennsylvania when this volume was published, has reached a point where it permits the accurate treatment of many theoretical problems which were once utterly beyond reach.

A final note: Early in life when we are struggling to articulate our innermost thoughts, the influence of others may be taken for granted. Later on, we realize that we owe great personal debts to all of our colleagues who have labored in their own way to unveil aspects of a marvelous structure. It is a privilege to acknowledge with deepfelt gratitude such debts to so many colleagues, both old and new, extending over more than half a century.

FREDERICK SEITZ.

NEW YORK,
December, 1986.

PREFACE TO THE FIRST EDITION

The present volume was written with a desire to satisfy the requirements of three types of possible reader: First, of course, students of physics and chemistry who desire to learn some details of a particular branch of physics that has general use; second, experimental physicists and chemists, and engineers and metallurgists with mathematical leanings who are interested in keeping an eye on a field of physics that is of possible value to them; and third, theoretical physicists at various stages of development who are interested in the present status of that phase of solid bodies that deals with electronic structure. The author fully realizes that the first two groups of reader do not wish to be concerned with all the intricacies of the theory, and for this reason he has attempted to edit the text by marking the more mathematical sections with an asterisk. It is recommended that readers not desiring to go through all this material read over it with an eye for the qualitative arguments and conclusions.

The author believes that an investigation of the table of contents will tell more concerning the scope of the book than a paragraph or two at this place. Although the book is a large one, it must be admitted that a number of very important topics are not treated. For example, the plastic properties of solids are only touched upon. The reason for this omission is, of course, that the theory involved in this field is not one that grows naturally out of modern quantum theory, and hence might better be treated under separate cover with some of the other structure-sensitive properties of solids. In addition, it was felt necessary to curtail the discussion of many interesting topics simply to avoid making the book much too long. In all such cases, which usually involve rather specialized subjects, an attempt has been made to give the reader reference material from which he may draw further information.

The author started to write this book in 1936 when he was at the University of Rochester and gratefully acknowledges the cooperation he received from Professor L. A. DuBridge in connection with it. The book was continued in spare time during the author's stay at the Research Laboratories of the General Electric Company in Schenectady from 1937 to 1939. He would like to say that the atmosphere of this organization proved very stimulating for writing, as well as for many other forms of research, and would like to express his gratitude to the directors for their interest and support. The book was completed at the University of Pennsylvania with the encouragement of Professor G. P. Harnwell. In addition, the author is obligated to the directors of the Westinghouse Research Laboratories for the privilege of spending a stimulating summer in East Pittsburgh in 1939.

The author also wishes to express his gratitude to the many friends and colleagues to whom he has turned for advice and discussions. Among these he particularly desires to mention Dr. R. P. Johnson, Dr. W. Shockley, Professor L. N. Ridenour, Dr. S. Dushman, Dr. E. U. Condon, Professor E. P. Wigner, and Professor J. H. Van Vleck.

Finally, he should like to acknowledge, though in a very inadequate manner, the constant help and encouragement furnished by his wife, Elizabeth Marshall Seitz, without whose aid this book probably never would have been written.

FREDERICK SEITZ.

PHILADELPHIA,
June, 1940.

CONTENTS

(NOTE: *Sections marked with an asterisk are more mathematical in nature, and may be read only for their qualitative content in a first reading.*)

CHAPTER V
QUANTUM MECHANICAL FOUNDATION
PART A

B. THE INTERACTION BETWEEN MATTER AND RADIATION★

CHAPTER VI
APPROXIMATE TREATMENT OF THE MANY-BODY PROBLEM

CHAPTER VII
MOLECULAR BINDING

CHAPTER VIII
THE BAND APPROXIMATION

CHAPTER IX

APPROXIMATIONAL METHODS

CHAPTER X

THE COHESIVE ENERGY

A. METALS

B. IONIC CRYSTALS

C. MOLECULAR CRYSTALS

CHAPTER XI

THE WORK FUNCTION AND THE SURFACE BARRIER

CHAPTER XII

THE EXCITED ELECTRONIC STATES OF SOLIDS

COMMONLY USED SYMBOLS

a_h	The Bohr radius of the hydrogen atom.
$\mathbf{A}$	The magnetic vector potential.
α_V	The coefficient of volume expansion.
β	The volume compressibility; also, the Bohr magneton.
c	The velocity of light; also, the velocity of elastic waves.
c_v	The specific heat at constant volume.
C_V	The molar or atomic heat at constant volume.
C_P	The molar heat at constant pressure.
γ	The damping frequency of an oscillator.
e	The electronic charge; also, the base of the natural system of logarithms.
$\mathbf{E}$	The electrostatic field intensity.
$\epsilon(\mathbf{k})$	The energy of an electron of wave number $\mathbf{k}$.
ϵ	The dielectric constant.
ϵ'	The energy of the uppermost electron in the filled band.
ϵ'_0	The energy of the uppermost electron in the filled band at absolute zero of temperature.
ϵ_0	The energy of the electron at the bottom of the filled band.
$\bar{\epsilon}$	The mean electronic energy.
$\bar{\epsilon}_0$	The mean electronic energy at absolute zero of temperature.
δ	The delta function.
Δ	The Laplacian operator.
f_E	The Einstein specific-heat function.
f_D	The Debye specific-heat function.
f	A partition function.
$g(r_s)$	The free-electron correlation energy.
h	Planck's constant.
$\hbar$	Planck's constant divided by 2π.
H	A Hamiltonian operator.
$\mathbf{H}$	The magnetic field intensity.
$\mathbf{J}, \mathbf{I}$	The current per unit area.
$\mathbf{k}$	The electronic wave-number vector; components k_x, k_y, k_z.
k	Boltzmann's constant; also the optical extinction coefficient.
k_0	The wave number of the electron at the top of the filled band.
$\mathbf{K}, \mathbf{K}_\alpha$	Principal vectors in the reciprocal lattice of a crystal.

l The mean free path between collisions (averaged for all scattering angles).

$l(\mathbf{k})$ The mean free path of the electron having wave number $\mathbf{k}$.

$\mathbf{L}$ Orbital angular momentum vector.

L The orbital angular momentum quantum number.

λ Wave length of an electron, light quantum, or lattice vibrational wave.

m The actual electronic mass.

m^* The effective electronic mass [determined from the $\epsilon(\mathbf{k})$ curve].

M The mass of an atom or ion.

$\mathbf{M}$ Orbital magnetic moment.

$\mathbf{M}_d$ Atomic electric dipole moment.

$\mathbf{M}_{ij}$ Matrix component of the atomic dipole moment.

M Magnetic moment per unit volume.

M_s The saturation magnetic moment.

$\boldsymbol{\mu}$ The electron spin magnetic moment vector.

μ The magnetic permeability.

n Index of refraction; exponent in Born's ion-ion repulsion function.

n_0 The number of particles per unit volume.

n_f The effective number of free electrons (usually the number of electrons in the band of width kT at the top of the filled region).

N The total number of atoms or electrons in a system; the complex index of refraction.

N_A Avogadro's number.

ν Frequency of vibration.

ν_m Maximum frequency in elastic vibrational spectrum of a lattice.

P The electric polarization.

φ The work function of a metal; also, the azimuthal angle.

r_s The radius of the sphere having the same volume as the atomic cell.

r_{12} The distance between two particles.

R The gas constant; also, the Hall constant; also, the reflection coefficient; also, the refractivity.

ρ Resistivity; also, the radiation density; also, the charge density.

S The total spin quantum number.

σ The electrical conductivity; also, the absolute value of the wave-number vector.

σ_s The saturation magnetization in units of Bohr magnetons per atom.

$\boldsymbol{\sigma}$ The wave-number vector for lattice vibrations; also, the spin angular momentum vector; the components are, σ_x, σ_y, σ_z in both cases.

σ_T The Thomson heat.

T The temperature, usually in degrees Kelvin.

Θ A characteristic temperature, such as that for the specific heat or the electrical conductivity; also, the effective cross section for collisions between electrons and lattice.

Θ_D The Debye characteristic temperature.

Θ_c The Curie temperature.

τ The mean time between collisions.

τ_i The primitive translation vectors of the lattice.

$v(\epsilon_0')$ The velocity of the uppermost electron in the filled electron band.

V The total volume of a system.

χ The magnetic susceptibility (per unit volume).

χ_m The molar magnetic susceptibility.

Z The atomic number.

THE MODERN THEORY OF SOLIDS

CHAPTER I

EMPIRICAL CLASSIFICATION OF SOLID TYPES

1. The Five Solid Types.—When using the term "solid" in this book, we shall refer to crystalline aggregates of atoms and molecules; that is, we shall have little to do with substances such as glasses that do not have definite lattice structure. In addition, we usually shall deal with crystals having relatively simple structures because they are most amenable to theoretical treatment. This limitation is not very important so long as we are interested only in general properties of solids, for substances with complicated structures can be classified in the same general way as simple ones. On the other hand, the restriction is very serious if the theory is looked upon as a tool for aid in making practical use of solids. There seems to be no way of removing this restriction other than to continue work along the lines that are developed here.

Although there is no unique way of classifying all the solids found in nature, the division that will be used here has enough natural advantages to make a discussion of alternatives unnecessary. It is based upon a survey of chemical, thermal, electrical, and magnetic characteristics. Briefly, the classification is as follows:

  a. Metals.
  b. Ionic crystals.
  c. Valence crystals.
  d. Semi-conductors.
  e. Molecular crystals.

Metals, which are distinguished primarily by their good electrical and thermal conductivity, are formed by the combination of the atoms of electropositive elements.

Ionic crystals are distinguished by good ionic conductivity at high temperatures, strong infrared absorption spectra, and good cleavage. They are formed by a combination of highly electropositive and highly electronegative elements, the salts, sodium chloride, magnesium oxide, etc., being the best examples.

1

Valence crystals, of which diamond and carborundum are examples, have poor electronic and ionic conductivity, great hardness, and poor cleavage. They are formed by combination of the lighter elements in the middle columns of the periodic chart.

Semi-conductors, of which zinc oxide and cuprous oxide are good examples, show a feeble electronic conductivity which increases with increasing temperature. It should be added that there is evidence that these substances are electronic conductors only when impure or when their composition is slightly different from that characteristic of ideal stoichiometric proportions, such as when there is an excess of zinc in zinc oxide. For this reason, semi-conductors are characterized by a tendency to favor addition of impurities and to disobey simple valence rules.

Finally, molecular crystals are the solids formed by inactive atoms such as the rare gases, and saturated molecules such as hydrogen and methane. They are characterized by low melting and boiling points, and they generally evaporate in the form of stable molecules.

As will be shown below, a large number of solids have properties that overlap those of two or more of these ideal groups. For this reason, the divisions should not be regarded as being clean-cut in the sense that a given solid belongs to only one of them.

We shall now give a more detailed discussion[1] of each of the five solid types.

2. Monatomic Metals.—Metals may be divided conveniently into two major classes, namely, monatomic metals and alloys. The literature relating to alloys naturally is much larger than that for monatomic metals. Since we shall not be interested in developing the theory of alloys beyond an elementary stage, we shall not give them a proportional amount of space.

We may recognize a further subdivision of metals into two groups depending upon whether the d shells[2] of the constituent atoms are filled or not. If the d shells are completely filled or completely empty, the properties of the metal usually are simpler than if they are not. The two cases will be discussed separately, the designation "simple metals" being used if the d shells are completely filled or completely empty, and "transition metals" in the alternative case.

[1] Since the methods used in obtaining most of this experimental material can be found in other places, they usually will not be discussed here; however, a few lesser known experiments concerning semi-conductors and ionic crystals are discussed in Secs. 4 and 6.

[2] Throughout this book, we shall use the conventional notation for the electronic orbital momentum quantum numbers (*cf.*, for example, H. E. White, *Introduction to Atomic Spectra* (McGraw-Hill Book Company, Inc., New York, 1934). In this notation the letters s, p, d, f, g, etc., designate the states having orbital angular momentum quantum numbers 0, 1, 2, 3, 4, etc., respectively.

a. Cohesion of Monatomic Metals.—The heat of sublimation, which is the energy required to dissociate a mol of substance into free atoms, is a convenient measure of the cohesion of a metal. Numerical values of the heats of sublimation that have been taken from the compilation by Bichowsky and Rossini[1] are given in Table I. The few values that appear in parenthesis do not occur in these tables and have been estimated by use of Trouton's rule, namely,

$$L = 0.0235 T_b$$

TABLE I.—THE HEATS OF SUBLIMATION OF MONATOMIC METALS
(In kg cal/mol at room temperature)

Monovalent Metals

Li	39.0	Cu	81.2
Na	25.9	Ag	68.0
K	19.8	Au	92.0
Rb	18.9		
Cs	18.8		

Divalent Metals

Be	75	Zn	27.4
Mg	36.3	Cd	26.8
Ca	47.8	Hg	14.6
Sr	47		
Ba	49		
Ra	(72.7)		

Trivalent Metals

Al	55	Ga	52
Sc	70	In	52
Yt	90	Tl	40
La	90		

Tetravalent Metals

Ti	100	Ge	85
Zr	110	Sn	78
Hf	(>72)	Pb	47.5
Th	177		

Pentavalent Metals

As	30.3
Sb	40.
Bi	47.8

Transition Element Metals

V	85	Nb	(>68)	Ta	(>97)
Cr	88	Mo	160	W	210
Mn	74	Ma		Re	
Fe	94.0	Ru	120	Os	125
Co	85.0	Rh	115	Ir	120
Ni	85.0	Pd	110	Pt	127
		U	220		

[1] F. R. BICHOWSKY and F. D. ROSSINI, *The Thermochemistry of the Chemical Substances,* Reinhold Publishing Corporation, New York, 1936.

where L is the heat of sublimation in kilogram calories per mol at the boiling point and T_b is the boiling temperature in degrees Kelvin. There are many interesting relationships among these values. One of the most

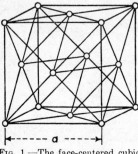

FIG. 1.—The face-centered cubic lattice.

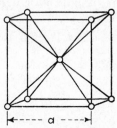

FIG. 2.—The body-centered cubic lattice.

striking ones is the fact that the atoms of transition metals on the whole are more tightly bound together than those of simple metals.

b. Crystal Structures.—Most of the monatomic metals crystallize in simple cubic or hexagonal structures. The three common types are shown in Figs. 1 to 3. More complex structures, which occur mainly among the atoms having higher valence, are shown in Figs. 4 to 11. Table II is a tabulation of the crystal parameter values for different

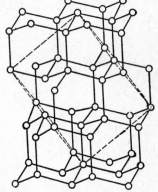

FIG. 4.—The diamond and gray tin lattice.

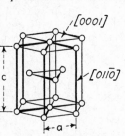

FIG. 3.—The close-packed hexagonal lattice showing two prominent crystallographic planes.

metals, including various allotropic modifications. These values have been taken from the three editions of *Strukturberichte.*[1]

---

[1] *Strukturberichte*, Leipzig (1931). Three supplements have appeared since the first volume.

Hume-Rothery[1] has pointed out that many complex structures, such as those of bismuth (Fig. 9), tin (Figs. 4 and 8), mercury (Fig. 6), and gallium (Fig. 7), are strikingly like those met among valence crystals, which are discussed below. For this reason, he would regard these substances as being

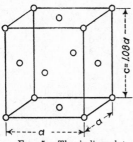

Fig. 5.—The indium lattice. The parameter values are given for indium.

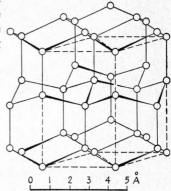

Fig. 6.—The mercury lattice.

intermediate between ideal metal and valence types, as we shall see below. This view is supported by observations on the conductivity and magnetic properties of these metals.

c. Allotropy.—Table II shows that at ordinary pressures the transition metals exhibit allotropy more commonly than do simple metals. Bridgman[2] has found, however, that many of the simpler metals change their structure at high pressures. Cesium, for example, has a close-packed modification which appears at 22,000 kg/cm² of pressure; similar changes

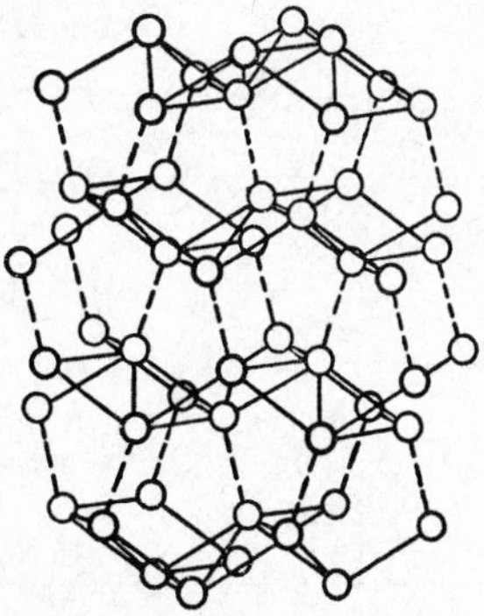

Fig. 7.—The gallium lattice, showing the layer structure in which each atom is surrounded by three neighbors.

Fig. 8.—The white tin lattice.

occur in magnesium. For this reason, it is doubtful whether polymorphism is a particular characteristic of any one group of metals.

[1] W. HUME-ROTHERY, *The Structure of Metals and Alloys* (Institute of Metals Monograph, London, 1936).

[2] P. W. BRIDGMAN, *Phys. Rev.*, **48**, 893 (1935); *Nat. Acad. Sci. Proc.*, **23**, 202 (1937).

TABLE II.—TABULATION OF CRYSTAL STRUCTURES AND PARAMETERS OF MONATOMIC METALS

(d is the nearest interatomic distance in angstroms and a is the edge length of the fundamental cube for cubic lattices. The parameters a and c for hexagonal close-packed lattices are shown in Fig. 3.)

The *alkali metals* form body-centered cubic lattices with the following parameter values.

	a	d
Li	3.46	3.00
Na	4.24	3.67
K	5.25	4.54
Rb	5.62	4.87
Cs	6.05	5.24

The *monovalent noble metals* have face-centered cubic lattices with the following parameter values.

	a	d
Cu	3.609	2.55
Ag	4.078	2.88
Au	4.070	2.87

Divalent Metals

	Type	a	c	d
Be	h.c.p.	2.28	3.59	a
Mg	h.c.p.	3.20	5.20	a
α Ca	f.c.c.	5.56	....	3.93
β Ca	h.c.p.	3.98	6.52	a
Sr	f.c.c.	6.06	....	4.28
Ba	b.c.c.	5.01	....	4.34
Zn	h.c.p.	2.65	4.930	a
Cd	h.c.p.	2.97	5.61	a
Hg	(see Fig. 6)			

Trivalent Metals

	Type	a	c	d
Al	f.c.c.	4.04	....	2.86
Sc				
Yt	h.c.p.	3.66	5.81	a
α La	h.c.p.	3.72	6.06	a
β La	f.c.c.	5.30	....	3.74
Ga	(see Fig. 7)			
In	tet.f.c.	4.59	4.94	(see Fig. 5)
α Tl	h.c.p.	3.45	5.52	a
β Tl	f.c.c.	4.84	....	3.42

TABLE II.—TABULATION OF CRYSTAL STRUCTURES AND PARAMETERS OF MONATOMIC METALS.—(*Continued*)

Tetravalent Metals

	Type	a	c	d
Ti	h.c.p.	2.953	4.73	2.892
α Zr	h.c.p.	3.23	5.14	a
β Zr	b.c.c.	3.61	....	3.13
Hf	h.c.p.	3.20	....	3.14
Ge	diam. str.	5.62	....	2.43
α Sn (gray)	diam. str.	6.46	....	2.80
β Sn (white)	(see Fig. 8)			
Pb	f.c.c.	4.93	....	3.48

Pentavalent Metals
As (see Fig. 9)
Sb As type; (see Fig. 9)
Bi As type; (see Fig. 9)

Transition Metals

	Type	a	c	d
V	b.c.c.	3.01	....	2.61
α Cr	b.c.c.	2.87	....	2.49
β Cr	h.c.p.	2.72	4.42	a
α Mn	(see Fig. 11)			
β Mn	......	12.58		
γ Mn	tet.f.c.	3.77	3.53	2.08
α, β, δ Fe (α is low-temperature magnetic form)	b.c.c.	2.86	....	2.58
γ Fe	f.c.c.	3.56	....	2.57
α Co	h.c.p.	2.51	4.11	a
β Co	f.c.c.	3.55	....	2.51
Ni	f.c.c.	3.51	....	2.48
Nb	b.c.c.	3.30	....	2.86
Mo	b.c.c.	3.14	....	2.96
Ma				
Ru	h.c.p.	2.765	4.470	
Rh	f.c.c.	3.78	....	2.67
Pd	f.c.c.	3.88	....	2.74
Ta	b.c.c.	3.29	....	2.72
α W	b.c.c.	3.16	....	2.73
β W	(see Fig. 11)			
Re	h.c.p.	2.76	4.45	a
Os	h.c.p.	2.71	4.32	a
Ir	f.c.c.	3.83	....	2.71
Pt	f.c.c.	3.92	....	2.71

There is, however, one type of allotropic change that is characteristic of transition metals. This may be illustrated by comparing the cases of tin and iron. The α, or gray, modification of tin is stable at very low temperatures, whereas the β, or white, modification is stable at high

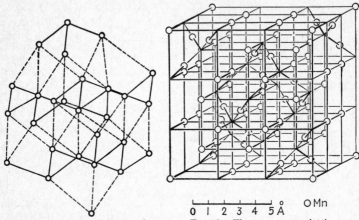

Fig. 9.—The bismuth lattice, showing the layer structure in which each atom is surrounded by three nearest neighbors.

Fig. 10.—The α manganese lattice.

temperatures (*cf.* Table II). The transition temperature, which is 18°C, was determined by Cohen and van Eijk[1] from measurements on the emf of an electrolytic cell that had one gray-tin electrode and one white-tin electrode. The case of tin is typical of the simple metals inasmuch as the α modification is not again stable in any range from 18°C to the melting point of the β phase. On the other hand, the body-centered modification of iron is stable[2] in two temperature ranges, namely, from 0° to 1179°K and from 1674°K to the melting point 1803°K. The face-centered, or γ, modification is stable in the intermediate range from 1179° to 1674°K. This "intrusion" of one phase into the range of another also occurs in cobalt.[3] In this case a face-centered cubic phase splits the stable range of a close-packed hexagonal phase into two parts.

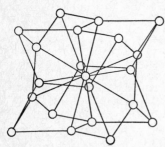

Fig. 11.—The β tungsten structure.

[1] E. Cohen and C. van Eijk, *Z. physik. Chem.*, **30**, 601 (1899).

[2] *Cf. Strukturberichte.*

[3] S. B. Hendricks, M. E. Jefferson, and J. F. Shultz, *Z. Krist.*, **73**, 376 (1930).

d. Atomic Radii.—It is often convenient to ascribe to each atom or ion a radius that is determined by the volume which the atom or ion occupies in a given compound. In monatomic metals the radius r is defined as half the distance between the centers of nearest neighbors, this definition being based upon the rigid-sphere concept of atoms, according to which the observed interatomic distance should be the distance for which neighboring spherical atoms come into contact, or twice the atomic radius. Since the nearest-neighbor distance is seldom precisely the same for two allotropic metal phases, it is clear that the rigid-sphere picture cannot be accurate. Nevertheless, the concept can be very valuable for semiquantitative work as will be seen when we discuss the

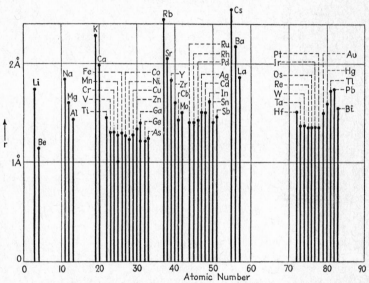

Fig. 12.—The radii of the metallic atoms as determined from the interatomic distance.

Hume-Rothery rules for alloys (*cf.* Sec. 3). Figure 12 shows the atomic radii as determined from the interatomic distances of Table II.

e. Electrical Conductivity.—The electrical resistivity ϱ of a substance is a tensor quantity that is defined in terms of the electrostatic field intensity $\mathbf{E}$ and the current per unit area $\mathbf{J}$ by the relation

$$\mathbf{E} = \varrho \cdot \mathbf{J}.$$

ϱ is a constant tensor for cubic crystals and may be represented by a single number in these cases. It has two independent values for hexagonal and tetragonal crystals. These may be determined by using fields parallel and perpendicular to the principal axes, since $\mathbf{E}$ and $\mathbf{J}$ are parallel to one another in these two cases. The two components of ϱ are respectively designated as the $\parallel$ and $\perp$ components.

TABLE III.—THE RESISTIVITIES OF METALS AT ROOM TEMPERATURE

(The resistivity ρ is expressed in units of 10^{-6} ohm-cm. $\parallel$ and $\perp$ designate, respectively, values in directions parallel and perpendicular to the principal axis in hexagonal and tetragonal crystals.)

Monovalent Metals

Li	8.75		Cu	1.56
Na	4.35		Ag	1.49
K	6.62		Au	2.04
Rb	12.0			
Cs	19.0			

Divalent Metals

Be	5.5		Zn	$\parallel 6.0$; $\perp 5.8$
Mg	$\parallel 3.50$; $\perp 4.22$		Cd	$\parallel 8.4$; $\perp 6.9$
Ca	9.80		Hg	($-45.5°$C) $\parallel 17.8$; $\perp 23.5$
Sr	32			
Ba	60			
Ra				

Trivalent Metals

Al	2.50		Ga	52.6
Sc			In	8.4
Yt			Tl	17.2
La	57.6			

Tetravalent Metals

Ti	47.5		Ge	89,000
Zr	41.0		Sn	$\parallel 13.1$; $\perp 9.1$
Hf	32.1		Pb	19.8
Th	18			

Pentavalent Metals

As	$\perp 28$
Sb	39
Bi	$\parallel 143$; $\perp 109$

Transition Metals

V	58.8	Nb	21	Ta	14
Cr	2.6	Mo	5.03	W	4.9
Mn $\begin{cases}\alpha \\ \beta \\ \gamma\end{cases}$	710(?) 91(?) 23	Ma Ru Rh	7.64 4.58	Re Os Ir	18.9 8.9 5.0
α Fe	8.71	Pd	10.2	Pt	9.81
α Co	6.2				
Ni	12.0				

Table III and Fig. 13 contain tabulations of the electrical resistivities[1] of monatomic metals at temperatures near 0°C. The lowest resistivities are those of the monovalent noble metals copper, silver and gold. In comparison, the alkali metals are only moderately good conductors.

[1] See, for example, the compilations of Landolt-Bornstein and the International Critical Tables.

Usually the resistivity decreases with increasing valence for the lighter elements and increases with increasing valence for the heavier elements, as may be seen by comparing the following two sequences:

Resistivity · 10⁶ ohm-cm		Resistivity · 10⁶ ohm-cm	
Na	4.35	Cu	1.56
Mg	3.50	Zn	5.8
Al	2.50	Ga	52.6

One of the most striking periodic properties of the resistivity is the large decrease that follows the completion of a *d* shell.   The change in

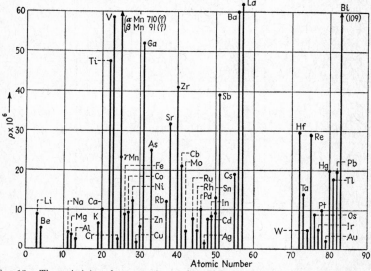

Fig. 13.—The resistivity of monatomic metals.   The ordinates are expressed in ohm-cm.

resistance from $12 \cdot 10^{-6}$ ohm-cm for nickel to $1.56 \cdot 10^{-6}$ for copper is the most prominent illustration of this.

The metals with the highest resistivities are those such as arsenic, antimony, bismuth, tin, mercury, and gallium which have complex structures, a fact lending additional support to Hume-Rothery's view that these are intermediate between ideal metals and insulating crystals.

The ratio of the resistance at temperature T to the resistance at zero degrees centigrade is shown in Fig. 14 for a number of metals.[1]   The fact that the curves are closely alike justifies our comparison, in the preceding paragraphs, of the room-temperature values.   The high-temperature resistivity of most metals varies linearly with temperature, whereas the low-temperature resistance varies with a higher power of T.   The

[1] *Ibid.*

most reliable measurements seem to show that the low-temperature variation is as T^5 for metals that are not superconducting. We shall discuss this topic more fully in Chap. XV.

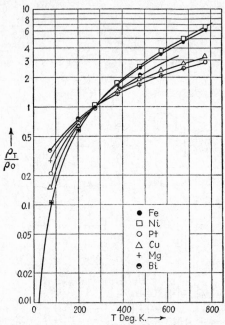

Fig. 14.—Temperature dependence of the relative resistivities of several metals. The ordinate is the ratio of the resistivity at temperature T to that at 0°C.

A large number of metals become superconducting below a temperature characteristic of the substance. These metals and their transition

TABLE IV.—THE SUPERCONDUCTING METALS AND THEIR TRANSITION TEMPERATURES

Metal	T_t, °K	Metal	T_t, °K
Zn	7.86	Ti	1.13
Cd	0.6	Zr	0.7
Hg	4.16	Hf	0.3
		Th	1.33
Al	1.14	Sn	3.72
Ga	1.07	Pb	7.2
In	3.38		
Tl	2.47	V	4.29
		Nb	9.22
		Ta	4.27

temperatures are listed in Table IV.   There seems to be no striking regularity beyond the fact that none of the monovalent metals is superconducting.   The resistances of nonsuperconducting metals at very low temperatures and the normal resistance of superconducting metals just above the transition temperature usually are dependent upon the previous history of the specimen on which measurements are made.   It is believed[1] that, at least in principle, one can divide the resistance into two parts, namely, a part that is characteristic of the pure substance and that extrapolates to zero at the absolute zero, and a part ρ_r, generally termed the residual resistance, that arises from imperfections and that presumably would be zero for a perfectly pure undistorted crystal.

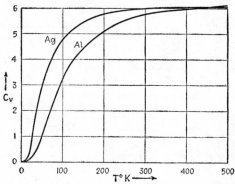

FIG. 15.—The atomic-heat curves of silver and aluminum.   The ordinates are cal/mol.

Since ρ_r does not seem to vary reversibly during temperature changes, it usually cannot be separated from a set of measured resistances in a precisely quantitative way.   It is known from fluctuations in resistance, however, that the residual resistance is of the same order of magnitude as the total resistance at 5°K.   Since the room-temperature value is about a thousand times larger than this, the fluctuations caused by residual resistance are of the order of 0.1 per cent at ordinary temperatures.

f. Specific Heats.—Figure 15 shows the temperature dependence of the atomic heats at constant volume C_v of silver and aluminum.[2]   These curves are typical of most of the simpler metals.   They are characterized by a monotonic rise from zero at absolute zero to a nearly constant

[1] *Cf.* E. GRÜNEISEN, *Handbuch der Physik*, Vol. XIII (1928).   More recent work, such as that of W. J. de Haas and G. J. van den Berg, *Physica*, **4**, 683 *ff.* (1937), seems to show that the residual resistance increases with decreasing temperature in the case of gold.

[2] Silver: A. EUCKEN, K. CLUSIUS, and H. WOLTINEK, *Z. anorg. Chem.*, **203**, 47 (1931).   Aluminum: C. G. MAIER and C. T. ANDERSON, *Jour. Chem. Phys.*, **2**, 513 (1934).

value at high temperatures.   This constant value should be about $3R$, according to the law of Dulong and Petit, where R is the gas constant. Actually, the measured values are slightly higher and rise with increasing temperature.   The part of the curves near $0°K$ may often be approximated closely by the expression $R(T/\Theta_D)^3$, where Θ_D is a constant, known

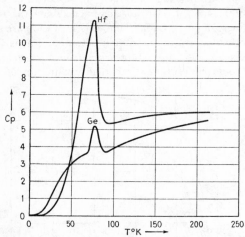

Fig. 16.—The atomic-heat curves of germanium and hafnium.   The ordinates are cal/mol.
(*After Simon and Cristescu.*)

as the Debye characteristic temperature.   Table V contains values of Θ_D for several metals that exhibit this type of specific-heat behavior.

Table V.—Characteristic Temperature of Several Simple Metals as Determined from the T^3 Law

(See Table XXXIV, Chap. III, for additional values.)

Metal	Θ_D
Ag	210
Ca	219
Zn	200
Tl	94
Sn	140
Bi	107

All the nontransition elements resemble silver and aluminum in that they have a limiting high-temperature atomic heat of about $3R$, but a number of them do not behave quite so simply at low temperatures. The differences vary from slight deviations from the T^3 law to greater ones represented by large peaks, such as those shown in the curves[1] for germanium and hafnium (Fig. 16).   The metals that exhibit anomalies

[1] S. Cristescu and F. Simon, *Z. physik. Chem.*, **25B**, 273 (1934).

of the extreme type generally have lattices in which more than one atom is contained in the unit cell.[1] For example, hafnium forms a close-packed hexagonal crystal, and germanium has the diamond lattice; both these types contain two atoms per unit cell.

The atomic-heat curves of the transition metals generally rise well above the Dulong and Petit value of $3R$ at high temperatures and increase linearly in this region.   Figure 17 shows the behavior[2] for γ iron, which is a typical case.   The ferromagnetic metals α iron and nickel show the same behavior but have additional peaks that accompany the decrease in their permanent magnetization.   Figures 17 and 18 illustrate[3] these two cases.

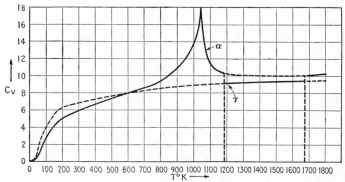

Fig. 17.—The atomic-heat curves of α and γ iron. The ordinates are cal/mol.   (*After Austin.*)

The specific heat of nickel does not follow the T^3 law in the region below 10°K but becomes linear in accordance with the equation[4]

$$C_v = 0.001744T \text{ cal/deg-mol (see Fig. 19).} \tag{1}$$

If the values given by this function are subtracted from the observed values, the residue is found to follow the T^3 law.   This fact indicates

---

[1] If τ_1, τ_2, τ_3 are the primitive translation vectors of the lattice, the unit cell is the unit of the lattice from which the entire lattice may be generated by translations of the type

$$\mathbf{T} = n_1\tau_1 + n_2\tau_2 + n_3\tau_3$$

in which n_1, n_2, and n_3 range over all integer values.   The volume of the unit cell is equal to the volume of the parallelepiped the edges of which are equal to τ_1, τ_2, τ_3, namely, $\tau_1 \cdot \tau_2 \times \tau_3$.

[2] *Cf.* the compilation of J. B. Austin, *Industrial Eng. Chem.*, **24**, 1225 (1932); **24**, 1388 (1932).

[3] See *ibid.* for iron and Landolt-Bornstein for a survey of the specific heats of nickel.

[4] W. H. Keesom and C. W. Clark, *Physica*, **2**, 513 (1935).

that the total specific heat of nickel is composed of two parts, one that has the same source as the specific heats of the nontransition metals and one that has another origin. The first part is believed to arise from thermal excitation of lattice vibrations; there is fairly conclusive evidence,

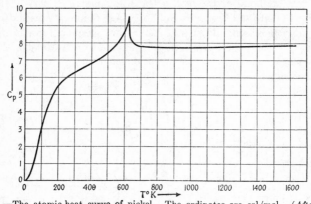

Fig. 18.—The atomic-heat curve of nickel. The ordinates are cal/mol. (*After Lapp.*)

which will be presented in Chap. IV, that the second is related to the excitation of the electrons in the unfilled d shells. In this connection, it is worth pointing out that at 1000°K the value of C_v in Eq. (1) is of the same order of magnitude as the difference between C_v and $3R$.

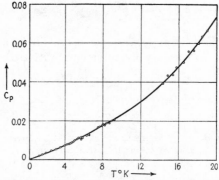

Fig. 19.—The atomic heat of nickel near absolute zero. The ordinates are cal/mol. (*After Keesom and Clark.*)

g. Magnetic Properties.—The magnetic susceptibility per unit volume χ, like the resistivity, is a tensor quantity. It is defined in terms of the magnetic field intensity H and the magnetization per unit volume M by the relation

$$M = \chi \cdot H.$$

We shall ordinarily use cgs units for these quantities, expressing H in gauss and M in terms of the cgs unit of dipole moment per unit volume. Another important magnetic quantity is the permeability ʊ which is defined in terms of H and the magnetic induction B, by the equation

$$B = \mu H.$$

ʊ and χ are related by the equation

$$\mu = 1 + 4\pi\chi.$$

χ is practically independent of both temperature and field strength for a large number of ʟontransition metals; however, it varies with

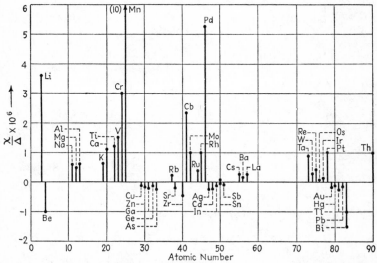

Fig. 20.—The specific susceptibilities of the monatomic metals at room temperature.

temperature in some cases.   The room-temperature values for the simple metals are given[1] in Table VI and in Fig. 20.   For convenience, the specific susceptibilities χ/Δ, where Δ is the density, rather than the susceptibilities are listed.   It may be seen that the metals in the short periods and those which precede the transition elements are paramagnetic, that is, have positive susceptibilities, whereas the metals which follow the transition metals are diamagnetic.   The susceptibilities usually are so small that traces of ferromagnetic impurity affect the measured values considerably and cause them to vary from specimen to specimen.

[1] See, for example, the compilations of Landolt-Bornstein and the International Critical Tables.   As in the case of the resistivity, the scalar components of the susceptibility are listed.

In addition, the susceptibility of a given specimen depends markedly on the mechanical treatment it receives. For example, Bitter[1] found that the susceptibility of a piece of copper wire could be varied by as much as 40 per cent by stretching. Similarly, Honda and Shimizu[2] found that cold working changed the susceptibility of a sample of copper

Table VI.—Room-temperature Specific Magnetic Susceptibilities of Monatomic Substances

(In cgs units)

Monovalent Metals

	$\chi \cdot 10^6$			$\chi \cdot 10^6$
Li	3.6		Cs	0.2
Na	0.6		Cu	−0.08
K	0.6		Ag	−0.19
Rb	0.2		Au	−0.14

Divalent Metals

Be	−1.0		Ba	0.15
Mg	0.5		Zn	−0.15
α Ca	1.1		Cd	−0.18
Sr	−0.2		Hg	−0.12

Trivalent Metals

Al	0.6		Ga	−0.24
Sc			In	−0.10
Yt			α Tl	−0.22
β La	20			

Tetravalent Metals

Ti	1.2		Ge	−0.12
Zr	−0.45		α Sn	−0.03
Hf			β Sn	0.03
Th	1.0		Pb	−0.12

Pentavalent Metals

As	−0.25
Sb	$\parallel$ −0.497; $\perp$ −1.38
Bi	$\parallel$ −1.0; $\perp$ −1.5

Transition Metals

V	1.5		Pd	5.2
α Cr	3.0		Ta	0.9
α Mn	10		α W	0.28
Nb	2.3		Re	0.4
Mo	1.0		Os	0.05
Ru	0.4		Ir	0.13
Rh	1.0		Pt	1.0

from negative to positive and that annealing after cold work restored the original diamagnetism. It is probably true that measurements on perfectly pure, unstrained specimens of the same metal would be closely alike. However, ordinary materials do not conform to these conditions.

[1] F. Bitter, *Phys. Rev.*, **36**, 978 (1930); also, *Introduction to Ferromagnetism* (McGraw-Hill Book Company, Inc., New York, 1937).

[2] K. Honda and Y. Shimizu, *Science Repts. Imp. Tôhoku Univ.*, **20**, 460 (1931); **22**, 915 (1933).

Figure 21 shows the temperature dependence of the measured[1] susceptibilities of a few nontransition elements.

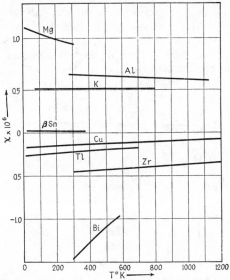

Fig. 21.—The temperature dependence of the specific susceptibilities of several metals.

The susceptibilities of the nonferromagnetic transition elements are all positive and are generally larger than those of the paramagnetic

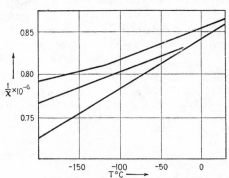

Fig. 22.—The temperature dependence of $1/\chi$ for several specimens of platinum.   (*After Collet and Foëx.*)

simple metals.   The temperature dependence[2] is shown in Fig. 22 for several specimens of platinum.

[1] See footnote 1, p. 17.
[2] P. Collet and G. Foëx, *Compt. rend.*, **192**, 1213 (1931).

The susceptibilities of ferromagnetic solids are so strongly dependent upon field strength that the magnetic properties are described most conveniently by giving M as a function of H. Figure 23 shows[1] M versus H curves at room temperature for three directions in a single crystal of nickel. M and H are parallel in each of the three cases. It may be observed that M is small when no field is present and that it rises very rapidly at first as H increases. It approaches a saturation value M_s in the [111] direction[2] at about fifty gauss and then remains practically constant. The intensity curves for the [110] and [100]

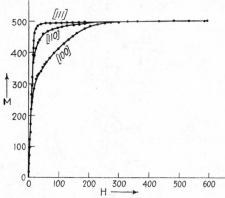

Fig. 23.—The magnetization curves of a single crystal of nickel. The abscissa is expressed in gauss. (*After Kaya.*)

directions bend over sharply at values of about $M_s \cos 30°$ and $M_s \cos 60°$, respectively. They then rise relatively slowly and approach the value M_s.

This behavior may be interpreted in terms of the domain concept of ferromagnetic materials which was first proposed by Weiss.[3] According

---

[1] S. Kaya, *Science Repts. Imp. Tôhoku Univ.*, **17**, 639 (1928).

[2] We shall commonly use this notation in which a crystallographic direction is specified by a set of integers (Miller indices) that are proportional to the direction cosines. In cubic crystals, the coordinate axes are usually taken as an ordinary Cartesian system; in a hexagonal crystal such as cobalt, however, one coordinate axis is taken in the direction of the hexagonal axis of the crystal, and three others, separated by 120 deg, are taken in the plane normal to the hexagonal axis. In the second case, directions are specified by four integers, the last being proportional to the direction cosine between the given direction and the hexagonal axis. We shall specify planes in a similar way by giving the integers that are proportional to the direction cosines of the normals to the planes.

[3] P. Weiss, *Jour. phys.*, **6**, 661 (1907). An excellent discussion of the present status of domain theory may be found in the book edited by R. Becker, *Probleme der technischen Magnetisierungskurve* (Julius Springer, Berlin, 1938). See also the more recent review article by W. F. Brown, *Jour. App. Phys.*, **11**, 160 (1940), and the book by R. Becker and W. Döring, *Ferromagnetismus* (Julius Springer, Berlin, 1939).

to this concept, ferromagnetic substances contain a large number of small domains that have an intrinsic value of magnetic intensity equal to M_s even in the absence of an external field. It is assumed that the direction of this intensity lies along one member of a prominent set of equivalent crystallographic directions, this set being the eight directions equivalent to [111] in the case of nickel, for example. The resultant magnetization of the entire crystal is zero when H is zero, for the domains have their magnetization distributed uniformly among the eight [111] directions. If a weak field is applied in the [111] direction, all the domains have their magnetization changed to this orientation and the crystal becomes magnetized to the saturation value M_s. This process of rotation is demonstrated very convincingly by the Barkhausen[1]

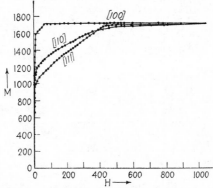

Fig. 24.—The magnetization curves of iron. The abscissa is expressed in gauss. (*After Honda and Kaya.*)

effect, which shows that magnetization takes place in very small discrete stages. The size of the domains may be estimated from the size of these steps if it is assumed that each jump represents the effect of one domain changing its direction. In this way, Bozorth and Dillinger[2] estimated that there are 10^9 domains per cubic centimeter.

It should be emphasized that the domains are not identical with crystalline units of the lattice, such as the grains in polycrystals. It now seems well established that the domain size, which may be larger or smaller than the grain size, is determined primarily by the magnetic interaction of different parts of a specimen and by variations in its internal stress. In addition, it should be mentioned that magnetization may take place by more or less continuous growth of properly oriented domains at their boundaries, much as crystals grow from nuclei.

[1] H. Barkhausen, *Physik. Z.*, **20**, 401 (1919).
[2] R. Bozorth and J. Dillinger, *Phys. Rev.*, **35**, 733 (1930).

If a weak field is applied in the [100] direction instead of the [111] direction, the domains are reoriented as nearly parallel to the [100] direction as possible without leaving the eight [111] directions. Since the angle between the [100] direction and the four nearest [111] axes is 55°, it follows that the largest value of |M| that can be obtained in this way is $M_s \cos 55°$. This actually is the value at the bend in the [100] curve of Fig. 23. Since M increases beyond this value as H is increased, it follows that the intensities of the domains can be bent away from the normal directions of magnetization and eventually become parallel to the direction of the applied field. The curve for the [110] direction supports the same picture.

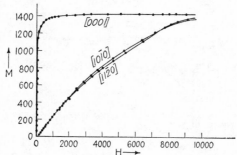

Fig. 25.—The magnetization curve of cobalt. The abscissa is expressed in gauss. (*After Kaya.*)

Figures 24 and 25 give corresponding curves for iron[1] and cobalt.[2] The [100] direction is the easy direction of magnetization in the first case, whereas the hexagonal axis is in the second. The [1010] and [1120] direction curves for cobalt show no sharply rising portions because they are orthogonal to the easy direction of magnetization.

Akulov[3] has shown that one may account for the form of the magnetization curves of iron by assuming that the energy E_m of the crystal varies with the direction of magnetization of the domain in the manner

$$E_m = K_1(S_1^2 S_2^2 + S_2^2 S_3^2 + S_3^2 S_1^2). \tag{2}$$

K_1 is a constant which ordinarily is determined experimentally, and S_1, S_2, S_3 are the direction cosines of the magnetic intensity. The total energy E_t of the crystal in the presence of a field H is then

$$E_t = E_m - HM \cos \varphi$$

[1] K. Honda and S. Kaya, *Science Repts. Imp. Tôhoku Univ.*, **15**, 721 (1926).

[2] S. Kaya, *Science Repts. Imp. Tôhoku Univ.*, **17**, 1157 (1928).

[3] N. S. Akulov, *Z. Physik*, **67**, 794 (1931).

TABLE VII.—MAGNETIC DATA OF THE FERROMAGNETIC MONATOMIC METALS
(The values of M_s and σ_s correspond to 0°K.)

	Fe	Co	Ni	Gd
M_s, cgs	1752	1446	512	1560
σ_s Bohr magnetons per atom	2.22	1.71	0.606	
Θ_c, °C	780	1075	365	16 ± 2
Θ_p, °C	774	1231	372	

where φ is the angle between H and M.   The value of φ corresponding to equilibrium for a given field intensity is determined by the condition

$$\frac{dE_t}{d\varphi} = 0.$$

This leads to a relationship between φ and H, from which the component of intensity in the field direction may be determined as a function of H that involves the constant K_1. Figure 26 shows the calculated and observed values of M for the [111] and [110] directions of iron; these were obtained by using

$$K_1 = 2.14 \cdot 10^5 \text{ ergs.}$$

This method of correlating experimental measurements with energy expressions of the type of Eq. (2) has been extended by Gans,[1] Bozorth,[2] and others.

The magnetization curves vary with temperature in two striking ways: (1) The value of M decreases with increasing temperature and eventually approaches zero at the ferromagnetic Curie point Θ_c. (2) The relative values of the magnetization curves for different directions change with temperature.   Figure 27 shows the ratio of $M_s(T)$ to $M_s(0)$, the value of M_s at 0°K, as a function of T/Θ_c for iron, cobalt, and nickel.

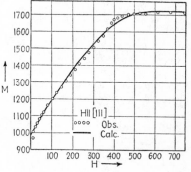

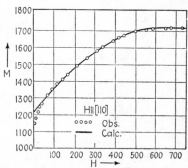

FIG. 26.—Calculated and observed magnetization curves for iron. (*After Akulov.*)

[1] R. GANS, *Physik. Z.*, **33**, 924 (1932).
[2] R. M. BOZORTH, *Phys. Rev.*, **50**, 1076 (1936).

Table VII contains[1] values of Θ_c for each of these substances as well as values of $M_s(0)$ and σ_s, the saturation moment per atom.

Above the Curie temperature, ferromagnetic crystals exhibit a paramagnetism that is the same order of magnitude as the paramagnet-

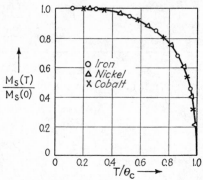

Fig. 27.—The temperature dependence of saturation magnetization for iron, cobalt and nickel. The abscissa is the ratio of the temperature to the Curie temperature θ_c and the ordinate is the ratio of the magnetization at temperature T to that at absolute zero. (*After Tyler.*)

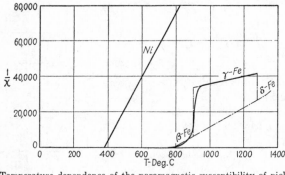

Fig. 28.—Temperature dependence of the paramagnetic susceptibility of nickel and iron above the Curie point.

ism of other transition metals. The susceptibility is highest at $T = \Theta_c$ and decreases with increasing temperature. The temperature dependence of the reciprocals of χ for iron and nickel is shown in Fig. 28. It may be noted that these curves are almost linear, a fact which shows that

$$\chi \cong \frac{C}{T - \Theta_p}$$

[1] After F. TYLER, *Phil. Mag.*, **9**, 1026 (1930); **11**, 596 (1931). See also E. C. STONER, *Magnetism and Matter* (Methuen & Company, Ltd., London, 1934).

where C and Θ_p are constants.   Some values of Θ_p, which is called the paramagnetic Curie point, are given in Table VII.

We have remarked in f that the specific heat of a ferromagnetic metal has a sharp peak in the neighborhood of the Curie point.   Peaks of this type appear in Figs. 17 and 18.   Figure 29 gives a more detailed plot of measured values for nickel.   This curve,[1] which is characteristic, also, of iron and cobalt, shows that the specific heat does not return to the normal $3R$ value above the Curie point.

It should be mentioned in passing that Urbain, Weiss, and Trombe[2] have found that metallic gadolinium is ferromagnetic, its ferromagnetic Curie point being 16°C.

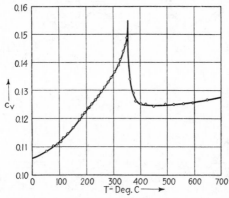

FIG. 29.—The specific heat of nickel in the vicinity of the Curie temperature.   (*After Moser.*)

We shall not discuss the many intricate facts about the properties of polycrystalline ferromagnetic materials.

3. Metal Alloys.—The metallurgist defines a metal alloy as a combination of two or more monatomic metals that has metallic properties. This definition does not require that the material should be a homogeneous phase, and, indeed, many useful alloys are not.   For simplicity, however, we shall restrict practically all of our discussion to single phases.

Alloys generally may be divided into two distinct classes, namely, substitutional alloys and interstitial alloys.   In the first type the different constituent atoms occupy the same type of lattice position.   Gold and silver form an alloy of this type in which both atoms occupy at random face-centered lattice positions.   As we shall see, one general requisite

---

[1] H. MOSER, *Physik. Z.*, **37**, 737 (1936).
[2] G. URBAIN, P. WEISS, and F. TROMBE, *Compt. rend.*, **200**, 2132 (1935).

for the formation of an alloy of this type is that the radii of the constituent atoms be nearly equal.

In interstitial alloys, one or more kinds of atom fit into the interstices of the lattice formed by another kind. Low-concentration carbon steels are alloys of this type. In these, carbon atoms probably occupy some of the face-centered positions of the ordinary body-centered structure of iron. The interstitial atom usually is much smaller than the atoms of the lattice into which it fits.

The properties of a large number of substitutional alloys have been investigated extensively, whereas information concerning interstitial alloys seems to be fragmentary. The main reason for this deficiency is

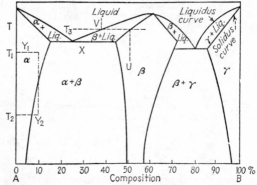

Fig. 30.—A symbolical phase diagram. The regions α, β and γ mark regions in which three different phases are stable. The phases α and β coexist at a point such as X in the region $\alpha + \beta$. The α phase Y_1 formed at the temperature T_1 may, in certain cases, be quenched to temperature T_2, where it is unstable, without actually reverting to the stable mixture of α and β.

that the small atoms, such as hydrogen, carbon, and nitrogen, of clear-cut interstitial cases are not good X-ray scatterers, so that structures of these alloys cannot be determined with certainty. We shall be concerned almost entirely with substitutional alloys in the following discussion.

a. The Phase Diagrams of Binary Alloys.—It is most convenient to discuss binary-alloy systems in terms of the conventional phase diagram. In this diagram the temperature-composition boundaries of the phases of the system are plotted as functions of composition. Figure 30 shows a typical case, the areas α, β, and γ marking regions in which three different phases exist. The composition and temperature may be varied within the limits of any one of these regions without changing the homogeneous structure of the material. If, however, one attempts to make an alloy corresponding to a point, such as X, that is not contained in one of these regions, the result is a mechanical mixture of two phases—the phases α and β in the case corresponding to the point X. It should be added that this statement is rigorously true only if we imply thermodynamically stable phases, for it is possible to prepare an unstable phase that corresponds to the temperature and composition of a point such as X. For example, it is possible that the α phase, formed at temperature and composition corresponding to the point Y_1 at the temperature T_1, where the α phase is thermodynamically stable, could be cooled to the temperature T_2, where it no longer is stable, without breaking into two phases in a measurable time. This procedure, known as quenching, has great practical importance and depends upon the fact that the time required to attain thermodynamical equilibrium may be very long at sufficiently low temperatures.

It may be proved by means of thermodynamics[1] that the boundaries of different phases usually are not continuous but are separated as in Fig. 30 (*cf.* Sec. 123).

The liquidus curve shown in the figure marks the temperature at which a solid phase begins to separate from the molten solution of two metals. This curve has significance only over a range of composition in which the molten metals are miscible. The solidus curve, on the other hand, marks the temperature at which a solid phase of given composition begins to melt. The two curves coincide only at special points such as at the ends of the diagram. At temperature T_3, the liquid and solid phases that may be in equilibrium with one another are given respectively by the two intercepts that the temperature line makes with the liquidus and solidus curves. Consider, for example, the case of Fig. 30 again. Starting with the solid of composition U, we find that this begins to melt at temperature T_3 and that the composition of the first sample of molten metal corresponds to the point (V, T_3). Conversely, if we start with the liquid of composition V and cool it to temperature T_3, the solid that forms has the composition U. It follows that the composition of the solid and melt changes as the process of melting or freezing proceeds in either of these two cases.

It is possible to derive a number of important and interesting relationships among liquidus, solidus, and solubility-limit curves by use of thermodynamics. We refer the reader to other sources[1] for the development of these topics.

b. Rules of Combination of Binary Substitutional Alloys of Simple Metals.—A systematic investigation of the phase diagrams of metal alloys has led to the formulation of a number of simple rules that correlate many facts. We shall include a brief summary of these rules for

---

[1] See, for example, G. TAMMANN, *The States of Aggregation* (translation by R. F. Mehl, D. Van Nostrand Company, Inc., New York, 1925); R. VOGEL, *Handbuch der Metallphysik*, Vol. II (Akademische Verlagsgesellschaft. Leipzig, 1937).

reference in later work.   They should not be accepted as though rigorous, for many exceptions exist.

1. *The rule of atomic size.*—This rule attempts to make more precise the qualitative notion that atoms must be nearly the same size if they form substitutional alloys over a wide range of composition.   It has been developed by many workers, but the most nearly quantitative formulation has been given by Hume-Rothery, Mabbott, and Channel-Evans.[1] These workers find that atoms the radii of which differ by more than about 15 per cent do not form extensive solid solutions.   If the difference is less than this, they are soluble over a wide range.   This rule is restricted by the condition that the radii must be derived from monatomic phases that have similar structures and that it should not be applied to systems

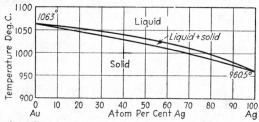

Fig. 31.—The phase diagram of the silver-gold system.   This is an example of a case in which the components are completely miscible.

in which one of the atoms has a tendency to exhibit valence characteristics, as do the atoms of arsenic, antimony, and bismuth.   These two conditions obviously are interrelated since atoms that have valence characteristics usually have complex lattice structures.   In Table VIII, the range of solubility of different metals in copper and the range of solubility of copper in these metals are compared with the atomic radii.

TABLE VIII.—Solubility Limits of the Primary Phases of Several Copper Alloys

(The solubilities are expressed in atom percentage of the solute.)

System	Size factor	Solubility in Cu	Solubility of Cu
Cu-Be	Favorable	16.5 Be	2.0 Cu
Cu-Mg	Unfavorable	6.5 Mg	0.01 Cu
Cu-Zn	Favorable	38.4 Zn	2.3 Cu
Cu-Cd	Unfavorable	1.7 Cd	0.12 Cu
Cu-Ga	Favorable	20.3 Ga	Very small
Cu-Tl	Unfavorable	Small	Small
Cu-Ge	Favorable	12.0 Ge	Small

[1] W. Hume-Rothery, G. W. Mabbott, and K. M. Channel-Evans, *Phil. Trans. Roy. Soc.*, **233**, 1 *ff.* (1934).   See also Hume-Rothery, *op. cit.*

The silver-gold system is one of the most favorable cases for high solubility, according to this rule, for the lattice of both constituents is face-centered cubic, the valences are the same, and the atomic radii are

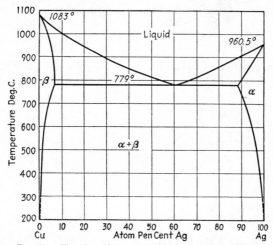

FIG. 32.—The phase diagram of the copper-silver system.

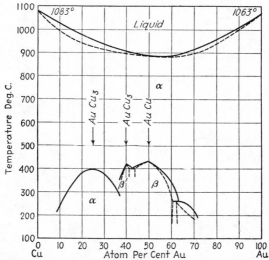

FIG. 33.—The phase diagram of the copper-gold system. These metals are completely miscible at all temperatures. The low-temperature curves correspond to ordered phases.

equal to within 2 per cent. Figure 31, which is the phase diagram of this system, shows that these metals form a single phase for the entire range of composition.

The radius of copper is about 13 per cent less than that of either gold or silver, and therefore the copper-gold and copper-silver systems should be borderline cases. The phase diagrams are shown[1] in Figs. 32 and 33. Whereas copper and silver do not mix, copper and gold are completely miscible except at low temperatures where more complex structures occur. These cases indicate that the rule of atomic sizes does not tell the entire story.

At the opposite extreme are lead and copper the radii of which differ by about 30 per cent and which do not mix in any proportion.

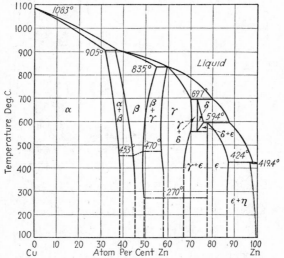

Fig. 34.—The brass (Cu-Zn) phase diagram. This is typical of substitutional alloys of atoms having different valences.

2. *The Hume-Rothery electron-atom ratio rules.*—The copper-rich and silver-rich portions of the phase diagrams of such systems as Cu-Zn, Cu-Cd, Cu-Al, Cu-In, Cu-Sn, Ag-Zn, Ag-Sn, in which the rule of favorable atomic sizes is satisfied, are strikingly similar. Figures 34 and 35 show the cases of Cu-Zn and Cu-Al which are typical examples. The β phase is body-centered cubic in both cases and appears immediately to the right of the primary face-centered phase of pure copper. The γ phase, which has a complex cubic structure, occurs next. The structure of this phase, which is shown in Fig. 36 for brass, is similar though not identical in all alloys; moreover, the γ phase generally has a high resistivity and a negative magnetic susceptibility and is brittle. The

[1] See M. Hansen, *Aufbau der Zweistofflegierung* (Julius Springer, Berlin, 1936) for references on the alloy systems discussed here.

ϵ phase of the brass system, which is close-packed hexagonal, also occurs in the copper-tin system.

Although these similar phases usually occur for different atomic concentrations, Hume-Rothery[1] has pointed out that they occur for about the same value of the ratio of valence electrons to atoms. In computing the number of valence electrons, he used the usual chemical valences, namely, one for copper, two for zinc,

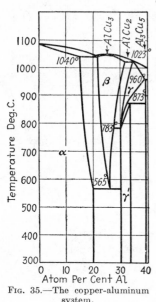

Fig. 35.—The copper-aluminum system.

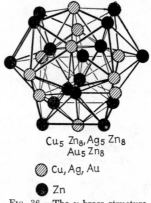

Cu₅ Zn₈, Ag₅ Zn₈
Au₅ Zn₈

⊘ Cu, Ag, Au

● Zn

Fig. 36.—The γ brass structure.

three for aluminum, and so on. Table IX gives a compilation of metals that form one or more of the three alloy phases mentioned above and

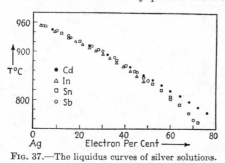

Fig. 37.—The liquidus curves of silver solutions.

that satisfy the Hume-Rothery rule. The electron-atom ratio characteristic of each structure is given at the head of each column.

[1] W. Hume-Rothery, *Jour. Inst. Metals*, **35**, 309 (1926). See also Hume-Rothery, *op. cit.*

TABLE IX.—PHASES THAT CONFORM TO HUME-ROTHERY'S ELECTRON-ATOM RATIO RULE

Electron-atom ratio	1.5	1.61	1.75
Structure.........................	β brass (b.c.c.)	γ brass type	h.c.p.
Nontransition cases.................	CuZn	Cu_5Zn_8	$CuZn_3$
	CuBe	Cu_5Cd_8	$CuCd_3$
	AgZn	Ag_5Zn_8	$AgZn_3$
	AgCd	......	$AgCd_3$
	AuZn	Au_5Zn_8	$AuZn_3$
	AuCd	......	$AuCd_3$
	Cu_3Al	Cu_9Al_4	Cu_3Sn
	Cu_3Ga	Cu_9Ga_4	Cu_3Ge
	Cu_5Sn	Cu_9In_4	Ag_3Sn
		$Cu_{31}Sn_8$	Au_5Al_3
Transition cases....................	CoAl	$CoZn_3$	
	NiAl		
	FeAl		

This rule is analogous to the ordinary rule of eight, being valid for substitutional alloys instead of for ionic and valence compounds.

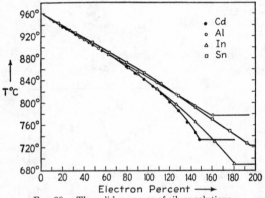

FIG. 38.—The solidus curves of silver solutions.

3. *The Hume-Rothery liquidus- and solidus-curve rules.*[1]—The liquidus curves of the primary solid solutions of elements such as zinc, cadmium, gallium, indium, tin, and antimony in copper and silver are similar in form and coincide if the electron percentage instead of the atomic

[1] W. HUME-ROTHERY, G. W. MABBOTT, and K. M. CHANNEL-EVANS, *Phil. Trans. Roy. Soc.*, **233**, 1 *ff.* (1934). See also HUME-ROTHERY, *op. cit.*

percentage is used as abscissa. In cases in which the solvent is monovalent, the electron percentage is obtained by multiplying the atom percentage with the valence of the solute. The liquidus curves of silver are given in Fig. 37 in order to show the extent to which the rule is valid. Figure 38 shows that the solidus curves obey the same principle.

c. *Alloys Involving Metals with Strong Valence Characteristics.*—The phase diagrams of systems in which one of the constituents is a metal of low valence, such as copper, silver, zinc, or magnesium, and the other is a less electropositive atom, such as arsenic, antimony, or bismuth, show that these substances do not combine to form extensive solid solutions, even when the size factors are favorable. Consider, for example, the phase diagram[1] of the magnesium-antimony system shown in Fig. 39.

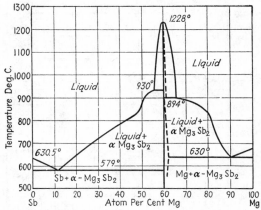

Fig. 39.—The magnesium-antimony system. These atoms are completely immiscible except for the compound Mg_3Sb_2.

The fact that the primary phases are narrow shows that neither atom is appreciably soluble in the lattice of the other. The intermediate phases occur over narrow regions of composition and for atomic ratios that are characteristic of ionic or valence compounds rather than of ideal metal alloys; moreover, the structures of these phases are often similar to those of ionic crystals. For example, the only intermediate phase in the magnesium-antimony system is the compound Mg_3Sb_2 in which the constituents are exhibiting their normal electropositive and electronegative valences. This compound exists in two phases, both of which have the structures of rare earth metal oxides.

We may conclude from evidence such as this that these alloys form part of a bridge between ideal metals and ionic crystals.

[1] See footnote 1, p. 30.

It should be added that arsenic and antimony are completely miscible in one another, a fact showing that they do not form ionic-like lattices unless they are combined with strongly electropositive elements.

d. Rules for Combination of Transition Metals.—When transition metals combine with simpler metals, they obey fairly closely the three

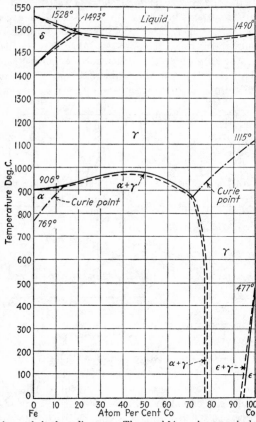

FIG. 40.—The iron-cobalt phase diagram. The α and δ iron phases are body-centered cubic, whereas the γ phase is face-centered. The ϵ phase is hexagonal close-packed.

rules that were presented in part *b*. In applying the rules, however, it is necessary to treat transition metals as though their valences were practically zero. For example, Table IX contains several alloys that have structures compatible with Hume-Rothery's electron-atom ratio rule (2 of part *b*) if this assumption is made.

Transition elements in the same row of the periodic chart have almost identical radii and combine over wide ranges of composition. Figure 40 shows the phase diagram of the iron-cobalt system which is a typical example.

Transition metals form an interesting sequence of interstitial alloys that has been studied extensively by Hägg.[1] He has found that the "metalloid" atoms hydrogen, nitrogen, carbon, and boron enter into the interstices of transition metals forming alloys of composition M_4X, M_2X, MX, and MX_2, where M is the metal atom and X is the metalloid, if the ratio of the radius of the metalloid atom to that of the metal is less than 0.59. This family of alloys usually forms lattices in which the metal atoms are arranged in cubic or hexagonal close-packed structures, although there are a few notable exceptions, such as tungsten carbide, WC, in which the tungsten atoms possess a simple hexagonal

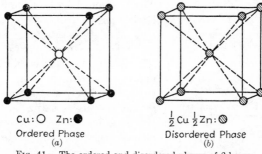

Cu:O   Zn:●
Ordered Phase
(a)

$\frac{1}{2}$ Cu $\frac{1}{2}$ Zn: ◍
Disordered Phase
(b)

Fig. 41.—The ordered and disordered phases of β brass.

arrangement. The iron-carbon system lies just outside the domain of applicability of Hägg's rules, for the radius ratio is about 0.63 in this case; however, the iron-nitrogen system is a typical one for which they are valid. The nontransition metals do not usually form genuinely metallic interstitial alloys when combined with the metalloid atoms, but rather tend to form more nearly ionic crystals, such as calcium carbide, presumably because they are more electropositive than the transition metals.

e. Ordered and Disordered Phases.—In an ideal substitutional alloy, different kinds of atom occupy a given set of lattice positions at random. Many alloys in which the atoms have this property at high temperatures change as the temperature is lowered. Consider, for example, the case of β brass,[2] which has the composition CuZn and the structure shown in Fig. 41b. At high temperatures, each site is occupied with equal prob-

---

[1] G. Hägg, *Z. phys. Chem. B,* **6,** 221 (1929); **7,** 339 (1930); **8,** 455 (1930).

[2] The possibility of order and disorder was first suggested by G. Tammann, *Z. anorg. Chem.,* **107,** 1 (1919). The structures of ordered and disordered β brass were first established by X-ray methods by F. W. Jones and C. Sykes, *Proc. Roy. Soc.,* **161,** 440 (1937).

ability by either type of atom. Below 480°C, however, the copper atoms prefer[1] body-centered positions and the zinc atoms prefer cube corners. These preferences increase as the temperature is lowered, and the structure becomes that of Fig. 41a at very low temperatures. The ordering process takes place continuously with decreasing temperature in this case and is completely reversible if the system is maintained at equilibrium. If we let p_{Cu} designate the probability that a body-centered position may be occupied by a copper atom and p_{Zn} the probability that it may be occupied by a zinc atom, we may conveniently define an order parameter S by the equation

$$S = p_{Cu} - p_{Zn}.$$

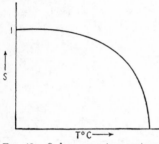

FIG. 42.—Order versus temperature for β brass (schematic).

This parameter varies from 0 to 1 as the lattice passes from the relatively disordered high-temperature phase to the ordered structure. Figure 42 shows schematically the way in which S depends upon temperature in β brass. This curve resembles closely the curve that shows the dependence of saturation magnetization of ferromagnetic materials upon temperature (*cf.* Fig. 27). The analogy with ferromagnetism becomes even more striking when one examines the specific heat curve[2] of β brass, which is shown in Fig. 43. It may be seen that there is a sharp peak, similar to the peak that occurs in nickel at the Curie temperature, at the temperature where ordering begins.

All changes between the ordered and disordered phases do not occur so gradually as that observed in β brass. For example, in the Cu_3Au system, a high degree of order occurs abruptly when the alloy is cooled below 380°C. This abrupt and reversible change is accompanied by the appearance of a latent heat.

Other substitutional phases that exhibit ordering are shown in Fig. 44.

f. Additional Properties of Substitutional Alloys of Nontransition Metals.—The thermal, electrical, and magnetic properties of metal alloys are, on the whole, much the same as those of monatomic metals. There are, however, a few striking differences that make additional discussion worth while. In this section, we shall consider alloys of nontransition metals.

---

[1] The body-centered and cube-corner positions are completely interchangeable in this case.

[2] MOSER, *op. cit.*

1. *Thermal properties.*—The difference between the heats of formation of alloys and of the metallic phases of their constituents has been

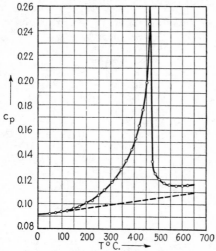

Fig. 43.—The specific heat of β brass during the transition from the ordered to a disordered structure.   (*After Moser.*)

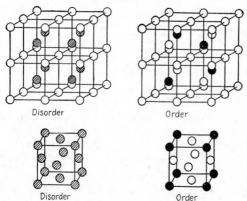

Fig. 44.—The upper pair of figures respectively represent the disordered and ordered arrangements of Fe₃Al.   O = Fe; ● = Al; ⊘ = ½Fe½Al.   The lower pair represent the disordered and ordered structures of CuPd.   O = Cu; ● = Pd.

determined in a number of cases.   Three methods are commonly used: (1) comparative measurements of the heat evolved when monatomic metals and alloys are dissolved in acids; (2) direct calorimetric measurements of the heats of reaction of monatomic metals; and (3) measurement of emfs of cells in which one of the electrodes is the alloy under investigation. The room-temperature values[1] of some results of this work, expressed in terms of kilogram calories per gram-atom, appear in Table X. The experimental errors are usually of the order of magnitude of 1 kg cal.

TABLE X.—ROOM-TEMPERATURE VALUES OF THE HEATS OF FORMATION OF ALLOYS
(In kg cal/gram-atom)

Cu-Zn System

β brass (unordered)	2.2
γ brass	2.6
$CuZn_3$	1.8

Ag-Cd System

β phase (unordered)	1.31
γ phase	1.42
$AgCd_3$	1.23

Au-Zn System

Au_3Zn	6.0
$AuZn$	5.5
$AuZn_3$	5.6

Miscellaneous Cases

Mg_2Sn	20	$MgLa$	2.9
$CaAl_3$	13	$CdSb$	1.8
Ca_2Zn_3	8	$AuSb_2$	1.2
Na_2Sn	7	Tl_7Bi	0.7
$NaHg$	5.4	Hg_5Tl_2	0.06

Cases Involving Transition Elements

Ni_3Sn	5.8
Ni_3Sn_2	7.5
$NiSn$	7.5
Al_5Co_2	12
Al_3Fe	6.3

The nearly ideal substitutional alloys, like the Cu-Zn and Ag-Cd systems, are not bound so tightly as compounds such as Mg_2Sn that evidently are transition cases. On the whole, however, it does not seem to be possible to draw any striking conclusions from this table.

The specific heats of nontransition metal alloys usually resemble those of monatomic metals in approaching zero at 0°K, in obeying Dulong and Petit's law at high temperature, and in increasing monotonically in the intervening region. The exceptions are those phases which undergo allotropic changes or which become ordered as described in part *e*. In the first case, there is a discontinuity in the specific heat curve and a latent heat, just as for any allotropic change. The behavior of the specific heat during ordering was discussed in part *e*.

Early investigation of the specific heats of alloys led to the formulation of the Kopp-Neumann law, which states that the molecular heat of any alloy is equal to the sum of the atomic heats of its constituent

[1] These are taken from the compilation of W. Biltz, *Z. Metallkunde*, **29, 73** (1937). See also W. Seith and O. Kubaschewski, *Z. f. Elektrochem.*, **43**, 743 (1937).

TABLE XI.—THE MOLECULAR HEAT OF Ag_3Au

Temperature, °C	Observed molar heat, cal	Sum of atomic heats, cal
100	24.926	24.942
200	25.415	25.475
300	26.005	26.000
400	26.599	26.513
500	27.195	27.012
600	27.789	27.500
700	28.384	27.979
800	28.937	28.463

monatomic metals.   More modern work has shown that this law is never precisely correct, although it is often correct to within 10 per cent.   Table

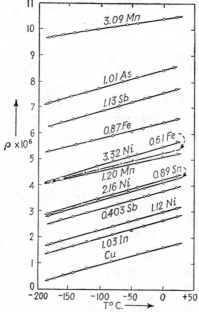

FIG. 45.—The resistivity versus temperature curves of a number of copper alloys in the range in which the resistivity of copper varies linearly with temperature.   It should be observed that the resistivities of the alloys are much higher at low temperatures, implying very large residual resistivities.   The resistivities are expressed in ohms-cm.   The numbers indicate the atomic per cent of the alloying metal.   (*After Linde.*)

XI gives a comparison[1] of the molecular specific heat of Ag_3Au and the sum of the atomic heats of the constituents over a 700° temperature

[1] J. A. BOTTEMA and F. M. JAEGER, *Proc. Roy. Soc. Amsterdam,* **35,** 928 (1932).

range. The close agreement in this case is partly a consequence of Dulong and Petit's law, since both the pure metals and the alloy obey it closely. It is evident that the Kopp-Neumann rule will fail badly in any temperature range in which the alloy becomes ordered.

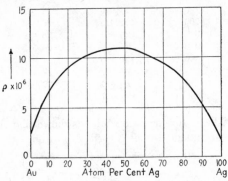

Fig. 46.—The resistivity of the silver-gold system at room temperature.

2. *Electrical resistivity of alloys.*—One of the most striking characteristics of the temperature-resistance curves of alloys is the fact that they do not extrapolate to zero at absolute zero so closely as those of monatomic metals do. In other words, their residual resistance usually is

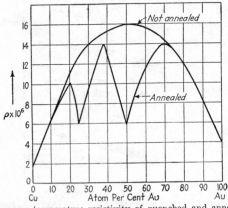

Fig. 47.—The room-temperature resistivity of quenched and annealed specimens of copper-gold alloys. The quenched specimens, which are not ordered, have the typical resistivity versus composition curve of perfect solid solutions. The annealed alloys show resistance minima at the compositions at which ordering occurs.

very large. This fact is shown by the curves[1] of Fig. 45 which are temperature-resistance plots for a number of copper alloys. One of the

---

[1] J. O. Linde, *Ann. Physik*, **15**, 219 (1932).

natural consequences of this large residual resistance is the fact that the low-temperature resistance of dilute primary solid solutions rises with increasing concentration of the solute. This behavior[1] is shown in

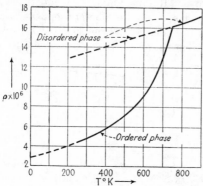

Fig. 48.—The temperature-resistivity curve of β brass during the ordering transition. The residual resistivity of the ordered alloy is much lower than that of the disordered phase.

Fig. 46 by the curve of resistance versus concentration for the Au-Ag system.

The resistance of an ordered phase invariably is lower than that of the disordered one, and there usually is a sharp kink in the curve of resistivity

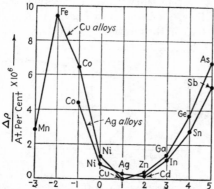

Fig. 49.—The increase in resistivity of copper and silver alloys per atom per cent of solute. The abscissa is the number of valence electrons of the solute atom relative to the closed d shell.

versus temperature at the temperature where ordering begins. These facts are illustrated[2] in Figs. 47 and 48, which show the dependence of

[1] See, for example, the references in Landolt-Bornstein.

[2] C. H. JOHANSSON and J. O. LINDE, *Ann. Physik*, **5**, 762 (1930); **25**, 1 (1936). G. BORELIUS, *Proc. Phys. Soc. (Sup.)*, **49**, 77 (1937).

resistivity on composition and on temperature for some alloys that were well enough annealed to allow ordering to take place.

Different atoms, dissolved in a given solvent metal, affect the resistivity in different ways. If, for example, one plots the increase in resistance per atom percentage of solute for different copper and silver solutions, one obtains the curves[1] of Fig. 49, which show that the resistance increases with the difference in valence of the solvent and solute atoms.

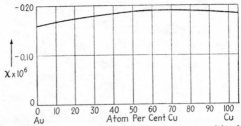

Fig. 50.—The dependence of magnetic susceptibility upon composition in the copper-gold system. (*After Shimizu.*)

3. *Magnetic susceptibilities.*—The specific magnetic susceptibilities of alloys of the nontransition metals are of the order of magnitude of 10^{-6}, just as are those of pure nontransition metals. If the constituents are soluble in all proportions and if the alloys do not form ordered phases, the variation of susceptibility with composition usually is uniform. Examples[2] of such cases are shown in Figs. 50 and 51. In other cases, particular phases have their own magnetic properties which may be

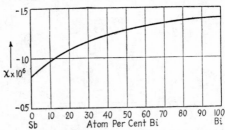

Fig. 51.—The dependence of the magnetic susceptibility upon composition in the antimony-bismuth system. (*After Shimizu.*)

considerably different from those of the pure metal. Thus, the γ brass type of phase usually is strongly diamagnetic, whereas the β brass type is usually normal. The susceptibility of the brass system[3] is shown in Fig. 52.

---

[1] See footnote 1, p. 40.

[2] Y. Shimizu, *Science Repts. Imp. Tôhoku Univ.*, **21**, 826 (1932).

[3] H. Endo, *Science Repts. Imp. Tôhoku Univ.*, **14**, 479 (1925).

g. Additional Properties of Transition-metal Alloys. 1. *Thermal properties.*—The heats of formation of only a few alloys that contain transition metals have been measured. Several of these values are contained in Table X. Generally speaking, they are of the same order of magnitude as those of phases of nontransition metals.

The specific heats of the transition-metal alloys generally show the same types of behavior as the specific heats of monatomic transition metals. Thus, they do not obey Dulong and Petit's law at high temperatures, if they are strongly paramagnetic or ferromagnetic; and if ferromagnetic, they exhibit "anomalous" peaks near the Curie point. There is a close correlation between the height of such peaks and the saturation moment of the ferromagnetic substance. For example, the addition of copper to nickel quenches the magnetization of the latter,

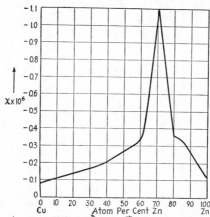

FIG. 52.—The magnetic susceptibility of the brass system, showing the large peak associated with the γ phase. (*After Endo.*)

and Fig. 53 shows[1] that the peak in the specific-heat curve disappears as the percentage of copper increases. Similar behavior has been observed in the chromium-nickel system.

2. *Electrical conductivity.*—The effect of temperature upon the electrical conductivity of alloys that contain transition metals has not been investigated so widely as has that upon the electrical conductivity of simpler alloys. The general facts, however, probably are about the same.[2] In disordered alloys, for example, there is a large residual resistance that decreases with increasing order. Typical composition-

---

[1] K. E. GREW, *Proc. Roy. Soc.*, **145**, 509 (1934).

[2] The resistance of several transition-metal alloys, such as constantan (Cu60 Ni40) and manganin (Cu84 Mn12 Ni4), are nearly temperature-independent over a temperature range of several hundred degrees.

resistance curves are enough like those of nontransition cases to require no additional comment.

3. *Magnetic properties.*—The magnetic properties of this group of alloys form a large and interesting body of material which we have only limited space to discuss.   Most of the alloys are paramagnetic and have

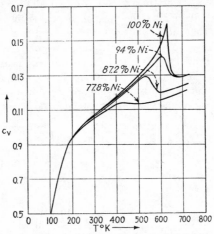

FIG. 53.—The quenching of the magnetic specific heat of nickel by addition of copper. (*After Grew.*)

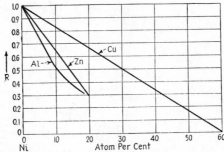

FIG. 54.—The quenching of the magnetization of nickel by the addition of simple metals. R is the ratio of the saturation magnetization at a given composition to that of pure nickel.   (*After Sadron.*)

susceptibilities that decrease with increasing temperature, just as do the susceptibilities of monatomic transition metals.   The alloys that contain one of the ferromagnetic metals, however, are ferromagnetic, at least for large concentrations of the ferromagnetic metal.   This ferromagnetism usually decreases with increasing dilution of the ferromagnetic constituent if the other constituent is not ferromagnetic.   For example,[1]

[1] C. SADRON, *Ann. Physik*, **17**, 371 (1932).

Fig. 54 shows the decrease in the saturation magnetization of a number of nickel alloys as their composition is varied.   These curves show the customary behavior, namely, the magnetization decreases uniformly as the concentration of solute is increased. The manganese-nickel system is an exception to this rule, for the magnetization passes through two peaks when manganese is added to nickel in gradually increasing amounts[1] (*cf.* Fig. 55).

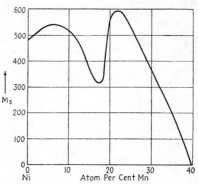

FIG. 55.—Dependence of the saturation magnetization of nickel-manganese alloys upon composition. (*After Kaya and Kussmann.*)

The alloys of ferromagnetic elements are all ferromagnetic. Figure 56 shows the behavior[2] of the saturation moment and the Curie point in the iron-cobalt system.

The alloys of copper and manganese[3] have the susceptibility curves shown in Fig. 57 at room and liquid-air temperatures.   The susceptibility of the phase that contains about 23 per cent of manganese is very high, indicating a strong

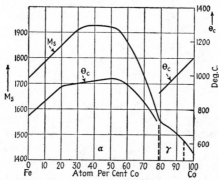

FIG. 56.—The saturation moment and the Curie temperature in the iron-cobalt system.

tendency toward ferromagnetism.   By adding aluminum or tin to this system, one obtains the Heusler alloys, of which some, such as the phase[4] of composition Cu_2AlMn, are ferromagnetic.   Ferromagnetic alloys also

[1] S. KAYA and A. KUSSMANN, *Z. Physik,* **72,** 293 (1931).

[2] Magnetization: A. KUSSMANN, B. SCHARNOW, and A. SCHULZE, *Z. tech. Physik* **13,** 449 (1932); P. WEISS and R. FORRER, *Compt. rend.,* **189,** 663 (1929).

[3] S. VALENTINER and G. BECKER, *Z. Physik,* **80,** 735 (1933).

[4] F. HEUSLER, *Verh. deut. physik. Ges.,* **5,** 219 (1903).

occur in the chromium-tellurium, manganese-arsenic, and platinum-chromium systems.

4. Ionic Crystals.—The salts produced by combining highly electropositive metals and highly electronegative elements such as the halogens, oxygen, and sulfur are the ideal ionic crystals. Other, more complex salts, such as metal carbonates and nitrates and ammonium halides, also may be classified as ionic crystals. We shall be interested principally in the diatomic types, however, since they are the easiest to handle theoretically.

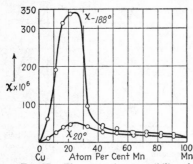

Fig. 57.—The magnetic susceptibility of the copper-manganese system. The abscissa is the atom per cent of manganese. (*After Valentiner and Becker.*)

Ionic crystals closely obey the ordinary rules of classical valency; in fact, most valence numbers are derived from investigations of the combining ratios of atoms in ionic compounds.

a. Cohesion.—The heats of formation[1] of a number of diatomic ionic crystals are given in Table XII. The standard state to which these values are referred is that of the monatomic gases of the constituents. It is noteworthy that the cohesive energy generally is larger for components containing atoms of higher valency than for compounds containing atoms of lower valency.

In many instances, it is convenient to refer the cohesive energies to a standard state of free ions rather than of free atoms. Thus, we shall have occasion to use the energy required to sublime sodium chloride into free Na^+ and Cl^- ions. These energies may be obtained from those of Table XII by adding the energy required to transfer valence electrons from the metal atoms to the electronegative atoms. In compounds of formula MX, this additional term obviously is a multiple of the difference between the ionization energy of the metal atoms and the electron affinity of the electronegative atoms. The first of these quantities has been determined very accurately, by spectroscopic means, for practically all metals. The second, however, has been measured only for the halogens. The most direct method of determining electron affinities has been developed by Mayer[2] and is based upon measurement of the equilibrium

[1] Bichowsky and Rossini, *op. cit.*

[2] J. E. Mayer, *Z. Physik*, **61**, 798 (1930). L. Helmholz and J. E. Mayer, *Jour. Chem. Phys.*, **2**, 245 (1934). P. P. Sutton and J. E. Mayer, *Jour. Chem. Phys.*, **2**, 146 (1934); **3**, 20 (1935).

density of atomic ions in heated alkali halide vapor.   From this quantity, it is possible to determine the heat of the reaction

$$MX \text{ (molecule)} \rightleftarrows M^+ \text{ (atom)} + X^- \text{ (atom)}$$

where M is the metal atom and X is the halogen atom.   By subtracting from this the heat of the reaction

$$MX \text{ (molecule)} \rightleftarrows M \text{ (atom)} + X \text{ (atom)}$$

and the ionization energy of the metal atom, one obtains the energy of the process

$$\text{Electron} + X \rightleftarrows X^-.$$

TABLE XII.—COHESIVE ENERGIES OF IONIC CRYSTALS RELATIVE TO THE MONATOMIC GASES OF THE CONSTITUENTS

(In kg cal/mol at room temperature)

Alkali Hydrides

LiH	112.5
NaH	91.8
KH	81.7
RbH	82.8
CsH	82.7

Alkali Halides

LiF	216.4		LiCl	162.6
NaF	193.7		NaCl	153.1
KF	186.1		KCl	153.1
RbF	183.9		RbCl	152.9
CsF	182.2		CsCl	154.0
LiBr	149.7		LiI	129.7
NaBr	139.5		NaI	120.8
KBr	140.8		KI	124.3
RbBr	141.6		RbI	125.5
CsBr	143.4		CsI	128.3

Other Monovalent Metal Halides

			CuCl	144.4
			$CuCl_2$	192.4
AgF	148.5		AgCl	127.2
			AuCl	129.2
			TlCl	117.5
CuBr	134.8		CuI	124.6
$CuBr_2$	169.0		CuI_2	137.2
AgBr	118.7		AgI	108.5
AuBr	122.3		AuI	117.4

Alkaline Earth Halides

			$BeCl_2$	245.4
MgF_2	363.7		$MgCl_2$	247.4
CaF_2	401.6		$CaCl_2$	296.2
SrF_2	399.6		$SrCl_2$	302.7
BaF_2	400.5		$BaCl_2$	312.1

TABLE XII.—COHESIVE ENERGIES OF IONIC CRYSTALS RELATIVE TO THE MONATOMIC
GASES OF THE CONSTITUENTS.—(*Continued*)

$BeBr_2$	208.2	BeI_2	165.6
$MgBr_2$	214.0	MgI_2	174.3
$CaBr_2$	263.8	CaI_2	227.5
$SrBr_2$	271.8	SrI_2	234.3
$BaBr_2$	283.2	BaI_2	244.8

Other Divalent Metal Halides

$ZnCl_2$	184.8	$ZnBr_2$	159.6
$CdCl_2$	177.6	$CdBr_2$	156.4
$HgCl_2$	125.8	$HgBr_2$	109.1
$PbCl_2$	191.0	$PbBr_2$	167.6

ZnI_2	128.4
CdI_2	126.4
HgI_2	91.1
PbI_2	140.5

Miscellaneous Cases

AlF_3	510	$AlCl_3$	308
SbF_3	352	$TlCl_3$	209
		$SbCl_3$	218
		$SbCl_5$	292
		$SnCl_2$	217
		$FeCl_2$	234
		$FeCl_3$	277

$AlBr_3$	262	AlI_3	209
$SbBr_3$	180	SbI_3	140
$SnBr_2$	193	SnI_2	168

Alkali Metal Oxides, Sulfides, and Selenides

Li_2O	279	Na_2S	208
Na_2O	210	K_2S	227
K_2O	185	Rb_2S	192
Cs_2O	179	Cs_2S	191

Li_2Se	224
Na_2Se	172
K_2Se	175

Other Cases

BeO	269		
MgO	242	MgS	185
CaO	258	CaS	228
SrO	247	SrS	226
BaO	241	BaS	226
Al_2O_3	667	Al_2S_3	449
TiO_2	436		

$CaSe$	191
$SrSe$	191
$BaSe$	191

Another scheme of the same type, also devised by Mayer, involves the use of halogen vapor instead of alkali halide vapor. Table XIII contains some values of the electron affinities of the halogens that were determined by Mayer, Helmholz, and Sutton, using these methods.

TABLE XIII.—ELECTRON AFFINITY OF HALOGEN ATOMS AS DETERMINED DIRECTLY BY MAYER'S METHODS
(In kg cal/mol)

Cl	88.3
Br	84.2
I	72.4

The electron affinities of the doubly charged negative ions O^{--}, S^{--}, and Se^{--}, etc. probably are negative; hence, these ions are unstable, and therefore their affinities cannot be determined by direct methods.

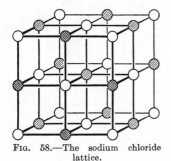

FIG. 58.—The sodium chloride lattice.   FIG. 59.—The cesium chloride lattice.

b. Crystal Structure.—Ideal ionic compounds usually crystallize in one of several simple structures.[1] The sodium chloride structure of Fig. 58 is characteristic of all the alkali halides[2] except the low-temperature modifications of cesium chloride, bromide, and iodide, which have the simple cubic structure of Fig. 59. Many divalent metal oxides, sulfides, selenides, and tellurides also have the sodium chloride lattice, although many others crystallize in the zincblende and wurtzite structures of Figs. 60 and 61. On the whole, it may be said that all four of these structures are characteristic of ionic crystals in which the constituent atoms have equal positive and negative valences. An interesting feature of these lattices is that they remain the same when metal and electronegative atoms are interchanged.

The fluorite lattice of Fig. 62 is typical of ionic compounds that have the formula M_2X or MX_2. This structure occurs among the alkaline earth halides such as calcium fluoride and barium fluoride and among the

---

[1] See *Strukturberichte.*

[2] Many of the alkali halides possess the cesium chloride structure at high pressures. See R. B. Jacobs, *Phys. Rev.,* **54,** 468 (1938).

alkali metal oxides and sulfides such as lithium oxide and sulfide. Many other compounds that have these formulae possess structures in which the chemical molecule shows a tendency to form an "island" in the lattice. This is evident, for example, in the rutile or titanium oxide lattice of Fig. 63 which may be regarded as being built of a body-centered arrangement of TiO_2 molecules. Since this behavior is typical of molecular crystals, it may be said that ionic crystals such as titanium oxide are mild transition cases between ionic and molecular types. Another

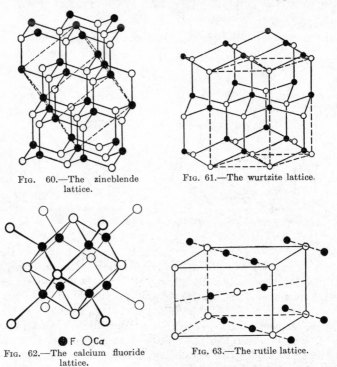

FIG. 60.—The zincblende lattice.

FIG. 61.—The wurtzite lattice.

● F ○ Ca

FIG. 62.—The calcium fluoride lattice.

FIG. 63.—The rutile lattice.

simple lattice that shows the same behavior is the pyrites or FeS_2 structure of Fig. 64. This is equivalent to a face-centered cubic arrangement of FeS_2 molecules. Carbon dioxide, which is a typical molecular compound, has a similar lattice.

The α-corundum structure of Fig. 65 is formed by several ionic compounds that have the form A_2B_3, such as Al_2O_3 and Fe_2O_3.

The structures of a number of ionic crystals are listed in Table XIV. This collection[1] also contains a few substances such as AlN and GaP,

[1] See *Strukturberichte.*

which are transition cases but which are of interest because they have typical ionic structures. Several ammonium salts also have been included in order to show that a radical may play the same role as an atomic ion.

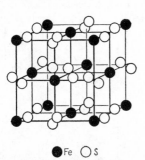

Fe ⬤   ◯ S

Fig. 64.—The pyrites lattice. Both ferric sulfide and carbon dioxide possess this structure.

⬤ Al   ◯ O

Fig. 65.—The α-corundum lattice.

The interatomic distances d of the alkali halides, show a regularity, namely, that the differences between the d values of NaF and KF, NaCl and KCl, NaBr and KBr, and NaI and KI are equal to within a few hundredths of an angstrom unit:

NaF-KF	NaCl-KCl	NaBr-KBr	NaI-KI
0.36 Å	0.33 Å	0.31 Å	0.30 Å

The same type of relationship is valid for the differences between the d values of the halides of other pairs of alkali metals and the d values of the alkali metal salts of pairs of halogens. It follows from this rule that we may associate with each ion a definite radius, the interatomic distance of each substance being given closely by the sum of the radii of the constituent ions. Thus, the crystals behave as though they were composed of rigid spherical ions that are in contact with one another. It obviously is not possible to determine the absolute values of the ionic radii from the d values alone, although it is possible to determine the differences of all the alkali metal ion radii and all the halogen-ion radii. Thus, it is necessary to know the absolute value of only one radius in order to determine all radii.

This additivity principle also is roughly valid for many other classes of ionic crystal, such as the copper and silver halides and the alkaline earth oxides, sulfides, and selenides.

TABLE XIV.—Some Crystal Constants of Ionic Solids
Monovalent Metal Halides
(See Table II, for notation)
The alkali metal hydrides probably have the NaCl type of lattice.

	Type	Parameters, Å		
		a	*c*	*d*
LiF	f.c.c.	4.02	....	2.01
NaF	f.c.c.	4.62	....	2.31
KF	f.c.c.	5.33	....	2.67
RbF	f.c.c.	5.63	....	2.82
CsF	f.c.c.	6.01	....	3.00
LiCl	f.c.c.	5.14	....	2.57
NaCl	f.c.c.	5.63	....	2.81
KCl	f.c.c.	6.28	....	3.14
RbCl	f.c.c.	6.54	....	3.27
CsCl	f.c.c.			
CsCl	s.c	4.11	....	3.56
LiBr	f.c.c.	5.49	....	2.75
NaBr	f.c.c.	5.96	....	2.98
KBr	f.c.c.	6.58	....	3.29
RbBr	f.c.c.	6.85	....	3.43
CsBr	s.c.	4.29	....	3.71
LiI	f.c.c.	6.00	....	3.00
NaI	f.c.c.	6.46	....	3.23
KI	f.c.c.	7.05	....	3.53
RbI	f.c.c.	7.33	....	3.66
CsI	s.c.	4.56	....	3.95
AgF	f.c.c.	4.92	....	2.46
CuCl	Zincblende	5.41	....	2.34
AgCl	f.c.c.	5.54	....	2.77
CuBr	Zincblende	5.68	....	2.46
AgBr	f.c.c.	5.76	....	2.88
CuI	Zincblende	6.05	....	2.62
α AgI	Wurtzite	4.59	7.53	
β AgI	Zincblende	6.49	....	2.81
TlCl	s.c.	3.84	....	3.33
TlBr	s.c.	3.97	....	3.44
TlI	s.c.	4.18	....	3.62

TABLE XIV.—SOME CRYSTAL CONSTANTS OF IONIC SOLIDS.—(*Continued*)
Alkaline Earth Halides
(Many of these have complex structures which we shall not discuss. The following are a few simple cases.)

	Type	Parameters, Å		
		a	*c*	*d*
MgF_2	Rutile (see Fig. 63)	4.64	3.06	2.05
CaF_2	Fluorite	5.45	....	2.36
BaF_2	Fluorite	6.19	....	2.68
ZnF_2	Rutile	4.72	3.14	2.10
CdF_2	Fluorite	5.40	....	2.34
Alkali Metal Oxides, Sulfides, and Selenides				
Li_2O	Fluorite	4.61	....	2.00
Li_2S	Fluorite	5.70	....	2.47
Na_2S	Fluorite	6.53	....	2.83
Monovalent Metal Oxides, Sulfides, Selenides				
Cu_2O	Complex cubic lattice	2.46	....	1.84
Cu_2S	Fluorite	5.59	....	2.42
Cu_2Se	Fluorite	5.75	....	2.49
Ag_2O	Same as Cu_2O	4.70	....	2.05
Bivalent Metal Oxides, Sulfides, Selenides				
BeO	Wurtzite	2.69	4.37	1.64
MgO	f.c.c.	4.21	....	2.10
CaO	f.c.c.	4.80	....	2.40
SrO	f.c.c.	5.15	....	2.58
BaO	f.c.c.	5.53	....	2.77
ZnO	Wurtzite	3.24	5.18	1.94
CdO	f.c.c.	4.70	....	2.35
BeS	Zincblende	4.86	....	2.10
MgS	f.c.c.	5.19	....	2.60
CaS	f.c.c.	5.68	....	2.84
SrS	f.c.c.	6.01	....	3.01
BaS	f.c.c.	6.37	....	3.19
α ZnS	Zincblende	5.42	....	2.35
β ZnS	Wurtzite	3.84	6.28	2.36
α CdS	Zincblende	5.82	....	2.52
β CdS	Wurtzite	4.14	6.72	2.52
HgS	Zincblende	5.84	....	2.53
BeSe	Zincblende	5.13	....	2.18
MgSe	f.c.c.	5.45	....	2.73
CaSe	f.c.c.	5.91	....	2.96
SrSe	f.c.c.	6.24	....	3.12
BaSe	f.c.c.	6.59	....	3.30

TABLE XIV.—SOME CRYSTAL CONSTANTS OF IONIC SOLIDS.—(*Continued*)

	Type	Parameters, Å		
		a	c	d

Bivalent Metal Oxides, Sulfides, Selenides

ZnSe	Zincblende	5.66	....	2.45
CdSe	Zincblende	6.05	....	2.62
HgSe	Zincblende	6.07	....	2.63
MgTe	Wurtzite	4.52	7.33	2.75
CaTe	f.c.c.	5.91	....	2.96
SrTe	f.c.c.	6.55	....	3.33
BaTe	f.c.c.	6.99	....	3.50
ZnTe	Zincblende	6.09	....	2.64
CdTe	Zincblende	6.46	....	2.79
HgTe	Zincblende	6.44	....	2.80

Oxides of Trivalent Metals

α Al_2O_3, Fe_2O_3, α Ga_2O_3 form crystals with the corundum structure of Fig. 65.

Miscellaneous Other Cases

Face-centered Cubic Type

	a	d
FeO	4.28	2.14
CoO	4.25	2.13
NiO	4.17	2.08
ScN	4.14	2.20

Simple Cubic Type

	a	d
NH_4Cl	3.86	3.34
NH_4Br	4.05	3.51
NH_4I	4.37	3.78

Zincblende Type

	a	d
AlP	5.45	2.36
GaP	5.44	2.36

Wurtzite Type

	a	c	d
NH_4F	4.39	7.02	2.63
AlN	3.11	4.98	1.87

c. Conductivity.—The halides and oxides of the simpler metals generally have an electrolytic conductivity that increases with increasing temperature. Figure 66 shows the specific conductivity of a number of very pure alkali halide crystals as determined by Lehfeldt.[1] The scale of abscissae is adjusted in order to be proportional to $1/T$, and the scale of ordinates is logarithmic. It may be observed that the low-temperature portions of the conductivity curves depend upon the history of the specimen, whereas the high-temperature portions are reproducible straight lines in this

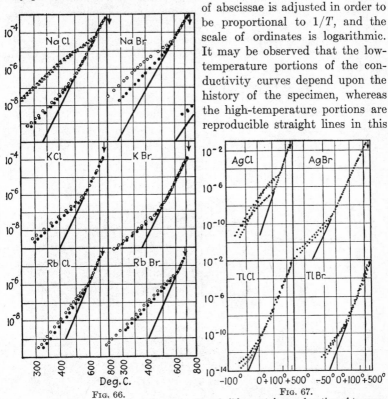

Fig. 66.

Fig. 67.

Fig. 66.—The ionic conductivity of the alkali halide crystals as a function of temperature. In all cases except that of NaCl the different sets of points refer to two artificial crystals. Additional measurements were made on natural crystals of sodium chloride. The ordinates are expressed in ohm⁻¹ cm⁻¹, the abscissas in degrees centigrade. (*After Lehfeldt.*)

Fig. 67.—The conductivity of silver and thallium halides. The ordinates are expressed in ohm⁻¹ cm⁻¹. (*After Lehfeldt.*)

type of plot. This fact shows that the high-temperature conductivity σ satisfies the relation

$$\sigma = A e^{-\frac{'\alpha}{T}}$$

where A and α are practically constant.

[1] W. Lehfeldt, *Z. Physik*, **85**, 717 (1933).

Similar curves are shown in Fig. 67 for silver and thallium halides.

It has been demonstrated[1] fairly conclusively that the conductivity of the halides is completely ionic. This fact is by no means obvious from the temperature dependence, for the electronic conductivity of semi-conductors usually follows the same law. Generally speaking, it is possible to establish the existence of ionic conductivity only by performing a number of indirect experiments, which will be discussed in connection with semi-conductors. Oxides, sulfides, and selenides usually exhibit some electronic conductivity.

Table XV gives the fractions n_+ and n_- of the current carried, respectively, by positive and negative ions in a number of halide crystals. These fractions are called the transport numbers of the corresponding ions.

TABLE XV.—THE TRANSPORT NUMBERS OF IONIC SOLIDS AT DIFFERENT
TEMPERATURES

Compound	Temperature, °C	n_+	n_-
NaF	500	1.000	0.000
	550	0.996	0.004
	600	0.916	0.084
	625	0.861	0.139
NaCl	400	1.000	0.000
	510	0.981	0.019
	560	0.946	0.054
	625	0.929	0.071
KCl	435	0.956	0.044
	500	0.941	0.059
	550	0.917	0.083
	600	0.884	0.166
AgCl	20–350	1.00	0.00
AgBr	20–300	1.00	0.00
BaF$_2$	500	0.00	1.00
BaCl$_2$	400–700	0.00	1.00
BeBr$_2$	350–450	0.00	1.00
PbF$_2$	200	0.00	1.00
PbCl$_2$	200–450	0.00	1.00
PbBr$_2$	250–365	0.00	1.00
PbI$_2$	255	0.39	0.61
	290	0.67	0.33

[1] See the survey article by C. Tubandt, *Handbuch der Experimental Physik*, Vol. XII, part 1.

d. Specific Heats.—The specific-heat curves of the diatomic salts of simple metals are normal in the sense that they obey Dulong and Petit's law at high temperatures and decrease monotonically with decreasing temperature. Figure 3, Chap. III, shows the atomic-heat curves of

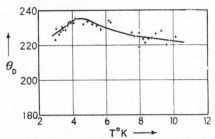

Fig. 68.—Variation of the characteristic temperature of potassium chloride with temperature near absolute zero. The characteristic temperature is defined by the T^3 law in cases of this type (see Sec. 19). (*After Keesom and Clark.*)

several alkali halides above 20°K. Keesom and Clark[1] have found that the specific heat of potassium chloride shows slight deviations from Debye's T^3 law at temperatures below 10°K (*cf.* Fig. 68). We shall discuss this effect in Chap. III.

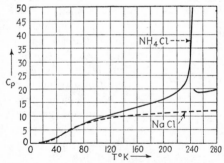

Fig. 69.—A comparison of the molar heats of ammonium chloride and sodium chloride. (*After Simon, v. Simson, and Ruhemann.*)

The specific heats of ammonium halides resemble those of the alkali halides at very low temperatures, but they have large anomalies in the region just below room temperature. Figure 69 gives a comparison of the specific-heat curves of ammonium chloride and sodium chloride[2] and shows the high peak that occurs at 250°K in the first case. As we

[1] Keesom and Clark, *Physica*, **2**, 698 (1935).
[2] F. Simon, O. v. Simson, and M. Ruhemann, *Z. physik. Chem.*, **129**, 339 (1927).

shall see later, this peak is connected with the reorientation of the NH_4^+ radical.[1]

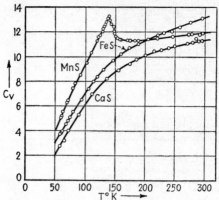

Fig. 70.—The specific-heat curves of manganous sulfide, ferrous sulfide, and calcium sulfide. The ordinate is cal/deg-mol. (*After Anderson.*)

Salts of transition metals usually do not obey Dulong and Petit's law at high temperatures; they show the same type of excess specific heat that is observed in metals that have unfilled d shells. Figure 70 shows the curves[2] for manganous sulfide and ferrous sulfide and, for comparison, the "normal" curve of calcium sulfide.

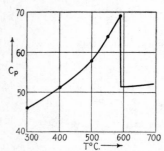

Fig. 71.—The specific-heat curve of magnetite, showing the peak that occurs at the Curie point. The ordinates are cal/deg-mol. (*After Weiss, Piccard and Carrard.*)

Several transition salts are ferromagnetic. Consequently, one might expect their specific-heat curves to have peaks near the ferromagnetic Curie point. Figure 71 shows the peak for magnetite,[3] which seems to be the only case that has been examined.

e. Magnetic Properties.—Most ideal ionic salts are diamagnetic, the exceptions being salts of transition metals, which usually are paramagnetic and sometimes are ferromagnetic.

[1] As a result of careful thermodynamical work in the vicinity of the transition temperature in ammonium chloride, A. Lawson [*Phys. Rev.*, **57**, 417 (1940)] has shown that the anomaly originates in a reorientation of the ammonium radicals rather than in onset of free rotation.

[2] C. T. Anderson, *Jour. Am. Chem. Soc.*, **53**, 476 (1931).

[3] P. Weiss, A. Piccard, and A. Carrard, *Arch. sci. phys. nat.*, **43**, 113 (1917).

The measured molar susceptibilities of a number of halides,[1] oxides, and sulfides of simple metals are given in Table XVI.   The experimental values fluctuate from specimen to specimen, as in simple metals; therefore, these values are not accurate to within more than a few per cent.

It may be verified that the susceptibilities of the alkali halides are additive to a comparatively high degree of accuracy.   This fact indicates that each ion preserves a characteristic diamagnetic susceptibility in each compound.   However, we cannot determine the absolute susceptibility of the ions from these data alone, just as we could not determine their ionic radii from crystallographic data alone.

The susceptibilities of other halides also seem to be additive, although the results fluctuate so widely in different experimental results that this statement cannot be made with certainty.

The copper salts furnish an interesting example.   The cuprous salts invariably are diamagnetic; however, cupric chloride and iodide are

Table XVI
Molar Susceptibilities of Salts of Nontransition Elements
(The unit used is 10^6 times the cgs unit.)

	F	Cl	Br	I
Li	......	− 25.4	− 37.3	− 55.8
Na	−19.6	− 30.8	− 43.2	− 60.3
K	−25.7	− 36.2	− 49.2	− 67.2
Rb	−31.9	− 46.4	− 56.7	− 67.1
Cs	......	....	....	− 92.5
CuX	......	− 38.1		
CuX$_2$	......	1243.	629.	
AgX	......	− 54.8	− 48.6	
AuX	......	− 67.0	− 60.8	− 91.0
TeX	......	− 58.2		
MgX$_2$	......	− 49.7	− 72.2	−111.4
CaX$_2$	−23.4	− 54.5		
SnX$_2$	......	− 63.0	− 85.3	−131.1
BaX$_2$	−22.8	− 74.0	−103.6	−160
ZnX$_2$	......	− 58.2	....	−112.7
CdX$_2$	......	....	....	−134.1
HgX$_2$	......	− 80.9		
SnX$_4$	......	− 114.1		

[1] See, for example, the compilations in Landolt-Bornstein and the International Critical Tables.

TABLE XVI.—(*Continued*)
Molar Susceptibilities of Transition-metal Salts
(Room-temperature values)

	$\chi_m \cdot 10^2$
FeCl$_2$	1.32
FeBr$_2$	1.36
FeI$_2$	1.36
FeSO$_4$	1.24
CoCl$_2$	1.22
CoBr$_2$	1.27
CoI$_2$	1.07
CoSe$_4$	1.05
CoO	10^5
NiCl$_2$	0.62
NiBr$_2$	0.55
NiI$_2$	0.38
NiO	10^4
PtCl$_3$	0.0136
CeF$_3$	0.220
CeCl$_3$	0.192
Ce$_2$S$_3$	0.492
Sm$_2$S$_3$	0.322
Sm$_2$(SO$_4$)$_3$	0.211

paramagnetic. This fact shows that the cuprous ion is a simple ion, whereas cupric ion is similar to the ions of transition metals.

The molar susceptibilities of a number of transition-metal salts also are listed in Table XVI. We shall describe the magnetic properties of some of these salts more fully in Chap. XVI.

The large susceptibilities of cobaltous oxide and nickel oxide suggest that these compounds are ferromagnetic. However, there do not seem to be any measurements on the magnetization curves of these substances. Magnetite, Fe$_3$O$_4$, and pyrrhotite Fe$_7$S$_8$, are the only salts that are ferromagnetic at room temperature and that have had their ferromagnetic properties measured.[1] The specific-heat curve of magnetite is shown in Fig. 71.

5. Valence Crystals. General Description.—Ideal valence crystals are monatomic nonconducting substances that have high cohesive energies and great hardness. Diamond is the prototype of this class, just as the alkali halides are the prototypes of ionic crystals. A characteristic of the diamond structure, which is shown in Fig. 4, is that the number of

---

[1] P. WEISS, *Jour. de Phys.*, **6**, 661 (1907). M. ZIEGLER, Thesis (Zurich, 1915). See also D. R. INGLIS, *Phys. Rev.*, **45**, 119 (1934).

nearest neighbors of each atom, namely, four, is equal to the ordinary valence of carbon. Table XVII contains some data[1] for diamond and other valence crystals. Boron is possibly another ideal valence crystal since it is also a very hard insulator; however, its structure does not seem to be known.

TABLE XVII.—SOME PROPERTIES OF SOLIDS HAVING VALENCE CHARACTERISTICS

Substance	Structure	Cohesive energy, kg cal/mol	Hardness (relative)	Resistivity, ohm-cm	Magnetic suscepti- bility $\cdot 10^6$
Diamond.....	Fig. 4 $d = 1.54$ Å	$(170.0 + a)^*$	10	10^{14}	-0.50
Graphite......	Fig. 72 $d = 1.42$	$(170.49 + a)^*$	0.5	$2 \cdot 10^{-5}$ at 0°C Decreases with decreasing temperature	-3.5
Boron........	....	115	9.5	10^{18}	-0.7
Silicon........	Diamond type $d = 2.35$	85	7	$8 \cdot 10^{-2}$	-0.13
Germanium...	Diamond type $d = 2.43$	85	...	$9 \cdot 10^{-2}$	-0.10
Gray tin......	Diamond type $d = 2.80$	78.6	....	....	-0.35
Silicon carbide	ZnS type $d = 1.89$, etc.	283	9		
Silicon dioxide	See text	405.7	7	10^{14}	-0.45
Boron nitride.	Fig. 72 $d = 1.45$	.....	....	....	0.0

* Two different methods of determining the heat of sublimation of graphite lead to values that differ by about 50 kg cal/mol.

Many substances may be classified between the valence type and one of the other types. For example, silicon, germanium, and gray tin crystallize in the diamond structure although they ordinarily have a much higher conductivity than diamond. Silicon and germanium may also be classified among semi-conductors, whereas gray tin may be classified among metals. Similarly, silicon carbide and silicon dioxide have some valence properties, such as great hardness, and some ionic characteristics, such as the ability to absorb infrared radiation strongly.

Silicon carbide crystallizes in several different lattice structures, which have in common the property that each atom is surrounded by four atoms of opposite type that are situated at the corners of a tetrahedron. One

[1] See, for example, the compilations of Landolt-Bornstein and the International Critical Tables.

form, for example, corresponds to the zincblende structure of Fig. 60. Similarly, silicon dioxide has several crystalline forms.   In each of these, a silicon atom is surrounded tetrahedrally by four oxygen atoms and each oxygen atom is joined to two silicon atoms.

The most stable solid form of carbon probably is graphite, which has a layer lattice structure similar to that of boron nitride, shown in Fig. 72.   This substance is not hard in spite of its high cohesive energy (*cf.* Table XVII), presumably because the planes of carbon atoms slide easily over one another.   Nevertheless, we shall classify graphite among the valence compounds since the forces between carbon atoms in the planes of graphite are believed to be similar to the forces between carbon atoms in diamond.   It may be seen from Table XVII, that graphite has a large conductivity which increases with decreasing temperature.   Since this conductivity is electronic, graphite may also be classified among metals.

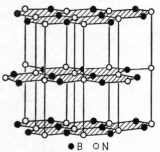

Let us consider the following two sequences of compounds:

LiF	NaF
BeO	MgO
BN	AlN.

Fig. 72.—The lattice of boron nitride.

● B   ○ N

In each of these sequences the valences of the electropositive and electronegative elements increase by unity as we move down a given column.   The top members are ideal ionic compounds.   The second members exhibit ionic conductivity and crystallize in typical ionic structure but are very hard.   Boron nitride has the structure shown in Fig. 72, which is similar to that of graphite, whereas aluminum nitride has the wurtzite structure, which is similar to that of diamond.   Evidently the properties of a compound of light elements become more nearly like those of typical valence crystals the nearer the center of the periodic chart its constituents lie.   This general rule is obeyed by many compounds of the lighter elements, the principal exceptions being molecular compounds.

In this connection, it may be recalled that the metals arsenic, antimony, and bismuth crystallize in a layer lattice structure in which the number of nearest neighbors of a given atom is equal to the electronegative valency of these elements.   This behavior, being analogous to that of ideal valence crystals, indicates that these substances should be classified between the metallic and the valence types.

6. Semi-conductors.   *a. General Properties.*—A number of solids have a small electronic conductivity that is negligible at very low temperatures and increases with increasing temperature. The experiments that distinguish this conductivity from ionic conductivity will be described in the next section. The electrical properties of these semi-conductors are so unique that it is convenient to introduce a separate classification for the group, even though it would be possible to place them among the other types. It will become apparent that this class does not possess the same degree of unity as the other four classes of solids.

Table XVIII contains a list of established semi-conductors and several substances that probably are semi-conductors. Most of these solids crystallize in the ionic type of structure and were discussed with ideal ionic crystals. Carborundum, on the other hand, was previously included under valence types. The semi-conducting specimens of most monatomic substances, such as silicon and tellurium, are usually impure.

TABLE XVIII
Semi-conductors
Monatomic Substances
Si(impure)
Te
Halides
AgI
CuI
Oxides

CuO	NiO	Cr_2O_3
ZnO	FeO	Fe_2O_3
BaO	WO_2	Fe_3O_4
CoO	UO_2	CU_2O

Sulfides and Selenides

PbS	
Ag_2S	Ag_2Se
CdS	
MoS_2	

Probable Semi-conductors
SiC(impure)
Ag_2Te

There are two methods of measuring the conductivities of semi-conductors. The first of these, which is used more commonly,[1] consists in placing a single crystal or a pressed powder specimen between two metal electrodes and measuring its resistance by some ordinary means such as a Wheatstone bridge. This direct method has a number of advantages; for example, the specimen may be heated or cooled easily, and it may be placed in any kind of atmosphere. Its main disadvantage is that contact resistance between the electrodes and the specimen, or

---

[1] TUBANDT, *op. cit.* See also the survey article on semi-conductors by B. Gudden, *Ergebnisse exakt. Natur.*, **13**, 223 (1934).

between granules of the powder, may affect the current-resistance curve. In some cases, these effects cause an apparent deviation from Ohm's law; in other cases, they simply give rise to a spurious value of the conductivity. For these reasons, it is always difficult to be certain that the conductivities obtained by the method actually are constants of the material under investigation.

An alternative method has been developed and employed by Gudden[1] and his coworkers Völkl and Guillery. They mix a quantity of the powdered semi-conducting material with a nonconducting dielectric, such as a heavy oil, and use the mixture as a dielectric medium in a condenser. The electrical conductivity of the semi-conductor is determined from an

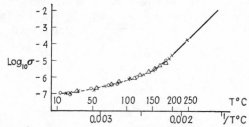

Fig. 73.—The ionic conductivity of silver chloride as determined by several methods. The values corresponding to the straight and dashed lines and to the crosses were measured by direct means. The values corresponding to the circles and triangles were measured by Völkl using the method described in the text.

investigation of the effective resistance of the condenser when it is part of a resonating high-frequency circuit. This procedure has the advantage that it eliminates contact resistance, for the current simply surges back and forth within the granules during the experiment. The principal disadvantages of the method are: (1) It does not allow a very wide choice of conditions under which measurements may be made. (2) It does not lead to very accurate results, since the experimental error usually is of the order of 10 per cent.

Results obtained by the two different methods agree in some cases and disagree widely in others. Guillery found, for example, that the pressed-powder and condenser methods give nearly identical results for most oxides, but widely different ones for stannic oxide and silicon carbide.

Figure 73 shows the conductivity of silver chloride, which is an ionic conductor, as determined by a number of workers,[2] using different methods (see legend to Fig. 73). In this case, the logarithm of the low-temperature conductivity is not linear when plotted as a function of $1/T$ (cf. Sec. 4).

[1] A Völkl, *Ann. Physik*, **14**, 193 (1932); P. Guillery, *ibid.*, **14**, 216 (1932).
[2] Völkl, *op. cit.*

Figure 74 shows the temperature dependence of the conductivity of cuprous oxide[1] as measured by the direct method. The conductivities of a number of semi-conductors give linear plots of this type, a fact showing that the conductivity varies with temperature in the manner

$$\sigma = A e^{-\frac{E}{kT}}$$

where E and A are practically constant for a given specimen. Meyer[2] has shown that A is about 1 ohm^{-1} cm^{-1} for many semi-conductors, whereas E varies considerably for different substances and for different specimens of the same substance. For example, values of E between 0.06 and 0.6 ev have been quoted for cuprous oxide.

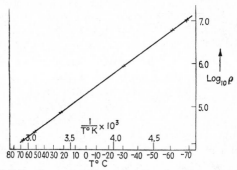

Fig. 74.—The resistivity of cuprous oxide as a function of temperature. The ordinates are ohm-cm. (*After Vogt.*)

b. Methods of Determining the Type of Conductivity.—One or more of the following three quantities are commonly measured in trying to determine whether or not the conductivity of a substance is electronic:

1. Transport numbers.
2. The Hall constant.
3. The thermoelectric coefficient.

We shall discuss the guiding principles that are used in each case.

1. *Measurement of transport numbers.*—An important characteristic of ionic conductivity is that electrolysis accompanies the flow of current, since this electrolysis should be absent in a substance the conductivity of which is entirely electronic. There is some electrolysis if the conductivity is partly ionic and partly electronic, but Faraday's transport law should not be valid in this case. Hence, in principle at least, one should be able to determine the fraction of electronic current by measur-

[1] W. Vogt, *Ann. Physik*, **7**, 190 (1930).
[2] W. Meyer, Thesis (Berlin, 1936); *Z. Physik*, **85**, 278 (1933).

ing the deviation from Faraday's law. Tubandt[1] and his coworkers have employed this method of determining semi-conductors with a great deal of success. Their procedure varies somewhat from case to case, but the underlying principles may be understood by considering the following hypothetical example.

Suppose it is suspected that a substance of formula MX, in which M is the metallic constituent and X is the electronegative constituent, conducts by both positive ions M^+ and electrons. In this case, Tubandt would place three pressed-powder or crystalline disks of the material in series with three disks of a substance MY, which is known to conduct only by means of M^+ ions, and would place these between two electrodes of the metal M (see Fig. 75). The disks adjoining the electrodes and the electrodes are weighed accurately. A current then is sent through the system, and the total quantity of electricity that passes is measured by means of a coulometer. If there is any electrolysis, the disks in contact with the electrodes usually become fastened to the electrodes during this procedure. The electrodes and the disks attached to them are weighed together in order to determine the amount of material lost by the anode and gained by the cathode. The positive-ion transport number may then be computed from all these measured quantities.

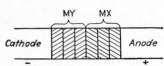

Fig. 75.—Arrangement of specimens and electrodes in measurement of transport numbers. In this case the specimen MX is a positive-ion conductor whose transport number is unknown, whereas MY is a conductor in which all of the current is carried by positive ions.

Tubandt uses the known ionic conductor MY in the circuit, partly to check the coulometer measurement and partly to make certain that negative ions do not leave the disk nearest the cathode. Three disks of each substance are employed in order that one disk may be in contact only with chemically similar substances. If the experiment is flawless, the weight of this disk should not change. A negative transport number can be determined in a similar way by placing three disks of a substance NX, which conducts only by negative ions X^-, between the anode and the three disks of the substance under test.

In practically all cases, this direct method yields results that agree with those determined by other methods. A set of transport numbers that Tubandt and other workers have obtained in this way and that are generally accepted at the present time are given in Table XIX. It should be noted that several semi-conductors have a small, but finite, ionic conductivity.

[1] See footnote 1, p. 56.

Tubandt's procedure seems to lead to incorrect results in the case of α silver sulfide, which we shall discuss briefly. In this case, Tubandt employed the scheme symbolized by Fig. 75, making the disks MX of silver sulfide, the disks MY of silver chloride, and the electrodes of silver. Since the α phase is stable only above 180°C, the system was maintained at an elevated temperature during the experiment. Tubandt found that the silver anode lost as much weight as the cathode gained, and he concluded that positive ions carry all of the current. This conclusion was contradicted by a large amount of subsequent work that indicated that the conductivity is mainly electronic. For example, the conductivity of silver sulfide at 180°C is about fifty times larger than that of any

TABLE XIX.—TRANSPORT NUMBERS OF SEMI-CONDUCTORS
(n_e = electronic transport number; n_+ = positive-ion transport number)

Substance	Temperature, °C	n_e	n_+
PbS	240	1.00	
α Ag$_2$S	Above 180	$\sim$0.99	$\sim$0.01
β Ag$_2$S	20	0.015	0.985
β Ag$_2$Se	20	$\sim$1.00	<0.01
β Ag$_2$Te	20	$\sim$1.00	<0.01
γ CuI	200	$\sim$1.00	$\sim$2.7 $\cdot$ 10^{-6}
	325	0.50	0.50
	400	0.00	1.00
Cu$_2$O	800	$\sim$1.00	$\sim$2 $\cdot$ 10^{-4}
	1000	$\sim$1.00	$\sim$5 $\cdot$ 10^{-4}

ionic conductor, and the Hall coefficient (*cf.* 2 below) is of the magnitude ordinarily associated with electronic conductivity.

At present, this contradiction is explained[1] in the following way. It is assumed that most of the conductivity of α silver sulfide is electronic, although a very small fraction is ionic. The halogen atoms that are released at the boundary between the silver chloride and silver sulfide disks by electrolysis of silver chloride reduce some of the sulfide and leave an equivalent amount of sulfur. This sulfur in turn reacts with an equivalent number of silver atoms that diffuse from the anode through the sulfide. This picture has many direct supports. For example, it is found that the anode does not lose an equivalent of weight if a very large current is passed through the silver sulfide. In this case, the diffusion of neutral silver atoms is not rapid enough to change all the liberated sulfur to sulfide.

---

[1] The work on silver sulfide is discussed by C. Wagner, *Z. physik. Chem.*, **B, 21**, 42 (1933); **23**, 469 (1933).

2. *Measurement of the Hall constant.*—In 1879, Hall[1] found that an emf may be produced across a strip of metal which is carrying a current by placing the strip in a magnetic field. For a cubic crystal, the direction and magnitude of the induced electrostatic field are given by the vector relationship

$$\mathsf{E} = R\mathsf{J} \times \mathsf{H}. \tag{1}$$

Here, E is the electrostatic field vector, J is the current per unit area, H is the magnetic field intensity, and R is the Hall constant of the material. The effect usually is measured by passing a large current I through a thin strip of thickness t that is placed in the magnetic field in such a way that H is normal to the surface. In this case, the emf $\mathsf{E_H}$ is

$$\mathsf{E_H} = R\frac{I\mathsf{H}}{t}.$$

In Sec. 37, we shall discuss a simple theory of the Hall effect in which the conductor is treated like a gas of free electrons. The results of this theory may be summarized by saying that they relate the Hall constant to the number of electrons per unit volume n and the value of the electronic charge $-e$ by means of the equation

$$R = -\frac{3\pi}{8nec}$$

in which c is the velocity of light. The sign of the charge on the carriers is thus the same as the sign of R. Strangely enough, this sign is positive for a number of metals, such as zinc and antimony, although their conductivity is undoubtedly electronic. The interpretation of this anomaly is one of the striking successes of the zone theory of solids, which will be developed in later chapters. In order of magnitude, the mobility[2] of the current carriers is given by the ratio of the Hall constant to the specific resistivity, which, in the vicinity of room temperature, is about 100 cm^2/volt-sec for most metals. These mobilities are about one hundred times larger than the mobilities of ions in the best solid ionic conductors. Incidentally, the Hall effect in ionic conductors is too small to be measured.

---

[1] E. H. Hall, *Am. Jour. Math.*, **2**, 287 (1879). A survey of early literature is given by L. L. Campbell in *Galvanomagnetic and Thermomagnetic Effects*, (Longmans, Green & Company, New York, 1923).

[2] The mobility μ of a current-carrying particle in a conductor is defined as the velocity with which the particle moves when placed in a unit electrostatic field. Thus, the conductivity σ is equal to $ne\mu$ where, as in the equation for the Hall constant, n is the number of conducting particles per unit volume and e is their charge.

Semi-conductors possess a measurable Hall effect, as Baedeker[1] first found in 1909, and the mobility of the carriers turns out to be of the same magnitude as the mobilities of electrons in metals. At the present time, the existence of a measurable Hall effect is accepted as proof that a given substance is a semi-conductor. The apparent sign of the charge of the current carriers is positive for about as many semi-conductors as it is negative. This anomaly is explained by the modern theory of solids in the same way as it is explained for metals.

The sign of the Hall coefficient apparently changes from positive to negative in some specimens of cuprous oxide[2] as the temperature is raised, being zero at about 500°C. The oxide remains an electronic conductor during the transition, however, a fact showing that the absence of a Hall effect does not furnish proof that the conductivity is ionic.

3. *Measurement of the thermoelectric effect.*—A current flows in two wires of different metals, which are joined together to form a closed circuit, if the two junctions are kept at different temperatures. This thermoelectric effect is usually described by giving the emf dE/dT that is developed for each degree difference in the temperatures of the junctions. These *thermoelectric coefficients* are additive in the sense that the value of dE/dT for two metals A and C is equal to the algebraic sum of the values for the metals A and B and the metals B and C. Hence, it is possible to find the value for any pair of metals if the value of each, relative to a standard, is known.

The thermoelectric effect is observed in semi-conductors but not in ionic conductors. For this reason, the effect is used as a test for electronic conductivity in the same way that the Hall effect is used.

c. *Factors That Influence the Conductivity.*—In addition to temperature, there are three factors that strongly influence the conductivity of a semi-conductor, namely, its impurity content, the mechanical treatment it has received, and the vapor pressures of the gases of its constituent atoms that are maintained in the surrounding atmosphere.

The first two of these influences have not been investigated in a systematic way. It is known, however, that different specimens of most monatomic semi-conductors, such as silicon, selenium, and tellurium, do not have the same conductivity at the same temperature, and it is concluded from this fact that impurities play an important role in determining the conductivity. Similarly, it is found that the conductivity of powders that are prepared from the same material by grinding varies with the amount of grinding, and the conclusion is drawn that

---

[1] K. BAEDEKER, *Ann. Physik*, **29**, 566 (1909); *Physik. Z.*, **13**, 1080 (1912).

[2] W. SCHOTTKY and F. WAIBEL, *Physik. Z.*, **34**, 858 (1934). See also *Physik. Z.*, **36**, 912 (1935) for correction in sign.

mechanical treatment influences the conductivity. It is possible that these phenomena are closely related, that is, that the distribution of impurity atoms in the semi-conductor may be partly connected with its mechanical history. Problems such as these await further investigation.

A large amount of work has been done on the effect of vapor pressure upon the conductivity. These investigations have thrown a great deal of light upon the origin of electronic conductivity in many of those semi-conductors which also may be catalogued under the heading of ionic crystals. The classical example of this type of work is Baedeker's experiment[1] on cuprous iodide, in which it was found that the conductiv-

TABLE XX

Substances in Which the Conductivity Increases with Increasing Vapor Pressure of the Electronegative Atom

Substance	Sign of Hall Coefficient
CuI	+
Cu_2S	
Cu_2O	+
NiO	
UO_2	
FeO	
CoO	

Substances in Which the Conductivity Decreases with Increasing Vapor Pressure of the Electronegative Atom

Substance	Sign of Hall Coefficient
Ag_2S	−
ZnO	−
CdO	−

Substances in Which the Conductivity Is Practically Unchanged with Increasing Vapor Pressure of the Electronegative Atom

CuO
Co_3O_4
Fe_3O_4
Fe_2O_3

ity of this substance increases with increasing iodine vapor pressure. The Hall coefficient, which is positive in this case, decreases at the same time, a fact indicating that the number of conducting particles increases with increasing iodine vapor pressure. Similar work[2] has been done on the effect of oxygen and sulfur vapor pressure upon the conductivity of oxide and sulfide semi-conductors. In some of these cases, the conductivity increases with increasing pressure of the electronegative element, and in others it decreases. Table XX contains a list of substances upon which investigations have been made and summarizes

[1] BAEDEKER, *op. cit.*

[2] A survey of work prior to 1935 is given by B. Gudden, *Ergebnisse exakt. Natur.*, **13**, 223 (1934). Additional references may be found in *Physik. Z.*, **36**, 717 (1935); **36**, 721 (1936).

qualitatively the results of this work.   Figures 76 and 77 show the dependence of conductivity[1] upon vapor pressure in the cases of cuprous oxide and cadmium oxide.   It may be observed that the conductivity is proportional to a power of the oxygen pressure in both these cases. This type of relationship occurs commonly.   It should be mentioned that we have chosen some of the "best" curves that are available in the

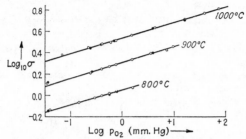

Fig. 76.—The dependence of the conductivity of cuprous oxide upon oxygen pressure
(After Dünwald and Wagner.)

literature and that all experimenters do not agree precisely upon the values of the conductivity of a given substance at a given temperature and vapor pressure.

The vapor pressure of the metallic constituent of a semi-conductor usually is not varied in these experiments.   Hilsch, Pohl,[2] and their coworkers, however, have placed alkali halide crystals in an atmosphere of alkali vapor for a long enough period of time to observe changes in conductivity.   The crystals become colored and exhibit a feeble electronic

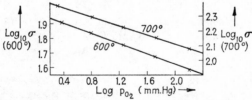

Fig. 77.—The dependence of the conductivity of cadmium oxide upon vapor pressure of
oxygen.   *(After Baumbach and Wagner.)*

conductivity at the end of this treatment.   This shows that even the most ideal ionic crystals may become semi-conductors under suitable conditions.

It should be added that the ionic conductivity of some semi-conductors (*cf.* Table XIX) seems to vary with the pressure of the electro-

---

[1] H. H. v. BAUMBACH and C. WAGNER, *Z. physik. Chem.*, **B22**, 208 (1933); H. DÜNWALD and C. WAGNER, *ibid.*, **22**, 214 *ff.* (1933).

[2] See the survey article by R. Pohl, *Proc. Phys. Soc. (Sup.)*, **49**, 3 (1937).

negative gas in the same way as the electronic conductivity. For example, Nagel and Wagner[1] have found evidence to support this in the case of cuprous oxide.

Schottky and Wagner have suggested that there is a correlation between the conductivity of semi-conductors that are sensitive to vapor pressure and the amount by which their composition deviates from ideal stoichiometric proportions. On this basis, they have developed a theory of semi-conductors that will be presented at appropriate places in the following chapters.

7. Molecular Crystals.—Of all the five solid types, we shall be least interested in molecular crystals. Since they are loosely bound aggregates of saturated atoms or molecules, many of their properties are determined primarily by the internal molecular structure, rather than by the solid binding, and thus they are outside the scope of this book. A number of substances that form molecular crystals are listed in Table XXI.[2] The prototypes of the class are the solids of the gaseous elements and of organic compounds that have low boiling points and low heats of sublimation. Several substances, such as sulfur, selenium, tellurium, phosphorus, and iodine, which are transition cases between molecular and valence types, are included as illustrations.

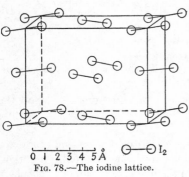

0 1 2 3 4 5Å   ⊖—⊖ I₂

FIG. 78.—The iodine lattice.

The rare gas solids crystallize in the face-centered cubic lattice, as may be seen in Table XXI. Helium forms a true solid only under a pressure of at least 25 atmospheres. The structure of this has not been determined.

Hydrogen, nitrogen, and oxygen have several phases which probably correspond to states in which the diatomic molecules have different relative orientations. All the high-temperature phases apparently are close-packed hexagonal structures of the diatomic molecules.

The solid phases of hydrochloric and hydrobromic acid are face-centered cubic lattices of the diatomic molecules above 98° and 110°K, respectively. Below these temperatures the structures are face-centered tetragonal. It is believed that the molecules are more randomly oriented in the high-temperature forms than in the low-temperature forms.

---

[1] See footnote 2, p. 70.

[2] See, for example, the compilations of data by Bichowsky and Rossini, *op. cit.*, and *Strukturberichte*.

The halogens, chlorine and iodine, have more complex structures in which the diatomic molecule behaves as a unit. The lattice of iodine is shown in Fig. 78.

Sulfur forms a lattice in which the units are S_8 molecules that have the ring structure shown in Fig. 79. The heat of sublimation relative to free S_8 molecules is about 20 kg cal/mol, a fact indicating that there are fairly large intermolecular forces. Selenium, tellurium, and phosphorus do not have typical molecular structures.

The crystals of organic molecules show a strong tendency to crystallize in simple structures. For example, carbon dioxide and methane

TABLE XXI.—DATA FOR MOLECULAR CRYSTALS

	Heat of sublimation, kg cal/mol	Structure
He	0.052	
Ne	0.52	
A	1.77	f.c.c.
Kr	2.67	
Xe	3.76	
H_2	2.44	There are several phases in each case. Apparently, the high-temperature phases are close-packed hexagonal arrangements of molecules
O_2	1.74	
N_2	1.50	
HCl	4.34	Two phases. The high-temperature phase is f.c.c. —the low-temperature, face-centered tetragonal
HBr	4.79	
Cl_2	6.0	Similar to Fig. 78
I_2	18.9	See Fig. 78
S_8	20	Lattice of S_8 molecules (see Fig. 79)
Se	30	Complex structures similar to those of arsenic, antimony, and bismuth
Te	25	
P	17.7	Complex valencelike structure
NH_3	6.29	Slightly distorted f.c.c. lattice
CO_2	8.24	Same as pyrites (Fig. 64)
CH_4	2.40	f.c.c. lattice

form face-centered cubic lattices of the constituent molecules. The ammonia lattice is very nearly the same, the difference being that the atoms at cube-corner and face-centered positions are slightly displaced relative to one another. This tendency toward comparatively simple arrangements extends even to very large molecules.

Practically all molecular crystals are diamagnetic.   This is in accord with the fact that the constituent molecules, being saturated, have no resultant spin.   Oxygen is an exception, since the normal state of the molecule is triplet.   Solid oxygen is strongly paramagnetic.

The specific-heat curves of many molecular crystals have large peaks.

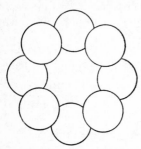

The typical cases of methane[1] and hydrochloric[2] acid are shown in Figs. 80 and 81.   It is believed that these peaks are associated with changes in degree of molecular orientation.

8. The Transition between the Solid Types. Figure 82 represents an attempt to show the interrelation of the solid types.   The metals are at the left, the two main classifications of these being monatomic metals and alloys. There is necessarily an abrupt transition be-

Fig. 79.—The unit ring molecule of sulfur. (*After Warren.*)

tween these two classes.   Valence and ionic types stand next to the right and are in one-to-one correspondence with the monatomic metals and the alloys.   The poorly conducting metals, such as bismuth, are transition cases between the ideal monatomic metals and the monatomic valence crystals diamond and boron.   Similarly, alloy

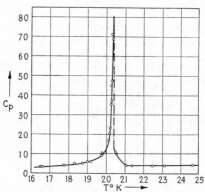

Fig. 80.—The specific heat of methane, showing the large peak at 20.4°K.   The ordinates are cal/deg-mol.   (*After Clusius and Perlick.*)

systems which have narrow phase boundaries, such as the antimony-magnesium system, are transition substances between alloys and ionic crystals.   In the same way, valence crystals, such as quartz

---

[1] K. Clusius and A. Perlick, *Z. physik. Chem.,* **24**, 313 (1924).

[2] A. Eucken and C. Karwat, *Z. physik. Chem.,* **112**, 467 (1924).   W. F. Giauque and R. Wiebe, *Jour. Am. Chem. Soc.,* **50**, 2193 (1928); **51**, 1441 (1929).

and carborundum, that have strong polar characteristics should be classed between ideal valence and ionic types. The molecular crystals are on the far right. The transition cases between these and valence and ionic crystals are substances such as sulfur and titanium oxide

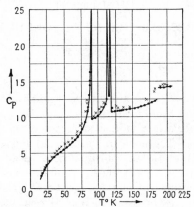

FIG. 81.—The specific-heat curve of hydrogen chloride. The ordinates are cal/deg-mol. (*After Giauque and Wiebe, and Eucken and Karwat.*)

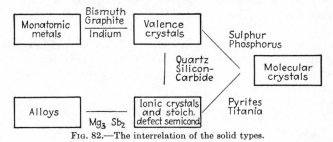

FIG. 82.—The interrelation of the solid types.

or pyrites, respectively, which are bound more tightly than molecular crystals but which show molecular coordination between atoms.

The several different types of semi-conductor cannot be fitted into this chart as a unit. Stoichiometrical defect or excess semi-conductors, such as zinc oxide, may be classed as ionic crystals. On the other hand, impurity semi-conductors, such as selenium, are transition solids between valence and molecular types that contain foreign atoms.

CHAPTER II

THE CLASSICAL THEORY OF IONIC CRYSTALS

9. Introduction.—The foundations of the classical theory of ionic crystals were laid about a quarter of a century ago by Madelung[1] and Born.[2] The basic concept of the theory is that the constituents of ionic crystals are positively charged metal atom ions and negatively charged electronegative atom ions. It is assumed that these ions are spherically symmetrical and that they interact with each other according to simple central force laws. The main interaction, according to the theory, is the ordinary electrostatic, or coulomb, force between the ions, which accounts for the large cohesive energies of the crystals. The electrostatic forces, which tend to contract the dimensions of the crystal, are balanced by repulsive forces which, from the classical viewpoint, have uncertain origin and which vary much more rapidly with interionic distance than do the coulomb forces between charges. Additional interactions are considered in the process of refinement and will be discussed later.

The repulsive term usually is chosen as a function of interionic distance that contains two adjustable parameters which are usually determined empirically by making the expression for the total energy satisfy the following two relations:

$$\left(\frac{dE}{dV}\right)_{V=V_0} = 0 \quad \text{and} \quad \left(\frac{d^2E}{dV^2}\right)_{V=V_0} = \frac{1}{V\beta}. \tag{1}$$

Here E is the energy of the crystal, V is its volume, and β is its compressibility.[3] These equations evidently express the conditions that the crystal should be at equilibrium under all forces and that the theoretical compressibility should be equal to the observed value.

As we shall see, the theory is remarkably successful in correlating many of the properties of ionic crystals. From a historical point of view, it may be said to form the basis for a quantitative understanding

---

[1] E. Madelung, *Gött. Nach.*, 100 (1909), 43 (1910); *Physik. Z.*, **11**, 898 (1910).

[2] A survey of Born's work may be found in the *Handbuch der Physik*, Vol. XXIV/2.

[3] The compressibility is defined by the relation

$$\beta = -\frac{1}{V}\left(\frac{\partial V}{\partial p}\right)_T.$$

Since $p = -\partial E/\partial V$, it follows that $1/V\beta = \partial^2 E/\partial V^2$.

of all solids, since it helped to distinguish between those facts which can be understood in terms of classical theory and those which cannot. We shall consider the theory here partly for this reason, and partly for the reason that the mathematical technique employed in it is of great value in more modern developments.

We shall now discuss in detail the various interaction terms employed in the classical theory. These will be considered under three headings: (1) electrostatic interaction, (2) repulsive interaction, and (3) multipole interaction.

10. Electrostatic Interaction Energy.—It is assumed in the classical ionic theory that the ions have charges corresponding to their normal chemical valence. Thus, sodium ions and chlorine ions in sodium chloride have, respectively, one electronic unit of positive and of negative charge, whereas magnesium ions and oxygen ions in magnesium oxide have two electronic units of positive and negative charge. According to electrostatic theory, the interaction energy of two nonoverlapping spherically symmetric charge distributions is

$$\frac{e_1 e_2}{r_{12}} \tag{1}$$

where e_1 and e_2 are the total charges on the distributions and r_{12} is the distance between their centers. Similarly, the total electrostatic energy E_e of n such charges of magnitude e_i $(i = 1, \cdots, n)$ is

$$E_e = \sum_{\text{pairs}} \frac{e_i e_j}{r_{ij}} \tag{2}$$

in which the summation extends over all pairs of charges, each pair being considered once. This also may be written in the form

$$E_e = \frac{1}{2} \sum_{i,j}' \frac{e_i e_j}{r_{ij}}, \tag{3}$$

where the summation is now a double sum over all charges and the superscript prime indicates that the cases $i = j$ are to be excluded. This convention will be used throughout this volume.

A detailed discussion of Ewald's[1] method of evaluating sums of type (3) for charges that are distributed in crystalline array may be found in the references of the footnote below. For two-atom crystals such as sodium chloride, cesium chloride, zinc sulfide, calcium fluoride, and aluminum oxide, the results may always be expressed in the simple form

$$E_e = -N_A A \frac{e_+ e_-}{l}. \tag{4}$$

---

[1] P. P. EWALD, Thesis (Munich, 1912); see also footnote 2, p. 76.

Here, E_e is the energy per mol; e_+ and e_- are the absolute values of the charges on the positive and negative ions, respectively; l is one of the characteristic crystal dimensions, such as the length of the cube edge, in cubic crystals, or the cation-anion distance; N_A is the Avogadro number; and A_l is the Madelung constant, which is characteristic of the type of crystal structure and is independent of the dimensions of the lattice. The numerical value of the Madelung constant evidently depends on the nature of the parameter l and on the units of charge, length, and energy that are used. Values of A_l are quoted in Table XXII for the types of ionic crystals that were discussed in Sec. 4. These values are given for three different choices of l: (1) when l is the closest cation-anion distance r_0; (2) when l is the cube root of the molecular volume δ_0; (3) when l is the length a_0 of the cube edge. The last case is significant only for cubic crystals.

TABLE XXII.—Madelung Constants for Some Typical Ionic Crystals
(The values in the three columns correspond to three different choices of lattice parameter:

r_0 cation-anion distance
δ_0 cube root of molecular volume
a_0 cube-edge length in cubic crystals.)

	A_{r_0}	A_{δ_0}	A_{a_0}
Sodium chloride	1.7476	2.2018	3.4951
Cesium chloride	1.7627	2.0354	2.0354
Zincblende (ZnS)	1.6381	2.3831	3.7829
Wurtzite (ZnS)	1.641	2.386	
Fluorite (CaF$_2$)	5.0388	7.3306	11.6366
Cuprite (CuO$_2$)	4.1155	6.5436	9.5044
Rutile (TiO$_2$)	4.816	7.70	
Anatase (TiO$_2$)	4.800	8.04	
Corundum (Al$_2$O$_3$)	25.0312	45.825	

It follows from Eq. (2) that the zero of energy is chosen in such a way that E_e vanishes when the ions are infinitely separated. Thus, the standard state is one of free ions. In the case of the alkali halides, $-E_e$ is about 10 per cent larger than the observed values of the cohesive energies U, referred to this standard state. For example, we have in the cases of sodium and cesium chloride

	E_e, kg cal/mol	U (obs.)
NaCl	205	182.8
CsCl	164	155.1

According to the classical picture, the energy associated with repulsive forces should account for the difference.

11. The Repulsive Term.—The repulsive force between ions is very small until the ions come into contact, whereupon it increases more rapidly than the electrostatic force. Two different forms of the repulsive interaction term have been considered in the course of the development of the theory. The first, which has no independent foundation, was introduced long before the advent of modern quantum theory; the second was introduced as a result of quantum theoretical treatments of ionic interaction. We shall now discuss both of these terms.

a. The b/r^n Interaction.—In his earlier work on ionic crystals, Born[1] assumed that the repulsive forces between ions gave rise to an interaction energy of the type

$$E_r = \frac{b}{r^n} \tag{1}$$

for the entire crystal. Here, b and n are constants and r is the distance between nearest unlike ions. If we assume that only nearest neighboring ions contribute to (1), this term implies that ions repel each other with a central force that varies as $1/r^{n+1}$.

In addition, Born assumed that the total energy of the crystal is simply the sum of the term (1) and an electrostatic term of the type (5) of the previous section. He then determined n and b by use of Eqs. (1) of the preceding section. We shall let E designate the energy of a mol of substance and shall express the molar volume V in terms of r by means of the equation

$$V = N_A \alpha r^3 \tag{2}$$

in which α is a constant that is characteristic of the type of lattice. Thus, the first two derivatives of E with respect to V are

$$\frac{dE}{dV} = \frac{1}{3N_A \alpha r^2} \frac{dE}{dr}, \tag{3}$$

$$\frac{d^2E}{dV^2} = \frac{1}{9N_A^2 \alpha^2 r^2} \frac{d}{dr}\left(\frac{1}{r^2} \frac{dE}{dr}\right). \tag{4}$$

Combining Eq. (1) with Eq. (4) of the previous section and solving Eq. (1) of Sec. 9 for b and n with the aid of Eqs. (3) and (4), we obtain

$$b = \frac{N_A A_{r_0} e_+ e_-}{n} r_0^{n-1}, \tag{5}$$

$$n = 1 + \frac{9\alpha r_0^4}{\beta e_+ e_- A_{r_0}}, \tag{6}$$

[1] See footnote 2, p. 76.

$$E = -\frac{N_A A_{r_0} e_+ e_-}{r_0}\left(1 - \frac{1}{n}\right). \qquad (7)$$

The values of n that are determined from Eq. (6) by use of experimental values of β generally lie in the vicinity of 9. They are somewhat smaller

TABLE XXIII.—VALUES OF n DERIVED FROM COMPRESSIBILITY MEASUREMENTS

Substance	n
LiF	5.9
LiCl	8.0
LiBr	8.7
NaCl	9.1
NaBr	9.5

TABLE XXIV.—THEORETICAL ENERGIES OF ALKALI HALIDE CRYSTALS DETERMINED BY USE OF THE b/r^n REPULSIVE POTENTIAL

Salt	n	U (calc.), kg cal/mol	U (exp.), kg cal/mol	Affinity, kg cal/mol
LiF	6.0	240.1	. . . . .	100.1
NaF	7.0	215.0	. . . . .	96.7
KF	8.0	190.4	. . . . .	95.3
RbF	8.5	181.8	. . . . .	98.0
CsF	9.5	172.8	. . . . .	98.8
LiCl	7.0	193.3	198.1	
NaCl	8.0	180.4	182.8	
KCl	9.0	164.4	164.4	
RbCl	9.5	158.9	160.5	
CsCl	10.5	148.9	155.1	
LiBr	7.5	183.1	189.3	
NaBr	8.5	171.7	173.3	
KBr	9.5	157.8	156.2	
RbBr	10.0	152.5	153.3	
CsBr	11.0	143.5	148.6	
LiI	8.5	170.7	181.1	
NaI	9.5	160.8	166.4	
KI	10.5	149.0	151.5	
RbI	11.0	144.2	149.0	
CsI	12.0	136.1	145.3	

than 9 for salts involving light ions and somewhat larger for salts involving heavy ions. Table XXIII contains several values that Slater[1] derived, using compressibilities that he obtained from his own measurements by extrapolation to absolute zero of temperature.

[1] J. C. SLATER, *Phys. Rev.*, **23**, 488 (1924).

Various workers have employed values of n that differ slightly from those derived from compressibilities. Thus, Lennard-Jones[1] has determined values from an investigation of the equation of state of rare gases, and Pauling[2] has used values obtained by rules that he derived from an approximate theoretical treatment of the interaction of closed-shell electronic configurations. It is clear from Eq. (7) that a change of n by unity affects E by 1 or 2 per cent. Since the various choices of n for a given substance differ by no more than 1, their differences are of minor consequence as far as the total energy is concerned.

Table XXIV contains a list of observed and calculated values of cohesive energies of alkali halide crystals. The theoretical values, which

TABLE XXV.—THEORETICAL ENERGIES OF OXIDES, SULFIDES, AND SELENIDES DETERMINED BY USE OF THE b/r^n REPULSIVE POTENTIAL

(The electron affinities of O, S, and Se have not been determined by direct experiment. The last column lists the values of these affinities which must be assumed if the calculated and observed results are to agree.)

Salt	n	U (calc.), kg cal/mol	Affinity, kg cal/mol
MgO	7.0	940.1	−175
CaO	8.0	842.1	−171
SrO	8.5	790.9	−160
BaO	9.5	747.0	−157
MgS	8.0	778.3	− 72
CaS	9.0	721.8	− 71
SrS	9.5	687.2	− 77
BaS	10.5	655.9	− 80
CaSe	9.5	698.8	− 94
SrSe	10.0	667.1	− 92
BaSe	11.0	637.1	− 97

were computed by Sherman,[3] include small corrections that allow them to be compared with room-temperature experimental values. Sherman used the values of n appearing in the second column that were determined by use of Pauling's rules. As we mentioned in the preceding paragraph, these differ only slightly from those of Table XXIII. Experimental values for the fluorides are not listed because the electron affinity of fluorine is not known. The last column, however, contains a list of

[1] J. E. LENNARD-JONES, Proc. Roy. Soc., **106**, 441, 463, 709 (1924); **109**, 476 (1925); **109**, 584 (1925).

[2] L. PAULING, Proc. Roy. Soc., **114**, 181 (1927); Jour. Am. Chem. Soc., **49**, 765 (1927); Z. Krist., **67**, 377 (1928).

[3] J. SHERMAN, Chem. Rev., **11**, 93 (1932).

electron affinities of fluorine that must be assumed in order to make the calculated and observed values agree. A similar compilation for several oxides, sulfides, and selenides is given in Table XXV. It should be noted that the values of affinities that must be assumed are negative in these cases.

Sherman used a set of affinities of negative ions that he derived from comparisons of theoretical and experimental cohesive energies of alkali halides and alkaline earth oxides, sulfides, and selenides in order to compare observed and computed energies of a number of other halides, oxides, sulfides, and selenides. Some of these cases are listed in Table XXVI. The agreement is very good for ideal ionic substances and is poor for substances such as cupric oxide.

b. The $ae^{-\frac{r}{\rho}}$ Interaction.—Investigations of interionic forces that have been carried out on the basis of quantum mechanics and that will be discussed in Chap. VII indicate that a repulsive term of the type b/r^n cannot be rigorously correct, although it may be a fair approximation for a short range of r. Born and Mayer[1] attempted to bring the older theory more nearly into accord with the quantum mechanical results by modifying it in two important ways: (1) They replaced the repulsion term (1) by an exponential expression of the type

$$\epsilon(r) = ae^{-\frac{r}{\rho}} \tag{8}$$

for the repulsive interaction of pairs of ions, in which a and ρ are constants. (2) They added another attractive term which is known as the van der Waals interaction. Mayer[2] subsequently modified this term and added another, as will be seen in the next section.

Born and Mayer found that they could take ρ in Eq. (8) as 0.345×10^{-8} cm for all types of ion if they determined a from the equation

$$a = b\left(1 + \frac{z_i}{n_i} + \frac{z_j}{n_j}\right)e^{\frac{r_i + r_j}{\rho}}. \tag{9}$$

Here, b is another fixed constant, z_i and z_j are the valences of two interacting ions, n_i and n_j are the numbers of valence electrons in the outer shells of the ions, and r_i and r_j are the ionic radii. The valences have negative signs for electronegative ions. n is equal to 8 for all simple ions except those which are isoelectronic with lithium, in which case n is 2. Born and Mayer used the ionic radii given in Table XXXII (page 93),

---

[1] M. BORN and J. E. MAYER, *Z. Physik*, **75**, 1 (1932); see paper following this as well.

[2] J. E. MAYER, *Jour. Chem. Phys.*, **1**, 270 (1933); see also J. E. MAYER and M. G. MAYER, *Phys. Rev.*, **43**, 605 (1933).

which were derived by Goldschmidt in the manner described in Sec. 14. Although Eq. (9) has some theoretical justification, its chief merit lies

Table XXVI.—Experimental and Theoretical Cohesive Energies of Halides Oxides, Sulfides, and Selenides Determined by Use of the b/r^n Repulsive Potential

(The experimental values are referred to a standard state of free ions. They involve the following electron affinities, obtained by Sherman in the manner described in the text:

F	98.5 kg cal/mol	O	−168 kg cal/mol
Cl	92.5 kg cal/mol	S	− 79.4 kg cal/mol
Br	87.1 kg cal/mol	Se	− 97 kg cal/mol
I	79.2 kg cal/mol		

The theoretical values are determined from the Born theory by use of the values of n which appear in the third column.)

Substance	Structure	n	U (exp.), kg cal/mol	U (theor.), kg cal/mol	Difference, kg cal/mol
AgF	NaCl	8.5	223.0	207.9	15.1
MgF$_2$	Rutile	7.0	688.3	696.8	− 8.5
CaF$_2$	Fluorite	8.0	618.0	617.7	+ 0.3
NiF$_2$	Rutile	8.0	713.2	697.1	16.1
CuCl	Zincblende	9.0	226.3	206.1	20.2
AgCl	NaCl	9.5	207.3	187.3	20.0
TlCl	CsCl	10.5	170.9	159.3	11.6
Li$_2$O	Fluorite	6.0	693	695	− 2
Cu$_2$O	Cuprite	8.0	788	682	106
Ag$_2$O	Cuprite	8.5	715	585	130
NiO	NaCl	8.0	966	968	− 2
ZnO	Wurtzite	8.0	972	977	− 5
PbO$_2$	Rutile	9.5	2,831	2,620	−211
Al$_2$O$_3$	Corundum	7.0	3,617	3,708	− 91
Na$_2$S	Fluorite	8.0	524	516	8
Cu$_2$S	Fluorite	9.0	683	612	71
ZnS	Wurtzite	9.0	851	816	35
ZnS	Sphalerite	9.0	851	819	32
PbS	NaCl	10.5	731	705	26
ZnSe	Sphalerite	9.5	845	790	55
PbSe	NaCl	11.0	735	684	51
Cu$_2$Se	Fluorite	9.5	685	599	86

in the fact that it leads to a good correlation between observed and calculated results without additional assumptions.

As an illustrative example, let us consider a crystal that contains two types of ion, each of which is surrounded by M like ions and M' unlike ions. If we consider only nearest like and unlike neighbors, the total repulsive interaction energy per mol is

$$E_r = N_A b \left[MC_{12}e^{\frac{r_1 + r_2 - r}{\rho}} + \frac{M'}{2}\left(C_{11}e^{\frac{2r_1}{\rho}} + C_{22}e^{\frac{2r_2}{\rho}}\right)e^{-\frac{a'r}{\rho}}\right] \qquad (10)$$

where C_{12}, C_{11}, and C_{22} are appropriate values of the quantity $[1 + (z_i/n_i) + (z_j/n_j)]$ in Eq. (9), r is the distance between like ions, and $a'r$ is the distance between unlike ones. If we desire to use an approximation corresponding to that of part (a), we should add the electrostatic term to this and apply Eqs. (1) of Sec. 9 to the sum. Instead, Born and Mayer included two additional terms which we shall discuss before proceeding further.

12. The Multipole Interactions and the Zero-point Energy. *a. Van der Waals Terms.*—It may be concluded[1] from the existence of solid and liquid phases of the rare gases that the constituent atoms have a small but finite attraction for one another. London investigated this attraction on the basis of quantum mechanics and was able to derive a simple approximate expression for the interaction energy. We shall use the results here and discuss the derivation in Sec. 58. It turns out that any two atoms or ions have, in addition to the terms of the type discussed in Secs. 10 and 11, an attractive energy term of the approximate form.

$$\epsilon_v = -\frac{3h}{2r^6} \cdot \frac{\nu_1\nu_2}{\nu_1 + \nu_2}\alpha_1\alpha_2 \qquad (1)$$

where ν_1 and ν_2 are the series limit frequencies of the discrete spectra of the two atoms or ions, and α_1 and α_2 are their polarizabilities. More accurate expressions have been developed for particular cases, but this expression is very useful for a general discussion. It will be shown in Sec. 58 that this attraction is connected with a synchronization of the motion of the electrons in the two atoms or ions.

The total van der Waals energy for a lattice of the same type as that which we considered in deriving Eq. (10) of the preceding section is

$$E_v = -\frac{N_A}{r^6}\left[S_v a_{12} + S_v'\frac{a_{11} + a_{22}}{2}\right] \qquad (2)$$

where a_{12}, a_{11}, and a_{22} are, respectively, the coefficients of $1/r^6$ in Eq. (1) for the positive-negative, positive-positive, and negative-negative ion pairs. S_v and S_v' are, respectively, the sums over all unlike and like ions of $1/R_i^6$ where R_i is the distance between a given ion and the ith ion in a lattice in which $r_0 = 1$. Numerical values of these rapidly convergent sums are given in Table XXVII for several types of structure.

<hr>

[1] LENNARD-JONES, *op. cit.*

In their original work, Born and Mayer evaluated the a that occur in (2) by means of London's equation (1), using values of α and ν that were determined from data on free ions. The results represent only about 1 per cent of the total energy and do not appreciably affect the order of magnitude of agreement between theoretical and experimental

TABLE XXVII.—COEFFICIENTS FOR THE VAN DER WAALS AND FOR THE DIPOLE-QUADRUPOLE ENERGIES

Lattice type	S_v	S'_v	S_M	S'_M
Sodium chloride...................	6.5952	1.8067	6.1457	0.8002
Cesium chloride...................	8.7088	3.5445	8.2007	2.1476
Zincblende.......................	4.354	0.762	4.104	0.253

values. These terms, however, are of considerable importance in determining the relative stability of two different lattice types, for they are comparable with the difference in energy of possible modifications. From a study of allotropy, Mayer[1] concluded that the α and ν values for free ions, when used in Eq. (1), do not account for the fact that CsCl, CsBr and CsI form simple cubic lattices instead of face-centered ones. He proposed that "better" values of these quantities should be determined from measurements of the optical properties of crystals by a method which we shall describe in Sec. 150.   These values nearly double

TABLE XXVIII.—COMPARISON OF VALUES OF THE CONSTANT C APPEARING IN THE DIPOLE-DIPOLE TERM FOR THE VAN DER WAALS ENERGY C/r^6
(In 10^{-60} erg-cm^6)

	Atomic data	Mayer's method
LiF	6	18
LiBr	102	183
NaCl	100	180
NaI	247	482
RbCl	384	691
CsBr	1,258	2,070

the van der Waals energy. Table XXVIII contains several values of the coefficient of $1/r^6$ that were determined by using the two types of data. We shall employ the energies given by Mayer's method in the following sections of this chapter. It should be emphasized, however, that in doing so we are departing somewhat from the original concepts of the ionic theory. It probably is true that the treatment of ionic crystals that would reproduce Mayer's values of α and ν from the funda-

---

[1] J. E. MAYER, *Jour. Chem. Phys.*, **1**, 270 (1933); **1**, 327 (1933).

mental equations of quantum mechanics would not substantiate all details of the classical picture.

b. *The Dipole-quadrupole Term.*—The energy term (2), which varies with the inverse sixth power of the distance, actually is the first term in an infinite series of terms that successively decrease in importance. The next term, which was first investigated by Margenau,[1] varies as $1/r^8$ and usually is about one-tenth as large as the van der Waals term at the observed interionic distance. We shall discuss the derivation of an expression for this interaction that is analogous to the derivation of (1) in Sec. 58. For present purposes, it is sufficient to point out that the contribution to the total energy of the crystal from this term is

$$E_M = -\frac{N_A}{r^8}\left(S_M d_{12} + S_M' \frac{d_{11} + d_{22}}{2}\right) \tag{3}$$

in which the d are analogous to the a in Eq. (2) and S_M and S_M' are sums of $1/R^8$ analogous to the sums in Eq. (2). Table XXVII contains values of S_M for several types of crystal. We shall use values of d that were derived by Mayer from empirical measurements [see part (a) of this section].

The energy term (3) is called the dipole-quadrupole interaction because it may be interpreted as arising from the interaction of a dipole moment on one atom with a quadrupole moment on the other. For the same reason, the van der Waals term is sometimes called the dipole-dipole term. The next term in the sequence, which varies as $1/r^{10}$, called the quadrupole-quadrupole term and is negligible in all cases in which we are interested.

c. *Zero-point Energy.*—The mechanical motion of the atoms or ions in a crystal containing N atoms may be treated as though the crystal were an assembly of $3N$ oscillators of various frequencies (*cf.* Sec. 22). According to quantum mechanics, an oscillator of frequency ν retains an energy of $h\nu/2$ at the absolute zero of temperature, which must be included along with the other energy terms when a comparison of experimental and theoretical cohesive energies is made. For present purposes, it is sufficiently accurate to use the Debye theory, according to which the frequencies are distributed between zero and a maximum ν_m in the manner described by the density function

$$f(\nu) = \frac{9N}{\nu_m^3}\nu^2. \tag{4}$$

Here, $f(\nu)d\nu$ is the number of oscillators having frequencies in the range from ν to $\nu + d\nu$. Thus, the total zero-point energy is

---

[1] H. Margenau, *Phys. Rev.*, **38**, 747 (1931). See also *Rev. Modern Phys.*, **11**, 1 (1939).

$$\int_0^{\nu_m} f(\nu)\frac{h\nu}{2}d\nu = \frac{9}{8}Nh\nu_m, \tag{5}$$

and the energy per mol of a diatomic crystal is

$$E_{h\nu} = N_A\frac{9}{4}h\nu_m. \tag{6}$$

The maximum frequency may be estimated to a sufficient degree of accuracy by using infrared absorption data, since the zero-point energy actually is small.

Following Mayer, we shall assume that the total energy of the crystal at absolute zero of temperature is given by the sum of the terms (2), (3), (6) of this section and Eqs. (4) and (10) of Secs. 10 and 11, respectively. This sum is

$$E_t = N_A\left\{-A\frac{e^2}{r} + b\left[MC_{12}e^{\frac{r_1+r_2-r}{\rho}} + \frac{M'}{2}(C_{11}e^{\frac{2r_1}{\rho}} + C_{22}e^{\frac{2r_2}{\rho}})e^{-\frac{a'r}{\rho}}\right] - \frac{1}{r^6}\left[S_v a_{12} + S'_v\frac{a_{11}+a_{22}}{2}\right] - \frac{1}{r^8}\left[S_M d_{12} + S'_M\frac{d_{11}+d_{22}}{2}\right] + \frac{9}{4}h\nu_m\right\}. \tag{7}$$

If we eliminate b by means of the first of Eqs. (1) of Sec. 9, we obtain

$$E_t = N_A\left[-A_{r_0}\frac{e^2}{r_0} + (1-k)\frac{\rho}{r_0}\left(A_{r_0}\frac{e^2}{r_0} + \frac{6C}{r_0^6} + \frac{8D}{r_0^8}\right) - \frac{C}{r_0^6} - \frac{D}{r_0^8} + \frac{9}{4}h\nu_m\right] \tag{8}$$

where

$$k = \frac{(a'-1)M'C_{22}\left(1 + \frac{C_{11}}{C_{22}}e^{-\frac{2\delta}{\rho}}\right)e^{\left(\frac{\delta}{r_0}+1-a'\right)\frac{r_0}{\rho}}}{2MC_{12} + a'M'C_{22}\left(1 + \frac{C_{11}}{C_{22}}e^{-\frac{2\delta}{\rho}}\right)e^{\left(\frac{\delta}{r_0}+1-a'\right)\frac{r_0}{\rho}}},$$

$$C = S_v a_{12} + S'_v\frac{a_{11}+a_{22}}{2},$$

$$D = S_M d_{12} + S'_M\frac{d_{11}+d_{22}}{2},$$

and δ is the difference of the ionic radii. The parameter ρ may be determined from E_t by means of the second of Eqs. (1), Sec. 9. Born and Mayer actually employed a slightly modified form of this equation which allowed them to use room-temperature values of β.

Table XXIX contains a compilation of the values of the five constituent terms of E_t for a number of halides. In addition, the sum of these terms is compared with the energies computed by Sherman (*cf.* Sec. 11). Born and Mayer did not use the dipole-quadrupole term E_M

and the empirical van der Waals term when they derived the values of the repulsive term for the alkali halides that appear in this table. They completely neglected the first term and used a smaller value for the second (*cf.* part *b*, Sec. 11). We have subsequently added the E_M and empirical E_v terms, but we have not corrected for the change that this induces in the repulsive term. This change is practically negligible as long as we are interested only in a comparison of E_t with Sherman's values.

TABLE XXIX.—CONTRIBUTIONS TO THE COHESIVE ENERGIES OF THE ALKALI HALIDES
AS GIVEN BY THE BORN-MAYER EQUATION
(In kg cal/mol)

	Madel-ung	Repul-sive	Dipole-dipole	Dipole-quad-rupole	Zero point	Total	Sherman
LiF	285.5	−44.1	3.9	0.6	−3.9	242.0	240.1
LiCl	223.5	−26.8	5.8	0.1	−2.4	200.2	193.3
LiBr	207.8	−22.5	5.9	0.1	−1.6	189.7	183.1
LiI	188.8	−18.3	6.8	0.1	−1.2	176.2	170.7
NaF	248.1	−35.3	4.5	0.1	−2.9	214.5	215.0
NaCl	204.3	−23.5	5.2	0.1	−1.7	184.4	180.4
NaBr	192.9	−20.6	5.5	0.1	−1.4	176.5	171.7
NaI	178.0	−17.1	6.3	0.1	−1.2	166.1	160.8
KF	215.1	−28.1	6.9	0.1	−2.2	191.8	190.4
KCl	183.2	−21.5	7.1	0.1	−1.4	167.5	164.4
KBr	174.5	−18.6	6.9	0.1	−1.2	161.7	157.8
KI	162.8	−15.9	7.1	0.1	−1.0	153.1	149.0
RbF	203.8	−26.2	7.9	0.1	−1.4	184.2	181.8
RbCl	175.8	−19.9	7.9	0.1	−1.2	162.7	158.9
RbBr	167.2	−17.6	7.9	0.1	−0.9	156.7	152.5
RbI	156.5	−15.4	7.9	0.1	−0.7	148.4	144.2
CsF	191.1	−23.9	9.7	0.1	−1.2	175.8	172.8
CsCl	162.5	−17.7	11.7	0.1	−1.0	155.6	148.9
CsBr	155.8	−16.4	11.4	0.1	−0.9	150.0	143.5
CsI	146.8	−14.6	11.1	0.1	−0.7	142.7	136.1

In computing the energy of the alkaline earth oxides and sulfides by means of Eq. (8), Mayer and Maltbie[1] used the mean of the values of ρ that were determined from the alkali halides. They estimated the van der Waals term from London's formula and omitted the dipole-quadrupole term. The results of this work are given in Table XXX and may

[1] J. E. MAYER and M. McC. MALTBIE, *Z. Physik,* **75,** 748 (1932).

be compared with Sherman's values in Table XXV. The affinities of oxygen and sulfur, determined from these newer computations, show a wider spread than those determined by Sherman.

TABLE XXX.—CRYSTAL ENERGIES DERIVED FROM THE BORN-MAYER EQUATION

Substance	Structure	$-E_l$, kg cal/mol	Affinity, kg cal/mol
MgO	NaCl	939	−190
CaO	NaCl	831	−165
SrO	NaCl	766	−144
BaO	NaCl	727	−147
MgS	NaCl	800	− 98
CaS	NaCl	737	−102
SrS	NaCl	686	− 85
BaS	NaCl	647	− 83

Bleick[1] has determined the lattice energy of the ammonium halides by the use of the Born-Mayer equation, treating the ammonium radical ion NH_4^+ as though it were a centrally symmetric unit. He obtained good agreement with measured cohesive energies in this way.

13. The Relative Stability of Different Lattice Types.—One of the important problems of crystal physics is that of determining the relative stability of different types of structure. This problem has a simple formal answer if questions of unstable equilibrium are neglected, for the stable lattice is that which has the lowest free energy. Unfortunately, the differences in free energies of different possible modifications may be extremely small; indeed, they are often less than the computational accuracy of the free energies. This fact is usually true, for example, of the differences in free energy of the four characteristic types of ionic structure (*cf.* Sec. 4). The free energies of different structures often are computed and compared in spite of this, the hope in such cases being that the computational errors lie in the same direction in each instance and cancel.

Mayer,[2] and Mayer and Levy,[3] using the Born-Mayer equation, have examined the relative stability of several halides in each of the structures listed in Table XXXI. They made no attempt to include temperature dependence, and so their discussion is valid only for the absolute zero of temperature. As they changed from the observed lattice type to a hypothetical one, they evaluated the differences of each of the constituent energy terms in Eq. (7) of Sec. 12, keeping r_0 fixed and equal to the

[1] W. E. BLEICK, *Jour. Chem. Phys.*, **2**, 160 (1934).
[2] J. E. MAYER, *Jour. Chem. Phys.*, **1**, 270 (1933); **1**, 327 (1933).
[3] J. E. MAYER and R. B. LEVY, *Jour. Chem. Phys.*, **1**, 647 (1933).

value for the observed stable lattice. They then added to the sum of these terms the change in energy of the hypothetical lattice as its interionic parameter changed from r_0 to the observed equilibrium value r_0'. This second term is equal to

$$\left(\frac{\partial^2 E_t'}{\partial r^2}\right)_{r=r_0'} \cdot \frac{(r_0 - r_0')^2}{2}$$

where E_t' is the energy of the hypothetical lattice. The zero-point energy was assumed to be the same for the different modifications. Table XXXI contains a list of the differences of the energy for the parameter r_0, the correction for change in density, and the sum of these two quantities. The sum would always be positive if the results agreed with experiment.

TABLE XXXI.—THE RELATIVE ENERGIES OF DIFFERENT LATTICE TYPES FOR SEVERAL HALIDES

(The first four columns of numbers are the changes in each of the contributions to the energy when the nearest-neighbor distance is held constant. The fifth column is the energy change δE obtained when the lattice expands or contracts to the true equilibrium distance. In kg cal/mol)

| Stable lattice | Hypothetical lattice | Change for fixed r_0 | | | | δE | Total change |
		Madelung	Dipole-dipole	Dipole-quadrupole	Repulsive		
AgF	Sodium chloride	Zincblende	+14.7	+ 7.9	+1.3	−13.7	−1.9	+ 8.3
AgCl	Sodium chloride	Zincblende	+13.0	+ 9.7	+1.9	−13.8	−1.9	+ 8.9
AgBr	Sodium chloride	Zincblende	+12.5	+ 9.7	+1.6	−13.4	−1.9	+ 8.4
AgI	Zincblende	Sodium chloride	−12.8	−17.7	−3.7	+30.0	−1.9	− 6.1
AgF	Sodium chloride	Cesium chloride	− 2.0	− 9.7	−1.7	+16.7	−0.9	+ 2.4
AgCl	Sodium chloride	Cesium chloride	− 1.7	−11.8	−2.3	+24.7	−0.9	+ 8.0
AgBr	Sodium chloride	Cesium chloride	− 1.7	−11.6	−2.2	+28.7	−0.9	+12.3
TlCl	Cesium chloride	Sodium chloride	+ 1.4	+ 8.1	+1.3	−10.5	−0.6	− 0.3
TlBr	Cesium chloride	Sodium chloride	+ 1.4	+ 7.9	+1.2	−10.1	−0.6	− 0.2
TlI	Cesium chloride	Sodium chloride	+ 1.3	+ 8.2	+1.6	−10.1	−0.6	+ 0.4
CuCl	Zincblende	Sodium chloride	−15.4	−10.5	−1.9	+33.4	−1.9	+ 3.7
CuBr	Zincblende	Sodium chloride	−14.7	−11.0	−2.0	+34.3	−1.9	+ 4.7
CuI	Zincblende	Sodium chloride	−13.6	−13.6	−2.6	+34.0	−1.9	+ 2.3

It is not positive for AgI, TlCl, and TlBr, for in these cases, the sodium chloride type of structure, instead of the observed one, has the lowest energy. It may be seen from the table that the attractive terms favor the stability of the following three types of structure in the order given: CsCl, NaCl, and ZnS. Conversely, the repulsive term favors them in the reverse order. It would require only a small increase in one of the attractive terms, such as the van der Waals term, to shift the calculated stable lattice of TlCl and TlBr from the sodium chloride type to the cesium chloride type. On the other hand, it would require a comparatively large change of an opposite kind to account for the stability of the zincblende structure in AgI. Mayer concluded from these facts that thallium salts and most silver salts conform closely to the assumptions of the ionic theory and that AgI has some nonionic characteristics.

May[1] has made a very thorough investigation of the problem of the relative stability of the cesium chloride and sodium chloride lattice for cesium chloride on the basis of the Born-Mayer equation. He found that this equation does not account for the stability of the cesium chloride lattice at absolute zero of temperature if a two-parameter repulsive term and Mayer's value of E_v are employed. In order to generalize the equation, he introduced two additional parameters. One of these was a multiplicative factor in the van der Waals term and the other was a factor in M' of Eq. (7), Sec. 12. In effect, the second parameter makes the constant b in the repulsive term different for like and unlike ions. These parameters were adjusted in order to make the cesium chloride lattice stable at absolute zero. The additional measured quantities that were used in doing this are the measured heat of the phase transition (about 1.34 kg cal/mol), and the observed change in lattice constant. The multiplicative constant of the attractive terms turns out to be 3.6, and the coefficient of M' to be 0.70. At the same time, the constant b is doubled, and ρ changes from 0.290 to 0.365Å. May suggested that part of the required increase in the attractive terms should be associated with a change in the purely electrostatic energy caused by distortion of the ions from spherical form. It is easy to show that the distortion of the charge on an ion in a cubic crystal may be described in the first approximation by fourth-order surface harmonics and that the associated potential, which is not centrally symmetrical, varies with r as $1/r^9$. There is, however, no conclusive quantitative evidence to support May's suggestion.

May also carried through a similar treatment of the stability of ammonium chloride and found further support for the conclusions drawn from the case of cesium chloride. In addition to this, he estimated the transition temperature for both cesium chloride and ammonium chloride and obtained qualitative agreement with the observed facts.

In contrast with May's work is that of Jacobs,[2] who correlated the phase transitions that occur in the alkali halides at high pressures with empirically determined constants that are more nearly like those used in the earlier work of Born and Mayer.

14. Ionic Radii.—We saw, in Sec. 4, Chap. I, that the nearest ion distances of the alkali halide crystals are additive. This fact, which was

---

[1] A. MAY, *Phys. Rev.*, **52**, 339 (1937).
[2] R. B. JACOBS, *Phys. Rev.*, **54**, 468 (1938).

first emphasized by Goldschmidt,[1] suggests that the ions of these crystals may be regarded as rigid spheres and that the equilibrium interionic distance is simply the distance at which these spheres come into contact. This view is justified in a rough way by the Born theory, which shows that the repulsive forces between ions vary about ten times more rapidly with interionic distance than do the electrostatic forces. The rigid sphere concept, however, cannot be used very widely for quantitative purposes; we shall discuss some of its limitations in a later paragraph of this section.

It is not possible to obtain the values of ionic radii from crystallographic data without knowing at least one radius at the start. Gold-

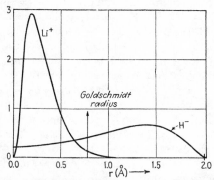

Fig. 1.—The charge distributions of Li$^+$ and H$^-$ ions as determined by Hylleraas and Bethe. The vertical line marks the point of contact of the Goldschmidt radii.

schmidt began his extensive work on the tabulation of radii with the following values of the radii of F$^-$ and O^{--}:

$$F^-: 1.33 \text{ A}; \qquad O^{--}: 1.32 \text{ Å}.$$

These values were determined by Wasastjernas[2] who correlated measurements on the optical properties of fluorides and oxides with a classical quantum theory of dispersion and thereby determined the radii of the outermost orbits of these ions. The theoretical basis of this work need not concern us here. We may, however, compare the value of the radius of Li$^+$ that is deduced from the value for F$^-$ with the electronic charge distribution of Li$^+$ as determined by modern quantum mechanics,[3] which is shown in Fig. 1. It is obvious that there is no precisely defined radius; however, the vertical line corresponding to Wasastjernas' radius is a reasonable point at which to say that the distribution stops. The right-

[1] V. M. Goldschmidt, *Skrifter det Norske Videnskaps* (1926, 1927).

[2] J. Wasastjernas, *Comm. Fenn.*, **1**, (1923); **6**, (1932).

[3] E. Hylleraas, *Z. Physik*, **54**, 347 (1929); H. Bethe, *Z. Physik*, **57**, 815 (1929).

hand part of Fig. 1 shows the distribution in H^-. The two curves are plotted in such a way that the distance between origins corresponds to the Li^+–H^- distance in lithium hydride. Table XXXII contains a

TABLE XXXII.—GOLDSCHMIDT'S IONIC RADII
(In Å)

H^-	1.27		O^{--}	1.32
F^-	1.33		S^{--}	1.74
Cl^-	1.81		Se^{--}	1.91
Br^-	1.96		Te^{--}	2.03
I^-	2.20			
Li^+	0.78		Mg^{++}	0.78
Na^+	0.98		Ca^{++}	1.06
K^+	1.33		Sr^{++}	1.27
Rb^+	1.49		Ba^{++}	1.43
Cs^+	1.65		Be^{++}	0.34
Tl^+	1.49		Zn^{++}	0.83
Cu^+	0.53		Cd^{++}	1.03
Ag^+	1.0		Hg^{++}	1.12
Mn^{++}	0.91		Al^{+++}	0.57
Fe^{++}	0.83		Se^{+++}	0.83
Co^{++}	0.82		Yt^{+++}	1.06
Ni^{++}	0.78		La^{+++}	1.22
Pb^{++}	1.32		Ga^{+++}	0.62
			In^{+++}	0.92
			Tl^{+++}	1.05

list of Goldschmidt's radii, which have been determined from additive systems. These values are self-consistent to within about 5 per cent.

Other radii have been obtained by other workers using different starting assumptions. For example, Huggins and Mayer[1] obtained the radii, given in Table XXXIII, for the alkali ions and halogen ions by adjusting

TABLE XXXIII.—THE RADII OF THE ALKALI METAL IONS AND THE HALOGEN IONS AS DETERMINED BY HUGGINS AND MAYER
(In Å)

Li^+	0.475		F^-	1.110
Na^+	0.875		Cl^-	1.475
K^+	1.185		Br^-	1.600
Rb^+	1.320		I^-	1.785
Cs^+	1.455			

the radii that appear in the Born-Mayer equation so that the theory would lead to the observed lattice constant; that is, b in Eq. (7) of Sec. 12 was given a fixed value and the ionic radii were adjusted so that Eq. (1) of Sec. 10 was satisfied for the observed lattice spacing. The values in Table XXXIII differ appreciably from Goldschmidt's values.

[1] M. L. HUGGINS and J. E. MAYER, *Jour. Chem. Phys.*, **1**, 643 (1933).

The cesium chloride type of structure has a larger electrostatic cohesive energy than the sodium chloride type for a given value of r_0, the nearest ion distance. In view of this fact, one might attempt to explain the stability of the sodium chloride type in most alkali halides on the basis of the rigid ion picture, by assuming that one type of ion is so large that like neighbors touch at greater values of r_0 in the cesium chloride structure than unlike ones do in the sodium chloride structure. A condition obviously necessary for this is that one set of ions should be so large that pairs touch in the cesium chloride structure for the observed interionic distance of the sodium chloride type of structure, that is, that the radii R_1 and R_2 should satisfy the conditions

$$R_1 \geq \frac{r_0}{\sqrt{3}}, \qquad R_2 \leq \left(1 - \frac{1}{\sqrt{3}}\right)r_0.$$

The ratios must then obey the inequality

$$\frac{R_2}{R_1} < \sqrt{3} - 1 = 0.73.$$

KF, RbF, RbCl, RbBr, and CsF do not satisfy this condition if Goldschmidt's radii are used; moreover, no reasonable and self-consistent change in radii would make all the ions in face-centered alkali halide crystals satisfy it.

In view of the work of the previous section, the failure of the simple rigid-sphere concept to cope with this problem is not surprising. The stability of an ionic crystal in any given structure is determined by a delicate balance of several types of force.

15. Implications of Deviations from the Cauchy-Poisson Relations.— In all the computational work that is discussed in the preceding sections of this chapter, it is explicitly assumed that the ions are spherically symmetrical and interact with spherically symmetrical forces. If this assumption were correct, the three elastic constants,[1] c_{11}, c_{12}, and c_{44}

---

[1] We shall use Voigt's notation for the elastic constants [see W. Voigt, *Lehrbuch der Kristallphysik* (Teubner, Leipzig, 1910; reissued 1928)]. If we designate the three tension components of the stress tensor by X_1, X_2, and X_3, the three shear components by X_4, X_5, and X_6, the three tension components of the strain tensor by x_1, x_2, and x_3, and the three shear components by x_4, x_5, and x_6, Hooke's stress-strain relation is given in terms of the thirty-six elastic constants c_{ij} ($i, j = 1, 2, \cdots, 6$) by the equations

$$X_i = \sum_{j=1}^{6} c_{ij} x_j.$$

If $E(x_1, \cdots, x_6)$ is the energy per unit volume of the crystal as a function of the

of a cubic crystal should satisfy the Cauchy-Poisson relation

$$c_{12} = c_{44}.$$

The measurements of Rose,[1] which are plotted in Fig. 2, show that this relation is not valid for sodium chloride at low temperatures, the

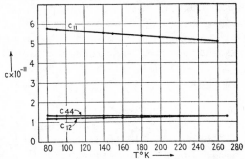

Fig. 2.—The three elastic constants of sodium chloride. c_{44} and c_{12} cross at 240°K. *(After Rose.)*

discrepancy being of the order of 10 per cent.   Similar measurements[2]

strains, it follows from the relation between force, work, and energy that

$$E(x_1, \cdots, x_6) = \frac{1}{2}\sum_{i,j} c_{ij}x_i x_j.$$

Thus,

$$c_{ij} = \frac{\partial^2 E}{\partial x_i \partial x_j}$$

and

$$c_{ij} = c_{ji};$$

hence, only twenty-one of the thirty-six c_{ij} are independent.   For cubic crystals, there are the additional relations.

$$c_{11} = c_{22} = c_{33}, \qquad c_{12} = c_{23} = c_{31}, \qquad c_{44} = c_{55} = c_{66}$$

and all other components are zero if the coordinate axes and cube axes are parallel; hence, there are only three independent components in this case.   In hexagonal crystals, there are the relations

$$c_{11} = c_{22}, \qquad c_{13} = c_{23}, \qquad c_{44} = c_{55}, \qquad c_{66} = \tfrac{1}{2}(c_{11} - c_{12}),$$

and all other components except c_{12} are zero, if the hexagonal axis is in the z direction. If the atoms of the crystal interact with central forces. there are the additional Cauchy-Poisson relations, which in the cases of cubic and hexagonal crystals are

$$c_{12} = c_{44}, \quad \text{and} \quad c_{12} = c_{44}, \quad c_{11} = \tfrac{3}{2}c_{12},$$

respectively.

[1] F. C. Rose, *Phys. Rev.*, **49**, 50 (1936).

[2] Data furnished by Quimby, Balamuth, and Rose on magnesium oxide appears in a paper by Barnes, Brattain, and Seitz, *Phys. Rev.*, **48**, 582 (1935).   See also the correction, *Phys. Rev.*, **49**, 405 (1936); M. A. Durand, *Phys. Rev.*, **50**, 449 (1936).

on magnesium oxide show a deviation between c_{12} and c_{44} of about 50 per cent. It is natural to ask how this discrepancy affects the Born theory.

The terms in the Born-Mayer equation may be divided into two classes, namely: the electrostatic term f_1, which varies slowly with r, and all others f_2, which vary rapidly in comparison. The first term is rigorously known, for although the charge distribution may not be precisely spherically symmetrical, the electrostatic-correction term arising from this distortion varies at least as rapidly as $1/r^9$ and may be classed in f_2. It follows that the deviations from the Cauchy-Poisson relations occur because some of the terms in f_2 should not be spherically symmetrical.

We have seen in the preceding work that the absolute magnitude of f_1 is about ten times that of f_2 for equilibrium values of r, even though the force terms $\partial f/\partial r$ are equal. The increase in importance of f_2 as we pass from a consideration of energies to one of forces is an immediate consequence of the fact that f_2 varies more rapidly with r than does f_1. This fact is shown in a striking way by the older form of the repulsive term b/r_0^n, where n is of the order of magnitude 10. The ratio of this term to the electrostatic term is

$$\frac{b}{A r^{n-1}} = \frac{1}{n},$$

whereas the ratio of the derivatives is n times this. Similarly, f_2'' is about ten times larger than f_1'', whence we may conclude that the elastic constants, which are related to second derivatives of the energy, are primarily determined by f_2. Thus, the experimental measurements of elastic constants show that f_2 is in error by about 10 per cent in sodium chloride and about 50 per cent in magnesium oxide. This error affects the cohesive energy by only 1 per cent in the first case and 5 per cent in the second, because f_2 accounts for only about 10 per cent of the cohesive energy. The energy differences of different crystallographic forms, however, are of the same order of magnitude as this error. Hence, it seems safe to say that these energies can be computed accurately only when the nonradial part of f_2 is properly included in the Born-Mayer equation.

16. Surface Energy.—One of the factors that determines whether or not a given crystallographic plane is a cleavage plane is its surface energy, that is, the energy σ per unit area required to separate the crystal along this plane. This energy usually is defined in such a way that the areas of the two halves of the crystal are counted separately. Surface energies of crystals with the sodium chloride lattice have been computed on the basis of the ionic theory by Madelung,[1] Born and Stern,[2] and Yamada.[3] Madelung took into account only electrostatic forces and found that the energy of a (100) surface is $0.520e^2/a^3$ per unit area, where a is the cube-edge distance of the unit cell.   Born and Stern improved this calculation by including repulsive forces and found the following values for the (100) and (110) planes:

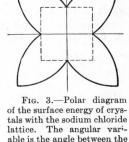

$$\left.\begin{aligned} \sigma_{100} &= 0.116\frac{e^2}{a^3}, \\ \sigma_{110} &= 0.315\frac{e^2}{a^3}. \end{aligned}\right\} \tag{1}$$

Since the second is about 2.7 times larger than the first, it follows that rock salt should split more easily along a (100) plane than along a (110) plane.   This result is in agreement with experiment, since only (100) planes show cleavage.   In both the preceding calculations, it was assumed that the ionic spacing near the surface is the same as in the interior of the lattice.   Madelung, however, has shown that, in the case of (100) surfaces, there should be a slight contraction of interionic distance in the direction normal to the (100) plane.

Fig. 3.—Polar diagram of the surface energy of crystals with the sodium chloride lattice. The angular variable is the angle between the normal to the surface plane and the (100) direction. Although this plot is valid only for planes whose normals lie in a principal plane, the minimum in the (100) direction is an absolute minimum. The dotted square represents the crystal form of lowest surface energy, as determined by Wulff's method. (*After Yamada.*)

Yamada extended Born and Stern's computations by calculating σ for all surface planes that have normals lying in the (100) plane. This includes as special cases planes that are equivalent to the (100) and (110) surfaces.   His values of σ are shown in the polar diagram of Fig. 3, in which the angular variable is the angle between the (100) direction and the normal to the plane.   This figure shows that the value of σ for the (100) plane is a relative minimum, whereas that for the (110) plane is a relative maximum.   Yamada also found that the surface energy of the (100) plane is smaller than that of any other plane, so that the (100) value is also an absolute minimum for the three-dimensional σ surface.

Wulff[4] has proved that the crystal form corresponding to the lowest surface energy may be obtained from a polar diagram of this type by taking the envelope of those planes that are orthogonal to lines passing through the origin at the points where these lines intersect the surface.

[1] E. Madelung, *Physik. Z.*, **19**, 524 (1918); **20**, 494 (1919).

[2] M. Born and O. Stern, *Sitzb. preuss. Akad. Wiss.*, 901 (1919).

[3] M. Yamada, *Physik. Z.*, **24**, 364 (1923); **25**, 52 (1924).

[4] G. Wulff, *Z. Krist.*, **34**, 449 (1901).

It is clear that this form is the dotted square of Fig. 3 in the two-dimensional case corresponding to Yamada's result for the (100) plane and is a cube in the three-dimensional case. In order to obtain the condition for thermodynamical equilibrium above absolute zero, Wulff's construction should be applied to the free-energy polar diagram instead of to the energy diagram.

If ionic crystals were perfect, it should be possible to estimate their breaking strength from the surface energy. Following Polanyi,[1] we shall assume that the interatomic forces between two separating crystal planes are important for a distance Δl of about ten angstrom units and shall determine a rough value of the breaking stress S by use of the energy equation

$$2\sigma \cong \Delta l S \tag{2}$$

where σ is the surface energy per unit area of the crystal. The value of σ for sodium chloride that is given by the first of the equations (1) is about 147 ergs cm^{-2}, which corresponds to a value of S of about 300×10^7 dynes cm^{-1}. The values ordinarily observed are about one hundred times smaller than this. Zwicky[2] carried through a more exact calculation of S on the basis of the ionic theory and found the value

$$S = 2{,}000 \times 10^7 \text{ dyne-cm}^{-2},$$

which is in even worse agreement with the measured values. Since measurements of the surface tension of molten salts lead to values of σ that are in comparative agreement with the theoretical values, we may not conclude that the computational methods for a perfect lattice are badly in error.

At the present time, it is believed that these discrepancies between observed and calculated breaking strengths arise from lattice imperfections. This type of explanation was first proposed by Griffith,[3] who suggested that the weak spots are tiny surface cracks. A more complete theory of plastic flow and related topics has been developed by Taylor and others. However, we shall refer the reader to other sources[5] for a discussion of this work.

[1] M. Polanyi, *Z. Physik*, **7**, 323 (1921).

[2] F. Zwicky, *Physik. Z.*, **24**, 131 (1923).

[3] A. A. Griffith, *Phil. Trans. Roy. Soc.*, **221**, 163 (1920).

[4] G. I. Taylor, *Trans. Faraday Soc.*, **24**, 121 (1928); *Proc. Roy. Soc.*, **145**, 362, 388 (1934).

[5] See E. Schmid and W. Boas, *Kristallplastizität* (Julius Springer, Berlin, 1935); C. F. Elam, *The Distortion of Metal Crystals* (Oxford University Press, 1935); also the report of the Bristol Conference on this topic, *Proc. Phys. Soc.*, January, 1940.

CHAPTER III

THE SPECIFIC HEATS OF SIMPLE SOLIDS

17. Introduction.—It was shown in Chap. I that the specific heats of practically all simple solids approach zero monotonically as the temperature approaches absolute zero. Since this fact cannot be explained satisfactorily on the basis of classical mechanics, Einstein's qualitative interpretation[1] on the basis of quantum theory may be regarded as one of the first triumphs of the theory. Einstein postulated, as had been done previously, that a simple crystal could be regarded as an aggregate of atomic oscillators, all of which vibrate with the same natural frequency. In addition, he assumed that the allowed energy states of these oscillators are integer multiples of $h\nu$, where ν is the frequency of oscillation and h is Planck's constant. In classical mechanics, it would have been assumed that the allowed energy states are continuous; and this assumption, according to classical statistical mechanics, would mean that Dulong and Petit's law should be valid at all temperatures. Einstein found upon applying Boltzmann's theorem to his postulated assemblage of quantum oscillators that the observed decrease in specific heat could be explained qualitatively.

A few years later, Debye[2] showed that most of the quantitative discrepancy between Einstein's results and observations could be removed by the introduction of a more accurate analysis of the modes of vibration of a simple solid. In short, he improved upon Einstein's assumption that all atoms oscillate with the same frequency. One outstanding result of Debye's work is the prediction that the specific heat should approach zero as T^3 near the absolute zero of temperature. This law is accurate for a large range of temperature in many cases; in others, it is not accurate.

Born and von Kármán[3] developed a method for computing the modes of vibration of a solid about the same time as did Debye. Although their method actually is more accurate than Debye's, since it is based upon fairly rigorous principles of atomic dynamics, Debye's results agreed so well with available experimental material that his theory was considered sufficiently accurate for practical purposes. In recent years,

---

[1] A. EINSTEIN, *Ann. Physik*, **22,** 180, 800 (1906); **34,** 170 (1911).

[2] P. DEBYE, *Ann. Physik*, **39,** 789 (1912).

[3] M. BORN and Th. VON KÁRMÁN, *Physik. Z.*, **13,** 297 (1912); **14,** 15 (1913).

however, a number of striking deviations from Debye's theory have been found in the low-temperature range where it might be expected to be most accurate.   For this reason, Blackman[1] reopened the question of determining the distribution of frequencies, and he found that Born and von Kármán's method gives something more nearly like the observed specific-heat curves.   Blackman's work is essentially qualitative, however, and it remains to be seen how well the low-temperature anomalies actually can be explained.

The three stages of development of the vibrational theory of specific heats, namely, the theory of Einstein, of Debye, and of Born, von Kármán, and Blackman, may be included in an approximation in which it is assumed that interatomic forces obey Hooke's law.   We shall treat specific heats of solids by this method in the present chapter.

It is believed that the details of specific-heat curves that cannot be explained by the Hooke's law approximation have two origins, namely, anharmonic interaction terms and thermal excitation of electrons. These topics will be discussed in later chapters.   Anharmonic interaction terms supposedly account for the following facts: (a) the anomalous peaks in the specific-heat curves of molecular crystals, such as solid methane, and of ionic crystals, such as ammonium chloride; and (b) a part of the high-temperature deviations from Dulong and Petit's law. Electronic interaction is believed to account for: (a) the linear temperature dependence of the specific heats of some metals near absolute zero, (b) the anomalous peaks in the specific heats of ferromagnetic metals and paramagnetic salts, and (c) a part of the high-temperature deviation from Dulong and Petit's law—particularly that of transition metals.

18. The Energy of an Assembly of Oscillators.—It will be shown in Sec. 22 that a crystal which contains N atoms that interact according to Hooke's law is mechanically equivalent to a set of $3N$ independent oscillators.   Thus, the mean total energy $\bar{E}$ of such a crystal is equal to the sum of the mean energies of the individual oscillators.   We shall derive the expression for the mean energy of an oscillator by assuming discrete energy levels in accordance with quantum theory.

The relative probability[2] of finding in its ith level a system that has levels of energy ϵ_i and degeneracy g_i is

$$r_i = g_i e^{-\frac{\epsilon_i}{kT}}. \tag{1}$$

[1] M. Blackman, *Z. Physik*, **86**, 421 (1933); *Proc. Roy. Soc.*, **148**, 384 (1935), **159**, 416 (1937); *Proc. Cambridge Phil. Soc.*, **33**, 94 (1937).

[2] For a discussion of Boltzmann's theorem, see various books on statistical mechanics such as: E. H. Kennard, *Kinetic Theory of Gases* (McGraw-Hill Book Company, Inc., New York, 1938); L. Brillouin, *Die Quantenstatistik* (Julius Springer, Berlin, 1931).

Using this, we find that the absolute probability p_i is

$$p_i = \frac{g_i e^{-\frac{\epsilon_i}{kT}}}{\sum_j g_j e^{-\frac{\epsilon_j}{kT}}} \tag{2}$$

where j is summed over all levels. The mean energy $\bar{\epsilon}$ of the system obviously is

$$\bar{\epsilon} = \sum_i p_i \epsilon_i = \frac{\sum_i \epsilon_i g_i e^{-\frac{\epsilon_i}{kT}}}{\sum_j g_j e^{-\frac{\epsilon_j}{kT}}}, \tag{3}$$

which is identical with the expression

$$\bar{\epsilon} = -\frac{d}{d(1/kT)} \log \left(\sum_i g_i e^{-\frac{\epsilon_i}{kT}} \right). \tag{4}$$

We shall call the sum

$$\sum_i g_i e^{-\frac{\epsilon_i}{kT}}$$

the partition function and shall designate it by f.

Let us evaluate Eq. (3) for a harmonic oscillator of which the energy levels, in accordance with quantum theory, are given by

$$\epsilon = nh\nu \tag{5}$$

where n takes all integer values and ν is the natural frequency of the oscillator. The levels are not degenerate in this case, whence g_i is unity. Thus, the partition function for the system is

$$f = \sum_{n=0}^{\infty} e^{-\frac{nh\nu}{kT}} = \sum_{n=0}^{\infty} (e^{-\frac{h\nu}{kT}})^n$$
$$= \frac{1}{1 - e^{-\frac{h\nu}{kT}}}. \tag{6}$$

According to Eq. (4), the mean energy is

$$\bar{\epsilon} = \frac{d}{d(1/kT)} \log (1 - e^{-\frac{h\nu}{kT}}) = \frac{h\nu e^{-\frac{h\nu}{kT}}}{1 - e^{-\frac{h\nu}{kT}}}$$
$$= \frac{h\nu}{e^{\frac{h\nu}{kT}} - 1}, \tag{7}$$

whereas the mean total energy of an assembly of $3N$ oscillators of different frequencies ν_i ($i = 1, 2, \cdots, 3N$) is

$$\bar{E} = \sum_{i=1}^{3N} \frac{h\nu_i}{e^{\frac{h\nu_i}{kT}} - 1}. \tag{8}$$

This sum may be replaced by an integral if the distribution of frequencies can be represented by an integrable function. In this case, which occurs commonly, as will be seen later,

$$\bar{E} = \int_0^{\nu_m} \frac{h\nu}{e^{\frac{h\nu}{kT}} - 1} q(\nu)d\nu \tag{9}$$

where $q(\nu)d\nu$ is the number of oscillators in the frequency range from ν to $\nu + d\nu$, and ν_m is the maximum frequency of any oscillator. Obviously, we may expect $q(\nu)$ to satisfy the equation

$$\int_0^{\nu_m} q(\nu)d\nu = 3N. \tag{10}$$

The derivative of Eq. (8) or (9) with respect to T is the molar heat C_V in the case in which N is the number of atoms in a mol of substance N_m. We have

$$C_V = \sum_{i=1}^{3N_m} k\left(\frac{h\nu_i}{kT}\right)^2 \frac{e^{\frac{h\nu_i}{kT}}}{(e^{\frac{h\nu_i}{kT}} - 1)^2} \tag{11}$$

in the general case, and we have

$$C_V = \int_0^{\nu_m} k\left(\frac{h\nu}{kT}\right)^2 \frac{e^{\frac{h\nu}{kT}}}{(e^{\frac{h\nu}{kT}} - 1)^2} q(\nu)d\nu, \tag{11a}$$

when Eq. (9) may be used. The summand in (11), namely,

$$k\left(\frac{h\nu}{kT}\right)^2 \frac{e^{\frac{h\nu}{kT}}}{(e^{\frac{h\nu}{kT}} - 1)^2}, \tag{12}$$

increases monotonically with temperature. It is zero at absolute zero; and as T approaches infinity, it approaches the constant value k in the manner indicated in Fig. 1. It should be clear that, with positive coefficients, any sum or integral of such a monotonically increasing function also must increase monotonically. Therefore, we may conclude that C_V is an increasing function of temperature regardless of the frequency distribution. This conclusion is valid only when the interatomic forces obey Hooke's law and when electrons do not contribute to the specific heat. Since (12) is equal to k when kT is much greater than $h\nu_m$, the sum (11) in this case reduces to

$$C_V = 3N_m k = 3nR \tag{13}$$

where R is the gas constant and n is the number of atoms per molecule of the crystal. Thus, the law of Dulong and Petit is valid at high temperatures for an assembly of oscillators. This limiting case also corresponds to the case in which quantum theory degenerates to classical theory, that is, when h approaches zero.

It is not possible to draw further important conclusions from this theory unless the form of the frequency distribution function $q(\nu)$ is known. In the next section, we shall discuss several different assumptions that have been made about this function.

19. Einstein's and Debye's Frequency Distributions. *a. Einstein's Distribution.*—Einstein[1] postulated, for simplicity, that the atoms of a solid oscillate independently and with the same frequency ν. The summand in Eq. (8) of Sec. 18 is constant under this assumption, whence,

$$\bar{E} = 3N \frac{h\nu}{e^{\frac{h\nu}{kT}} - 1}, \tag{1}$$

and

$$C_V = 3nR \left(\frac{h\nu}{kT}\right)^2 \frac{e^{\frac{h\nu}{kT}}}{(e^{\frac{h\nu}{kT}} - 1)^2}$$

$$= 3nR f_E \left(\frac{h\nu}{kT}\right). \tag{2}$$

f_E will be called the Einstein specific-heat function (*cf.* Fig. 1). The parameter ν in Eq. (2) is usually adjusted so as to obtain the best possible fit of theoretical and observed specific-heat curves. Its magnitude usually is of the order of an infrared frequency. Although the parts of specific-heat curves for which C_v is greater than $3nR/2$ usually can be fitted closely with an Einstein specific-heat function, the low-temperature portions usually cannot. It may be recalled from the discussion of Chap. I that the observed specific heat usually drops to zero as T^3, whereas Eq. (2) varies as

$$3nR \left(\frac{h\nu}{kT}\right)^2 e^{-\frac{h\nu}{kT}}$$

[1] EINSTEIN, *op. cit.*

which approaches zero more rapidly. We shall discuss the comparison of observed and theoretical values in more detail after presenting Debye's theory.

b. Debye's Distribution.—Debye[1] pointed out that the possible oscillation frequencies of a solid body range from very low values, corresponding to sound waves, to very high ones, corresponding to infrared absorption peaks of solids. Since the contribution of a low-frequency oscillator to the low-temperature specific heat is larger than that of a high-frequency one, we might have anticipated that Einstein's assumption, which neglects the long waves, would not give a large enough specific heat at very low temperatures.

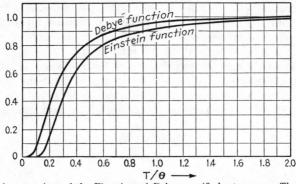

Fig. 1.—A comparison of the Einstein and Debye specific-heat curves. The frequency associated with the Einstein curve is equal to the maximum Debye frequency.

Debye suggested that a more accurate distribution might be found by treating the crystal as though it were a continuous elastic medium which has elastic constants that are independent of frequency. It has become conventional to follow Debye's original simplified treatment of the problem in which the medium is assumed to be isotropic. Instead, one might use the actual elastic constants. The factor that makes the more precise procedure difficult to apply is not so much the problem of finding the normal modes of vibration as the problem of expressing the frequency density in a tractable analytical form. In the isotropic case the frequency is independent of the direction of propagation, so that the distribution function may be expressed simply. We shall proceed with the simplified treatment and discuss the alternative one later.

The following point should be kept in mind. The vibrational frequencies of a continuous medium range from zero to infinity, corresponding to wave lengths that range from infinity to zero, whereas, in an actual solid, wave lengths shorter than interatomic distances are not independ-

[1] DEBYE, *op. cit.*

ent of longer ones.   This interdependence is shown for a simple example in Fig. 2.   The transverse displacements of the particles in this figure may be described equally well by the function *a* or *b*, or by any one of an infinite number of shorter waves that have the same amplitudes as *a* or *b* at the atomic positions.

The same conclusion may be drawn from the fact that a system which contains N atoms has only $3N$ degrees of freedom, whereas a strictly continuous medium has an infinite number of degrees of freedom.   This interdependence of different modes is not exhibited as long as the medium is treated as though it were continuous.   Hence, it is necessary to introduce a condition that limits the number of modes used in determining the specific heat.   This may be done conveniently by neglecting all

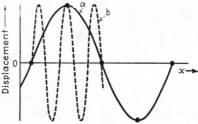

Fig. 2.—An example of the way in which a given periodic displacement of a lattice may be described by different harmonic functions *a* and *b*.   The wave numbers σ_a and σ_b of the two functions bear the relation $\sigma_b = \sigma_a + 1/l$, where l is the lattice distance.

modes having a frequency greater than the value ν_m defined by the equation

$$\int_0^{\nu_m} f(\nu)d\nu = 3N, \tag{3}$$

in which $f(\nu)$ is the distribution function for a continuous medium.   This, however, is not a unique way of selecting the modes of vibration.   We could, for example, choose them by placing a restriction upon the range of wave lengths rather than upon the range of frequencies.   The question of choosing modes does not arise when they are derived by a detailed atomic treatment like that of Secs. 20, 21, and 22 for this method automatically gives us a unique set of $3N$ modes.

Let us consider a rectangular parallelepiped of an isotropic medium, bounded by the six planes $x = 0$, $x = L_x$, $y = 0$, $y = L_y$, $z = 0$, $z = L_z$. The equations for the propagation of longitudinal and transverse vibrations in this medium are

$$\Delta u_l = \frac{1}{c_l^2} \frac{\partial^2 u_l}{\partial t^2}, \tag{4a}$$

$$\Delta u_t = \frac{1}{c_t^2} \frac{\partial^2 u_t}{\partial t^2}, \tag{4b}$$

respectively, where u_l and u_t are the amplitudes of the vibrations and c_l and c_t are their velocities. These velocities may be related to the two elastic constants for isotropic media by means of the equations[1]

$$c_l = \sqrt{\frac{m}{c_{11}}}, \qquad c_t = \sqrt{\frac{2m}{c_{11} + c_{12}}}$$

in which m is the density.

The boundary conditions employed when one searches for the normal modes of vibration depend upon the specific physical conditions under which the body vibrates. Since the frequency distribution turns out to be insensitive to the choice of boundary conditions, we shall assume for simplicity that the surface planes are rigidly held. In this case, the boundary condition is that u should be zero at these planes. With this condition, the normal modes are

$$u_l = A_l \sin \frac{\pi n_x x}{L_x} \sin \frac{\pi n_y y}{L_y} \sin \frac{\pi n_z z}{L_z} \sin 2\pi \nu_l t, \qquad (5a)$$

$$u_t = A_t \sin \frac{\pi n_x x}{L_x} \sin \frac{\pi n_y y}{L_y} \sin \frac{\pi n_z z}{L_z} \sin 2\pi \nu_t t, \qquad (5b)$$

where the n_i are arbitrary integers and the A_i are constants. The frequencies ν_l and ν_t are related to the n_i by means of the equations

$$\nu_l^2 = \frac{c_l^2}{4}\left(\frac{n_x^2}{L_x^2} + \frac{n_y^2}{L_y^2} + \frac{n_z^2}{L_z^2}\right), \qquad (6a)$$

$$\nu_t^2 = \frac{c_t^2}{4}\left(\frac{n_x^2}{L_x^2} + \frac{n_y^2}{L_y^2} + \frac{n_z^2}{L_z^2}\right). \qquad (6b)$$

There are three independent modes of vibration associated with each set of integers n_x, n_y, n_z, namely, two transverse waves of frequency ν_t and one longitudinal wave of frequency ν_l.

The quantities

$$\lambda_x = \frac{2L_x}{n_x}, \qquad \lambda_y = \frac{2L_y}{n_y}, \qquad \lambda_z = \frac{2L_z}{n_z} \qquad (7)$$

---

[1] In an isotropic medium, the elastic constants c_{ij} (*cf.* footnote 1, p. 94) are related by the equations.

$$c_{11} = c_{22} = c_{33},$$
$$c_{12} = c_{23} = c_{31},$$
$$c_{44} = c_{55} = c_{66} = \frac{(c_{11} - c_{12})}{2}$$

and all others are zero. Thus only two constants are independent. The equations for propagation of a wave in an elastic medium degenerate to (4a) and (4b) in this case. See, for example, W. Voigt, *Lehrbuch der Kristallphysik*, p. 587 (Teubner, Leipzig, 1928); G. Joos, *Theoretical Physics* (G. E. Stechert & Company, New York, 1934).

are the wave lengths of the modes (5a) and (5b) measured along the x, y, and z axes, respectively. The reciprocals of these quantities give the number of waves per unit length, or *the wave-number components* σ_x, σ_y, and σ_z:

$$\sigma_x = \frac{n_x}{2L_x}, \qquad \sigma_y = \frac{n_y}{2L_y}, \qquad \sigma_z = \frac{n_z}{2L_z}. \tag{8}$$

In terms of these quantities, the frequencies (6a) and (6b) become

$$\nu_l = c_l \sigma,$$
$$\nu_t = c_t \sigma, \tag{6c}$$

where

$$\sigma = \sqrt{\sigma_x^2 + \sigma_y^2 + \sigma_z^2}. \tag{9}$$

The standing wave (5) may be viewed as the sum of a set of traveling waves, for which the prototype function is the exponential function

$$A e^{i2\pi(\sigma \cdot \mathbf{r} - \nu t)} \tag{10}$$

where A is a constant, σ is a constant vector having components σ_x, σ_y, and σ_z, ν is the frequency of the wave, and $\mathbf{r}$ is the position vector whose components are the Cartesian coordinates of space x, y, and z. At a given time the phase of (10) is constant at all points that satisfy the equation

$$\sigma \cdot \mathbf{r} = \text{constant},$$

that is, on planes orthogonal to the direction σ. Thus these planes are the wave fronts. The normal distance between planes that differ in phase by 2π clearly is $1/|\sigma|$. Hence, $|\sigma|$ is the reciprocal wave length or the wave number, that is, the number of waves per unit distance normal to the wave front. For this reason, σ is called the wave-number vector. The phase of (10) is constant at points that satisfy the equation

$$\sigma \cdot \mathbf{r} - \nu t = \text{constant},$$

which describes the motion of the wave front planes in the direction of σ with a velocity equal to $\nu/|\sigma|$. The standing wave (5) may be obtained by adding together the sixteen functions that can be derived from (10) by exchanging the signs of ν and of the three components of σ.

The number of modes of vibration having values of σ less than the value σ' may be computed conveniently in the following way:

Let us represent the continuous variables σ_x, σ_y, and σ_z in a three-dimensional coordinate system. The points defined by Eqs. (8) are then distributed throughout the positive octant of this reference system and are arranged at the mesh points of a simple rectangular lattice. The equation $\sigma = \sigma'$ defines a sphere in the same reference system, and the volume of the part of this sphere lying in the positive octant is $\frac{1}{6}\pi\sigma'^3$. The total number of points (8) that lie in this octant is

$$L_x L_y L_z \frac{4}{3}\pi\sigma'^3, \tag{11}$$

since the average volume occupied by any one point is $1/8L_x L_y L_z$. Equation (11) may be written as

$$V\frac{4}{3}\pi\sigma'^3 \tag{12}$$

where V is the volume of the crystal.

There are three modes of vibration associated with each point (8); therefore, the total number of modes for which σ is less than σ' is

$$N(\sigma') = V4\pi\sigma'^3. \tag{13}$$

The equation $\nu = \nu'$ defines two spheres in σ space, one corresponding to longitudinal modes of vibration, and the other to transverse modes. The radii of these spheres are given by the equations

$$\sigma'_l = \frac{\nu'}{c_l}, \qquad \sigma'_t = \frac{\nu'}{c_t}, \tag{14}$$

respectively. Using Eq. (12), we find that the number of points in the first sphere is $\frac{4}{3}\pi\sigma_l'^3$, and the number in the second is $\frac{4}{3}\pi\sigma_t'^3$. Hence, the total number of modes having frequency less than ν' is

$$N(\nu') = V\frac{4\pi}{3}\left(\frac{1}{c_l^3} + \frac{2}{c_t^3}\right)\nu'^3, \tag{15}$$

since one longitudinal mode and two transverse modes correspond to each value of σ. We may define a mean velocity c by means of the equation

$$\frac{3}{c^3} = \frac{1}{c_l^3} + \frac{2}{c_t^3}, \tag{16}$$

and we may write Eq. (15) in the form

$$N(\nu') = V\frac{4\pi}{c^3}\nu'^3. \tag{17}$$

The corresponding expression for the case of a nonisotropic medium cannot be derived so simply, for the dependence of ν on the variables σ_x, σ_y, and σ_z usually will not be so simple as in (6a) and (6b).

Debye determined the highest allowed frequency ν_m, by setting (17) equal to $3N$ where N is the number of atoms in the solid. In addition, he assumed that the number of modes of vibration in the frequency range from ν' to $\nu' + d\nu'$ is given by the differential of (17), which is equivalent to assuming that the distribution function $f(\nu)$ in Eq. (9) of Sec. 18 is

$$f(\nu) = 3V\frac{4\pi}{c^3}\nu^2.$$ (18)

Under these conditions, Eq. (9) of Sec. 18 becomes

$$\bar{E} = \frac{V12\pi h}{c^3}\int_0^{\nu_m}\frac{\nu^3 d\nu}{e^{\frac{h\nu}{kT}}-1}.$$ (19)

We may eliminate c by means of Eq. (17), after setting $N(\nu') = 3N$. $\bar{E}$ then is

$$\bar{E} = \frac{9Nh}{\nu_m^3}\int_0^{\nu_m}\frac{\nu^3 d\nu}{e^{\frac{h\nu}{kT}}-1}.$$ (20)

Next, we may set $x = h\nu/kT$ and write (20) in the form

$$\bar{E} = 9Nh\nu_m\left(\frac{kT}{h\nu_m}\right)^4\int_0^{\frac{h\nu_m}{kT}}\frac{x^3 dx}{e^x-1}.$$ (21)

Thus, the molar heat is equal to

$$C_V = 9nR\left(\frac{kT}{h\nu_m}\right)^3\int_0^{\frac{h\nu_m}{kT}}\frac{e^x x^4 dx}{(e^x-1)^2}.$$ (22)

Following Debye, we shall define a characteristic temperature Θ_D by the

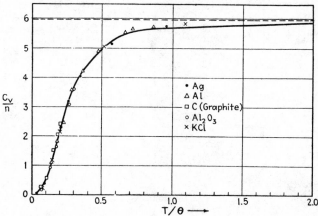

FIG. 3.—Comparison of the Debye specific-heat curve and the observed specific heats of a number of simple substances.

equation $k\Theta_D = h\nu_m$ and shall write (22) in the form

$$C_V = 3nRf_D\left(\frac{\Theta_D}{T}\right)$$ (23)

where

$$f_D\left(\frac{\Theta_D}{T}\right) = 3\left(\frac{T}{\Theta_D}\right)^3 \int_0^{\frac{\Theta_D}{T}} \frac{e^x x^4}{(e^x - 1)^2} dx. \tag{24}$$

The function $f_D(x)$ is shown in Fig. 1, along with Einstein's function. It

Table XXXIV.—The Debye Characteristic Temperatures of Solids

Substance	Θ_D	Substance	Θ_D
Metals			
Na	150	Al	390
K	100	Ga	125
Cu	315	In	100
Ag	215	Tl	100
Au	170	La	150
Be	1,000	Ti	350
Mg	290	Zr	280
Ca	230	Hf	213
Sr	170	Ge	290
Zn	250	Sn	260
Cd	172	Pb	88
Hg	96		
		Sb	140
		Bi	100
Cr	485	Ta	245
Mn	350	W	310
Fe	420	Re	300
Co	385	Os	250
Ni	375	Ir	285
Mo	380	Pt	225
Ru	400		
Rh	370		
Pd	275		
Ionic Crystals			
KCl	227	CaF$_2$	474
NaCl	281	FeS$_2$	645
KBr	177		
AgCl	183		
AgBr	144		

should be noted that f_D approaches zero as $(T/\Theta_D)^3$.

Figure 3 shows the manner[1] in which the atomic-heat curves of a number of simple solids may be fitted with a Debye function. It is obvious

[1] See, for example, the compilation of data in Landolt-Bornstein.

from Fig. 1 that the Einstein function would not apply nearly so well at low temperatures. Table XXXIV contains values of Θ_D that were determined from curves of this type.

The mean velocity c and the characteristic temperature are related by means of the equation

$$\Theta_D = \frac{h\nu_m}{k} = \frac{h}{k}c\left(\frac{3N}{4\pi V}\right)^{\frac{1}{3}}. \tag{25}$$

Born and von Kármán,[1] employing room-temperature values of the elastic constants, found excellent agreement between the observed and calculated temperatures of aluminum, copper, silver, lead, sodium chloride, and potassium chloride for temperatures above 25°K. Similar agreement has been found by Hopf and Lechner[2] for calcium fluoride and iron sulfide and by Schrödinger[3] for iron and carbon. Eucken,[4] however, showed that this close agreement usually disappears if the elastic constants for absolute zero are used instead. For example, Table XXXV contains a comparison of observed and calculated values

TABLE XXXV.—Comparison of Characteristic Temperatures Determined from the T^3 Law with Those Computed from the Low-temperature Elastic Constants

Substance	Θ_D (T^3 law)	Θ_D (calc.)
Cu	329	353
Ag	212	241
Al	399	502

of Θ_D for copper, silver, and aluminum. It should be noted that the Θ_D calculated from absolute-zero data are larger than the observed values. We may invariably expect this kind of discrepancy if the observed Θ_D agrees with the value computed from room-temperature data, since the elastic constants increase with decreasing temperature.

In addition to the discrepancy pointed out by Eucken, Grüneisen and Goens[5] have found that the Θ_D for zinc and cadmium determined from room-temperature data are larger than the experimental values. The disagreement would be emphasized even more in this case by using elastic constants corrected to absolute zero.

---

[1] Born and von Kármán, *op. cit.*
[2] L. Hopf and G. Lechner, *Verh. deut. physik. Ges.*, **16**, 643 (1914).
[3] E. Schrödinger, *Handbuch der Physik*, Vol. X.
[4] A. Eucken, *Verh. deut. physik. Ges.*, **15**, 571 (1913).
[5] E. Grüneisen and E. Goens, *Z. Physik*, **26**, 250 (1924).

Thus, we may say that at least part of the excellent agreement between the observed specific-heat curves and Debye's curve is fortuitous.

c. *Modifications of Debye's Equation Based upon the Continuum Hypothesis.*—Several modifications of Debye's equation were suggested in the period following his paper. We shall discuss briefly the work of Born,[1] the theoretical basis of which is examined in Sec. 21.

In treating molecular solids, Born postulated that only those $3N$ degrees of freedom that correspond to *intermolecular* vibrations of a system of N molecules should be treated by the continuum method. The specific heat associated with the remaining $3(n - 1)$ sets of N frequencies that correspond to intramolecular vibrations of the n atoms

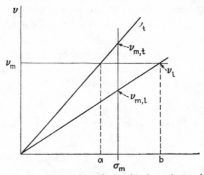

FIG. 4.—Frequency versus wave-number relationship for an isotropic continuous medium. The two lines represent curves for the two types of polarization. In the Debye scheme the $3N$ modes are selected by choosing all waves having frequency less than ν_m. Thus the transverse modes corresponding to the wave-number range a–b are excluded, whereas the longitudinal modes in the same range are retained. In the Born method all modes having wave numbers less than σ_m are retained, all others are discarded.

in a molecule should be approximated by $3(n - 1)$ Einstein functions. In addition, he suggested that the continuum frequencies that are used should be selected by taking all modes of vibration associated with a region of wave-number space instead of restricting the range by means of Eq. (17). The difference between Born's and Debye's methods may be understood by considering an isotropic medium. The two lines of Fig. 4 represent the dependence of frequency on wave number for the transverse and longitudinal waves. In Debye's scheme, the modes are selected by taking all frequencies less than ν_m. Thus, waves having only one type of polarization are used in the range of σ extending from a to b. Born suggested that it would be preferable to use all waves for which σ is less than the value σ_m that is defined by the condition that $N(\sigma')$ in Eq. (13) be equal to $3\bar{N}$; that is,

[1] M. BORN, *Dynamik der Kristallgitter* (Teubner, Leipzig, 1915).

$$\sigma_m = \left(\frac{3n_0}{4\pi}\right)^{\frac{1}{3}} \tag{26}$$

where n_0 is the molecular density. The limiting frequencies of the longitudinal and transverse waves then are

$$\begin{aligned} \nu_{m,l} &= c_l\sigma_m, \\ \nu_{m,t} &= c_t\sigma_m, \end{aligned} \tag{27}$$

and the density of each type of vibration is

$$\left.\begin{aligned} f_l(\nu) &= V\frac{4\pi\nu^2}{c_l^3} \qquad 0 \le \nu \le \nu_{m,l}, \\ f_t(\nu) &= 2V\frac{4\pi\nu^2}{c_t^3} \qquad 0 \le \nu \le \nu_{m,t}. \end{aligned}\right\} \tag{28}$$

The contribution to the specific heat from these modes is

$$R\left[f_D\left(\frac{\Theta_l}{T}\right) + 2f_D\left(\frac{\Theta_t}{T}\right)\right]$$

where $\Theta_l = h\nu_{m,l}/k$ and $\Theta_t = h\nu_{m,t}/k$. To this, Born would add $3(n-1)$ Einstein terms in order to include the intramolecular vibrations.

Born suggested that the same method of selecting frequencies could also be used in good approximation in anisotropic solids. In this case, the three waves of different polarization have different velocities for each direction of propagation. Thus, the limiting frequencies $\nu_{m,1}$, $\nu_{m,2}$, and $\nu_{m,3}$ are functions of the direction of propagation. If the direction variables are taken to be the polar angles θ and φ, the specific heat associated with the continuum modes of vibration now is

$$C_B = \frac{R}{4\pi}\sum_{j=1}^{3}\int_\Omega f_D\left(\frac{\Theta_j(\theta,\varphi)}{T}\right)d\Omega \tag{29}$$

where

$$\Theta_j(\theta,\varphi) = \frac{h\nu_{m,i}(\theta,\varphi)}{k} = \frac{hc_i(\theta,\varphi)}{k}\left(\frac{3n_0}{4\pi}\right)^{\frac{1}{3}} \tag{30}$$

and $d\Omega$ is the differential of solid angle. If we add to this the Einstein term associated with intramolecular vibrations, we obtain

$$C_V = C_B + \sum_{i=1}^{3(n-1)} Rf_E\left(\frac{h\nu_i}{kT}\right). \tag{31}$$

Försterling[1] computed C_B for the cubic salts sodium chloride, potassium chloride, calcium fluoride, and iron sulfide, using the room-temper-

---

[1] K. Försterling, *Z. Physik*, **3**, 9 (1920); *Ann. Physik*, **61**, 549 (1920); *Z. Physik*, **8**, 251 (1922).

ature elastic constants.   As we shall see in Sec. 23, it is doubtful whether Born's method of dividing the specific heat into Debye and Einstein terms can be justified in the case of these ideal ionic crystals.   Försterling employed an ingenious expansion method for evaluating the integrals in Eq. (29) and found that the result could be approximated closely by means of three Debye functions.   The three Einstein frequencies were taken to be the same in the diatomic case, whereas two sets of three equal values were used in the triatomic cases.   These frequencies were chosen in order to give the best fit with the observed results.   The three computed characteristic temperatures and the assumed vibrational frequencies are listed in Table XXXVI.   Table XXXVII contains a

TABLE XXXVI

Substance	Calculated characteristic temperatures			Assumed frequencies	
	Θ_1	Θ_2	Θ_3	$h\nu_1/k$	$h\nu_2/k$
NaCl	354	216	194	218	
KCl	289	170	150	186	
CaF$_2$	558	348	274	306	502
FeS$_2$	631	439	404	406	620

TABLE XXXVII.—COMPARISON OF OBSERVED AND CALCULATED MOLAR HEAT OF POTASSIUM CHLORIDE

(In cal/mol)

T, °K	Observed	Calculated
70	7.54	7.52
87	8.66	8.63
137	10.36	10.38
235	11.46	11.38

comparison of the observed and calculated values of C_V for potassium chloride above 70°K.   The accuracy is about the same in the other cases.   The contributions from the Einstein terms decrease very rapidly at low temperatures and are negligible below 10°K.   Hence, only the C_B term need be considered in this range.   This term, however, varies as $T^3/\overline{\Theta}^3$ where $1/\overline{\Theta}^3$ is the mean of $1/\Theta^3(\theta, \varphi)$.   Since Keesom and Clark have found that potassium chloride does not satisfy the T^3 law below 10°K, it follows that Born's modification of Debye's theory is subject to some of the same objections as the original theory.

More recently, Lord, Ahlberg, and Andrews[1] have applied a modification of Born's theory to crystalline benzene, which is an ideal molecular

[1] R. C. LORD, J. E. AHLBERG, and D. H. ANDREWS, *Jour. Chem. Phys.*, **5**, 649 (1937).

substance. In addition to using the continuum theory for the $3N$ intermolecular vibrations, they employed the theory for the $3N$ modes of motion in which the molecules undergo coupled torsional oscillations.[1] Einstein functions were employed for the remaining $27N$ intramolecular modes of vibration, the observed vibrational frequencies of the free benzene molecules being used in these terms. Since the intramolecular frequencies are comparatively high, the Einstein terms are negligible at 60°K and only the $6N$ continuum modes need be considered below

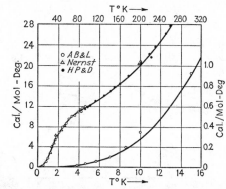

FIG. 5.—The observed and calculated molar heats of benzene. The lower curve represents the low-temperature heat, the upper curve corresponds to the higher temperature values. (*After Lord, Ahlberg and Andrews.*)

this temperature. These workers found that the molecular heat in this region could be fitted closely by the single Debye term

$$6Rf_D\left(\frac{\Theta_D}{T}\right)$$

in which $\Theta_D = 150°$. The agreement is shown in the lower half of Fig. 5. The observed and calculated molar heats for a wider range are given in the upper half of the same figure. Unfortunately, a rather questionable $C_P - C_V$ correction must be added above 150°K, so that the agreement in the higher temperature range is not so significant as that in the lower. It may be observed that the Einstein terms play an important role between 60° and 150°K.

The preceding method cannot be applied to a molecular crystal in a temperature range in which torsional oscillation changes to free rotation. We shall discuss the theory for this case in Chap. XIV.

[1] In a molecular crystal in which the inner molecular forces are much larger than the intermolecular forces, it may be expected that there are a set of $3N$ low-frequency torsional modes of vibration as well as the $3N$ low-frequency translational ones.

A striking example of a substance for which Born's equation (29) is more successful in predicting the specific heat than Debye's is the alkali metal lithium.   The elastic constants of this substance have been computed by Fuchs,[1] using a method that will be described in Chap. X. The elastic constants of lithium have not been measured, but those of sodium have been measured by Quimby and Siegel.[2]   The agreement between observed and calculated results in this case may be found in Table LXII, Chap. X.   It should be observed that the relation

$$\frac{2c_{44}}{c_{11} - c_{12}} = 1$$

for isotropic media is far from being satisfied, showing that these crystals are much less isotropic than the alkali halides.   Fuchs used the elastic

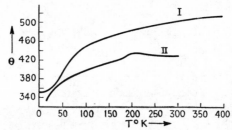

Fig. 6.—The observed and calculated $\Theta_D(T)$ curves of lithium.   The theoretical curve I was calculated by Fuchs using Born's modification of Debye's theory and the theoretical elastic constants of lithium.   The experimental curve II was determined by Simon and Swain.

constants of lithium to compute the three characteristic temperatures for each direction of propagation [*cf.* Eqs. (29) and (30)] and then computed the specific heat from these.   Figure 6 gives a comparison of the observed and computed values of $\Theta(T)$ obtained by equating the observed and calculated specific-heat curves to a single Debye function.   It is evident from curve II that Debye's law is inaccurate over a range of temperature comparable with the mean Debye temperature.   The theoretical curve duplicates the general trend of the experimental curve and, hence, is in better agreement with it than the straight line corresponding to Debye's theory would be.   However, the agreement is still far from exact.   It is possible that Blackman's method of computing the specific-heat curve, which is discussed in Sec. 23, would further improve the theoretical values.[3]

[1] K. Fuchs, *Proc. Roy. Soc.*, **153**, 622 (1936); **157**, 444 (1936).

[2] S. L. Quimby and S. Siegel, *Phys. Rev.*, **54**, 293 (1938).

[3] R. C. Lord has pointed out to the writer that the specific heat of lithium may be fitted closely by a combination of Einstein and Debye functions, which suggests that the lattice may be somewhat molecular.   This suggestion awaits X-ray confirmation.

20. Observed Deviations from Debye's Law of Vibrational Specific Heats.—The specific heats of a number of simple solids exhibit deviations from Debye's law that imply that Debye's theory requires modification. We shall exclude molecular crystals from consideration at the present time and shall discuss only the cases that involve metals or simple ionic crystals. The discrepancies may be placed under three headings, as follows:

a. The Linear Term in Metals.—Figure 19, Chap. I, shows that nickel has a linear specific-heat term in addition to the Debye term. A number of other metals show a similar anomaly near absolute zero that is now ascribed to the specific heat of their free electrons. We shall discuss a simple theory of this term in the next chapter.

b. Low-temperature Anomalies of Monotonically Increasing Curves.—Figure 68, Chap. I, shows the variation with temperature of the Debye characteristic temperature of potassium chloride. This curve was obtained by Keesom and Clark by equating the observed molecular heat to Debye's expression

$$C_V = 464.5\left(\frac{T}{\Theta_D}\right)^3$$

and solving for Θ_D. It may be observed that, instead of being a constant, this quantity shows a distinct peak about $4.3°K$. As we mentioned in the previous section, Born's modification of Debye's theory cannot explain this because the Einstein terms in Eq. (29) are negligible in this temperature range. Deviations of this type have been explained qualitatively by Blackman. We shall discuss his work in Sec. 23.

c. The Anomalous Peaks of Germanium and Hafnium.—These peaks, which are shown in Fig. 16, Chap. I, cannot be explained by any theory that assumes Hooke's-law forces, for no superposition of Einstein functions leads to a curve with a maximum. Although the peaks in these metals are similar to the peaks occurring in molecular crystals, the explanation in terms of molecular rotation evidently cannot be applied.

21. The Vibrational Modes of One-dimensional Systems. *a. Monatomic Lattice.*—Suppose that we have a one-dimensional lattice of atoms extending along the x axis. We shall assume that the atoms are spaced at distances a from one another and that they interact with Hooke's-law forces. Consider the longitudinal modes of vibration from the following two standpoints: first, the Debye standpoint in which the modes are determined by solving the equation

$$\frac{1}{c_l^2}\frac{d^2\psi(x)}{dt^2} = \frac{d^2\psi(x)}{dx^2} \tag{1}$$

where $\psi(x)$ is the longitudinal displacement and c_l is the constant velocity of propagation of very long waves; and, the atomic standpoint in which the normal modes are determined by solving simultaneously the equations of motion for each atom.

The independent harmonic solutions of (1) for a string of length L fixed at both ends are

$$\psi(x) = A_n \sin \frac{2\pi n_x}{2L} \sin 2\pi \nu t \tag{2}$$

where n is an integer and the frequency ν is related to it by the equation

$$\nu = c_l \frac{n}{2L} = c_l \sigma. \tag{3}$$

It is obvious that $n/2L$ is the wave number σ of the standing wave. Since there are an infinite number of independent modes of vibration for a continuous string, we must limit the frequencies if we want to take account of the fact that there are a finite number of atoms in the string. This may be done by excluding those modes for which n is greater than N, the total number of atoms, that is, by neglecting modes with wave lengths less than $2a$.

Next, let us derive the equations of motion for the atoms of the string. We shall number the atoms consecutively from 1 to N, starting at the origin of coordinates, and we shall let x_n be the variable that measures the displacement of the nth atom from its equilibrium position. In addition, we shall assume that each atom interacts only with its immediate neighbors. Under this condition, the equation of motion of the nth atom is

$$m\frac{d^2x_n}{dt^2} = -\mu[(x_n - x_{n+1}) - (x_{n-1} - x_n)] \tag{4}$$

where μ is Hooke's constant for a pair of atoms and m is the mass of an atom.

The equations for the two end atoms obviously are different from the equations for interior atoms, for the former lack neighbors on one side. This fact could complicate the procedure of finding solutions of (4); however, we shall avoid the difficulty by using a method that was employed first by Born and von Kármán.[1] We shall assume that there are additional atoms on both sides of the string, in order that the equations for the end atoms may be the same as those for interior atoms. This assumption cannot affect the nature of the frequency-distribution function in any important way, as long as the number of atoms is sufficiently large. In addition, we shall assume that the phase of x_1 is the same as

[1] Born and von Kármán, *op. cit.*

the phase of the hypothetical $N + 1$st atom. As an alternative, we might assume that the end atoms are fixed. The "periodic" boundary condition is more advantageous, however, since it allows us to find elemental running-wave solutions without dealing with an infinite string and, at the same time, takes proper account of the number of degrees of freedom.

The function

$$x_n(l, \nu) = A e^{2\pi i \left(\frac{l}{N} n - \nu t \right)}, \tag{5}$$

where A is a constant and l is an integer, satisfies the periodic boundary condition and reduces all of Eqs. (4) to the same form, namely,

$$-4\pi^2 m \nu^2 = -\mu 2 \left(1 - \cos \frac{2\pi l}{N} \right). \tag{6}$$

Hence, the real and imaginary parts of (5) are physically interesting solutions of Eq. (4). Each of these parts defines a real traveling wave of wave length $\lambda = Na/l$, or wave number $\sigma = l/Na$, for which the end atoms move in phase. We may also construct real standing-wave solutions of fixed frequency ν' by properly combining the four functions of type (5) for which $l = \pm l'$ and $\nu = \pm \nu'$. The two independent waves of this type are

$$x_n = A \sin \frac{2\pi l'}{N} n \sin (\nu' t + \delta) \tag{7a}$$

and

$$x_n = A \cos \frac{2\pi l'}{N} n \sin (\nu' t + \delta') \tag{7b}$$

where δ and δ' are arbitrary phases.

The set of independent modes of different wave number are those corresponding to values of l for which the quantities (5) are different functions of n. Thus, the modes belonging to $l = l'$ and $l = l' + N$ are not independent, since $x_n(l', \nu)$ is equal to $x_n(l' + N, \nu)$ for all values of n. It is clear that there are only N independent values of l and that these may be chosen to be those lying in the range from zero to $N - 1$. The sine and cosine functions of $2\pi l'n/N$ appearing in (7a) and (7b) are different only for half this range of l, because of the relation

$$\sin x = - \sin (2\pi - x), \tag{8a}$$
$$\cos x = \cos (2\pi - x). \tag{8b}$$

Thus, if we choose to represent the normal modes by means of the functions (5), the independent range of l extends from zero to $N - 1$, whereas, if we choose the two functions (7a) and (7b) instead, the range extends from zero to $N/2$. In either case, however, there are just N independent modes, in agreement with the fact that there are only N degrees of freedom in the system.

The relationship between frequency and wave number, given by Eq. (6), may be reduced to

$$\nu = \frac{1}{\pi}\sqrt{\frac{\mu}{m}}\left|\sin\frac{\pi l}{N}\right| = \frac{1}{\pi}\sqrt{\frac{\mu}{m}}\left|\sin\pi\sigma a\right|. \tag{9}$$

This equation becomes identical with Eq. (3) for values of σ small enough to allow us to replace the sine by its argument, for then

$$\nu = a\sqrt{\frac{\mu}{m}}\sigma.$$

Figure 7 illustrates the expression (9) for the independent range of the σ. This curve shows that it is not generally permissible to assume that the

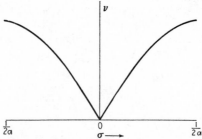

Fig. 7.—The $\nu(\sigma)$ curve for a monatomic linear lattice. The independent range of σ is taken to extend from $-1/2a$ to $1/2a$ corresponding to values of l extending from $-N/2$ to $N/2$.

velocity of propagation of elastic waves is independent of frequency, as one does in deriving Debye's frequency distribution.

Grüneisen and Goens[1] first suggested that one might explain the discrepancy between the characteristic temperatures obtained from the T^3 law and those computed from elastic data on the basis of differences in velocity between long and short waves. The fact that short waves travel more slowly in the linear lattice suggests that in computing Θ_D we should use elastic constants somewhat smaller than those obtained from ordinary measurements. Blackman has investigated more fully the significance of this suggestion (*cf.* Sec. 23).

We shall next find the frequency distribution corresponding to Eq. (9). Equal ranges of σ contain equal numbers of modes, since σ is proportional to l. Moreover,

[1] Grüneisen and Goens, *op. cit.*

$$d\sigma = \frac{dl}{Na}.$$

Hence, the density $dl/d\nu$ is

$$\frac{dl}{d\nu} = \frac{1}{d\nu/dl} = \frac{Na}{d\nu/d\sigma} = \frac{N}{\sqrt{\mu/m} \, \cos \pi\sigma a}$$

$$= \frac{N\sqrt{m/\mu}}{\sqrt{1 - (\nu^2\pi^2 m/\mu)}}.$$

Thus, the density is constant for small values of ν, just as in the continuous case, but it becomes infinite when $\sigma = 1/2a$.

　　b. Diatomic Lattices.—Let us extend the preceding problem by adding particles of mass M at points midway between the particles of mass m so that the distance between neighbors now is $a/2$. We shall label the masses with integers extending from 1 to $2N$ in this case so that the odd integers correspond to the masses m and the even ones to the masses M. The equations of motion are then

$$\left.\begin{aligned}
m\frac{d^2x_{2n+1}}{dt^2} &= -\mu[(x_{2n+1} - x_{2n}) - (x_{2n+2} - x_{2n+1})], \\
M\frac{d^2x_{2n}}{dt^2} &= -\mu[(x_{2n} - x_{2n-1}) - (x_{2n+1} - x_{2n})],
\end{aligned}\right\} \tag{10}$$

where μ is the force constant between unlike neighbors.

　　If the Born-von Kármán boundary conditions are used again, a normal coordinate substitution is

$$\left.\begin{aligned}
x_{2n+1} &= Ae^{2\pi i\left(l\frac{2n+1}{2N} - \nu t\right)}, \\
x_{2n} &= Be^{2\pi i\left(l\frac{2n}{2N} - \nu t\right)},
\end{aligned}\right\} \tag{11}$$

for integer l. The wave number σ is l/Na in this case. A and B satisfy the homogeneous linear equations

$$\left.\begin{aligned}
(4\pi^2 m\nu^2 - 2\mu)A + 2\mu \cos\left(2\pi\frac{l}{2N}\right)B &= 0, \\
2\mu \cos\left(2\pi\frac{l}{2N}\right)A + (4\pi^2 M\nu^2 - 2\mu)B &= 0,
\end{aligned}\right\} \tag{12}$$

which have a solution only for those values of ν that satisfy the compatibility equation

$$\begin{vmatrix}
(4\pi^2 m\nu^2 - 2\mu) & 2\mu \cos\left(2\pi\frac{l}{2N}\right) \\
2\mu \cos\left(2\pi\frac{l}{2N}\right) & (4\pi^2 M\nu^2 - 2\mu)
\end{vmatrix} = 0, \tag{13}$$

the roots of which are

$$4\pi^2\nu^2 = \frac{\mu}{mM}\left(M + m \pm \sqrt{M^2 + m^2 + 2Mm \cos 2\pi\frac{l}{N}}\right) \quad (14)$$

It is easy to show that all independent modes occur in the range of l extending from zero to $2N$ and that there are two modes for each value

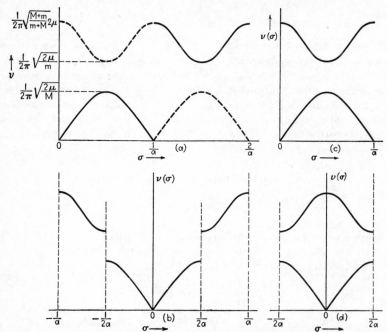

Fig. 8.—$\nu(\sigma)$ curves for a diatomic linear lattice. In a the independent range extends from 0 to $2/a$ and the function is single-valued. In b the independent range extends from $-1/a$ to $1/a$ so that the curve is symmetrical and single-valued. In c and d the function is double-valued, being unsymmetrical in the first case and symmetrical in the second. Case d is the reduced-zone scheme which will be used frequently in subsequent work.

of l, corresponding to the two roots of (14). Only half of these $4N$ modes are independent, however, since the factor $\cos 2\pi l/2N$ in Eqs. (12) repeats its values in the range of l extending from zero to $2N$.

We may obtain a one-to-one relationship between the modes of vibration and the $2N$ values of σ by arbitrarily restricting ν to be a single-valued function of σ, in the manner illustrated in Fig. 8a. The dotted curves show the discarded branches. It is sometimes convenient to choose the domain of σ to extend from $-\dfrac{1}{a}$ to $+\dfrac{1}{a}$ instead of from zero to $2/a$. In this case, the single-valued $\nu(\sigma)$ curve is symmetrical, as in Fig. 8b.   On the other hand, we may choose the range of σ to extend from zero to $1/a$, and we may use both branches of the frequency curve in the manner shown in Fig. 8c.   It is also possible to obtain a symmetrical $\nu(\sigma)$ curve in this case by choosing σ so that it ranges from $-1/2a$ to $+1/2a$, as in Fig. 8d.   We shall usually use this mode of description, which is called the "reduced-zone scheme."

The relationship between A and B may be determined by substituting ν^2 from Eq. (14) in either of the equations (12).   The ratio A/B is

$$\frac{A}{B} = \frac{2M \cos 2\pi(l/2N)}{[M - m \pm \sqrt{M^2 + m^2 + 2Mm \cos (2\pi l/N)}]}. \tag{15}$$

It is interesting to note that the two normal modes which correspond to the points of discontinuity at $\sigma = \pm 1/2a$ are those in which one of the two types of mass is stationary.   This fact may be shown by setting $l = N/2$ in Eq. (12).   In addition, it should be noted that ν approaches zero linearly near the origin, just as in the monatomic case, a fact showing that acoustical waves travel with constant velocity.

The number of modes of vibration per unit range of σ is constant for each branch of the $\nu(\sigma)$ curve and is equal to Na.   Hence, the distribution as a function of frequency is

$$f(\nu) = Na\frac{d\sigma}{d\nu},$$

which may be evaluated from Eq. (14).   This function, which is

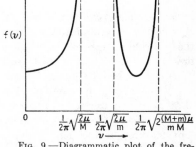

Fig. 9.—Diagrammatic plot of the frequency distribution of modes of vibration for the diatomic linear lattice. The density approaches infinity at the three points indicated by vertical dashed lines.

shown in Fig. 9, has nonvanishing values over two ranges of frequency corresponding to the two branches of the $\nu(\sigma)$ curve.

It may be seen that we obtain a new equation of type (12) for each particle added to the unit cell of the lattice and that a new root is simultaneously added to the determinantal equation that determines the frequency.   Hence, if we want to retain ν as a single-valued function of σ, we must extend the domain of σ by $1/a$ for each particle added.

Another interesting one-dimensional case is that in which the two particles in the unit cell of a one-dimensional lattice have equal masses but interact more strongly with one another than with their neighbors. This case is analogous to that of a molecular crystal in which intramolecular forces are stronger than intermolecular ones.   If α is the

Hooke's constant for interaction between the pairs in the unit cell, β the constant for interaction of one of these atoms with the neighbor in a different cell, and m the atomic mass, the equation analogous to (14) is

$$4\pi^2\nu^2 = \frac{2(\alpha + \beta) \pm \sqrt{4(\alpha + \beta)^2 - 16\alpha\beta \sin^2(2\pi l/2N)}}{2m}.$$ In the reduced-

zone scheme, l is an integer extending from $-\frac{N}{2}$ to $\frac{N}{2}$, and the wave number is $\sigma = l/Na$, as before. When α becomes very large in comparison with β, the upper branch of this function reduces to the constant value

$$\nu = \frac{1}{\pi}\sqrt{\frac{\alpha}{2m}},$$

which is the natural frequency of two atoms having mass m and force constant β. The lower curve reduces to

$$\nu = \frac{1}{\pi}\sqrt{\frac{\beta}{2m}}\left|\sin\frac{\pi l}{N}\right|$$

under the same conditions. This expression is almost the same as (9); however, $2m$ appears instead of m. It is clear that the pair of atoms in the unit cell move as one in the modes associated with the low-frequency branch of the $\nu(\sigma)$ curve and oscillate as though in free space in the modes of the high-frequency branch. This condition obtains, of course, only when α is much larger than β. If α and β are comparable, the pair of atoms in the unit cell will behave less as an independent unit.

These simple one-dimensional examples illustrate most of the important features of the general three-dimensional case, which we shall discuss next. In particular, it is possible to see the origin of Born's postulates which were presented in part c of Sec. 19.

In connection with Born's first postulate, it should be pointed out that only the modes associated with the lowest branch of the $\nu(\sigma)$ curve approach acoustical vibrations at long wave lengths. Hence, this is the only branch we may reasonably expect to approximate by means of the $\nu(\sigma)$ curve for a continuous medium. There are no other branches for a monatomic lattice (case a), but there are others in polyatomic cases. The number of degrees of freedom in the lowest branch or in any single branch is equal to the number of unit cells in the lattice in the linear case and is equal to three times this number in the three-dimensional case. From this fact, Born concluded that it is not permissible to obtain the distribution of more than $3N$ modes by treating the solid as though it were continuous. He effectively assumed that the other branches of the $\nu(\sigma)$ curve are constants when he assumed that their contribution to the specific heat could be expressed in terms of Einstein functions. This assumption is reasonable for molecular crystals, in which the intramolecular forces are much stronger than intermolecular ones (*cf.* the last one-dimensional example); but it cannot apply to crystals such as potassium chloride, in which there is high coordination between the atoms in the lattice.

Born's second assumption follows from the fact that in three dimensions each of the three branches of $\nu(\sigma)$, corresponding to the three directions of polarization, extends over the same range of wave numbers, just as the different branches do in the one-dimensional case.

22. General Three-dimensional Case.*—Let us now consider a general three-dimensional lattice[1] of atoms that interact with Hooke's-law forces. We shall discuss a crystal having translational vectors τ_1, τ_2, and τ_3 that are not necessarily orthogonal to each other. The unit cell may be taken as a rhombohedral parallelepiped the edges of which are determined by the translational vectors. The crystal specimen may have an arbitrary shape; we shall conveniently assume, however, that it is a rhombohedral parallelepiped of which the faces are parallel to the faces of the unit cell and the edges L_1, L_2, and L_3 are large integer multiples of the lengths of the edges of a unit cell; that is,

$$L_1 = N_1|\tau_1|, \qquad L_2 = N_2|\tau_2|, \qquad L_3 = N_3|\tau_3|. \tag{1}$$

The crystal obviously contains $N_1N_2N_3$ unit cells. A given cell may be specified relative to one corner of the crystal by the vector

$$\mathbf{T}(p_1,p_2,p_3) = p_1\tau_1 + p_2\tau_2 + p_3\tau_3, \tag{2}$$

which extends from the corner of the crystal to the corner of the cell in question. We shall call the cell that is specified by the integers p_1', p_2', p_3' the p'th cell in the lattice. In addition, we shall assume that there are n atoms in the unit cell and that the positions of these relative to the corner nearest the origin of coordinates are specified by the n position vectors $\varrho_\alpha(\alpha = 1, 2, \cdots, n)$. The position vector $\mathbf{r}_\alpha(p_1,p_2,p_3)$, of the αth atom in the pth unit cell relative to the origin is then given by the sum

$$\mathbf{r}_\alpha(p_1,p_2,p_3) = \mathbf{T}(p_1,p_2,p_3) + \varrho_\alpha.$$

---

* This section may be omitted in a first reading. It is demonstrated that the general conclusions that were drawn from the one-dimensional case concerning the $\nu(\sigma)$ curves also apply to the three-dimensional one with the difference that the wave number is now a three-dimensional vector whose independent values range throughout a polyhedron.

[1] For additional details concerning the problem of determining the vibrational modes of lattices, see R. B. Barnes, R. R. Brattain, and F. Seitz, *Phys. Rev.*, **48**, 582 (1935); R. R. Lyddane and K. F. Herzfeld, *Phys. Rev.*, **54**, 846 (1938).

We shall designate the coordinates of the αth atom in the pth unit cell relative to its equilibrium position by $x_\alpha{}^i(p_1,p_2,p_3)$ where i takes values 1, 2, and 3 corresponding to each of three orthogonal Cartesian coordinates. The equations of motion for this atom have the form

$$m_\alpha \frac{d^2 x_\alpha{}^i(p_1,p_2,p_3)}{dt^2} = f_\alpha{}^i(p) \qquad (3)$$

where m_α is the mass of the αth atom and $f_\alpha{}^i(p)$ is a homogeneous linear function of the relative displacements of the α-pth atom and all other atoms:

$$f_\alpha{}^i = \sum_{j,\beta,p'} \mu_{\alpha\beta}{}^{ij}(p,p')[x_\beta{}^j(p') - x_\alpha{}^j(p)]. \qquad (4)$$

We shall consider the interaction with all neighbors rather than, as in the one-dimensional case, merely the interaction with nearest neighbors. The coefficients $\mu_{\alpha\beta}{}^{ij}(p,p')$, which obviously are the interatomic Hooke constants, depend only on the relative values of p' and p for given α, β, i, j, that is, only on the relative equilibrium positions of two atoms. Thus, the equations of motion for the atoms in the pth cell differ from those for equivalent atoms in the qth cell only by the translation $\mathbf{T}(p_1,p_2,p_3) - \mathbf{T}(q_1,q_2,q_3)$. We shall make no assumptions about the number of physically important terms that appear in the right-hand side of (4), although we may expect that nearest neighbors will have the largest coefficients.

Since we shall generally be interested in forces that may be expressed as the gradient of a potential, we may require that

$$f_\alpha{}^i(p) = - \frac{\partial V}{\partial x_\alpha{}^i(p)}. \qquad (5)$$

The necessary and sufficient conditions for these equations to be satisfied are

$$\frac{\partial f_\alpha{}^i(p)}{\partial x_\beta{}^j(p')} = \frac{\partial f_\beta{}^j(p')}{\partial x_\alpha{}^i(p)},$$

or, as one may readily see,

$$\mu_{\alpha\beta}{}^{ij}(p,p') = \mu_{\beta\alpha}{}^{ji}(p',p), \qquad (6a)$$

$$\sum_{\beta,p'} \mu_{\alpha\beta}{}^{ij}(p,p') = \sum_{\gamma,p''} \mu_{\alpha\gamma}{}^{ji}(p,p''). \qquad (6b)$$

In order to introduce the Born-von Kármán boundary conditions, we shall assume that surface atoms of the crystal have the same equations of motion as the interior atoms. We are then able to show that the

$3nN_1N_2N_3$ equations (3) may be reduced to $3n$ independent equations. The appropriate substitution that satisfies the Born-von Kármán boundary conditions is, in fact,

$$x_\alpha{}^i(p_1,p_2,p_3) = \frac{\xi_\alpha{}^i(\mathbf{d})}{\sqrt{m_\alpha}} e^{2\pi i(\sigma \cdot \mathbf{r}_\alpha(p_1,p_2,p_3) - \nu t)} \tag{7}$$

where the $\xi_\alpha{}^i$ are constants and $\mathbf{d}$ is a vector that satisfies the conditions

$$\begin{aligned}
\mathbf{d} \cdot \mathbf{T}(N_1,0,0) &\equiv N_1 \mathbf{d} \cdot \boldsymbol{\tau}_1 = l_1, \\
\mathbf{d} \cdot \mathbf{T}(0,N_2,0) &\equiv N_2 \mathbf{d} \cdot \boldsymbol{\tau}_2 = l_2, \\
\mathbf{d} \cdot \mathbf{T}(0,0,N_3) &\equiv N_3 \mathbf{d} \cdot \boldsymbol{\tau}_3 = l_3,
\end{aligned} \right\} \tag{8}$$

where l_1, l_2, and l_3 are integers. The solution of Eqs. (8), which obviously express the Born-von Kármán boundary conditions, is

$$\mathbf{d} = \frac{l_1}{N_1} \frac{\boldsymbol{\tau}_2 \times \boldsymbol{\tau}_3}{|\boldsymbol{\tau}_1 \boldsymbol{\tau}_2 \boldsymbol{\tau}_3|} + \frac{l_2}{N_2} \frac{\boldsymbol{\tau}_3 \times \boldsymbol{\tau}_1}{|\boldsymbol{\tau}_1 \boldsymbol{\tau}_2 \boldsymbol{\tau}_3|} + \frac{l_3}{N_3} \frac{\boldsymbol{\tau}_1 \times \boldsymbol{\tau}_2}{|\boldsymbol{\tau}_1 \boldsymbol{\tau}_2 \boldsymbol{\tau}_3|} \tag{8a}$$

where the cross designates the conventional vector product and $|\boldsymbol{\tau}_1 \boldsymbol{\tau}_2 \boldsymbol{\tau}_3|$ is the determinant

$$\begin{vmatrix} t_{11} & t_{21} & t_{31} \\ t_{12} & t_{22} & t_{32} \\ t_{13} & t_{23} & t_{33} \end{vmatrix}.$$

Equation (8a), like Eq. (2), defines the mesh points of a lattice having primitive translation vectors $\mathbf{s}_1$, $\mathbf{s}_2$, and $\mathbf{s}_3$, where

$$\mathbf{s}_1 = \frac{\boldsymbol{\tau}_2 \times \boldsymbol{\tau}_3}{N_1 |\boldsymbol{\tau}_1 \boldsymbol{\tau}_2 \boldsymbol{\tau}_3|}, \qquad \mathbf{s}_2 = \frac{\boldsymbol{\tau}_3 \times \boldsymbol{\tau}_1}{N_2 |\boldsymbol{\tau}_1 \boldsymbol{\tau}_2 \boldsymbol{\tau}_3|}, \qquad \mathbf{s}_3 = \frac{\boldsymbol{\tau}_1 \times \boldsymbol{\tau}_2}{N_3 |\boldsymbol{\tau}_1 \boldsymbol{\tau}_2 \boldsymbol{\tau}_3|}. \tag{9}$$

This lattice has been called the reciprocal lattice by Gibbs, who first used it.

Let us consider for a moment the simple lattice consisting of one type of atom, say the αth in the unit cell. When the $\xi_\alpha{}^i$ ($i = 1, 2, 3$) are fixed, expression (6) describes a running wave for which the wave-number vector is $\mathbf{d}$. There are $N_1N_2N_3$ ways of choosing the l_i in (8) to give independent waves, and these may be selected to lie in the range

$$-\frac{N_1}{2} \le l_1 \le \frac{N_1}{2}, \qquad -\frac{N_2}{2} \le l_2 \le \frac{N_2}{2}, \qquad -\frac{N_3}{2} \le l_3 \le \frac{N_3}{2}.$$

This domain obviously corresponds to values of the wave-number vector $\mathbf{d}$ that lie within the rhombohedral parallelepiped having corners at the eight points

$$\mathbf{d} = \pm \frac{N_1}{2}\mathbf{s}_1 \pm \frac{N_2}{2}\mathbf{s}_2 \pm \frac{N_3}{2}\mathbf{s}_3. \tag{10}$$

The waves (7) are independent for a larger domain of $\mathfrak{d}$ if we consider all atoms at once; in fact, it is easy to see that the domain may be chosen to be n times larger. However, the difference between modes for which $\mathfrak{d}$ lies inside (10) and those for which it does not is simply a matter of the relative phases of the motions of different atoms in the unit cell. Hence, we may condense the larger domain of $\mathfrak{d}$ to the domain determined by (10) by absorbing these phases in the ξ^i and introducing n sets of ξ^i instead of one. This procedure is analogous to the one used in passing from the description of Fig. 8b to the reduced-zone scheme of Fig. 8d in the one-dimensional case.

We shall frequently use the relation

$$\sum_p e^{2\pi i (\sigma - \sigma') \cdot \mathbf{r}_\alpha(p)} = N_1 N_2 N_3 \delta_{\sigma,\sigma'} \tag{11}$$

where the summation extends over all $N_1 N_2 N_3$ values of p, and $\delta_{\sigma,\sigma'}$, the Kronecker delta function, is zero when $\mathfrak{d} \neq \mathfrak{d}'$ and is equal to 1 when $\mathfrak{d} = \mathfrak{d}'$.

Substituting (7) in Eq. (3) and multiplying the result by $e^{-2\pi i [\sigma \cdot \mathbf{r}_\alpha(p') - \nu t]}$, we obtain the following $3n$ equations for the $\xi_\alpha{}^i$ after summing over p':

$$-4\pi^2 \nu^2 \xi_\alpha{}^i = g_\alpha{}^i (\xi_1^1, \xi_1^2, \cdots, \xi_n^3, \mathfrak{d}). \tag{12}$$

$g_\alpha{}^i$ clearly is a homogeneous linear function of the $\xi_\alpha{}^i$ and of the differences $\mathbf{r}_\alpha(p) - \mathbf{r}_\beta(q)$. We have, in fact,

$$g_\alpha{}^i = \sum_{j,\beta,p'} \frac{\mu_{\alpha\beta}{}^{ij}}{\sqrt{m_\alpha}} \left(\frac{\xi_\beta{}^j}{\sqrt{m_\beta}} e^{2\pi i \sigma \cdot [\mathbf{r}_\beta(p') - \mathbf{r}_\alpha(p)]} - \frac{\xi_\alpha{}^j}{\sqrt{m_\alpha}} \right) \tag{13}$$

$$\equiv \sum_{j,\beta} \lambda_{\alpha\beta}{}^{ij}(\mathfrak{d}) \xi_\beta{}^j$$

where

$$\lambda_{\alpha\beta}{}^{ij}(\mathfrak{d}) = \sum_{p'} \frac{\mu_{\alpha\beta}{}^{ij}(p,p')}{\sqrt{m_\alpha m_\beta}} e^{2\pi i \sigma \cdot [\mathbf{r}_\beta(p') - \mathbf{r}_\alpha(p)]} \tag{14a}$$

for $\alpha \neq \beta$ and $p \neq p'$, and

$$\lambda_{\alpha\alpha}{}^{ij}(\mathfrak{d}) = -\frac{1}{m_\alpha} \sum_{\beta \neq \alpha, p' \neq p} \mu_{\alpha\beta}{}^{ij}(p,p'). \tag{14b}$$

The $\lambda_{\alpha\beta}$ are real only in the special case in which each atom is a center of symmetry, that is, in the case in which there is an atom of type β at the position $-[\mathbf{r}_\beta(p') - \mathbf{r}_\alpha(p)]$, relative to atom α in the p'th cell, for each atom at the position $+[\mathbf{r}_\beta(p') - \mathbf{r}_\alpha(p)]$. This condition is satisfied in some of the simplest lattices, but the $\lambda_{\alpha\beta}$ are usually complex. We see, however, that

$$\lambda_{\alpha\beta}{}^{ij}(\mathfrak{d}) = \lambda_{\beta\alpha}{}^{ji*}(\mathfrak{d}), \tag{15}$$

because of Eqs. (6a) and (6b). Hence, the matrix of the $\lambda_{\alpha\beta}$ is Hermitian.

The homogeneous linear equations (12), namely,

$$-4\pi\nu^2\xi_\alpha{}^i(\mathfrak{d}) = \sum_{j,\beta}\lambda_{\alpha\beta}{}^{ij}(\mathfrak{d})\xi_\beta{}^j(\mathfrak{d}), \tag{16}$$

will have solutions only for those values of ν which satisfy the usual determinantal condition. This secular equation has $3n$ real roots ν_t^2 ($t = 1, \cdots, 3n$), since $\lambda_{\alpha\beta}$ is Hermitian. The $3n$ independent sets of $\xi_\alpha{}^i$, which we may distinguish by a subscript t, satisfy the orthogonality relation

$$\sum_{i,\alpha}\xi_{\alpha,t}{}^i(\mathfrak{d})\xi_{\alpha,t'}{}^{i*}(\mathfrak{d}) = 0 \tag{17}$$

when $t \neq t'$. These coefficients may be normalized in such a way that this sum is unity when $t = t'$; that is,

$$\sum_{t'}\xi_{\alpha,t}{}^i(\mathfrak{d})\xi_{\alpha,t'}{}^{i*}(\mathfrak{d}) = \delta_{t,t'}. \tag{17a}$$

Thus, we see that in all there are $3nN_1N_2N_3$ independent functions of type (7) when we include the $N_1N_2N_3$ independent values of $\mathfrak{d}$.

The $\xi_\alpha{}^i$ are usually complex when the $\lambda_{\alpha\beta}$ are complex. We may obtain physically interesting real functions, however, by taking the real and imaginary parts of the quantities $\xi_{\alpha,t}{}^i(\mathfrak{d})e^{2\pi i[\sigma\cdot r_\alpha(p) - \nu(\sigma)t]}$. This procedure does not double the modes of vibration, since spatial parts of the $3n$ complex functions associated with $-\mathfrak{d}$ are the complex conjugates of those associated with $+\mathfrak{d}$, as may be seen by taking the complex conjugate of Eq. (16). The real functions have the form

$$x_\alpha{}^i(p) = a_\alpha{}^i \sin 2\pi[\mathfrak{d}\cdot r_\alpha(p) - \nu t] + b_\alpha{}^i \cos 2\pi[\mathfrak{d}\cdot r_\alpha(p) - \nu t]$$

where $a_\alpha{}^i$ and $b_\alpha{}^i$ are now real. Thus the motion of the αth atom is described by the vector

$$\mathbf{A}_\alpha \sin 2\pi[\mathfrak{d}\cdot r_\alpha(p) - \nu t] + \mathbf{B}_\alpha \cos 2\pi[\mathfrak{d}\cdot r_\alpha(p) - \nu t], \tag{18}$$

where $\mathbf{A}_\alpha$ and $\mathbf{B}_\alpha$ are the vectors

$$\mathbf{A}_\alpha = \begin{pmatrix} a_\alpha{}^1 \\ a_\alpha{}^2 \\ a_\alpha{}^3 \end{pmatrix}, \qquad \mathbf{B}_\alpha = \begin{pmatrix} b_\alpha{}^1 \\ b_\alpha{}^2 \\ b_\alpha{}^3 \end{pmatrix}. \tag{19}$$

If **A** and **B** are not equal and if neither is zero, the quantity (18) describes harmonic motion in an elliptical path, the plane of the ellipse being the plane containing the two vectors **A** and **B**. The direction of the major and minor axes may be found in the following way: Let us choose Cartesian coordinates x and y in the plane determined by **A** and **B**. Then, the motion of the particle is given by

$$x = A_x \sin \alpha + B_x \cos \alpha = \sqrt{A_x^2 + B_x^2} \sin \left(\alpha + \tan^{-1} \frac{B_x}{A_x} \right),$$

$$y = A_y \sin \alpha + B_y \cos \alpha = \sqrt{A_y^2 + B_y^2} \cos \left(\alpha - \tan^{-1} \frac{A_y}{B_y} \right),$$

where $\alpha = 2\pi[\mathfrak{d} \cdot \mathbf{r}_\alpha(p) - \nu t]$ and A_x, A_y and B_x, B_y are the components of **A** and **B**. The condition that the ellipse should be in normal form is that

$$\frac{B_x}{A_x} = -\frac{A_y}{B_y}. \tag{20}$$

If θ is the angle between **B** and **A** and if ϕ is the angle that **A** makes with the x axis, we have

$$
\begin{aligned}
A_x &= |\mathbf{A}| \cos \phi, \\
A_y &= |\mathbf{A}| \sin \phi, \\
B_x &= |\mathbf{B}| \cos (\theta + \phi), \\
B_y &= |\mathbf{B}| \sin (\theta + \phi),
\end{aligned}
$$

and (20) becomes

$$|\mathbf{B}|^2 \sin 2(\theta + \phi) = -|\mathbf{A}|^2 \sin 2\phi,$$

which determines ϕ.

Thus, we see that the real normal modes usually describe elliptically polarized elastic waves. They are plane-polarized in the special case in which **A** or **B** is zero, that is, when ξ_α^i is real.

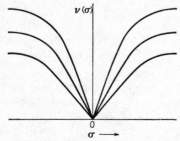

FIG. 10.—Schematic diagram of the $\nu(\sigma)$ curves of a monatomic three-dimensional lattice. Actually the values of σ range over a three-dimensional zone rather than a one-dimensional one so that this corresponds to the $\nu(\sigma)$ relation for a line in σ space that passes through the origin. It should be noted that the $\nu(\sigma)$ relation is linear near the origin.

The $\lambda_{\alpha\beta}$ that appear in (16) are continuous functions of $\mathfrak{d}$. Hence, it may be expected that the frequencies $\nu(\mathfrak{d})$ are continuous functions of $\mathfrak{d}$ in the domain (10). In general, there are $3n$ continuous branches of the frequency curve. These branches may coincide for certain values of $\mathfrak{d}$, particularly if the lattice has a high degree of symmetry. In analogy with the one-dimensional case, there are three branches of the $\nu(\mathfrak{d})$ curve that approach zero linearly as $\mathfrak{d}$ approaches zero. The long wave-length modes of this type correspond to ordinary acoustical waves which travel with constant velocity. Thus, we may expect that the three branches of $\nu(\mathbf{\sigma})$ will behave schematically, as illustrated in Fig. 10, if there is one atom per unit cell. Similarly, the six branches will behave as illustrated in Fig. 11, if there are two atoms in the unit cell, etc.

Before leaving this treatment of the three-dimensional problem, we shall derive expressions for the total energy and for the Lagrangian and Hamiltonian functions. In terms of the $x_\alpha{}^i(p)$, the kinetic energy is

$$T = \sum_{i,\alpha,p} \frac{m_\alpha}{2}[\dot{x}_\alpha{}^i(p)]^2. \qquad (21)$$

The potential energy, which is defined by Eqs. (5), is given by the expression

$$-V = \tfrac{1}{2} \sum_{i,\alpha,p} \sum_{j,\beta,p'} \mu_{\alpha\beta}{}^{ii}(p,p')[x_\beta{}^j(p') - x_\alpha{}^i(p)]x_\alpha{}^i(p)$$

$$\equiv \tfrac{1}{2} \sum_{i,\alpha,p} f_\alpha{}^i(p)x_\alpha{}^i(p). \qquad (22)$$

Let us now express $x_\alpha{}^i(p)$ as a series of the form

$$x_\alpha{}^i(p) = \sum_{t,\sigma} a_t(\mathbf{\sigma}) \frac{\xi_{\alpha,t}{}^i(\mathbf{\sigma})}{\sqrt{m_\alpha N}} e^{2\pi i \sigma \cdot \mathbf{r}_\alpha(p)} \qquad (23)$$

FIG. 11.—Schematic diagram, analogous to Fig. 10 for the diatomic case.

where N is the total number of unit cells and $a_t(\mathbf{\sigma})$ is the time-dependent amplitude of the tth mode having wave number $\mathbf{\sigma}$. We shall assume that the ξ are normalized in the sense of Eq. (17a). The amplitude $a_t(\mathbf{\sigma})$ may then be expressed in terms of the $x_\alpha{}^i(p)$ in the following way. If Eq. (23) is multiplied by

$$\sqrt{\frac{m_\alpha}{N}} \xi_{\beta,t'}{}^i{}^*(\mathbf{\sigma}')e^{-2\pi i \sigma' \cdot \mathbf{r}_\alpha(p)} \qquad (24)$$

and the result is summed over i, α, and p, the right-hand side reduces to $3na_{t'}(\mathbf{\sigma}')$ because of the orthogonality relations. Hence,

$$a_{t'}(\mathbf{\sigma}') = \frac{1}{3n} \sum_{i,\alpha,p} \sqrt{\frac{m_\alpha}{N}} x_\alpha{}^i(p) \xi_{\alpha,t'}{}^i{}^*(\mathbf{\sigma}')e^{-2\pi i \sigma' \cdot \mathbf{r}_\alpha(p)}. \qquad (25)$$

Since the secular equations for $\mathbf{\sigma}$ and $-\mathbf{\sigma}$ are identical, we may choose the index t in such a way that

$$\xi_{\alpha,t}{}^i(-\mathbf{\sigma}) = \xi_{\alpha,t}{}^i{}^*(\mathbf{\sigma}) \qquad (26)$$

and write Eq. (23) in the form

$$x_\alpha{}^i(p) = \frac{1}{\sqrt{m_\alpha N}}\sum_{\sigma,t}{}'' [a_t(\mathbf{\sigma})\xi_{\alpha,t}{}^i(\mathbf{\sigma})e^{2\pi i\sigma\cdot\mathbf{r}_\alpha(p)} + a_t(-\mathbf{\sigma})\xi_{\alpha,t}{}^{i*}(\mathbf{\sigma})e^{-2\pi i\sigma\cdot\mathbf{r}_\alpha(p)}]$$

$$(27)$$

where Σ'' implies that the summation of $\mathbf{\sigma}$ is carried out in such a way that only one of the two points $\pm\mathbf{\sigma}$ is counted. We may now specify that

$$a_t{}^*(\mathbf{\sigma}) = a_t(-\mathbf{\sigma}) \tag{28}$$

in order that (27) may be real. If (27) is substituted into Eqs. (21) and (22), it is found that these equations reduce to

$$\left.\begin{aligned}
T &= \sum_{\sigma,t}{}'' \dot{a}_t(\mathbf{\sigma})\dot{a}_t{}^*(\mathbf{\sigma}), \\
V &= \sum_{\sigma,t}{}'' 4\pi^2 \nu_t^2(\mathbf{\sigma})a_t(\mathbf{\sigma})a_t{}^*(\mathbf{\sigma}),
\end{aligned}\right\} \tag{29}$$

because of the orthogonality conditions. Now it is convenient to replace the complex variable $a_t(\mathbf{\sigma})$ by the two real variables

$$[a_t(\mathbf{\sigma}) + a_t{}^*(\mathbf{\sigma})]\sqrt{2}, \tag{30a}$$

$$[a_t(\mathbf{\sigma}) - a_t{}^*(\mathbf{\sigma})]\frac{\sqrt{2}}{i}. \tag{30b}$$

We shall do this in such a way that the $3n$ variables (30a) are associated with the point $\mathbf{\sigma}$ and the $3n$ variables (30b) are associated with the point $-\mathbf{\sigma}$; that is, we shall introduce real variables $\alpha_t(\mathbf{\sigma})$ defined by the relations

$$\begin{aligned}
\alpha_t(\mathbf{\sigma}) &= \frac{[a_t(\mathbf{\sigma}) + a_t{}^*(\mathbf{\sigma})]}{\sqrt{2}}, \\
\alpha_t(-\mathbf{\sigma}) &= \frac{[a_t(\mathbf{\sigma}) - a_t{}^*(\mathbf{\sigma})]}{i\sqrt{2}}.
\end{aligned} \tag{31}$$

Since the inverses of the equations are

$$\begin{aligned}
a_t(\mathbf{\sigma}) &= \frac{\alpha_t(\mathbf{\sigma}) + i\alpha_t(-\mathbf{\sigma})}{\sqrt{2}}, \\
a_t{}^*(\mathbf{\sigma}) &= \frac{\alpha_t(\mathbf{\sigma}) - i\alpha_t(-\mathbf{\sigma})}{\sqrt{2}},
\end{aligned} \tag{32}$$

Eqs. (29) may be transformed to

$$\left.\begin{aligned}
T &= \frac{1}{2}\sum_{t,\sigma} \dot{\alpha}_t^2(\mathbf{\sigma}), \\
V &= \frac{1}{2}\sum_{t,\sigma} 4\pi^2 \nu_t^2(\mathbf{\sigma})\alpha_t^2(\mathbf{\sigma}),
\end{aligned}\right\} \tag{33}$$

where δ is now summed over the entire zone of wave numbers. Thus, the Lagrangian function of the system is

$$L = T - V = \frac{1}{2}\sum_{t,\sigma}[\dot{\alpha}_t^2(\delta) - 4\pi^2\nu_t^2(\delta)\alpha_t^2(\delta)],\tag{34}$$

whereas the Lagrangian equations are

$$\ddot{\alpha}_t(\delta) = -4\pi^2\nu_t^2(\delta)\alpha_t(\delta),\tag{35}$$

which are the same as those for a linear harmonic oscillator, as we should expect.

The momentum variables $p_t(\delta)$ are defined in terms of the Lagrangian function by the equations

$$p_t(\delta) = \frac{\partial L}{\partial \dot{\alpha}_t(\delta)} \equiv \dot{\alpha}_t(\delta).\tag{36}$$

Thus, the Hamiltonian function is

$$\left.\begin{aligned}H &= \sum_{t,\sigma}p_t(\delta)\dot{\alpha}_t(\delta) - L\\&= \frac{1}{2}\sum_{t,\sigma}[p_t^2(\delta) + 4\pi^2\nu_t^2(\delta)\alpha_t^2(\delta)].\end{aligned}\right\}\tag{37}$$

23. Blackman's Computations.—Blackman[1] has determined by direct computation the frequency-distribution function of the normal modes of several simple lattices and has used the results to determine specific heats. The lattices he treated do not correspond to actual cases; however, the consequences of these computations make it seem reasonable that the discrepancies between experiment and the Debye theory which we listed under *b* in Sec. 20 may be explained, at least in part, by an extension of his work.

We shall discuss the results of two of Blackman's computations, namely, those for the linear lattice with two different atoms and those for a three-dimensional simple cubic lattice.

a. The Linear Lattice.—In Sec. 21, we derived the expression for the energy of the vibrational modes of a one-dimensional lattice containing two atoms. The result is

$$4\pi^2\nu^2(\delta) = \frac{\mu}{mM}(M + m \pm \sqrt{M^2 + m^2 + 2mM\cos 2\pi\delta a})\tag{1}$$

where δ is the wave number, which may be taken to extend from $-1/2a$ to $1/2a$. The distribution of modes as a function of frequency is shown in

[1] BLACKMAN, *op. cit.*

Fig. 9. We may compare the specific heat derived by the use of this distribution function $f(\nu)$ with that derived by the use of the distribution for a continuous string. It is easy to show from the results of the first part of a in Sec. 21 that the distribution function in the second case is

$$f_c(\nu) = \frac{2Na}{c_l}.$$

If we restrict the domain of ν so that the number of modes is $2N$, then

$$f_c(\nu) = \frac{2N}{\nu_m} \qquad 0 \leq \nu \leq \nu_m \tag{2}$$

from which we obtain a one-dimensional Debye specific heat

$$C_V\left(\frac{\Theta}{T}\right) = 2Nk\left(\frac{T}{\Theta}\right)\int_0^{\frac{\theta}{T}} \frac{x^2 e^x}{(e^x - 1)^2}dx \tag{3}$$

where $\Theta = h\nu_m/k$.

Blackman evaluated the specific heat, using the actual distribution

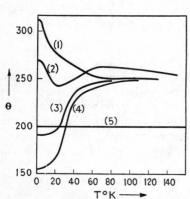

function for the one-dimensional lattice, and then equated this to (3) at each temperature and obtained a value of Θ. The maximum frequency of the lattice was chosen in each case so that $h\nu_m/k = 200°$. The dependence of Θ on temperature for several values of the ratio m/M is shown in Fig. 12. It may be observed that the curves approach a nearly constant value at high temperatures but deviate considerably in the range below 40°K. The details near $T = 0$ are not given exactly in this figure; more precise computations show that the $\Theta(T)$ curve for $m/M = 8$ has a minimum near 10°K.

Fig. 12.—$\Theta(T)$ curves for a one-dimensional lattice. (1) $m/M = 1$, (2) $m/M = 3$, (3) $m/M = 8$, (4) $m/M = 13$, (5) continuum. (*After Blackman.*)

These curves leave little room for doubt that the deviations from Debye's distribution function can be important.

b. Simple Cubic Lattice.—Blackman carried through a similar computation for a monatomic simple cubic lattice (*cf.* Fig. 13). He assumed that each atom interacts only with its six nearest neighbors and twelve next nearest neighbors. For simplicity, he fixed the ratio γ/α of the force constants at 0.05, where α is the Hooke's constant for nearest neighbors and γ is that for next nearest neighbors.

There is one particle per unit cell in this case; hence, the secular equation is of third degree, the three roots corresponding to the three directions of polarization of vibrational waves. It turns out that the three branches of the $\nu(\mathbf{\sigma})$ curve meet at $\mathbf{\sigma} = 0$ and at the eight corners of the cube defined by

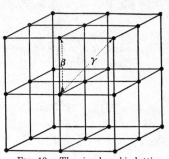

$$\sigma_x = \pm 1/2a,\, \sigma_y = \pm 1/2a,\, \sigma_z = \pm 1/2a,$$

where a is the distance between nearest neighbors. For this reason, the range of frequency happens to be the same for each branch of $\nu(\mathbf{\sigma})$.

Figure 14 shows the relative number of modes as a function of ν for each of the branches of $\nu(\mathbf{\sigma})$. The unit of

FIG. 13.—The simple cubic lattice. The Hooke's constant for nearest neighbors is β. That for next nearest is γ.

frequency has been chosen arbitrarily to make $\nu_m = 1.55$. The fourth continuous curve is the relative distribution of all modes. Actually,

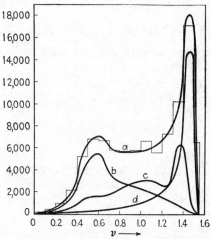

FIG. 14.—Relative scale plot of the frequency distribution of modes of vibration for the simple cubic lattice. Curves b, c, and d correspond to separate directions of polarization; curve a is the sum. (*After Blackman.*)

these curves were determined by approximate means; the stepwise curve illustrates the total distribution function as Blackman computed it.

One striking difference between the three-dimensional and one-dimensional cases is that the peak in $f(\nu)$ does not occur at ν_m in the former. Although the gradient of $\nu(\mathfrak{d})$ does vanish at the corners of the cube in $\mathfrak{d}$ space within which all independent modes lie, this fact does not lead to a peak in $f(\nu)$, for the volume of $\mathfrak{d}$ space in which the gradient of $\nu(\mathfrak{d})$ vanishes is an infinitesimal of higher order. Blackman believes that the three-dimensional and one-dimensional cases usually differ in this respect.

Figure 15 is a plot of the Θ versus T curve that Blackman obtained by comparing the specific heat of the simple cubic lattice with that for the Debye continuum. The absolute units of ν_m were fixed arbitrarily in order that the high temperature Θ should be 144°. It may be seen that Θ varies considerably in the range below 40° and that it does not approach the high-temperature value at 0°K. Thus, if this were an actual crystal, Debye's law would appear to be valid experimentally above 40°K, but the value of Θ_D that would be obtained would differ

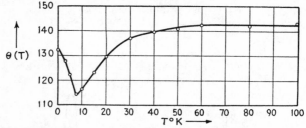

Fig. 15.—The $\Theta(T)$ curve for the simple cubic lattice. (*After Blackman.*)

from the value obtained from observations near absolute zero. In addition, the T^3 law would not be valid below 10°K, as one might expect from Debye's theory.

The substances that conform most closely to Blackman's model are the alkali halides, such as potassium chloride, in which the masses of positive and negative ions are nearly equal. The low-temperature behavior of Θ for KCl has been measured by Keesom and Clark and is described in part b, Sec. 20. Blackman's results do not agree with their experimental results very closely, since Θ passes through a minimum at low temperatures in his model, whereas a maximum actually is observed. It is possible that more extensive assumptions about the interaction forces between neighboring atoms would give better agreement with experiment.

24. The $C_P - C_V$ Correction.—The molar heat ordinarily measured is C_P, the heat at constant pressure. However, the theories we have been discussing in this chapter are based on the assumption that the interatomic distance is kept constant as the temperature changes; hence, they refer to C_V, the molar heat at constant volume. C_P and C_V are connected by the thermodynamical equations[1]

$$C_P - C_V = -T\frac{(\partial V/\partial T)_P^2}{(\partial V/\partial P)_T} \tag{1}$$

where V is the molar volume and P is the pressure. If we set

$$\alpha_V = \frac{1}{V}\left(\frac{\partial V}{\partial T}\right)_P, \qquad \beta = -\frac{1}{V}\left(\frac{\partial V}{\partial P}\right)_T, \tag{2}$$

where α_V is the coefficient of volume expansion and β is the compressibility, Eq. (1) becomes

$$C_P - C_V = TV\frac{\alpha_V^2}{\beta}. \tag{3}$$

The coefficient of volume expansion is practically equal to three times the coefficient of linear expansion; hence,

$$C_P - C_V = TV\frac{9\alpha_l^2}{\beta}. \tag{4}$$

Although α_l is comparatively easy to measure at any temperature, β is usually measured only in the vicinity of room temperature. For this reason, it is necessary to obtain values of β at other temperatures by an extrapolation method of some kind.

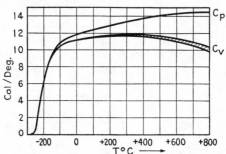

Fig. 16.—The C_P and C_V curves for sodium chloride. The ordinate is in cal/mol-deg.

Figures 16 and 17 show C_P and C_V for sodium chloride and for lead as determined by Eucken and Dannöhl.[2] They used their own values of α_l and values of β that were extrapolated linearly from room-temperature values of Slater and Bridgman. One of the most interesting features of these results is the fact that C_V seems to drop below the Dulong and

[1] See, for example, G. Birtwistle, *The Principles of Thermodynamics*, pp. 71 *ff.* (Cambridge University Press, 1925).

[2] A. Eucken and W. Dannöhl, *Z. Elektrochem.*, **40**, 814 (1934).

Petit value at high temperatures in the case of sodium chloride. The same workers have found a similar drop in silver. Since the rise at high temperatures in the case of lead may be explained in terms of an electronic specific heat (*cf.* Chap. IV), they conclude that the drop observed in the other cases is a common property of the contribution to C_V from lattice vibrations.

Unfortunately, this conclusion rests upon the assumption that the linearly extrapolated values of β are correct. Eucken and Dannöhl

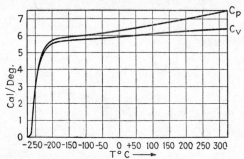

FIG. 17.—The C_P and C_V curves for lead. The ordinate is in cal/mol-deg.

believe that this assumption is justified by the fact that the quantity γ defined by the equation

$$\gamma = \frac{3\alpha_l V}{C_V \beta},\tag{5}$$

in which C_V is their value of the specific heat at constant volume, is practically independent of temperature, the reason for this being that Grüneisen,[1] in developing an equation of state for metals on the assumption of central-force interaction of atoms, found that γ should be temperature-independent. It does not seem entirely safe to accept this result of Grüneisen's theory, however, since the interatomic forces are far from central. If Grüneisen's relation were correct, the determination of C_V would be greatly simplified, for then β could be replaced by γ and we should have

$$\frac{C_P}{C_V} = 1 + 3\gamma\alpha_l T.\tag{6}$$

Thus, γ could be determined from room-temperature measurements, and values of α_l alone would need to be measured at other temperatures.

It should be added that the deviations from Dulong and Petit's law implied by Fig. 17 are not unreasonable, for Born and Brody[2] have shown that potential interaction terms that are cubic in atomic displacements have an effect of this kind at high temperatures.

[1] E. GRÜNEISEN (see *Handbuch der Physik*, Vol. X, for a survey of this work).

[2] M. BORN and E. BRODY, *Z. Physik*, **6**, 132 (1921).

CHAPTER IV

THE FREE-ELECTRON THEORY OF METALS AND SEMI-CONDUCTORS

25. Introduction.—Drude[1] first suggested that the electrical and thermal properties of metals might be correlated by assuming that metals contain free electrons in thermal equilibrium with the atoms of the solid. This hypothesis has passed through several stages of development and remains the cornerstone of the theory of metals. Drude employed the hypothesis to derive approximate expressions for electrical and thermal conductivity. In this work, he introduced the concept of a mean free path for collision of the free electrons, which has also been retained in a modified form in subsequent developments.

Lorentz[2] carried Drude's postulates to their logical conclusion in a more accurate and extensive treatment of the problem. He assumed that the electron velocities in a metal that is in field free space at constant temperature obey the Maxwell-Boltzmann distribution laws, and he determined by an ingenious method the appropriate modification of this distribution when electric fields and temperature gradients are present. Using these results, he was able to make more precise computations of the conductivities than Drude had made. In addition, he was able to treat various thermoelectric effects. As sometimes happens in such cases, Drude's results were in somewhat better agreement with experiment than Lorentz's results. These differences are of minor importance, however, when compared with two major objections to the theory, namely: (1) The manner in which Maxwell-Boltzmann statistics were employed implies that the electrons contribute a larger share of the specific heat of metals than is possible if the Einstein-Debye theory is applicable to atomic vibrations in metals. (2) It was necessary to assume that the electronic mean free path becomes infinite at the absolute zero of temperature in order to explain the vanishing of resistance at absolute zero. The theory presented no plausible reason for this fact.

The theory remained in this unsatisfactory state until after the discovery of the Pauli principle and the development of Fermi-Dirac

---

[1] P. Drude, *Ann. Physik*, **1**, 566 (1900). See H. A. Lorentz, *The Theory of Electrons* (Teubner, Leipzig, 1909 and G. P. Stechert & Co., New York, 1923) for a discussion of this early work.

[2] H. A. Lorentz, *Amsterdam Proc.*, 1904–1905; see also Lorentz, *op. cit.*

statistics.   Then Sommerfeld[1] modified Lorentz's treatment by employing quantum statistics instead of classical statistics.   This procedure, as we shall see, removed practically all the difficulties except those relating to the behavior of the mean free path at low temperatures.   Houston[2] and Bloch,[3] however, were able to justify the occurrence of large mean free paths on the basis of a quantum mechanical investigation of the interaction between electrons and lattice ions.

On the whole, then, Drude's original idea has withstood the test of time.   The free-electron theory merits a thorough discussion, partly for this reason and partly because it furnishes us with a clear semiquantitative picture of some of the most useful properties of metals.   We may point out, however, that the free-electron picture, as we shall present it in this chapter, does not include an interpretation of the cohesive properties of metals and does not explain why some substances are metals and others are not.   These topics can be understood clearly only when solids are treated on the basis of quantum mechanics.

A. METALS

26. Distribution of Electron Velocities.—Following Drude, we shall employ the following simple model of a metal.   We shall assume that

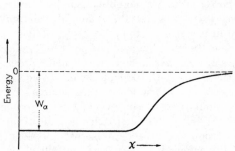

Fig. 1.—Schematic diagram of the potential of a metal.   The value of the potential inside is $-W_a$, that outside is zero.   The major part of the variation between $-W_a$ and zero takes place near the surface.

the electronic potential energy is constant in the interior of the metal and is equal to $-W_a$, relative to an arbitrary zero of potential at infinite distance.   The precise way in which the potential energy varies in the vicinity of the surface need not concern us at present; a schematic diagram of the variation is shown in Fig. 1.   The total energy of a moving

[1] A. Sommerfeld, Z. Physik, **47**, 1 (1928).   See also the review articles: A. Sommerfeld and H. Bethe, Handbuch der Physik, Vol. XXIV/2 (1934).   A. Sommerfeld and N. H. Frank, Rev. Modern Phys., **3**, 1 (1931).

[2] W. V. Houston, Z. Physik, **48**, 449 (1928).

[3] F. Bloch, Z. Physik, **52**, 555 (1928).

electron in the interior of the metal then is

$$E = \frac{1}{2m^*}\mathbf{p}^2 - W_a \tag{1}$$

where $\mathbf{p}$ is the momentum vector. If the electronic potential actually were constant, m^* would be the electronic mass. It happens, however, that it often is possible to approximate the electronic energy by an equation of type (1), even for fairly complex internal potential fields, by allowing m^* to take values different from the electronic mass. The reason for this fact will be made clear in later sections of the book that deal with methods of computing the energies of electrons in solids.

We shall now develop the expressions for the velocity distribution of electrons in a metal in those cases in which classical and Fermi-Dirac statistics[1] are valid. Actually, it is permissible to use only the second type of statistics, but it is interesting to compare the differences between the two forms.

Before proceeding, we shall determine the number of states having energy E, since this quantity is involved in the expressions for the distribution functions. In a rigorous treatment, the degeneracy should be determined by solving the Schrödinger equation; however, the following simple method leads to correct results. We may associate with each electron a six-dimensional phase space of which the six coordinates are the three positional coordinates x, y, z and the three components of momenta p_x, p_y, p_z of an electron. If we arbitrarily divide this phase space into cells of volume h^3, we may obtain the proper density of states by associating two states with each cell. These two states correspond to electrons that move in the same orbit with opposite orientation of electron spin. This procedure may be justified roughly by use of the

---

[1] For a discussion of the differences between Fermi-Dirac and classical statistics, see, for example, G. Joos, *Theoretical Physics*, Chap. 37 (G. E. Stechert & Company, New York, 1934); also Sommerfeld and Bethe, *op. cit.*, and L. Brillouin, *Die Quantenstatistik* (Julius Springer, Berlin, 1930).

In classical statistics, the number of particles in an assembly that are in the level of degeneracy g_i and energy ϵ_i is

$$n_i = g_i A e^{-\frac{\epsilon_i}{kT}}$$

where A is a constant that is to be adjusted so that the sum of n_i over all levels is equal to the total number of particles.

In Fermi-Dirac statistics, which is valid for electrons,

$$n_i = \frac{g_i}{e^{\frac{\epsilon_i - \epsilon'}{kT}} + 1}$$

in which ϵ' is the adjustable parameter analogous to A in the classical case. It may be shown that ϵ' is the free energy per particle.

phase-integral condition of classical quantum mechanics, which states that the volume of phase space associated with each level is h for each positional coordinate. Thus, the volume is h^3 for a particle in ordinary three-dimensional space. A more rigorous justification will be given in a later chapter.

In the present case, in which the electrons are allowed to wander freely throughout the volume V of the metal, we may assume that each cell in phase space extends throughout the volume V associated with the positional coordinates. Then the different cells may be completely specified by giving the domain of momentum space they cover. The average number of cells in the parallelepiped that has edges extending from p_x to $p_x + \Delta p_x$, p_y to $p_y + \Delta p_y$, and p_z to $p_z + \Delta p_z$, etc., is

$$\Delta G = V \frac{\Delta p_x \Delta p_y \Delta p_z}{h^3}, \tag{2}$$

since $V \Delta p_x \Delta p_y \Delta p_z$ is the volume of phase space occupied by the parallelepiped. In a crystal of ordinary size, for which V is greater than 10^{-12} cc, the values of Δp_x, Δp_y, Δp_z that are associated with one cell are infinitesimally small for all practical purposes. Hence, we may replace the discrete distribution of cells by a continuous one and say that the number of cells dG, in the volume $V dp_x dp_y dp_z$ of phase space, is

$$dG = V \frac{dp_x dp_y dp_z}{h^3}. \tag{3}$$

The number dG_s of states of both kinds of spin, associated with the same volume, is twice dG; that is,

$$dG_s = 2V \frac{dp_x dp_y dp_z}{h^3}. \tag{4}$$

We may now derive the expression for the number of levels lying in the range from E to $E + dE$. According to Eq. (1), the relation

$$E = \text{constant}$$

defines a sphere in momentum space of radius $\sqrt{(E + W_a)2m^*}$. The volume dP between concentric spheres the radii of which differ by dE is clearly

$$dP = 4\pi (E + W_a)2m^* d\sqrt{(E + W_a)2m^*}$$
$$= 2\pi \sqrt{(E + W_a)(2m^*)^3} dE. \tag{5}$$

This may be simplified by setting

$$\epsilon = E + W_a, \tag{6}$$

for then

$$dP = 2\pi(2m^*)^{\frac{3}{2}}\sqrt{\epsilon}d\epsilon. \tag{7}$$

The variable ϵ evidently measures the electronic energy relative to $-W_a$. According to Eq. (4) the number of states associated with this volume of momentum space is

$$g(\epsilon)d\epsilon = \frac{4\pi V(2m^*)^{\frac{3}{2}}}{h^3}\sqrt{\epsilon}d\epsilon = C\sqrt{\epsilon}d\epsilon \tag{8}$$

where

$$C = \frac{4\pi V(2m^*)^{\frac{3}{2}}}{h^3} \tag{9}$$

and $g(\epsilon)$ is the density of states. It should be emphasized that Eq. (8) is valid only when (1) is true.

We are now prepared to discuss the distribution function for classical and Fermi-Dirac statistics.

a. Classical Distribution.—The classical, or Maxwell-Boltzmann, distribution function, is[1]

$$n_i = g_i A e^{-\frac{\epsilon_i}{kT}}, \tag{10}$$

where n_i is the number of particles in the level of energy ϵ_i which is g_i-fold degenerate, and A is the normalizing parameter. Thus, according to Eq. (8), the number dn of electrons in the energy range $d\epsilon$ is

$$dn = CAe^{-\frac{\epsilon}{kT}}\sqrt{\epsilon}d\epsilon. \tag{11}$$

If the total number of particles is N, we find, upon integrating (11) over all values of ϵ from zero to infinity, that A is related to N by the equation

$$A = \frac{2}{\sqrt{\pi}}\frac{N}{(kT)^{\frac{3}{2}}}\frac{1}{C} = \frac{h^3 n_0}{2(2\pi m^*kT)^{\frac{3}{2}}} \tag{12}$$

where $n_0 = N/V$ is the number of electrons per unit volume. Hence,

$$dn = \frac{2}{\sqrt{\pi}}\frac{N}{(kT)^{\frac{3}{2}}}e^{-\frac{\epsilon}{kT}}\sqrt{\epsilon}d\epsilon. \tag{13}$$

This result, which is independent of h and of V, shows that the size of the cell in phase space is not important so long as the cell is small enough to permit the use of a continuous distribution of levels. Incidentally, Eq. (13) is identical with the corresponding equation for the Maxwell distribution of kinetic energies of gas molecules.

[1] See footnote 1, p. 141.

The mean energy per electron $\bar{\epsilon}$ is equal to $1/N$ times the integral of ϵdn over all values of ϵ; that is,

$$\bar{\epsilon} = \frac{2}{\sqrt{\pi}} \frac{1}{(kT)^{\frac{3}{2}}} \int_0^\infty e^{-\frac{\epsilon}{kT}} \epsilon^{\frac{3}{2}} d\epsilon$$
$$= \tfrac{3}{2}kT. \tag{14}$$

Thus, according to classical statistics, the total electronic energy per mol of a monatomic metal is

$$E = \tfrac{3}{2}zRT$$

where z is the number of free electrons per atom. This result is contradicted by experiment, if we assume that z is equal to the number of valence electrons per atom, for it predicts an electronic heat per gram atom of $3zR/2$. We have seen in Chap. III, however, that practically all the specific heats of most metals can be ascribed to lattice vibrations. This contradiction is sufficient to rule out the use of classical statistics for describing the distribution of the free electrons in metals.

b. *Fermi-Dirac Distribution.*—The quantum statistical distribution function that should replace (10) is[1]

$$n_i = \frac{g_i}{e^{-\alpha + \frac{\epsilon_i}{kT}} + 1} = \frac{g_i}{e^{\frac{(\epsilon_i - \epsilon')}{kT}} + 1} \tag{15}$$

where, for convenience, α has been replaced by ϵ'/kT. This equation becomes

$$dn = C \frac{\sqrt{\epsilon} d\epsilon}{e^{\frac{(\epsilon - \epsilon')}{kT}} + 1} \tag{16}$$

when the continuous distribution method is used.

The parameter ϵ', which must be fixed in such a way that the integral of (16) is equal to the total number of electrons, cannot be determined so easily as the corresponding parameter in the classical case. We shall evaluate ϵ' for several limiting cases, using different methods in each one.

1. *Absolute zero.*—The form of

$$f\left(\frac{\epsilon - \epsilon'}{kT}\right) = \frac{1}{e^{\frac{\epsilon - \epsilon'}{kT}} + 1} \tag{17}$$

as a function of ϵ is illustrated in Fig. 2 for various relations between T and ϵ'. In the limiting case of absolute zero, (17) is unity when ϵ is less

---

[1] See footnote 1, p. 141.

than ϵ_0', the value of ϵ' at 0°K, and is zero for greater values of ϵ.   Hence, Eq. (16) may be written

$$dn = \begin{cases} C\sqrt{\epsilon}\,d\epsilon & 0 \le \epsilon \le \epsilon_0' \\ 0 & \epsilon_0' < \epsilon \end{cases} \tag{18}$$

That is, at absolute zero all the energy levels below ϵ_0' are completely filled with electrons, whereas all above ϵ_0' are completely empty (*cf.* Fig. 3).   Also, all cells in momentum space for which p is less than

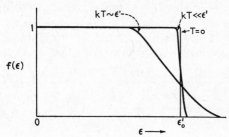

FIG. 2.—The Fermi-Dirac distribution function for several relationships between T and ϵ'.

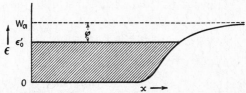

FIG. 3.—The filling of the lowest energy levels of the metal at absolute zero.   All levels below $\epsilon = \epsilon_0'$ are completely occupied; all above are empty.   The work function φ is the difference between the top of the filled region and the potential at $x = \infty$.

$\sqrt{2m\epsilon_0'}$ are entirely filled, whereas those that have greater momenta are completely empty.

The integral of (18) is

$$N = C \int_0^{\epsilon_0'} \sqrt{\epsilon}\,d\epsilon = \frac{2}{3}C\epsilon_0'^{\frac{3}{2}}. \tag{19}$$

Hence,

$$\epsilon_0' = \frac{h^2}{2m^*}\left(\frac{3n_0}{8\pi}\right)^{\frac{2}{3}} \tag{20}$$

where $n_0 = N/V$, as previously.   It is interesting to note that ϵ_0' depends only upon the density of particles.

Using Eq. (20) we find that the mean energy is

$$\bar{\epsilon}_0 = \frac{1}{N}\int_0^{\epsilon_0'} C\epsilon\sqrt{\epsilon}\,d\epsilon = \frac{3}{5}\epsilon_0' = \frac{3}{10}\frac{h^2}{m^*}\left(\frac{3n_0}{8\pi}\right)^{\frac{2}{3}}. \tag{21}$$

The value of ϵ_0' for a number of metals is given in Table XXXVIII, in which we have assumed that m^* is the actual electronic mass. Since these numbers are of the magnitude of several electron volts, we see that the results of quantum statistics are appreciably different from those of classical statistics, in which $\bar{\epsilon}$ is zero at absolute zero. Thus, in classical statistics, an energy W_a would be required to remove an electron from a metal, whereas, in quantum statistics, the minimum energy required is $\phi = W_A - \epsilon_0'$ (*cf.* Fig. 3). The quantity ϕ, which is called the work function, may be evaluated experimentally by determining the light quantum of lowest energy that will eject electrons from a metal. Photoelectrically determined[1] values of ϕ are listed in Table XXXVIII. ϕ is greatly dependent on the condition of the surface through which elec-

TABLE XXXVIII

Metal	Valence	ϵ_0', ev	ϕ, ev	$W_a = \epsilon_0' + \phi$
Li	1	4.72	2.2	6.9
Na	1	3.12	1.9	5.0
K	1	2.14	1.8	3.9
Cu	1	7.04	4.1	11.1
Ag	1	5.51	4.7	10.2
Au	1	5.54	4.8	10.3
Be	2	14.3		
Ca	2	4.26	3.2	7.5
Al	1	5.63	3.0	8.6
Al	3	11.7	3.0	14.7

trons are ejected, and the values in Table XXXVIII are averages for the cleanest surfaces that have been obtained.

We shall see below that at low temperatures $\bar{\epsilon}$ is equal to ϵ_0' plus a term that varies with temperature as $(kT)^2/\epsilon_0'$. Since this term ordinarily is very small compared with unity, the complete filling of the lowest energy levels that are shown in Fig. 3 is not appreciably altered at ordinary temperatures.

2. *The case in which kT is small in comparison with ϵ_0'.*—This case is a very useful one since ϵ_0' is much larger than kT below the melting point of most metals.

The equation for determining ϵ' in this case, as well as the general one, is

$$N = C \int_0^\infty \frac{\sqrt{\epsilon}\,d\epsilon}{e^{\frac{\epsilon - \epsilon'}{kT}} + 1}. \qquad (22)$$

[1] See, for example, the compilations by J. A. Becker, *Rev. Modern Phys.*, **7**, 95 (1935), and A. L. Hughes and L. A. DuBridge, *Photoelectric Phenomena* (McGraw-Hill Book Company, Inc., New York, 1932).

Sommerfeld and Bethe[1] have developed a convenient method for evaluating integrals of this type.   Let us write Eq. (22) in the form

$$N = C\int_0^\infty f\sqrt{\epsilon}d\epsilon, \tag{23}$$

using the notation of Eq. (17).   Integrating this expression by parts, we obtain

$$N = -C\int_0^\infty \frac{df}{d\epsilon}\frac{2}{3}\epsilon^{\frac{3}{2}}d\epsilon. \tag{24}$$

It will be found that a number of important integrals may be expressed in an analogous form, namely,

$$a = C\int_0^\infty \frac{df}{d\epsilon}\alpha(\epsilon)d\epsilon \tag{25}$$

where $\alpha(\epsilon)$ is a continuous function of ϵ.

Following Sommerfeld and Bethe, let us transform the integration variable in (25) from ϵ to η defined by

$$\eta = \frac{\epsilon - \epsilon'}{kT}.$$

This is equivalent to choosing ϵ' to be the origin of the energy scale and expressing the energy in units of kT.   In addition, let us write

$$\alpha(\epsilon) = \beta(\eta).$$

Then, Eq. (25) becomes

$$a = C\int_{-\frac{\epsilon'}{kT}}^\infty \beta(\eta)\frac{df(\eta)}{d\eta}d\eta. \tag{26}$$

For small values of kT, ϵ'/kT is so large that the lower limit may be replaced by $-\infty$, whence

$$a \cong C\int_{-\infty}^\infty \beta(\eta)\frac{df(\eta)}{d\eta}d\eta. \tag{27}$$

Let us assume that β may be expanded into a Taylor series.   Then,

$$\beta(\eta) = \beta(0) + \left[\frac{\partial\beta}{\partial\eta}\right]_0\eta + \frac{1}{2}\left[\frac{\partial^2\beta}{\partial\eta^2}\right]_0\eta^2 + \cdots,$$

---

[1] SOMMERFELD and BETHE, *op. cit.*

and Eq. (27) is

$$a = C \int_{-\infty}^{\infty} \frac{\partial f}{\partial \eta} \left(\beta(0) + \left[\frac{\partial \beta}{\partial \eta} \right]_0 \eta + \frac{1}{2} \left[\frac{\partial^2 \beta}{\partial \eta^2} \right]_0 \eta^2 + \cdots \right) d\eta.$$

The first integral is

$$\beta(0) C \int_{-\infty}^{\infty} \frac{\partial f}{\partial \eta} d\eta = \beta(0) C [f(\eta)]_{-\infty}^{\infty} = -\beta(0) C.$$

The second integral vanishes because $\partial f / \partial \eta$ is an even function of η.

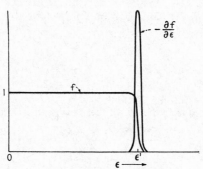

FIG. 4.—The functions f and $-\dfrac{\partial f}{\partial \epsilon}$ as functions of ϵ (schematic).

The third term, which is the only other one that need be considered in the low-temperature approximation, is

$$-\frac{1}{2} \left[\frac{\partial^2 \beta}{\partial \eta^2} \right]_0 C \int_{-\infty}^{\infty} \frac{\eta^2 e^{-\eta}}{(e^{-\eta} + 1)^2} d\eta. \tag{28}$$

This may be evaluated by expanding the integrand in powers of $e^{-\eta}$. The integral then is

$$2 \int_0^{\infty} \eta^2 (e^{-\eta} - 2e^{-2\eta} + 3e^{-3\eta} - \cdots) d\eta = 4 \left(1 - \frac{1}{2^2} + \frac{1}{3^2} - \frac{1}{4^2} + \cdots \right)$$

$$= \frac{\pi^2}{3}.$$

Hence,

$$a \cong -C \left(\beta(0) + \frac{\pi^2}{6} \left[\frac{\partial^2 \beta}{\partial \eta^2} \right]_0 \right).$$

Transforming back to the variable ϵ, we have

$$a \cong -C \left(\alpha(\epsilon') + \frac{\pi^2}{6} (kT)^2 \left[\frac{\partial^2 \alpha}{\partial \epsilon^2} \right]_{\epsilon = \epsilon'} \right). \tag{29}$$

It is easy to see that the infinite series of which the terms in (29) are the first few members converges very rapidly when kT is much less than ϵ'. The form of the function $df/d\epsilon$ for this case is illustrated in Fig. 4. It is obvious that this function vanishes everywhere except in a region of width kT at $\epsilon = \epsilon'$, where it has a steep maximum. Since

$$\int_{-\infty}^{\infty} \frac{df}{d\epsilon} d\epsilon = -1,$$

the main term in the expansion of a is simply $-C\alpha(\epsilon')$, the value of the integrand at ϵ'. The additional terms are corrections for the finite width of the function $\partial f/\partial\epsilon$, and are small as long as the width is small.

Let us return to Eq. (24). Using Eq. (29), we find that

$$N = C\left(\frac{2}{3}\epsilon'^{\frac{3}{2}} + \frac{\pi^2}{12}\frac{(kT)^2}{\epsilon'^{\frac{1}{2}}}\right), \tag{30}$$

which reduces to Eq. (19) when T becomes zero. We may solve this for ϵ' when T is not zero by replacing ϵ' in the denominator of the second term by ϵ_0'. It is found in this way that

$$\epsilon' \cong \epsilon_0'\left[1 - \frac{\pi^2}{8}\left(\frac{kT}{\epsilon_0'}\right)^2\right]^{\frac{2}{3}}$$

$$\cong \epsilon_0'\left[1 - \frac{\pi^2}{12}\left(\frac{kT}{\epsilon_0'}\right)^2\right]. \tag{31}$$

The mean energy per electron is

$$\bar{\epsilon} = \frac{C}{N}\int_0^{\infty} f\epsilon^{\frac{3}{2}}d\epsilon = -\frac{C}{N}\int_0^{\infty}\frac{df}{d\epsilon}\frac{2}{5}\epsilon^{\frac{5}{2}}d\epsilon. \tag{32}$$

When Eqs. (19) and (29) are used, this reduces to

$$\bar{\epsilon} = \frac{C}{N}\left(\frac{2}{5}\epsilon'^{\frac{5}{2}} + \frac{\pi^2}{4}\epsilon'^{\frac{1}{2}}(kT)^2\right)$$

$$= \frac{3}{2}\frac{1}{\epsilon_0'^{\frac{3}{2}}}\left(\frac{2}{5}\epsilon'^{\frac{5}{2}} + \frac{\pi^2}{4}\epsilon'^{\frac{1}{2}}(kT)^2\right). \tag{33}$$

Upon simplifying this equation by means of Eq. (31), we find

$$\bar{\epsilon} = \frac{3}{5}\epsilon_0'\left[1 + \frac{5}{12}\pi^2\left(\frac{kT}{\epsilon_0'}\right)^2\right]$$

$$= \bar{\epsilon}_0\left[1 + \frac{5}{12}\pi^2\left(\frac{kT}{\epsilon_0'}\right)^2\right] \tag{34}$$

where $\bar{\epsilon}_0$ is the value of $\bar{\epsilon}$ for absolute zero [*cf.* Eq. (21)].

The derivative of Eq. (34) with respect to T is the electronic heat γ_V; that is,

$$\gamma_V = k \frac{\pi^2}{2} \frac{kT}{\epsilon_0}. \tag{35}$$

A linear term, which ordinarily is much smaller than the classical value of $3k/2$, has been observed in a number of metals and will be discussed in Secs. 27 and 28. It is easy to see from the method we have used to arrive at Eq. (34) that only the part of the electronic distribution near $\epsilon = \epsilon'$ contributes to the temperature-dependent part of $\bar{\epsilon}$. The electrons of lower energy are hemmed in by filled cells to such an extent that ordinarily they are not excited. Thus, only those electrons that are in the energy range of width kT near the top of the occupied levels are free in the classical sense. In fact, one may obtain a specific heat of the same order of magnitude as (35) by assuming that a fraction kT/ϵ' of the electrons are free and have the classical electronic heat $3k/2$.

Equation (29) may be used to derive expressions for the mean value of various quantities in cases more general than that in which $g(\epsilon)$ has the form (8). Suppose that $g(\epsilon)$ is an arbitrary function of ϵ. Then, the integrals for N and $\bar{\epsilon}$ are

$$N = \int_0^\infty g(\epsilon) f d\epsilon = -\int_0^\infty \left[\int_0^\epsilon g(x) dx \right] \frac{df}{d\epsilon} d\epsilon \tag{36}$$

$$N\bar{\epsilon} = \int_0^\infty \epsilon g(\epsilon) f d\epsilon = -\int_0^\infty \left[\int_0^\epsilon x g(x) dx \right] \frac{df}{d\epsilon} d\epsilon. \tag{37}$$

Using Eq. (29), we find that these expressions may be reduced to

$$N = \int_0^{\epsilon'} g(\epsilon) d\epsilon + \frac{\pi^2}{6} (kT)^2 g'(\epsilon'), \tag{36a}$$

$$N\bar{\epsilon} = \int_0^{\epsilon'} \epsilon g(\epsilon) d\epsilon + \frac{\pi^2 (kT)^2}{6} [\epsilon' g'(\epsilon') + g(\epsilon')]. \tag{37a}$$

The total electronic heat may be derived from (37a) by differentiating this equation with respect to T and is

$$N\gamma_V = \left\{ \epsilon' g(\epsilon') + \frac{\pi^2 k^2 T^2}{6} [2g'(\epsilon') + \epsilon' g''(\epsilon')] \right\} \frac{d\epsilon'}{dT} + \frac{\pi^2 k^2 T}{3} [\epsilon' g'(\epsilon') + g(\epsilon')]. \tag{38}$$

An expression for the derivative of ϵ' with respect to temperature may be obtained by taking the temperature derivative of Eq. (36a). The result is

$$\frac{d\epsilon'}{dT} \simeq -\frac{\pi^2}{3} k^2 T \frac{g'(\epsilon')}{g(\epsilon')}.$$

Substituting this in Eq. (38), we find

$$\gamma_V \cong \frac{\pi^2 k^2 T}{3} \frac{g(\epsilon')}{N}. \tag{39}$$

Equation (35) is the special case of this equation in which g has the value $C\sqrt{\epsilon}$.

3. *The case in which ϵ' is large and negative.*—We shall see that this case is of interest at very high temperatures. When ϵ' is negative, the quantity

$$e^{\frac{\epsilon'}{kT}} \tag{40}$$

is less than unity. Hence, we may expand the expressions for N and $\bar{\epsilon}$ in terms of it. The results, to terms in the first power of (40), are

$$N \cong C e^{\frac{\epsilon'}{kT}} \int_0^\infty e^{-\frac{\epsilon}{kT}} \sqrt{\epsilon} \, d\epsilon, \tag{41a}$$

$$\bar{\epsilon} \cong \frac{C e^{\frac{\epsilon'}{kT}}}{N} \int_0^\infty e^{-\frac{\epsilon}{kT}} \epsilon^{\frac{3}{2}} d\epsilon. \tag{41b}$$

These equations become identical with those for the case of classical statistics, which was discussed in part a, when we set

$$A = e^{\frac{\epsilon'}{kT}}.$$

This result shows that classical statistics are valid under the conditions in which (40) is small in comparison with unity. Substituting A from Eq. (12), we find that

$$\frac{\epsilon'}{kT} = \log \frac{h^3 n_0}{2(2\pi m k T)^{\frac{3}{2}}}.$$

The condition that must be satisfied if this is to be negative and large is

$$kT \gg \left(\frac{16}{9\pi}\right)^{\frac{1}{3}} \frac{h^2}{2m} \left(\frac{3n_0}{8\pi}\right)^{\frac{2}{3}} = \left(\frac{16}{9\pi}\right)^{\frac{1}{3}} \epsilon_0'.$$

For most metals, this condition is satisfied only at temperatures far above the melting point.

4. *Intermediate case.*—The evaluation of integrals in the intermediate case, in which kT is comparable with ϵ_0', involves a relatively large amount of computation and will not be discussed here. Some of the more important results have been derived by Mott[1] and by Stoner.[2]

---

[1] N. F. Mott, *Proc. Roy. Soc.*, **152**, 42 (1935).
[2] E. C. Stoner, *Phil. Mag.*, **21**, 145 (1936).

For example, the behavior of the electronic heat, as a function of kT/ϵ_0', is shown in Fig. 5.

27. The Specific Heats of Nontransition Metals.—The expression that was derived in the preceding section for the heat of free electrons in a metal, namely,

$$\gamma_V = \frac{k}{2}\pi^2\frac{kT}{\epsilon_0'} \tag{1}$$

per electron, is small compared with the contribution from lattice vibra-

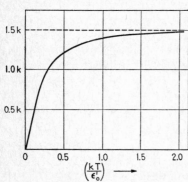

tions at ordinary temperatures. At low temperatures, however, it should become comparable with the Debye value of

$$k\frac{12\pi^4}{5}\left(\frac{T}{\Theta_D}\right)^3 \tag{2}$$

per atom, since the quantity (2) decreases more rapidly with decreasing temperature than (1). The ratio of (1) to (2), namely,

$$\frac{5}{24\pi^2}\frac{kT}{\epsilon_0'}\left(\frac{\Theta_D}{T}\right)^3,$$

approaches unity in the neighborhood of 1°K when Θ_D is of the order of 100° and ϵ_0' is of the order of 1 electron volt. Thus, according to this result, an appreciable part of the specific heat of simple metals in the neighborhood of 1°K should have electronic origin.

Keesom[1] and Kok have observed an electronic specific heat of this type in the simple metals silver, zinc, copper, and aluminum. The most accurate measurements have been made on copper and aluminum, for which the molar heats are

$$\text{Cu: } C_V = 0.888 \cdot 10^{-4}RT + 3Rf_D\left(\frac{335}{T}\right), \tag{3}$$

$$\text{Al: } C_V = 1.742 \cdot 10^{-4}RT + 3Rf_D\left(\frac{419}{T}\right), \tag{4}$$

where $f_D(\Theta/T)$ is the Debye function. Equation (4) is valid only above 1.13°K, for the metal changes to the superconducting phase at this temperature. It is possible to determine an "experimental" ϵ_0' from α,

[1] Ag, Zn: W. H. Keesom and J. A. Kok, *Physica*, **1,** 770 (1934). Cu: J. A. Kok and W. H. Keesom, *Physica*, **3,** 1035 (1936). Al: J. A. Kok and W. H. Keesom, *Physica*, **4,** 835 (1937).

the coefficient of T in the linear term in (3) and (4). In fact, we have from Eq. (1)

$$\epsilon_0'(\text{exp.}) = \frac{\pi^2 kR}{2\alpha}z \tag{5}$$

where z is the number of valence electrons per atom. Values of this quantity are given in Table XXXIX. In the case of copper, it is assumed that there is 1 free electron per atom, whereas with aluminum values are given for the cases in which it is assumed that there are 1 and 3 electrons per atom. Evidence that will be discussed in Chap. XIII indicates that there actually are 3 free electrons per atom. Dividing the computed ϵ_0' of Table XXXIX by the experimental ones, we obtain the ratio of the effective electronic mass to the real mass. These ratios also appear in Table XXXIX.

Equations so precise as (3) and (4) are not given for silver and zinc. The experimenters estimate, however, that m^*/m is of the order of unity in both cases.

TABLE XXXIX

	ϵ_0' (exp.), ev	z	m^*/m
Cu	4.78	1	1.47
Al	2.44	1	2.30
	7.30	3	1.61

28. The Electronic Specific Heats of Transition Metals at Low Temperatures.—Experimental investigations that we shall discuss presently show that transition metals often have a much larger electronic specific heat than do simple metals. In order to treat this topic at the present point, it is necessary to accept some simple consequences of the band theory of solids; these will be justified in later chapters.

According to the band theory,[1] the ten states of an atomic d shell contribute a quasi-continuous band of $10N$ electronic levels to the metal, where N is the total number of atoms. Of these levels, $5N$ correspond to one orientation of spin, and the other $5N$ correspond to the opposite orientation. Figure 6 shows the position and width of the d electron band relative to the levels of the ordinary valence electrons. At present, we shall not discuss the quantum mechanical principles that determine the width ΔE of the d band, and the relative position of the valence and d levels. The density of valence-electron levels is so much less than the density of d levels that the point below which there are $11N$ levels of both valence and d type is above the top of the d band. In other words, the

---

[1] This d electron-band model was first proposed by N. F. Mott, *Proc. Phys. Soc.*, **47**, 571 (1935). and has been extended by J. C. Slater, *Phys. Rev.*, **49**, 537 (1936).

ordinary valence levels are filled above the top of the d band if N electrons are added to the valence band (*cf.* Fig. 6).   Thus, the d band is completely filled in metals such as copper, silver, and gold, which have $11N$ electrons outside the rare gas shells.   Conversely, the d band is not completely filled in the transition metals that immediately precede copper, silver, or gold in the periodic chart.

We shall let $g_v(\epsilon)$ and $g_d(\epsilon)$, respectively, designate the density of levels in the valence and d bands, as functions of energy.   The origin of ϵ will be taken to be at the bottom of the valence band.   For present

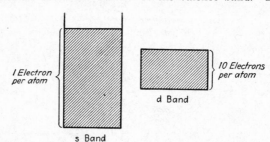

Fig. 6.—The relative widths and the density of levels in the s and d bands.   The shaded area of the s band represents the width of energy levels occupied by one electron per atom. There is room for ten electrons per atom in the d band.   If there are eleven valence electrons per atom, as in copper, silver and gold, both bands are filled as shown in this figure. For the case in which there are ten electrons per atom see Fig. 7.

purposes, we shall assume that $g_v(\epsilon)$ has the form of $g(\epsilon)$ in Eq. (8) of Sec. 26, namely,

$$g_v(\epsilon) = C\sqrt{\epsilon} \tag{1}$$

where the mass appearing in C does not differ from the true electronic mass by a factor larger than 2 or 3.   As we shall see, there is good evidence that near the bottom of the d band $g_d(\epsilon)$ has the same form as the expression (1) for free electrons, namely,

$$g_d(\epsilon) = C_b\sqrt{\epsilon - E_d} \qquad \epsilon \geq E_d. \tag{2}$$

where

$$C_b = \frac{4\pi V (2m_b)^{\frac{3}{2}}}{h^3} \tag{3}$$

in which m_b is the effective mass at the bottom of the band.   We shall assume that near the top of the band

$$g_d(\epsilon) = C_t\sqrt{(E_d + \Delta E - \epsilon)} \qquad \epsilon \lesssim E_d + \Delta E \tag{4}$$

where

$$C_t = \frac{4\pi V (2m_t)^{\frac{3}{2}}}{h^3} \tag{5}$$

in which m_t is the effective mass near the top of the band.

It may be expected that m_b and m_t will turn out to be much larger than the electronic mass, since the density of levels in the d band is much larger than the density of levels in the valence-electron band.

According to Eq. (39), which is valid when kT is much less than ϵ_0', the electronic heat is

$$\gamma_v = \frac{\pi^2 k^2 T}{3} \frac{g(\epsilon_0')}{N} \tag{6}$$

per electron, where $g(\epsilon_0')$ is the total density of levels at the top of the filled region.   If there are more than $10N$ electrons,

$$g(\epsilon_0') = g_v(\epsilon_0'), \tag{7}$$

since then the d band is completely filled and the valence band is filled beyond the top of the d band.   Hence, we may expect that in this case the electronic heat is of the same magnitude as that of other simple metals.   This was shown to be true in the previous section, for it was found there that the electronic heats of copper, silver, and zinc are the same order of magnitude as that of aluminum.   On the other hand, $g(\epsilon_0')$ is

$$g(\epsilon_0') = g_v(\epsilon_0') + g_d(\epsilon_0') \tag{8}$$

when there are fewer than $11N$ electrons, for then the d band is only partly filled.   We may expect (8) to be much greater than (7), since we have seen that the density of d levels is much greater than the density of valence levels.   Consequently we may expect a larger electronic specific heat for transition metals.

Before presenting the experimental facts for nickel, we shall find it convenient to discuss the implications of ferromagnetism from the energy-band picture.   Let us divide the d band into two bands, each having density $g_d(\epsilon)/2$, corresponding to two opposite orientations of electron spin (*cf.* Fig. 7).   In a paramagnetic transition metal, these two bands are filled to exactly the same level when no external magnetic field is present (*cf.* Fig. 7a).   Thus, the intrinsic magnetic moment is zero in this case.   We may interpret many of the properties of ferromagnetic substances by assuming that at the absolute zero of temperature one of the two d bands is completely filled, whereas the other is filled to the same height as the valence band (*cf.* Fig. 7b).   The metal then possesses an intrinsic magnetic moment per unit volume, since there is an excess of electrons with spin oriented in the direction associated with the filled band.   If β is the magnetic moment per electron, the magnetic moment M per unit volume is seen to be

$$\mathsf{M} = \beta \Delta n_d \tag{9}$$

where Δn_d is the difference between the numbers of electrons per unit volume in the d band having each orientation.   This occurrence of the freezing of electrons in one-half of the d band is accompanied by a decrease in the total energy of the metal.   We shall discuss this in a later chapter. Since all ferromagnetic metals become paramagnetic at sufficiently high temperatures, after a continuous decrease of the intrinsic magnetization, we may conclude that the frozen electronic structure gradually melts as the temperature is raised.

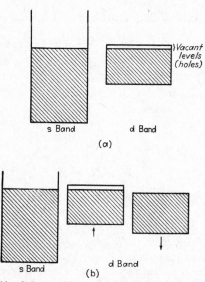

Fig. 7.—Filling of levels in transition metals, which have ten or fewer valence electrons per atom.   Case a, paramagnetic metal in which levels of both spins in the d band are filled to the same height.   Case b, ferromagnetic metal in which the levels in the d band of one spin are preferentially filled.   The levels in the other half of the d band and in the s band are filled to the same height.

The number Δn_d may be determined for ferromagnetic metals from Eq. (9) by using experimental values of M and assuming that β is the Bohr magneton

$$\beta = \frac{eh}{4\pi mc}. \tag{10}$$

This number, which is given for nickel, cobalt, and iron in Table VII, Chap. I, is also equal to the number of electrons per unit volume missing from the unfilled half of the d band.   For this reason, it is often called the number of holes in the d band.   Thus, there is 0.6 hole per atom in the d band of nickel, etc.   Since the nickel d band is nearly filled, we shall assume that the density $g_{d/2}$ of vacant levels in the partly filled half band is half of (4), or that

$$g_{d/2}(\epsilon) = \frac{C_t \sqrt{E_d + \Delta E - \epsilon}}{2}.$$

The parameter ϵ' may be determined from this by means of the following equation which states that the number of holes is $0.6N$, where N is the total number of atoms:

$$0.6 = \frac{C_t}{2N} \int_{\epsilon'}^{E_d + \Delta E} \sqrt{E_d + \Delta E - \epsilon} \, d\epsilon. \tag{11}$$

Integrating this, we find that

$$\delta\epsilon' \equiv E_d + \Delta E - \epsilon_d' = \left(\frac{1.8N}{C_t}\right)^{\frac{2}{3}} \tag{12}$$

where $\delta\epsilon'$ is the width of the unfilled region. Hence, the density of levels at the point ϵ' in the partly filled half of the d band is

$$g_{d/2}(\epsilon') = \frac{C_t^{\frac{2}{3}}}{2}(1.8N)^{\frac{1}{3}}. \tag{13}$$

According to Eq. (6), the contribution from the d electrons to the electronic heat is

$$\gamma_V = \frac{\pi^2 k^2 T}{3} \frac{C_t^{\frac{2}{3}}}{2N^{\frac{1}{3}}}(1.8)^{\frac{1}{3}} \tag{14}$$

per electron. If we use Eq. (12), we may replace C_t by $\delta\epsilon'$. The result is

$$\gamma_V = 0.3\frac{\pi^2 k^2 T}{\delta\epsilon'}, \tag{15}$$

or the molar heat is

$$C_V = 0.3R\pi^2\frac{kT}{\delta\epsilon'}. \tag{16}$$

Keesom and Clark[1] have found that the molar heat of nickel at very low temperatures can be expressed by the equation

$$C_V = 8.72 \cdot 10^{-4}RT + 3Rf_D\left(\frac{413}{T}\right). \tag{17}$$

The electronic term is about ten times larger than that for simple metals

[1] W. H. KEESOM and C W. CLARK, *Physica*, **2**, 513 (1935).

such as copper or aluminum.   Equating (16) to the first term in (17), we find

$$\frac{\delta\epsilon'}{k} \cong 3400°\text{K},$$

which is about 0.29 ev.   This should be corrected slightly to include the small contribution of the valence electrons to the specific heat.   The result, however, would not be appreciably different.

The ratio of m_t to the true electronic mass may be determined easily from the preceding equations.   The result is

$$\frac{m_t}{m} \sim 28,$$

which shows that the d electrons behave as though they were relatively heavy particles.

Keesom and Kurrelmeyer[1] have measured the specific heat of α iron at low temperatures and find

$$C_v = 0.60 \cdot 10^{-3}RT + 2.36 \cdot 10^{-6}RT^3.$$

In this case, the number of holes in the d band is of the order of 2.2 per atom, if we assume that we may interpret the data of Table VII of Chap. I (page 23) in the same way as for nickel.   The corresponding values of $\delta\epsilon'$ and m_t/m are listed in Table XL.   The fact that m_t/m is smaller than for nickel is partly connected with the fact that Eq. (4) is not valid for the holes in the d band of iron, as will be seen in Chap. XIII.

It has become conventional to associate the specific heat of the d electrons in transition metals with the holes in the d bands.   This procedure is convenient because the use of quantities such as Δn in (9) and $\delta\epsilon'$ in (12) and (16), which are, respectively, the number of holes in the unfilled region and the width of this region, reduces the expression for the specific heat to its simplest form.   We shall see later that this convention has many other advantages.

Large electronic specific heats have been observed in palladium and platinum.[2]   However, it is not possible to determine $\delta\epsilon'$ for these metals from the observed data, for they are not ferromagnetic.   Since these metals occupy positions in the periodic table similar to nickel, we shall assume that they also have 0.6 hole per atom.   These holes are distributed equally among levels of both kinds of spin so that the equation relating C_t and $\delta\epsilon'$ is now different from (12).   Instead of (12), we find

$$\delta\epsilon' = \left(\frac{0.9N}{C_t}\right)^{\frac{2}{3}};$$

[1] W. H. Keesom and B. Kurrelmeyer, *Physica*, **6**, 364 (1939).
[2] Pd: G. L. Pickard, *Nature*, **138**, 123 (1936).
   Pt: J. A. Kok and W. H. Keesom, *Physica*, **3**, 1035 (1936).

however, the equation connection C_V and $\delta\epsilon'$ turns out to be the same as (16). Using the observed electronic heats per mol of metal, namely,

$$Pd: C_V = 1.6 \cdot 10^{-3} RT,$$
$$Pt: C_V = 0.804 \cdot 10^{-3} RT,$$

we obtain the values of $\delta\epsilon'$ and m_t/m given in Table XL.

TABLE XL.—THE EFFECTIVE MASSES OF THE HOLES IN THE d BANDS OF SEVERAL TRANSITION METALS

(Derived from the observed electronic heats on the assumption that the holes are perfectly free)

Metal	$\delta\epsilon'$, ev	m_t/m
Ni	0.29	28
α Fe	1.58	12
Pd	0.16	43
Pt	0.32	22

29. The Pauli Theory of the Paramagnetism of Simple Metals.—It may be seen from Fig. 20, Chap. I, that the metals that follow the rare gases in the periodic chart are weakly paramagnetic. Since it may be shown that the inner closed-shell electrons of these substances give a diamagnetic contribution to the total susceptibility, we may conclude that the paramagnetic susceptibility is associated with the valence electrons. Pauli[1] proposed the following simple semiquantitative interpretation of this paramagnetic term.

Let us divide the quasi-continuous band of levels shown in Fig. 3 into two bands, one for electrons of a given spin and one for electrons of the opposite spin, just as we did for the d band in the preceding section. The density of levels in each of these two bands obviously is just half the density in the band of Fig. 3. In the absence of a magnetic field, each band is

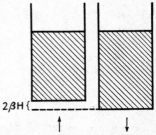

FIG. 8.—The relative displacement of the levels of different spin. The levels with magnetic moment parallel to the field are lowered by βH, those of opposite spin are raised by the same amount. The energy difference $2\beta H$ is exaggerated.

filled to exactly the same value of ϵ. If the metal is placed in a homogeneous magnetic field of intensity H, the band of levels associated with electrons having spin parallel to the field is lowered by an amount βH, and the other band is raised by the same amount. Here, β is the magnetic moment of the electron which, according to the theory of

[1] W. PAULI, *Z. Physik*, **41**, 81 (1927).

electron spin, is given by Eq. (10) of the preceding section. Figure 8 shows schematically the behavior of the levels. Evidently, the equilibrium distribution of electrons is that in which both types of level are filled to exactly the same point on an energy scale, for otherwise we could gain energy by removing electrons from the highest filled band to the lowest. Hence, more electrons have their magnetic moment parallel to the field than antiparallel to it. The number Δn of electrons that leave the antiparallel band and enter the other is equal to the number of electrons in the energy range of width βH at the top of one of the bands, that is,

$$\Delta n = \beta H g_s(\epsilon') \tag{1}$$

where $g_s(\epsilon')$ is the density of levels of given spin at the top of the filled region.   The relation (1) is valid only for fields that satisfy the condition

$$\beta H \ll \epsilon'.$$

All ordinary fields are included in this condition, since βH is of the order of 10^{-3} ev for the strongest attainable fields and ϵ' is of the order of 1 volt. The difference between the number of electrons in the two bands, namely $2\Delta n$, is

$$2\Delta n = 2\beta H g_s(\epsilon'), \tag{2}$$

whence the magnetic moment per unit volume M is

$$M = \frac{2\beta^2 H g_s(\epsilon')}{V}$$

and the susceptibility is

$$\chi = \frac{2\beta^2 g_s(\epsilon')}{V}. \tag{3}$$

We shall assume that the electrons are perfectly free and shall substitute $g_e(\epsilon')$ from Eq. (8) of Sec. 28; that is,

$$g_e(\epsilon') = \frac{g(\epsilon')}{2} = \frac{C}{2}\sqrt{\epsilon'} \cong \frac{3}{4}N\frac{1}{\epsilon_0}.$$

Then,

$$\chi = \frac{3}{2}n_0\beta^2\frac{1}{\epsilon_0'}.$$

The relationship between χ, ϵ_0', the number of valence electrons per atom, Z, and the atomic volume a, is

$$\chi = \frac{81.0Z}{a\epsilon_0'}10^{-6} \tag{4}$$

where χ is now expressed in cgs units, a is expressed in cubic angstroms, and ϵ_0' is expressed in electron volts.

Table XLI contains values of χ, computed from Eq. (4), for the alkali metals and several alkaline earth metals. It should be observed that the computed susceptibilities are often more positive than the observed ones, a fact indicating that a diamagnetic correction is frequently needed. One of the principal sources of this correction is the closed-shell ion cores

TABLE XLI.—A COMPARISON OF THE OBSERVED SUSCEPTIBILITIES OF SEVERAL METALS WITH THOSE COMPUTED ON THE BASIS OF THE FREE-ELECTRON THEORY
(In cgs units)

Metal	$\chi \cdot 10^6$	
	Observed	Calculated
Li	2.0	0.80
Na	0.63	0.65
K	0.58	0.53
Be	−1.85	1.38
Mg	0.87	0.98
Ca	1.70	0.89

of the atoms. In the cases of copper, silver, and gold, the diamagnetic contribution from the newly filled d shells is large enough to cancel the paramagnetic term and to make the metal diamagnetic.

The transition metals such as platinum and palladium are strongly paramagnetic. Since the free-electron theory may be used to explain the electronic heat (*cf.* Sec. 28), we naturally should attempt to apply to these metals the analogue of Eq. (4), namely,

$$\chi = \frac{81.0 Z}{a \delta \epsilon_0'} 10^{-6}$$

where Z is the number of holes per atom. It was first shown by Mott and Jones,[1] and may be verified by simple calculations, that the values of $\delta \epsilon_0'$ that are required to explain the observed values of χ are four or five times smaller than those derived from the experimental values of the electronic heat. This discrepancy shows that the free-electron model is much too simple for the d-shell electrons.

30. Thermionic and Schottky Emission.—Up to this point, we have had no cause to consider the way in which the electronic potential of a metal varies near the surface. The form of this curve is important, however, when we consider any process in which electrons pass through

[1] N. F. MOTT and H. JONES, *Theory of the Properties of Metals and Alloys*, pp. 194 *ff*. (Oxford University Press, 1936).

the surface.   We shall discuss two phenomena of this type in the present section, namely, thermionic emission and Schottky emission.   Thermionic emission[1] is the phenomenon in which an electronic current evaporates from a heated metal in the absence of an external electric field, whereas Schottky emission refers to the evaporation that occurs when the metal is at a negative potential.   Both these emission phenomena are strongly temperature-dependent.   An additional temperature-independent emission occurs when the potential of the metal becomes suffi-

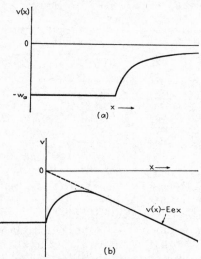

Fig. 9.—*a*, the image-force potential barrier [*cf.* Eq. (3)]; *b*, the effect of a field on the image-force barrier [*cf.* Eq. (5)].

ciently negative.   This *field emission*, which can be explained in terms of the quantum mechanical process of penetration through a barrier, will not concern us in the present chapter.

Suppose that we have an electron at a distance x from the surface of an uncharged metal, where x is large compared with an interatomic distance and small compared with the dimensions of the surface.   Then the only force that acts upon the electron is the classical attractive image force F_i, which is given by the equation

$$F_i = -\frac{e^2}{4x^2} \tag{1}$$

[1] See the following review articles: S. Dushman, *Rev. Modern Phys.*, **2**, 381 (1930); Becker, *op. cit.* See also J. H. de Boer, *Electron Emission and Adsorption Phenomena* (The Macmillan Company, New York, 1935) and A. L. Reimann, *Thermionic Phenomena* (Chapman and Hall, London, 1934).

in which e is the electronic charge.   The potential energy associated
with this force is clearly

$$V_i = -\frac{e^2}{4x}. \tag{2}$$

The form of the potential may be expected to deviate from this when the
surface is not homogeneous, when x is comparable with the interatomic
distance, or when x is comparable with the linear dimensions of the
surface.

We shall assume for simplicity that the actual potential has the form

$$V(x) = \begin{cases} -\dfrac{e^2}{4x + (e^2/W_a)} & x \geq 0, \\ -W_a & x \leq 0, \end{cases} \tag{3}$$

where the region in which x is negative corresponds to the interior of the
metal.   It is clear that this function, which is illustrated in Fig. 9,
changes continuously from $-W_a$ to zero as we pass from the interior
of the metal to infinity and that it approaches (2) for large distances.   In
an actual case, we might expect dV/dx to be continuous; however, the
function (3) does not satisfy this condition.

If the metal is charged negatively, so that the repulsive field is E, the
potential

$$V_e = -\mathsf{E}ex \tag{4}$$

must be added to (3).   The total field V_t then is

$$V_t = \begin{cases} -\dfrac{e^2}{4x + (e^2/W_a)} - \mathsf{E}ex & x \geq 0 \\ -W_a & x \leq 0. \end{cases} \tag{5}$$

The additional field term in (5) has the effect of lowering the height of the
potential barrier at the surface of the metal.   The maximum value of
V_t is

$$V_m \cong -\sqrt{\mathsf{E}}e^{\frac{3}{2}}, \tag{6}$$

as may be proved by solving the equation $dV/dx = 0$.

Hence, when there is a field present, the effective work function φ_E is

$$\varphi_\mathsf{E} = \varphi - \sqrt{\mathsf{E}}e^{\frac{3}{2}}, \tag{7}$$

where φ is the work function defined in Sec. 26.

Let us now compute the number of electrons that evaporate from a
unit area of the metal in unit time.   We shall assume that the energy ϵ
of the electrons inside the metal may be written as a function of the

components of momentum, and we shall choose the x axis to be normal to the surface. Then, as long as the potential is a function of x alone, the y and z components of the momentum of any electron that passes through the surface are preserved. Hence, if a given electron moving toward the surface is to surmount the surface barrier, its total energy ϵ must be greater than the barrier height $W_a - \sqrt{\mathsf{E}}e^{\frac{1}{2}} = \epsilon' + \varphi_{\mathsf{E}}$, by an amount $(p_y^2 + p_z^2)/2m$. Thus, we must have

$$\epsilon \geq \epsilon' + \varphi_{\mathsf{E}} + \frac{1}{2m}(p_y^2 + p_z^2) = \epsilon''. \tag{8}$$

The total number of electrons with momenta in the range p_x to $p_x + dp_x$, etc., striking a unit area of the surface in unit time, is

$$n(p_x,p_y,p_z)v_x dp_x dp_y dp_z = n(p_x,p_y,p_z)\frac{\partial \epsilon}{\partial p_x}dp_x dp_y dp_z, \tag{9}$$

since

$$v_x = \frac{\partial \epsilon}{\partial p_x}. \tag{9a}$$

According to Eqs. (4) and (15), Sec. 26, $n(p_x,p_y,p_z)$, which is the number of electrons per unit volume of phase space, is

$$n(p_x,p_y,p_z) = \frac{2}{h^3}\frac{1}{e^{\epsilon(p_x,p_y,p_z)-\epsilon'} + 1}. \tag{10}$$

The total number of electrons ν that strike a unit area in unit time is equal to the integral of (9) over all values of p_x, p_y, and p_z that satisfy the relation (8). The integration over dp_x may be replaced by an integration over the variable ϵ, since

$$\frac{\partial \epsilon}{\partial p_x}dp_x = d\epsilon.$$

The resulting integral is

$$\nu = \frac{2}{h^3}\int_{-\infty}^{\infty}\int_{-\infty}^{\infty}\int_{\epsilon''}^{\infty}\frac{d\epsilon dp_y dp_z}{e^{\frac{\epsilon-\epsilon'}{kT}} + 1}$$

$$= \frac{2kT}{h^3}\int_{-\infty}^{\infty}\int_{-\infty}^{\infty}\log\left[1 + e^{-\frac{\varphi_{\mathsf{E}} + \frac{(p_y^2 + p_z^2)}{2m}}{kT}}\right]dp_y dp_z. \tag{11}$$

The exponential function appearing in the integrand ordinarily is very small, since φ_{E} is much greater than kT. Hence, we may expand the logarithm and keep only the first terms. The result is

$$\nu \cong \left(\frac{2kT}{h^3}\right)e^{-\frac{\varphi E}{kT}}\int_{-\infty}^{\infty}\int_{-\infty}^{\infty}e^{-\frac{p_y{}^2+p_z{}^2}{2mkT}}dp_y dp_z$$

$$= \frac{8\pi m(kT)^2}{h^3}e^{-\frac{\varphi E}{kT}}\int_0^{\infty}e^{-x^2}x\,dx = \frac{4\pi m(kT)^2}{h^3}e^{-\frac{\varphi E}{kT}}. \tag{12}$$

Equation (12) is the Richardson-Dushman[1] equation which was first derived on the basis of thermodynamics.

If we let r designate the probability that the electrons which have sufficient energy to get over the barrier are reflected back, we find that the current per unit area is

$$I = AT^2(1-r)e^{-\frac{\varphi}{kT}}e^{-\sqrt{E}\frac{e^{\frac{3}{2}}}{kT}} \tag{13}$$

where

$$A = \frac{4\pi mek^2}{h^3} = 120\frac{\text{amp}}{\text{cm}^2\text{deg}^2}. \tag{14}$$

Had we used the classical distribution function

$$n(p_x,p_y,p_z) = \frac{n_0}{(2\pi mkT)^{\frac{3}{2}}}e^{-\frac{\epsilon}{kT}}, \tag{15}$$

where n_0 is the number of electrons per unit volume, we should have found that

$$I = en_0\sqrt{\frac{kT}{2\pi m}}(1-r)e^{-\frac{W_a}{kT}}. \tag{16}$$

This becomes identical with Eq. (13) at high temperatures if we set

$$W_a = \epsilon' + \varphi$$

and if we recall that

$$e^{-\frac{\epsilon'}{kT}} = \frac{2(2\pi mkT)^{\frac{3}{2}}}{h^3 n_0}.$$

On the other hand, the emission corresponding to Eq. (16) is much larger than that corresponding to (13) if W_a in (16) is regarded as the measured work function. As a matter of fact, Eq. (16) gives the same result as Eq. (13) in this case only if a fraction

$$\frac{3\sqrt{\pi}}{4}\left(\frac{kT}{\epsilon_0'}\right)^{\frac{3}{2}}$$

of all electrons is assumed to be free.

The reflection coefficient r is much less than unity if the barrier is described fairly well by the image-force potential. This fact was shown

[1] See footnote 1, p. 162.

first by Nordheim[1] on the basis of wave mechanics. With this simplification, Eq. (13) may be written

$$I = AT^2 e^{-\frac{\varphi}{kT}} e^{-\sqrt{\mathsf{E}}\frac{e^{\frac{3}{2}}}{kT}}. \tag{17}$$

The relative dependence of I upon field strength has been found to obey this equation very closely. Table XLII contains values of A and of φ that have been determined by experimental measurements[2] on thermionic emission. It should be observed that A actually does not have the theoretical value (14) in any case. The observed values do not depart from the calculated ones by a very large factor in most cases, but there are large deviations in a few. The possible interpretations of these deviations may be understood by examining the assumptions upon

Table XLII.—Thermionic Data for Several Metals

Metal	A, amp/cm^2-deg^2	φ, ev
Ca	~ 60	3.2
Cs	~ 160	1.8
Mo	~ 60	4.3
Ni	~ 27	~ 5.0
Pt	$\sim 10^4$	5.0
Ta	~ 50	4.1
Th	~ 60	3.4
W	~ 60	4.5

which the derivation of Eq. (13) was based. We shall present these assumptions categorically.

a. Apparently, we were assuming that the electrons are strictly free inside the metal when we set the mass m that appears in Eq. (8) equal to the electronic mass. Actually, this assumption[3] has not been made, for the momenta and mass in Eq. (8) may be regarded as the values *outside* the metal. It is necessary to assume, however, that the components of momenta in the plane of the surface are preserved as the electrons pass through the surface, if the manner in which Eq. (8) was used in deciding the limits of integration of (11) is to be correct. This assumption is justifiable only if the potential gradient parallel to the surface is zero.

b. We were implicitly assuming that the electrons do not interact appreciably when we used the Fermi distribution function for each

---

[1] L. W. Nordheim, *Proc. Roy. Soc.*, **121**, 626 (1928).

[2] These values of A and φ have been taken from the compilations referred to in footnote 1, p. 162.

[3] *Cf.* Sommerfeld and Bethe, *op. cit.*, p. 436.

electron without including interaction terms. The density of electrons emitted through the surface is so small that it is easy to justify[1] this assumption.

c. The fact that the integration over p_x could be replaced by an integration over ϵ is valid under broader conditions than those contained in our model. In wave mechanics, the integration variable that replaces the x component of momentum is h times the x component of electronic wave number k_x. It turns out that the group velocity in the x direction v_x, is related to the energy by the equation[2]

$$v_x = \frac{1}{h}\frac{\partial \epsilon}{\partial k_x},$$

which is equivalent to the relation

$$v_x = \frac{\partial \epsilon}{\partial p_x}$$

that we used previously.

d. We also assumed that the surface is plane. There seems to be little doubt that even the smoothest surfaces are rough in a submicroscopic sense. Metal surfaces are apparently made[3] of many different crystallographic planes which are inclined relative to one another. The form of the potential function before each plane depends both upon the crystallographic orientation of the plane and upon the nature of any contamination that may be present on it. Although the total potential difference between any point inside the metal and a point at infinity is W_a when there is no external field, the barrier before some of the surfaces may rise to values higher than W_a and drop to W_a at larger distances. Since the highest point of the barrier determines the work function, we should expect the entire surface to behave as though composed of many surfaces which emit more or less in accordance with Eq. (13) but which have different work functions. The variations in surface potential from point to point also imply that there is a tangential force. The existence of this force makes the process of separating the integral (11) into independent integrals over p_x and over p_y and p_z an approximation.

e. The assumption that the reflection coefficient r is zero seems to be justifiable for any reasonably clean surface. It is not justifiable,[4] how-

---

[1] This does not mean that the electrons will not congregate outside the metal and give rise to space-charge effects.

[2] See Chap. VIII.

[3] Direct evidence for this has been given by R. P. Johnson and W. Shockley, *Phys. Rev.*, **49**, 436 (1936).

[4] Evidence for other barriers has been presented by W. A. Nottingham, *Phys. Rev.*, **49**, 78 (1936).

ever, if the potential function has a large peak at a distance comparable with the electronic wave length before it approaches the image-force value. There is no reason for expecting such peaks for clean metal surfaces; they may occur, however, when the surface has been oxidized or when layers of other foreign atoms have been absorbed.

f. Finally, it should be pointed out that in deriving Eq. (13) we have assumed that the work function is independent of temperature. Since metals are heated to very high temperatures during thermionic experiments, this assumption is not justifiable.[1] This topic will be discussed further in Chap. XI.

By way of summary, it may be said that the deviations from the Richardson-Dushman equation probably arise from a combination of effects that are connected with the composite nature of metal surfaces and the temperature dependence of the work function.

31. Boltzmann's Equation of State; Lorentz's Solution*.—Let us consider a system of particles that is in dynamic equilibrium under external forces. For example, the system may consist of the electrons in a metal that is acted upon by stationary external electric and magnetic fields. When the steady-state current is flowing, this system is in a state of dynamic equilibrium of the type we wish to consider.

Let

$$f_n(x,y,z,v_x,v_y,v_z)dxdydzdv_xdv_ydv_z \tag{1}$$

be the number of particles having position coordinates in the range from x to $x + dx$, etc., and velocity coordinates in the range from v_x to $v_x + dv_x$, etc. We may obtain a condition on f_n, whenever the system is in a steady state, by asking that it should be independent of time. Now, f may vary with time in two independent ways: (1) It may vary because particles are moving from one region of space to another and are accelerated by the external field during this motion. This variation, which takes place continuously, is called the "drift variation" and may be evaluated in the following way. The number of particles that, at time $t + dt$, have drifted to the cell of phase space corresponding to the coordinates x, y, z, v_x, v_y, v_z must be equal to the number that were in the cell located at $x - v_x dt$, $y - v_y dt$, $z - v_z dt$, $v_x - \alpha_x dt$, $v_y - \alpha_y dt$, $v_z - \alpha_z dt$ at time t, where α_x, α_y, α_z are the components of acceleration. This relationship holds only for a time interval dt so short that collisions have not had a large effect on the distribution. Thus, the change due to drift in the number of particles having coordinates x, y, and z and velocity v_x, v_y, v_z in a time dt is

---

[1] See J. A. Becker and W. H. Brattain, *Phys. Rev.*, **45**, 694 (1934).

$$(\Delta f)_d = f_n(x - v_x dt, y - v_y dt, z - v_z dt, v_x - \alpha_x dt, v_y - \alpha_y dt, v_z - \alpha_z dt, t) - f_n(x,y,z,v_x,v_y,v_z,t)$$

$$= -\left(\frac{\partial f_n}{\partial x}v_x + \frac{\partial f_n}{\partial y}v_y + \frac{\partial f_n}{\partial z}v_z + \frac{\partial f_n}{\partial v_x}\alpha_x + \frac{\partial f_n}{\partial v_y}\alpha_y + \frac{\partial f_n}{\partial v_z}\alpha_z\right)dt.$$

Consequently the rate of change of f caused by drift is

$$\left(\frac{df_n}{dt}\right)_d = -\frac{\partial f_n}{\partial x}v_x - \frac{\partial f_n}{\partial y}v_y - \frac{\partial f_n}{\partial z}v_z - \frac{\partial f_n}{\partial v_x}\alpha_x - \frac{\partial f_n}{\partial v_y}\alpha_y - \frac{\partial f_n}{\partial v_z}\alpha_z. \qquad (2)$$

(2) f may vary because of the relatively discontinuous changes in velocity that accompany collisions. If

$$\Theta(v_x,v_y,v_z;v_x',v_y',v_z')dv_x'dv_y'dv_z' \qquad (3)$$

is the probability per unit time that a particle will change its velocity from v_x, v_y, v_z to a value having components in the ranges extending from v_z' to $v_z' + dv_z'$, etc., the total number the velocity of which alters from v_x, v_y, v_z to some other value is

$$a = f_n(x,y,z,v_x,v_y,v_z)\int \Theta(v_x,v_y,v_z;v_x',v_y',v_z')dv_x'dv_y'dv_z'. \qquad (4)$$

Similarly, the number the velocity of which changes to v_x, v_y, v_z from another value is

$$b = \int f_n(v_x'',v_y'',v_z'')\Theta(v_x'',v_y'',v_z'';v_x,v_y,v_z)dv_x''dv_y''dv_z''. \qquad (5)$$

Thus, the rate of change of f_n caused by collisions is

$$\left(\frac{df_n}{dt}\right)_c = b - a. \qquad (6)$$

The total rate of change of f_n is the sum of (2) and (6). The condition for equilibrium is that this sum should vanish or that

$$\frac{\partial f_n}{\partial x}v_x + \frac{\partial f_n}{\partial y}v_y + \frac{\partial f_n}{\partial z}v_z + \frac{\partial f_n}{\partial v_x}\alpha_x + \frac{\partial f_n}{\partial v_y}\alpha_y + \frac{\partial f_n}{\partial v_z}\alpha_z = b - a, \qquad (7)$$

which is Boltzmann's equation[1] of state.

In a homogeneous specimen of metal that is at constant temperature in a field-free space, the components of the gradient of f_n, $\partial f_n/\partial x$, $\partial f_n/\partial y$, $\partial f_n/\partial z$, and the components of acceleration α_x, α_y, α_z vanish. Equation (7) then reduces to

$$a = b$$

which states that the numbers of particles that leave and enter a given volume of momentum space as a result of collisions are equal. On the

[1] L. BOLTZMANN, *Vorlesung über Gastheorie* (J. A. Barth, Leipzig, 1923).

other hand, the left-hand side of Eq. (7) does not vanish if there is a temperature gradient in the metal or if there is an external field. Hence, a is not equal to b in this case.

Suppose that we have homogeneous electrical fields in the x and y directions and a homogeneous magnetic field in the z direction. According to classical mechanics, the acceleration $\boldsymbol{\alpha}$ is given by the equation

$$m\boldsymbol{\alpha} = -e\mathsf{E} - \frac{e}{c}\mathbf{v} \times \mathsf{H} \tag{8}$$

where E and H are the electric and magnetic fields, respectively, and e is the absolute value of the electronic charge. For the field assumed above, we have

$$\left.\begin{aligned}
m\alpha_x &= -\left(e\mathsf{E}_x + \frac{e}{c}v_y\mathsf{H}_z\right), \\
m\alpha_y &= -\left(e\mathsf{E}_y - \frac{e}{c}v_x\mathsf{H}_z\right), \\
m\alpha_z &= 0.
\end{aligned}\right\} \tag{9}$$

Hence, Eq. (7) becomes

$$\frac{\partial f_n}{\partial x}v_x + \frac{\partial f_n}{\partial y}v_y + \frac{\partial f_n}{\partial z}v_z - \frac{\partial f_n}{\partial v_x}\left(\frac{e\mathsf{E}_x}{m} + \frac{e}{c}\frac{v_y}{m}\mathsf{H}_z\right) - \frac{\partial f_n}{\partial v_y}\left(\frac{e}{m}\mathsf{E}_y - \frac{e}{c}\frac{v_x}{m}\mathsf{H}_z\right) = b - a. \tag{10}$$

We shall discuss this equation in the case first treated by Lorentz,[1] namely, when the medium is homogeneous and isotropic, no magnetic field is present, and the electrical field is in the x direction. Equation (10) then reduces to

$$\frac{\partial f_n}{\partial x}v_x - \frac{\partial f_n}{\partial v_x}\frac{e\mathsf{E}_x}{m} = b - a \tag{11}$$

since f_n does not depend upon y and z.

Lorentz simplified the collision terms by making the following three assumptions.

1. The electrons undergo only elastic collisions. This assumption seemed to be reasonable at the time it was made since it had been postulated that the electrons were deflected principally by direct collisions with the ions. It may be shown that electrons, being relatively light, would lose little energy in such processes. This interpretation of the electronic collisions is not accepted as completely rigorous at the present time. Nevertheless we shall employ Lorentz's assumption since the results that may be derived from it have semiquantitative value.

[1] H. A. LORENTZ, *The Theory of Electrons* (Teubner, Leipzig, 1909, and G. E. Stechert & Company, New York, 1923).

2. The electronic scattering is isotropic; that is, Θ in (3) is independent of the relative directions of the velocity before and after collisions.

3. The distribution function f_n, for the case when a field is present, is related to the function f_n^0, for the case when no field is present, by the equation

$$f_n = f_n^0 + v_x \chi(v) \tag{12}$$

where χ is a small undetermined function that depends upon the velocity only through the speed $v = \sqrt{v_x^2 + v_y^2 + v_z^2}$. Equation (12) may be regarded as expressing the form of f_n when it is expanded as a series in powers of E_x and only first-power terms are retained.

According to assumption 1, the probability Θ is zero unless

$$v = v'.$$

It is important to note that Θ should be infinite when this condition is satisfied if the total probability of a collision, namely,

$$\int \Theta(v_x, v_y, v_z; v_x', v_y', v_z') dv_x' dv_y' dv_z',$$

is different from zero. In order to avoid the mathematical difficulties that accompany the use of a discontinuous function, we shall introduce a new collision function

$$\eta(v; \theta, \varphi; \theta', \varphi') \sin \theta' d\theta' d\varphi' \tag{13}$$

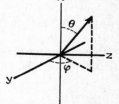

Fig. 10.—The polar angles φ and θ.

which gives the probability that a particle that is traveling with speed v in the direction described by the polar angles θ, φ (*cf.* Fig. 10) is deflected into the solid angle $\sin \theta' d\theta' d\varphi'$ in the direction θ', φ' without a change in speed. Θ and η are obviously connected by the equation

$$\eta(v; \theta, \varphi; \theta' \varphi') = \int_0^\infty \Theta(v_x, v_y, v_z; v_x', v_y', v_z') v'^2 dv'. \tag{14}$$

According to assumption 2, η is isotropic; hence, we see that

$$a = f_n \eta 4\pi, \tag{15}$$

or according to Eq. (12),

$$a = [f_n^0(v) + v_x \chi(v)]4\pi\eta. \tag{16}$$

Substituting (12) in the expression (5), we find

$$b = f_n^0 4\pi\eta + v\chi(v)\eta \int \frac{v_x'}{v'} \sin \theta' d\theta' d\varphi'.$$

The second integral vanishes since $v'_x/v' = \cos \theta'$. Hence,

$$b = f_n^0(v)4\pi\eta, \tag{17}$$

and

$$b - a = -v_x\chi(v)4\pi\eta. \tag{18}$$

We might have foreseen that $b - a$ would not contain f_n^0 from the fact that a is equal to b when E_x is zero.

Since the factor $4\pi\eta$ is the total number of collisions an electron makes in a second, $4\pi\eta/v$ is the total number that it makes in traveling 1 cm, and the reciprocal of this quantity is the mean free path l. Thus,

$$l = \frac{v}{4\pi\eta}. \tag{19}$$

We may now simplify the drift terms by using Eq. (12). Substituting (12) in the left-hand side of Eq. (11), we find

$$\left(\frac{df_n^0}{dt}\right)_{\text{drift}} = v_x\frac{\partial f_n^0}{\partial x} + v_x^2\frac{\partial\chi}{\partial x} - \frac{e}{m}E_x\frac{\partial f_n^0}{\partial v_x} - \frac{e}{m}E_xv_x\frac{\partial\chi}{\partial v_x} - \frac{e}{m}E_x\chi. \tag{20}$$

We may drop the terms in χ since, by assumption, they are considerably smaller than those in f_n^0. Hence, Eq. (11) becomes

$$v_x\frac{\partial f_n^0}{\partial x} - \frac{e}{m}E_x\frac{\partial f_n^0}{\partial v_x} = -\frac{v_xv}{l}\chi(v). \tag{21}$$

We may replace the derivative with respect to v_x by one with respect to ϵ if we use the relation

$$\epsilon = \frac{m}{2}v^2 = \frac{m}{2}(v_x^2 + v_y^2 + v_z^2).$$

Equation (21) then becomes

$$\frac{\partial f_n^0}{\partial x} - eE_x\frac{\partial f_n^0}{\partial \epsilon} = -\frac{v}{l}\chi(v). \tag{22}$$

Solving this equation for χ we obtain

$$\chi = -\frac{l}{v}\left(\frac{\partial f_n^0}{\partial x} - eE_x\frac{\partial f_n^0}{\partial \epsilon}\right), \tag{23}$$

which relates the unknown function χ to the properties of the known distribution function f_n^0.

The process of reasoning that leads to Eq. (23) may be summarized as follows: In using (12), we assume that the distribution function is modified by the addition of a small term if an electric field is present. The form of this term implies that there are more electrons moving in the direction of the force than in the opposite direction, since this term is negative for negative values of v_x if it is positive for positive ones. Since the collision probability is unaffected by the field, the contribution of f_0 to the collision terms is zero in the presence of a field, just as in the absence of one, and the collision terms depend only upon χ. The drift terms are calculated by considering the effect of the field on the unperturbed distribution function. It is assumed that the small additional term in χ does not appreciably affect the drift terms.

The equation for χ reduces to the simple form (23) only as a result of the simplifying assumptions that were made concerning the collision probability Θ. It would not be possible to remove χ from under the integral sign in the expression for b if the particle speeds were not conserved during collisions. Hence, we should arrive at an integral equation in place of (23) in a more general case. This occurs, for example in the quantum mechanical treatment[1] of the equation of state for electrons in metals, since the electrons are found to be inelastically scattered by the lattice when the collision process is examined in the light of quantum mechanical laws.

Gans[2] has generalized the Lorentz equation to include the case in which there are both electric and magnetic fields present. We shall assume that the electric field is in the x and the y directions and that the magnetic field is in the z direction, so that the equation of state is (10). In place of (12), Gans assumes that

$$f_n = f_n^0 + v_x\chi_1 + v_y\chi_2 \tag{24}$$

where χ_1 and χ_2 depend upon the velocity components only through the speed v. In place of Eq. (18) he then obtains

$$b - a = -\frac{v}{l}(v_x\chi_1 + v_y\chi_2). \tag{25}$$

We may substitute Eq. (24) in the drift term of Eq. (9). The expression may be simplified by setting

$$\frac{\partial}{\partial v_x} = mv_x\frac{\partial}{\partial \epsilon} \quad \text{and} \quad \frac{\partial}{\partial v_y} = mv_y\frac{\partial}{\partial \epsilon}.$$

Many terms then cancel, and the result is

$$-\left(\frac{\partial f_n}{\partial t}\right)_d = v_x\frac{\partial f_n^0}{\partial x} + v_y\frac{\partial f_n^0}{\partial y} - \frac{\partial f_n^0}{\partial \epsilon}(v_xe\mathsf{E}_x + v_ye\mathsf{E}_y) - \frac{e}{c}\frac{v_y}{m}\mathsf{H}_z\chi_1 + $$
$$\frac{e}{c}\frac{v_x}{m}\mathsf{H}_z\chi_2. \tag{26}$$

[1] See Chap. XV.
[2] R. GANS, *Ann. Physik*, **20**, 293 (1906).

Terms that involve products of E_x or E_y and χ_1, χ_2 or involve derivatives of the χ have been regarded as negligible in comparison with the terms in E_x and E_y that are retained.

Equating coefficients of v_x and of v_y in Eqs. (25) and (26), we obtain

$$\left.\begin{aligned}
\frac{\partial f_n^0}{\partial x} - e E_x \frac{\partial f_n^0}{\partial \epsilon} &= -\frac{v}{l}\chi_1 - \frac{e}{mc}H_z\chi_2, \\
\frac{\partial f_n^0}{\partial y} - e E_y \frac{\partial f_n^0}{\partial \epsilon} &= \frac{e}{mc}H_z\chi_1 - \frac{v}{l}\chi_2.
\end{aligned}\right\} \tag{27}$$

The result of solving these equations for χ_1 and χ_2 is

$$\left.\begin{aligned}
\chi_1 &= -\frac{l}{v}\frac{f_1 - sf_2}{s^2 + 1}, \\
\chi_2 &= -\frac{l}{v}\frac{f_2 + sf_1}{s^2 + 1},
\end{aligned}\right\} \tag{28}$$

where

$$s = \frac{el}{mvc}H_z = \frac{l}{v}k, \qquad k = \frac{eH_z}{mc}, \tag{29}$$

and

$$\left.\begin{aligned}
f_1 &= \frac{\partial f_n^0}{\partial x} - e E_x \frac{\partial f_n^0}{\partial \epsilon}, \\
f_2 &= \frac{\partial f_n^0}{\partial y} - e E_y \frac{\partial f_n^0}{\partial \epsilon}.
\end{aligned}\right\} \tag{30}$$

32. Electrical and Thermal Conductivity*.—The electrical and thermal currents[1] **i** and **c** that are associated with an electron having velocity **v** are, respectively,

$$\mathbf{i} = -e\mathbf{v} \qquad \text{and} \qquad \mathbf{c} = \mathbf{v}\frac{mv^2}{2}. \tag{1}$$

The total electrical and thermal currents I_x and C_x that pass through a unit area at x, y, z, normal to the x direction of a metal, may be expressed by the integrals

$$I_x = -\int e v_x f_n d\sigma, \tag{2a}$$

$$C_x = \int v_x \frac{1}{2}mv^2 f_n d\sigma, \tag{2b}$$

where f_n is the distribution function of electron velocities, $d\sigma$ is $dv_x dv_y dv_z$, and the integral extends over all values of v_x, v_y, and v_z. I_x and C_x depend upon x, y, and z whenever f_n is dependent upon these variables.

[1] First treated by Sommerfeld, *op. cit.*

The integrals in (2a) and (2b) vanish when f_n is dependent on the components of velocity only through the scalar velocity, for then the current associated with electrons traveling in one direction is exactly compensated by the current associated with electrons traveling in the opposite direction. Thus, the contribution to the currents from f_n^0 in Eqs. (12) and (24) in Sec. 31 is zero.

Let us consider the form of Eqs. (2a) and (2b) in the case in which there are an electric field and a thermal gradient in the x direction. The distribution function then is given by Eq. (12) of the preceding section where χ satisfies Eq. (23) of the same section. Consequently Eqs. (2a) and (2b) become

$$I_x = e \int \frac{v_x^2 l}{v}\left(\frac{\partial f_n^0}{\partial x} - e\mathsf{E}_x\frac{\partial f_n^0}{\partial \epsilon}\right)d\sigma, \tag{3a}$$

$$C_x = -\frac{m}{2} \int v_x^2 lv\left(\frac{\partial f_n^0}{\partial x} - e\mathsf{E}_x\frac{\partial f_n^0}{\partial \epsilon}\right)d\sigma. \tag{3b}$$

The quantities in parentheses in each of these integrals involve v_x, v_y, and v_z only through the scalar velocity v. Hence, we may replace v_x^2 by $v^2/3$, $d\sigma$ by $4\pi v^2 dv$, and the triple integration by a single integral of v extending from zero to infinity. Thus,

$$I_x = \frac{4\pi e}{3} \int_0^\infty v^3 l\left(\frac{\partial f_n^0}{\partial x} - e\mathsf{E}_x\frac{\partial f_n^0}{\partial \epsilon}\right)dv, \tag{4a}$$

$$C_x = -\frac{4\pi m}{6} \int_0^\infty v^5 l\left(\frac{\partial f_n^0}{\partial x} - e\mathsf{E}_x\frac{\partial f_n^0}{\partial \epsilon}\right)dv. \tag{4b}$$

a. Electrical Conductivity.—The simplest case to consider is that in which the temperature of the metal is constant. Then, f_n^0 is independent of x, and the current is

$$I_x = -\frac{4\pi e^2}{3}\mathsf{E}_x \int_0^\infty v^3 l\frac{\partial f_n^0}{\partial \epsilon}dv. \tag{5}$$

We shall use the Fermi-Dirac distribution function for metals. Thus,

$$f_n^0(v_x,v_y,v_z) = \frac{2m^3}{h^3}f(\epsilon) \tag{6}$$

where

$$f(\epsilon) = \frac{1}{e^{\frac{(\epsilon-\epsilon')}{kT}} + 1}. \tag{7}$$

At ordinary temperatures,

$$\epsilon' \cong \epsilon_0' = \frac{h^2}{2m}\left(\frac{3n_0}{8\pi}\right)^{\frac{2}{3}} \tag{8}$$

where n_0 is the number of electrons per unit volume (*cf.* Sec. 26).

It is convenient to change the variable of integration from v to ϵ. Equation (5) then becomes

$$I_x = -\mathsf{E}_x \frac{e^2}{3} \frac{C}{V} \sqrt{\frac{2}{m}} \int \epsilon l \frac{\partial f(\epsilon)}{\partial \epsilon} d\epsilon \tag{9}$$

where [cf. Eq. (9), Sec. 26]

$$\frac{C}{V} = \frac{4\pi(2m)^{\frac{3}{2}}}{h^3} = \frac{3}{2} \frac{n_0}{\epsilon_0'^{\frac{3}{2}}}. \tag{10}$$

We may now employ the approximation expressed by Eq. (29), Sec. 26, and write

$$\int_0^\infty \epsilon l \frac{\partial f(\epsilon)}{\partial \epsilon} d\epsilon = -\epsilon_0' l(\epsilon_0') \tag{11}$$

where $l(\epsilon_0')$ is the value of l for $\epsilon = \epsilon_0'$. This approximation is very accurate for ordinary valence electrons. Then, I_x becomes

$$I_x = \mathsf{E}_x \frac{e^2}{3} \frac{C}{V} \sqrt{\frac{2}{m}} \epsilon_0' l(\epsilon_0') \tag{12}$$

$$= \mathsf{E}_x \frac{e^2}{2} \frac{n_0}{\epsilon_0'^{\frac{3}{2}}} \sqrt{\frac{2}{m}} l(\epsilon_0')$$

$$= \mathsf{E}_x \frac{e^2 n_0 l(\epsilon_0')}{mv(\epsilon_0')}.$$

Hence, the electrical conductivity is

$$\sigma = \frac{I_x}{\mathsf{E}_x} = \frac{e^2 n_0 l(\epsilon_0')}{mv(\epsilon_0')}. \tag{13}$$

According to this equation, the conductivity depends upon temperature through the factor l alone, since all other quantities in Eq. (13) should be practically constant.

b. Thermal Conductivity.—The electronic thermal conductivity may be obtained by solving Eqs. (4a) and (4b) for C_x in the case in which there is a uniform temperature gradient dT/dx in the x direction and in which there is no electrical current flowing. The thermal conductivity κ is defined by the ratio

$$\kappa = -\frac{C_x}{dT/dx}. \tag{14}$$

$\partial f_n^0/\partial x$ is not zero when there is a temperature gradient. In fact, if we recall that f_n^0 is a function of $\alpha = (\epsilon - \epsilon')/kT$ [cf. Eqs. (6) and (7)], we see that

$$\frac{\partial f_n^0}{\partial x} = \frac{\partial f_n^0}{\partial T}\frac{dT}{dx} = kT\frac{\partial f_n^0}{\partial \epsilon}\frac{d\alpha}{dT}\frac{dT}{dx} = -\frac{\partial f_n^0}{\partial \epsilon}\left[\frac{\epsilon}{T} + T\frac{d}{dT}\left(\frac{\epsilon'}{T}\right)\right]\frac{dT}{dx}. \quad (15)$$

Equations (4a) and (4b) now become

$$I_x = -\frac{e}{3}\sqrt{\frac{2}{m}}\left\{K_1\left[e\mathsf{E}_x + T\frac{d}{dT}\left(\frac{\epsilon'}{T}\right)\frac{dT}{dx}\right] + \frac{K_2}{T}\frac{dT}{dx}\right\}, \quad (16a)$$

$$C_x = \frac{1}{3}\sqrt{\frac{2}{m}}\left\{K_2\left[e\mathsf{E}_x + T\frac{d}{dT}\left(\frac{\epsilon'}{T}\right)\frac{dT}{dx}\right] + \frac{K_3}{T}\frac{dT}{dx}\right\}, \quad (16b)$$

where

$$K_i = \frac{C}{V}\int_0^\infty \epsilon^i l\frac{df}{d\epsilon}d\epsilon. \quad (17)$$

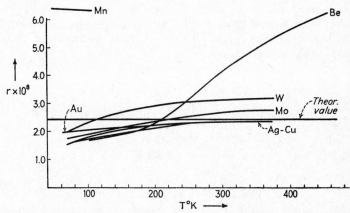

Fig. 11.—The Wiedemann-Franz ratio for several metals. The ratio $r(r = \kappa/T\sigma)$ is given in units in which σ is measured in ohms^{-1} cm^{-1}, κ is measured in watts deg^{-1} cm^{-1}, and T is the absolute temperature. The theoretical value is 2.45×10^{-8} watt-ohm/deg^2

We may eliminate E_x from (16b) by substituting this quantity from (16a), after setting $I_x = 0$. The result is

$$\kappa = -\frac{C_x}{dT/dx} = -\frac{1}{3}\sqrt{\frac{2}{m}}\left(\frac{K_3K_1 - K_2^2}{K_1T}\right). \quad (18)$$

According to Eq. (29), Sec. 26,

$$K_i = -\frac{C}{V}\left\{\epsilon_0'^i l(\epsilon_0') + \frac{\pi^2}{6}k^2T^2\left[\frac{d^2(l\epsilon^i)}{d\epsilon^2}\right]_{\epsilon=\epsilon_0'}\right\}. \quad (19)$$

It is obvious that the total contribution to (18) from the first term of Eq. (19) is zero. The principal contribution from the second term may be reduced to

$$\kappa = \frac{\pi^2}{3}\frac{k^2 n_0 l(\epsilon_0')}{mv(\epsilon_0')}T \quad (20)$$

by straightforward manipulation. We find that

$$\frac{\kappa}{T\sigma} = \frac{\pi^2}{3}\left(\frac{k}{e}\right)^2,$$ (21)

upon comparing this expression with the one for the conductivity. In other words, the free-electron theory predicts that $\kappa/T\sigma$ should be a universal constant. A relationship of this type was observed first by Wiedemann and Franz;[1] for this reason, the ratio $\kappa/T\sigma$ is called the Wiedemann-Franz ratio. Experimental values usually differ somewhat from the theoretical ones. Figure 11 shows plots of the observed[2] ratio for several metals. Silver, copper, and gold obey the theoretical relation fairly well at high temperatures but do not at low temperatures. On the other hand, beryllium and β manganese show large deviations.[3]

33. Electrothermal Effects*.—The rate at which heat accumulates in a unit volume of a wire that is carrying both electrical and thermal currents in the x direction is

$$\frac{dH}{dt} = I_x\mathsf{E}_x - \frac{\partial C_x}{\partial x}.$$ (1)

The first term in this equation is the electrical work done, and the second is the divergence of the heat current which, according to the equation of continuity, is the rate at which heat flows into the volume. We may solve Eq. (16a) of the previous section for E_x thus:

$$\mathsf{E}_x = -\frac{3}{e^2}\sqrt{\frac{m}{2}}\frac{1}{\mathrm{K}_1}I_x - \frac{T}{e}\frac{d}{dT}\left(\frac{\epsilon'}{T}\right)\frac{dT}{dx} - \frac{\mathrm{K}_2}{\mathrm{K}_1}\frac{1}{Te}\frac{dT}{dx}.$$ (2)

The expression (16b) for C_x then becomes

$$C_x = -\frac{\mathrm{K}_2}{\mathrm{K}_1}\frac{1}{e}I_x + \frac{1}{3}\sqrt{\frac{2}{m}}\left(\frac{\mathrm{K}_3\mathrm{K}_1 - \mathrm{K}_2^2}{T\mathrm{K}_1}\right)\frac{dT}{dx}$$ (3)

when E_x is replaced by (2). If we now substitute Eqs. (2) and (3) in Eq. (1), we obtain

$$\frac{dH}{dt} = \frac{I_x^2}{\sigma} + \frac{I_x}{e}T\frac{d}{dx}\left(\frac{\mathrm{K}_2}{T\mathrm{K}_1} - \frac{\epsilon'}{T}\right) + \frac{d}{dx}\left(\kappa\frac{dT}{dx}\right)$$ (4)

where σ and κ are the electrical and thermal conductivities, which are given, respectively, by Eqs. (13) and (20) of the preceding section.

[1] WIEDEMANN and FRANZ, *Ann. Physik*, **89**, 497 (1853).

[2] Taken from Landolt-Bornstein. The measurements were made principally by W. Meissner.

[3] These deviations are most probably associated with the assumption that the electrons are perfectly free.

The first term in Eq. (4) is the Joule heat, and the third is that arising from ordinary flow of heat. Both these terms are independent of the relative directions of the electrical current and the thermal gradient. The second term, on the other hand, represents an accumulation of heat that depends upon both the direction of the current and the direction of the temperature gradient. The influence of this term is usually considered in three separate cases which we shall now discuss.

a. The Thomson Effect.—Suppose that we have a wire carrying a current I_x and having a uniform temperature gradient dT/dx. Then, in accordance with the second term in (4), heat will be produced at a rate

$$\frac{dH}{dt} = \frac{I_x}{e}T\frac{d}{dT}\left(\frac{K_2}{TK_1} - \frac{\epsilon'}{T}\right)\frac{dT}{dx}. \tag{5}$$

This is known as the Thomson heat, and the negative of the coefficient of $I_x dT/dx$, namely,

$$\sigma_T = -\frac{T}{e}\frac{d}{dT}\left(\frac{K_2}{TK_1} - \frac{\epsilon'}{T}\right), \tag{6}$$

is called the Thomson coefficient. The quantity in the parenthesis of (6) may be evaluated by use of Eq. (19) of the preceding section. The first-order terms cancel and the second-order terms lead to the result

$$\frac{1}{T}\left(\frac{K_2}{K_1} - \epsilon'\right) = k^2T\frac{\pi^2}{6}\left[\frac{1}{l\epsilon'}\frac{d^2(l\epsilon'^2)}{d\epsilon'^2} - \frac{1}{l}\frac{d^2(l\epsilon')}{d\epsilon'^2}\right] = k^2T\frac{\pi^2}{3}\left[\frac{1}{\epsilon'_0} + \frac{l'(\epsilon'_0)}{l(\epsilon'_0)}\right]. \tag{7}$$

Hence, σ_T is

$$\sigma_T = -\frac{k^2T}{e}\frac{\pi^2}{3}\left[\frac{1}{\epsilon'_0} + \frac{l'(\epsilon'_0)}{l(\epsilon'_0)}\right]. \tag{8}$$

The quantity $l'(\epsilon'_0)/l(\epsilon'_0)$ cannot be determined without a theory of the mean free path. We shall treat this as an unknown and shall use measured values of σ_T to evaluate it. Values of $\epsilon'_0 l'(\epsilon'_0)/l(\epsilon'_0)$ and of σ_T/T are given in Table XLIII for the alkali metals.[1] It should be noticed that the observed values are similar for all cases except Li which has a different sign and is extremely large. Whether this difference is real or is due to experimental error remains to be seen. The negative value of σ_T for most of the alkalies shows that the electrons carry heat from hot to cold regions of the metal.

b. Peltier Effect.—When a current passes from one metal to another that is at the same temperature (for example, from 1 to 2), some heat usually is evolved or absorbed at the junction. This is known as the Peltier heat: the Peltier coefficient $\pi_{1\to2}$ is defined by the relation

[1] See SOMMERFELD and FRANK, *op. cit.*

$$\pi_{1 \to 2} = -\frac{dH/dt}{I_x}$$

where dH/dt is the heat evolved per unit area of the junction. According to the second term of Eq. (4),

$$\pi_{1 \to 2} = -\frac{T}{e} \int_1^2 \frac{d}{dx}\left(\frac{K_2}{TK_1} - \frac{\epsilon'}{T}\right) dx$$
$$= -\frac{T}{e}\left[\left(\frac{K_2}{TK_1} - \frac{\epsilon'}{T}\right)_2 - \left(\frac{K_2}{TK_1} - \frac{\epsilon'}{T}\right)_1\right]. \tag{9}$$

Hence, comparing this with Eq. (6), we see that

$$\frac{d}{dT}\left(\frac{\pi_{1 \to 2}}{T}\right) = \frac{\sigma_{T,2} - \sigma_{T,1}}{T}. \tag{10}$$

This equation may be derived on the basis of purely thermodynamical

TABLE XLIII

σ_T/T observed, micro-volts/deg^2	$\epsilon_0' l'(\epsilon_0')/l(\epsilon_0')$
Li	+0.40	−76
Na	−0.0282	2.69
K	−0.0275	1.29
Rb	−0.069	4.05
Cs	−0.062	2.88

reasoning.[1] Thus, the agreement between the observed and computed values of (9) should be neither better nor worse than the agreement between observed and computed values of σ_T.

c. *The Seebeck Effect.*—When the junctions of two metals 1 and 2 that are connected in series to form a closed loop are kept at different temperatures T' and T'', an emf acts within the circuit. The Seebeck emf F_S is defined as the value of the total force when the current is zero. F_S may be computed by integrating the expression (2) for E_x around the circuit

$$F_S = -\oint E_x dx = \frac{1}{e}\oint\left[T\frac{d}{dx}\left(\frac{\epsilon'}{T}\right) + \frac{K_2}{K_1 T}\frac{dT}{dx}\right]dx$$
$$= \frac{1}{e}\oint\left(-\frac{\epsilon'}{T}\frac{dT}{dx} + \frac{K_2}{K_1 T}\frac{dT}{dx}\right)dx. \tag{11}$$

---

[1] These effects and others are discussed from the thermodynamical viewpoint by P. W. Bridgman, *The Thermodynamics of Electrical Phenomena in Metals* (The Macmillan Company, New York, 1934).

We may now change the variable of integration from x to T and divide the integral (11) into integrals over each metal.   Then,

$$F_S = -\frac{1}{e}\int_{T'}^{T''}\left[\left(\frac{K_2}{K_1 T} - \frac{\epsilon'}{T}\right)_2 - \left(\frac{K_2}{K_1 T} - \frac{\epsilon'}{T}\right)_1\right]dT.$$

Hence, according to (9),

$$F_S = \int_{T'}^{T''} \frac{\pi_{1\to 2}}{T}\,dT. \tag{12}$$

This equation also may be derived on a purely thermodynamical basis.

34. The Isothermal Hall Effect*.—Let us consider a metal that has electric fields in the x and y directions and a magnetic field in the z direction.   The equation of state then is Eq. (24), Sec. 31, in which χ_1 and χ_2 satisfy Eqs. (28) of the same section.   It may be seen that the electrical and thermal currents in the x and y directions are

$$I_x = -\frac{e}{3}\int v^2\chi_1 d\sigma = \frac{4\pi e}{3}\int vl\frac{f_1 - sf_2}{s^2 + 1}v^2 dv, \tag{1a}$$

$$I_y = -\frac{e}{3}\int v^2\chi_2 d\sigma = \frac{4\pi e}{3}\int vl\frac{f_2 + sf_1}{s^2 + 1}v^2 dv, \tag{1b}$$

$$C_x = \frac{m}{6}\int v^4\chi_1 d\sigma = -\frac{4\pi m}{6}\int v^3l\frac{f_1 - sf_2}{s^2 + 1}v^2 dv, \tag{1c}$$

$$C_y = \frac{m}{6}\int v^4\chi_2 d\sigma = -\frac{4\pi m}{6}\int v^3l\frac{f_2 + sf_1}{s^2 + 1}v^2 dv. \tag{1d}$$

We shall consider the emf induced in the y direction, when there is a current flowing in the x direction and when the metal is at constant temperature.   This transverse field is known as the Hall emf.   The conditions describing this physical situation are

$$\frac{\partial f_n^0}{\partial x} = \frac{\partial f_n^0}{\partial y} = 0 \qquad \text{and} \qquad I_y = 0. \tag{2}$$

Then Eqs. (1a) and (1b) are

$$\left.\begin{aligned}
I_x &= -\frac{4\pi e^2}{3}\left(\mathsf{E}_x\int vl\frac{1}{s^2 + 1}\frac{\partial f_n^0}{\partial\epsilon}v^2 dv - \mathsf{E}_y\int vl\frac{s}{s^2 + 1}\frac{\partial f_n^0}{\partial\epsilon}v^2 dv\right), \\
0 &= -\frac{4\pi e^2}{3}\left(\mathsf{E}_y\int vl\frac{1}{s^2 + 1}\frac{\partial f_n^0}{\partial\epsilon}v^2 dv + \mathsf{E}_x\int vl\frac{s}{s^2 + 1}\frac{\partial f_n^0}{\partial\epsilon}v^2 dv\right).
\end{aligned}\right\} \tag{3}$$

We shall change the variable of integration to ϵ and replace f_n^0 by the Fermi-Dirac function.   These equations then become

$$I_x = -\frac{e^2}{3}\sqrt{\frac{2}{m}}(E_x L_1 - E_y L_2),$$
$$0 = -\frac{e^2}{3}\sqrt{\frac{2}{m}}(E_y L_1 + E_x L_2),$$

(4)

where

$$L_1 = \frac{C}{V}\int_0^\infty \frac{\epsilon l}{s^2 + 1}\frac{\partial f}{\partial \epsilon}d\epsilon, \qquad L_2 = \frac{C}{V}\int_0^\infty \frac{\epsilon l s}{s^2 + 1}\frac{\partial f}{\partial \epsilon}d\epsilon,$$

(5)

and f is the Fermi-Dirac function.

Solving Eqs. (4) for E_x and E_y, we find

$$E_x = -\frac{3}{e^2}\sqrt{\frac{m}{2}}\frac{L_1}{L_1^2 + L_2^2}I_x,$$
$$E_y = \frac{3}{e^2}\sqrt{\frac{m}{2}}\frac{L_2}{L_1^2 + L_2^2}I_x.$$

(6)

The first equation shows that the electrical conductivity in a magnetic field is

$$\sigma(H_z) = -\frac{e^2}{3}\sqrt{\frac{2}{m}}\frac{L_1^2 + L_2^2}{L_1}$$

(7)

Now,

$$L_2 = \frac{el(\epsilon_0')}{mv(\epsilon_0')c}H_z L_1,$$

(8)

whence the second of Eqs. (6) may be written

$$E_y = -\frac{el(\epsilon_0')}{mv(\epsilon_0')c}\frac{H_z I_x}{\sigma(H_z)}.$$

The coefficient of $H_z I_x$ in this is the Hall constant R:

$$R = -\frac{el(\epsilon_0')}{mv(\epsilon_0')c\sigma(H_z)},$$

(9)

which, when $\sigma(H_z)$ is equal to the expression for zero magnetic field, reduces to

$$R = -\frac{1}{n_0 ec}$$

(10)

[cf. Eq. (13) of Sec. 32] where n_0 is the number of electrons per unit volume.[1] It may be shown by tracing through the preceding computation that the negative sign of Eq. (10) arises because the conductivity is related to an electronic current. The sign would be reversed if the carriers were positively charged.

[1] We may see from Eq. (9) that the product $R\sigma$ is the mobility (see footnote 2, p. 68).

TABLE XLIV.—COMPARISON OF OBSERVED HALL CONSTANTS WITH THOSE COMPUTED
ON THE BASIS OF THE FREE-ELECTRON THEORY
(In volts/cm-abamp-gauss.   Experimental values refer to room temperature.)

Metal	$R \cdot 10^{12}$	
	Observed	Calculated
Cu	− 5.5	− 7.4
Ag	− 8.4	−10.4
Au	− 7.2	−10.5
Li	− 17.0	−13.1
Na	− 25.0	−24.4
Be	+ 24.4	− 2.5
Zn	+ 3.3	− 4.6
Cd	+ 6.0	− 6.5
Al	− 3.0	− 3.4
Fe	+100	
Co	+ 24	
Ni	− 60	
Bi	H ⊥ ∼ −1,000	∼ − 4.1
	H ∥ ∼ + 300	

Experimental and theoretical values[1] of R appear in Table XLIV.
The computed numbers are the same order of magnitude as the observed
ones for the monovalent metals.   The signs are opposite for the divalent
metals beryllium, zinc, and cadmium and for iron and cobalt.   The
magnitude of the observed coefficient is one hundred times larger than

TABLE XLV.—THE ELECTRON MOBILITIES OF SEVERAL METALS AS DERIVED FROM
THE PRODUCT OF THE CONDUCTIVITY AND THE HALL CONSTANT
(In cm²/volt-sec)

Metal	μ	Metal	μ
Cu	34.8	Zn	5.8
Ag	56.3	Cd	7.9
Au	29.7	Al	10.1
Li	19.1	Bi	⊥ 9.1
Na	48.0		∥ 2.1
Be	44.4		

[1] See, for example, the compilations of Landolt-Bornstein and the International
Critical Tables.

the theoretical one for bismuth. Both these types of anomaly have been explained semiquantitatively by the band theory of solids.

The product of the Hall constant and the electrical conductivity is the electron mobility μ. Values of this quantity for several metals are given in Table XLV. It may be seen that the mobility differs less from metal to metal than do the conductivity and the Hall constant. Using Eq. (13) of Sec. 32 and these mobilities, we find that $l \sim 5 \cdot 10^{-7}$ cm for the best conductors.

By a straightforward manipulation with Eqs. (29), Sec. 31, we may reduce the expression (7) for $\sigma(\mathsf{H}_z)$ to

$$\sigma(\mathsf{H}_z) = \sigma(0)\left(1 - \frac{B\mathsf{H}_z^2}{1 + C\mathsf{H}_z^2}\right) \tag{11}$$

where

$$B = \frac{\pi^3}{3}\left[\frac{ekTl(\epsilon_0')}{m^2v^3(\epsilon_0')}\right]^2 \tag{12}$$

and

$$C = \sigma^2 R^2, \tag{13}$$

in which R is the Hall constant (10).

Equation (11) predicts that σ should decrease quadratically with H_z

Table XLVI.—A Comparison of the Measured Hall Constants with Those Determined from the Conductivity in Magnetic Fields

(In volts/cm-abamp-gauss)

Metal	$R \cdot 10^{12}$	
	Observed	From $\sigma(\mathsf{H}_z)$
Cu	−5.5	−7.5
Ag	−8.4	−12.5
Au	−7.2	−17.5
Zn	3.3	∼35
Cd	6.0	∼100
Al	−3.0	13
Sb	∼2,000	∼1,000
Bi	⊥ −1,000	∼ −8,000

for weak fields and should asymptotically approach a constant value $\sigma_0\left(1 - \dfrac{B}{C}\right)$ for large field strengths. The observed values of C usually agree in order of magnitude with those determined from directly measured values of the Hall coefficients. Table XLVI contains values of R for copper, silver, and gold, determined directly as well as from measured

values of $\sigma(H_z)$.[1] Figure 12 illustrates the observed dependence of
resistance upon H_z for a specimen of copper. The saturation effect
predicted by Eq. (11) does not appear for the range of field strengths
that has been employed in these metals.   Table XLVI shows values of
R for zinc, cadmium, aluminum, antimony, and bismuth which were
also determined from $\sigma(H_z)$ as well as by direct measurement.   The
agreement is not so good as for the monovalent metals.   The computed
values of B are about 10^4 times smaller than the observed ones at ordinary
temperatures, a fact showing that a more exact treatment of the problem
is necessary.

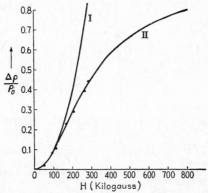

Fig. 12.—Variation of the resistivity of copper in a magnetic field.   The ordinate is
$\Delta\rho/\rho_0$ where ρ_0 is the resistivity in the absence of a field and $\Delta\rho = \rho(H) - \rho_0$.   Curve I is
obtained by fitting the points in the quadratic region near the origin with Eq. (11) when
C is zero.   Curve II is obtained by adjusting both B and C.   The value of C obtained in
this way agrees closely with the theoretical value, whereas the observed B is about 10^4
times larger than the theoretical value.   *(After Sommerfeld and Frank.)*

Qualitative explanations of this discrepancy have been given by
Sommerfeld and Frank and by Sommerfeld and Bethe,[2] and more
accurate treatments have been developed by Jones and Zener,[3] and by
Davis.[4]   It may be shown that B depends upon the fluctuations in
velocity of an electron in the metal.   In the derivation of the relation
(12), it is explicitly assumed that the electronic energy is an isotropic
function of velocity, so that the only fluctuations that occur result from
the fact that the Fermi-Dirac distribution function has a tail of width kT.
In an actual metal it turns out, as we shall see in the following chapters,
that the energy versus velocity function is not usually isotropic, because
of the interactions between electrons and the lattice.   The reader is

[1] Measurements by P. Kapitza, *Proc. Roy. Soc.,* **123**, 292, 342 (1929).
[2] SOMMERFELD and FRANK, *op. cit.*   SOMMERFELD and BETHE, *op. cit.*
[3] H. JONES and C. ZENER, *Proc. Roy. Soc.,* **145**, 268 (1934).
[4] L. DAVIS, *Phys. Rev.,* **56**, 93 (1939).

referred to the paper by Davis for a more complete discussion of this topic.

B. SEMI-CONDUCTORS

35. A Simple Model of a Semi-conductor.—We shall discuss the theory of semi-conductors on the basis of a simple model[1] that is adequate for understanding most of the characteristic features of these substances. We shall assume that there are n_b bound electronic states per unit volume having energy $-\Delta\epsilon$ and that these levels are completely occupied at absolute-zero temperature by n_e electrons. In addition, we shall assume that above these bound states there is a conduction band of levels that

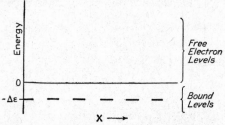

Fig. 13.—Model of a semi-conductor. The ordinate is energy and the abscissa is a symbolical positional coordinate x. The region above zero energy is quasi-continuously dense with free-electron levels. The levels at $-\Delta\epsilon$ correspond to electrons that are bound at particular positions in the lattice.

are completely unoccupied at absolute zero (see Fig. 13). The density $g_c(\epsilon)$ of conduction levels will be taken as

$$g_c(\epsilon)d\epsilon = C\sqrt{\epsilon}d\epsilon \qquad \epsilon > 0 \qquad (1)$$

where for a unit volume,

$$C = \frac{4\pi(2m^*)^{\frac{3}{2}}}{h^3} \qquad (2)$$

[cf. Eq. (9), Sec. 26]. We shall assume that it is permissible to speak of velocity, momentum, etc., for electrons in the conduction band in the conventional way.

Some electrons are thermally excited from the bound states to the conduction band at temperatures above the absolute zero. The number of electrons $n(\epsilon)$ per unit volume having energy ϵ is

$$n(\epsilon) = \frac{g(\epsilon)}{e^{\frac{\epsilon-\epsilon'}{kT}} + 1} \qquad (3)$$

[1] A model similar to this was proposed by A. H. Wilson, *Proc. Roy. Soc.*, **133**, 458 (1931); **134**, 277 (1931).

where $g(\epsilon)$ is the density of levels and ϵ' should be determined from the usual condition

$$\int_{-\infty}^{\infty} n(\epsilon)d\epsilon = n_b. \tag{4}$$

The density function $g(\epsilon)$ is given by Eq. (1) when ϵ is greater than zero, and it is zero for all negative values of ϵ except $-\Delta\epsilon$. According to our assumptions about the density of bound states, the density must be so great at $\epsilon = -\Delta\epsilon$ that

$$\int_{-\infty}^{0} g(\epsilon)d\epsilon = n_b. \tag{5}$$

The discontinuous function defined in this way may be approximated as closely as we please by a continuous function. The only one of its properties we shall use is the relation

$$\int_{-\infty}^{0} f(\epsilon)g(\epsilon)d\epsilon = n_b f(-\Delta\epsilon) \tag{6}$$

where $f(\epsilon)$ is any continuous function of ϵ.

We shall divide the integral in Eq. (4) into two integrals, one extending from $-\infty$ to 0 and the other extending from 0 to ∞. The first integral is

$$\int_{-\infty}^{0} g(\epsilon)\frac{1}{e^{\frac{\epsilon-\epsilon'}{kT}} + 1}d\epsilon = n_b\frac{1}{e^{\frac{-\Delta\epsilon-\epsilon'}{kT}} + 1}, \tag{7}$$

as we may see from Eq. (6). The second integral is

$$C\int_{0}^{\infty}\frac{\sqrt{\epsilon}d\epsilon}{e^{\frac{\epsilon-\epsilon'}{kT}} + 1}. \tag{8}$$

This evidently is equal to the total number of excited electrons, which, to begin with, we shall assume small compared with n_b. It is obvious from the form of the Fermi-Dirac distribution function that the quantity (8) is small only when $-\epsilon' \gg kT$, that is, when $e^{\frac{\epsilon'}{kT}} \ll 1$. We may use this fact to expand (8) in terms of $e^{\frac{\epsilon'}{kT}}$. If we retain only the first term, the result is

$$C\int_{0}^{\infty}\frac{\sqrt{\epsilon}d\epsilon}{e^{\frac{\epsilon-\epsilon'}{kT}} + 1} \cong Ce^{\frac{\epsilon'}{kT}}\int_{0}^{\infty}e^{-\frac{\epsilon}{kT}}\sqrt{\epsilon}d\epsilon = Ce^{\frac{\epsilon'}{kT}}(kT)^{\frac{3}{2}}\frac{\sqrt{\pi}}{2}. \tag{9}$$

Thus, Eq. (4) is

$$Ce^{\frac{\epsilon'}{kT}}(kT)^{\frac{3}{2}}\frac{\sqrt{\pi}}{2} + n_b\frac{1}{e^{\frac{-\Delta\epsilon+\epsilon'}{kT}} + 1} = n_b,$$

or

$$Ce^{\frac{\epsilon}{kT}}(kT)^{\frac{3}{2}}\frac{\sqrt{\pi}}{2} = n_b \frac{e^{-\frac{\Delta\epsilon+\epsilon'}{kT}}}{e^{-\frac{\Delta\epsilon+\epsilon'}{kT}} + 1}. \tag{10}$$

The quantity on the right-hand side of this equation is the number of electrons that have been excited from the bound states. The condition that must be satisfied if this is to be small compared with n_b is that ϵ' should be much larger than $-\Delta\epsilon$, so that the exponential in the denominator may be dropped. We have, as a result,

$$e^{\frac{2\epsilon'}{kT}}C(kT)^{\frac{3}{2}}\frac{\sqrt{\pi}}{2} = n_b e^{-\frac{\Delta\epsilon}{kT}}, \tag{11}$$

or

$$\epsilon' = -\frac{\Delta\epsilon}{2} + \frac{kT}{2}\log\frac{n_b 2}{C(kT)^{\frac{3}{2}}\sqrt{\pi}}. \tag{12}$$

The second quantity on the right is of the order of magnitude kT for ordinary densities of bound electrons. Thus ϵ' is very nearly equal to $-\Delta\epsilon/2$ when $\Delta\epsilon$ is greater than kT (Fig. 14).

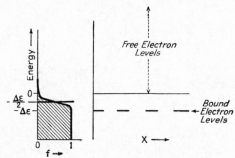

Fig. 14.—The position of the distribution function $f(\epsilon)$ relative to the energy level diagram of Fig. 13. The point $\epsilon = \epsilon_0$ of the distribution function occurs at the point $-\Delta\epsilon/2$ when $\Delta\epsilon >> kT$.

Substituting (12) into Eq. (3) and neglecting small quantities, we find that

$$n(\epsilon) = n_b^{\frac{1}{2}}\left[\frac{2C}{(kT)^{\frac{3}{2}}\sqrt{\pi}}\right]^{\frac{1}{2}}e^{-\frac{\Delta\epsilon}{2kT}}e^{-\frac{\epsilon}{kT}}\sqrt{\epsilon}. \tag{13}$$

This becomes identical with the classical distribution (13), Sec. 26, if we say that the number of free electrons per unit volume n_f is given by the temperature-dependent quantity

$$n_f = n_b^{\frac{1}{2}}\left[\frac{\sqrt{\pi}}{2}(kT)^{\frac{3}{2}}C\right]^{\frac{1}{2}}e^{-\frac{\Delta\epsilon}{2kT}}. \tag{14}$$

If kT is large compared with $\Delta\epsilon$, it is evident that practically all the electrons will have evaporated from the bound levels and that n_f may then be set equal to n_b.

We shall proceed to discuss the conductivity, the thermoelectric effects, and the Hall effect in semi-conductors and shall use Eq. (10) in the same way that we used the Fermi-Dirac distribution in metals.   For convenience, we shall write Eq. (13) in the form

$$n(\epsilon) = ae^{-\frac{\epsilon}{kT}}\sqrt{\epsilon} \tag{15}$$

where

$$a = n_f \frac{2}{\sqrt{\pi}} \frac{1}{(kT)^{\frac{3}{2}}}.$$

36. Electrical Conductivity*.—According to Eq. (5), Sec. **32**, the electrical conductivity σ is

$$\sigma = \frac{I_x}{E_x} = -\frac{4\pi e^2}{3}\int_0^\infty v^3 l\frac{\partial f_n^0}{\partial\epsilon}dv \tag{1}$$

where $f_n^0 dv_x dv_y dv_z$ is the number of free electrons per unit volume having velocity components in the range from v_x to $v_x + dv_x$, etc.   The relation between ϵ and v is

$$\epsilon = \frac{m^*v^2}{2}$$

in our model of a semi-conductor.   Hence, f_n^0 is related to $n(\epsilon)$ in Eq. (12) of the preceding section by the equation

$$4\pi f_n^0 v^2 dv = n(\epsilon)d\epsilon. \tag{2}$$

Thus,

$$\begin{aligned}
4\pi f_n^0 &= \frac{n(\epsilon)}{\sqrt{\epsilon}}2\left(\frac{m^*}{2}\right)^{\frac{3}{2}} \\
&= 2\left(\frac{m^*}{2}\right)^{\frac{3}{2}} ae^{-\frac{\epsilon}{kT}},
\end{aligned} \tag{3}$$

and

$$\sigma_c = -\frac{e^2}{3}\sqrt{\frac{2}{m^*}}a\int_0^\infty l\epsilon\frac{\partial}{\partial\epsilon}(e^{-\frac{\epsilon}{kT}})d\epsilon. \tag{4}$$

A simpler way of obtaining the same result is to recall that the distribution function in the present case is the limit of the Fermi-Dirac function when ϵ' is negative and is equal to the quantity in Eq. (12) of the preceding section.   Then the equation for the conductivity (and for the thermoelectric effects) may be derived by replacing the Fermi-Dirac function f

in all the equations of Secs. 32, 33, and 34 by the following approximate value:

$$f \cong e^{\frac{\epsilon'}{kT}} e^{-\frac{\epsilon}{kT}} = \left[\frac{n_b 2}{C(kT)^{\frac{3}{2}}\sqrt{\pi}} \right]^{\frac{1}{2}} e^{-\frac{\Delta\epsilon}{2kT}} e^{-\frac{\epsilon}{kT}}$$

$$= \frac{a}{C} e^{-\frac{\epsilon}{kT}}. \tag{4a}$$

Equation (9), Sec. 32, then leads to Eq. (4).

We shall assume that l is a constant l_0 for the range of velocities over which the integrand is appreciable. Then,

$$\sigma_c = \frac{e^2}{3}\sqrt{\frac{2}{m^*}} a l_0 kT = \frac{4 n_f l_0 e^2}{3\sqrt{2\pi m^* kT}}. \tag{5}$$

This equation was derived first by Lorentz,[1] who used it for metals under the assumption that n_f is the number of free electrons per unit volume. In passing, let us compare Eq. (5) with σ in Eq. (13), Sec. 32. We shall call the latter σ_q and shall set $n_f = n_0$. The ratio of the two conductivities is

$$\frac{\sigma_q}{\sigma_c} = \frac{3}{4} \frac{\sqrt{2\pi m^* kT}}{mv(\epsilon_0')} \frac{l(\epsilon_0')}{l_0}. \tag{6}$$

If we assume that $l(\epsilon_0')/l_0$ is of the order of unity, this ratio is of the order $\sqrt{kT/\epsilon_0'}$, which is about 10^{-1} at room temperature for ordinary metals. Thus, σ_c agrees with experiment at room temperature only if we assume that either n_f or l_0 is about 10^{-1} times as large as the corresponding quantities that appear in σ_q. Thus, the classical mean free path must be of the order of 10^{-8} cm if n_f is assumed to be equal to n_0. Then, contrary to experiment, the contribution to the electronic heat must be taken as $3R/2$ per mol. On the other hand, if we assume that the classical mean free path is the same as the quantum mean free path, we must assume that

$$n_f \cong n_0 \sqrt{\frac{kT}{\epsilon_0'}}. \tag{7}$$

There is no a priori reason for making this assumption on purely classical grounds. Actually, the use of quantum statistics is equivalent to eliminating all but a fraction of the free electrons, as we have pointed out in previous sections.

---

[1] See footnote 2, p. 139.

Returning to semi-conductors, for which n_f is expressed by Eq. (14) of the preceding section, we see that

$$\sigma = n_b^{\frac{1}{2}} \frac{4\sqrt{2}}{3} \frac{e^2 l_0}{h^{\frac{3}{2}}} (2\pi m^* kT)^{\frac{1}{4}} e^{-\frac{\Delta\epsilon}{2kT}}. \tag{8}$$

In the range of temperature in which $kT \ll \Delta\epsilon/2$, the coefficient of $e^{-\frac{\Delta\epsilon}{2kT}}$ varies so slowly with temperature, compared with the exponential $e^{-\frac{\Delta\epsilon}{2kT}}$ that (8) agrees within experimental error with the observed law

$$\sigma = A e^{-\frac{E}{kT}} \tag{9}$$

(*cf.* Sec. 6). We conclude that the observed E and A are related to quantities in Eq. (9) by the equations

$$E = \frac{\Delta\epsilon}{2},$$

$$A = n_b^{\frac{1}{2}} \frac{4\sqrt{2}}{3} \frac{e^2 l_0}{h^{\frac{3}{2}}} (2\pi m^* kT)^{\frac{1}{4}}$$

$$= 0.024 l_0 n_b^{\frac{1}{2}} T^{\frac{1}{4}} \text{ ohm}^{-1} \text{ cm}^{-1}. \tag{10}$$

The numerical value of A has been determined by setting m^* equal to the actual electronic mass.

37. Thermoelectric Effects and the Hall Effect in Semi-conductors.— According to Eq. (6), Sec. 33, the Thomson coefficient σ_T is

$$\sigma_T = -\frac{T}{e} \frac{d}{dT}\left(\frac{K_2}{TK_1} - \frac{\epsilon'}{T} \right) \tag{1}$$

where the quantities K_2 and K_1 must be evaluated by use of the classical value of f in the manner described in Sec. 26. We find

$$\frac{K_2}{TK_1} = k \frac{\int_0^\infty x^2 e^{-x} dx}{\int_0^\infty x e^{-x} dx} = 2k \tag{2}$$

if we assume that l is constant. The expression (2) contributes nothing to (1), since it is independent of temperature. Hence,

$$\sigma_T = \frac{T}{e} \frac{d}{dT}\left(\frac{\epsilon'}{T} \right) = \frac{\Delta\epsilon}{2T} - \frac{3k}{4} \log T. \tag{3}$$

The Peltier coefficient $\pi_{1\to2}$ and the Seebeck emf F_s are related to σ_T by Eqs. (10) and (12) of Sec. 33, namely:

$$\frac{d}{dT}\left(\frac{\pi_{1\to2}}{T} \right) = \frac{\sigma_{T,2} - \sigma_{T,1}}{T}, \tag{4}$$

$$F_s = \int_{T'}^{T''} \frac{\pi_{1\to2}}{T} dT. \tag{5}$$

The value of the Hall constant R is

$$R = -\frac{L_2}{L_1}\frac{1}{\mathsf{H}_z\sigma(\mathsf{H}_z)} \tag{6}$$

[*cf.* Eq. (6), Sec. 34], where the ratio L_2/L_1 now is

$$\frac{L_2}{L_1} = \frac{e\mathsf{H}_z}{mc}\,l_0\,\frac{\displaystyle\int_0^\infty \frac{\epsilon}{v}\frac{\partial f}{\partial \epsilon}d\epsilon}{\displaystyle\int_0^\infty \epsilon\frac{\partial f}{\partial \epsilon}d\epsilon} \tag{7}$$

$$= \frac{e\mathsf{H}_z}{4mc}\,l_0\sqrt{\frac{2\pi m}{kT}}.$$

Hence,

$$R\sigma(\mathsf{H}_z) = -\frac{e}{\sqrt{2\pi mkT}}\frac{\pi l_0}{2c}. \tag{8}$$

As in Sec. 32, we shall assume that $\sigma(\mathsf{H})$ is equal to $\sigma(0)$ for ordinary fields. Then σ is given by Eq. (5), Sec. 36, and

$$R = -\frac{3\pi}{8}\frac{1}{n_f ec}. \tag{9}$$

All the electrical quantities in these equations are expressed in electrostatic units.

Let us apply these equations to the case of zinc oxide, which is a typical semi-conductor. Its properties, which have been measured by Fritsch,[1] are strongly dependent upon such factors as thermal treatment and oxygen vapor pressure. This behavior is characteristic of most semiconductors (*cf.* Chap. I). The temperature dependence of the conductivity of a particular specimen is shown in Fig. 15a. These results may be fitted by the function

$$\sigma = Ae^{-\frac{E}{kT}}$$

where A is 3.72 ohm^{-1} cm^{-1} and E is 0.013 ev. This specimen was then kept at 900°C for 30 hours in an atmosphere of oxygen at 120 atmospheres. As a result of this treatment, A changed to 2.1 ohm^{-1} cm^{-1}, and E changed to 0.38 ev. Thus, the room-temperature conductivity fell by a factor of about 10^6. The fact that A was nearly the same before and after the heat treatment indicates, according to Eq. (10), that the product $ln_b^{\frac{1}{3}}$ did not change by a large factor during heating. The Hall constant of this specimen was not measured, but it was measured on another specimen before and after heating in the high-pressure furnace. In this

[1] O. Fritsch, *Ann. Physik*, **22**, 375 (1935).

case, A changed from 5.63 to 0.62 ohm^{-1} cm^{-1}, and E changed from 0.012 to 0.063 ev after 120 hours of heating. It is interesting to note that the change induced in E in this case was not nearly so large as in the case cited above, although the heating in oxygen was extended about four times as long. Apparently erratic results of this kind are characteristic of semi-conductors. The room-temperature Hall constants, before and after heating, were -9×10^{-8} and -380×10^{-8} volt/cm-amp-gauss, respectively. From these results and Eq. (9), we may conclude that n_f dropped from $8 \cdot 10^{17}$ to $2 \cdot 10^{16}$ electrons per cubic centimeter as a

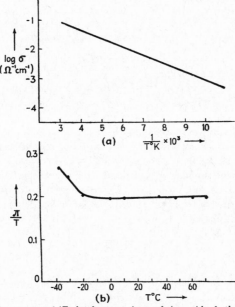

FIG. 15.—a, the log σ versus $1/T$ plot for a specimen of zinc oxide; b, the Seebeck emf per degree for the same specimen. The ordinates are in units of millivolt/deg.

result of heating. The product $R\sigma$ at room temperature, was $-30 \cdot 10^{-8}$ gauss^{-1} before heating and remained practically unchanged. This, according to Eq. (8), implies that l_0 was about $1.7 \cdot 10^{-7}$ cm.

Figure 15b shows the Seebeck emf per degree as a function of temperature for a copper-zinc oxide system. These measurements were made on the second specimen discussed in the previous paragraph after it was heated. The Seebeck emf per degree should be equal to the derivative of (5) with respect to T'''; that is, it should be equal to

$$\frac{\pi_{1\rightarrow2}}{T}. \tag{10}$$

This expression may be related to the Thomson coefficients of copper and of zinc oxide by means of Eq. (4). We shall assume that the coefficient for copper is negligible compared with that for zinc oxide, basing this assumption on the fact that the Seebeck emf for metals is about 10^{-3} times as large as that for semi-conductors. The quantity (10) is then

$$\frac{\pi_{1\to2}}{T} \cong \frac{\epsilon'}{Te} \tag{11}$$

because of (3) and (4). Substituting the value of ϵ' derived in Sec. 35 [*cf.* Eqs. (12) and (14)], we find

$$\frac{\pi_{1\to2}}{T} \cong -\frac{1}{e}\left(\frac{\Delta\epsilon}{T} + k \log \frac{n_f}{n_b}\right). \tag{12}$$

Since the measured value of $\Delta\epsilon$ is $2 \cdot 0.063$ ev, Eq. (12) predicts an effect of the order of magnitude $0.4 \cdot 10^{-3}$ volt per degree, which agrees roughly with the observed effect. The precise variation, however, does not seem to lie within the descriptive power of the simple theory.

This manner of correlating the properties of a semi-conductor may be applied to a large number of other cases. It should be emphasized that the simple theory we have used applies only to those substances for which the Hall effect is negative. There are, however, a large number of substances, such as copper iodide, for which the Hall constant has a positive sign, as though positive charges carry the current. We shall return to a discussion of these substances after developing the band theory of solids.

CHAPTER V

QUANTUM MECHANICAL FOUNDATION

In Part A of this chapter we shall consider the principles and theorems of quantum mechanics that have particular use in the theory of the solid state. Although some of this material may be well known to the reader,[1] we shall present it here in order to place it in a form that is consistent with the treatment of later chapters. In Part B, we shall discuss the theory of radiation.

PART A

38. Elementary Postulates of the Theory.—The development of the nonrelativistic form of quantum mechanics, with which we shall be solely concerned, brought with it a revision of the logical and mathematical discipline of mechanics. This revision was necessary in order to include principles that are applicable in a larger domain of the physical world than that in which the classical laws are valid. There is a correspondence between the classical and the quantum mechanical laws, for the domain of the former is contained in the domain of the latter. Thus, there is a quantity in quantum mechanics corresponding to each quantity in classical mechanics, and the quantum laws reduce to the classical ones when Planck's constant may be regarded as a very small quantity.

One of the primary features of quantum mechanics is the introduction of a state function which is said to describe a given dynamical system completely when the system is in a given state of motion. This function, as we shall use it, is an ordinary function of Cartesian coordinates. All available information may be derived from the state function by the proper use of certain operators that correspond, individually, to measurable quantities, such as position coordinates, momentum, energy, etc. Neither state functions nor the operators have immediate physical significance; only certain quantities derived by proper juxtaposition of the two are measurable numbers.

---

[1] General references: P. A. M. DIRAC, *The Principles of Quantum Mechanics* (Oxford University Press, New York, 1935). S. DUSHMAN, *Elements of Quantum Mechanics* (John Wiley & Sons, Inc., New York, 1938). E. C. KEMBLE, *The Fundamental Principles of Quantum Mechanics* (McGraw-Hill Book Company, Inc., New York, 1937). V. ROJANSKY, *Introductory Quantum Mechanics* (Prentice-Hall, Inc., New York, 1938).

The most convenient definition of an operator, from our standpoint, is the following: An operator is a quantity symbolizing a process in which a given function is changed into another function. For example, the process of taking the square root of a function defines an operator. If we designate this operator by $\sqrt{}$ and the operating process by a dot, we have

$$\sqrt{} \cdot f = \sqrt{f};$$

that is, $\sqrt{}$ operating on f gives $\sqrt{f}$. Similarly, we may regard the ordinary differential symbol $\partial/\partial x$ as an operator for which

$$\frac{\partial}{\partial x} \cdot f(x) = \frac{\partial f}{\partial x}.$$

Likewise, the multiplication of the functions $f(x)$ and $v(x)$ defines an operator since the product $v(x)f(x)$ is a new function of x.

Complete description in quantum theory does not imply precise knowledge of all measurable quantities at all instants of time as it does in classical mechanics. Quantum mechanics is primarily a statistical theory; its results tell us the mean or expectation values of measurements. Thus repetition of the procedure for making a precise measurement of a given quantity usually should not lead to a repetition of results even when the system on which measurements are made is in the same state at the beginning of each measurement. This principle is to be contrasted with the principles of classical theory, according to which one should expect precisely repeatable results under identical experimental conditions.

The statistical formulation of quantum theory is believed to be ultimate, in contrast with the formulation of classical statistical mechanics in which probability is introduced only as a convenient tool. Thus, it is believed that the limitation of description contained in quantum theory can be verified by direct experiment. The electron-diffraction experiments[1] of Davisson and Germer, Thomson, and Rupp and the molecular-beam experiments of Rabi[2] and his coworkers have gone a long way toward providing this verification. Even without this direct experimental check of the uncertainty relations, however, there is overwhelming experimental evidence of other kinds to justify the use of quantum mechanics for the types of problem that are considered in this book.

[1] C. DAVISSON and L. H. GERMER, *Phys. Rev.*, **30**, 705 (1927). G. P. THOMSON, *Proc. Roy. Soc.*, **117**, 600 (1928); **119**, 651 (1928). E. RUPP, *Ann. Physik* **85**, 981 (1928).

[2] I. RABI, *Phys. Rev.*, **49**, 324 *ff.* (1936).

Although experimental results usually should not be precisely repeatable, there is one case in which duplicate observations should give identical results. Suppose that α is the operator corresponding to a given observable quantity and that f is the state function of a system for a particular state of motion. Precisely the same values of the measured quantity associated with α should be obtained when the system is in the state f if

$$\alpha \cdot f = af \tag{1}$$

where a is constant. Moreover, the number a should be the result of each measurement. A function f that satisfies an equation of this type is by definition an *eigenfunction* of the operator α. a is called the eigenvalue of f.

From a practical standpoint, the problems in which we are primarily interested are to determine (1) the possible forms that may be given to the state function and to the dynamical operators and (2) the dynamical laws of the theory. The solutions may be placed in many possible forms, just as in classical theory. A serviceable form for our purposes is the one based on Schrödinger's scheme. It may be summarized in the following way:

a. The operators that correspond to the Cartesian coordinates x_i, y_i, z_i of the ith particle of a system and to the time variable t are taken to be the variables themselves. The state function f is then chosen as a function of these variables. Then,

$$f = f(x_1, y_1, z_1, \cdots, x_n, y_n, z_n, t).$$

b. The operators of the variables that are conjugate to these variables in the classical sense, namely, the corresponding momenta p_{x_1}, p_{y_1}, p_{z_1}, $\cdots$, p_{z_n}, and the negative of the Hamiltonian function H are taken in the form

$$\frac{\hbar}{i}\frac{\partial}{\partial x_1}, \quad \frac{\hbar}{i}\frac{\partial}{\partial y_1}, \quad \cdots, \quad \frac{\hbar}{i}\frac{\partial}{\partial z_n}, \quad \frac{\hbar}{i}\frac{\partial}{\partial t}, \tag{2}$$

respectively. In classical dynamics, $-H$ is the conjugate of t only in the special case in which it is independent of time. The assignment of the operator $-\dfrac{\hbar}{i}\dfrac{\partial}{\partial t}$ to H is assumed to hold, however, even when H is dependent on time.

c. The operator corresponding to any classical dynamical variable that is a function of the x, the p, and t may be obtained by replacing the p by the differential operator introduced in part *b*. In particular, the operator corresponding to the Hamiltonian function, which generally has the form,

$$H(x_1, \cdots, z_n, p_{x_1}, \cdots, p_{z_n}, t)$$

is

$$H\left(x_1, \cdots, z_n, \frac{\hbar}{i}\frac{\partial}{\partial x_1}, \cdots, \frac{\hbar}{i}\frac{\partial}{\partial z_n}, t\right). \tag{3}$$

This operator must be equal to $-\dfrac{\hbar}{i}\dfrac{\partial}{\partial t}$, if the identification presented in part b is also correct. Since this relation is not an identity, it is necessary to assume that the theory is concerned only with those state functions that satisfy the relation

$$H\left(x_1, \cdots, z_n, \frac{\hbar}{i}\frac{\partial}{\partial x_1}, \cdots, \frac{\hbar}{i}\frac{\partial}{\partial z_n}, t\right)f = -\frac{\hbar}{i}\frac{\partial f}{\partial t}. \tag{4}$$

This is the fundamental dynamical equation of quantum mechanics, which is used to determine the state function. We shall refer to it as the first Schrödinger equation.

d. The states that are eigenfunctions of the energy operator satisfy the equation

$$Hf = -\frac{\hbar}{i}\frac{\partial f}{\partial t} = Ef. \tag{5}$$

Hence, they must have the form

$$f = \Psi(x_1, \cdots, z_n)e^{-i\frac{E}{\hbar}t} \tag{6}$$

where Ψ satisfies the equation

$$H\Psi = E\Psi. \tag{7}$$

This is the second Schrödinger equation.

e. It was mentioned in connection with Eq. (1) that the observed value of α is always a if a system is in a state f that satisfies the equation

$$\alpha f = af.$$

If the state function is not an eigenfunction of α, the mean value $\bar{a}$ of the measured value is

$$\bar{a} = \frac{\int f^*\alpha \cdot f d\tau}{\int f^* f d\tau} \tag{8}$$

where the integration is to be extended over all values of the x and t; f^* is the complex conjugate of f; and $d\tau$ is the product of dt and the volume element in $3n$-dimensional configuration space. Whenever $\int f^* f d\tau$ is infinite, $\bar{a}$ is the limit to which the ratio (8) approaches as the volume of integration is gradually increased to include the entire space. Since all measured quantities are real, the allowed operators α must satisfy the relation

$$\int f^*(\alpha \cdot f)d\tau = \int f(\alpha \cdot f)^* d\tau. \tag{9}$$

Such operators are called Hermitian.

f. One may legitimately ask for the operator that corresponds to the measurement of whether or not the system has the coordinates $x_1', \cdots, z_n', t'$. It is difficult to formulate this operator in a mathematically rigorous fashion in the Schrödinger theory. This difficulty may be avoided by defining the operator to be one for which the integral (8) has the value

$$f^*f(x_1', \cdots, z_n', t'). \tag{10}$$

Thus, $f^*f(x_1', \cdots, z_n', t')$ is interpreted as the probability that the system has the coordinates $x_1', \cdots, z_n', t'$. If f is an eigenfunction of the Hamiltonian operator, it has the form (6), and the quantity $f^*f = \Psi^*\Psi$ is independent of time. States of this type are said to be stationary.

g. It is natural to expect that f^*f should be an integrable function if f^*f is interpreted as the relative probability that the system should occupy the coordinates $x_1', \cdots, z_n', t'$. This is a general restriction on the state function. Ordinarily, this condition is fulfilled by demanding that f should be finite everywhere; the more accurate condition, however, is that the integral of f^*f over any finite volume should exist and be finite.

It may easily be shown[1] that classical mechanics and quantum mechanics lead to identical results in the ordinary large-scale domain in which classical mechanics is ordinarily applied.

39. Auxiliary Theorems.—There are several additional theorems that are frequently used in conjunction with the preceding principles because they are useful in applying the theory to specific problems. They may be listed in the following way.

a. The purely spatial part of an eigenfunction (6) of a time-independent Hamiltonian operator may be used without the time-dependent factor $e^{-\frac{iEt}{\hbar}}$ when the mean value of any time-independent operator is computed, since these factors cancel out of the integrands in Eq. (8). Since the eigenfunctions of other operators usually do not satisfy the time-dependent Schrödinger equation (4) and hence cannot be written in the form (6), their purely spatial eigenfunctions usually cannot be used in the same way. We shall mention two useful theorems concerning the spatial eigenfunctions of Hermitian operators.

1. The space eigenfunctions $\psi_1, \psi_2, \cdots$ of any time-independent Hermitian operator form a complete orthogonal set; therefore, any space

[1] See KEMBLE, *op. cit.*, p. 49.

function φ, such as an eigenfunction of a Hamiltonian operator, may be expanded in terms of them in a Fourier-series fashion. Thus,

$$\varphi = \sum_i a_i \psi_i \tag{1}$$

where

$$a_i = \frac{\int \varphi \psi_i^* d\tau(x_1, \cdots, z_n)}{\int |\psi_i|^2 d\tau}. \tag{2}$$

2. The condition that must be satisfied in order that two operators α and β may have all eigenfunctions in common is that they should commute, that is, that they must satisfy the condition

$$(\alpha\beta - \beta\alpha)\psi = 0 \tag{3}$$

for arbitrary ψ. This theorem is particularly useful when we have operators that commute with the Hamiltonian operator of a system, for then we may choose the stationary states to be eigenfunctions of all these operators.

b. Normalizing Conditions.—Since it is possible to interpret $|f(x_1, \cdots, z_n, t)|^2$ as the relative probability that the system has the coordinates $x_1, \cdots, z_n$, at time t, it is reasonable to ask that f should satisfy the equation

$$\int |f(x_1, \cdots, z_n, t)|^2 d\tau(x_1, \cdots, z_n) = 1 \tag{4}$$

at each instant of time, where the integration extends over all space. This is known as the normalizing condition. Since the relation

$$\frac{d}{dt} \int |f|^2 d\tau(x_1, \cdots, z_n) = 0 \tag{5}$$

may be proved by use of the Schrödinger equation and the Hermitian condition on H, a state function that is normalized at one instant of time remains normalized at all later times. As a rule, normalized functions are used because they give absolute rather than relative probabilities.

c. The Variational Theorem.—One of the most powerful tools for obtaining solutions of the Schrödinger equation by means of approximate methods is furnished by the variational theorem of quantum mechanics. This theorem states that functions $\Psi(x_1, \cdots, z_n)$, for which the variation of the mean value

$$\bar{a} = \frac{\int \Psi^* \alpha \Psi d\tau(x_1, \cdots, z_n)}{\int \Psi^* \Psi d\tau} \tag{6}$$

is zero, satisfy the relation

$$\alpha\Psi = a\Psi. \tag{7}$$

The converse of this is also true.

We shall prove this theorem for the case in which α is a time-independent Hermitian operator and in which integration extends over all space.

We have, as the condition on $\bar{a}$,

$$\delta\bar{a} = \delta\frac{\int\Psi^*\alpha\Psi d\tau}{\int\Psi^*\Psi d\tau} = 0 \tag{8}$$

where $\delta\bar{a}$ indicates the variation in $\bar{a}$ that is to be associated with a variation $\delta\Psi$ in Ψ. If we set

$$A = \int\Psi^*\alpha\Psi d\tau, \tag{9}$$
$$B = \int\Psi^*\Psi d\tau,$$

$\delta\bar{a}$ becomes

$$\delta\bar{a} = \delta\left(\frac{A}{B}\right) = \frac{B\delta A - A\delta B}{B^2} = 0. \tag{10}$$

Hence, we must have

$$\frac{A}{B}\delta B - \delta A = 0,$$

or

$$\bar{a}\delta B - \delta A = 0, \tag{11}$$

since

$$\bar{a} = \frac{A}{B}.$$

Now,

$$\delta A = \int\delta\Psi^*\alpha\Psi d\tau + \int\Psi^*\alpha\delta\Psi d\tau,$$
$$\delta B = \int\delta\Psi^*\Psi d\tau + \int\Psi^*\delta\Psi d\tau,$$

whence (11) may be written in the form

$$\int\delta\Psi^*(\bar{a} - \alpha)\Psi d\tau + \int\Psi^*(\bar{a} - \alpha)\delta\Psi d\tau = 0$$

or in the form

$$\int\delta\Psi^*(\bar{a} - \alpha)\Psi d\tau + \int\delta\Psi(\bar{a} - \alpha)\Psi^* d\tau = 0, \tag{12}$$

since α is Hermitian. If we express Ψ in terms of its real and imaginary parts, Ψ_r and Ψ_i, respectively, (11) becomes

$$\int\delta\Psi_r(\bar{a} - \alpha)\Psi_r d\tau + \int\delta\Psi_i(\bar{a} - \alpha)\Psi_i d\tau = 0.$$

The necessary and sufficient conditions for this equality are that

$$(\bar{a} - \alpha)\Psi_r = 0,$$
$$(\bar{a} - \alpha)\Psi_i = 0,$$

since both $\delta\Psi_r$ and $\delta\Psi_i$ are arbitrary. Hence, the necessary and sufficient condition for the validity of Eq. (8) is that Ψ should satisfy the equation

$$\alpha\Psi = \bar{a}\Psi. \tag{13}$$

It is not possible without further investigation to say whether a particular Ψ satisfying (13) gives $\bar{a}$ a maximum value, a minimum value, or just an inflection point.

If we apply this theorem to the Hamiltonian operator H, there generally is a lowest value of the mean value corresponding to the stationary state of lowest energy. States of higher energy are invariably orthogonal to this, as we mentioned in part *a*. Hence, the state Ψ_2, just above the lowest Ψ_1, may be specified by the two conditions

$$\delta\frac{\int\Psi_2{}^*H\Psi_2 d\tau}{\int\Psi_2{}^*\Psi_2 d\tau} = 0, \tag{14}$$

$$\int\Psi_2{}^*\Psi_1 d\tau = 0. \tag{15}$$

The necessary and sufficient condition for these equations is that Ψ_2 should satisfy the relation

$$H\Psi_2 = (E - \lambda)\Psi_2 = E'\Psi_2$$

where the Lagrangian parameter λ may be determined by the condition (15). Higher discrete states may be defined in a similar way by the condition (14) and by additional conditions of the type (15) which express the fact that the higher state is orthogonal to all lower ones.

The variational theorem shows that the accuracy of the mean value of H for a given approximate function f is usually greater than the accuracy of the mean value of other quantities. This fact may be shown directly as follows. Suppose that f is expressed in the form

$$f = \Psi + \alpha\Phi \tag{16}$$

where Ψ is the exact eigenfunction, which f represents approximately, and $\alpha\Phi$ is the part of f orthogonal to Ψ. We shall assume that Ψ and Φ are normalized and that α is a small number. The mean value of H for the function (16) is

$$\frac{\int(\Psi^* + \alpha\Phi^*)H(\Psi + \alpha\Phi)d\tau}{(1 + \alpha^2)} = \frac{(E + \alpha^2\int\Phi^*H\Phi d\tau)}{(1 + \alpha^2)}$$

where E is the eigenvalue of Ψ. Thus, the fractional accuracy of Ψ is the square root of that of E. Since the mean values of other operators involve terms in the first power of α, their fractional accuracy is also of the order of the square root of the accuracy of the energy.

40. Electron Spin*.[1]—A significant feature of the preceding formulation of quantum mechanics is that it contains an operator for each classical variable. It has been found necessary to introduce other operators that do not correspond to classically measurable quantities, in order to explain certain experimental observations. Most important among these operators are those associated with electron spin. Historically, they find their origin in an attempt of Goudsmit and Uhlenbeck to explain certain features of atomic spectra that had not been previously interpreted. These workers were led to assume that an electron possesses a spin about an axis passing through its center and that the total spin angular momentum is equal to $\hbar/2$. This condition on the angular momentum may be expressed in the form

$$|\mathbf{d}| = \sqrt{\sigma_x^2 + \sigma_y^2 + \sigma_z^2} = \frac{\hbar}{2} \tag{1}$$

where σ_x, σ_y, and σ_z are, respectively, the x, y, and z components of spin angular momentum. In addition, they found it necessary to assume that the magnetic moment $\mathbf{\mu}$, associated with electron spin, is related to the mechanical moment $\mathbf{d}$, by means of the equation

$$\mathbf{\mu} = -\frac{e}{mc}\mathbf{d}. \tag{2}$$

In contrast with this, the relationship between the orbital angular momentum $\mathbf{L}$ (that is, the angular momentum of an electron moving about an axis that does not pass through it) and the orbital magnetic moment $\mathbf{M}_0$ is

$$\mathbf{M}_0 = -\frac{e}{2mc}\mathbf{L}; \tag{3}$$

that is, there is an additional factor of 2 in the right-hand side of Eq. (2). There have been several attempts, based on classical electromagnetic theory, to prove that the mechanical and magnetic moments of a spinning spherical charge distribution actually do satisfy the relation (2). At present, this work is generally regarded as inapplicable, aside from being inconclusive, for the phenomenon of electron spin is viewed as lying outside the domain in which classical concepts have meaning.

The most complete theory of spin yet devised was discovered by Dirac[2] in a search for a form of quantum mechanics that satisfies the principles of relativity. Equation (2) arises as a by-product of other assumptions in Dirac's theory. We shall not be concerned with the

[1] See *ibid.*, p. 510.
[2] See ROJANSKY, *op. cit.*

details of this theory, for we are interested only in the nonrelativistic approximation. In this case, spin may be handled by a scheme developed by Pauli to which Dirac's treatment reduces when the velocities of electrons are small compared with the velocity of light.

The significance of spin in Pauli's theory is connected with the statement that electrons are particles for which the complete state function has two components instead of one. The two components are not scalar functions in the sense that they are independent of the choice of coordinate system in three-dimensional space, for they transform between one another in a complex way when coordinate axes are transformed. Thus, it is customary to introduce the concept of a two-dimensional spin space in which the two-component functions of the state function are represented as orthogonal components of a spinor. A transformation of coordinate axes in ordinary space induces a corresponding transformation in spin space. We shall not consider the details of spinor transformation theory because we shall have no explicit use for the transformation equations.[1] It should be mentioned, however, that the transformation characteristics of two-dimensional spinors are considerably different from those of two-dimensional vectors.

The introduction of spin in the quantum theory of electrons is analogous to the introduction of the concept of polarization into the theory of light. Suppose that we were acquainted with none of the polarization phenomena of optics. Then it would be possible to describe many optical experiments, such as those of interference or of energy transport, by assuming that the amplitude of a light wave is a scalar quantity, just as the amplitude of sound. As soon as experiments on polarization are performed, however, we are compelled to say that there is a vector character associated with the amplitude of a light wave. This vector character may be described by taking components of the amplitude in two orthogonal directions of polarization. The analogy between the electron and the light wave does not hold in a quantitative way, however, for the two independent directions of electron polarization are separated by 180 degrees rather than by 90 degrees, as they are for light. This fact marks the difference between vector character and spinor character.

We shall represent the coordinate variables in spin space by ζ_1 and ζ_2, respectively. A unit spinor in the direction of the ζ_1 axis has components $\zeta_1 = 1$, $\zeta_2 = 0$ and may be represented by the column matrix

$$\mathbf{n}(1,0) = \begin{pmatrix} 1 \\ 0 \end{pmatrix}; \tag{4}$$

---

[1] See, for example, E. P. Wigner, *Gruppentheorie* (Vieweg, Braunschweig, Germany, 1931).

a unit spinor along the ζ_2 axis may be represented by

$$\mathbf{n}(0,1) = \begin{pmatrix} 0 \\ 1 \end{pmatrix}. \tag{5}$$

Pauli assumed that the components of spin angular momentum σ_x, σ_y, and σ_z could be represented in spin space by the matrices

$$\sigma_x = \frac{\hbar}{2}\begin{pmatrix} 0 & 1 \\ 1 & 0 \end{pmatrix}, \qquad \sigma_y = \frac{\hbar}{2}\begin{pmatrix} 0 & -i \\ i & 0 \end{pmatrix}, \qquad \sigma_z = \frac{\hbar}{2}\begin{pmatrix} 1 & 0 \\ 0 & -1 \end{pmatrix}. \tag{6}$$

The spinor $\mathbf{n}(1,0)$ of Eq. (4) satisfies the relation

$$\sigma_z \cdot \mathbf{n}(1,0) = \frac{\hbar}{2}\mathbf{n}(1,0), \tag{7}$$

and $\mathbf{n}(0,1)$ satisfies

$$\sigma_z \cdot \mathbf{n}(0,1) = -\frac{\hbar}{2}\mathbf{n}(0,1). \tag{8}$$

Hence, these two spinors are said to correspond respectively to the precise values $+\frac{1}{2}\hbar$ and $-\frac{1}{2}\hbar$ of the z component of spin angular momentum. A direct computation of the matrix

$$\mathfrak{d}^2 = \sigma_x^2 + \sigma_y^2 + \sigma_z^2, \tag{9}$$

which corresponds to the total spin angular momentum, shows that

$$\mathfrak{d}^2 = \tfrac{3}{4}\hbar^2\begin{pmatrix} 1 & 0 \\ 0 & 1 \end{pmatrix}. \tag{10}$$

Thus any spinor $\mathbf{n}(\zeta_1, \zeta_2)$ satisfies the equation

$$\mathfrak{d}^2 \cdot \mathbf{n}(\zeta_1, \zeta_2) = \tfrac{3}{4}\hbar^2 \cdot \mathbf{n}(\zeta_1, \zeta_2), \tag{11}$$

and the precise value of the square of the total spin angular momentum is $3\hbar^2/4$.

In the coordinate system in spin space for which the matrices of σ_x, σ_y, and σ_z have the form (6), the two components of the state function may be labeled by a variable ζ_z which takes two values, namely, $+1$ for the coordinate going with the diagonal element $\hbar/2$ of σ_z and -1 for the coordinate going with the value $-\hbar/2$. In other words, we may label the axes in spin space with the eigenvalues of $2\sigma_z/\hbar$, instead of by the indices 1 and 2. The state function then has the form $f(x,y,z,\zeta_z)$, it being understood that $f(x,y,z,1)$ and $f(x,y,z,-1)$ are the components of a spinor

$$\begin{pmatrix} f(x,y,z,1) \\ f(x,y,z,-1) \end{pmatrix}. \tag{12}$$

All state functions (12) are eigenfunctions of the total square spin angular momentum (10). The reader may readily verify that the state functions

$$\begin{pmatrix} f(x,y,z) \\ 0 \end{pmatrix} \quad \text{and} \quad \begin{pmatrix} 0 \\ f(x,y,z) \end{pmatrix}$$

are eigenfunctions of σ_z and that

$$\begin{pmatrix} f(x,y,z) \\ \pm f(x,y,z) \end{pmatrix} \quad \text{and} \quad \begin{pmatrix} f(x,y,z) \\ \pm if(x,y,z) \end{pmatrix}$$

are eigenfunctions of σ_x and σ_y, respectively.

Classically, the components of angular momentum of a particle, relative to the origin of coordinates, are given by

$$\left.\begin{aligned} m_x &= yp_z - zp_y, \\ m_y &= zp_x - xp_z, \\ m_z &= xp_y - yp_x, \end{aligned}\right\} \tag{13}$$

where x, y, z are the spatial coordinates of the particle and p_x, p_y, p_z are the components of angular momentum. The quantum operators that correspond to (13) are

$$\left.\begin{aligned} m_x &= \frac{\hbar}{i}\left(y\frac{\partial}{\partial z} - z\frac{\partial}{\partial y} \right), \\ m_y &= \frac{\hbar}{i}\left(z\frac{\partial}{\partial x} - x\frac{\partial}{\partial z} \right), \\ m_z &= \frac{\hbar}{i}\left(x\frac{\partial}{\partial y} - y\frac{\partial}{\partial x} \right). \end{aligned}\right\} \tag{14}$$

It is easily shown that these obey the commutation rules

$$\left.\begin{aligned} m_y m_z - m_z m_y &= \hbar i m_x, \\ m_z m_x - m_x m_z &= \hbar i m_y, \\ m_x m_y - m_y m_x &= \hbar i m_z. \end{aligned}\right\} \tag{15}$$

Since the spin matrices (6) obey exactly the same commutation rules, they may be viewed as quantum mechanical angular-momentum operators even though this angular momentum cannot be measured directly, even in an idealized experiment.

Along with the spin operators, we may introduce the following operators:

$$\left.\begin{aligned} \mu_x &= \frac{e}{mc}\sigma_x, \\ \mu_y &= \frac{e}{mc}\sigma_y, \\ \mu_z &= \frac{e}{mc}\sigma_z, \end{aligned}\right\} \tag{16}$$

which we shall call the components of spin magnetic moment. The choice of coefficient in these relations is justified both by Dirac's theory and by experiment, as we have said previously.

When the mechanical system contains n electrons instead of 1, the state function must be viewed as a spinor in a 2^n-dimensional spin space. We may choose the coordinate system of this in such a way that the components of the state function are labeled by n variables $\zeta_{z_1}, \zeta_{z_2}, \cdots,$ ζ_{z_n}, each of which takes the two values ± 1; that is, we may write the state function in the form

$$f(x_1, \cdots, z_n, \zeta_{z_1}, \cdots, \zeta_{z_n}). \tag{17}$$

The function of $x_1, \cdots, z_n$ associated with each of the 2^n possible values of the ζ_z is a component of a 2^n-dimensional spinor. It is implied that the operators $\sigma_{z_i}(i = 1, \cdots, n)$, corresponding to the z components of spin of the n electrons, are 2^n-dimensional diagonal matrices in this coordinate system and that the diagonal element of $2\sigma_{z_i}/\hbar$ is $+1$ or -1, respectively, for the coordinate axes for which ζ_{z_i} is $+1$ or -1. Similarly, the matrices σ_{x_i} and σ_{y_i} are represented by 2^n-dimensional matrices that have the form (6) in the two-dimensional subspace that is associated with the two axes labeled by $\zeta_1', \zeta_2', \cdots, \zeta_{i-1}', \pm 1, \zeta_{i+1}', \cdots, \zeta_n'$. All other components are zero. It is easy to show that the σ satisfy the commutation rules

$$\left. \begin{aligned} \sigma_{y_i}\sigma_{z_j} - \sigma_{z_j}\sigma_{y_i} &= \delta_{ij}\hbar i\sigma_{x_i}, \\ \sigma_{z_i}\sigma_{x_j} - \sigma_{x_j}\sigma_{z_i} &= \delta_{ij}\hbar i\sigma_{y_i}, \\ \sigma_{x_i}\sigma_{y_j} - \sigma_{y_j}\sigma_{x_i} &= \delta_{ij}\hbar i\sigma_{z_i}. \end{aligned} \right\} \tag{18}$$

The matrices $\mathfrak{d}_i^2 = \sigma_{x_i}^2 + \sigma_{y_i}^2 + \sigma_{z_i}^2$ are diagonal matrices, all diagonal elements of which are equal to $3\hbar^2/4$ [*cf.* Eq. (10)].

The three matrices, defined by the equations

$$\left. \begin{aligned} \Sigma_x &= \sum_{i=1}^n \sigma_{x_i} \\ \Sigma_y &= \sum_{i=1}^n \sigma_{y_i} \\ \Sigma_z &= \sum_{i=1}^n \sigma_{z_i} \end{aligned} \right\} \tag{19}$$

are called the operators of the components of the total spin angular momentum, and

$$\Sigma^2 = \Sigma_x^2 + \Sigma_y^2 + \Sigma_z^2 \tag{20}$$

is called the square of the total spin angular momentum. $\Sigma_x,\ \Sigma_y,\ \Sigma_z$

commute with Σ^2 and obviously satisfy the commutation rules

$$\left.\begin{aligned}
\Sigma_y\Sigma_z - \Sigma_z\Sigma_y &= i\hbar\Sigma_x, \\
\Sigma_z\Sigma_x - \Sigma_x\Sigma_z &= i\hbar\Sigma_y, \\
\Sigma_x\Sigma_y - \Sigma_y\Sigma_x &= i\hbar\Sigma_z,
\end{aligned}\right\} \tag{21}$$

because of Eqs. (18). The one-electron operators σ_{x_i}, σ_{y_i}, and σ_{z_i} commute with Σ_x, Σ_y, Σ_z, respectively; but the pairs σ_{x_i} and Σ_y do not commute with one another. Moreover, the one-electron operators do not commute with Σ^2 when there are two or more electrons in the system. Hence, the one-electron operators will not be in diagonal form if the coordinate axes are selected so that Σ_z and Σ^2 are diagonal.

If the Hamiltonian operator does not contain any spin terms, as happens in many actual cases, it commutes with all spin operators. Then the stationary states of the system may be chosen to be eigenfunctions of any set of commuting spin operators, such as the set $\Sigma_z, \sigma_{z_i}, \cdots,$ σ_{z_n} or the set Σ^2, Σ_z. The second set is particularly useful when the Pauli principle, which we shall describe in the next section, is properly taken into account.

41. The Pauli Principle and Related Restrictions.—The Hamiltonian operator of any system that contains at least two particles of a given kind remains invariant in form if the coordinates of like particles are permuted among themselves. It may be shown that for this reason the set of stationary states of the system that are associated with a given eigenvalue of the Hamiltonian transform among themselves in one of several different ways when the coordinate variables are permuted. The theory of groups of transformations is particularly concerned with this property of eigenfunctions. This topic need not interest us at present, however, because of the Pauli exclusion principle which states that the physically permissible solutions of Schrödinger's equations behave in definite ways when the coordinates of particles are permuted. In particular, the Pauli principle requires that the state function be antisymmetric under electron permutations; that is, the state function must transform into its negative under odd permutations of electrons and into itself under even ones. It is also true that the state function must be symmetric under the permutation of the coordinates of light quanta; that is, it must transform into itself under both even and odd permutations in this case.

The fundamental reason for these requirements is not completely understood at present. It seems reasonable, however, to expect that the exclusion principle will appear as the natural consequence of some general invariance principle, possibly unformulated as yet, just as the concept of spin arises out of the requirement of relativistic invariance in Dirac's theory.

Statistical theories, corresponding to classical statistical mechanics, that take the exclusion principle into account have been developed[1] by Fermi and Dirac for the antisymmetric case and by Bose and Einstein for the symmetric one.

It seems to be a general rule in nature that all elementary particles, except photons, obey Fermi-Dirac statistics. The statistical behavior of nuclei, as is determined from the analysis of band spectra, is explained completely by assuming that nuclei are composed of protons and neutrons. In order to remove the single exception, photons, it has been postulated that light quanta are composed of two elementary particles which are not observed separately in standard optical experiments. The evidence for this postulate, however, is not very conclusive.

The antisymmetric states are the most important ones from the standpoint of the electron theory of solids. Although the actual process of selecting such states will be discussed in detail in sections that deal with the approximate solutions of the Schrödinger equation, there are several points that should be brought out here.

Suppose that we have a system that contains n electrons. Let us designate them by integers ranging from 1 to n in order to establish a normal arrangement. There are $n!$ possible different permutations of these n indices, each of which relabels the electrons in different ways. The Hamiltonian operator is invariant under these permutations of indices. For each permutation, say the νth of the set of $n!$, we may introduce a permutation operator P_ν, which is defined in such a way that it permutes electrons in the manner described by the νth permutation. Any operator α that is invariant under the νth permutation satisfies the relation

$$P_\nu \alpha g = \alpha P_\nu g$$

for an arbitrary function of g. We may write the operator equation corresponding to this

$$P_\nu \alpha = \alpha P_\nu$$

and may say that α commutes with P_ν. It is quite clear that two permutation operators generally do not satisfy the relationship

$$P_\nu P_{\nu'} g = P_{\nu'} P_\nu g$$

for arbitrary g. The Pauli restriction on the allowable state functions, namely,

$$P_\nu f = (-1)^{p(\nu)} f,$$
$$P_{\nu'} f = (-1)^{p(\nu')} f,$$

where $p(\nu)$ is the order of the νth permutation, implies, however, that

$$P_\nu P_{\nu'} f = P_{\nu'} P_\nu f$$

[1] See, for example, L. BRILLOUIN, *Quantenstatistik* (Julius Springer, Berlin, 1930).

for all allowable state functions.   Thus, for our purposes, all permutation operators may be said to commute.

The one-electron spin operators do not commute with the permutation operators, since each spin operator refers to a specific electron.   Hence, we cannot expect to find functions that are simultaneous eigenfunctions of H, of $P_{\nu_1}, \cdots, P_{\nu_n!}$, and of the one-electron spin quantities.   On the other hand $\mathbf{\Sigma}^2$ and Σ_z, which commute with one another, do commute with the P_ν since they contain all electron variables symmetrically. Hence, when H is independent of spin, we may expect to find stationary states that satisfy the Pauli principle and are eigenfunctions of $\mathbf{\Sigma}^2$ and Σ_z.   For this reason, the physically interesting states are eigenfunctions of $\mathbf{\Sigma}^2$ and Σ_z rather than eigenfunctions of the one-electron operators.

We shall accept without proof the following theorems[1] concerning the eigenvalues of $\mathbf{\Sigma}^2$ and Σ_z.

a. For a system of n electrons, the eigenvalues of $\mathbf{\Sigma}^2$ have the value

$$S(S + 1)\hbar^2$$

where S may range from $n/2$ down to 0 or $\frac{1}{2}$, depending respectively upon whether n is even or odd. S is called the total spin quantum number, and the eigenfunctions of $\mathbf{\Sigma}^2$ are said to be states of definite multiplicity.

b. There are $2S + 1$ degenerate states associated with each value of S.   These states may be chosen in such a way that they are eigenfunctions of Σ_z and have eigenvalues ranging from $+S$ to $-S$, by integer steps.   The number $2S + 1$ is called the multiplicity of the degenerate level associated with S.

Rules for constructing eigenfunctions of definite multiplicity may be found in the references in footnote.[2]

PART B. THE INTERACTION BETWEEN MATTER AND RADIATION*

We shall develop the theory of radiation[3] in this part of the present chapter for use in discussing the optical properties of solids.   The

---

[1] See, for example, E. U. Condon and G. H. Shortley, *The Theory of Atomic Spectra* (Cambridge University Press, 1935).

[2] Eigenfunctions of definite multiplicity are used extensively in the theory of molecular valence, in which they are called "bond functions."   A discussion of the theory of these functions may be found in the following papers: H. Eyring and G. E. Kimball, *Jour. Chem. Phys.*, **1**, 239 (1933); G. Rumer, *Nachr. Gott.*, M. P. Klasse, 337 (1932); J. H. Van Vleck and A. Sherman, *Rev. Modern Phys.*, **7**, 167 (1935).

* This part is used primarily in Chap. XVII, which deals with the optical properties of solids, and may be omitted by a reader not immediately interested in this topic.

[3] General references: G. Breit, *Rev. Modern Phys.*, **4**, 504 (1932); E. Fermi, *Rev. Modern Phys.*, **4**, 87 (1932); W. Heitler, *The Quantum Theory of Radiation* (Oxford University Press, New York, 1936).

topics of principal interest are the theory of light quanta and the theory of the interaction between matter and light.   We shall begin with a brief discussion of the classical theory and shall use this to develop the quantum equations.

42. The Classical Electromagnetic Equations.[1] *a. The Radiation Field.*—Maxwell's equations for free space are

$$\text{div } \mathsf{E} = 0, \qquad\qquad \text{div } \mathsf{H} = 0,$$
$$\text{curl } \mathsf{E} = -\frac{1}{c}\frac{\partial \mathsf{H}}{\partial t}, \qquad \text{curl } \mathsf{H} = \frac{1}{c}\frac{\partial \mathsf{E}}{\partial t}, \qquad (1)$$

where E and H are the electric and magnetic field intensities.   The time-dependent solutions of these equations that correspond to light waves have the form

$$\left.\begin{array}{l} \mathsf{E} = \mathsf{E}_0 e^{2\pi i(\eta\cdot\mathbf{r}-\nu t)}, \\ \mathsf{H} = \mathsf{H}_0 e^{2\pi i(\eta\cdot\mathbf{r}-\nu t)}, \end{array}\right\} \qquad (2)$$

where E_0 and H_0 are constant vectors, $\mathbf{n}$ is the wave number vector, and ν is the frequency of the wave.   These quantities are interrelated by the equations

$$\left.\begin{array}{ll} \mathsf{E}_0\cdot\mathbf{n} = 0, & \mathsf{H}_0\cdot\mathbf{n} = 0, \\ \dfrac{\mathbf{n}^2}{\nu^2} = \dfrac{1}{c^2}, & \mathsf{H}_0\cdot\mathsf{E}_0 = 0, \\ \mathsf{H}_0^2 = \mathsf{E}_0^2, & \end{array}\right\} \qquad (3)$$

from which it may be concluded that E_0 and H_0 are orthogonal to one another and to the direction of propagation of the light wave.

It is convenient to express Maxwell's equations and the solution (2) in terms of the vector and scalar potentials A and φ, which are related to E and H by the equations

$$\left.\begin{array}{l} \mathsf{E} = -\dfrac{1}{c}\dfrac{\partial \mathsf{A}}{\partial t} - \text{grad } \varphi, \\ \mathsf{H} = \text{curl } \mathsf{A}. \end{array}\right\} \qquad (4)$$

The equations for A and φ are

$$\Delta\varphi - \frac{1}{c^2}\frac{\partial^2\varphi}{\partial t^2} = 0, \qquad \Delta\mathsf{A} - \frac{1}{c^2}\frac{\partial^2\mathsf{A}}{\partial t^2} = 0,$$
$$\text{div } \mathsf{A} = \frac{1}{c}\frac{\partial\varphi}{\partial t}. \qquad (5)$$

[1] M. Abraham and R. Becker, *Classical Electricity and Magnetism* (Blackie & Son, Ltd., London, 1932); R. Becker, *Theorie der Elektrizität*, Vol. II (Julius Springer, Berlin, 1932).

In the case of light waves, we may set φ equal to zero and take A to be

$$\mathsf{A} = \frac{c}{2\pi i \nu} \mathsf{E}_0 e^{2\pi i (\eta \cdot \mathbf{r} - \nu t)}. \tag{6}$$

It may be seen that Eqs. (4) then lead to Eqs. (2) and (3).

If we have a number of waves of different frequencies in a cubical space of volume V, the real vector potential of the system may be expressed in the form of a series of traveling waves

$$\mathsf{A} = \sum_{\eta, s} \mathbf{f}_s(\mathbf{n}) \frac{c}{[\nu_s(\mathbf{n}) V]^{\frac{1}{2}}} \{ A_s(\mathbf{n}) e^{2\pi i (\eta \cdot \mathbf{r} - \nu_s(\eta) t)} + A_s{}^+(\mathbf{n}) e^{-2\pi i (\eta \cdot \mathbf{r} - \nu_s(\eta) t)} \}. \tag{7}$$

The quantities that appear in this equation are very similar to those that appear in the Fourier resolutions of the atomic motions in lattice theory (*cf.* Sec. 22). The quantities $\mathbf{f}_s(\mathbf{n})(s = 1, 2)$ are real polarization vectors that satisfy the equations

$$\begin{aligned} \mathbf{f}_s(\mathbf{n}) \cdot \mathbf{n} = 0, \qquad \mathbf{f}_1(\mathbf{n}) \cdot \mathbf{f}_2(\mathbf{n}) = 0, \\ \mathbf{f}_s \cdot \mathbf{f}_s = 1, \end{aligned} \Bigg\} \tag{8}$$

and the $A_s(\mathbf{n})$ are complex constants that are proportional to the amplitudes of the wave of wave number $\mathbf{n}$ having E in the direction $\mathbf{f}_s(\mathbf{n})$. Since we shall deal with real quantities, we shall assume that

$$A_s{}^*(\mathbf{n}) = A_s{}^+(\mathbf{n}). \tag{9}$$

The summation of $\mathbf{n}$ extends over the allowed values of this variable, which are given by the equations

$$\eta_x = \frac{n_x}{L}, \qquad \eta_y = \frac{n_y}{L}, \qquad \eta_z = \frac{n_z}{L} \tag{10}$$

where n_x, n_y, and n_z are integers and L is the length of an edge of the box. It may be seen from this that the density of the allowed values of $\mathbf{n}$ in wave-number space is $L^3 = V$. The factor $c/(\nu V)^{\frac{1}{2}}$ is introduced into (7) in order to simplify results that we shall obtain below.

We shall now derive a Hamiltonian function for this radiation field using Eq. (7). In material systems, the Hamiltonian, when independent of time, is equal to the total energy. Hence, we should expect to obtain a Hamiltonian for the radiation field by computing the energy E_r, namely,

$$E_r = \frac{1}{8\pi} \int (\mathsf{E}^2 + \mathsf{H}^2) dV. \tag{11}$$

If Eqs. (4) are used to compute E and H from the vector potential, Eq. (11) becomes

$$E_r = \sum_{\eta, s} 2\pi \nu_s(\mathbf{n}) a_s(\mathbf{n}) a_s{}^*(\mathbf{n}) \tag{12}$$

where

$$a_s(\mathbf{n}) = A_s(\mathbf{n})e^{-2\pi i \nu_s(\eta)t} \Big\}$$
$$a_s{}^*(\mathbf{n}) = A_s{}^*(\mathbf{n})e^{+2\pi i \nu_s(\eta)t}. \Big\} \tag{13}$$

If we now regard $a_s(\mathbf{n})$ and $ia_s{}^*(\mathbf{n})$ as conjugate variables and (12) as the Hamiltonian of the system, we find that the Hamiltonian equations are

$$\dot{a}_s(\mathbf{n}) = \frac{\partial H_R}{\partial i a_s{}^*(\mathbf{n})} = -2\pi i \nu_s(\mathbf{n}) a_s(\mathbf{n}),$$

$$\dot{a}_s{}^*(\mathbf{n}) = \frac{\partial H_R}{\partial a_s(\mathbf{n})} = 2\pi i \nu_s(\mathbf{n}) a_s{}^*(\mathbf{n}). \tag{14}$$

The solutions of these equations lead to the time dependence expressed by Eqs. (13). Hence, (11) actually is the Hamiltonian function of the system.

Since the variables $a_s(\mathbf{n})$ and $a_s{}^*(\mathbf{n})$ are not real, it is convenient to replace them by the real quantities

$$p_s(\mathbf{n}) = \frac{1}{\sqrt{2}}(a_s + a_s{}^*),$$

$$q_s(\mathbf{n}) = \frac{i}{\sqrt{2}}(a_s - a_s{}^*). \tag{15}$$

The Hamiltonian, when expressed in terms of these variables, is

$$H_R = \frac{1}{2}\sum_{s,\eta} 2\pi \nu_s(\mathbf{n})[p_s^2(\mathbf{n}) + q_s^2(\mathbf{n})], \tag{16}$$

which is the same as for a system of harmonic oscillators. It is this fact that justifies the statement that a radiation field is equivalent to a system of harmonic oscillators. The Hamiltonian (16) may be used as the starting point for a discussion of the quantum theory of the radiation field.

If we express the vector potential $\mathbf{A}$ in terms of the a and the p and q, we find

$$\mathbf{A} = \sum_{\eta,s} \mathbf{f}_s(\mathbf{n}) \frac{c}{[\nu_s(\mathbf{n})V]^{\frac{1}{2}}}[a_s(\mathbf{n})e^{2\pi i \eta \cdot \mathbf{r}} + a_s{}^*(\mathbf{n})e^{-2\pi i \eta \cdot \mathbf{r}}] \tag{17}$$

$$= \sum_{\eta,s} \mathbf{f}_s(\mathbf{n}) \frac{c}{\sqrt{2}[\nu_s(\mathbf{n})V]^{\frac{1}{2}}}(p_s \cos 2\pi \mathbf{n} \cdot \mathbf{r} + q_s \sin 2\pi \mathbf{n} \cdot \mathbf{r}).$$

b. The Interaction between Matter and Radiation.[1]—The force that acts on an electron of charge $-e$ that is moving with velocity $\dot{\mathbf{r}}$ in an electro-

---

[1] M. Abraham and R. Becker, *Classical Electricity and Magnetism* (Blackie & Son, Ltd., London, 1932); R. Becker, *Theorie der Elektrizität*, Vol. II (Julius Springer, Berlin 1932).

magnetic field is given by Lorentz's equation

$$\mathbf{f} = -e\mathbf{E} - \frac{e}{c}\dot{\mathbf{r}} \times \mathbf{H}.$$

Hence, the Newtonian equation of motion is

$$m\ddot{\mathbf{r}} = -e\mathbf{E} - \frac{e}{c}\dot{\mathbf{r}} \times \mathbf{H}.$$

This equation may be derived from the Lagrangian function

$$L = \frac{1}{2}m\dot{\mathbf{r}}^2 - \frac{e}{c}\mathbf{A}\cdot\dot{\mathbf{r}} + e\varphi, \tag{18}$$

as may be verified by writing out the Eulerian equations associated with this function. The components of momenta are

$$p_x = \frac{\partial L}{\partial \dot{x}} = m\dot{x} - \frac{e}{c}\mathbf{A}_x,$$

etc., so that the Hamiltonian $H = \mathbf{p}\cdot\dot{\mathbf{r}} - L$ is

$$H = \frac{1}{2m}\left(\mathbf{p} + \frac{e}{c}\mathbf{A}\right)^2 - e\varphi$$
$$= \frac{1}{2m}\mathbf{p}^2 + \frac{e}{mc}\mathbf{p}\cdot\mathbf{A} + \frac{e^2}{2mc^2}\mathbf{A}^2 - e\varphi. \tag{19}$$

The terms

$$H_I = \frac{e}{mc}\mathbf{p}\cdot\mathbf{A} + \frac{e^2}{2mc^2}\mathbf{A}^2 \tag{20}$$

are of interest when the charge is in a radiation field because they give the interaction between the field and the particle. The other two terms constitute the Hamiltonian in the absence of a field.

It should now be clear that the total Hamiltonian of a system of n electrons in a radiation field is

$$H = H_M + \sum_i \left[\frac{e}{mc}\mathbf{p}_i \cdot \mathbf{A}(\mathbf{r}_i) + \frac{e^2}{2mc^2}\mathbf{A}^2(\mathbf{r}_i) \right]$$
$$= H_M + H_I \tag{21}$$

where H_M is the Hamiltonian in the absence of a radiation field and H_I is the interaction term. If we desire to include the radiation field in the Hamiltonian, we must add to (21) the function H_R that was derived in Part A. The total Hamiltonian then is

$$H_T = H_M + H_I + H_R. \tag{22}$$

43. The Semiclassical Method of Treating Radiation.—Previous to the development of Dirac's theory of radiation,[1] which, at present, is the most accurate method of treating radiation problems, Schrödinger,[2] Gordon,[3] and Klein[4] developed a simpler theory which is still useful for obtaining some of the important equations in a simple way. In this scheme, the classical interaction term H_I in the Hamiltonian (21) of the preceding section is treated as a time-dependent perturbation in the Schrödinger equation. If the radiation field is zero in the classical sense that the system is in the dark, the system is unperturbed and cannot change its state. If, on the other hand, the field is finite, the system may change its state by emission, absorption, or scattering of light. The fact that the state can change only if radiation is present shows a fundamental weakness in the theory, for it cannot be used to treat the problem of spontaneous emission of light by an excited atom. In spite of this weakness, the method leads to many correct results relatively simply. An important reason for this simplicity is the fact that the Schrödinger-Gordon-Klein method can treat absorption, emission, and dispersion in an approximation in which the $e^2 A^2 / 2mc^2$ terms of Eq. (21) are neglected as small quantities, whereas the Dirac theory can treat dispersion only in an approximation in which all the terms of H_I are retained.

The semiclassical method will now be used to develop the equations for absorption, emission, and dispersion of radiation by an atom. We shall assume that the radiation field extends continuously over a finite range of frequencies and that the vector potential of the wave of wave number $\mathbf{n}$ is

$$\mathsf{A}(\mathbf{n}) = \frac{c}{2\pi i \nu} \mathsf{E}_0(\mathbf{n})[e^{2\pi i(\eta \cdot \mathbf{r} - \nu t)} - e^{-2\pi i(\eta \cdot \mathbf{r} - \nu t)}] \tag{1}$$

where E_0 is the amplitude of the electrostatic field. The interaction term in the Hamiltonian arising from this vector potential is

$$H_I' = \sum_i \frac{e}{2\pi i \nu m} \mathbf{p}_i \cdot \mathsf{E}_0[e^{2\pi i(\eta \cdot \mathbf{r}_i - \nu t)} - e^{-2\pi i(\eta \cdot \mathbf{r}_i - \nu t)}]. \tag{2}$$

We shall designate the Hamiltonian of the atom by H_M and the stationary states by Ψ_i where

$$H_M \Psi_i = E_i \Psi_i. \tag{3}$$

[1] See footnote 3, p. 210.
[2] E. SCHRÖDINGER, *Ann. Physik*, **81**, 134 (1926).
[3] W. GORDON, *Z. Physik*, **40**, 117 (1926).
[4] O. KLEIN, *Z. Physik*, **37**, 895 (1926).

It is now of interest to search for a solution of the time-dependent Schrödinger equation

$$(H_M + H_I')f = -\frac{\hbar}{i}\frac{\partial f}{\partial t} \tag{4}$$

that is equal to Ψ_0 at time $t = 0$. This solution may be expressed in terms of the Ψ_i in the following way:

$$f = a_0\Psi_0 e^{-\frac{i}{\hbar}E_0 t} + \sum_{i\neq 0} a_i\Psi_i e^{-\frac{i}{\hbar}E_i t} \tag{5}$$

where a_0 is unity at time zero and the a_i are small quantities that are zero at the same time. Substituting this in Eq. (4), we find after a simple reduction that employs Eq. (3)

$$-\frac{\hbar}{i}\sum_i \frac{\partial a_i}{\partial t}\Psi_i e^{-\frac{i}{\hbar}E_i t} = a_0 H_I'\Psi_0 e^{-\frac{i}{\hbar}E_0 t} \tag{6}$$

in which small terms involving the product of a_i and H_I' have been neglected. If this equation is multiplied by $\Psi_j{}^*$ and the result is integrated over the electronic coordinates, it is found that

$$-\frac{\hbar}{i}\frac{\partial a_j}{\partial t} = a_0 H_{j0} e^{-\frac{i}{\hbar}(E_0 - E_j)t} \tag{7}$$

where

$$H_{j0} = \int \Psi_j{}^* H_I'\Psi_0 d\tau. \tag{8}$$

In deriving (7), we have used the relation

$$\int \Psi_j{}^*\Psi_i d\tau = \delta_{ij}. \tag{9}$$

We shall assume in the following work that a_0 is close to unity at all times in which we are interested. Since H_{j0} involves time, it is convenient to show this dependence explicitly by using Eq. (2). Equation (8) then becomes

$$H_{j0} = \mathsf{E}_0 \cdot (\mathsf{C}_{j0} e^{-2\pi i\nu t} - \mathsf{C}_{j0}{}^+ e^{2\pi i\nu t}) \tag{10}$$

where

$$\mathsf{C}_{j0} = \frac{e}{2\pi i\nu m}\int \Psi_j{}^*\left(\sum_i \mathbf{p}_i e^{2\pi i\eta\cdot\mathbf{r}_i}\right)\Psi_0 d\tau \tag{11}$$

and $\mathsf{C}_{j0}{}^+$ is the same quantity with the sign of $\mathbf{n}$ reversed. Thus, Eq. (7) is

$$-\frac{\hbar}{i}\frac{\partial a_j}{\partial t} = \mathsf{E}_0 \cdot [\mathsf{C}_{j0} e^{-\frac{i}{\hbar}(E_0 - E_j + h\nu)t} - \mathsf{C}_{j0}{}^+ e^{-\frac{i}{\hbar}(E_0 - E_j - h\nu)t}].$$

The integral of this equation that vanishes at $t = 0$ is

$$a_j = -\mathbf{E}_0 \cdot \left[\mathbf{C}_{j0} \frac{1 - e^{-\frac{i}{\hbar}(E_0 - E_j + h\nu)t}}{E_0 - E_j + h\nu} - \mathbf{C}_{j0}{}^+ \frac{1 - e^{-\frac{i}{\hbar}(E_0 - E_j - h\nu)t}}{E_0 - E_j - h\nu} \right]. \quad (12)$$

When either one of the relations

$$E_0 - E_j \pm h\nu = 0 \quad (13)$$

is not closely satisfied, a_j oscillates very rapidly with time about the value zero. In this case, it may be said that the atom remains in the state Ψ_0 and behaves like a system undergoing forced oscillations far from resonance. On the other hand, when either one of the relations (13) is satisfied we may say that the atom resonates and changes its state. We shall first discuss the case of resonance and return to the other later.

In the case of resonance, we may interpret $|a_j|^2$ as the probability that at time t the system is found in the state Ψ_j. Since $h\nu$ is positive, the case

$$E_0 - E_j = h\nu \quad (14)$$

can occur only when $E_j < E_0$, so that it corresponds to induced emission. The case

$$E_j - E_0 = h\nu, \quad (15)$$

on the other hand, corresponds to induced absorption. The probability that either process occurs in time t is, respectively,

$$P_\nu(t) = |\mathbf{E}_0 \cdot \mathbf{C}_{j0}|^2 \omega(E_0 - E_j \pm h\nu) \quad (16)$$

where the negative sign corresponds to (14) and the positive to (15), and

$$\omega(\epsilon) = \frac{2\left(1 - \cos \frac{\epsilon t}{\hbar}\right)}{\epsilon^2} = \frac{4 \sin^2 \frac{\epsilon t}{2\hbar}}{\epsilon^2}. \quad (17)$$

This varies quadratically with time at small time for any given value of ν. If the radiation is continuous and extends over a sufficiently broad range of the spectrum, however, the total probability varies linearly with time. Let us suppose that the energy of the radiation that lies in the range of frequency from ν to $\nu + d\nu$ and is polarized in the direction $\mathbf{n}_0$ is $\rho(\nu)d\nu$, where ρ is practically constant near the resonance frequency. Since the mean density is related to the amplitude $\mathbf{E}_0$ of (1) by the equation

$$\rho = \frac{\mathbf{E}_0^2}{2\pi},$$

we may rewrite Eq. (16) in the form

$$P_\nu(t) = 2\pi\rho|\mathbf{n}_0 \cdot \mathbf{C}_{j0}|^2\omega(E_0 - E_j \pm h\nu). \tag{18}$$

The total probability of the transition then is

$$P(t) = \int_0^\infty P_\nu(t)d\nu = 2\pi\rho\int_0^\infty |\mathbf{n}_0 \cdot \mathbf{C}_{j0}|^2\omega(E_0 - E_j \pm h\nu)d\nu. \tag{19}$$

The function ω has a peak of half width $\Delta\nu = 1/t$ at the resonance frequency. For the optical absorption and emission time in which we shall be interested, t is of the order of 10^{-8} sec, so that $h\Delta\nu$ is of the order of 10^{-8} ev, which is very small compared with ordinary values of ν. Hence, we may replace ω by a delta function of the same area. Now,

$$\int_{-\infty}^\infty 4\frac{\sin^2\dfrac{\epsilon t}{2\hbar}}{\epsilon^2}d\epsilon = \frac{2t}{\hbar}\int_{-\infty}^\infty \frac{\sin^2 x}{x^2}dx = \frac{2\pi t}{\hbar}.$$

Hence, $\omega(\epsilon)$ may be replaced by

$$\frac{2\pi t}{\hbar}\delta(\epsilon) \tag{19a}$$

where $\delta(\epsilon)$ is a delta function that satisfies the equation

$$\int_{-\Delta\epsilon}^{\Delta\epsilon} \delta(\epsilon)d\epsilon = 1.$$

Thus, $P(t)$ is

$$P(t) = \frac{8\pi^3}{h^2}|\mathbf{n}_0 \cdot \mathbf{C}_{j0}|^2\rho_\nu t \tag{20}$$

where ρ_ν is the density at the frequency defined by Eq. (14) or (15).

Equation (20) shows that the induced emission and absorption probabilities are proportional to the radiation density and that the selection rules for transitions are determined by the matrix components of the operator $\sum_i \mathbf{p}_i e^{2\pi i \eta \cdot \mathbf{r}_i}$. These matrix components can easily be expressed in terms of the matrix components of the atomic dipole moment

$$\mathbf{M} = -\sum_i e\mathbf{r}_i \tag{21}$$

in the case in which $\mathbf{n} \cdot \mathbf{r}$ varies so little over the atomic system that $e^{2\pi i \eta \cdot \mathbf{r}}$ can be replaced by unity. If the Schrödinger equation

$$H\Psi_k = E_k\Psi_k$$

is multiplied by $\left(\sum_i \mathbf{r}_i\right)\Psi_j{}^*$ and the result is subtracted from the equation

$$H\Psi_j{}^* = E_j\Psi_j{}^*,$$

after the latter is multiplied by $\left(\sum_i \mathbf{r}_i\right)\Psi_k$, it is found that

$$(E_j - E_k)\int \Psi_j{}^*\left(\sum_i \mathbf{r}_i\right)\Psi_k d\tau = -\frac{\hbar^2}{2m}\int\left(\sum_i \mathbf{r}_i\right)\left[\Psi_k\left(\sum_i \Delta_i\right)\Psi_j{}^* - \Psi_j{}^*\left(\sum_i \Delta_i\right)\Psi_k\right]d\tau. \quad (22)$$

The right-hand side may, by use of Green's theorem, be transformed to

$$\frac{\hbar^2}{2m}\int \sum_i (\Psi_k \operatorname{grad}_i \Psi_j{}^* - \Psi_j{}^* \operatorname{grad}_i \Psi_k)d\tau. \quad (23)$$

It is assumed that Ψ_k vanishes at large distances, so that the surface integrals in Green's formula can be dropped. In a similar way, the integral of the second term in (23) may be shown to be equal to the first. Hence,

$$\sum_i \int \Psi_j{}^* \operatorname{grad}_i \Psi_k d\tau = -\frac{m}{\hbar^2}(E_j - E_k)\int\left(\sum_i \mathbf{r}_i\right)\Psi_j{}^*\Psi_k d\tau. \quad (24)$$

According to Eq. (11),

$$\mathbf{C}_{j0} = -\frac{e\hbar}{2\pi\nu m}\sum_i \int \Psi_j{}^*(\operatorname{grad}_i \Psi_0)d\tau$$

when $e^{2\pi i \eta \cdot \mathbf{r}}$ is unity, so that we find with the use of Eq. (24)

$$-\mathbf{C}_{j0} = \frac{E_j - E_0}{2\pi\nu\hbar}\int \Psi_j{}^*\left(\sum_i e\mathbf{r}_i\right)\Psi_0 d\tau$$

$$= \pm\mathbf{M}_{j0}. \quad (25)$$

Here

$$\mathbf{M}_{j0} = -\int \Psi_j{}^*\left(\sum_i e\mathbf{r}_i\right)\Psi_0 d\tau, \quad (26)$$

and opposite signs are valid for the cases of Eqs. (14) and (15), respectively.

Equation (20) may now be written

$$P(t) = \frac{8\pi^3}{h^2}|\mathbf{M}_{j0} \cdot \mathbf{n}_0|^2\rho_\nu t. \quad (27)$$

The average value of the coefficient of $\rho_\nu t$ in this equation, namely,

$$\frac{8\pi^3}{3h^2}|\mathbf{M}_{j0}|^2, \tag{28}$$

is the Einstein B coefficient for induced transitions between two states Ψ_0 and Ψ_j.   Here

$$|\mathbf{M}_{j0}|^2 = |M_{x,j0}|^2 + |M_{y,j0}|^2 + |M_{z,j0}|^2. \tag{29}$$

If the lowest atomic state is an S state,[1] which is spherically symmetrical in electron coordinates, the dipole matrix components are finite only if the excited state is a P function, which has the symmetry of a first-order surface harmonic.   If the three degenerate P functions are chosen to have the symmetry of the functions $\sum_i x_i$, $\sum_i y_i$, and $\sum_i z_i$, respectively, and are labeled with indices x, y, z, it may be shown that

$$M_{x,x0} = M_{y,y0} = M_{z,z0},$$
$$M_{x,y0} = M_{x,z0} = \cdots = M_{y,x0} = \cdots = M_{z,y0} = 0.$$

Thus, light polarized in either the x, y, or z direction may induce a transition to one of the triply degenerate states.

In order to discuss by the semiclassical method the scattering of light when the frequency is not near the resonance frequency, it is necessary to compute the mean value of the atomic dipole moment $\mathbf{M}$ for the state f of Eq. (5).   The result is a time-dependent function that contains terms which vary harmonically.   We shall assume that the real coefficient of the term that varies as $e^{2\pi i \nu t}$ is the amplitude of a forced atomic oscillation of frequency ν.   We shall also assume that $e^{2\pi i \eta \cdot \mathbf{r}}$ may be replaced by unity.

The mean value of $\mathbf{M}$ for the state f is

$$\mathbf{M}' = \int f^* \mathbf{M} f d\tau$$
$$= \int \Psi_0^* \mathbf{M} \Psi_0 d\tau + \sum_i{}' \left[a_i \mathbf{M}_{i0}^* e^{\frac{i}{\hbar}(E_0 - E_i)t} + a_i^* \mathbf{M}_{i0} e^{-\frac{i}{\hbar}(E_0 - E_i)t} \right] \tag{30}$$

in which terms involving squares of the a have been neglected.   If it is recalled that a_i has the form

$$a_i = \mathbf{E}_0 \cdot \mathbf{M}_{i0} \frac{E_i - E_0}{h\nu} \left[\frac{1 - e^{-\frac{2\pi i(E_0 - E_i + h\nu)t}{h}}}{E_0 - E_i + h\nu} - \frac{1 - e^{-\frac{2\pi i(E_0 - E_i - h\nu)t}{h}}}{E_0 - E_i - h\nu} \right], \tag{31}$$

[1] We shall use the conventional notation of atomic spectra in which the states having total angular momentum 0, 1, 2, 3, 4, etc., in units of $\hbar$, are designated respectively by S, P, D, F, G, etc.   See, for example, H. E. White, *Introduction to Atomic Spectra* (McGraw-Hill Book Company, Inc., New York, 1934).

it is found that the terms that oscillate with frequency ν are

$$\sum_k{}' \mathsf{E}_0 \cdot \mathbf{M}_{k0}\mathbf{M}_{k0}{}^* \frac{E_k - E_0}{h\nu}\left(\frac{1}{E_0 - E_k - h\nu} - \frac{1}{E_0 - E_k + h\nu}\right)(e^{-2\pi i\nu t} + e^{2\pi i\nu t})$$

$$= \sum_k \frac{\mathsf{E}_0 \cdot \mathbf{M}_{k0}\mathbf{M}_{k0}{}^*}{h} \frac{2\nu_{k0}}{\nu_{k0}{}^2 - \nu^2}(e^{-2\pi i\nu t} + e^{2\pi i\nu t}) \quad (32)$$

where

$$\nu_{k0} = \frac{(E_k - E_0)}{h}.$$

Since the electrostatic field intensity is

$$\mathsf{E} = \mathsf{E}_0(e^{-2\pi i\nu t} + e^{2\pi i\nu t}),$$

the atomic polarizability tensor $\boldsymbol{\alpha}$ for unmodified scattering is

$$\boldsymbol{\alpha} = \sum_k{}' \frac{\mathbf{M}_{k0}{}^*\mathbf{M}_{k0}}{h} \frac{2\nu_{k0}}{\nu_{k0}{}^2 - \nu^2}. \quad (33)$$

This tensor is a constant with the value

$$\sum_k{}' |\mathbf{M}_{0k}|^2 \frac{2\nu_{k0}}{\nu_{k0}{}^2 - \nu^2} \quad (34)$$

if the state Ψ_0 is an S state. The remaining terms in the mean value of $\mathbf{M}$ depend upon time through functions of the type

$$e^{-\frac{i}{h}(E_0 - E_i)t}$$

These terms do not have significance as induced scattering terms and will not be considered further.

In order to discuss Raman, or modified, scattering of light, it is necessary to extend the preceding computation by considering matrix components of $\mathbf{M}$ between states f_0 and f_i. By a correspondence-principle argument[1] that is not very satisfactory, one may then arrive at the equation

$$\alpha_{\nu + \nu_{0n}} = \sum_k \left(\frac{\mathbf{M}_{0k}\mathbf{M}_{kn}}{\nu_{kn} + \nu} - \frac{\mathbf{M}_{kn}\mathbf{M}_{0k}}{\nu_{0k} + \nu}\right) \quad (35)$$

for the polarizability tensor associated with the absorption of frequency ν and the emission of frequency $\nu + \nu_{0n}$. The final frequency must, of course, be positive.

44. The Current Operator.—It is necessary to use the matrix components of the current operator in applying the semiclassical method to

[1] O. KLEIN, *Z. Physik*, **37**, 895 (1926).

solids, as we shall do in Chap. XVII.   This operator may be derived in the following simple way.   If we multiply the Schrödinger equation

$$-\frac{\hbar}{i}\frac{\partial}{\partial t}\Psi_k = -\sum_i \frac{\hbar^2}{2m}\Delta_i\Psi_k + \frac{\hbar e}{imc}\sum_i \mathbf{A}(\mathbf{r}_i)\cdot\mathrm{grad}_i\,\Psi_k + V\Psi_k \qquad (1)$$

by $\Psi_k{}^*$ and subtract from it the equation for $\Psi_k{}^*$ multiplied by Ψ_k, we obtain

$$\frac{\partial|\Psi_k|^2}{\partial t} = \frac{i\hbar}{2m}\sum_i (\Psi_k{}^*\Delta_i\Psi_k - \Psi_k\Delta_i\Psi_k{}^*) - \frac{e}{mc}\sum_i \mathrm{div}_i\,(\mathbf{A}(\mathbf{r}_i)\Psi_k{}^*\Psi_k), \qquad (2)$$

since div $\mathbf{A} = 0$ for the fields in which we shall be interested.   Now,

$$\Psi_k{}^*\Delta_i\Psi_k - \Psi_k\Delta_i\Psi_k{}^* = \mathrm{div}_i\,(\Psi_k{}^*\,\mathrm{grad}_i\,\Psi_k - \Psi_k\,\mathrm{grad}_i\,\Psi_k{}^*),$$

so that Eq. (2) may be written

$$-e\frac{\partial|\Psi_k|^2}{\partial t} = -\sum_i \mathrm{div}_i\left[-\frac{e\hbar}{2mi}(\Psi_k{}^*\,\mathrm{grad}_i\,\Psi_k - \Psi_k\,\mathrm{grad}_i\,\Psi_k{}^*) - \right.$$
$$\left. \frac{e^2}{mc}\mathbf{A}(\mathbf{r}_i)\Psi_k{}^*\Psi_k\right]. \qquad (3)$$

The quantity $-e|\Psi_k|^2$ may be regarded as the charge density ρ in $3N$-dimensional space.   Hence, if we compare Eq. (3) with the equation of continuity, namely,

$$\frac{\partial\rho}{\partial t} = -\mathrm{div}\,\mathbf{J},$$

where $\mathbf{J}$ is the current, we obtain

$$\mathbf{J}_{kk} = \sum_i\left[-\frac{e\hbar}{2mi}(\Psi_k{}^*\,\mathrm{grad}_i\,\Psi_k - \Psi_k\,\mathrm{grad}_i\,\Psi_k{}^*) - \frac{e^2}{mc}\mathbf{A}(\mathbf{r}_i)\Psi_k{}^*\Psi_k\right]. \qquad (4)$$

We shall regard this as the diagonal matrix component of the $3N$-dimensional current operator of which the general element is

$$\mathbf{J}_{jk} = \sum_i\left[-\frac{e\hbar}{2mi}(\Psi_j{}^*\,\mathrm{grad}_i\,\Psi_k - \Psi_k\,\mathrm{grad}_i\,\Psi_j{}^*) - \frac{e^2}{mc}\mathbf{A}(\mathbf{r}_i)\Psi_j{}^*\Psi_k\right]. \qquad (5)$$

Now, the charge density $\rho(\mathbf{r}_1)$ of the first electron is

$$\rho(\mathbf{r}_1) = \int' \rho(\mathbf{r}_1, \cdots, \mathbf{r}_n)d\tau' \qquad (6)$$

where the primed integral extends over the variables of all electrons except the first. Thus, the current associated with the first electron is

$$\mathbf{i}_{jk}(\mathbf{r}_1) = \int' \Big[-\frac{e\hbar}{2mi}(\Psi_j{}^* \operatorname{grad}_1 \Psi_k - \Psi_k \operatorname{grad}_1 \Psi_j{}^*) -$$
$$\frac{e^2}{mc}\mathbf{A}(\mathbf{r}_1)\Psi_j\Psi_k{}^* \Big] d\tau', \quad (7)$$

and the mean total current at the point $\mathbf{r}_1$ is

$$\mathbf{J}_{kk}(\mathbf{r}_1) = N\int' \Big[-\frac{e\hbar}{2mi}(\Psi_k{}^* \operatorname{grad}_1 \Psi_k - \Psi_k \operatorname{grad}_1 \Psi_k{}^*) -$$
$$\frac{e^2}{mc}\mathbf{A}(\mathbf{r}_1)\Psi_k\Psi_k{}^* \Big] d\tau'. \quad (8)$$

45. Line Breadth.[1]—The absorption and emission lines of an atomic system that has discrete levels obviously should be infinitely narrow in the delta-function approximation of Eq. (19a), Sec. 43. The same result is obtained even if $\omega(\epsilon)$ is not replaced by a delta function, for the function

$$\omega(\epsilon) = \frac{\sin^2 \dfrac{\epsilon t}{2\hbar}}{\epsilon^2} \quad (1)$$

may be made as narrow as we please by making t large enough. Thus, after a long time the frequency distribution of the total radiation from a system of excited atoms should be very narrow. This result is a consequence of the approximations that were used in solving Eqs. (7), Sec. 43, and does not occur if the equations are solved more accurately by taking into account the fact that the coefficient $a_0(t)$ for the initial state is not a constant but varies with time.

It is possible to obtain a nearly self-consistent solution of Eq (6) Sec. 43, in a number of important cases by assuming that

$$a_0 = e^{-2\pi\Gamma t} \quad (2)$$

where Γ is a constant. With this assumption, the equation for a_j is

$$-\frac{\hbar}{i}\frac{\partial a_j}{\partial t} = H_{j0}e^{-\frac{i}{\hbar}(E_0 - E_j - ih\Gamma)t} \quad (3)$$

in the approximation in which terms other than a_0 may be neglected in the right-hand side. The solution of this equation is

$$a_j = -\mathbf{E}_0 \cdot \Big[\mathbf{C}_{j0}\frac{1 - e^{-\frac{i}{\hbar}(E_0 - E_j - ih\Gamma + h\nu)t}}{E_0 - E_j - ih\Gamma + h\nu} - \mathbf{C}_{j0}^+\frac{1 - e^{-\frac{i}{\hbar}(E_0 - E_j - ih\Gamma - h\nu)t}}{E_0 - E_j - ih\Gamma - h\nu} \Big]. \quad (4)$$

[1] See the survey article by V. Weisskopf, *Phys. Zeits.*, **34**, 1 (1933).

A value of Γ may be obtained by substituting (2) into the equation for a_0, namely,

$$-\frac{\hbar}{i} \frac{\partial a_0}{\partial t} = \sum_j H_{0j} a_j e^{-\frac{i(E_j - E_0)t}{\hbar}}. \tag{5}$$

In the semiclassical theory, Γ is proportional to the energy density of radiation and is, as a consequence, very small when the radiation density is small.   In the Dirac theory, however, it contains an additional constant term that is related to the probability coefficient for spontaneous emission from the upper to the lower member of the pair of levels between which the optical transitions are occurring.

The function $a_j(t)$ given by Eq. (4) reduces to $a_j(t)$ of Eq. (12), Sec. 43, when t is much smaller than $1/\Gamma$.   Hence, the results of Sec. 43 are valid only for comparatively short times.   At times long compared with $1/\Gamma$, the square of the absolute value of one of the terms in (4) reduces to

$$\frac{|\mathbf{E}_0 \cdot \mathbf{C}_{j0}|^2}{[(E_0 - E_j) \pm h\nu]^2 + h^2\Gamma^2} \tag{6}$$

a fact showing that the distribution of transition probability for different frequencies is governed by a function of the form

$$\frac{1}{[(E_0 - E_j) \pm h\nu]^2 + h^2\Gamma^2}, \tag{7}$$

which, as a function of ν, has a peak at

$$\nu = \frac{|E_0 - E_j|}{h} \tag{8}$$

of which the width at half maximum is

$$\Delta\nu = 2\Gamma. \tag{9}$$

It may be shown that Γ is of the order of magnitude 10^8 sec^{-1} for ordinary atomic transitions so that this *natural width* ordinarily is small compared with emission frequencies.

The fact that an emission line or an absorption line has a finite natural breadth does not imply that energy is not conserved.   This breadth finds its origin in the fact that the energy of the excited atomic state is uncertain because the interaction between the atom and the radiation field, expressed by the term H_I, is uncertain.   Thus the half width could be made as small as we please by making H_I sufficiently small.

Natural broadening is usually masked by one or more kinds of broadening that have a completely different origin.   In the case of gases, for example, lines are broadened by the Doppler effect, since the atoms move with different speeds, and by the interaction between atoms. Doppler broadening is not important in crystals, but the analogue of interaction broadening is important. We shall discuss it briefly here and in more detail in Chap. XVII.

If we have a large number of free atoms that are infinitely separated from one another and are stationary, their electronic levels are discrete

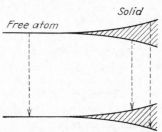

FIG. 1.—The discrete atomic energy levels may be broadened in the solid. If transitions between many levels in each band are allowed, the resulting emission "line" will be broad. This occurs in metals.

so that the energy levels of the entire system are discrete. Thus, the width of the emission lines is determined entirely by natural broadening. The atoms interact, however, if they are brought within a finite distance of one another, and this interaction broadens those levels of the entire system which were degenerate when the atoms were infinitely separated,

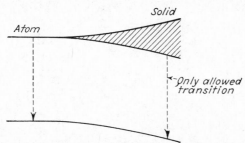

FIG. 2.—In this case the lowest level remains nondegenerate in passing from the system of free atoms to the solid. If selection rules forbid transitions from more than one of the excited states to the lower level, the emission line should be sharp. In actual cases of this type the line is broadened because of coupling between the electrons and the lattice.

as we shall see in Sec. 66. If transitions are allowed between many of the levels in two bands, the corresponding emission "line" should now have a breadth that is determined by the selection rules and the width of the bands. Thus the lines may broaden because the atomic levels are spread into bands (*cf.* Fig. 1).

It frequently happens, in the case of solids, that selection rules forbid transitions from more than one level of a band to a nondegenerate level of the entire system (*cf.* Fig. 2).   The corresponding line would have only the natural breadth, in this case, if only the electronic states of the system had to be taken into account.   An assembly of atoms, however, has, in addition to its electronic states, a system of energy states that are associated with internuclear motion.   In the case of solids, these states may be described in terms of lattice oscillations, as we have seen in Chap. III.   It turns out that the vibrational states may be excited during transitions between electronic states.   Since the range of allowed vibrational frequencies is continuous, the amounts of energy that can be given to the vibrational system are spread over a continuous range. Now, the energy that is not given to the vibrational system during spontaneous emission is radiated as light; hence, the emission lines have vibrational broadening.   This vibrational broadening is one of the most important of the factors that determine the shape of absorption and emission bands of insulating solids at temperatures above absolute zero.

CHAPTER VI

APPROXIMATE TREATMENTS OF THE MANY-BODY PROBLEM

46. Introduction.—This chapter deals with some of the methods that have been devised to handle the Schrödinger equation for a mechanical system in which there are at least two interacting particles. In practically all these schemes, an attempt is made to select one member of a given set of admittedly approximate wave functions by use of the variational theorem (*cf.* Chap. V, Sec. 39). In one method—the variational method—the set of approximate state functions is obtained by writing down a definite function of the electronic variables that contains a number of parameters. The best function of the family is chosen by fixing these parameters so that the mean value of the Hamiltonian is a minimum. In another scheme—that of Hartree, Fock, and Slater—the starting set of functions is chosen as a combination of one-particle functions, that is, functions that involve the coordinates of only one particle. The one-particle functions are then determined by use of the variational theorem.

Since the exact solutions of the Schrödinger equation for any many-body system usually are very intricate functions of all variables and since the functions that may be manipulated are restricted to fairly simple types, even the best function obtained by one of the approximate treatments usually leads to an energy that differs appreciably from the experimental value. There are exceptions to this statement, such as the case of the normal state of helium, which we shall find useful for locating the cause of error in other problems.

We shall begin the discussion with a few remarks concerning the Hamiltonian operator that will be used in solids. These will be followed by a presentation of the two schemes that were described above.

47. The Hamiltonian Function and Its Mean Value.—In order to place the problem of determining the stationary states of solids upon a working basis, it is necessary to overlook certain terms in the Hamiltonian operator that are of secondary importance. First, we shall neglect the effects that arise from the motion of nuclei, assuming that the nuclei are at rest. The nuclear coordinates then enter into the Hamiltonian as parameters. In later chapters that deal with phase-changes, conductivity, and optical properties, we shall be interested both in the motion of nuclei and in the effect of nuclear motion on electrons. Second, we shall

227

assume that there is no radiation present so that we shall not be concerned with radiation interaction terms. Neither of these neglected effects introduces appreciable errors, as long as we are interested only in the eigenvalues of the electronic system.

In addition to the foregoing assumptions, it is expedient to assume that all extranuclear electrons may be divided into two classes: (1) the inner electrons, which belong to closed shells that are rigidly attached to the nuclei and are not affected appreciably by changes in interatomic distances; (2) the outer, or valence, electrons, which are affected by changes in interatomic distance. The latter are responsible for most of the ordinary properties of solids. We shall assume that the effect of the rigidly bound electrons on the valence electrons may be described by means of a potential term of the same type as that which takes into account the effect of the nuclei. In other words, it will be assumed that the valence-electron wave function may be determined by use of a Hamiltonian operator in which the effect of the rigid-shell electrons is taken into account by the presence of an ordinary potential function. The validity of this method must be investigated for each solid and will be discussed in particular cases later. It will be seen that this procedure is usually satisfactory for simple substances.

According to these assumptions, the Hamiltonian operator for n valence electrons is

$$H = \sum_{i=1}^{n}\left(-\frac{\hbar^2}{2m}\Delta_i + V_i\right) + \frac{1}{2}\sum_{i,j=1}^{n}{}' \frac{e^2}{r_{ij}} + I \qquad (1)$$

where the indices i and j are to be summed over all n electrons, $-\hbar^2\Delta_i/2m$ is the kinetic energy operator of the ith electron, and V_i, which is the same for all valence electrons, is the potential energy of the ith electron in the field of the nuclei and bound electrons. e^2/r_{ij} is the coulomb interaction potential between the ith and jth electrons, where e is the absolute value of the electronic charge and r_{ij} the distance between electrons. It should be noted that the cases $i = j$ are excluded in the last summation. Finally, I is a constant representing the interaction between the nuclei and between rigid-shell electrons on different atoms. Although this term does not enter into the determination of the valence-electron wave functions, it is included here to indicate that it may not be neglected when the cohesive energy of the solid is computed. It should be added that both V_i and I contain the internuclear distances parametrically.

Now, the stationary state of lowest energy minimizes the mean-value integral

$$E = \int\Psi^*H\Psi d\tau \qquad (2)$$

according to the mean-value theorem.   For this reason, the mean-value integral is of importance and merits considerable discussion.   Let us break the Hamiltonian (1) into two parts, namely,

$$T = -\sum_i \frac{\hbar^2}{2m}\Delta_i, \tag{3a}$$

$$V = \sum_i V_i + \frac{1}{2}\sum_{i,j}' \frac{e^2}{r_{ij}} + I, \tag{3b}$$

where T is the total kinetic energy operator and V is the total potential energy operator.   In terms of T and V, Eq. (2) is

$$E = \int\Psi^*T\Psi d\tau + \int\Psi^*V\Psi d\tau. \tag{4}$$

We may write the first integral in the form

$$\frac{\hbar^2}{2m}\sum_i \int |\text{grad}_i \Psi|^2 d\tau \tag{5}$$

if we apply Green's theorem and assume that the surface integral vanishes, as is true in the cases in which we shall be interested.   We may conclude that the mean value of T is a positive quantity since the integrand in (5) is positive.   The second integral in (4) may be written in the form

$$\int|\Psi|^2 V d\tau, \tag{6}$$

for V is simply a function of the variables of integration.   This integral may be regarded as the classical potential energy of the charge distribution

$$\rho = e|\Psi|^2 \tag{7}$$

in the potential field V/e, both the charge distribution and the potential function being static in the $3n$-dimensional space of all electronic coordinates.   Thus, *the mean value of the potential energy may be given a straightforward classical interpretation in terms of $3n$-dimensional charge and potential distributions.*[1]

Since the mean value of T is positive, it follows that ordinary atomic systems are energetically stable only because the mean value of V can be negative.   Evidently the first term of (3b), which contains the energy of attractive nuclear-electron forces, is entirely responsible for the fact that (6) may be negative.   At first sight, one might suppose that the

---

[1] Although Ψ is a function of the spin variables as well as the space variables, it is possible to average over spin without affecting this conclusion, since neither T nor V depends upon spin.   Thus ρ in (7) may be taken as a simple function of the space variables.

wave function Ψ that minimizes E would distribute the electronic charge in such a way that the integral (6) should be as negative as possible. It is easy to see, however, that this wave function would have a very large kinetic energy associated with it. To minimize (6), it would be necessary to choose Ψ in such a way that ρ is small in regions where V is positive and large in regions where V is negative. This means, however, that Ψ would vary rapidly from point to point or that its gradient would be large. The mean value of (5) would then be large, and it is easy to show (*cf.* next paragraph) that E would not be as low as possible. On the other hand, suppose that we attempt to minimize E by making the mean kinetic energy as small as possible. In this case, we should choose Ψ to be a very slowly varying function, so that its gradient might be very small. A function of this type would not give a charge distribution ρ that preferentially localizes the electrons in the regions of negative V, so that E again would not be as small as possible. Thus we see that the mean values of T and V tend to counterbalance one another. If the mean value of V is large and negative, the mean value of T is large and positive; if the mean value of T is nearly zero, the mean value of V is also nearly zero. Hence, the function that minimizes E must effect a compromise between both terms.

This competition between T and V is indicated by the uncertainty relation

$$\Delta p_i \Delta q_i \geqq \hbar.$$

Any increase in the localization of electrons implies that the Δq_i are decreased, a fact that in turn implies an increase in Δp_i. Since the mean value of T is roughly $\sum_i (\Delta p_i)^2/2m$, the kinetic energy increases as the Δq_i decrease. Suppose, for example, that we have an electron moving in the potential field e/r of a proton. The potential energy is roughly $-e^2/\Delta r$, where Δr is the uncertainty in r, whereas the kinetic energy is roughly $\hbar^2/2m(\Delta r)^2$. The sum of these two energies does not have a minimum when Δr is zero or infinite, which would respectively minimize the two terms, but it has a minimum for $\Delta r = \hbar^2/me^2 \cong 0.5 \cdot 10^{-8}$ cm. Incidentally, this example shows that the scale of atomic size, namely, $\hbar^2/me^2$, is determined by the compromise between kinetic and potential energy.

Although the competition between T and V tempers the wave function in such a way that neither of the terms is minimized, the wave function always favors both terms to some extent. Thus, it is found that the wave function for the lowest energy state is generally smooth and yet localizes the charge in suitable regions of potential. Since the potential (3b) becomes very large and positive in regions of the $3n$-dimensional coordinate space where the variables r_{ij} are very small, we may expect that ρ will generally exhibit a smooth minimum of the type shown in Fig. 1 in the neighborhood of such regions.

Upon these effects, which are related to the variational principle, are superimposed those which arise from the condition that Ψ must be antisymmetric. Roughly, the lowest energy members of the family of antisymmetric functions usually have a higher mean kinetic energy than some of the functions that might be allowed were it not for the Pauli principle. The change in sign of Ψ under permutation of the variables of any two electrons implies that Ψ usually changes its sign in certain regions of configuration space. The word usually is employed because in some simple exceptional cases, such as that of the lowest state of atomic helium, the antisymmetric condition affects only the spin variables.

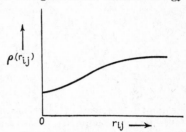

Fig. 1.—It may be expected that the electron density function possesses a smooth minimum of this type because of electron repulsion.

The change in sign of Ψ implies in turn that there is a nonvanishing gradient in the same regions, and this implies a contribution to the kinetic energy (5). Unless this contribution is compensated by a drop in the mean potential energy, the best antisymmetric function does not give E an absolute minimum. Thus, the lowest states of all free atoms that involve more than two electrons (that is, atoms beyond helium) are raised because of the Pauli principle. On the other hand, we shall see that the potential energy of antisymmetric valence-electron wave functions in solids is usually lower than the energy of other functions and that the decrease in potential energy sometimes counterbalances the gain in kinetic energy. The origin of this decrease in V may be understood in the following way. As we remarked above, the antisymmetry of Ψ implies that the wave function may change its sign at regions of configuration space where r_{ij} is zero. Hence, the Pauli principle may imply that ρ has a minimum at regions where r_{ij} is zero. This condition, however, is just that which must be satisfied if the positive contribution to the potential from the e^2/r_{ij} terms is to be small.

48. The Helium-atom Problem.—We shall discuss only one of the important problems solved by the variational method, namely, that of helium. The Hamiltonian for this two-electron system is

$$H = -\frac{\hbar^2}{2\mu}\Delta_1 - \frac{\hbar^2}{2\mu}\Delta_2 + \frac{e^2}{r_{12}} - \frac{Ze^2}{r_1} - \frac{Ze^2}{r_2} \tag{1}$$

where r_1 and r_2 are the distances of the two electrons from the nucleus of charge Ze ($Z = 2$ for He) and μ is the reduced electronic mass, namely,

$$\mu = \frac{mM}{m + M}$$

in which m is the electronic mass and M is the nuclear mass. The other quantities in (1) have been defined previously. In the approximation in which (1) is valid, the wave functions may be written in the form[1]

$$\Psi = \Phi(\mathbf{r}_1,\mathbf{r}_2)\sigma(\zeta_1,\zeta_2) \tag{2}$$

where Φ is a function of the space variables, symbolized by $\mathbf{r}_1$ and $\mathbf{r}_2$, and σ is a function of the spin variables. It may be shown from the symmetry of (1) under permutations of electrons that the only forms of Φ that lead to stationary values of $\int\Phi^*H\Phi d\tau$ are either symmetric or antisymmetric under interchange of $\mathbf{r}_1$ and $\mathbf{r}_2$. This characteristic is peculiar to the two-electron problem. We shall designate a symmetrical Φ by Φ_S and an antisymmetrical one by Φ_A. According to the Pauli principle, σ must be antisymmetric when Φ is symmetric and symmetric when Φ is antisymmetric, in order that Ψ may be antisymmetric. There are three symmetric spin functions for two electrons, namely,[2]

$$\left.\begin{aligned}
\sigma_S^1 &= \eta_1(1)\eta_2(1), \\
\sigma_S^0 &= \eta_1(1)\eta_2(-1) + \eta_1(-1)\eta_2(1), \\
\sigma_S^{-1} &= \eta_1(-1)\eta_2(-1),
\end{aligned}\right\} \tag{3}$$

and there is one antisymmetric function, namely,

$$\sigma_A^0 = \eta_1(1)\eta_2(-1) - \eta_1(-1)\eta_2(1). \tag{4}$$

These functions are eigenfunctions of the spin operators

$$\mathbf{\Sigma}^2 = (\sigma_{x_1} + \sigma_{x_2})^2 + (\sigma_{y_1} + \sigma_{y_2})^2 + (\sigma_{z_1} + \sigma_{z_2})^2 \tag{5}$$

and

$$\Sigma_z = (\sigma_{z_1} + \sigma_{z_2}). \tag{6}$$

The functions (3) correspond to the eigenvalue $2\hbar^2$ of (5) and to the eigenvalues $\hbar$, 0 and $-\hbar$ of (6), respectively, whereas (5) corresponds to the eigenvalue 0 of both (5) and (6). We may expect that a *singlet state* $\Phi_S\sigma_A^0$ has the lowest energy, since the presence of Φ_S implies a low kinetic energy (*cf.* Sec. 47).

The most extensive and systematic investigation of the mean value

$$E = \int\Phi_S^*H\Phi_S d\tau \tag{7}$$

---

[1] For simplicity, we shall designate the spin variable of the ith electron by ζ_i instead of by ζ_{z_i}, as in the preceding chapter.

[2] The function $\eta_i(\pm 1)$ is identical with the function $\eta(\zeta_{z_i}, \pm 1)$ of the preceding chapter.

has been carried through by Hylleraas,[1] who used functions containing an increasing number of parameters in successive stages of approximation. In the case of helium, his method leads to a total binding energy that agrees with the experimental value to within a few hundredths of a per cent. We shall discuss the various types of function that he employed and shall attempt to interpret his results in terms of the general remarks of the preceding section. In this discussion, it is convenient to replace the six Cartesian coordinates of the electron by the variables

$$\left.\begin{array}{l} s = r_1 + r_2, \\ t = r_1 - r_2, \\ u = r_{12}, \end{array}\right\} \tag{8}$$

and by three other angular variables upon which the wave function of the lowest state does not depend.

a. First Approximation.—The simplest function used by Hylleraas has the form

$$\Phi_S = e^{-\alpha s} = e^{-\alpha r_1} e^{-\alpha r_2}, \tag{9}$$

which contains only one parameter, namely, α. The use of this function is equivalent to assuming that both electrons move in a coulomb field and have hydrogen-like wave functions corresponding to the $1s$ state. Since $\alpha = Z/a_h$ for one-electron atoms, where Z is the atomic number of the nucleus and a_h is the radius of the first Bohr orbit in hydrogen ($a_h = 0.531$ Å), the value of $\alpha a_h = Z'$ that gives $\int \Phi_S{}^* H \Phi_S d\tau$ its minimum value furnishes a value of the "effective nuclear charge" in which each electron moves. It is found that $Z' = Z - \frac{5}{16}$, which is $\frac{27}{16}$ for helium. The mean value of H is then $\dfrac{-(Z - \frac{5}{16})^2 e^2}{a_h}$, or 5.695 Rydberg units (1 Rydberg unit equals 13.54 ev), which should be compared with the observed value of 5.810 Rydberg units.

A part of this difference of 0.7 ev per electron may be removed by using a better eigenfunction of the one-electron type. The Hartree-Fock procedure, which will be discussed more fully in the following sections, starts from the assumption

$$\Phi_S = \phi(\mathbf{r}_1)\phi(\mathbf{r}_2) \tag{10}$$

and determines ϕ in such a way as to minimize (7); that is, it determines the best Φ that may be written in the form (10). The energy associated[2] with the Hartree wave function is 5.734 Rydberg units, which leaves

[1] E. HYLLERAAS, *Z. Physik*, **54**, 347 (1929). See also H. BETHE, *Handbuch der Physik*, Vol. XXIV/1, p. 324.

[2] BETHE, *op. cit.*, p. 368.

about two-thirds of the difference, or about 0.45 ev per electron. This difference arises from the fact that electrons really do not move independently of one another but are *correlated*. The wave function not only is a function of the variables $s = r_1 + r_2$ and $t = r_1 - r_2$ but is also a function of $u = r_{12}$. This dependence has been discussed in Sec. 47 and is related to the fact that a potential term of the type e^2/u appears in the Hamiltonian.

b. *Second Approximation.*—In the next approximation, Hylleraas chose the function

$$\Phi_S = e^{-\alpha s}(1 + a_1 u + b_1 t^2) \tag{11}$$

and found

$$E = -5.805$$

for

$$\alpha = \frac{1.82}{a_h}.$$

$$a_1 = \frac{0.29}{a_h},$$

$$b_1 = \frac{0.13}{a_h^2}.$$

This leaves a difference of only 0.03 ev per electron between the computed and observed values. The function (11) shows the expected correlation between electrons since the variable u enters in it.

c. *Higher Approximations.*—Hylleraas extended these computations by using a power series in u, s, and t^2 as the coefficient of $e^{-\alpha s}$ and arrived at a limiting energy of 5.80749 Rydberg units which actually is greater than the observed value by about 0.0002 Rydberg unit. This discrepancy does not imply any flaw in the variation method but is due to the fact that the Hamiltonian (1) neglects relativistic effects which would introduce a correction of this order of magnitude.

One important fact that may be gained from this investigation is that the method of one-electron functions yields an energy which is in error by about 0.5 ev per electron, because it does not involve the necessary correlations. Since the binding energy of many solids is of the order of 1 ev per electron, we may expect that the cohesive energies derived from one-electron functions will often have a relatively high percentage of error.

49. The One-electron Approximation.[1]—The one-electron scheme has proved to be the most fruitful of several approximate methods that

---

[1] General discussions of this topic may be found in the tract by L. Brillouin *Actualités Scientifiques* iv (Hermann et Cie., Paris, 1934), and in E. U. Condon and G. H. Shortley, *The Theory of Atomic Spectra* (Cambridge University Press, 1935).

have been developed for obtaining qualitative and semiquantitative solutions of the Schrödinger equation when many electrons are present. As was mentioned in the introduction to the present chapter, this scheme is based upon a plan for constructing wave equations for an n-electron system from n one-electron wave functions.

It was assumed in the introductory development of the one-electron approximation that the total state function can be represented by the product of n one-electron functions.   Thus,

$$\Psi(x_1, y_1, z_1, \cdots, x_n, y_n, z_n) = \psi_1(x_1, y_1, z_1)\psi_2(x_2, y_2, z_2) \cdots$$
$$\psi_n(x_n, y_n, z_n). \quad (1)$$

Here the ψ_i are the normalized one-electron functions and x_i, y_i, z_i are the spatial coordinates of the ith electron. Spin variables were not explicitly introduced. In accordance with the Pauli principle, as it had been introduced into the Bohr theory, it was assumed that no more than two electrons can have identical functions and that any two electrons that do have identical ψ_i have opposite spins. The function (1) does not include electronic correlations explicitly since the state function for the ith electron is ψ_i, regardless of the position of the other electrons.

Hartree[1] suggested, on the basis of plausibility, that each one-electron function in (1) should satisfy a one-electron Schrödinger equation in which the potential includes a term that takes into account the coulomb field of the other electrons as well as the fields arising from nuclei and other charged particles. He chose this term as the classical electrostatic potential of the $n - 1$ normalized charge distributions $|\psi_i|^2$. In other words, his equation for ψ_j is[2]

$$-\frac{\hbar^2}{2m}\Delta_j\psi_j + \left[V_j(x_j, y_j, z_j) + \sum_i{}' e^2 \int \frac{|\psi_i|^2}{r_{ij}}d\tau_i \right]\psi_j = \epsilon_j\psi_j \quad (2)$$

where V_j is the field arising from nuclei, etc.   Hartree developed a practical method, now known as the method of the self-consistent field, for solving the set of simultaneous equations, and applied this procedure to a number of atoms in a large-scale program of investigating the periodic chart.   This program is still under way, although modified equations, which we shall discuss below, are now used in place of (2).

---

[1] D. R. HARTREE, *Cambridge Phil. Soc.*, **24**, 89 (1928).

[2] In the rest of this chapter, we shall designate the volume element for the ith electron by $d\tau_i$ when spin is not included and by $d\tau_i'$ when spin is included.   In cases in which ambiguity may occur, however, we shall use the notation of the preceding chapter, namely, $d\tau(x_i, y_i, z_i)$, etc.   Similarly, the volume element for the coordinates of the ith and jth electrons will sometimes be written as $d\tau_i d\tau_j$ and at other times as $d\tau_{ij}$.

The mean value of the Hamiltonian

$$H = \sum_i \left(-\frac{\hbar^2}{2m}\Delta_i + V_i \right) + \frac{1}{2}\sum_{i,j}' \frac{e^2}{r_{ij}} \tag{3}$$

for the function (1) is seen to be simply

$$E = \sum_{i=1}^n \left(-\frac{\hbar^2}{2m}\int \psi_i{}^*\Delta\psi_i d\tau_i + \int \psi_i{}^* V_i\psi_i d\tau_i \right) +$$
$$\frac{1}{2}\sum_{i,j}' e^2 \int\int \frac{|\psi_i|^2|\psi_j|^2}{r_{ij}} d\tau_i d\tau_j, \tag{4}$$

which reduces to

$$E = \sum_{i=1}^n \epsilon_i - \frac{1}{2}\sum_{i,j}' e^2 \int\int \frac{|\psi_i|^2|\psi_j|^2}{r_{ij}} d\tau_i d\tau_j \tag{5}$$

when the ψ_i satisfy Hartree's equations.

There is only one equation of the type (2) for the normal state of helium since the two ψ are identical in this case. The energy (5) that was derived by using the solution of this equation was discussed in part a of the previous section.

It was recognized during the period after Hartree's first work that the Pauli principle has a more natural position in the new quantum theory than in the old. In accordance with the discussion of Chap. V, the principle is taken to imply that all wave functions must be antisymmetric under permutation of all electronic coordinates, including the spin variables. The function (1) is not a satisfactory wave function from this standpoint, for it is not antisymmetric. An allowable antisymmetric function may be constructed from the same set of one-electron functions, however, if each ψ_i is replaced by a function φ_i that is the product of ψ_i and a spin function $\eta_i(\zeta_i)$, thus:

$$\varphi_i(\mathbf{r}_j) = \psi_i(x_i, y_i, z_i)\eta_i(\zeta_i) \tag{6}$$

where $\mathbf{r}_j$ in the left-hand side of (6) stands for the four variables x_j, y_j, z_j, ζ_j. We shall assume in the following discussion that the η_i are eigenfunctions of σ_z, the z component of spin. Hence, they may be labeled as $\eta_i(+1)$ or $\eta_i(-1)$, in the cases going with the eigenvalues $\hbar/2$ and $-\hbar/2$, respectively. The antisymmetric function that may be constructed from the φ_i is

$$\Psi = \sum_P (-1)^p P \cdot [\varphi_1(\mathbf{r}_1)\varphi_2(\mathbf{r}_2) \cdots \varphi_n(\mathbf{r}_n)] \tag{7}$$

where P runs over the $n!$ permutations of the n variables and p is the parity[1] of the pth permutation. This sum has the property that it may be written in the form of a determinant

$$\Psi = \begin{vmatrix} \varphi_1(\mathbf{r}_1)\varphi_1(\mathbf{r}_2) & \cdots & \varphi_1(\mathbf{r}_n) \\ \varphi_2(\mathbf{r}_1)\varphi_2(\mathbf{r}_2) & \cdots & \varphi_2(\mathbf{r}_n) \\ \cdot & & \cdot \\ \cdot & & \cdot \\ \cdot & & \cdot \\ \varphi_n(\mathbf{r}_1)\varphi_n(\mathbf{r}_2) & \cdots & \varphi_n(\mathbf{r}_n) \end{vmatrix}, \tag{8}$$

in which the φ_i are elements. The antisymmetric properties of Ψ are now evident from the properties of determinants, for the process of interchanging two columns, which reverses the sign of Ψ, corresponds to a permutation of the corresponding variables. Moreover, Ψ vanishes identically when two φ_i are equal. Hence, the Pauli principle, as it was employed in the Bohr theory, is automatically satisfied.

It may be shown from elementary principles of group theory[2] that (7) is the only antisymmetric combination of the φ_i. This function is not usually the only antisymmetric combination of the ψ_i, however, for it may be possible to assign spin functions to the ψ_i in more than one way. This possibility will be discussed further in the following sections.

The electrons do not move independently so long as there is more than one independent term in (7); that is, electronic motion is correlated in the antisymmetric wave function even though this function is composed of one-electron functions. These correlations are more or less accidental, for they arise from the Pauli principle rather than from the requirement that the electrons should keep away from one another. It turns out that these accidental correlations sometimes favor cohesion by keeping the electrons apart and sometimes hinder it by piling the electrons together. We shall have occasion to examine particular cases in detail in later chapters.

The function (8) is not normalized when the φ_i are normalized, so that (8) must be multiplied by a constant. Although this constant usually depends upon the choice of φ_i, it is simply $1/\sqrt{n!}$ in the particular case in which the φ_i are orthogonal, that is, when

$$\int \varphi_i{}^*\varphi_j d\tau(x,y,z,\zeta) = \delta_{ij}. \tag{9}$$

The integration in (9) implies a summation over the two values of spin

---

[1] The parity of a permutation is the number of interchanges that must be made in order to obtain the permutation from the standard arrangement. Thus, the parity of the permutation 2143 of the integers 1234 is 2, since the permutation may be obtained by interchanging 1 and 2, and 3 and 4, respectively.

[2] E. P. WIGNER, *Gruppentheorie* (Vieweg, Braunschweig, Germany, 1931).

variable ζ.   If Ψ is given by (7),

$$\int \Psi^*\Psi d\tau' =$$
$$\int \sum_P \sum_{P'} (-1)^{p+p'} P[\varphi_1^*(\mathbf{r}_1) \cdots \varphi_n^*(\mathbf{r}_n)] P'[\varphi_1(\mathbf{r}_1) \cdots \varphi_n(\mathbf{r}_n)] d\tau', \quad (10)$$

which is equal to

$$\sum_P \int P|\psi_1(\mathbf{r}_1)|^2 |\psi_2(\mathbf{r}_2)|^2 \cdots |\psi_n(\mathbf{r}_n)|^2 d\tau = n!$$

because of (9).

We shall assume hereafter that the condition (9) is satisfied.   This does not place any important restriction upon the φ_i since those which have opposite spins are automatically orthogonal and those which have parallel spins may be made orthogonal by the Schmidt method.[1]   This orthogonalization process does not affect (8), for a determinant remains unchanged if a constant multiple of the elements in one row is added to the corresponding elements in another, and application of the Schmidt method is equivalent to doing this.

Let us split the Hamiltonian into kinetic and potential energy parts, as in Sec. 47.   The mean value of the former, namely,

$$T = -\frac{\hbar^2}{2m} \sum_1^n \Delta_i,$$

is

$$-\sum_{i=1}^n \frac{\hbar^2}{2m} \int \frac{1}{n!} \left[\sum_P (-1)^p P(\varphi_1^*(\mathbf{r}_1)\varphi_2^*(\mathbf{r}_2) \cdots \varphi_n^*(\mathbf{r}_n)) \right] \Delta_i \cdot$$

$$\left[\sum_{P'} (-1)^{p'} P'(\varphi_1(\mathbf{r}_1) \cdots \varphi_n(\mathbf{r}_n)) \right] d\tau(x_1, \cdots, z_n, \zeta_1, \cdots, \zeta_n)$$

$$= -\sum_{i=1}^n \frac{\hbar^2}{2m} \int \varphi_i^* \Delta \varphi_i d\tau(x,y,z,\zeta) = -\sum_{i=1}^n \frac{\hbar^2}{2m} \int \psi_i^* \Delta \psi_i d\tau(x,y,z) \quad (11)$$

when the eigenfunction is (7).   This result is exactly the same as that derived by use of (1).   Hence, the mean value of T is unaffected by the use of a determinantal eigenfunction.   The same statement is true for that part of the potential energy which can be written in the form $\sum_{i=1}^n V_i$,

[1] See, for example, R. COURANT and D. HILBERT, *Methoden der mathematische Physik* (Julius Springer, Berlin, 1924).

where V_i depends only upon the coordinates of the ith electron and is the same function of these as V_j is of the coordinates of the jth electron, as can be seen from the fact that the form of (11), as a sum of one-electron integrals, depends only upon the property that T is a sum of one-electron operators. Hence,

$$\sum_i \bar{V}_i = \sum_i \int \varphi_i^* V(\mathbf{r}) \varphi_i d\tau(x,y,z,\zeta). \tag{12}$$

If V is independent of spin, this is

$$\sum_i \int \psi_i^* V(\mathbf{r}) \psi_i d\tau(x,y,z). \tag{13}$$

The mean value of those terms of the Hamiltonian, such as

$$\frac{1}{2} \sum_{i,j}' \frac{e^2}{r_{ij}}, \tag{14}$$

in which the elements of summation depend upon the coordinates of two electrons, is affected by the use of an antisymmetric function. The mean value of a typical term of (14), e^2/r_{12}, is

$$\frac{1}{n!} \int \left[\sum_P (-1)^p P(\varphi_1^*(\mathbf{r}_1) \, \cdots \, \varphi_n^*(\mathbf{r}_n)) \right] \frac{e^2}{r_{12}} \cdot$$
$$\left[\sum_{P'} (-1)^{p'} P'(\varphi_1(\mathbf{r}_1) \, \cdots \, \varphi_n(\mathbf{r}_n)) \right] d\tau' =$$
$$\frac{1}{n!} \sum_{P,P'} (-1)^{p+p'} \int P[\varphi_1^*(\mathbf{r}_1) \, \cdots \, \varphi_n^*(\mathbf{r}_n)] \frac{e^2}{r_{12}} P'[\varphi_1(\mathbf{r}_1) \, \cdots \, \varphi_n(\mathbf{r}_n)] d\tau'. \tag{15}$$

If a given permutation P sends $\mathbf{r}_i$ into $\mathbf{r}_1$ and $\mathbf{r}_j$ into $\mathbf{r}_2$ and the remaining $n-2$ variables into $\mathbf{r}_3, \cdots, \mathbf{r}_n$ in some particular way, the integral in (15) that contains this P vanishes unless the P' in the same integral sends the same set into $\mathbf{r}_3, \cdots, \mathbf{r}_n$ and either $\mathbf{r}_i$ into $\mathbf{r}_1$ and $\mathbf{r}_j$ into $\mathbf{r}_2$ or $\mathbf{r}_i$ into $\mathbf{r}_2$ and $\mathbf{r}_j$ into $\mathbf{r}_1$. When P' does satisfy this condition, the integration over the variables $\mathbf{r}_3, \cdots, \mathbf{r}_n$ reduces (15) to

$$\frac{e^2}{n(n-1)} \sum_{i,j} \left[\int \frac{\varphi_i^*(\mathbf{r}_1)\varphi_j^*(\mathbf{r}_2)\varphi_i(\mathbf{r}_1)\varphi_j(\mathbf{r}_2)}{r_{12}} d\tau_{12}' - \right.$$
$$\left. \int \frac{\varphi_i^*(\mathbf{r}_1)\varphi_j^*(\mathbf{r}_2)\varphi_i(\mathbf{r}_2)\varphi_j(\mathbf{r}_1)}{r_{12}} d\tau_{12}' \right]$$

because of (9). Since the result is the same for the other $n(n-1)$ terms in the summation (14), the mean value of this quantity is

$$\frac{e^2}{2}\sum_{i,j}'\left[\int\frac{|\varphi_i(\mathbf{r}_1)|^2|\varphi_j(\mathbf{r}_2)|^2}{r_{12}}d\tau'_{12} - \int\frac{\varphi_i{}^*(\mathbf{r}_1)\varphi_j{}^*(\mathbf{r}_2)\varphi_i(\mathbf{r}_2)\varphi_j(\mathbf{r}_1)}{r_{12}}d\tau'_{12}\right]. \quad (16$$

The summation over the spins in the first terms may be performed a
once, giving

$$\frac{e^2}{2}\sum_{i,j}'\int\frac{|\psi_i(\mathbf{r}_1)|^2|\psi_j(\mathbf{r}_2)|^2}{r_{12}}d\tau(x_1, \cdots, z_n). \quad (17$$

The corresponding sum in the second term is zero whenever $\eta_i \neq \eta_j$
however, we shall leave this result in the form given in (16) for th
present.

The expression (17) is exactly the same as the last term in (4), whicl
was derived by taking the mean value of (14) for the simple produc
function (1). Hence, the only effect on the mean value of H of usin
the determinantal form of Ψ is the introduction of the *exchange energy*

$$-\frac{e^2}{2}\sum_{i,j}'\int\frac{\varphi_i{}^*(\mathbf{r}_1)\varphi_j{}^*(\mathbf{r}_2)\varphi_i(\mathbf{r}_2)\varphi_j(\mathbf{r}_1)}{r_{12}}d\tau'_{12}. \quad (18$$

This term results, not from any unusual nonclassical force betwee
electrons, but because we have used the determinantal eigenfunctio
instead of (1). As we said previously, the product function (1) contain
no interelectronic correlations, whereas the determinantal function (7
does contain them. The exchange term (18) is simply the contributio
that these correlations make to the energy. If the exchange energy i
negative, the charge distribution $e|\Psi|^2$, corresponding to (7), has a lowe
self-potential than the charge distribution corresponding to (1), for th
accidental correlations in (7) keep the electrons apart. On the othe
hand, if (18) is positive, the accidental correlations in (7) raise the sel
energy of the charge distribution, because they push electrons togethe
We cannot predict the sign of (18) without knowing the form of the φ
each case must be investigated separately.

In order to illustrate the ways in which exchange terms are related t
correlations, we shall derive the expression for the probability densit
of two electrons in a set of n. Let $\mathbf{r}_1$ and $\mathbf{r}_2$ be the coordinates of th
electrons and $P(\mathbf{r}_1,\mathbf{r}_2)$ the probability density. To find $P(\mathbf{r}_1,\mathbf{r}_2)$ we mus
integrate $|\Psi|^2$ over all variables except $\mathbf{r}_1$ and $\mathbf{r}_2$. The result, which ma
be derived easily by the methods that were used in obtaining (16), is

$$P(\mathbf{r}_1,\mathbf{r}_2) = \frac{1}{n(n-1)}\left[\sum_{i,j}'|\varphi_i(\mathbf{r}_1)|^2|\varphi_j(\mathbf{r}_2)|^2 - \right.$$
$$\left.\sum_{\substack{i,j \\ \|\,\text{spins}}}'\varphi_i{}^*(\mathbf{r}_1)\varphi_j{}^*(\mathbf{r}_2)\varphi_i(\mathbf{r}_2)\varphi_j(\mathbf{r}_1)\right] \quad (19$$

where the second sum extends only over pairs of φ_i with parallel spin. It follows from the manner of deriving $P(\mathbf{r}_1,\mathbf{r}_2)$ that the coulomb and exchange energy is simply $n(n-1)/2$ times the integral

$$e^2 \int \frac{P(\mathbf{r}_1,\mathbf{r}_2)}{r_{12}} d\tau_1 d\tau_2, \tag{20}$$

which is the self-energy of the charge distribution $eP(\mathbf{r}_1,\mathbf{r}_2)$.

Let us evaluate (19) for the special case of perfectly free electrons in a large cubical box.   If we use periodic boundary conditions, the wave functions are

$$\psi_{\mathbf{k}} = \frac{1}{\sqrt{V}} e^{2\pi i \mathbf{k}\cdot\mathbf{r}}. \tag{21}$$

Here, V is the volume of the box and $\mathbf{k}$ is the electronic wave-number vector, the components of which take the discrete values

$$k_x = \frac{n_x}{L}, \qquad k_y = \frac{n_y}{L}, \qquad k_z = \frac{n_z}{L} \tag{22}$$

where n_x, n_y, n_z are arbitrary integers and L is the length of an edge of the box.   We shall assume that all values of $\mathbf{k}$ that lie within a sphere of radius k_0 appear in (19) with both spins.   This system of free electrons evidently is equivalent to the system used in the simple Sommerfeld model of a metal.

Since $|\psi_{\mathbf{k}}|^2 = 1/V$, the first sum in (19) is equal to the constant value $1/V^2$.   The second sum may be written

$$\sum_{\mathbf{k}_1,\mathbf{k}_2}' e^{2\pi i(\mathbf{k}_1 - \mathbf{k}_2)\cdot(\mathbf{r}_1 - \mathbf{r}_2)} \tag{23}$$

where $\mathbf{k}_1$ and $\mathbf{k}_2$ are to be summed over all values in the sphere of radius k_0.   If the number of electrons is sufficiently large and the levels are sufficiently dense, this sum may be replaced[1] by an integral and then reduced to

$$\frac{9}{2}\left[\frac{2\pi k_0 r \cos 2\pi k_0 r - \sin 2\pi k_0 r}{(2\pi k_0 r)^3} \right]^2$$

where $r = |\mathbf{r}_1 - \mathbf{r}_2|$.   Hence,

$$P(\mathbf{r}_1,\mathbf{r}_2) = \frac{1}{V^2}\left\{ 1 - \frac{9}{2}\left[\frac{2\pi k_0 r \cos 2\pi k_0 r - \sin 2\pi k_0 r}{(2\pi k_0 r)^3} \right]^2 \right\}.$$

The coefficient of $1/V^2$ is plotted in Fig. 2 as a function of r.   The

[1] E. P. WIGNER and F. SEITZ, *Phys. Rev.*, **43**, 804 (1933).

probability that the electrons will be at the same point is just half the probability that they will be far apart and varies smoothly in between. The second term in (19) is entirely responsible for the correlation term which, in turn, gives rise to the exchange energy when (20) is computed. In cases in which the electrons are not entirely free, some correlation effect is provided by the first terms in (19).

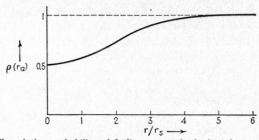

Fig. 2.—The relative probability of finding two perfectly free electrons at a distance r from one another. The parameter r_s is the radius of the sphere whose volume is equal to the mean volume per electron. In this case the correlation effect corresponding to the minimum near $r = 0$ arises from exchange.

The introduction of antisymmetric wave functions detracts considerably from the plausibility of Hartree's equations (2) since they do not take into account the correlations that give rise to the exchange energy. For this reason, Fock[1] and Slater[2] suggested independently that the variational theorem should be used to derive a set of equations for the best one-electron functions. We shall discuss these equations in Sec. 51 after investigating the question of multiplicity in the next section. We shall see that Hartree's equations are satisfactory when Ψ has the form (1) but that additional terms must be added to these when the wave function has the form (7).

50. Eigenfunctions of Definite Multiplicity.—It was pointed out in the discussion of Sec. 41, Chap. V, that the antisymmetric stationary states of any system can be chosen as eigenfunctions of both Σ^2 and Σ_z as well as of H, as long as the Hamiltonian is independent of spin. It was also pointed out that the functions that have the same eigenvalue $\hbar^2 S(S + 1)$ of Σ^2 can be divided into groups of $2S + 1$ which have the same energy. This $(2S + 1)$-fold degenerate level is said to have multiplicity $2S + 1$. The constituent states may be chosen as eigenfunctions of Σ_z, and there is then one state for each of the $2S + 1$ possible eigenvalues of Σ_z which range from S to $-S$ with integer differences. States of different multiplicity usually have different energies except in the special case in which accidental degeneracy occurs. For these

[1] V. Fock, *Z. Physik*, **61**, 126 (1930). See also footnote 1, p. 234.

[2] J. C. Slater, *Phys. Rev.*, **35**, 210 (1930).

reasons, the eigenfunctions of H, that is, the functions that minimize the integral

$$\int \Psi^* H \Psi d\tau'$$

are usually eigenfunctions of Σ^2. The eigenfunctions may be chosen as states of definite multiplicity, however, even when there is accidental degeneracy. Hence, we shall consider this to be a general condition on Ψ.

The antisymmetric function (7) of the preceding section usually is not an eigenfunction of Σ^2 when the ψ_i are different and the η_i are selected at random. The determinant has unit multiplicity (that is, $S = 0$), however, in the special case in which the ψ_i are equal in pairs and the spins of the members of equal pairs are opposite.[1] Hence, the determinant is a satisfactory function from the standpoint of multiplicity in this particular case. Fortunately, this case is an important one for all simple solids, since they have unit multiplicity in the normal state. As a result, the form of Fock's equations, which is discussed in the next section and is based upon a determinantal wave function, is valid for the normal state of simple solids. The exceptional solids are ferromagnetic and strongly paramagnetic substances.

51. Fock's Equations.—The results that are obtained by following Fock and Slater's plan for determining the best one-electron functions from the variational theorem are derived in Appendix I. It is found that these results depend upon the initial choice of the complete wave function. If the function has the form (1), Sec. 49, the equations that the ψ_i must satisfy turn out to be Hartree's equations (2). If, in addition, we specify that the different ψ are to be orthogonal or that

$$\int \psi_i^* \psi_j d\tau = \delta_{ij}, \tag{1}$$

the best equations are

$$-\frac{\hbar^2}{2m}\Delta\psi_i + \left(V_i + \sum_j{}' e^2 \int \frac{|\psi_j|^2}{r_{ij}} d\tau_j\right)\psi_i = \epsilon_i\psi_i + \sum_j{}' \lambda_{ij}\psi_j. \tag{2}$$

These equations have the same form as Hartree's except for the presence of the term

$$\sum_j{}' \lambda_{ij}\psi_j,$$

which arises from the condition (1). The λ_{ij} are the Lagrangian parameters associated with the orthogonality stipulation.

The Pauli principle is not properly included in either of the total wave functions on which Hartree's equations and the equations (2) are based. For reasons discussed in Chap. V, it is necessary to use an anti-

[1] See footnote 2, p. 210.

symmetric function of the electron coordinates. When this is done by employing a determinantal function of the type (7) of Sec. 49 with the condition

$$\int \varphi_i \varphi_j d\tau(x,y,z,\zeta) = \delta_{ij}, \tag{3}$$

the best equations are found to be (*cf.* Appendix I)

$$-\frac{\hbar^2}{2m}\Delta\varphi_i(\mathbf{r}_1) + \left[V(\mathbf{r}_1) + \sum_j{}' e^2 \int \frac{|\varphi_j(\mathbf{r}_2)|^2}{r_{12}}d\tau_2' \right]\varphi_i(\mathbf{r}_1) -$$

$$\sum_j{}' \left[e^2 \int \frac{\varphi_i(\mathbf{r}_2)\varphi_j{}^*(\mathbf{r}_2)}{r_{12}}d\tau_2' \right]\varphi_j(\mathbf{r}_1) = \epsilon_i\varphi_i(\mathbf{r}_1) + \sum_j{}' \lambda_{ij}\varphi_j(\mathbf{r}_1). \tag{4}$$

The parameters λ have the same significance as those in (2), whereas the essentially new terms, namely,

$$-\sum_j{}' \left[e^2 \int \frac{\varphi_i(\mathbf{r}_2)\varphi_j{}^*(\mathbf{r}_2)}{r_{12}}d\tau_2' \right]\varphi_j(\mathbf{r}_1), \tag{5}$$

which we shall call "exchange terms," arise from the use of a determinantal eigenfunction. It should be noted that the integrals in (5) are functions of $\mathbf{r}_1$. The exchange terms may be regarded as "non-conventional" potential integrals that take into account the accidental correlations of the determinantal wave function, just as the exchange integrals of Sec. 49 may be regarded as taking into account the change in self-energy caused by these correlations.

Equations (4), which we shall henceforth call "Fock's equations," have many symmetrical properties not possessed by Hartree's equations. If we add the expression

$$\left[e^2 \int \frac{|\varphi_i(\mathbf{r}_2)|^2}{r_{12}}d\tau_2 \right]\varphi_i(\mathbf{r}_1)$$

to the first summation on the left-hand side of (4) and subtract the same expression from the second summation, we obtain

$$-\frac{\hbar^2}{2m}\Delta\varphi_i(\mathbf{r}_1) + \left[V + \sum_j e^2 \int \frac{|\varphi_j(\mathbf{r}_2)|^2}{r_{12}}d\tau_2' \right]\varphi_i(\mathbf{r}_1) -$$

$$\sum_j \left[e^2 \int \frac{\varphi_i{}^*(\mathbf{r}_2)\varphi_i(\mathbf{r}_2)}{r_{12}}d\tau_2' \right]\varphi_j(\mathbf{r}_1) = \sum_j \lambda_{ij}\varphi_j(\mathbf{r}_1), \tag{6}$$

in which $\lambda_{ii} = \epsilon_i$ and none of the sums is primed. Next, let us set

$$\rho(\mathbf{r}_1,\mathbf{r}_2) = \sum_j \varphi_j{}^*(\mathbf{r}_1)\varphi_j(\mathbf{r}_2), \tag{7}$$

which is known as "Dirac's density matrix." In terms of this function, the first summation in (6) is simply

$$\left[e^2 \int \frac{\rho(\mathbf{r}_2, \mathbf{r}_2)}{r_{12}} d\tau_2' \right] \varphi_i(\mathbf{r}_1), \tag{8}$$

and the second is

$$-e^2 \int \frac{\rho(\mathbf{r}_2, \mathbf{r}_1) \varphi_i(\mathbf{r}_2)}{r_{12}} d\tau_2'. \tag{9}$$

Following Dirac,[1] we shall define an operator A in terms of (9) by the relation

$$A \varphi_i(\mathbf{r}_1) = -e^2 \int \frac{\rho(\mathbf{r}_2, \mathbf{r}_1) \varphi_i(\mathbf{r}_2)}{r_{12}} d\tau'. \tag{10}$$

Thus, the operator A, acting upon $\varphi_i(\mathbf{r}_1)$, multiplies this function by $\rho(\mathbf{r}_2, \mathbf{r}_1)$, changes the variable $\mathbf{r}_1$ to $\mathbf{r}_2$, and integrates over $\mathbf{r}_2$. It may be readily verified that A is both linear and Hermitian. The integral that appears in (8) we shall designate by U; it is simply the coulomb potential of the three-dimensional charge distribution. If we use the foregoing terminology, (6) reduces to

$$H^F \varphi_i(\mathbf{r}_1) = \sum_j \lambda_{ij} \varphi_j(\mathbf{r}_1) \tag{11}$$

where H^F is the *Fock Hamiltonian operator*, namely,

$$H^F = -\frac{\hbar^2}{2m} \Delta + V + U + A, \tag{12}$$

which is the same for all electrons in the system.

The parameters λ_{ij} in (11), aside from λ_{ii}, should be selected in such a way as to ensure the orthogonality of the φ_i. A possible choice of these parameters when the spectrum is non-degenerate is $\lambda_{ij} = 0 (i \neq j)$, since functions that satisfy the equation

$$H^F \varphi_i = \epsilon_i \varphi_i$$

are orthogonal because H^F is Hermitian. This is not the only possible choice of λ_{ij}, however, and others have been used on occasion; but we shall find that this choice is a convenient one in a large part of the following discussion.

The difficulties encountered in solving Hartree's equations also arise with Fock's equations, for it is usually necessary to obtain a solution by some method of successive approximations, such as Hartree's method of the self-consistent field. This procedure is more difficult to apply to Fock's equations because the exchange terms introduce many complications.

[1] P. A. M. Dirac, *Cambridge Phil. Soc.*, **26**, 376 (1930).

In the next section, we shall give a brief summary of the solutions of Hartree's and Fock's equations for free atoms. The actual technique does not interest us so much as the results and their deviation from experimental results, for these give us an estimate of the error that may be expected when the equations are solved for solids.

52. The Solutions of Hartree's and Fock's Equations for Single Atoms.—The simplest nontrivial problem to which the methods of Hartree and Fock-Slater are applicable is that of the normal state of helium, which we have discussed in Sec. 48. Hartree's and Fock's equations are identical in this case since the electrons have antiparallel spins and the exchange terms are zero. The total energy of the atom, as given[1] by this approximation, is found to be 0.076 Rydberg unit higher than the observed value of 5.810 Rydberg units, a fact indicating that electronic correlations are important to the extent of 0.45 ev per electron. Henceforth, we shall call an energy difference such as this 0.45 ev, which measures the error in the energy derived from a one-electron approximation, a "correlation energy." The connotation of this term is evident from the discussion in preceding sections.

The method employed to determine the one-electron function ψ in the case of helium is characteristic of the self-consistent-field computations of Hartree and his school. This procedure is completely described by Condon and Shortley[2] and will not be thoroughly discussed here. It need only be mentioned that the procedure consists, essentially, in assuming a starting function for each electron, determining the potential integrals appearing in Hartree's equation from these, solving the equations for the new wave functions, and comparing these functions with the ones originally assumed. If the two sets agree, the system is said to be self-consistent and the equations solved; if not, the procedure is repeated until the initial and final functions do agree. Naturally, there is no fixed plan that ensures that this procedure will converge rapidly, since a great deal depends upon a good choice of starting functions. Other workers, such as Brown[3] and Torrance,[4] have developed variations of the scheme originally used by Hartree. All these methods involve practically the same steps of approximation and will be regarded as the same here.

We are fundamentally more interested in the solutions of Fock's equations than in those of Hartree's equations, since the former should lead to more accurate results. Solutions of one or both of these equations have been obtained for a number of atoms listed at the end of this

[1] Hartree, *op. cit.*

[2] Condon and Shortley, *op. cit.*

[3] F. W. Brown, *Phys. Rev.*, **44**, 214 (1933).

[4] C. C. Torrance, *Phys. Rev.*, **46**, 388 (1934).

section. Among these, those for beryllium and carbon, determined respectively by Hartree and Hartree[1] and by Torrance, are of principal interest since the absolute binding energies of these atoms are known.

We shall use the results of the atomic computations to study two topics, namely, the accuracy of the energy states of a given atom containing n electrons relative to the lowest energy state of that atom with $n - 1$ electrons, and the error in the absolute energy of an entire atom. The first of the two quantities gives a rough estimate of the relative accuracy of Hartree's and Fock's equations, whereas the second indicates the absolute error and will give a correlation energy. To date, beryllium and carbon are the only cases, other than helium, in which the absolute energy has been completely investigated from the theoretical standpoint.

When an atomic configuration involves only closed shells, it may easily be proved[2] that Fock's equations possess a self-consistent solution for which

$$\psi = R_i(r)\Theta_{l_m}(\theta)\Phi_m(\varphi)$$

where R, Θ, Φ are functions of each of the spherical coordinates, respectively, and the spherical harmonics $\Theta_{l_m}(\theta)$, $\Phi_m(\varphi)$ are chosen to agree with the conventions of the one-electron approximation for atomic spectra. This theorem is not valid for Hartree's equations, for they do not possess the same symmetry as Fock's equations. The effective potential for an electron in a closed shell arising from the same closed shell is not spherically symmetrical in Hartree's equations, for the summations in these equations are primed. However, when dealing with closed shells, Hartree generally takes only the spherically symmetrical part of the potential in order that the equations may be separated in spherical polar coordinates. For this reason, his results are not exact solutions except in special cases, such as the normal states of helium and beryllium, in which the configurations are closed shells of s functions. We shall now proceed to discuss particular cases.

a. Beryllium.[3]—Both Fock's and Hartree's equations have been solved for the normal $1s^2 2s^2$ state of beryllium to a high degree of accuracy. The resulting $1s$ functions are very nearly alike, but the $2s$ functions show considerable difference. The two types of $2s$ function are difficult to compare with one another because the solution of Fock's equation is orthogonal to the $1s$ function, whereas the solution of Hartree's is not. The latter may be made orthogonal to the $1s$, however, and the results are given in Fig. 3.

The total energies of Be and of Be++, as determined by different methods, appear in Table XLVII.

[1] D. R. HARTREE and W. HARTREE, *Proc. Roy. Soc.*, **150A**, 9 (1935).

[2] CONDON and SHORTLEY, *op. cit.*

[3] HARTREE and HARTREE, *op. cit.*

The difference between the total energy of beryllium as computed from Fock's equations and the observed value is about 0.19 Rydberg unit, or about 2.56 ev, whereas the difference between the two values for Be^{++} is 0.07 Rydberg unit, or 0.9 ev.   In order to obtain an estimate of the correlation energy per electron, we shall divide the first of these by 4 and the second by 2, obtaining 0.65 ev and 0.45 ev, respectively.   It

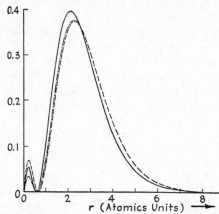

Fig. 3.—A comparison of the square of the 2s functions of beryllium obtained by solving Hartree's and Fock's equations.   The full curve represents the solution of Fock's equations; the broken curves represent the orthogonalized and non-orthogonalized solutions of Hartree's equations.

might seem at first sight that it would be more proper to divide the first by 4! and the second by 2!, since there are 4! and 2! interacting pairs in each case.   This procedure would not be so reasonable, however, since the correlation effect is larger for electrons in the same shell than for electrons in different shells.   According to these results, the mean

TABLE XLVII

	Hartree	Fock	Experimental
E(Be), R.u.	-29.115	-29.140	-29.331 ± 0.008
E(Be^{++})	-27.235	-27.235	-27.307 ± 0.008

correlation energy increases slightly as the number of electrons increases, a fact showing that the one-electron approximation becomes less accurate. The correlation energy of 0.07 Rydberg unit for Be^{++} is almost exactly the same as the value of 0.077 for He.

If we assume that the error in E(Be^{++}) $-$ E(Be) arises purely from the correlations between the 2s electrons, this correlation energy is found to be 0.119 Rydberg unit, or about 0.81 ev per electron. Part of this error actually arises from correlations of the $1s$ and the $2s$ electrons, but this is probably a small fraction.

b. Carbon.—Calculations of the energy levels of carbon have been carried out by Torrance[1] and Ufford.[2] Torrance solved Fock's equations for the $1s^2 2s^2 2p^2$ configuration. The energies of the ^3P, ^1D, and ^1S states were evaluated by Ufford using the functions obtained from the solution. Since this is not a closed-shell configuration, Torrance had to replace the asymmetric fields that arise from p electrons by spherically symmetrical ones. In addition, he used the form of Fock's equations discussed in Sec. 51, which is valid for a single determinantal eigenfunction, although the wave functions of the lowest states do not actually have this form.[3] For these reasons, Ufford's results should be somewhat higher than those that would be derived from more accurate one-electron functions. The computed cohesive energy of the normal atom is 1,019.66 ev, which may be compared with the observed one of 1,024.84 ev. The difference gives a total correlation energy of 5.18 ev, which is about 0.86 ev per electron.

Torrance also computed one-electron functions for the $1s^2 2s 2p^3$ configuration from which Ufford was able to derive energy values for excited states of the carbon atom. In addition, Ufford computed matrix components of the Hamiltonian between the two configurations and determined new energy levels by taking into account the perturbing effect that the configurations exert on one another. This procedure is equivalent to using new wave functions that are linear combinations of the unperturbed wave functions for the normal and excited states. The relative positions of the levels in both approximations are shown in Fig. 4 and are compared with the observed values. The computed cohesive energy of the atom is changed from 1,019.66 ev to 1,020.09 ev by the interaction between configurations,

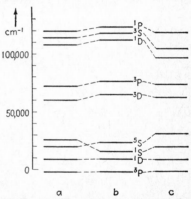

Fig. 4.—The relative positions of the levels of the $2s^2 2p^2$ and $2s 2p^3$ configurations of carbon in different approximations. *a*, strict one-electron approximation; *b*, approximation in which interaction between configurations is taken into account; *c*, observed term values. It should be noted that the positions of the ^5S and ^1S levels are inverted in going from *a* to *b*.

[1] Torrance, *op. cit.*

[2] C. W. Ufford, *Phys. Rev.*, **53**, 568 (1938).

[3] Modified forms of Fock's equations have been developed by G. H. Shortley, *Phys. Rev.*, **50**, 1072 (1936).

so that this perturbation does not account for a very large part of the error in cohesive energy.

c. *Oxygen.*—Hartree and Black[1] have solved Hartree's equation for oxygen in various states of ionization, using the approximation described above, in which only the radial part of the field arising from p functions is employed. For this reason and because of the fact that Hartree's rather than Fock's equations were solved, the results are not so significant as in the case of beryllium and carbon. One point that may give us some confidence in them, however, is the fact that the energies derived from Hartree's and Fock's equations do not differ appreciably in the case of beryllium. The energies of several states of O^{++}, O^+, and O, relative to the ground state of the atom having one less electron, are listed in Table XLVIII.

TABLE XLVIII

		Calculated, Rydberg units	Observed, Rydberg units	Difference
O^{++}	^3P	3.976	4.050	0.074
	^1D	3.778	3.868	0.090
	^1S	3.482	3.659	0.176
O^+	^4S	2.516	2.602	0.086
	^2D	2.258	2.334	0.076
	^2P	2.084	2.210	0.126
O	^3P	0.832	1.00	0.168
	^1D	0.686	0.856	0.170
	^1S	0.468	0.694	0.226

If the equations that were solved had been Fock's and if they had been solved exactly, we might regard the energy difference on the right of Table XLVIII as the correlation energy of the electron that is removed to give the ground state. Actually, these values are only approximately equal to the correlation energies. It should be noted that the correlation energy is greater for the excited states, particularly for those having low values of multiplicity and angular momentum. The dependence on multiplicity is probably related to the fact that states with low multiplicity are least affected by the Pauli principle so that the effects of "accidental correlations" are not so prominent as in the other cases; hence, other correlations that are not provided by the Pauli principle are more important.

d. *Other Cases.*—A large number of other atoms have been investigated by Hartree and additional workers. Although the Hartree potentials arising from inner-shell electrons prove to be very valuable for the process of determining wave functions of valence electrons in solids, we shall not discuss these cases here, since they would carry us too far afield. We shall refer to some of the results in later sections, however, and so we shall list those atoms for which Hartree fields have been obtained.

Ag^+	M. M. Black, *Mem. and Proc. Manchester Lit. Phil. Soc.* (1934–1935).
Al^{+3}	D. R. Hartree, *Proc. Roy. Soc.*, **151**, 96 (1935).
A	D. R. Hartree and W. Hartree, *Proc. Roy. Soc.*, **166**, 450 (1938).
B	F. W. Brown, J. H. Bartlett, and C. G. Dunn, *Phys. Rev.*, **44**, 296 (1933).
Be, Be^+	D. R. Hartree and W. Hartree, *Proc. Roy. Soc.*, **150**, 9 (1935); **154**, 588 (1936).
Ca	D. R. Hartree and W. Hartree, *Proc. Roy. Soc.*, **149**, 210 (1935); **164**, 167 (1938).
Cs^+	D. R. Hartree, *Proc. Roy. Soc.*, **143**, 506 (1934).
Cl^-	D. R. Hartree, *Proc. Roy. Soc.*, **141**, 281 (1933); **156**, 45 (1936).
Cu^+	D. R. Hartree, *Proc. Roy. Soc.*, **141**, 281 (1933); **157**, 490 (1936).
F, F^-	D. R. Hartree, *Proc. Roy. Soc.*, **151**, 96 (1935); F. W. Brown, *Phys. Rev.*, **44**, 214 (1933).
He	D. R. Hartree, *Cambridge Phil. Soc.*, **24**, 89 and 111 (1928).
Hg	D. R. Hartree and W. Hartree, *Proc. Roy. Soc.*, **149**, 210 (1935).
K	D. R. Hartree, *Proc. Roy. Soc.*, **143**, 506 (1934), **166**, 450 (1938); *Proc. Cambridge Phil. Soc.*, **34**, 550 (1938).
Li	J. Hargreaves, *Proc. Cambridge Phil. Soc.*, **25**, 75 (1928).
Na	V. Fock and Mary Petrashen, *Physik. Z.*, **6**, 368 (1934); D. R. Hartree and W. Hartree, *Proc. Cambridge Phil. Soc.*, **34**, 550 (1938).
Ne	F. W. Brown, *Phys. Rev.*, **44**, 214 (1933).
O, O^+, O^{++}, O^{+3}	D. R. Hartree and M. M. Black, *Proc. Roy. Soc.*, **139**, 311 (1933).
Rb^+	D. R. Hartree, *Proc. Roy. Soc.*, **151**, 96 (1935).
Si^{+4}	J. McDougall, *Proc. Roy. Soc.*, **138**, 550 (1932).
W	M. F. Manning and J. Millman, *Phys. Rev.*, **49**, 848 (1936).

53. Types of Solution of Fock's Equations for Multiatomic Systems.—Two independent types[1] of solution of Fock's equations have been widely used in multiatomic systems, namely: the Heitler-London, or atomic, type; and the Hund-Mulliken-Bloch, or molecular, type. In the Heitler-London scheme, it is assumed that the ψ_i are large only about single atoms or ions. Thus, in H_2, the two one-electron functions have the form shown roughly in Fig. 5a. This type of solution is accurate when the atoms of the multiatomic system are far from one another and the atomic or ionic properties of the constituent atoms are pronounced. In the Hund-Mulliken-Bloch scheme, on the other hand, it is

---

[1] The workers after which these schemes are named actually did not use them in connection with Fock's equations but simply used one-electron functions of the corresponding form. In the case of solids, we shall call the second scheme simply the "Bloch scheme" or the "band scheme."

assumed that each ψ extends over the entire system of atoms and has equal amplitude at equivalent atoms. In H_2, for example, both electrons have the type of wave function that is illustrated in Fig. 5b. The use of this type of function is equivalent to assuming that the atoms of the system are affected by combination to such an extent that the

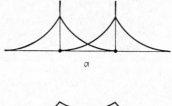

a

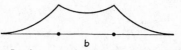

b

Fig. 5.—In the Heitler-London approximation the electrons of H_2 have separate wave functions of the type shown in *a*. In the Hund-Mulliken approximation both have the function *b*, which is distributed symmetrically between both atoms.

valence electrons belong to the entire molecule rather than to a single atom.

It has not been shown whether or not both types of solution always exist and whether or not they are the only solutions of Fock's equations. We shall show, however, that both types can exist.

Let us consider Fock's equations in the form

$$-\frac{\hbar^2}{2m}\Delta\psi_i(\mathbf{r}_1) + \left[V_i(\mathbf{r}_1) + \sum_j{}' e^2 \int \frac{|\psi_j|^2}{r_{12}} d\tau_2 \right]\psi_i(\mathbf{r}_1) -$$
$$\sum_j{}' \left(e^2 \int \frac{\psi_j{}^*\psi_i d\tau_2}{r_{12}} \right)\psi_j(\mathbf{r}_1) = \epsilon\psi_j(\mathbf{r}_1). \quad (1)$$

We shall discuss the case of atomic functions first. The coulomb terms

$$\sum_j{}' e^2 \int \frac{|\psi_j|^2}{r_{12}} d\tau_2$$

obtained from Heitler-London functions screen part of the contribution to the ionic potential V from all atoms except the ith and make the attractive coulomb field largest at the ith atom. Thus, if we neglect the exchange term, we may expect the solutions of Eqs. (1) which are obtained when the potentials have been derived from atomic functions to be localized about individual atoms, that is, to have the form of atomic functions. Hence, we may expect to find a self-consistent solution of atomic functions in Hartree's case. This conclusion remains valid when the exchange terms are included if the atoms are far apart, for then these terms are small. Since it has not yet proved possible to discuss the influence of exchange terms in any general way when the atoms are close together, it has not been demonstrated that atomic types of solution always occur in this case.

It is easy to see that the molecular types of solution always exist, for if we write Fock's equations in the form of Eq. (11), Sec. 61, it follows that no atom is preferred as long as the ψ_i prefer no atom. Hence, if a starting set of functions of the molecular type is used in applying a self-consistent scheme, all subsequent solutions, including the self-consistent one, will be of this type.

We shall show in later chapters (*cf.* Chaps. VII and VIII) that the two types of solution lead to identical antisymmetric total wave functions in an important case, namely, that of the normal state of a molecule the constituents of which have rare gas structure.

The molecular types of wave function, particularly those of lowest energy which have few nodes, are smoother than the atomic wave functions. Hence, we may expect that the mean kinetic energy usually will be lower when molecular functions are used. This advantage of the Hund-Mulliken-Bloch scheme is balanced by the fact that the scheme relies upon the accidental correlations introduced by the Pauli principle to reduce the energy of electron repulsion. The Heitler-London scheme, on the other hand, reduces this energy by keeping the electrons on separate atoms. The results for the problem of molecular hydrogen, which we shall discuss in the next chapter, show that in this case the advantages and disadvantages of the two schemes are about equal. Incidentally, the cohesive energy obtained by both schemes is in error by about 0.5 ev per electron, which shows that the solutions of Fock's equations are far from exact.

Both these one-electron schemes have been used extensively in solids, since each has its advantage for different types of problem. For example, the Bloch scheme is preferable in discussing metallic conductivity, whereas the Heitler-London scheme is preferable in discussing cohesion in ionic crystals. We shall develop both approximations in the following chapters, letting physical reasonableness be our guide in their application.

CHAPTER VII

MOLECULAR BINDING

54. Introduction.—There is an intimate connection between the binding properties of molecules and of solids as far as quantum mechanical principles are concerned. For this reason, we shall discuss some of the features of molecular binding. Since there are many important topics in the theory of molecules upon which we shall not touch, the following discussion should not be regarded as complete.

A fairly complete investigation along exact lines has been carried out for many of the simpler molecules such as H_2^+, H_2, and Li_2. The Hartree-Fock scheme, discussed in the last chapter, plays a very minor role in this work, for the variational scheme has been employed directly. However, we shall be able to interpret some of the results in terms of the Hartree-Fock scheme. In cases such as H_2^+ and H_2, in which the final results are almost as accurate as those obtained by Hylleraas for atomic helium, it is possible to gain an abundance of valuable information.

In addition to these quantitative investigations, there have been a number of qualitative discussions of more complicated molecules based on the Heitler-London and the Hund-Mulliken schemes. This work has proved to be extremely useful in the hands of the physical chemist who is willing to introduce an ample amount of empirical knowledge into any scheme he uses.

The Hamiltonian operator used in discussing the electronic structure of simple molecules is generally the same as the operator (1) of Sec. 47, Chap. VI, in which nuclear coordinates appear parametrically in V_i and I and in which spin interactions are neglected. For this reason, we may take over all the general remarks of the last chapter.

55. The Hydrogen-molecule Ion.—The simple molecule H_2^+ has the Hamiltonian operator

$$-\frac{\hbar^2}{2m}\Delta - \left(\frac{e^2}{r_a} + \frac{e^2}{r_b} - \frac{e^2}{r_{ab}}\right) \tag{1}$$

where r_a and r_b are the distances of the electron from the two protons, which are separated by a distance r_{ab}. The corresponding Schrödinger equation

$$-\frac{\hbar^2}{2m}\Delta\psi - \left(\frac{e^2}{r_a} + \frac{e^2}{r_b} - \frac{e^2}{r_{ab}}\right)\psi = E\psi \tag{2}$$

254

is separable when expressed in terms of the elliptical variables

$$\xi = \frac{r_a + r_b}{r_{ab}}, \qquad \eta = \frac{r_a - r_b}{r_{ab}}, \qquad \phi, \tag{3}$$

where ϕ is the angle that the plane containing the electron and the two nuclei makes with a fixed plane passing through the nuclei. In fact, if we set

$$\psi = \Xi(\xi) H(\eta) \Phi(\phi), \tag{4}$$

the separated equations are

$$\frac{d}{d\xi}\left((\xi^2 - 1)\frac{d\Xi}{d\xi}\right) + \left(-\lambda\xi^2 + 2D\xi - \frac{\mu^2}{\xi^2 - 1} + \tau\right)\Xi = 0, \tag{5a}$$

$$\frac{d}{d\eta}\left((1 - \eta^2)\frac{dH}{d\eta}\right) + \left(\lambda\eta^2 - \frac{\mu^2}{1 - \eta^2} - \tau\right)H = 0, \tag{5b}$$

$$\frac{d^2\Phi}{d\phi^2} = -\mu^2\Phi, \tag{5c}$$

where

$$\lambda = \frac{-mr_{ab}^2\left(E - \frac{e^2}{r_{ab}}\right)}{2\hbar^2}, \qquad D = \frac{r_{ab}me^2}{\hbar^2}, \tag{5d}$$

and μ and τ are separational parameters.

From (5c), it is clear that $\Phi = e^{i\mu\phi}$, whence μ may take on only integer values. In addition, we know that the two states for which $\mu = \pm\mu'$ will have identical energies, for only μ^2 appears in the other two equations. In other words, all levels except those for which μ is zero are two-fold degenerate. Since the angular momentum about the axis joining the nuclei is simply $\mu\hbar$, it is conventional to designate the states for which $\mu = 0, 1, 2 \cdots$ by σ, π, δ, respectively, in analogy with the atomic designation involving the angular-momentum quantum number.

On general grounds, we should expect the lowest state to be one for which $\mu = 0$ since this type of function does not have an angular nodal plane. This state was investigated first by Burrau,[1] but a more accurate treatment has been given by Teller[2] who, in addition, carried through an investigation of higher states. We shall not discuss Teller's work in detail, except to say that he solved (5b) by means of power series and found the eigenfunctions of (5a) by use of the variational equation going with this self-adjoint differential equation. If the number of nodes in Ξ and H is designated by n_ξ and n_η, respectively, it should be possible to label all the states by these two quantum numbers and μ. The

[1] O. BURRAU, *Danske Videnskab. Selskab*, **7**, 14 (1927).

[2] E. TELLER, *Z. Physik*, **61**, 458 (1930).

convention usually adopted, however, is to employ the quantum numbers n and l of the hydrogen-like state into which a given molecular eigenfunction degenerates when r_{ab} approaches zero. A simple investigation reveals that n_ξ goes into the radial quantum number and that n_η goes into the polar quantum number $l - \mu$, as may be seen from the fact that ξ becomes a radial variable and η becomes the variable $\cos \theta$, where θ is the polar angle in spherical polar coordinates. In other words

$$\left. \begin{aligned} n &= n_\xi + l + 1; \\ l &= n_\eta + \mu. \end{aligned} \right\} \tag{6}$$

The energies of several states are shown as functions of r_{ab} in Figs. 1 and 2. In Fig. 1, the abscissa is the internuclear distance expressed in Bohr units and the ordinate is the purely electronic part of the energy

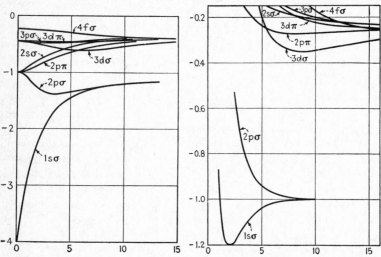

FIG. 1.—The electronic energy of several states of $H_2{}^+$ as a function of interatomic distance. (*After Teller.*) The ordinate is in Rydberg units.

FIG. 2.—The total energy of several states of $H_2{}^+$ as a function of interatomic distance. These curves are derived from those of Fig. 3 by adding the energy of nuclear repulsion. (*After Teller.*)

(that is, the nuclear repulsion is neglected). In Fig. 2, the abscissae are the same as in Fig. 1, whereas the ordinates represent the total energy. The stable states are those of Fig. 2 which possess a minimum for finite values of r_{ab}. It may be seen that the only members of the computed set that possess this property are the $1s\sigma$, $3d\sigma$ and $2p\pi$ states. The energies of the minima relative to a zero in which all particles are separated from each other by an infinite distance are given in Table XLIX.

TABLE XLIX

State	Nuclear separation, a_h	Energy, Rydberg units	Energy of dissociation, ev
$1s\sigma$	2.00	-1.20537	2.781
$3d\sigma$	11.5	-0.350	1.35
$2p\pi$	8	-0.265	0.20

The last column in this table is the energy required to dissociate the molecule into an atom and a proton.

When the parameter r_{ab} becomes very large, the amplitudes of the wave functions become negligibly small at distances midway between the nuclei. The wave function then reduces, for all practical purposes, to two hydrogen wave functions, each centered about one of the protons. Only one of these two states should be regarded as the final state if the separation is sufficiently large, for there is only a negligible probability of the electron jumping from one atom to another. The quantum numbers of this atomic state, which may be determined by a simple analysis involving parabolic coordinates, are connected with those of the molecule in the following way: The total quantum number n' of the final state is related to n_ξ, n_η, and μ by the equations

$$
n' = \begin{cases} n_\xi + \dfrac{n_\eta}{2} + \mu + 1 & (n_\eta \text{ even}) \\[2mm] n_\xi + \dfrac{n_\eta + 1}{2} + \mu + 1 & (n_\eta \text{ odd}). \end{cases} \tag{7}
$$

No definite l value may be assigned to the final state, for the hydrogen-atom wave functions obtained by removing a proton from H_2^+ are not eigenfunctions of angular momentum. The values of l entering into these wave functions, when they are expressed as a linear combination of eigenfunctions of angular momentum, range over the allowed values that may be associated with n', that is, from $n'-1$ to 0. The value of m for the final state is the same as the value of μ before separation. It may be seen from (6) that the value of n obtained by coalescing the two nuclei is either greater than or equal to the value (7) obtained by separation.

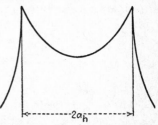

FIG. 3.—The charge distribution associated with the normal wave function of H_2^+. (*After Burrau.*)

The electronic charge distribution of the lowest state is shown in Fig. 3, for equilibrium separation of protons. It should be noted that the amplitude is large and nearly constant at distances midway between the protons. The disadvantage of having two repelling protons is more than compensated by the following two facts: (1) There is a larger region in which the electron may have negative potential energy when there are two protons instead of one. (2) The wave function may be smooth over a larger region of space. The first fact has associated with it a drop in potential energy; the second, a drop in kinetic energy. The $2p\sigma$ state, which combines with the $1s\sigma$ at large internuclear distances, has a nodal plane midway between protons. Thus, neither is it smooth, nor does it allow the electron to have an appreciable probability of being at midway regions where the potential field favors binding. Figure 2 shows that it is entirely repulsive, as we might expect.

56. The Hydrogen Molecule.—The hydrogen molecule has been treated approximately by a large number of workers. We shall consider first the solution obtained by James and Coolidge,[1] since it is considerably more accurate than any other. Their procedure is patterned after the one Hylleraas followed when working with helium (*cf.* Sec. 48, Chap. VI). We shall see that there is a close correspondence between the conclusions that may be drawn from the solutions of both problems.

a. James and Coolidge expressed the Hamiltonian operator

$$-\frac{\hbar^2}{2m}(\Delta_1 + \Delta_2) - \frac{e^2}{r_{1a}} - \frac{e^2}{r_{1b}} - \frac{e^2}{r_{2a}} - \frac{e^2}{r_{2b}} + \frac{e^2}{r_{12}} + \frac{e^2}{r_{ab}} \qquad (1)$$

in terms of the four elliptical variables

$$\xi_1 = \frac{r_{1a} + r_{1b}}{r_{ab}}, \qquad \xi_2 = \frac{r_{2a} + r_{2b}}{r_{ab}},$$

$$\eta_1 = \frac{r_{1a} - r_{1b}}{r_{ab}}, \qquad \eta_2 = \frac{r_{2a} - r_{2b}}{r_{ab}}, \qquad (2a)$$

which are analogous to the set (3) of the previous section. Instead of using ϕ_1 and ϕ_2 as the remaining pair of electron variables, they chose the set

$$\theta = \phi_1 + \phi_2 \qquad \text{and} \qquad \rho = \frac{2r_{12}}{r_{ab}}. \qquad (2b)$$

The interelectronic distance was used explicitly in order that electronic correlation, discussed in the last chapter, might play a role appropriate to its importance. The variable θ does not enter into the wave function of the lowest state for the same reason that ϕ does not enter into the wave function of the lowest state of H_2^+.

The starting wave function for the lowest state was taken in the power-series form

$$e^{-\delta(\xi_1 + \xi_2)} \sum C_{mnjkp} (\xi_1^m \xi_2^n \eta_1^j \eta_2^k \rho^p + \xi_1^n \xi_2^m \eta_1^k \eta_2^j \rho^p) \qquad (3)$$

[1] H. M. James and A. S. Coolidge, *Jour. Chem. Phys.*, **1**, 825, (1933).

where δ and the C are parameters that must be determined.   Naturally, only the first few terms of the series were used.   The various steps that were taken in extending the computation may be summarized in the following way.

1. Only the exponential term was retained in the first approximation. The best value of δ depends upon r_{ab} and is 1.696 for the observed separation of 1.40 Bohr units.   In this case, the binding energy of the molecule is 2.56 ev, which should be compared with the experimental value of 4.73 ± 0.04 ev.

The wave function

$$e^{-\delta(\xi_1 + \xi_2)}$$

may be written in the form

$$e^{-\frac{\delta(r_{1a}+r_{1b})}{r_{ab}}} e^{-\frac{\delta(r_{2a}+r_{2b})}{r_{ab}}}, \tag{4}$$

which corresponds to the product of two one-electron wave functions of the Hund-Mulliken type.   Hence, the results for this case give us a lower limit to the accuracy that should be expected if the Fock-Hartree procedure based on the Hund-Mulliken scheme were employed.   From this result, it is difficult to say to what extent a rigorous solution of Fock's equations would improve the calculated energy.   A more accurate solution of the Hund-Mulliken type will be discussed in part *b*.

2. Neglecting all terms in (3) that depend upon ρ, James and Coolidge found that the best energy they could obtain for $r_{ab} = 1.40a_h$ is about 4.27 ev, which differs from the observed value by approximately $\frac{1}{2}$ ev. Since this approximation is the best possible one in which correlations are not included, it gives an upper limit to the possible accuracy of the Fock-Hartree procedure.   It is doubtful whether the energy of the Fock-Hartree approximation would be nearly as good as this, however, since (3) is much more intricate than a product of one-electron functions even when terms in ρ are neglected.   Thus, the correlation energy correction would be at least $\frac{1}{4}$ ev per electron, if the Hund-Mulliken scheme were used.

3. The simplest function employed in which the variable ρ occurs is the five-parameter expression

$$e^{-\delta(\xi_1+\xi_2)}[a_1(\eta_1^2 + \eta_2^2) + a_2\eta_1\eta_2 + a_3(\xi_1 + \xi_2) + a_4\rho]. \tag{5}$$

This leads to an energy of 4.507 ev with $r_{ab} = 1.40a_h$.   The parameter values are listed in Table L.   It should be noted that δ has a value considerably different from that discussed in case 1.   Thus, just as in the case of helium, the inclusion of a linear term in ρ leads to a better energy.

4. As a final step, a thirteen-parameter expression involving quadratic terms in ξ_1, ξ_2, and ρ, as well as those appearing in (5), was employed.

The parameters C_{mnjkp} are tabulated in Table L, and it may be seen that in some cases they differ greatly from the values discussed in case 3. The final energy is about 4.69 ev, which lies within the experimental error of the observed value. The computed internuclear distance is identical with the observed one.

TABLE L

	Case 1	Case 3	Case 4
δ	1.69609	2.23779	2.2350
C_{00020}	....	0.80483	1.19279
C_{00110}	....	−0.27997	−0.45805
C_{10000}	....	−0.60985	−0.82767
C_{10200}	....	....	−0.17134
C_{10020}	....	....	−0.12101
C_{10110}	....	....	0.12394
C_{20000}	....	....	0.08323
C_{00001}	....	0.19917	0.35076
C_{00021}	....	....	0.07090
C_{00111}	....	....	−0.01143
C_{10001}	....	....	−0.03987
C_{00002}	....	....	−0.01197

b. We shall now compare the accuracy of the Heitler-London and Hund-Mulliken schemes in so far as computations on H_2 allow us to make a comparison. It is easy to show that Fock's equations for the two schemes are essentially different in this case, as they are in most multiatomic systems (an exceptional case will be discussed in Sec. 58).

For a two-electron system, the singlet eigenfunction based upon one-electron functions is

$$\frac{1}{\sqrt{2}}[\psi_1(\mathbf{r}_1)\psi_2(\mathbf{r}_2) + \psi_1(\mathbf{r}_2)\psi_2(\mathbf{r}_1)][\eta_1(1)\eta_2(-1) - \eta_2(1)\eta_1(-1)]. \qquad (6)$$

Here, ψ_1 and ψ_2 are the one-electron functions that, in the Heitler-London scheme, are centered about different nuclei. Fock's equation for ψ_1 is

$$-\frac{\hbar^2}{2m}\Delta_1\psi_1(\mathbf{r}_1) + \left(-\frac{e^2}{r_{1a}} - \frac{e^2}{r_{1b}} + e^2\int\frac{|\psi_2(\mathbf{r}_2)|^2}{r_{12}}d\tau_2\right)\psi_1(\mathbf{r}_1) +$$
$$\left(e^2\int\frac{\psi_1(\mathbf{r}_2)\psi_2(\mathbf{r}_2)}{r_{12}}d\tau_2\right)\psi_2(\mathbf{r}_1) = \epsilon\psi_1(\mathbf{r}_1) + \lambda\psi_2(\mathbf{r}_1), \qquad (7)$$

and there is a similar equation for ψ_2.

The space part of the total wave function (6) reduces to $\psi(\mathbf{r}_1)\psi(\mathbf{r}_2)$ when ψ_1 and ψ_2 are equal, as in the Hund-Mulliken scheme. In this case, Fock's equation for $\psi(\mathbf{r})$ is

$$-\frac{\hbar^2}{2m}\Delta\psi + \left(-\frac{e^2}{r_{1a}} - \frac{e^2}{r_{2a}} + e^2\int\frac{|\psi(\mathbf{r}_2)|^2}{r_{12}}d\tau_2\right)\psi = \epsilon\psi. \tag{8}$$

Equation (7) reduces not to (8) when we set $\psi_1 = \psi_2$, but instead to

$$-\frac{\hbar^2}{2m}\Delta\psi + \left(-\frac{e^2}{r_{1a}} - \frac{e^2}{r_{2a}} + 2e^2\int\frac{|\psi(\mathbf{r}_2)|^2}{r_{12}}d\tau_2\right)\psi = \epsilon\psi, \tag{9}$$

in which the coulomb integral contains a factor 2. Thus the different assumptions of the two cases lead to different systems of equations. The two schemes should be regarded as different approximate solutions of the same problem rather than as different solutions of the same set of one-electron equations.

The most accurate attempt to apply the Heitler-London scheme to the hydrogen molecule was made by Wang,[1] who used hydrogen-like atomic wave functions of the form $e^{-\alpha r}$, where r is the distance of an electron from the nucleus and α is an adjustable screening parameter. The best value of the cohesive energy obtained by this procedure is 3.76 ev corresponding to $r_{ab} = 1.41a_h$. The effective proton charge Ze is related to α by the equation

$$Ze = e\alpha a_h.$$

For the best value of α, one finds that $Z = 1.17$. At first sight, it may seem surprising that this is greater than unity. It should be recalled, however, that Z must increase from unity to the value of $\frac{27}{16}$ for atomic helium (*cf.* Sec. 48) as r_{ab} decreases from infinity to zero.

On the other hand, the most accurate attempt to apply the Hund-Mulliken scheme to the hydrogen molecule was made by Hylleraas.[2] He constructed a determinantal eigenfunction from one-electron solutions of a two-center system, similar to H_2^+. He assumed that the charge on the centers was $e/2$, rather than e as in H_2^+, in order to compensate for the nuclear screening effect that one electron exerts on the other. The cohesive energy obtained by taking the mean value of the Hamiltonian of H_2 with this determinantal eigenfunction is 3.6 ev for an internuclear distance of $1.40a_h$. This fact shows that Fock's equation would lead to the correct energy to within at least 0.55 ev per electron if the Hund-Mulliken scheme were used. In view of the remarks made under 2, part *a*, we can say that the correlation energy of H_2 for the Hund-Mulliken scheme lies between 0.55 and 0.25 ev per electron.

We may conclude[3] from these two cases that the Heitler-London and Hund-Mulliken schemes are about equally successful as far as the

---

[1] S. C. WANG, *Phys. Rev.*, **31**, 579 (1928).

[2] E. HYLLERAAS, *Z. Physik*, **71**, 741 (1931).

[3] Other methods of treating H_2 are surveyed by J. H. Van Vleck and A. Sherman, *Rev. Modern Phys.*, **7**, 167 (1935).

problem of minimizing the energy is concerned.   The error in each case is of the order of 0.5 ev per electron.

57. Molecular Lithium.—Next to hydrogen, diatomic lithium is the simplest stable molecule that contains only one type of atom.   We shall outline briefly the computations that deal with it.

In early work, several attempts were made to compute the binding energy of Li_2 by taking into account only the two $2s$ valence electrons and treating $1s$ shells as though rigidly fixed.   The interactions between the closed shells on different atoms were neglected.   It was first pointed out by James[1] that the relatively simple methods used in treating the two valence electrons give a binding energy that agrees with the experimental value of 1.14 ev only because the closed-shell interactions are neglected.   He carried out several types of calculation that show the following facts.

a. If the closed-shell interactions are neglected, a binding energy greater than the observed value may be obtained by using essentially the same procedure that James and Coolidge employed for H_2.   There are two reasons for this fact: (1) The ion cores repel one another more strongly when the shells are taken into account than when they are not. The origin of this additional repulsion will be discussed in the next section.   (2) The wave function obtained by a variational procedure usually violates the Pauli principle unless the closed-shell wave functions are explicitly included in the varied wave function.

b. A binding energy of 0.62 ev may be obtained by an involved variational computation in which closed-shell wave functions are included in the varied function.   The interelectronic distance variables were not introduced into this wave function since they would have made the computations prohibitively complicated.

James's work on Li_2 is interesting from the standpoint of computations dealing with monatomic solids, for many features of the two cases are identical.   Since closed inner shells are present in all the interesting solids, it is important to know the extent to which they can be neglected. The preceding discussion shows that the problem of closed shells must be approached with care if relative binding energies are to have much significance.   It will become apparent later that there are several redeeming features in the case of solids.   Most important among these is the fact that equilibrium distances in solids are usually much larger than in molecules.   For example, the closest distance of approach of lithium atoms in the metal is $5.65a_h$, whereas it is $5.05a_h$ in the molecule.

58. Closed-shell Interaction and van der Waals Forces.[2]—There is one case in which the Heitler-London and Hund-Mulliken schemes are

---

[1] H. M. James, *Jour. Chem. Phys.*, **2**, 794 (1934).

[2] See review article by H. Margenau, *Rev. Modern Phys.*, **11**, 1 (1939), for a summary of the development of the theory of van der Waals forces.

identical, namely, the case of electron configurations that correspond to the interaction of closed shells.   In this case, the singlet eigenfunction is a simple determinant, for all electronic wave functions appear in pairs with opposite spin. The determinant formed from Heitler-London wave functions may be rearranged so as to satisfy the Hund-Mulliken conditions. We shall demonstrate this theorem by considering the interaction of two normal helium atoms.   It will be evident that the principles involved in this particular case are generally applicable.

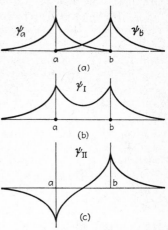

We shall designate the wave functions centered about one of the atoms by ψ_a and those centered about the other by ψ_b. We may assume that ψ_a and ψ_b are symmetrical in the sense that they become interchanged if the two nuclei are interchanged (*cf.* Fig. 4).   The wave function for the molecule is then

Fig. 4.—*a*, the two Heitler-London functions of H_2. Their sum *b* and their difference *c* are Hund-Mulliken functions.

$$\begin{vmatrix} \psi_a(\mathbf{r}_1)\eta_1(1) & \psi_a(\mathbf{r}_2)\eta_2(1) & \psi_a(\mathbf{r}_3)\eta_3(1) & \psi_a(\mathbf{r}_4)\eta_4(1) \\ \psi_a(\mathbf{r}_1)\eta_1(-1) & \psi_a(\mathbf{r}_2)\eta_2(-1) & \psi_a(\mathbf{r}_3)\eta_3(-1) & \psi_a(\mathbf{r}_4)\eta_4(-1) \\ \psi_b(\mathbf{r}_1)\eta_1(1) & \psi_b(\mathbf{r}_2)\eta_2(1) & \psi_b(\mathbf{r}_3)\eta_3(1) & \psi_b(\mathbf{r}_4)\eta_4(1) \\ \psi_b(\mathbf{r}_1)\eta_1(-1) & \psi_b(\mathbf{r}_2)\eta_2(-1) & \psi_b(\mathbf{r}_3)\eta_4(-1) & \psi_b(\mathbf{r}_4)\eta_4(-1) \end{vmatrix}. \tag{1}$$

It should be noted that corresponding elements in the first and third rows have identical spin functions, as do those in the second and fourth rows.   Let us now add the third row to the first, subtract the first row from the third, and repeat this procedure with the second and fourth rows.   We then obtain the determinant

$$\begin{vmatrix} \psi_I(\mathbf{r}_1)\eta_1(1) & \psi_I(\mathbf{r}_2)\eta_2(1) & \psi_I(\mathbf{r}_3)\eta_3(1) & \psi_I(\mathbf{r}_4)\eta_4(1) \\ \psi_I(\mathbf{r}_1)\eta_1(-1) & \psi_I(\mathbf{r}_2)\eta_2(-1) & \psi_I(\mathbf{r}_3)\eta_3(-1) & \psi_I(\mathbf{r}_4)\eta_4(-1) \\ \psi_{II}(\mathbf{r}_1)\eta_1(1) & \psi_{II}(\mathbf{r}_2)\eta_2(1) & \psi_{II}(\mathbf{r}_3)\eta_3(1) & \psi_{II}(\mathbf{r}_4)\eta_4(1) \\ \psi_{II}(\mathbf{r}_1)\eta_1(-1) & \psi_{II}(\mathbf{r}_2)\eta_2(-1) & \psi_{II}(\mathbf{r}_3)\eta_3(-1) & \psi_{II}(\mathbf{r}_4)\eta_4(-1) \end{vmatrix} \tag{2}$$

where

$$\psi_I = \psi_a + \psi_b, \qquad \psi_{II} = \psi_a - \psi_b. \tag{3}$$

Since (2) has been obtained by adding other rows to the rows of (1), the two determinants are equivalent. ψ_I and ψ_{II}, however, are Hund-Mulliken wave functions, (*cf.* Fig. 4). Hence, in this case, the two schemes are equivalent. In more general cases, the determinant corresponding to (1) has more rows and columns, but alternate rows have equal spin functions and may again be combined to form Hund-Mulliken wave functions. We shall discuss the crystalline case in Chap. VIII.

The Heitler-London scheme has been used to compute the energy of interaction between closed shells. For convenience, the one-electron functions employed in these computations were based upon approximate atomic functions rather than upon solutions of Fock's equations. We shall discuss several examples in detail.

a. Helium Interaction.—The most accurate treatment of the repulsive interaction of two helium atoms has been given by Slater[1] who computed the mean value of the interaction potential H_I of the two molecules, using a total wave function that was constructed by taking an appropriate linear combination of atomic eigenfunctions. His work is not a strict application of the Heitler-London method, since the atomic wave functions that he used were not constructed of one-electron functions alone but contained the interelectronic distance variable. In analytical form, his atomic wave function $\Phi(1,2)$ is

$$\Phi(1,2) = \begin{cases} 1.392e^{-2(r_1+r_2)+0.5r_{12}+0.0107(r_1^2+r_2^2)}; & (r_1, r_2 < 3), \\ 1.241e^{-2r_2-1.344r_1}r_1^{-0.255}\left(1 + \dfrac{0.0707}{r_1}\right); & (r_1 > 3; r_2 < 3), \quad (4) \\ 1.241e^{-2r_1-1.344r_2}r_2^{-0.255}\left(1 + \dfrac{0.0707}{r_2}\right); & (r_2 > 3; r_1 < 3). \end{cases}$$

Here r_1, r_2, and r_{12}, are the radial distances between the electrons and the nucleus and between the electrons, expressed in units of the Bohr radius. If we designate the two nuclei by a and b and let a subscript on Φ indicate the nucleus about which the electrons are centered, the complete antisymmetric wave function for He_2 that is obtainable from (4) is

$$\Psi(1,2,3,4) = \Phi_a(1,2)\Phi_b(3,4) - \Phi_a(1,4)\Phi_b(2,3) - \Phi_a(2,3)\Phi_b(1,4) + \Phi_a(3,4)\Phi_b(1,2), \quad (5)$$

in which it is implied that each Φ has an appropriate spin factor. The function (5) reduces to (3) when $\Phi(1,2)$ is simplified to a product of one-electron functions.

The repulsive interaction $V(R)$ that is derived from (5) is a fairly complicated function of the internuclear distance R, but it may be accurately approximated by means of the analytical expression

$$V(R) = 481e^{-\frac{R}{0.412}}, \quad (6)$$

in units of ev, for R greater than $2a_h$.

[1] J. C. SLATER, *Phys. Rev.*, **32**, 339 (1928).

b. Neon Interaction.—Bleick and Mayer[1] have determined the interaction energy in the case of two neon atoms, using one-electron atomic functions that were obtained by approximating Brown's results[2] analytically. The radial parts of the $2s$ and $2p$ functions are

$$R_{2s} = 13.6e^{-8.22r} - r(14.7e^{-3.69r} + 4.76e^{-2.15r}), \left.\vphantom{\begin{matrix}a\\b\end{matrix}}\right\} \tag{7}$$
$$R_{2p} = r(17.9e^{-3.80r} + 2.30e^{-1.69r}).$$

The internuclear distances used in their work were so large that the $1s$ functions did not overlap appreciably. The results of a straightforward, though laborious, computation of the interaction energy are given in Table LI for three values of the internuclear distance.

TABLE LI

R, Bohr Units	$V(R)$, ev.
3.41	2.15
4.35	0.22
6.05	0.003

These three values may be fitted by the simple function

$$1.18 \cdot 10^4 e^{-\frac{R}{0.395}} \tag{8}$$

with an error of about 10 per cent.

The fact that $V(R)$ is approximately exponential in both the cases a and b provides a rough justification for the exponential term that Born and Mayer used to express the repulsive interaction of ions in crystals.

These repulsive terms do not describe the interaction energy properly at large distances since the tendency of electrons to avoid one another has not been taken into account. An additional attractive term is found when this is done. Since the additional term was implicitly postulated by van der Waals when he proposed his equation of state for gases, it is called the "van der Waals interaction."

Let us consider the van der Waals energy[3] for two atoms a and b that have m and n electrons, respectively, and are separated by a distance R. We shall assume that the atoms lie along the x axis and shall designate the Cartesian coordinates of the electrons relative to the nucleus of the atom on which they reside by (x_{ai}, y_{ai}, z_{ai}) and (x_{bj}, y_{bj}, z_{bj}), respectively, where i ranges from 1 to m and j ranges from 1 to n. The interaction potential H_I of the two atoms may be computed by straightforward principles of electrostatic theory and may be expanded in terms of the Cartesian variables. The result is

[1] W. E. Bleick and J. Mayer, *Jour. Chem. Phys.*, **2**, 252 (1934).

[2] W. G. Brown (*cf.* Sec. 52).

[3] F. London, *Z. Physik*, **63**, 245 (1930).

$$H_I = \sum_{i=1}^{m} \sum_{j=1}^{n} \left\{ -\frac{e^2}{R^3}(2x_{ai}x_{bj} - y_{ai}y_{bj} - z_{ai}z_{bj}) + \frac{3}{2}\frac{e^2}{R^4}[r_{ai}^2 x_{bj} - x_{ai}r_{bj}^2 + \right.$$

$$(2y_{ai}y_{bj} + 2z_{ai}z_{bj} - 3x_{ai}x_{bj})(x_{ai} - x_{bj})] + \frac{3}{4}\frac{e^2}{R^5}[r_{ai}^2 r_{bj}^2 - 5r_{ai}^2 x_{bj}^2 -$$

$$\left. 5r_{bj}^2 x_{ai}^2 - 15x_{ai}^2 x_{bj}^2 + 2(4x_{ai}x_{bj} + y_{ai}y_{bj} + z_{ai}z_{bj})^2] + \cdots \right\}. \quad (9)$$

The terms in $1/R^3$, $1/R^4$ and $1/R^5$ are called, respectively, the dipole-dipole, dipole-quadrupole, and quadrupole-quadrupole interaction terms because they are similar to the interaction energy of the corresponding types of multipole.

The expressions (6) and (8) are mean values of H_I for the approximate wave functions constructed from atomic wave functions, and they may be looked upon as the first-order terms in the perturbation formula

$$E_I(R) = \int \Psi_0^* H_I \Psi_0 d\tau - \sum_\alpha \frac{|\int \Psi_0^* H_I \Psi_\alpha d\tau|^2}{E_0 - E_\alpha} + \cdots \quad (10)$$

where Ψ_0 is the lowest state and the Ψ_α are higher states. Hence, the second term is the van der Waals energy, if we may assume that higher terms in the expansion (10) are negligible.

London has derived a somewhat rough but general expression for the contribution to the van der Waals energy from the dipole-dipole potential in (9). For simplicity, he assumed that the unperturbed wave functions of the normal and excited states may be represented by products of functions of the separate atoms, thus,

$$\Psi = \Phi_a{}^\mu \Phi_b{}^\nu \quad (11)$$

in which μ and ν correspond to different atomic states. This approximation is poor when the atoms are very close together, for it violates the Pauli principle. It is accurate, however, for large atomic separations when the wave functions of different atoms do not overlap. On this assumption the integrals in the numerator of the second term in (10) degenerate into sums of products of integrals over separate atoms, for each of the terms in (9) is a product of terms involving variables of electrons on different atoms. These one-atom integrals, which have the form

$$\int \Phi_a{}^{*0}\left(\sum_a x_{ai}\right)\Phi_a{}^\nu d\tau_a, \quad (12)$$

play an important role in the theory of optical radiation (*cf.* Chap. V). They vanish unless a dipole transition is allowed between the states Φ_a^0

and $\Phi_a{}^\nu$. Moreover, there is a close connection between the integrals which involve $\sum_i x_{ai}$ and those which involve $\sum_i y_{ai}$ and $\sum_i z_{ai}$, for the Φ are wave functions for spherically symmetrical atoms. For these reasons, the numerator of the second term in (10) may be simplified considerably, and the sum may be reduced to

$$E_{d\text{-}d}(R) = 6\frac{e^4}{R^6}\sum_{\mu\nu}\frac{\left|\left(\sum_i z_{ai}\right)_{0\mu}\right|^2 \left|\left(\sum_j z_{bj}\right)_{0\nu}\right|^2}{E_a^0 + E_b^0 - E_a{}^\mu - E_b{}^\nu} \tag{13}$$

where

$$\left.\begin{aligned}\left(\sum_i z_{ai}\right)_{0\mu} &= \int \Phi_a{}^{0*}\left(\sum_i z_{ai}\right)\Phi_a{}^\mu d\tau,\\ \left(\sum_j z_{bj}\right)_{0\nu} &= \int \Phi_b{}^{0*}\left(\sum_j z_{bj}\right)\Phi_b{}^\nu d\tau,\end{aligned}\right\} \tag{14}$$

and the E are the energy values of the wave functions that appear in these integrals.

$E_a{}^\mu$ and $E_b{}^\nu$ are sometimes replaced by constant mean values $\bar E_a$ and $\bar E_b$ in order that the numerator may be summed alone. These mean values, which are not defined independently of Eq. (13), are usually assumed to be approximately equal to the ionization energy of the atoms. By the use of this approximation, Eq. (13) becomes

$$E_{d\text{-}d}(R) = 6\frac{e^4}{R^6}\frac{\left[\left(\sum_i z_{ai}\right)^2\right]_{00}\left[\left(\sum_j z_{bj}\right)^2\right]_{00}}{E_a^0 + E_b^0 - \bar E_a - \bar E_b}, \tag{15}$$

since

$$\sum_\mu\left|\left(\sum_i z_{ai}\right)_{0\mu}\right|^2 = \left[\left(\sum_i z_{ai}\right)^2\right]_{00}. \tag{16}$$

The polarizability α of an atom is related to its energy $E(\mathsf{E})$, in a field of intensity E, by means of the equation

$$E(\mathsf{E}) = E_0 + \tfrac{1}{2}\alpha\mathsf{E}^2. \tag{17}$$

Thus, according to perturbation theory, we have

$$\alpha = -2e^2\sum_\mu\frac{\left|\left(\sum_i z_{ai}\right)_{0\mu}\right|^2}{E_0 - E_\mu}, \tag{18}$$

which is approximately equal to

$$-2e^2\frac{\left[\left(\sum_i z_{ai}\right)^2\right]_{00}}{E_0 - \bar E_a}.$$

Hence, (15) is approximately equal to

$$E_{d\text{-}d} \cong -\frac{3}{2}\frac{1}{R^6}\frac{(E_0 - \bar{E}_a)(E_0 - \bar{E}_b)}{(E_a^0 + E_b^0 - \bar{E}_a - \bar{E}_b)}\alpha_a\alpha_b \tag{19}$$

where α_a and α_b are, respectively, the polarizabilities of atoms a and b.

Margenau[1] and Mayer[2] have applied similar methods in order to obtain the contribution to the van der Waals energy from the dipole-quadrupole and quadrupole-quadrupole terms. The end result for the dipole-quadrupole term that corresponds to (19) is

$$E_{d\text{-}q} = -\frac{3}{2e^2}\frac{1}{R^8}\frac{[\alpha_a\alpha_b^2(E_a^0 - \bar{E}_a)(E_b^0 - \bar{E}_b)^2 + \alpha_a^2\alpha_b(E_a^0 - \bar{E}_a)^2(E_b^0 - \bar{E}_b)]}{(E_a^0 + E_b^0 - \bar{E}_a - \bar{E}_b)}. \tag{20}$$

The quadrupole-quadrupole term may be developed in a similar way, as is described in Margenau's paper.

We shall now discuss the results of computations for hydrogen and helium.

a. Hydrogen.—For two hydrogen atoms, (19) becomes

$$-\frac{6}{R^6}\frac{e^2}{a_h}, \tag{21}$$

in which R is expressed in units of a_h, when $\bar{E} - E_0$ is set equal to the ionization energy $e^2/2a_h$. In the same approximation, the complete expression for the van der Waals energy, through quadrupole-quadrupole terms, is

$$-6\frac{e^2}{a_h}\left(\frac{1}{R^6} + \frac{22.5}{R^8} + \frac{236}{R^{10}}\right). \tag{22}$$

This may be used to estimate the relative magnitude of the different terms.

More accurate expressions for the dipole-dipole term have been derived by Eisenschitz and London[3] and by Slater and Kirkwood.[4] The first workers summed (13) directly and obtained

$$-\frac{6.47}{R^6}\frac{e^2}{a_h} \tag{23}$$

whereas the second employed a variational method and found

$$-\frac{6.49}{R^6}\frac{e^2}{a_h}. \tag{24}$$

[1] H. MARGENAU, *Phys. Rev.*, **38**, 747 (1931); **40**, 387 (1932); see also *op. cit.*
[2] J. E. MAYER, *Jour. Chem. Phys.*, **1**, 270 (1933).
[3] R. EISENSCHITZ and F. LONDON, *Z. Physik*, **60**, 491 (1930).
[4] J. C. SLATER and J. G. KIRKWOOD, *Phys. Rev.*, **37**, 682 (1931).

b. *Helium.*—Margenau[1] has derived an expression equivalent to (22) for helium.   The result is

$$-\frac{1.62}{R^6}\frac{e^2}{a_h}\left(\frac{1}{R^6} + \frac{7.9}{R^8} + \frac{30}{R^{10}}\right). \tag{25}$$

A more exact expression for the dipole-dipole term, which was derived by Slater and Kirkwood, is[2]

$$-\frac{1.59}{R^6}\frac{e^2}{a_h}. \tag{26}$$

The sum of (26) and (6) has a shallow minimum with a depth of 0.75×10^{-3} ev at $5.5a_h$.   This minimum accounts for the weak cohesive energy of liquid helium (see Fig. 5).

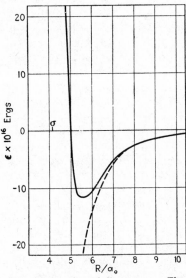

FIG. 5.—The total interaction energy of two helium atoms.   The dotted curve is the van der Waals interaction.   (*After Slater and Kirkwood.*)

A similar minimum should be expected for all rare gas atoms.   Bleick and Mayer find that the total interaction energy for neon, obtained by adding (8) and (19), has a minimum of $1.3 \cdot 10^{-3}$ ev when R is equal to $6.5a_h$.   These values agree roughly with those obtained from empirical data by Lennard-Jones.[3]

[1] Margenau has recently developed a revised form of Eq. (25) which he regards as a more accurate representation of the true van der Waals energy than either Eq. (25) or (26) (see *Phys. Rev.*, **55**, 1137 (1939)).

[2] SLATER and KIRKWOOD, *op. cit.*

[3] See survey by J. E. Lennard-Jones, *Physica*, **4**, 941 (1937).

59. Molecular Valence.[1]—The problem of providing a reason fo
the inertness of the rare gas atoms in their interaction with one anothe
is solved by the computations discussed in the last section, for they sho
that closed-shell structures have very weak attractive forces.   A simila
problem arises in connection with molecules such as H_2, N_2, CH_4, C_2H_6
since they also form very stable units that do not interact strongly wit
one another.   This problem is of importance to us because these mole
cules are constituents of an important class of solids.

The origin of the weak intermolecular forces is believed to be largely
the same as that in the case of the rare gases, namely, the van der Waal
interaction.   Computations that support this are discussed in Sec. 88
Chap. X.   The problem of understanding the internal stability of these
molecules, however, is not so easy to answer in a quantitative way
Since the internal binding energy of the stablest molecules is of the
order of magnitude of 1 ev per electron, this problem can be solved
quantitatively only by solving the Schrödinger equation for molecular
systems to a higher degree of accuracy than is generally feasible at present.

Physical chemists have attempted to avoid some of the difficulties
associated with solving the Schrödinger equation accurately by introducing semiempirical schemes.   These are usually patterned after one of the
one-electron schemes, the matrix components that enter into the theoretical results being judiciously replaced by quantities derived from
empirical data.   From what is known of the accuracy of one-electron
approximations, it is doubtful whether actual computations based on
these one-electron schemes would yield results that agree with experimental results as well as those of the semiempirical schemes do.   The
latter are such a distinct improvement over older valence theories, however, that they have great value in discussing many properties of molecules.   We shall present some of the qualitative results of these schemes
in the sections of Chap. XIII that deal with valence crystals.

[1] See the review article by J. H. Van Vleck and A. Sherman, *Rev. Modern Phys.*,
7, 167 (1935).

CHAPTER VIII

THE BAND APPROXIMATION

60. Qualitative Importance of the Band Scheme.—Prior to the introduction of quantum mechanics, it had been believed that insulators have low electronic conductivities because their valence electrons are localized on definite atoms or molecules and cannot jump from one atom to another. The electrons in metals, on the other hand, were considered to be free to roam throughout the lattice, and the high conductivity was believed to arise from this freedom of motion. If we attempt to use these qualitative notions in order to understand both the conductivities and the cohesive energies of solids, we face considerable difficulty unless we are willing to assume that the binding forces arise from essentially different sources in each type of solid.

Suppose, for example, we assume that there are only two kinds of interatomic force, namely: (1) electrostatic forces between bound charge distributions, and (2) undefined forces that are ultimately connected with the presence of free electrons and are of primary importance in metals. It is possible to account for the cohesion and insulating properties of ionic and molecular crystals in terms of (1). We may assume, as is done in the Madelung-Born theory, that the constituents of ionic crystals are ions and that the main part of the cohesion arises from the electrostatic attraction between these. Similarly, we may assume that the molecular constituents of molecular crystals are electrostatically neutral and that the cohesive energy arises from multipole forces of an electrostatic type. Since these forces should be weaker than the forces between ions, we are able to understand the relatively smaller cohesive energies of molecular crystals.

We meet with difficulty in discussing insulating valence crystals such as diamond. In this case, the atoms are electrostatically neutral, as in molecular crystals, and yet the cohesion is as great as in ionic crystals and metals. This difficulty was removed in classical theory by assuming that in addition to (1) and (2) there are valence forces which are responsible for the large cohesion of diamond and quartz.

As in many other cases in which classical views led to complication, the introduction of quantum mechanics produces order in a relatively simple way. In particular, the band concept of solids, which is based upon the Bloch scheme and which has been developed by many workers,[1] has been very useful in coordinating many of the properties of solids that could not be adequately understood before. It may be recalled that the Bloch scheme is based on a one-electron approximation in which the one-electron functions have the same amplitude at equivalent positions in each unit cell. We shall see in the next section that these functions have the form

$$\psi_k = \chi_k(\mathbf{r})e^{2\pi i k \cdot \mathbf{r}}, \tag{1}$$

where $\mathbf{r}$ is the position vector, whose components are x, y, z; $\chi_k(\mathbf{r})$ has the translational periodicity of the lattice; and $\mathbf{k}$ is a wave-number vector. $\mathbf{k}$ may be defined in terms of the reciprocal lattice of the crystal (*cf.* Sec. 22) if the Born-von Kármán boundary conditions are used in determining the functions (1).

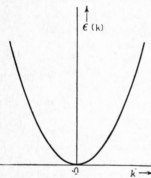

In the simplest case, χ_k is a constant so that ψ_k is a free-electron wave function for which the dependence of energy on $\mathbf{k}$ is

$$\epsilon(\mathbf{k}) = \frac{h^2}{2m}\mathbf{k}^2 \tag{2}$$

Fig. 1.—The $\epsilon(\mathbf{k})$ curve for perfectly free electrons.

(see Fig. 1). This approximation corresponds to that of the Sommerfeld theory of metals, which was discussed in Chap. IV. We shall see that its use is equivalent to assuming that the Fock-Hartree field for the electrons is constant—a condition that is nearly satisfied in many simple metals.

In the opposite extreme, corresponding to tightly bound inner-shell electrons, χ_k is zero everywhere in the unit cell except in the immediate vicinity of the particular atom the inner shells of which are being described. It turns out, in this case, that the portion of χ_k near the atom is identical with the inner-shell wave function of the free atom. Moreover, the energy $\epsilon(\mathbf{k})$ is practically independent of $\mathbf{k}$ and has a different value for each type of inner-shell electron.

[1] The qualitative existence of bands was first pointed out by M. J. O. Strutt, *Ann. Physik*, **84**, 485 (1927); **85**, 129 (1928). The band picture was extended by: L. Brillouin, *Compt. rend.*, **191**, 198, 292 (1930); *Jour. phys.*, (VII), **1**, 377 (1930) [*cf.* also *Quantenstatistik* (Julius Springer, Berlin, 1931)]; P. M. Morse, *Phys. Rev.*, **35**, 1310 (1930); R. Peierls, *Ann. Physik*, **4**, 121 (1930) [*cf.* also *Ergebnisse exakt. Natur.*, **11**, 264 (1932)]; R. de L. Kronig and W. G. Penney, *Proc. Roy. Soc.*, **130**, 499 (1931).

In intermediate cases in which χ_k is neither completely constant nor localized, the energy $\epsilon(\mathbf{k})$, as a function of $\mathbf{k}$, is not completely quasi-continuous, as in the free-electron case. Instead, $\epsilon(\mathbf{k})$ exhibits the property of *banding;* that is, it is quasi-continuous for large ranges of $\mathbf{k}$ but is discontinuous for certain values of $\mathbf{k}$. The regions of continuity lie between a set of concentric polyhedra which are centered about the origin of $\mathbf{k}$ space; the points where the discontinuities occur are at the surfaces of the polyhedra. In the following sections, we shall investigate the relationships from which the form of these polyhedra may be deduced.

The regions between the polyhedra are called "zones" and the polyhedra are called "zone boundaries." The magnitude of the discontinuities at the zone boundaries depends upon the extent to which χ_k deviates from a constant value and is zero for perfectly free electrons. Figure 2 illustrates the discontinuities for a typical case in which the $\epsilon(\mathbf{k})$ curve is plotted as a function of the points on a line that passes through the point $\mathbf{k} = 0$. The discontinuities occur at points where this line intersects the different polyhedra.

The transition from the free-electron type of wave function to the rigidly bound electron type may be regarded as taking place when χ_k changes from a constant to a highly localized function. During this transition $\epsilon(\mathbf{k})$ develops discontinuities which become larger and larger until $\epsilon(\mathbf{k})$ is constant within each zone.

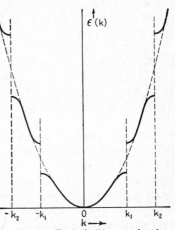

Fig. 2.—Typical $\epsilon(\mathbf{k})$ curve for electrons that are not perfectly free. This corresponds to values of $\mathbf{k}$ that lie on a line passing through the origin of $\mathbf{k}$ space. The discontinuities occur at points k_1 and k_2 at which the line cuts the zone-boundary polyhedra. These points are different for different directions in $\mathbf{k}$ space. The dashed parabola represents the $\epsilon(\mathbf{k})$ curve for perfectly free electrons (*cf.* Fig. 1).

When constructing an antisymmetric wave function from the Bloch functions, we are not allowed to assign a function of given $\mathbf{k}$ to more than two electrons because of the Pauli principle. Hence, we must use a large number of different values of $\mathbf{k}$ in order to assign functions to all electrons in a solid. We shall say that levels are occupied when the wave functions corresponding to them have been assigned to electrons. As in treating atoms and molecules, we shall assume that, in the normal state of the system, the lowest one-electron energy levels are occupied as far as possible. It turns out that the number of states in any zone is an integer multiple of the number of unit cells in the crystal. Hence, it may happen that a low-lying set of zones is completely occupied and that the levels in zones of higher energy are completely empty. The conditions that must be satisfied if this is to occur in the lowest state are as follows:

a. The number of electrons present must be exactly equal to the number of states in an integer number of zones. It turns out that this condition is satisfied in all insulators and in the alkaline earth metals. It is not satisfied in the alkali metals, for in these there are twice as many states per zone as there are electrons.

b. The highest filled zone must not have any energy levels that lie above the lowest energy levels in the next highest zone. If these energy bands overlap, the energy of the system could be lowered by transferring electrons from the highest levels of the last filled zone to the lower levels of the next zone. Complete filling of zones is prohibited in the alkaline earth metals because condition b is not satisfied.

Let us consider the difference between the properties of a substance in which occupied zones are completely filled and those of one in which they are not. In both cases, the electrons normally are paired in such a way that for each electron moving in a given direction there is another moving in the opposite direction with the same speed. Hence, the current carried by each electron cancels that carried by the other, and the resultant current of the entire solid is zero. This statistical balance may be disturbed easily in a substance that does not have completely filled bands; for, by the application of a weak electrostatic field, some of the electrons may be made to jump to the near-lying unoccupied levels, thus changing the average velocity from zero to a finite value. This type of shift in statistical balance was described in the sections of Chap. IV that deal with the Lorentz-Sommerfeld theory of conductivity. On the other hand, the highest occupied levels may be separated from the unoccupied ones by several electron volts if the solid has completely filled bands. In this case, a very strong electrical field would be required to induce the electrons to jump from occupied to unoccupied levels. Hence, the crystal with completely filled bands is an insulator even though its electrons are wandering throughout the lattice.

Thus, we see that the electrical properties of two substances that have similar one-electron functions may be vastly different. In this connection, we may anticipate that the one-electron functions of diamond and of metals are similar so that the cohesive forces have similar origin in both cases.

Since the introduction of the band scheme permits us to modify the classical concept of "bound electrons" in valence crystals, it is natural to ask whether or not we need retain the classical concept when discussing ionic and molecular crystals. This question may be answered fairly unambiguously in both cases.   In ionic crystals, the classical picture is a fair approximation but is not rigorously correct since the valence-electron wave functions are not entirely localized about the cations.   This means that the Bloch functions of ionic crystals possess properties similar to those for metals.   Conductivity is absent because of the way in which zones are filled rather than because the electrons are not free to move from one atom to another.   Since the amplitude of the wave functions unquestionably is small at regions midway between molecules in molecular crystals, we should expect small electronic conductivity even if the zone structure did not exist.   Nevertheless, the presence of filled zones plays a predominant role in entirely prohibiting electronic conductivity.

It should be kept in mind that the zone scheme rests upon an approximation.   Although it serves a useful purpose in providing a model of a solid that is adequate for describing many important properties in a simple, straightforward way, the picture is not a perfect one, and it may lead to incorrect results if it is not used with sufficient care.   In particular, it should not be applied without reserve to problems that involve excited states of insulators, for reasons which are discussed in Chap. XII.

61. The Connection between Zone Structure and Crystal Symmetry. Since the concept of zone structure is based upon a particular type of one-electron approximation, it is natural to ask for those properties of the Fock-Hartree equations that in this case lead to the existence of zones.   This question has the relatively simple answer that zone structure is characteristic of any eigenvalue equation in which the operator remains unchanged under the primitive translations of the lattice.   Thus the eigenvalues E of any equation that has the form

$$H\psi = E\psi, \tag{1}$$

where H has crystallographic symmetry, are practically continuous except for certain unallowed regions.   This topic may be made the basis of a very elegant and practical group-theoretical discussion[1] in which we shall not indulge at this point.   We shall consider, instead, several examples of equations of type (1) in which the symmetry conditions that are required for zone structure occur.   From these, we may derive general conditions from which the precise form of zone structure may be determined in any case.

Case a.   The One-dimensional Oscillating Lattice in Classical Mechanics.—One of the simplest problems in which zone properties occur is that of determining the vibrational modes of a long one-dimensional

---

[1] The group-theoretical side of the existence of zones is discussed in the following papers: F. Seitz, *Ann. Math.*, **37**, 17 (1936); L. P. Bouckaert, R. Smoluchowski, and E. Wigner, *Phys. Rev.*, **50**, 58 (1936); C. Herring, *Phys. Rev.*, **52**, 361 (1937), **52**, 365 (1937); M. I. Chodorow and M. F. Manning, *Phys. Rev.*, **52**, 731 (1937).

lattice of particles that interact with harmonic forces. Several special cases of this problem were discussed in Sec. 21, Chap. III. We shall review these now.

First, we considered the case in which the particles have equal mass and are separated by a distance a. The equations of motion for this case are

$$m\frac{d^2x_n}{dt^2} = -\mu[(x_n - x_{n-1}) - (x_{n+1} - x_n)] \tag{2}$$

where x_n is the displacement of the nth atom from its equilibrium position, μ is Hooke's constant for interaction between nearest neighbors, and m is the mass of a particle. Since we are searching for solutions that are periodic in time, we may replace Eq. (2) by

$$-m(2\pi\nu)^2x_n = -\mu[(x_n - x_{n-1}) - (x_{n+1} - x_n)]. \tag{3}$$

This equation is of the type (1) since we may regard the x_n as components of a vector $\mathbf{X}$. Thus,

$$\mathbf{X} = \begin{pmatrix} x_1 \\ x_2 \\ \cdot \\ \cdot \\ \cdot \\ x_{n-1} \\ x_n \\ x_{n+1} \\ \cdot \\ \cdot \\ \cdot \end{pmatrix}. \tag{4}$$

Using (4) we may write Eq. (3) in the form

$$\sum_m M_{nm}X_n = -\nu^2X_n, \tag{5}$$

or

$$\mathbf{M} \cdot \mathbf{X} = -\nu^2\mathbf{X} \tag{5a}$$

where $\mathbf{M}$ is a tensor or matrix the components of which are

$$M_{n,n} = -\frac{2\mu}{4\pi^2m},$$
$$M_{n,n+1} = M_{n-1,n} = \frac{\mu}{4\pi^2m}, \qquad M_{n,m} = 0, \qquad m \neq \begin{cases} n \\ \text{or} \\ n \pm 1 \end{cases}.$$

The matrix $\mathbf{M}$ clearly has the symmetry of the lattice since Eqs. (2) are the same for all masses. The solutions of these equations were found to have the form

$$x_n = A e^{2\pi i \sigma n a}. \tag{6}$$

Here σ is the wave number l/Na, where l is an arbitrary integer and N is the number of cells in the lattice. The frequency associated with (6) is

$$\nu_l = \frac{1}{2\pi}\sqrt{\frac{\mu}{m}}\,|\sin \pi\sigma a|. \tag{7}$$

The independent values of l may be chosen to range from $-N/2$ to $+N/2$. Equation (7), for the corresponding range of σ, is shown in Fig. 7, Sec. 21. This curve does not exhibit the discontinuities characteristic of zone structure, for only one zone occurs in the present problem.

Next, we shall consider the extended problem in which two different masses occur in the unit cell. If the particles are separated uniformly by a distance $a/2$ and are labeled by integers extending from zero to $2N$, the normal modes are

$$x_{2n} = B e^{2\pi i \sigma\left(\frac{2n}{2}a\right)},$$
$$x_{2n+1} = A e^{2\pi i \sigma\left(\frac{2n+1}{2}a\right)}, \tag{8}$$

where the wave number σ is again equal to l/Na. The frequencies and the constants A and B may be determined by solving an appropriate second-order secular equation which was discussed in Sec. 21. We there found the result

$$4\pi^2\nu^2 = \frac{\mu}{mM}(M + m \pm \sqrt{M^2 + m^2 + 2Mm \cos 2\pi\sigma a}),$$

in which we may choose the independent range of σ to extend from $-1/a$ to $1/a$ and obtain the result shown in Fig. 8, Sec. 21. Discontinuities characteristic of zone structure occur at $\sigma = \pm 1/2a$, so that there are two zones in this case.

If another mass is added to the unit cell in such a way that neighboring masses are separated by a distance $a/3$, the new normal coordinates are

$$x_{3n} = A e^{2\pi i \sigma \frac{3n}{3}a},$$
$$x_{3n+1} = B e^{2\pi i \sigma \frac{3n+1}{3}a},$$
$$x_{3n+2} = C e^{2\pi i \sigma \frac{3n+2}{3}a}, \tag{9}$$

and so forth, if the masses are at points labeled by integers extending from zero to $3N$. There are three zones in this case, since discontinuities occur when σ is equal to $\pm 1/2a$ and $\pm 1/a$.

It should be noted that we may write Eqs. (6), (8), and (9) in the form

$$x(\xi) = \chi(\xi)e^{2\pi i\sigma\xi} \tag{10}$$

where ξ is the positional coordinate of a given mass, $x(\xi)$ is the displacement of the mass from its equilibrium position, and $\chi(\xi)$ is a discontinuous function having periodicity a that takes values different from zero only at points where atoms are situated. Thus, in the case corresponding to Eq. (9), χ is equal to A at $\xi = 3na/3$, to B at $\xi = (3n + 1)\dfrac{a}{3}$, and to C at $\xi = (3n + 2)\dfrac{a}{3}$ and is zero everywhere else. The form of

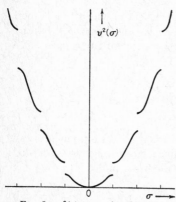

the function (10) evidently is the same as that of the function (1) in the preceding section.

If we continue to add masses at equivalent points within the unit cell, we eventually find it convenient to use a density function $\rho(\xi)$ and a force function $\mu(\xi)$, both of which are continuous and have periodicity a. The normal coordinate should still be expressible in the form (10) but $\chi(\xi)$ should now be a continuous function of ξ with periodicity a. The independent modes of vibration then correspond to values of σ extending from $-\infty$ to $+\infty$, and ν^2 is discontinuous whenever σ is equal to $\pm r/2a$, where r is an arbitrary integer. This case, in which there evidently is an infinite number of zones, is represented schematically in

FIG. 3.—$\nu^2(\sigma)$ curve for the normal modes of a continuous string having periodically varying density and Hooke's constant (extended zone scheme). If the string were perfectly uniform the $\nu^2(\sigma)$ curve would be a parabola. This curve should be compared with that of Fig. 2.

Fig. 3. The differential equation that is satisfied by $x(\xi)$ is the wave equation

$$-4\pi^2\nu^2 x = \frac{\mu}{\rho}\frac{d^2x}{d\xi^2}$$

which has the eigenvalue form (1).

Case b. The One-dimensional Schrödinger Equation. 1. Kramers' general treatment.—From among the many methods that have been developed to show that the eigenvalues of the Schrödinger equation exhibit band structure when the potential function is periodic, we shall select a particularly good one that is due to Kramers.[1]

[1] H. A. KRAMERS, *Physica*, **2**, 483 (1935).

Suppose that we have an equation of the form

$$\frac{d^2\psi}{dx^2} + [E - V(x)]\psi = 0 \tag{11}$$

where

$$V(x + a) = V(x). \tag{12}$$

Then, if $\psi_1(x)$ and $\psi_2(x)$ are solutions of this equation, the functions $\psi_1(x + a)$ and $\psi_2(x + a)$ also are solutions. Since a second-order equation possesses only two independent solutions, we must have the relationships

$$\begin{aligned}\psi_1(x + a) &= a_{11}\psi_1(x) + a_{12}\psi_2(x),\\ \psi_2(x + a) &= a_{21}\psi_1(x) + a_{22}\psi_2(x),\end{aligned} \tag{13}$$

if $\psi_1(x)$ and $\psi_2(x)$ are independent. From these equations, we may derive the relationship

$$\begin{vmatrix}\psi_1(x + a) & \psi_2(x + a)\\ \psi_1'(x + a) & \psi_2'(x + a)\end{vmatrix} = \begin{vmatrix}\psi_1(x) & \psi_2(x)\\ \psi_1'(x) & \psi_2'(x)\end{vmatrix} \begin{vmatrix}a_{11} & a_{12}\\ a_{21} & a_{22}\end{vmatrix} \tag{14}$$

where $\psi' = d\psi/dx$. Now, the quantity

$$\begin{vmatrix}\psi_1(x) & \psi_2(x)\\ \psi_1'(x) & \psi_2'(x)\end{vmatrix}, \tag{15}$$

which is called the "*Wronskian*," is a constant in the present case.[1] Hence, we may conclude that

$$\begin{vmatrix}a_{11} & a_{12}\\ a_{21} & a_{22}\end{vmatrix} = 1. \tag{16}$$

We may now choose linear combinations φ_1 and φ_2 of ψ_1 and ψ_2 that have the property

$$\begin{aligned}\varphi_1(x + a) &= \lambda_1\varphi_1(x)\\ \varphi_2(x + a) &= \lambda_2\varphi_2(x).\end{aligned} \tag{17}$$

The coefficients λ_1 and λ_2 may be determined from the a in Eq. (13) by means of the equation

$$\begin{vmatrix}a_{11} - \lambda & a_{12}\\ a_{21} & a_{22} - \lambda\end{vmatrix} = \lambda^2 - (a_{11} + a_{22})\lambda + 1 = 0. \tag{18}$$

The quantity $\mu = (a_{11} + a_{22})$ is real since ψ_1 and ψ_2 may always be chosen to be real functions.

Kramers distinguishes between the three cases $|\mu| > 2, |\mu| < 2, |\mu| = 2$, which we shall discuss categorically.

---

[1] See, for example, WHITTAKER and WATSON, *Modern Analysis* (Cambridge University Press, 1935).

i. When $|\mu|$ is greater than 2, Eq. (18) has unequal real roots so that φ_1 and φ_2 satisfy the equations

$$\varphi_1(x + a) = \lambda_1\varphi_1(x), \qquad \varphi_2(x + a) = \frac{1}{\lambda_1}\varphi_2(x). \qquad (19)$$

ii. When $|\mu|$ is less than 2, the roots are complex conjugates of one another, whence

$$\begin{aligned}\varphi_1(x + a) &= e^{i\alpha}\varphi_1(x), \\ \varphi_2(x + a) &= e^{-i\alpha}\varphi_2(x).\end{aligned} \qquad (20)$$

It may be shown very easily that in this case φ_1 and φ_2 are complex conjugates of one another.

iii. When $|\mu|$ is equal to 2, λ is ± 1, both roots are equal, and Eq. (6) is replaced by

$$\begin{aligned}\varphi_1(x + a) &= \pm\varphi_1(x), \\ \varphi_2(x + a) &= \pm\varphi_2(x) + b\varphi_2(x).\end{aligned} \qquad (21)$$

Thus, both functions φ_1 and φ_2 satisfy the equation

$$\varphi(x + a) = \pm\varphi(x) \qquad (21a)$$

only when b vanishes.

In case *i*, the ratio

$$\frac{\varphi(x + na)}{\varphi(x)} \qquad (22)$$

certainly becomes infinite when n approaches either $+\infty$ or $-\infty$. Hence, this type of eigenfunction must be excluded if the periodic field extends over the entire range of x between $+\infty$ and $-\infty$.

In case *ii*, both eigenfunctions are periodic and satisfy the relation

$$|\varphi(x + a)| = |\varphi(x)|.$$

These solutions are allowable as long as they remain finite in any unit cell.

In case *iii*, there is at least one solution of type (21a) and possibly two, depending upon whether b is zero or not.

The solutions for case *i* correspond to the unallowed regions of energy, whereas those for cases *ii* and *iii* correspond to the allowed regions. We shall see that the functions corresponding to case *iii* are solutions associated either with points in wave-number space at the zone boundaries or at points such as $\delta = 0$ for which the solutions have periodicity a.

It should be mentioned that the solutions for case *i* need not always be excluded if the periodic field does not extend to infinity; for in the finite case, which corresponds to an actual crystal, the wave functions associated with some values of E in the allowed region may not diverge. In these cases, which were first pointed out by Tamm, the functions have their maximum absolute values near the boundary of the lattice and decrease rapidly on both sides of this point. We shall discuss these solutions further in Sec. 70.

Kramers was able to express the quantities μ, $d\mu/dE$, and $d^2\mu/dE^2$ in terms of integrals that involve the functions φ_1 and φ_2. From these expressions, he deduced that $\mu(E)$ has the form illustrated in Fig. 4,

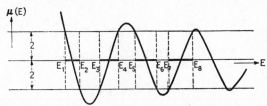

Fig. 4.—The $\mu(E)$ curve. The allowed values of E correspond to the ranges in which μ lies between 1 and −1. (*After Kramers.*)

which approaches $2 \cos a\sqrt{E}$ when E approaches ∞ and approaches $e^{a\sqrt{-E}}$ when E approaches $-\infty$. In intermediate regions, it oscillates between values greater than 2 and less than −2, in the manner shown. Thus, there are alternate continuous bands of allowed and unallowed levels. The unallowed bands become vanishingly small when E is large and positive. All values of E that are sufficiently negative are unallowed because of the monotonic increase of μ.

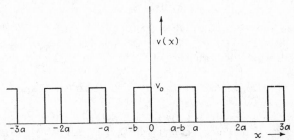

Fig. 5.—The one-dimensional periodic potential of Kronig and Penney.

2. *The case of Kronig and Penney.*—One of the simplest examples of a one-dimensional periodic field has been treated by Kronig and Penney.[1] This example merits attention because of the direct way in which it yields the general features of zone structure. Let us consider the periodic potential illustrated in Fig. 5, for which

[1] KRONIG and PENNEY, *op. cit.*, 499 (1931). See also V. ROJANSKY, *Introductory Quantum Mechanics*, Sec. 49 (Prentice-Hall, Inc., New York, 1938).

$$V = V_0, \qquad -b \leq x \leq 0,$$
$$V = 0, \qquad 0 \leq x \leq a - b,$$
$$V(x + a) = V(x). \tag{23}$$

Thus, V has the constant value V_0 for a range of length b in each unit cell. In the regions where V is zero, the general solution of the Schrödinger equation is

$$\psi_1 = Ae^{i\alpha x} + Be^{-i\alpha x} \tag{24}$$

where

$$\alpha = \frac{\sqrt{2mE}}{\hbar}.$$

In the other region, we have

$$\psi_2 = Ce^{\beta x} + De^{-\beta x} \tag{25}$$

where

$$\beta = \frac{\sqrt{2m(V_0 - E)}}{\hbar}.$$

Since we are searching for solutions of Kramers' class ii, we must have

$$\psi_2(-b) = e^{-i\lambda a}\psi_1(a - b),$$
$$\psi_2'(-b) = e^{-i\lambda a}\psi_1'(a - b), \tag{26}$$

where $\lambda/2\pi$ is the wave number. In addition, we must have

$$\psi_2(0) = \psi_1(0),$$
$$\psi_2'(0) = \psi_1'(0), \tag{27}$$

because of the continuity requirement at $x = 0$. We find, after substituting (24) and (25) in (26) and (27) and solving the determinantal compatibility equation, the condition

$$\cos \lambda a = \frac{(\beta^2 - \alpha^2)}{2\alpha\beta} \sinh \beta b \sin \alpha(a - b) + \cosh \beta b \cos \alpha(a - b), \tag{28}$$

which may be used to determine the allowed values of E.

Following Kronig and Penney, we may, at this point, introduce the simplifying conditions

$$b \to 0,$$
$$V_0 \to \infty,$$

and we may stipulate that these limiting values are approached in such a way that the quantity

$$\frac{mV_0}{\hbar^2}b(a - b) \tag{29}$$

remains constant. This restriction assures us that the "potential area" $V_0 b$ is finite. With these conditions, the quantity βb in Eq. (28) is equal to

$$\sqrt{\frac{2cb}{a-b}}$$

where c is the limiting value of (29). Equation (28) approaches the value

$$\cos \lambda a = \frac{c}{\alpha a} \sin \alpha a + \cos \alpha a \tag{30}$$

in the limit as $b \to 0$. This equation may be satisfied whenever the quantity on the right-hand side lies in the interval from -1 to $+1$ since λ may then take real values. As is shown in Fig. 6, we obtain allowed

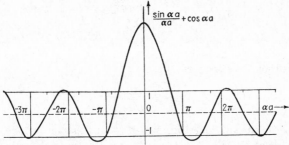

Fig. 6.—The function $\{c \sin \alpha a / \alpha a + \cos \alpha a\}$. The allowed values of E are given by those ranges of $\alpha = \sqrt{2mE}/\hbar$ in which this function lies between -1 and $+1$. It may be seen that the unallowed ranges become smaller as E increases. This curve is analogous to that of Fig. 4. (*After Kronig and Penney.*)

bands of energy that become closer and closer as E approaches infinity. It may be verified that the coefficients A and B in (24) have the ratio

$$-\frac{A}{B} = \frac{1 - e^{-i(\lambda+\alpha)a}}{1 - e^{-i(\lambda-\alpha)a}} \tag{31}$$

in the range of x that extends from 0 to a. The values of A and B in any other range, such as that extending from na to $(n+1)a$, may be obtained from the values in the range from 0 to a by multiplying by the factor $e^{i\lambda(n+1)a}$.

Case c. The Two-dimensional Schrödinger Equation.—By discussing three types of two-dimensional cases, we may obtain some of the most important principles of zone-structure theory. The first case, which was investigated by Brillouin,[1] deals with electrons that are practically free. The periodic potential then may be treated as a small perturbation. In

---

[1] *Cf. Quantenstatistik, op. cit.*

the second case, the periodic potential is larger than in the first but may be expressed in the form

$$V(x,y) = V_x(x) + V_y(y) + V_p(x,y) \tag{32}$$

where $V_x(x)$ and $V_y(y)$ are large compared with $V_p(x,y)$. The Schrödinger equation is separable in this case if $V_p(x,y)$ is neglected. The zones that are obtained when V_p is included by perturbation theory usually do not have the same form as in Brillouin's case. We shall see however, that the two sets may be made identical by suitable rearrangement. The third case to be discussed is the more general one in which perturbation methods cannot be employed.

1. *Brillouin's case.*—We shall assume that the Schrödinger equation has the form

$$-\frac{\hbar^2}{2m}\left(\frac{\partial^2\psi}{\partial x^2} + \frac{\partial^2\psi}{\partial y^2}\right) + [V(x,y) - E]\psi = 0.$$

For simplicity, we shall discuss the case in which $V(x,y)$ has the periodicity a in both the x and y directions. The normalized unperturbed eigenfunctions then have the form

$$\psi_{\mathbf{k}}^0 = \frac{1}{\sqrt{S}}e^{2\pi i \mathbf{k}\cdot\mathbf{r}} \tag{33}$$

where S is the area of the lattice, which we shall assume is a square having the edge length Na. If we adopt the Born-von Kármán boundary conditions, the permissible components of $\mathbf{k}$ are

$$k_x = \frac{n_x}{Na}, \qquad k_y = \frac{n_y}{Na} \tag{34}$$

where n_x and n_y are arbitrary integers. The unperturbed energy of the wave function (33) is

$$E_{\mathbf{k}}^0 = \frac{h^2}{2m}\mathbf{k}^2, \tag{35}$$

and the entire energy spectrum above zero is quasi-continuous.

To the first order in perturbed quantities, the eigenfunctions of Eq. (32) are

$$\psi_{\mathbf{k}}(x,y) = \frac{1}{\sqrt{S}}e^{2\pi i \mathbf{k}\cdot\mathbf{r}} + \sum_{\mathbf{k}'}\frac{\left[\dfrac{1}{S^{\frac{3}{2}}}\displaystyle\int e^{-2\pi i \mathbf{k}'\cdot\mathbf{r}'}V(x',y')e^{2\pi i \mathbf{k}\cdot\mathbf{r}'}dx'dy'\right]e^{2\pi i \mathbf{k}'\cdot\mathbf{r}}}{E_{\mathbf{k}}^0 - E_{\mathbf{k}'}^0} \tag{36}$$

where the summation extends over all values of k_x' and k_y' and the integration extends over the entire lattice. The integrals in Eq. (36) may be simplified by use of the translational symmetry of $V(x,y)$. We have in fact

$$\int_S e^{2\pi i(\mathbf{k}-\mathbf{k'})\cdot\mathbf{r}} V(x,y)dxdy = \left[\int_A e^{2\pi i(\mathbf{k}-\mathbf{k'})\cdot\mathbf{r}} V(x,y)dxdy\right]\sum_{\mathbf{d}} e^{2\pi i(\mathbf{k}-\mathbf{k'})\cdot\mathbf{d}} \quad (37)$$

where $\mathbf{d}$ is the vector

$$\mathbf{d} = \begin{pmatrix} na \\ ma \end{pmatrix}, \qquad (m,\ n = 0,\ 1,\ 2,\ \cdots,\ N)$$

and the integral in the right-hand term extends over the unit cell, that is, over the ranges of x and y lying between zero and a. The summation in (37) vanishes identically unless the vector $(\mathbf{k} - \mathbf{k'})$ satisfies the relation

$$\begin{aligned} (k_x - k'_x)a &= p_x, \\ (k_y - k'_y)a &= p_y, \end{aligned} \right\} \quad (38)$$

where p_x and p_y are integers. Hereafter, we shall reserve the letter $\mathbf{K}$ for wave-number vectors of which the scalar products with the primitive translation vectors of the lattice are integers. Thus, we may write Eq. (38) in the form

$$\mathbf{k'} - \mathbf{k} = \mathbf{K}. \quad (39)$$

Vectors of type $\mathbf{K}$ possess the important property that the functions $e^{2\pi i\mathbf{K}\cdot\mathbf{r}}$ have the same periodicity as the lattice. Using (38), we may now write Eq. (36) as

$$\psi_{\mathbf{k}} = \chi_{\mathbf{k}}(x,y)e^{2\pi i\mathbf{k}\cdot\mathbf{r}} \quad (40)$$

where the function

$$\chi_{\mathbf{k}} = \frac{1}{\sqrt{S}}\left(1 + \sum_{\mathbf{K}} \frac{\dfrac{e^{2\pi i\mathbf{K}\cdot\mathbf{r}}}{S}\displaystyle\int_S e^{-2\pi i\mathbf{K}\cdot\mathbf{r'}} V dx' dy'}{E_{\mathbf{k}}^0 - E^0_{\mathbf{k}+\mathbf{K}}}\right)$$

has the periodicity of the lattice. As we have remarked previously, the form of (40) may be deduced rigorously on the basis of symmetry.

Turning now to the energy, we have in the second approximation

$$E_{\mathbf{k}} = \frac{h^2}{2m}\mathbf{k}^2 + \frac{1}{S}\int V dxdy + \frac{1}{S^2}\sum_{\mathbf{k'}} \frac{|\int e^{2\pi i(\mathbf{k}-\mathbf{k'})\cdot\mathbf{r}} V dxdy|^2}{E_{\mathbf{k}}^0 - E_{\mathbf{k'}}^0}. \quad (41)$$

The first two terms on the right do not affect the continuity of E as a function of $\mathbf{k}$. It may be seen, however, that the third term becomes infinite when $\mathbf{k}$ and $\mathbf{k'}$ satisfy Eq. (39) and the relation

$$E_{\mathbf{k}}^0 = E_{\mathbf{k'}}^0 \quad (42)$$

is fulfilled. It is precisely in this case that the perturbation method we have used needs revision. Instead of using the unperturbed wave functions (33), we should select those "proper linear combinations" of the degenerate functions that make the integral

$$V^0{}_{k'k} = \int \psi_{k'}{}^{0*} V(x,y) \psi_k{}^0 dx dy \tag{43}$$

vanish. Then the numerators of the troublesome terms vanish, and the degeneracy (42) is removed, so that there is a discontinuity in E at the points where Eqs. (39) and (42) are satisfied.

The origin of this discontinuity may be illustrated by approximate considerations of the following type. We shall assume that the values of $\mathbf{k}$ and $\mathbf{k}'$ for which degeneracy nearly occurs and for which Eq. (39) is satisfied occur in pairs. If $\psi_k{}^0$ and $\psi^0{}_{k+K}$ are the functions associated with these pairs, the proper linear combinations $\psi_k{}^{0'}$ and $\psi_{k'}{}^{0'}$ have the form

$$\psi_k{}^{0'} = a\psi_k{}^0 + b\psi^0{}_{k+K}, \Big\}$$
$$\psi_{k'}{}^{0'} = c\psi_k{}^0 + d\psi^0{}_{k+K}, \Big\}$$

in which a, b, c, and d must be chosen so that the off-diagonal components of the energy matrix

$$\begin{pmatrix} E_k{}^0 + V^0{}_{kk} & V^0{}_{k,k+K} \\ V^0{}_{k+K,k} & E^0{}_{K+k} + V^0{}_{k+K,k+K} \end{pmatrix}$$

vanish. We shall assume that the V^0 are continuous functions of $\mathbf{k}$. The equations corresponding to this condition are

$$\epsilon_k a + V^0{}_{k,k'} b = \lambda a, \Big\}$$
$$V^0{}_{k',k} a + \epsilon_{k'} b = \lambda b, \Big\} \tag{44}$$

in which we have set $\mathbf{k}' = \mathbf{k} + \mathbf{K}$ and

$$\epsilon_k = E_k{}^0 + V^0{}_{kk}.$$

The λ for which Eqs. (44) have solutions are the new unperturbed energies and satisfy the equation

$$\lambda = \frac{(\epsilon_k + \epsilon_{k'}) \pm \sqrt{(\epsilon_k - \epsilon_{k'})^2 + 4|V^0{}_{kk'}|^2}}{2}. \tag{45}$$

When $|V^0{}_{kk'}|$ is negligible in comparison with $\epsilon_k - \epsilon_{k'}$, the roots of this equation are ϵ_k and $\epsilon_{k'}$. When ϵ_k and $\epsilon_{k'}$ become nearly alike, however, the roots do not become closer than $2|V^0{}_{kk'}|$, which implies a discontinuity in the energy versus $\mathbf{k}$ curve. This is illustrated schematically in Fig. 7.

Now, Eq. (42) is satisfied whenever

$$|\mathbf{k}| = |\mathbf{k}'|, \tag{46}$$

because of Eq. (35).   Hence, (39) and (46) are the only equations that must be satisfied if there is to be a discontinuity.   It may be seen that these conditions determine a set of lines which satisfy the equations

$$\mathbf{k} \cdot \frac{\mathbf{K}}{2} = \frac{\mathbf{K}^2}{4} \tag{47}$$

where $\mathbf{K}$ may be any one of the $\mathbf{K}$-type vectors.   These lines are illustrated in Fig. 8 for the lattice under discussion.

Brillouin has pointed out that all similarly shaded regions in Fig. 8 may be pieced together in a unique way to form a square that is identical with the central zone.   When this is done, points in any two squares that overlap when the squares are placed on top of one another differ by a vector of type $\mathbf{K}$.   For this reason, similarly shaded regions are said to belong to the same zone.   It may be observed that this piecing process requires only that the sections be *translated* to the central zone by a vector of type $\mathbf{K}$; that is, the sections do not need to be folded or rotated. If it is recalled that $e^{2\pi i \mathbf{K} \cdot \mathbf{r}}$ has the translational periodicity of the lattice, it may be seen that the functions that go with a given zone may be written in the form

$$\chi_{\mathbf{k}} e^{2\pi i \mathbf{k} \cdot \mathbf{r}},$$

where $\mathbf{k}$ ranges over all points in the inner zone and $\chi_{\mathbf{k}}$ has translational symmetry.

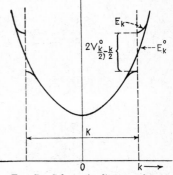

FIG. 7.—Schematic diagram showing the effect of the perturbing potential V on the energy levels.   The unperturbed energy curve $E_{\mathbf{k}}^0$ is continuous and the energy associated with the state $\mathbf{k}$ is equal to that of the state $-\mathbf{k}$.   The matrix components of V vanish unless the two states satisfy the relation $\mathbf{k}' = \mathbf{k} + \mathbf{K}$, where $\mathbf{K}$ is a principal lattice vector. If the matrix of V connecting these two states is diagonalized, the new $E_{\mathbf{k}}$ curve possesses a discontinuity at $\mathbf{k} = \pm \mathbf{K}/2$.

In addition, it may be seen from Eq. (45) that $E_{\mathbf{k}}$ is continuous when regarded only as a function of the values of $\mathbf{k}$ in the first zone.   In other words, we may represent $E_{\mathbf{k}}$ as a multiple-valued function of the points in the first zone instead of as a single-valued, discontinuous function of the points in the entire $\mathbf{k}$ plane.   The correspondence between the two types of representation may be made clear by cutting up each of the surfaces in the multiple-valued representation and by placing them over the outer zones.   The reduced-zone scheme, which uses only the inner zone, has the advantage that it requires knowing the geometrical form of only one zone.

It should be emphasized that the symmetry of the zone structure shown in Fig. 8 arises from the high degree of symmetry of $E_{\mathbf{k}}^0$, the unperturbed energy surface. The zone structure does not have the same form when this symmetry is absent. We shall illustrate other possibilities in case II.

The important zone relations

$$\mathbf{k'} - \mathbf{k} = \mathbf{K}, \tag{48a}$$
$$\mathbf{k}^2 = \mathbf{k'}^2, \tag{48b}$$

occur in the theory[1] of diffraction of X rays by crystals in which they are called Laue's equations. An X ray having wave-number vector $\mathbf{k}$ that impinges on a crystal can have its wave-number vector changed to $\mathbf{k'}$ by diffraction only if these equations are satisfied. Thus, we may

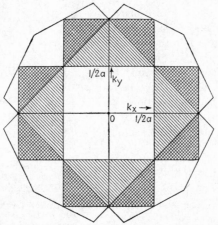

Fig. 8.—The first four Brillouin zones for a square, two-dimensional lattice. The similarly shaded areas may be translated into the first zone by vectors of type $\mathbf{K}$ and will then exactly cover this zone. The heavy dots indicate principal vectors in $\mathbf{k}$ space.

determine the X-ray diffraction pattern of a crystal from the Brillouin zone pattern, and vice versa. This identification of Eqs. (48a) and (48b) with Laue's equations shows that the occurrence of zones in the electronic problem is intimately connected with the wave properties of electrons.

2. *Case* II.—We shall now consider the Schrödinger equation when the potential has the form (32). If V_p is neglected, we know that the unperturbed equation may be separated into the ordinary equations

$$\left. \begin{array}{l} -\dfrac{\hbar^2}{2m}\dfrac{d^2\xi(x)}{dx^2} + (V_x - E_x)\xi(x) = 0, \\[2mm] -\dfrac{\hbar^2}{2m}\dfrac{d^2\eta(y)}{dy^2} + (V_y - E_y)\eta(y) = 0. \end{array} \right\} \tag{49}$$

[1] *Cf.* A. H. Compton and S. K. Allison, *X-Rays* (D. Van Nostrand Company, Inc., New York, 1934).

The complete unperturbed wave function is

$$\psi^0(x,y) = \xi(x)\eta(y), \tag{50}$$

and the unperturbed energy is

$$E^0 = E_x + E_y. \tag{51}$$

From parts a and b of this section, it follows that

$$\left.\begin{array}{l} \xi = \chi_{k_x}(x)e^{2\pi i k_x x}, \\ \eta = \chi_{k_y}(y)e^{2\pi i k_y y}, \end{array}\right\} \tag{52}$$

where k_x and k_y satisfy Eqs. (34). Hence,

$$\psi_\mathbf{k}^0 = \chi_\mathbf{k}^0 e^{2\pi i \mathbf{k} \cdot \mathbf{r}}, \tag{53}$$

just as in Eq. (40), where

$$\chi_\mathbf{k}^0 = \chi_{k_x}\chi_{k_y}.$$

Moreover, both E_x and E_y possess discontinuities at points that satisfy the relations

$$\begin{array}{l} k_x = \dfrac{n_x}{2a} \\[2mm] k_y = \dfrac{n_y}{2a} \end{array} \quad (n_x, \, n_y = 0, \, \pm 1, \, \pm 2, \, \cdots).$$

Since E^0 has the same discontinuities as E_x and E_y, we know that discontinuities appear in the unperturbed problem along the lines shown in Fig. 9, which illustrates the energy contours of the unperturbed energy function in a typical case.

When applying the perturbation theory, we may use the same simplifications that were used in simplifying Eq. (36). This procedure is permissible because χ in Eq. (53) has the same translational periodicity as V_p. The perturbed functions now are

$$\psi_\mathbf{k} = e^{2\pi i \mathbf{k} \cdot \mathbf{r}}\left(\chi_\mathbf{k}^0 + \sum_\mathbf{K} \chi^0_{\mathbf{k}+\mathbf{K}} e^{2\pi i \mathbf{K} \cdot \mathbf{r}} \frac{\int_S \psi_\mathbf{k}^0 V \psi^{*0}_{\mathbf{k}+\mathbf{K}} dx'dy'}{E_\mathbf{k}^0 - E^0_{\mathbf{k}+\mathbf{K}}}\right). \tag{54}$$

Hence, the form of Eq. (40) is maintained, and many of the remarks made in connection with $E_\mathbf{k}^0$ in the preceding case are valid again. Thus, $E_\mathbf{k}$, the perturbed energy, is discontinuous whenever

$$E_\mathbf{k}^0 = E^0_{\mathbf{k}+\mathbf{K}}. \tag{55}$$

This fact may be derived from an equation analogous to (41). One difference between this case and the last is that Eq. (55) is not necessarily satisfied when $|\mathbf{k}| = |\mathbf{k} + \mathbf{K}|$, as it was in Brillouin's problem, for the energy contours usually are not circles. Instead, they have lower symmetry which depends upon the symmetry of $V_x + V_y$. Let us assume for simplicity that the unperturbed potential has the symmetry of a square, which is the highest possible. We may then use one of the elementary results of the theory of symmetry,[1] which states that Eq. (55) is satisfied whenever $\psi_{\mathbf{k}}^0$ is sent into $\psi^0{}_{\mathbf{k}+\mathbf{K}}$ by the symmetry operators under which the Schrödinger equation is invariant. From the form of (54), it follows that this happens in the present highly symmetric case

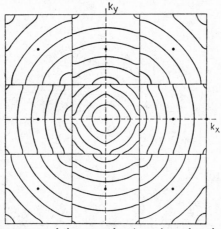

Fig. 9.—Energy contours of the zero-order eigenvalues when the potential has the form (32). Discontinuities occur only at the lines corresponding to the zones of one-dimensional lattices.

whenever $\mathbf{k}$ is sent into $\mathbf{k} + \mathbf{K}$ by one of the eight symmetry operations of a square, that is, whenever

$$\left.\begin{matrix} k_x = \pm(k_x + K_x) \\ k_y = \pm(k_y + K_y) \end{matrix}\right\} \quad \text{or} \quad \left\{\begin{matrix} k_x = \pm(k_y + K_y) \\ k_y = \pm(k_x + K_x) \end{matrix}\right. . \qquad (56)$$

The $\pm$ signs in the two sets of cases may be taken in arbitrary combinations. The conditions (56) include all the lines shown in Fig. 9 and all the additional straight lines shown in Fig. 10. Many of the lines that appear in Brillouin's pattern are absent, however, because (56) is not so stringent as the condition $|\mathbf{k}| = |\mathbf{k}'|$.

There is a large amount of accidental degeneracy in addition to the symmetrical degeneracy that occurs at the points satisfying (56). It is not difficult to see that this accidental degeneracy adds just enough zones to make up for the difference between the pattern of Fig. 10 and Bril-

[1] *Cf.* E. Wigner, *Gruppentheorie* (Vieweg, Braunschweig, Germany, 1931).

louin's pattern. The additional zone boundaries, however, are usually curved rather than straight lines. The amount by which these curves deviate from straight lines depends upon the amount by which the energy surface $E_k{}^0 = E_x + E_y$ deviates from a parabola of revolution. An additional curved zone is illustrated in Fig. 10 for the case corresponding to Fig. 9. The resulting pattern possesses square symmetry.

The sections having similar shading in Fig. 10 have the same total area as the central zone and fit exactly into the first zone if cut along the lines of the Brillouin pattern. Moreover, overlapping points of the two squares still correspond to values of **k** that differ by a vector of type **K**.

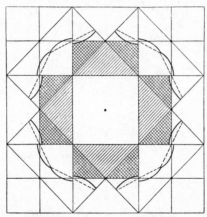

Fig. 10.—The lines of discontinuity in the case in which the non-diagonal matrix components of the perturbing term in (11) are taken into account. Only the straight lines are the same as for Brillouin's case, in which the zero-order eigenfunctions are free-electron functions (cf. Fig. 8). The additional straight lines of Brillouin's pattern are replaced by curved lines, of which one set is shown. The corresponding Brillouin zone is represented by dotted lines.

We shall present these statements without proof since they are easy to demonstrate. It follows that the differences between Brillouin's pattern and Fig. 10 have only superficial importance, for by properly cutting and translating the zones of Fig. 10 it is possible to piece them together to form the Brillouin pattern. As a result of this process, some of the starting eigenfunctions are relabeled, since a ψ previously associated with a point **k** may be associated with a point **k** + **K** after the redistribution. These facts show that knowledge of the central zone is sufficient to provide a complete description of zone structure, for we may always regard the energy surface as a multiple-valued function in this domain.

If the potential function (32) has no symmetry, aside from translational periodicity, we may anticipate that the symmetry of $E_k{}^0$ is much lower than it was previously. It is true that the separation into Eqs. (49)

proceeds as before and that the unperturbed energy (51) has the discontinuity pattern of Fig. 9, but $E_\mathbf{k}$ does not have equal values at points that are connected by Eqs. (56). The remaining symmetry, which is called the "natural symmetry of the Schrödinger equation," is expressed by the theorem that ψ and ψ^* are independent solutions of the Schrödinger equation having the same energy, if ψ is a complex solution and the potential function is real. Now, the conjugates of the solutions ξ_{k_x} and η_{k_y} of Eq. (49) are ξ_{-k_x} and η_{-k_y}, respectively. Thus, the symmetry of the one-dimensional energy curves relative to the origin of $\mathbf{k}$ space is due to the natural symmetry[1] of the Schrödinger equation. This symmetry

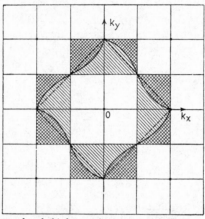

FIG. 11.—First, second and third zones in a case in which the potential function (32) has no symmetry other than translational symmetry. The only zones that are straight lines, as in Brillouin's scheme, are the vertical and horizontal lines of Fig. 9. All other zones are curved lines. Only one curved zone, namely the second, is shown in this figure. The fourth zone is curved as in Fig. 10.

carries over to the two-dimensional energy-level diagram and accounts for the straight lines of Fig. 9. Since all other degeneracy that occurs is accidental, the other zone boundaries usually are irregular curves, as in Fig. 11, which have the residual symmetry expressed by the equations

$$E_{k_x,k_y} = E_{-k_x,-k_y}, \quad \cdots \cdots$$

As before, the similarly shaded regions may be cut and pieced together to form the Brillouin pattern.

3. *The general case.*—It follows from the discussion of the two preceding cases that Brillouin's zone pattern usually does not occur unless a perturbation method in which the unperturbed wave functions are free-electron functions is used. The only unambiguous way in which to

[1] *Cf.* E. WIGNER, *Gött. Nachr.*, 546 (1932).

arrive at Brillouin's pattern in a general case is to use the reduced-zone scheme, in which E_k is determined as a many-valued function in the first Brillouin zone, and to cut and distribute the energy surfaces among the outer zones.   The distribution must take place in such a way that an energy value associated with the point **k** in the inner zone goes to a point **k** + **K** in an outer one.   If the single-valued representation is obtained by another method, the zone structure usually depends upon the method employed.   Certain zone boundaries, which may be established by an investigation of symmetry degeneracy, always appear in cases of high symmetry.   Other boundaries are not unique and may be altered by choice.   There will be certain fixed points through which the zones must pass, however, even in the case of lowest symmetry.   These points are determined by the relation

$$E_k = E_{-k}, \tag{57}$$

which expresses the natural symmetry of the Schrödinger equation. Thus, the zone lines must pass through the points

$$\mathbf{k} = \frac{\mathbf{K}}{2}. \tag{58}$$

The natural advantages of the Brillouin zone scheme are (1) that it determines zones by the Laue conditions, which have significance in the theory of X-ray diffraction, and (2) that it preserves a correspondence between the **k** vectors for perfectly free electrons and those in the lattice. If the wave functions for electrons in the crystal are labeled according to Brillouin's scheme and if the crystalline potential field is adiabatically decreased to zero, the wave-number values will be preserved and will agree with those for the resulting free electrons.

In spite of these advantages, the Brillouin scheme is not always the simplest one to use, particularly when one is dealing with cases in which primitive translations are not equal or are not orthogonal to one another. Let us consider a case in which the primitive translations are orthogonal but not equal.   The lines defined by Eq. (58), which are shown in Fig. 12, do not form zones that are as simple as those of Fig. 8.   It is easy, however, to determine a set that is almost as simple.   Such a set, for example, is shown in Fig. 13.   These lines are parallel to the fundamental **K** vectors instead of orthogonal to them, as Brillouin's zone boundaries are. The relative simplicity of this scheme is illustrated in an even more striking manner in an oblique case.   It is still true, of course, that both methods of construction may be brought into coincidence by appropriately cutting and rearranging zones.

Case d.   The Three-dimensional Schrödinger Equation.—The three-dimensional case is a straightforward generalization of the two-dimensional one.   All the remarks in case c may be extended at once to cover the case in which **k** is a three-dimensional vector.   If we introduce the Born-von Kármán boundary conditions, **k** takes on the discrete, though dense, values defined by the equations

$$N_i \mathbf{k} \cdot \boldsymbol{\tau}_i = n_i \qquad (n_i = 0, \pm 1, \pm 2, \pm \cdots), \qquad (59)$$

which are identical with Eqs. (8), Sec. 22, Chap. III.[1]   Here the n_i are arbitrary integers, the $\boldsymbol{\tau}_i (i = 1, 2, 3)$ are the three primitive translations

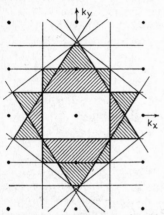

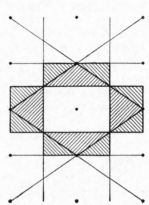

Fig. 12.—Brillouin zone scheme for a two-dimensional rectangular lattice. The zone boundaries, which are defined by Eq. (27), are more complex than those of Fig. 8 because they are orthogonal to the **K** vectors.

Fig. 13.—Alternative to Fig. 12 in which the zone boundaries are drawn parallel to the **K** vectors. This simplifies the pattern.

of the lattice, and the N_i are the number of cells that extend along an axis of the crystal parallel to $\boldsymbol{\tau}_i$.

The three-dimensional Brillouin zones are determined by planes that satisfy the equations

$$\mathbf{k} \cdot \frac{\mathbf{K}}{2} = \frac{\mathbf{K}^2}{4}, \qquad (60)$$

---

[1] It may be noted at this point for future reference that the density of points in **k** space, as determined from the solutions

$$\mathbf{k} = \frac{n_1}{N_1} \frac{\tau_2 \times \tau_3}{|\tau_1 \tau_2 \tau_3|} + \frac{n_2}{N_2} \frac{\tau_3 \times \tau_1}{|\tau_1 \tau_2 \tau_3|} + \frac{n_3}{N_3} \frac{\tau_1 \times \tau_2}{|\tau_1 \tau_2 \tau_3|}$$

of Eq. (34), is V, where V is the volume of crystal.   This may be shown by computing the reciprocal of the volume of the unit cell of the reciprocal lattice.   If spin degeneracy is included, the density of states is $2V$.

which are analogous to (48).   The vectors $\mathbf{K}$ are defined by the relations

$$\mathbf{K} \cdot \boldsymbol{\tau}_i = l_i \tag{61}$$

where the l are arbitrary integers.   These equations, which generalize (38), are special cases of (59) and their solutions [*cf.* Eq. (8*a*), Sec. 22, Chap. III] are:

$$\mathbf{K} = l_1 \frac{\boldsymbol{\tau}_2 \times \boldsymbol{\tau}_3}{|\boldsymbol{\tau}_1 \boldsymbol{\tau}_2 \boldsymbol{\tau}_3|} + l_2 \frac{\boldsymbol{\tau}_3 \times \boldsymbol{\tau}_1}{|\boldsymbol{\tau}_1 \boldsymbol{\tau}_2 \boldsymbol{\tau}_3|} + l_3 \frac{\boldsymbol{\tau}_1 \times \boldsymbol{\tau}_2}{|\boldsymbol{\tau}_1 \boldsymbol{\tau}_2 \boldsymbol{\tau}_3|}. \tag{62}$$

The three-dimensional eigenfunctions always have the form

$$\psi_{\mathbf{k}} = \chi_{\mathbf{k}} e^{2\pi i \mathbf{k} \cdot \mathbf{r}}.$$

where $\chi_{\mathbf{k}}$ has the translational periodicity of the lattice.   This fact may be proved either by perturbation theory, as was done in case *c*, or by use of group theory.

We may obtain zone patterns that are different from Brillouin's by basing the determination of zones on some scheme that does not start from free-electron waves.   Some of the simpler zones determined by other methods are the same as those obtained by use of (61) in cases of high symmetry.   As before, the differences in zone pattern are superficial, since any pattern may be made to coincide with the Brillouin pattern by appropriately rearranging points in $\mathbf{k}$ space.   We shall discuss further details of the three-dimensional case in the next section.

Although the discussion of *c* and *d* has been restricted to the Schrödinger equation, the conclusions are valid for the solutions of any eigenvalue problem in which there is translational symmetry.   The problem of determining the normal modes of vibration of a crystal is a case of this type.   We have seen in Chap. III that zone theory plays an important role in the classification of solutions of this problem.

62. Survey of Rules and Principles Concerning Three-dimensional Zones.—In this section, we shall amplify the remarks of part *c* of the previous section by tabulating rules for constructing three-dimensional zones.   Some of these rules were discussed in Sec. 22; others may be derived by applying the principles introduced there.   The rules are as follows:

a. All zones have equal volume in wave-number space.   This rule follows from the fact that all zones may be mapped in one another.

b. It is possible to neglect all zones except the first if[1] $\epsilon(\mathbf{k})$ is regarded as a multiple-valued function in this zone.   This reduced-zone scheme is most useful when one is determining surfaces by a direct solution of

---

[1] We shall usually designate the eigenvalues of the three-dimensional periodic one-electron functions by $\epsilon(\mathbf{k})$.

Fock's equations, for then one may derive the energy bands without becoming involved in many of the geometrical complexities of zone structure.

c. Each zone contains $2N$ states, where N is the number of cells in the lattice. The factor 2 arises from the two possible orientations of spin, and the factor N from the fact that N values of $\mathbf{k}$ are associated with each zone. It follows that the maximum number of electrons that may occupy any zone is $2N$.

d. There is a close correspondence between the laws for determining Brillouin zones and those for determining X-ray diffraction. This correspondence is indicated in part by the fact that the equations of Brillouin zone boundaries are the same as Laue's equations for X-ray diffraction; however, the correlation may be much closer. Mott and Jones[1] have pointed out, for example, that in any monatomic solid in which the electronic potential may be approximated by a sum $\sum_{S} V_S$ of contributions V_S from each atom, the energy discontinuities are small for values of wave numbers for which the X-ray structure factor is small and are large when the structure factor is large. This type of correlation usually does not occur in polyatomic solids because the scattering powers of atoms for low-energy electrons and for X rays are not necessarily proportional for different atoms.

e. Lattices that have the same type of translational symmetry have equivalent zone patterns since zone structure is determined by the $\mathbf{K}$ vectors and these are determined by the primitive translation vectors. The magnitude of the gaps, which is determined by the distribution of potential in the unit cell, however, may be entirely different for crystals that have the same zone structure. Hence, the zone boundaries at which the largest gaps occur may be completely different in translationally similar crystals. This fact is analogous to the fact that the intensities of X-ray diffraction spots may be very different in crystals that have the same translational symmetry.

f. If Brillouin's zones are used to describe the states of all electrons in the solid, it may be convenient to regard the K-shell electrons as filling the first set, the L-shell electrons the next, etc. The valence electrons then occupy zones at the outer fringe of the filled region. It is usually more convenient, however, to disregard inner-shell electrons when one is discussing properties of the solid that do not involve them explicitly. The valence electrons then may be regarded as occupying the central zones.

---

[1] N. F. Mott and H. Jones, *The Theory of Metals and Alloys*, Chap. V (Oxford University Press, New York, 1936).

g. Two neighboring zones may have overlapping energy levels even though the energy associated with the outer zone is higher than that associated with the inner at any particular point of the boundary. An example of a case in which overlapping occurs is shown in Fig. 14.

h. All occupied zones cannot be completely filled in a substance that has an odd number of electrons per unit cell. Hence, occupied and unoccupied levels are immediately adjacent in a substance of this type. Since this is the condition for metallic conductivity, these substances are

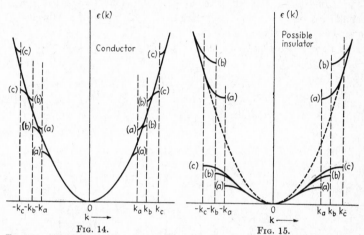

Fig. 14.   Fig. 15.

Fig. 14.—$\epsilon(\mathbf{k})$ curves for the case in which two zones overlap. This figure gives superposition of energy curves for values of **k** on lines that pass through the origin of **k** space and extend in three prominent directions. The intercepts of these lines with the zone boundary occur, respectively, at $\mathbf{k}_a$, $\mathbf{k}_b$, and $\mathbf{k}_c$. The highest point at $\mathbf{k}_a$, which belongs to the second zone, lies below the lowest point of $\mathbf{k}_b$ which belongs to the first zone, etc. (compare with Fig. 15).

Fig. 15.—$\epsilon(\mathbf{k})$ curves for the case in which two zones do not overlap. This figure is the same as Fig. 14 except that the gaps are so large that the uppermost levels of the lower band are always below the lowest levels of the upper band. A substance having $\epsilon(\mathbf{k})$ curves of this type is an insulator if there is an even number of electrons.

metals (*cf.* Sec. 60). If the substance has an even number of electrons per unit cell, the type of conductivity depends upon the nature of the gaps at the boundary of the filled region. Suppose that there are $2m$ electrons per unit cell. The mth zone is not completely filled if the mth and the $m + 1$st zones overlap, for then some electrons prefer the lowest levels of the $m + 1$st zone to the highest levels of the mth zone. Thus, the substance should be a metal in this case. On the other hand, it should be an insulator if they do not overlap. Cases corresponding to both these types are shown in Figs. 14 and 15.

We shall not discuss those properties of zones and energy surfaces which may be derived by application of group theory, for to do so would carry us too far afield. The development of this topic may be found in the references of footnote 1, page 275.

63. Examples of Zone Structure.—For illustrative purposes, we shall describe in this section the simplest zones of several lattices. The precise description of zones is important, at present, only for a few simple crystals, such as the monovalent and divalent metals and the alkali halides. More complex solids can be handled only in rough approximation. Thus, for γ brass all that is important is to know the shape of one or two zones near the limit of the filled region and whether the gaps at the boundaries of these zones are large or small. We shall discuss these cases as we need them in later sections.

a. Face-centered Cubic Lattice.—The primitive translations of the face-centered cubic lattice may be chosen as

$$\tau_1 = \begin{pmatrix} a \\ a \\ 0 \end{pmatrix}, \qquad \tau_2 = \begin{pmatrix} a \\ 0 \\ a \end{pmatrix}, \qquad \tau_3 = \begin{pmatrix} 0 \\ a \\ a \end{pmatrix}. \qquad (1)$$

The components of these vectors are expressed in Cartesian coordinates. Solving the equations

$$\mathbf{k} \cdot N_i \tau_i = n_i \qquad (i = 1, 2, 3), \qquad (2)$$

which define the values of $\mathbf{k}$, we find

$$\mathbf{k} = \frac{n_1}{2N_1 a} \begin{pmatrix} -1 \\ -1 \\ 1 \end{pmatrix} + \frac{n_2}{2N_2 a} \begin{pmatrix} -1 \\ 1 \\ -1 \end{pmatrix} + \frac{n_3}{2N_3 a} \begin{pmatrix} 1 \\ -1 \\ -1 \end{pmatrix}. \qquad (3)$$

Hence, the reciprocal lattice is body-centered.

The $\mathbf{K}$ vectors are given by those values of (3) for which n_1, n_2, and n_3 are integer multiples of N_1, N_2, and N_3, respectively. The first three sets of these vectors are:

$$\frac{1}{2a} \begin{pmatrix} -1 \\ -1 \\ 1 \end{pmatrix}, \qquad \frac{1}{2a} \begin{pmatrix} -1 \\ 1 \\ -1 \end{pmatrix}, \qquad \frac{1}{2a} \begin{pmatrix} 1 \\ -1 \\ -1 \end{pmatrix}, \qquad \frac{1}{2a} \begin{pmatrix} 1 \\ 1 \\ 1 \end{pmatrix}. \qquad (I)$$

$$\frac{1}{2a} \begin{pmatrix} 2 \\ 0 \\ 0 \end{pmatrix}, \qquad \frac{1}{2a} \begin{pmatrix} 0 \\ 2 \\ 0 \end{pmatrix}, \qquad \frac{1}{2a} \begin{pmatrix} 0 \\ 0 \\ 2 \end{pmatrix}. \qquad (II)$$

$$\left. \begin{array}{ccc} \frac{1}{2a} \begin{pmatrix} 2 \\ 2 \\ 0 \end{pmatrix}, & \frac{1}{2a} \begin{pmatrix} 0 \\ 2 \\ 2 \end{pmatrix}, & \frac{1}{2a} \begin{pmatrix} 2 \\ 0 \\ 2 \end{pmatrix}. \\[2em] \frac{1}{2a} \begin{pmatrix} 2 \\ -2 \\ 0 \end{pmatrix}, & \frac{1}{2a} \begin{pmatrix} 0 \\ 2 \\ -2 \end{pmatrix}, & \frac{1}{2a} \begin{pmatrix} 2 \\ 0 \\ -2 \end{pmatrix}. \end{array} \right\} \qquad (III)$$

The first two zones are shown in Figs. 16a and b and are bounded by planes orthogonal to (I) and (II). The higher zones are more complicated and involve other **K** vectors. It may be seen that the illustrated zones are uniquely specified by symmetry conditions.

   b. Body-centered Lattice.—The primitive translations in this case are

$$\tau_1 = \begin{pmatrix} a \\ a \\ a \end{pmatrix}, \qquad \tau_2 = \begin{pmatrix} -a \\ a \\ a \end{pmatrix}, \qquad \tau_3 = \begin{pmatrix} -a \\ -a \\ a \end{pmatrix}, \tag{4}$$

whence

$$\mathbf{k}_1 = \frac{n_1}{2N_1 a}\begin{pmatrix} 1 \\ 0 \\ -1 \end{pmatrix} + \frac{n_2}{2N_2 a}\begin{pmatrix} 1 \\ 1 \\ 0 \end{pmatrix} + \frac{n_3}{2N_3 a}\begin{pmatrix} 0 \\ -1 \\ 1 \end{pmatrix}. \tag{5}$$

In other words, the inverse lattice in this case is face-centered.

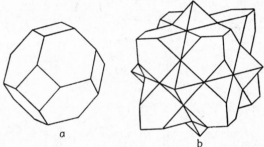

$$\begin{array}{cc} a & \\ & b \end{array}$$

Fig. 16.—The first and second zones for a face-centered cubic lattice. The first has half the volume of the cube that is determined by extending the six square faces. The second has the same volume as this cube.

   The first two sets of **K** vectors are:

$$\frac{1}{2a}\begin{pmatrix} 1 \\ 1 \\ 0 \end{pmatrix}, \quad \frac{1}{2a}\begin{pmatrix} 1 \\ -1 \\ 0 \end{pmatrix}, \quad \frac{1}{2a}\begin{pmatrix} 1 \\ 0 \\ 1 \end{pmatrix}, \quad \frac{1}{2a}\begin{pmatrix} 1 \\ 0 \\ -1 \end{pmatrix}, \quad \frac{1}{2a}\begin{pmatrix} 0 \\ 1 \\ 1 \end{pmatrix}, \quad \frac{1}{2a}\begin{pmatrix} 0 \\ 1 \\ -1 \end{pmatrix} \tag{I}$$

$$\frac{1}{2a}\begin{pmatrix} 2 \\ 0 \\ 0 \end{pmatrix}, \quad \frac{1}{2a}\begin{pmatrix} 0 \\ 2 \\ 0 \end{pmatrix}, \quad \frac{1}{2a}\begin{pmatrix} 0 \\ 0 \\ 2a \end{pmatrix}. \tag{II}$$

The first two zones appear in Fig. 17.

   c. Close-packed Hexagonal Lattice.—The primitive translations of the close-packed hexagonal lattice are

$$\tau_1 = \begin{pmatrix} a \\ 0 \\ 0 \end{pmatrix}, \qquad \tau_2 = \begin{pmatrix} 0 \\ b/2 \\ \sqrt{3}b/2 \end{pmatrix}, \qquad \tau_3 = \begin{pmatrix} 0 \\ -b/2 \\ \sqrt{3}b/2 \end{pmatrix}, \tag{6}$$

for which

$$\mathbf{k}_1 = \frac{n_1}{N_1 a}\begin{pmatrix}1\\0\\0\end{pmatrix} + \frac{2n_2}{N_2 b\sqrt{3}}\begin{pmatrix}0\\\sqrt{3}/2\\1/2\end{pmatrix} + \frac{2n_3}{N_3 b\sqrt{3}}\begin{pmatrix}0\\-\sqrt{3}/2\\1/2\end{pmatrix}. \qquad (7)$$

Thus, this lattice is its own reciprocal. The first zone is the hexagonal

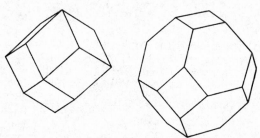

FIG. 17.—The first two zones for the body-centered cubic lattice. The surfaces of the first are normal to **K** vectors (I) of the text, whereas the surfaces of the second are normal to **K** vectors (II) and **K** vectors that lie in the (111) direction.

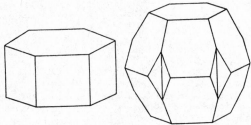

FIG. 18.—The first two Brillouin-type zones for a close-packed hexagonal crystal. The second zone is not uniquely defined by symmetry and may be drawn in many different ways.

prism shown in Fig. 18, which is determined by the following two sets of **K** vectors:

$$\frac{1}{a}\begin{pmatrix}1\\0\\0\end{pmatrix}. \qquad (I)$$

$$\frac{2}{\sqrt{3}b}\begin{pmatrix}0\\\sqrt{3}/2\\1/2\end{pmatrix}, \qquad \frac{2}{\sqrt{3}b}\begin{pmatrix}0\\-\sqrt{3}/2\\1/2\end{pmatrix}, \qquad \frac{2}{\sqrt{3}b}\begin{pmatrix}0\\0\\1\end{pmatrix}. \qquad (II)$$

The form of the second zone, which is determined by the sets (I) and (II) and the set

$$\frac{2}{\sqrt{3b}}\begin{pmatrix}\sqrt{3}b/2a\\\sqrt{3}/2\\1/2\end{pmatrix}, \quad \frac{2}{\sqrt{3b}}\begin{pmatrix}\sqrt{3}b/2a\\-\sqrt{3}/2\\-1/2\end{pmatrix}, \quad \frac{2}{\sqrt{3b}}\begin{pmatrix}\sqrt{3}b/2a\\-\sqrt{3}/2\\1/2\end{pmatrix},$$

$$\frac{2}{\sqrt{3b}}\begin{pmatrix}\sqrt{3}b/2a\\\sqrt{3}/2\\-1/2\end{pmatrix}, \quad \frac{2}{\sqrt{3b}}\begin{pmatrix}\sqrt{3}b/2a\\0\\1\end{pmatrix}, \quad \frac{2}{\sqrt{3b}}\begin{pmatrix}\sqrt{3}b/2a\\0\\-1\end{pmatrix}, \qquad \text{(III)}$$

depends upon the axial ratio $\sqrt{3}b/2a$. The form for an ideally close-packed lattice is illustrated in Fig. 18.

64. Cases of Coalescence of the Heitler-London and Bloch Schemes*.

We saw in Sec. 58, Chap. VII, that the Heitler-London and Hund-Mulliken schemes are equivalent when applied to the state of a molecule whose constituent atoms or ions have only closed-shell configurations. In this case, the total wave function is a single determinant in both the one-electron schemes, and it is possible to transform one determinant into the other by adding to each row an appropriate linear combination of other rows. A similar theorem is valid for solids. The Heitler-London and Bloch schemes are identical whenever the Heitler-London scheme leads to completely filled shells. To prove this, all that is necessary is to show that a complete zone of N Bloch functions may be constructed by taking linear combinations of the N Heitler-London functions associated with translationally equivalent atoms in the N unit cells of the lattice, for each of these Heitler-London functions appears once with each spin in the determinantal eigenfunction.

Let $\psi(\mathbf{r} - \mathbf{r}(n))$ be the closed-shell function that is centered about the point $\mathbf{r}(n)$ in the nth unit cell. The functions formed from these by taking the linear combinations

$$\psi_{\mathbf{k}} = \sum_n e^{-2\pi i \mathbf{k} \cdot \mathbf{r}(n)}\psi(\mathbf{r} - \mathbf{r}(n)), \qquad (1)$$

where $\mathbf{r}(n)$ is summed over all unit cells in the lattice and $\mathbf{k}$ is a vector in the reciprocal lattice of the crystal, are Bloch functions. This may be seen[1] by changing $\mathbf{r}$ to $\mathbf{r} - \boldsymbol{\tau}$ where $\boldsymbol{\tau}$ is any primitive translation of the lattice, for then (1) becomes

$$\sum_{\mathbf{r}(n)+\tau} e^{2\pi i \mathbf{k} \cdot \boldsymbol{\tau}} e^{-2\pi i \mathbf{k} \cdot [\mathbf{r}(n)+\tau]}\psi(\mathbf{r} - (\mathbf{r}(n) + \boldsymbol{\tau})) = e^{2\pi i \mathbf{k} \cdot \boldsymbol{\tau}}\psi_{\mathbf{k}},$$

since in an ideal unbounded lattice the summation over $\mathbf{r}(n) + \boldsymbol{\tau}$ is equivalent to a summation over $\mathbf{r}(n)$. Thus, the function in (1) is multiplied by a factor $e^{2\pi i \mathbf{k} \cdot \boldsymbol{\tau}}$ when the crystal is translated by an amount

---

[1] F. Bloch, *Z. Physik*, **52**, 555 (1928), pointed out first that the sum (1) is the combination of atomic functions which satisfies the periodic boundary conditions.

τ. N independent functions of type (1) may be constructed from th $\psi(\mathbf{r} - \mathbf{r}(n))$, and this number is just sufficient to fill a zone.

The converse would be a theorem stating that the Heitler-London an Bloch schemes are equivalent whenever the Bloch scheme leads to a se of completely filled zones. It does not seem to be possible to prove thi theorem generally. It is probably valid, however, in many special cases Consider the N functions

$$\psi(\mathbf{r} - \mathbf{r}(n)) = \sum_{\mathbf{k}} \chi_{\mathbf{k}} e^{-2\pi i \mathbf{k} \cdot [\mathbf{r} - \mathbf{r}(n)]} \tag{2}$$

where $\chi_{\mathbf{k}} e^{2\pi i \mathbf{k} \cdot \mathbf{r}}$ is a Bloch function, $\mathbf{k}$ is summed over all the N values in single zone, and $\mathbf{r}(n)$ ranges over N translationally equivalent position in the N cells of the lattice. The function (2) is localized in the nth unit cell if $\chi_{\mathbf{k}}$ varies sufficiently slowly with $\mathbf{k}$; hence, in this case, it is function of the Heitler-London type. Suppose, for simplicity, that χ is independent of $\mathbf{k}$. Then the function (2) degenerates to

$$\chi(\mathbf{r}) \sum_{\mathbf{k}} e^{-2\pi i \mathbf{k} \cdot [\mathbf{r} - \mathbf{r}(n)]}. \tag{3}$$

The phases of the exponent in the summands of this function rang roughly between the values $\pm 2\pi |\mathbf{r} - \mathbf{r}(n)||\mathbf{k}(\text{max.})|$, where $\mathbf{k}(\text{max.})$ is th distance from $k = 0$ to the farthest point in the zone. Hence, the sur in (3) has its maximum value when $\mathbf{r}$ is in the nth unit cell and is ver small when $|\mathbf{r} - \mathbf{r}(n)|$ is large. Although this property is affected χ depends upon $\mathbf{k}$, we may expect (2) to be localized as long as th dependence is not strong.

Thus, in some cases, we may transform a determinantal wave functio that is based upon a filled zone of Bloch functions into a determinant c Heitler-London functions that are localized about any set of points in th unit cell we choose. It may happen, however, that the Heitler-Londo wave functions which are obtained in this way do not have desirabl symmetry characteristics and that a better set can be found.

Before leaving this topic, we shall discuss the connection betwee the energy parameters in Fock's equations for the Bloch scheme and th Heitler-London scheme when the two are equivalent. For the Bloc scheme, the equations may be chosen in the form

$$H\psi_{\mathbf{k}} = \epsilon(\mathbf{k})\psi_{\mathbf{k}} \tag{4}$$

since the $\psi_{\mathbf{k}}$ are automatically orthogonal (*cf.* Sec. 51). The corresponc ing equations for the Heitler-London functions are

$$H\psi_r = \epsilon_0 \psi_n + \sum_{m}{}' \alpha_{nm} \psi_m \tag{5}$$

in which

$$\psi_n = \psi(\mathbf{r} - \mathbf{r}(n))$$

is localized about the atom at the position $\mathbf{r}(n)$ and the α_{nm} are normalization parameters that must be introduced because the ψ_n form a highly degenerate set of functions not automatically orthogonal.

The $\psi_\mathbf{k}$ and ψ_n are connected by the equations

$$\psi_\mathbf{k} = \frac{1}{\sqrt{N}} \sum_m e^{2\pi i \mathbf{k} \cdot \mathbf{r}(m)} \psi_m. \qquad (6)$$

It may be verified that the operators H in Eqs. (4) and (5) are the same because of Eqs. (6). The relation between $\epsilon(\mathbf{k})$ and ϵ_0 may be found by multiplying Eq. (5) by $e^{2\pi i \mathbf{k} \cdot \mathbf{r}(n)}$, summing over n, and subtracting Eq. (4). The result is

$$(\epsilon_0 - \epsilon(\mathbf{k}))\psi_\mathbf{k} + \frac{1}{\sqrt{N}} \sum_{n,m}{}' e^{2\pi i \mathbf{k} \cdot \mathbf{r}(n)} \alpha_{nm}\psi_m = 0. \qquad (7)$$

If we multiply this by $\psi_\mathbf{k}^*$ and integrate, we find

$$\epsilon(\mathbf{k}) = \epsilon_0 + \frac{1}{N} \sum_{n,m}{}' \alpha_{nm} e^{2\pi i \mathbf{k} \cdot [\mathbf{r}(n) - \mathbf{r}(m)]}. \qquad (8)$$

We shall discuss relationships of this type in more detail in the next section.

65. Approximate Bloch Functions for the Case of Narrow Bands*.—

It is possible to use Eq. (6) of the preceding section to obtain Bloch type functions from atomic functions even when the Heitler-London and Bloch schemes are not equivalent; indeed, these functions are sometimes very useful for semiquantitative discussions. The relationship between the energies in this case may be derived in the following way: We shall assume, for simplicity, that we have a monatomic lattice of atoms that possess one valence electron each and that the Schrödinger equation for the wave function $\psi(\mathbf{r} - \mathbf{r}(n))$ of the electron in the nth atom is

$$\left\{ -\frac{\hbar^2}{2m} \Delta + V(\mathbf{r} - \mathbf{r}(n)) \right\} \psi(\mathbf{r} - \mathbf{r}(n)) = \epsilon \psi(\mathbf{r} - \mathbf{r}(n)) \qquad (1)$$

when the atom is free. The function $V(\mathbf{r} - \mathbf{r}(n))$ is the electronic potential, which we may expect to be negative. When the atoms are brought together, the potential at the nth atom is changed by the addition of the term

$$\sum_m{}' V(\mathbf{r} - \mathbf{r}(m))$$

in which m is summed over all atoms except the nth. In order to determine[1] the energy of the Bloch wave function

$$\psi_k = \frac{1}{\sqrt{N}} \sum_n e^{2\pi i k \cdot r(n)} \psi(r - r(n)),$$

it is necessary to evaluate the integral

$$\epsilon(k) = \int \psi_k^* \left\{ -\frac{\hbar^2}{2m}\Delta + \sum_m V(r - r(m)) \right\} \psi_k d\tau$$

$$= \frac{1}{N} \Bigg(\sum_{n,m} e^{2\pi i k \cdot [r(n) - r(m)]} \int \psi^*(r - r(m)) \left\{ -\frac{\hbar^2}{2m}\Delta + \right.$$

$$\left. V(r - r(n)) \right\} \psi(r - r(n)) d\tau +$$

$$\sum_{n,m} e^{2\pi i k \cdot [r(n) - r(m)]} \int \psi^*(r - r(m)) \sum_{l \neq n} V(r - r(l)) \psi(r - r(n)) d\tau. \Bigg)$$

We shall assume that the atomic functions satisfy the relations

$$\int \psi^*(r - r(m)) \psi(r - r(n)) d\tau = \delta_{m,n}$$

and that

$$\int \psi^*(r - r(m)) V(r - r(l)) \psi(r - r(n)) d\tau$$

is zero unless one of the conditions

$$n = l, \qquad n = m, \qquad l = m$$

is satisfied. Equation (3) may then be simplified to

$$\epsilon(k) = \epsilon' + \frac{1}{N} \sum_{n,m} \alpha'_{nm} e^{2\pi i k \cdot [r(n) - r(m)]}$$

where

$$\epsilon' = \epsilon + \sum_l \int |\psi(r)|^2 V(r - r(l)) d\tau$$

and

$$\alpha'_{nm} = \int \psi^*(r - r(m)) V(r - r(n)) \psi(r - r(n)) d\tau.$$

Both α'_{nm} in this equation and α_{nm} in Eq. (8) of the preceding section are larger when the atoms overlap a great deal than when they are far apart; moreover, in the limit as the atoms become free, (4) approaches the energy ϵ' of the electron in the free atom for all values of k. We may

[1] *Ibid.*

conclude that in the solid there is a band of Bloch levels corresponding to one zone of $\mathbf{k}$ space for each electronic state of the free atom (*cf.* Fig. 19).

Since the lattice has translational symmetry, it follows that $\alpha'_{n+l,n}$ is independent of n. Thus, Eq. (4) may be written in the form

$$\epsilon(\mathbf{k}) = \epsilon' + \sum_l \alpha'_l e^{2\pi i \mathbf{k} \cdot \mathbf{r}(l)} \tag{7}$$

in which

$$\alpha'_l = \alpha'_{n+l,n} \tag{8}$$

and the summation extends over all values of l. Equation (8) of the preceding section evidently can be placed in the same form. Now, atomic functions may always be chosen as real if there is no magnetic field present. Thus, the α'_l may be regarded as real, and Eq. (7) may be placed in the form

$$\epsilon(\mathbf{k}) = \epsilon' + \sum_l \alpha'_l \cos 2\pi \mathbf{k} \cdot \mathbf{r}(l)$$

$$= \epsilon' + \sum_l \alpha' - 2\sum_l \alpha'_l \sin^2 \pi \mathbf{k} \cdot \mathbf{r}(l). \tag{9}$$

Only the α_l corresponding to very near neighbors are important when the bands are comparatively narrow. Hence, only the first few terms in the series (8) need be retained in this case. The α corresponding to immediate neighbors are equal in the three cubic lattices and in the ideally close-packed hexagonal crystal when the atomic functions are s type. Hence, in this case the nearest-neighbor terms in (9) may be written as

$$\epsilon(\mathbf{k}) = \epsilon'' - 2\alpha \sum_{\mathbf{r}_n} \sin^2 \pi \mathbf{k} \cdot \mathbf{r}_n \tag{10}$$

where $\mathbf{r}_n$ ranges over the vectors connecting an atom with its nearest neighbors and

$$\epsilon'' = \epsilon' + z\alpha$$

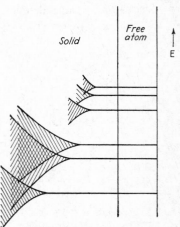

FIG. 19.—The one-electron levels of the free atom split into bands when the atom enters the solid.

in which z is the number of nearest neighbors. Now, if $V(\mathbf{r} - \mathbf{r}(n))$ in the integral (6) is negative, as the electronic potential of an atom should be, and if the portions of the ψ that overlap have the same sign, as we may expect for well-separated s functions, the α in (9) are negative. Thus, we may conclude that before appreciable overlapping occurs the $\epsilon(\mathbf{k})$ curves for bands that arise from s electrons increase with increasing values of $|\mathbf{k}|$ in any direction starting from the origin of $\mathbf{k}$ space (*cf.* Fig. 20).

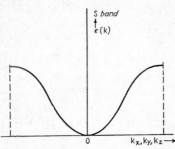

FIG. 20.—The $\epsilon(\mathbf{k})$ curve for an s electron band is concave upwards for all directions in $\mathbf{k}$ space.

The same conclusion cannot be drawn in cases in which the atomic functions are not s type. Suppose, for example, that the atomic ψ are p functions of the form

$$\frac{x}{r}f(r). \tag{11}$$

Then the overlapping portions of the ψ of atoms that have different x coordinates have opposite signs, and their α are positive (*cf.* Fig. 21). Moreover, although the overlapping portions for atoms that lie in the same x plane have the same sign, the product is small because (11) has a node in this plane. Thus, in this case the function $\epsilon(\mathbf{k})$ should decrease with increasing values of $\mathbf{k}$ in the x direction of $\mathbf{k}$ space and should increase in the y and z directions (*cf.* Fig. 22). Cases in which the functions are higher than s or p type must be considered individually.

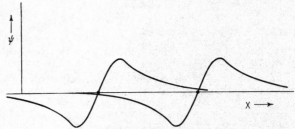

FIG. 21.—The overlapping parts of p functions of type (11) that have different x coordinates have opposite sign.

It is readily found that, in monatomic crystals having the three simplest cubic structures, Eq. (10) reduces to

$$\epsilon(\mathbf{k}) \cong \epsilon'' - 4\pi^2\alpha a^2 \mathbf{k}^2 \tag{12}$$

in the neighborhood of the origin of $\mathbf{k}$ space, where a is the edge length of the fundamental cube. This may be made identical with the expression for effectively free electrons, namely,

$$\epsilon(\mathbf{k}) = \frac{h^2}{2m^*}\mathbf{k}^2,$$

by setting

$$\frac{1}{m^*} = -\frac{8\pi^2}{h^2}\alpha a^2.$$

Since α is negative for s electrons, they have a positive effective mass. On the other hand, it is clear from what has been said in the preceding paragraph that the effective mass of electrons near the origin of a p band is negative for motion in the x direction. The significance of negative masses will be discussed in Sec. 68.

Mott and Jones[1] have used $\epsilon(\mathbf{k})$ functions of type (10) to compute the function $dN/d\epsilon$, which gives the number of states per unit energy range.[2] Since the density of points in $\mathbf{k}$ space is uniform and is equal[3] to $2V$, it follows that

$$\delta N = 2V d\tau_\mathbf{k},$$

where $d\tau_\mathbf{k}$ is the differential of volume in $\mathbf{k}$ space and δN is the number of states in this volume. Now, if we choose coordinates so that the differential of volume is bounded on two faces by surfaces of constant energy,

$$d\tau_\mathbf{k} = dS\left(\frac{dk}{d\epsilon}\right)d\epsilon,$$

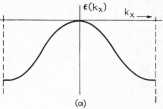

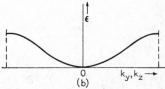

FIG. 22.—$\epsilon(\mathbf{k})$ curves for a p band. The $\epsilon(k_x)$ curve is concave downward, whereas the $\epsilon(k_y)$ and $\epsilon(k_z)$ curves are concave upward. The curvature is less in the second case because the overlapping is smaller.

where dS is the differential area of the surface of constant energy, $d\epsilon$ is the energy difference of the two bounding surfaces, and $dk/d\epsilon$ is the change in k in going from one energy surface to the other. Evidently,

$$\left(\frac{dk}{d\epsilon}\right) = \frac{1}{|\mathrm{grad}_\mathbf{k}\,\epsilon(\mathbf{k})|}, \tag{13}$$

whence

$$n(\epsilon') \equiv \left(\frac{dN}{d\epsilon}\right)_{\epsilon=\epsilon'} = \iint \frac{2V dS}{|\mathrm{grad}_\mathbf{k}\,\epsilon(\mathbf{k})|} \tag{14}$$

in which the integral extends over the surface of energy ϵ'.

[1] *Cf.* MOTT and JONES, *op. cit.*

[2] In the following sections of this book, we shall usually designate $dN/d\epsilon$ by $n(\epsilon)$.

[3] See footnote 1, p. 294.

Figure 23 gives a comparison of the $n(\epsilon)$ curves for the first zone of a body-centered lattice that are obtained with the assumptions that $\epsilon(\mathbf{k})$ is given by Eq. (10) and by the free-electron equation

$$\epsilon(\mathbf{k}) = \frac{h^2}{2m^*}\mathbf{k}^2,$$

respectively. In the second case, the function varies as $\sqrt{\epsilon}$ until the spherical contours intersect the zone boundary, whereupon the density of levels decreases.

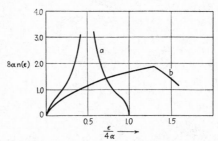

Fig. 23.—$dN/d\epsilon$ curves for a body-centered cubic lattice. Curve b is derived by assuming the electrons are perfectly free; curve a is obtained from an $\epsilon(\mathbf{k})$ relation of the type (10).

66. The Total Electronic Wave Function of the Solid.—Let us digress momentarily from the one-particle approximation and discuss the wave functions of the solid as a whole. To begin with, we shall assume that the constituents are far apart. If the lowest states of these atoms, ions, or molecules are nondegenerate as in rare gas solids, ordinary molecular crystals, or ionic crystals that are referred to a normal state of ions with rare gas configurations, the lowest state of the entire assembly also is nondegenerate. On the other hand, if the constituents are twofold degenerate, as in the alkali metals, the total degeneracy of the system of N separated atoms is 2^N. More generally, the degeneracy of the lowest state of the entire assembly is g^N if the lowest state of the constituent is g-fold degenerate. The first excited state of the nondegenerate assembly is Nf-fold degenerate if the first excited state of the constituents is f-fold degenerate. The factor N appears because any one of the N constituents may be excited.

Let us now bring the atoms or molecules closer together in order to form a solid. We may expect some of the levels of the entire solid that are highly degenerate at infinite separation to split into levels of lower degeneracy as the interatomic distance decreases to the point where the charge distributions of different atoms overlap. This splitting may occur in any one of several ways. (1) The components of the splitting level may be separated from one another by finite distances (*cf.* Fig. 24*a*). (2) The level may split into a quasi-continuous band (Fig. 24*b*). (3) It may break into separate continuous bands (Fig. 24*c*). (4) It may break into bands and discrete levels (Fig. 24*d*). A nondegenerate level remains nondegenerate, of course, although it may cross or mingle with the progeny of degenerate levels (Figs. 25*a* and 25*b*).

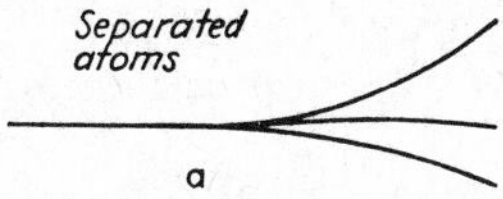

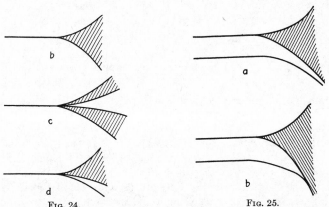

The splitting of degenerate levels may be understood from a general standpoint in the following way. It can be shown on the basis of

FIG. 24.   FIG. 25.

FIG. 24.—Possible behavior of a degenerate level of a system of atoms as the atoms are combined to form a solid. In *a* the level splits but remains discrete; in *b* it splits into a quasi-continuous band; in *c* it splits into two bands; in *d* it splits into a band and a discrete level. The discrete levels in *a* and *d* may be highly degenerate.

FIG. 25.—The upper level is highly degenerate whereas the lower is nondegenerate. In case *a* the lower level does not merge with the continuum; in case *b* it does.

symmetry theory that the wave functions of the entire crystal have the form

$$\chi_{k_1, \ldots, k_n}(r_1, \cdots, r_n)e^{2\pi i\left(\sum_i k_i \cdot r_i\right)} \tag{1}$$

Here, r_i is the position vector of the ith electron; $\chi_{k_1, \ldots, k_n}$ is a periodic function in the sense that

$$\chi_{k_1, \ldots, k_n}(r_1 + \tau, r_2 + \tau, \cdots, r_n + \tau) = \chi_{k_1, \ldots, k_n}(r_1, \cdots, r_n) \tag{2}$$

where τ is any translational vector of the lattice. If periodic boundary conditions are used, the k_i satisfy the relations

$$\left(\sum_i k_i\right) \cdot T_\alpha = l_\alpha \qquad (\alpha = 1, 2, 3),$$

where T_α is one of the three vectors corresponding to the edge lengths of the crystal and l_α is an integer. The function (1) evidently is a

$3n$-dimensional generalization of a Bloch function, and $\sum_i \mathbf{k}_i$ is a generalized wave-number vector. For each set of integers l_1, l_2, and l_3, there is an independent value of $\sum_i \mathbf{k}_i$. When the atoms are widely separated, we may expect $\chi_{\mathbf{k}_1, \ldots, \mathbf{k}_n}$, as in the one-dimensional case, to be a function that has a relatively large amplitude only when the electrons are near the atoms. The modulation factor $e^{2\pi i \left(\sum_i \mathbf{k}_i \cdot \mathbf{r}_i \right)}$ then has relatively minor importance; and, as in the case of bound electrons in the Bloch scheme, the degeneracies may be high since many wave functions that have different values of $\sum_i \mathbf{k}_i$ may have the same energy. On the other hand, we may expect χ to have an appreciable amplitude for a much wider range of the $\mathbf{r}_i$ when the atoms are close together. The value of $\sum_i \mathbf{k}_i$ then affects the energy, and the degeneracy is at least partly removed.

If the lowest level is not degenerate and if, as in Fig. 25a, it does not mingle with a continuum, the solid in its normal state is an insulator. In this case, we may regard the effect of an electrostatic field as a perturbation and may express the perturbed state as a linear combination of the unperturbed functions. The amplitude with which the excited states appear in this function is proportional to the field intensity and is small as long as the field is not strong. Hence, we should expect a finite electronic polarizability for small electrostatic fields, just as for ordinary atoms and molecules in which the lowest level is discrete.

Consider next the cases illustrated in Fig. 26 in which there are low-lying continuous bands. The lowest state now has a large number of other states arbitrarily near to it. When this solid is placed in an electrostatic field, it may be possible to construct from the lowest levels of the continuum a stable perturbed wave function which takes maximum advantage of the applied field by neutralizing this field within the solid. This property corresponds to infinite polarizability and is characteristic of metals. A continuum of levels is a necessary but not a sufficient condition for metallic polarizability, as we shall see presently. We may conclude that Fig. 26 is representative of the energy-level diagram of a metal. Figure 26b then corresponds to the case of an alkaline earth metal. The normal state at infinite separation is not degenerate in this case because the atoms have closed-shell (s^2) configurations.

It is not possible to describe a continuum of the type required for metallic behavior by means of the Heitler-London approximation.

Since the Hartree-Fock field for the Heitler-London functions must have a trough in the region where the electrons are localized, just as in the case of atoms, we may expect the lowest energy levels that are associated with this trough to be discrete. Hence, the lowest and first excited levels of the system should be separated by a finite energy.

The Bloch scheme, on the other hand, leads to a continuum of the required type, as we have seen in the preceding sections of this chapter. Thus the Bloch scheme is preferable to the Heitler-London for a purely qualitative description of the properties of metals.

Although the lowest state of insulators may be described equally well by either approximation, the Bloch scheme may be inferior for the lowest excited states. Consider, for example, the following system. We shall start with a set of N infinitely separated atoms having nondegenerate normal states. One excited state of this system is that in which an electron is removed from one atom and is placed on another, a positive and a negative ion thus being formed. We shall assume for simplicity that this is the first excited state. Its degeneracy is $g_p g_n N(N - 1)$, where g_p and g_n are the degeneracies of the positive and negative ions, respectively. The factor $N(N - 1)$ occurs because an electron may be taken from any one of the N ions and placed on any one of the $N - 1$ remaining atoms. Some of this degeneracy disappears as the atoms are brought within a finite distance of one another, because of the interaction between positive and negative ions. The energy of a pair of spherically

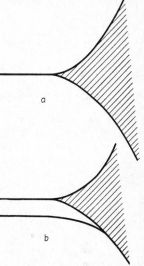

Fig. 26.—Cases in which the lowest levels form a quasi-continuous band. In case a there is no nondegenerate level and the low-lying degenerate level breaks into a band. In case b the nondegenerate level is absorbed into the continuum of the band arising from an excited level. Case a is characteristic of the alkali metals, case b of the alkaline earth metals.

symmetrical nonoverlapping ions decreases by an amount $-e^2/r_s$ as the ions approach one another, where r_s is the distance between their centers. Thus, the degeneracy of the new levels is $n_s g_p g_n N$, where n_s is the number of neighbors at distance r_s from a given atom. This remaining degeneracy is partly removed as the atoms overlap, and each level may be expected to split into a band. Suppose that the splitting has reached the point corresponding to the line A, Fig. 27, when the system attains equilibrium relative to interatomic displacements. The first band of excited states then differ from the normal state in that an electron has

been transferred from one atom to an immediate neighbor. Since the
negative and positive ions that are produced remain at a fixed distance
relative to one another, the crystal does not possess electronic conductivity when in any one of the levels of this band of excited states. It is
not possible to describe nonconducting excited states of this type by
means of the Bloch approximation. A crystal in which all occupied
zones are completely filled is an insulator, but it becomes an electronic
conductor as soon as one electron has been removed to an unoccupied
zone. Thus, the use of the Bloch approximation for an insulator is
equivalent to assuming that a continuum which is similar to the con-
tinua of metals lies above the lowest nondegenerate level of the crystal.

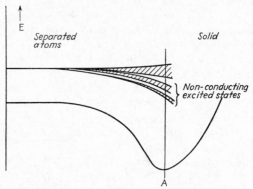

Fig. 27.—Case in which the lowest level is nondegenerate and the excited degenerate
level splits into several bands, the lowest of which is nonconducting. The nonconducting
states cannot be described by the Bloch approximation.

The Heitler-London approximation is somewhat more suitable than
the Bloch scheme for describing the nonconducting excited states.
However, it is open to the objection that in using it the assumption must
be made that a particular atom is ionized and that the electron is localized
on the neighbors of this atom. Since all atoms in the model discussed
above are equivalent, the excitation would be described more appropriately if it extended throughout the lattice. In Sec. 96, we shall discuss
an approximational scheme that satisfies this condition.

The Bloch scheme is not always invalid when the lowest state of the
solid is not degenerate for large separations. Free alkaline-earth metal
atoms are normally in a $^1S_0$ state, and yet they combine to form metals.
In this case, the degenerate level that corresponds to all atoms being in
the first excited state apparently drops below the nondegenerate state
and broadens into a continuous band as the atoms are brought together
(*cf.* Fig. 26b). This band presumably is similar to the low-lying band
of the alkali metals and gives the alkaline earths their metallic charac-

teristics.   It follows, from this example, that the first excited state of an insulator may be conducting for actual interatomic separations even though this state is nonconducting for large distances.   A continuum that arises from degenerate states in which many atoms are excited or ionized can cross or become very close to the nonconducting levels, leaving the solid in the condition corresponding to that shown in Fig. 28.

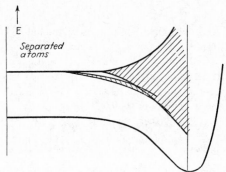

Fig. 28.—Case in which the conducting levels arising from the first excited state of an insulator overlap or come very close to the nonconducting levels at the actual interatomic distance.   In this case, unlike that of Fig. 27, the first excited state is conducting.

67. Koopmans's Theorem*.—We shall now prove a theorem, due to Koopmans,[1] which states that the energy parameter ϵ_j in Fock's equations for a solid, namely,

$$H^F \varphi_i = \epsilon_i \varphi_i, \tag{1}$$

is the negative of the energy required to remove the electron in the state φ_j from the solid when the space part of the φ are Bloch type functions.

Suppose that the Hamiltonian for the entire solid is designated by H. Then when the jth electron is removed, the new Hamiltonian is

$$H' = H - \left(-\frac{\hbar^2}{2m}\Delta_j + \sum_i{}' \frac{e^2}{r_{ij}} + V_j \right) \tag{2}$$

where V_j is the ion-core potential.   The wave function of the initial state is

$$\Psi = \frac{1}{\sqrt{n!}} \begin{vmatrix} \varphi_1(\mathbf{r}_1) \ldots \ldots \ldots \ldots \varphi_{j-1}(\mathbf{r}_1) \ \varphi_j(\mathbf{r}_1) \ \varphi_{j+1}(\mathbf{r}_1) \ldots \ldots \ldots \ldots \varphi_n(\mathbf{r}_1) \\ \cdot \qquad\qquad\qquad\qquad\qquad\qquad\qquad\qquad\qquad\qquad\qquad\qquad \cdot \\ \cdot \qquad\qquad\qquad\qquad\qquad\qquad\qquad\qquad\qquad\qquad\qquad\qquad\qquad \cdot \\ \varphi_1(\mathbf{r}_n) \ldots \ldots \ldots \ldots \ldots \ldots \ldots \ldots \ldots \ldots \ldots \ldots \ldots \ldots \varphi_n(\mathbf{r}_n) \end{vmatrix}, \tag{3}$$

[1] T. Koopmans, *Physica*, **1**, 104 (1933).

whereas the wave function for the state in which the electron having the function φ_j has been removed is

$$\Psi_j = \frac{1}{\sqrt{(n-1)!}} \begin{vmatrix} \varphi_1(\mathbf{r}_1) \ldots \ldots \ldots \varphi_{j-1}(\mathbf{r}_1) \; \varphi_{j+1}(\mathbf{r}_1) \ldots \ldots \ldots \varphi_n(\mathbf{r}_1) \\ \cdot \qquad\qquad\qquad\qquad\qquad\qquad\qquad \cdot \\ \cdot \qquad\qquad\qquad\qquad\qquad\qquad\qquad \cdot \\ \varphi_1(\mathbf{r}_n) \ldots \ldots \ldots \ldots \ldots \ldots \ldots \ldots \ldots \ldots \ldots \varphi_n(\mathbf{r}_n) \end{vmatrix} . \quad (4)$$

Since the Fock Hamiltonian is practically unchanged when a Bloch type function is removed, because the electron charge is spread throughout the entire crystal, the φ in (3) and (4) are practically identical. This conclusion clearly is valid only when the electronic system is very large and when the one-electron functions are of the extended type.

The work done in removing the electron is simply

$$\int \Psi_j {}^* H' \Psi_j d\tau' - \int \Psi^* H \Psi d\tau \qquad (5)$$

in which $d\tau'$ excludes the variables of the jth electron. Now,

$$\int \Psi_j {}^* H' \Psi_j d\tau' = \sum_{i \neq j} \int \varphi_i {}^*(\mathbf{r}_1) \left[-\frac{\hbar^2}{2m} \Delta_1 + V(\mathbf{r}_1) \right] \varphi_i(\mathbf{r}_1) d\tau_1 +$$

$$\frac{e^2}{2} \sum_{i,k \neq j}' \left[\int \frac{|\varphi_i(\mathbf{r}_1)|^2 |\varphi_k(\mathbf{r}_2)|^2}{r_{12}} d\tau_{12} - \int \frac{\varphi_i {}^*(\mathbf{r}_1) \varphi_k {}^*(\mathbf{r}_2) \varphi_i(\mathbf{r}_2) \varphi_k(\mathbf{r}_1)}{r_{12}} d\tau_{12} \right]. \quad (6)$$

$\int \Psi^* H \Psi d\tau$ is equal to the same expression with the difference that the summations include the index j. Thus, (5) is

$$-\left\{ \int \varphi_i {}^*(\mathbf{r}_1) \left[-\frac{\hbar^2}{2m} \Delta_1 + V(\mathbf{r}_1) + \sum_i e^2 \int \frac{|\varphi_i(\mathbf{r}_2)|^2}{r_{12}} d\tau_2 \right] \varphi_j(\mathbf{r}_1) d\tau_1 - \right.$$

$$\left. \sum_i e^2 \int \frac{\varphi_i {}^*(\mathbf{r}_1) \varphi_j {}^*(\mathbf{r}_2) \varphi_i(\mathbf{r}_2) \varphi_j(\mathbf{r}_1)}{r_{12}} d\tau_{12} \right\} =$$

$$-\int \varphi_i {}^*(\mathbf{r}_1) H^F \varphi_j(\mathbf{r}_1) d\tau = -\epsilon_j$$

(*cf.* Sec. 51), which proves Koopmans's theorem.

It should be noted that Koopmans's theorem also tells us that, in the Fock approximation, the energy required to take an electron from the state $\varphi_{\mathbf{k}}$ to the state $\varphi_{\mathbf{k}'}$ is $\epsilon(\mathbf{k}') - \epsilon(\mathbf{k})$ when the one-electron functions are of the Bloch type.

68. Velocity and Acceleration in the Bloch Scheme.—We shall now develop several theorems[1] concerning the behavior of electrons in the

---

[1] These theorems have been proved independently by several workers. See, for example, A. Sommerfeld and H. Bethe, *Handbuch der Physik*, Vol. XXIV/2; H. Jones

Bloch approximation. In particular, we shall be interested in the relation between electron velocity and energy and that between acceleration and external fields. In quantum theory, the quantities corresponding to velocity, acceleration, and force are represented by operators. The ordinary measured values of these quantities for a given state of a system are the mean values of the corresponding operators. Hence, in order to proceed in a rigorous fashion, we should compute these mean values and should derive relationships among them from the quantum laws. Actually, the correct final results may be obtained much more simply by using wave-packet methods. We shall employ this procedure.

a. The Relation between Velocity and Energy.—In order to determine the velocity of an electron that has a given energy $\epsilon(\mathbf{k}')$ and has the wave-number vector $\mathbf{k}'$, we shall construct a packet by adding together wave functions associated with values of $\mathbf{k}$ in the vicinity of $\mathbf{k}'$ and by determining the group velocity of this packet. This group velocity corresponds to the measured or mean value of the velocity of an electron having wave number $\mathbf{k}'$. When the proper time dependence is included, the Bloch wave functions are

$$\psi_{\mathbf{k}} = \chi_{\mathbf{k}} e^{2\pi i \mathbf{k} \cdot \mathbf{r}} e^{-2\pi i \frac{\epsilon(\mathbf{k})}{h} t} \tag{1}$$

where $\chi_{\mathbf{k}} e^{2\pi i \mathbf{k} \cdot \mathbf{r}}$ is the space-dependent part and $e^{-2\pi i \frac{\epsilon(\mathbf{k})}{h} t}$ is the time-dependent part. The packet, whose constituent functions are centered about $\mathbf{k}'$, is

$$f(\mathbf{r},t) = \int a(\mathbf{k}) \chi_{\mathbf{k}} e^{2\pi i \left(\mathbf{k} \cdot \mathbf{r} - \frac{\epsilon}{h} t \right)} d\tau(\mathbf{k}) \tag{2}$$

where $|a(\mathbf{k})|$ has a maximum value at $\mathbf{k} = \mathbf{k}'$ and decreases rapidly on either side. The integration extends over the range of values of $\mathbf{k}$ about $\mathbf{k}'$ in which $|a(\mathbf{k})|^2$ is appreciably different from zero and which may be made as small as we please. If we set $\mathbf{k} = \mathbf{k}' + \Delta\mathbf{k}$ and expand $\epsilon(\mathbf{k})$ in a Taylor series about $\mathbf{k}'$, we obtain

$$\epsilon(\mathbf{k}) = \epsilon(\mathbf{k}') + \text{grad}_{\mathbf{k}'}\, \epsilon(\mathbf{k}') \cdot \Delta\mathbf{k} + \cdots . \tag{3}$$

We shall assume that the range in which $\Delta\mathbf{k}$ is important is so small that we need retain only those terms indicated in (3). Equation (2) then becomes

$$f(\mathbf{r},t) \cong e^{2\pi i \left[\mathbf{r} \cdot \mathbf{k}' - \frac{\epsilon(\mathbf{k}')}{h} t \right]} \int a(\Delta\mathbf{k}) \chi_{\mathbf{k}} e^{2\pi i \left(\mathbf{r} - \frac{\text{grad}_{\mathbf{k}} \epsilon}{h} t \right) \cdot \Delta\mathbf{k}} d\tau(\Delta\mathbf{k})$$

and C. Zener, *Proc. Roy. Soc.*, **144**, 101 (1934); H. Fröhlich, *Theorie der Metalle* (Julius Springer, Berlin, 1937).

where the integration now extends over a range of $\Delta \mathbf{k}$ about the value $\Delta \mathbf{k} = 0$. $\chi_{\mathbf{k}}$ varies so slowly with $\mathbf{k}$ that we may take it outside the integral and write

$$f(\mathbf{r},t) \cong \chi_{\mathbf{k}'}(\mathbf{r}) e^{2\pi i \left[\mathbf{r} \cdot \mathbf{k}' - \frac{\epsilon(\mathbf{k}')}{h} t \right]} \int a(\Delta \mathbf{k}) e^{2\pi i \left(\mathbf{r} - \frac{\text{grad}_{\mathbf{k}'}\epsilon}{h} t \right) \cdot \Delta \mathbf{k}} d\tau(\Delta \mathbf{k}). \quad (4)$$

Equation (4) describes a wave packet whose phase is determined by the function

$$\chi_{\mathbf{k}'} e^{2\pi i \left[\mathbf{k}' \cdot \mathbf{r} - \frac{\epsilon(\mathbf{k}')}{h} t \right]}, \quad (5)$$

which is the Bloch function corresponding to $\mathbf{k} = \mathbf{k}'$. Its group behavior is determined by the function

$$\int a(\Delta \mathbf{k}) e^{2\pi i \left(\mathbf{r} - \frac{\text{grad}_{\mathbf{k}'}\epsilon}{h} t \right) \cdot \Delta \mathbf{k}} d\tau(\Delta \mathbf{k}), \quad (6)$$

of which the argument is $\mathbf{r} - \dfrac{\text{grad}_{\mathbf{k}} \epsilon}{h} t$. Thus, the function (6) has constant amplitude at those points for which $\mathbf{r} - t \, \text{grad}_{\mathbf{k}} \epsilon$ is constant, that is, at points that move with the constant velocity

$$\mathbf{v}(\mathbf{k}') = \frac{1}{h} \text{grad}_{\mathbf{k}'} \epsilon(\mathbf{k}'), \quad (7)$$

which is the group velocity of the packet. The classical expression $|\mathbf{v}| = \sqrt{2\epsilon/m}$ for the speed is valid only in the special case in which

$$\epsilon = \frac{h^2}{2m} \mathbf{k}^2,$$

that is, for perfectly free electrons. The relation between speed and ϵ is different in all other cases. For example, near the top of a band where the slope of ϵ decreases, the speed decreases with increasing energy and actually may become zero if the slope of ϵ happens to vanish at the boundary of the zone.

The current $\mathbf{i}$ carried by an electron having velocity $\mathbf{v}$ is $-e\mathbf{v}$. Hence, we obtain from (7)

$$\mathbf{i} = -e \frac{\text{grad}_{\mathbf{k}} \epsilon(\mathbf{k})}{h}. \quad (8)$$

b. *The Effective Electron Mass.*—We shall now derive the relation between acceleration and applied force. It follows from Eq. (7) that

$$\frac{d\mathbf{v}}{dt} = \frac{d}{dt} \left[\frac{\text{grad}_{\mathbf{k}} \epsilon(\mathbf{k})}{h} \right] = \frac{\text{grad}_{\mathbf{k}} [d\epsilon(\mathbf{k})/dt]}{h}. \quad (9)$$

The classical relation between force and energy is

$$\frac{d\epsilon}{dt} = \mathbf{F} \cdot \mathbf{v},\tag{10}$$

which should remain valid for the mean values of quantum theory. Substituting (10) in (9), we find

$$\frac{d\mathbf{v}}{dt} = \frac{\text{grad}_k \ \mathbf{F} \cdot \mathbf{v}}{h} = \frac{\mathbf{F} \cdot [\text{grad}_k \ \text{grad}_k \ \epsilon(\mathbf{k})]}{h^2}\tag{11}$$

where $\text{grad}_k \ \text{grad}_k \ \epsilon(\mathbf{k})$ is a tensor the nine components of which are

$$\begin{pmatrix} \dfrac{\partial^2 \epsilon}{\partial k_x^2} & \dfrac{\partial^2 \epsilon}{\partial k_y \partial k_x} & \dfrac{\partial^2 \epsilon}{\partial k_z \partial k_x} \\[2mm] \dfrac{\partial^2 \epsilon}{\partial k_x \partial k_y} & \dfrac{\partial^2 \epsilon}{\partial k_y^2} & \dfrac{\partial^2 \epsilon}{\partial k_z \partial k_y} \\[2mm] \dfrac{\partial^2 \epsilon}{\partial k_x \partial k_z} & \dfrac{\partial^2 \epsilon}{\partial k_y \partial k_z} & \dfrac{\partial^2 \epsilon}{\partial k_z^2} \end{pmatrix}.\tag{12}$$

Equation (11) is analogous to the classical equation

$$\frac{d\mathbf{v}}{dt} = \frac{1}{m}\mathbf{F}$$

and shows that in the Bloch approximation an electron in an energy band behaves as though it had an effective mass $\mathbf{m}^*$ represented by the tensor

$$\frac{1}{\mathbf{m}^*} = \frac{1}{h^2} \text{grad}_k \ \text{grad}_k \ \epsilon(\mathbf{k}).\tag{13}$$

Thus, the force and acceleration usually are not in the same direction.

Suppose that the electron is moving in the direction of a principal axis of the tensor (12) and that this direction is chosen to lie along the x axis. Then, the effective mass for acceleration in the x direction is

$$m^* = \frac{h^2}{d^2\epsilon(\mathbf{k})/dk_x^2}.\tag{14}$$

As we may see from Fig. 2, this is usually negative when the electron is near the top of the band. Hence, an electron in this position behaves as though it had a negative mass. The type of force that we shall usually consider is the combination of electrostatic and Lorentz force:

$$\mathbf{F} = -e\mathbf{E} - \frac{e}{c}\mathbf{v} \times \mathbf{H}.$$

Since e appears linearly in this, an electron with a negative effective mass is equivalent to a particle with a positive charge.

It was found while developing the free-electron theory of metals that the electrons that reside in the energy band of width kT at the top of the filled region are principally responsible for the electrical properties of metals. If this band lies above the inflection point of $\epsilon(\mathbf{k})$, all these electrons behave like positive charges. This fact evidently offers an explanation of the anomalous Hall effect: metals and semi-conductors that have a positive Hall constant contain nearly filled bands.

It sometimes is convenient to ascribe all the "anomalous" properties of metals and semi-conductors that are associated with the electrons of negative mass to the vacant levels or holes at the top of the nearly filled band. Thus, these holes may be treated as though they were positively charged particles having effective mass

$$\frac{1}{\mathbf{m}^*} = - \operatorname{grad}_\mathbf{k} \operatorname{grad}_\mathbf{k} \epsilon(\mathbf{k}).$$

This convention is not simply an analogy, for Heisenberg[1] has shown that it is possible to replace the wave equation for the electrons in a nearly filled band by an approximate wave equation for the holes in the band. In this equation, the holes play the same role as positive charges. Heisenberg's theorem may also be applied to nearly filled atomic shells.

In addition to the preceding relations, we shall find the following one useful. By combining Eqs. (10) and (7), we find

$$\mathbf{F} \cdot \mathbf{v} = \frac{d\epsilon}{dt} = \frac{d\mathbf{k}}{dt} \cdot \operatorname{grad} \epsilon = h \frac{d\mathbf{k}}{dt} \cdot \mathbf{v},$$

whence

$$\frac{d\mathbf{k}}{dt} = \frac{\mathbf{F}}{h}. \tag{15}$$

In this connection, Houston[2] has made an interesting investigation of the motion of an electron in the presence of an electrostatic field for the band approximation. He has shown that when the field is small, or when the electronic wave-number vector is not near the zone boundary, a good approximation for the solution of the time-dependent Schrödinger equation is

$$\chi_\mu e^{2\pi i \mu \cdot \mathbf{r}} e^{-\frac{2\pi i}{h} \int \epsilon(\mu) dt}$$

where $\mathbf{\mu} = \mathbf{k} - \dfrac{e\mathbf{E}t}{h}$, $\mathbf{E}$ being the field intensity. This function degenerates to $\psi_\mathbf{k} e^{-\frac{2\pi i \epsilon(\mathbf{k})t}{h}}$ when $\mathbf{E}$ is zero. In this approximation the electronic

[1] W. Heisenberg, *Ann. Physik*, **10**, 888 (1931).
[2] W. V. Houston, *Phys. Rev.*, **57**, 184 (1940).

wave-number vector moves through $\mathbf{k}$ space at a uniform rate eE/h, or, in the reduced-zone scheme, the wave-number vector moves uniformly to the boundary of the zone in the direction of E, suffers a Bragg reflection, that is, jumps to the opposite face of the zone, and starts back along the same path, repeating the process indefinitely. During these cycles, the system remains in the same energy surface $\epsilon(\mathbf{k})$, so that there is no transition between zones. In higher approximation, the electron may jump between zones for sufficiently strong fields. Houston has shown that these transitions occur when the wave-number vector is near the zone boundary and that in a one-dimensional case the probability of a transition in a single cycle is

$$\frac{4\pi^2 e^{-\alpha}}{(1 - e^{-\alpha})^2}$$

where $\alpha = mV^2d/eE\hbar$, in which m is the electronic mass, $2V$ is the energy gap, and d is the lattice distance. A similar expression had previously been derived by Zener[1] and applied to a discussion of the problem of dielectric breakdown in insulating crystals. It is believed at present that dielectric breakdown in simple crystals, such as the alkali halides, occurs for fields weaker than those required to cause transitions from the filled to the unoccupied levels by this process (see Sec. 133).

69. Modification of Boltzmann's Equation for the Bloch Scheme.— We are now in a position to modify Boltzmann's equation to suit the quantum mechanical one-electron mode of description instead of the classical one (*cf.* Sec. 31, Chap. IV). In place of the distribution function $f(x,y,z,v_x,v_y,v_z)$, which gives the number of particles at x, y, and z with velocity components v_x, v_y, and v_z, we shall introduce the function

$$f(x,y,z,k_x,k_y,k_z) \equiv f(\mathbf{r},\mathbf{k}), \tag{1}$$

which gives the number of particles at x, y, z with wave-number components k_x, k_y, k_z.

The drift term, analogous to (2), Sec. 31, Chap. IV, is

$$\left(\frac{df}{dt}\right)_{\text{drift}} = -\frac{\partial f}{\partial x}v_x - \frac{\partial f}{\partial y}v_y - \frac{\partial f}{\partial z}v_z - \frac{\partial f}{\partial k_x}\dot{k}_x - \frac{\partial f}{\partial k_y}\dot{k}_y - \frac{\partial f}{\partial k_z}\dot{k}_z, \tag{2}$$

which may be expressed in terms of $\text{grad}_\mathbf{k}\,\epsilon(\mathbf{k})$ and $\mathbf{F}$ by means of Eqs. (7) and (15) of the preceding section. The result is

$$\left(\frac{df}{dt}\right)_{\text{drift}} = -\frac{\text{grad}_\mathbf{k}\,\epsilon}{h} \cdot \text{grad}_\mathbf{r}\,f - \frac{\mathbf{F}}{h} \cdot \text{grad}_\mathbf{k}\,f. \tag{3}$$

[1] C. Zener, *Proc. Roy. Soc.*, **145**, 523 (1934).

The collision terms may now be expressed in terms of the function $\Theta(k_x,k_y,k_z;k'_x,k'_y,k'_z)dk'_xdk'_ydk'_z$, which gives the probability that a particle changes its wave-number components from k_x, k_y, k_z to values in the range from k'_x to $k'_x + dk'_x$, etc. The quantities that correspond to a and b, Sec. 31, Chap. IV, are then

$$a = f(x,y,z,k_x,k_y,k_z)\int\Theta(k_x,k_y,k_z;k'_x,k'_y,k'_z)dk'_xdk'_ydk'_z, \tag{4}$$
$$b = \int f(x,y,z,k''_x,k''_y,k''_z)\Theta(k''_x,k''_y,k''_z;k_x,k_y,k_z)dk''_xdk''_ydk''_z. \tag{5}$$

The equation of equilibrium analogous to Boltzmann's equation, is

$$\frac{\text{grad}_\mathbf{k}\,\epsilon}{h}\cdot\text{grad}_\mathbf{r}\,f + \frac{\mathbf{F}}{h}\cdot\text{grad}_\mathbf{k}\,f = b - a, \tag{6}$$

which may be solved when $b - a$ is known.

In general, a and b may be determined from (4) and (5) when Θ has been determined by means of a quantum mechanical analysis. We shall discuss particular cases of this procedure in Chap. XV.

It may be concluded from the results of the present discussion that the developments of Chap. IV can be modified in two important respects, namely:

a. Although the electrons may behave as though effectively free, so that the relation

$$\epsilon = \frac{h^2}{2m^*}\mathbf{k}^2 = \frac{\mathbf{p}^2}{2m^*} \tag{7}$$

is valid, as assumed in Chap. IV, it should be recognized that the effective mass m^* may be negative in cases in which the bands are nearly filled. As we have seen in the preceding section, this situation may be treated by assuming that both m^* and the electronic charge are positive. Thus, the results of Chap. IV can be employed in these cases by reversing the sign of e.

b. It may be necessary to use the tensor character of the effective mass in order to take into account the anisotropy even of cubic metals. This procedure probably is necessary in discussing the dependence of resistivity on magnetic field strength; for as we have seen in Sec. 34, the results derived on the basis of (7) are badly in error but would be improved if the velocity at the top of the filled region were an anisotropic function.

70. Additional Energy States.—It was mentioned in part *b*, Sec. 61, that some of the nonperiodic states that are labeled under Kramers' class *i* may be permitted if the crystal is not infinitely large. If the crystal does extend to infinity in all directions, the amplitude of these states diverges in at least one direction. The possibilities for the finite lattice were first pointed out by Tamm[1] who showed that the divergences

[1] I. Tamm, *Physik. Z. Sowj.*, **1**, 733 (1932).

can be avoided in special cases.   We shall discuss his computations for a simple model in the following paragraphs.   Before doing so, however, we should add that a computation that is based upon the use of a static one-electron field, as Tamm's is, is not completely adequate for establishing the general conditions under which nonperiodic states are permitted, since the actual field in which an electron moves is determined in part by the electron itself.

Tamm considers the simple potential field that is shown in Fig. 29, which, for positive values of x, is the same as that used by Kronig and Penney and has the constant value W for negative values.   As before, we shall allow V_0 to approach infinity and shall assume that $mV_0b(a - b)/\hbar^2$

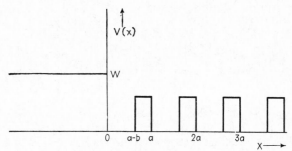

FIG. 29.—The potential function used by Tamm.

approaches the constant value c in the limit.   One restriction on the solutions then is

$$\cos \lambda a = \frac{c}{\alpha a} \sin \alpha a + \cos \alpha a \tag{1}$$

[*cf.* Eq. (30), Sec. 61], where

$$\alpha = \frac{\sqrt{2mE}}{\hbar}$$

and $e^{i\lambda a}$ is the factor by which the wave function is multiplied when x changes to $x + a$.   We found previously that λ is necessarily real, for otherwise the wave functions would diverge at large values of x.   Hence, the allowed values of E were determined by the condition that the right-hand side of (1) should lie between the values -1 and $+1$.   These conclusions are no longer valid since the wave functions are different from Kronig and Penney's in the negative region of x.   Instead, they are exponential functions of the type

$$\psi_\eta = c'e^{\eta x} \tag{2}$$

where

$$\eta = \frac{\sqrt{2m(W - E)}}{\hbar}.$$

These functions should be joined to Kronig and Penney's at the origin of x.

It turns out that all of Kronig and Penney's solutions going with real values of λ may be joined to functions of type (2), if the phases are chosen properly. Hence, the continuous spectrum of periodic states is not altered. We must consider the case in which λ is imaginary.

The solution that decreases by a factor $\epsilon e^{-\mu a}$ in successive cells of the lattice along the positive x axis is

$$\psi_\mu = \left(e^{-i\alpha x} - \frac{1 - \epsilon e^{(\mu - i\alpha)a}}{1 - \epsilon e^{(\mu + i\alpha)a}} e^{i\alpha x} \right), \tag{3}$$

where ϵ is $+1$ when the right-hand side of (1) is greater than $+1$ and is -1 when the right-hand side is less than -1. The condition (1) is replaced by

$$\epsilon \cosh \mu a = \frac{c}{\alpha a} \sin \alpha a + \cos \alpha a. \tag{4}$$

The additional condition that must be satisfied if (2) and (3) are to join smoothly at the origin is

$$\eta \frac{\sin \alpha a}{\alpha} + \cos \alpha a = \epsilon e^{-\mu a} \tag{5}$$

for the case in which $E < W$. Eliminating μ from (4) and (5), we find that E must satisfy the equation

$$a\alpha \cotan a\alpha = \frac{a^2 \gamma^2}{c} - \sqrt{a^2 \gamma^2 - a^2 \alpha^2} \tag{6}$$

where

$$\gamma^2 = \frac{2mW}{\hbar^2}.$$

This equation possesses one root in each interval of $a\alpha$ extending from $n\pi$ to $(n + 1)\pi$. Since there is one continuous band in each of these intervals, we may conclude that there is one "surface" state in the energy region between each pair of neighboring bands, when E is less than W.

When E is greater than W, the functions (2) are periodic and may be joined to (3) for arbitrary values of E. The resulting solutions are periodic outside the lattice and exponentially damped inside.

We may conclude that the type of surface barrier which is considered here permits states which lie in the "unallowed" energy region.

These states are exponentially damped on both sides of the surface when E is less than W and are periodic outside the lattice and damped inside when E is greater than W. The first type represents electrons that are localized at the surface, whereas the second represents electrons that impinge on the surface from outside and are reflected back.

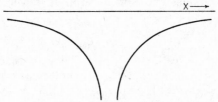

FIG. 30.—A simple potential trough.

Fowler[1] has pointed out that in a finite one-dimensional crystal the surface states occur in pairs, one state being associated with each end of the crystal.

In addition, Shockley[2] examined the origin of the surface levels more thoroughly on the basis of a more general one-dimensional model and considered the dependence of the levels on lattice parameter. He found that if the potential of the normal lattice may be expressed as a periodic sum of simple troughs of the type shown in Fig. 30 the surface levels can

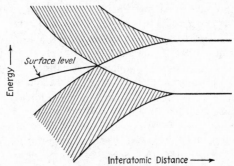

FIG. 31.—Schematic diagram showing the manner in which the surface levels occur in a case in which the potential is a periodic sum of troughs of the type shown in Fig. 30.

occur only if there is a separate potential trough at the surface or if the energy bands arising from separate atomic levels overlap (see Fig. 31). The second case evidently cannot occur at large interatomic separations. It turns out that the bands do not overlap in the Kronig-Penney model and that Tamm obtained surface levels only because he implicitly intro-

[1] R. H. FOWLER, *Proc. Roy. Soc.*, **141**, 56 (1933).

[2] W. SHOCKLEY, *Phys. Rev.*, **56**, 317 (1939). See also the similar discussion by W. G. POLLARD, *Phys. Rev.*, **56**, 324 (1939).

duced an additional potential trough at the surface in cutting off the potential function in the manner shown in Fig. 29.

Shockley also showed that one level is removed from the lower band and one is removed from the upper band for each pair of surface levels that occurs. Thus, if the lower band were completely filled and the upper band were completely empty at large interatomic separations, two electrons of opposite spin would be forced from the lower band into the surface levels when overlapping occurs.

The generalization for three dimensions is apparently straightforward. Suppose that we have a crystal that is bounded by the plane $x = 0$, the crystal being on the positive side of the x axis. We may expect some wave functions that have energies lying in the unallowed regions and that have the form

$$\psi = \chi e^{-\mu x} e^{2\pi i(k_y y + k_z z)} \tag{7}$$

inside the lattice and the form

$$\psi = e^{-\eta x} e^{2\pi i(k_y y + k_z z)} \tag{8}$$

outside. In (7), χ is a function possessing the periodicity of the lattice; η is real when E is less than W, the potential outside of the lattice, and is imaginary when E is greater than W. As before, the second type of solution is permitted for all values of E in the unallowed regions. We may expect more than one solution of the first type in the unallowed regions in which they may occur, however, for k_x and k_y may take all values associated with a two-dimensional zone system that is determined by the translational symmetry of the surface. Roughly speaking, we may expect as many states in these unallowed regions as there are atoms on the surface. Since this number ordinarily is about one million times smaller than the total number of atoms in a crystal, we should not expect these surface levels to affect the bulk properties of a substance.

There does not seem to be any direct experimental evidence for the existence of surface states, although Tamm suggests that charges on the surfaces of some charged insulators may be bound in states of this type.

Shockley has made a qualitative generalization of the results of his investigation of the one-dimensional model. His conclusions are as follows:

a. Surface levels will not occur between the ordinary X-ray levels or in the forbidden region of most simple insulating salts such as sodium chloride, for neither of these cases corresponds to overlapping-band systems. The basis for these conclusions is discussed in Chap. XIII.

b. Surface states should occur in the forbidden region between the highest filled band and lowest vacant band of diamond, for this gap occurs as a result of the overlapping of an s band and a p band (see Sec. 109).

Since the number of surface states is approximately twice the number of electrons that are forced from the filled band, the surface band should not be completely filled.   In an ideal case, this statement would imply that the surface should be conducting; however, various effects, such as surface cracks or adsorbed atoms, could easily impair this type of conductivity, which does not seem to be observed.

c. Surface states should occur near the overlapping bands of all metals.   In the monovalent metals, such as the alkali metals, in which the occupied band is usually not filled to the zone boundary, these levels would be completely empty.   They should be partly filled in the divalent metals, however.

More important than the existence of surface states is the fact, suggested by these computations, that bound electronic states may be

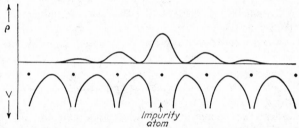

Fig. 32.—Schematic representation of a localized impurity level.   The lower curve is the lattice potential which is distorted by an impurity atom.   The upper curve represents the localized charge distribution associated with an impurity level that lies in the forbidden region.

associated with any flaw or discontinuity in an otherwise perfectly periodic lattice in the manner illustrated in Fig. 32.   Suppose, for example, that we have an infinite one-dimensional lattice, such as that of Kronig and Penney, and that we alter the potential in a single cell by lowering it from zero to $-W'$.   It is easy to show by the method used above that in the forbidden energy regions there are then states corresponding to electrons which are localized in the vicinity of the singular cell.   If the cell extends from $-a$ to 0, the allowed forms of the wave function within this range are

$$\sin \gamma'\left(x + \frac{a}{2}\right) \quad \text{or} \quad \cos \gamma'\left(x + \frac{a}{2}\right) \tag{9}$$

where

$$\gamma' = \frac{\sqrt{2m(E + W')}}{\hbar}. \tag{10}$$

The localized functions correspond to cases in which the functions (9) are joined smoothly to functions of type (3) at $x = 0$ and $x = a$.   The conditions for this in the two cases (9) are, respectively,

$$\frac{\gamma'}{\alpha} \cotan \frac{\gamma'a}{2} \sin \alpha a + \cos \alpha a = \epsilon e^{-\mu a} \tag{11}$$

and

$$-\frac{\gamma'}{\alpha} \tan \frac{\gamma'a}{2} \sin \alpha a + \cos \alpha a = \epsilon e^{-\mu a}. \tag{11a}$$

Either one of these conditions and the condition (4) must be satisfied simultaneously. The resulting equations are more complicated than (6) and will not be discussed in detail here. They show that when E is negative we may expect approximately as many trapped electron states as there are discrete states for a simple barrier of width a and depth $-V'$, whereas, for values of E greater than zero, there usually is one state in each forbidden region. However, there may be more or less than one in particular regions. These cases depend in a complex way upon the relative values of αa and $\gamma'a$.

Thus, we may conclude that impurity atoms or lattice imperfections induce additional energy states which correspond to electrons localized in the vicinity of the impurity or defect. These states lie in the regions between continuous bands. We shall find that they can play a very important role, particularly in semi-conductors.

71. Optical Transitions in the Zone Approximation.—Before leaving the zone approximation, we shall find it convenient to discuss optical transition probabilities. We found in Sec. 43 that the probability of an optical transition between two states Ψ_α and Ψ_β contains the following integral

$$\int \Psi_\alpha^* \sum_i \mathbf{p}_i e^{2\pi i \eta \cdot \mathbf{r}_i} \Psi_\beta d\tau(x_1, \cdots, z_n, \zeta_1, \cdots, \zeta_n) \tag{1}$$

in which $\mathbf{p}_i$ is the momentum operator for the ith electron, $\mathbf{n}$ is the wave-number vector of the light quantum that is absorbed or emitted, and the integration extends over all of the electronic coordinates. In the band approximation, Ψ_α and Ψ_β are determinantal wave functions that are constructed of Bloch one-electron functions. Since the operator

$$\sum_i \mathbf{p}_i e^{2\pi i \eta \cdot \mathbf{r}_i} \tag{2}$$

is a sum of one-electron operators, the integral in (1) vanishes if Ψ_α and Ψ_β differ with respect to more than one Bloch function; moreover, (1) vanishes even in this case, unless the electron spins associated with the two different Bloch functions are the same. Hence, only one electron can change its state during the absorption or emission of a single light quantum.

If ψ_k and $\psi_{k'}$ are, respectively, the different Bloch functions in the functions Ψ_α and Ψ_β, (1) may be reduced to the form

$$\int \psi_k^* \mathbf{p} e^{2\pi i \eta \cdot \mathbf{r}} \psi_{k'} d\tau \, (x,y,z) \tag{3}$$

where the integral now extends over the coordinates of one electron. Let us write ψ_k and $\psi_{k'}$ in the typical Bloch function forms

$$\begin{aligned} \psi_k &= \chi_k e^{2\pi i \mathbf{k} \cdot \mathbf{r}}, \\ \psi_{k'} &= \chi_{k'} e^{2\pi i \mathbf{k'} \cdot \mathbf{r}}. \end{aligned} \tag{4}$$

The integral (3) may then be written as

$$\int e^{2\pi i (\mathbf{k'} + \eta - \mathbf{k}) \cdot \mathbf{r}} \chi_k^* \left[\frac{\hbar}{i} \operatorname{grad} \chi_{k'} - \hbar 2\pi (\mathbf{n} + \mathbf{k'}) \chi_{k'} \right] d\tau. \tag{5}$$

The quantity

$$f = \chi_k^* \left[\frac{\hbar}{i} \operatorname{grad} \chi_{k'} - \hbar 2\pi (\mathbf{n} + \mathbf{k'}) \chi_{k'} \right] \tag{6}$$

has the periodicity of the unit cell, so that (5) may be written in the form

$$\sum_{\mathbf{r}_i} e^{2\pi i (\mathbf{k'} + \eta - \mathbf{k}) \cdot \mathbf{r}_i} \int_i e^{2\pi i (\mathbf{k'} + \eta - \mathbf{k}) \cdot (\mathbf{r} - \mathbf{r}_i)} f d\tau_i \tag{7}$$

where the integral $\int_i$ extends over the ith unit cell of the lattice and τ_i is a vector extending from a corner of this cell to the origin of coordinates. Since each integral in the sum (7) is the same, this series is

$$A \sum_{\mathbf{r}_i} e^{2\pi i (\mathbf{k'} + \eta - \mathbf{k}) \cdot \mathbf{r}_i} \tag{8}$$

where

$$A = \int_0 e^{2\pi i (\mathbf{k'} + \eta - \mathbf{k}) \cdot \mathbf{r}} f d\tau_0$$

in which the integral extends over the unit cell at the origin. Now, (8) vanishes unless

$$\mathbf{k'} + \mathbf{n} - \mathbf{k} = \mathbf{K} \tag{9}$$

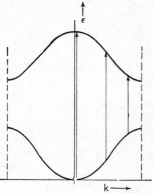

Fig. 33.—Diagram showing the allowed transitions in the reduced-zone scheme.

where $\mathbf{K}$ is a principal vector in the inverse lattice. Hence, transitions are allowed only between one-electron states the wave-number vectors of which satisfy the relation (9). The wave length of a light quantum ordinarily is large compared with the wave lengths of an electron of comparable energy. Hence, $\mathbf{n}$ is usually much smaller than $\mathbf{k}$ or $\mathbf{k'}$ and may be neglected. Equation (9) then simplifies to

$$\mathbf{k'} = \mathbf{k} + \mathbf{K},$$

which states that electrons may make only vertical transitions in the reduced-zone scheme (*cf.* Fig. 33).

CHAPTER IX

APPROXIMATIONAL METHODS

72. Introduction.—One of the most useful methods for obtaining approximate solutions of the Schrödinger equation for solids will be discussed briefly in this chapter. This discussion is supplementary to that of Chap. VI, for the one-electron schemes described there form the basis for the method described here. This method begins by replacing Fock's equations, which usually cannot be separated into one-variable equations, by central field equations that are separable. When accurate one-electron functions have been derived in this way, they are used to compute coulomb and exchange energies. Following this, an attempt is made to estimate the correlation effect and correlation energy. It is difficult to treat these quantities either accurately or concisely; however, they have been handled in a few special cases that will be discussed near the end of the chapter.

73. The Cellular Method.—The primary requirement of a practical plan for solving Fock's equations is that it should replace them by accurate separable equations. Hartree's procedure in the case of free atoms (*cf.* Chap. VI) is a good illustrative example of such a plan. Hartree's equations are not separable when they are applied to an electron configuration that involves an incompletely filled shell of p or d electrons. If the nonspherical part of the coulomb potential of p or d electrons is dropped, however, the equations become separable and may be solved by the methods used for ordinary differential equations. The error made in dropping the nonspherical terms lies within the limits of natural error of the Hartree field method and may be conveniently corrected by perturbation methods.

A similar procedure is possible in solids.[1] Let us restrict the discussion, for the present, to the case of Hartree's equations and overlook the exchange terms. These equations are

$$-\frac{\hbar^2}{2m}\Delta\psi_k(\mathbf{r}_1) + \left[V(\mathbf{r}_1) + \sum_{k'}' e^2 \int \frac{|\psi_{k'}(\mathbf{r}_2)|^2}{r_{12}}d\tau_2 \right]\psi_k(\mathbf{r}_1) = \epsilon\psi_k(\mathbf{r}_1) \quad (1)$$

where $V(\mathbf{r}_1)$ is the total ion-core potential and the sum in the second term extends over all electrons except the kth. The wave function near the

[1] E. WIGNER and F. SEITZ. *Phys. Rev.*, **43**, 804 (1933); **46**, 509 (1934).

nucleus of any atom is determined by the ion-core field of this atom, for this field becomes very large compared with the other potential terms in (1). In simple solids in which the ion cores are closed shells, this field is spherically symmetrical. The potential arising from other parts of the lattice is comparable with that of the ion core of this atom at distances from the nucleus that are of the same order as interatomic dimensions. If the crystal has a high degree of rotational symmetry relative to the nucleus, the potential of the rest of the lattice is nearly spherically symmetrical. Thus, it may be expected that there is a large domain about each nucleus in which the field may be replaced by a spherically symmetrical one. This principle may be used in a wide range of cases, although it is less accurate in crystals having low symmetry than in cubic lattices, such as the alkali metals and alkali halides.

The preceding observation on the symmetry of the field in the neighborhood of each nucleus suggests that the lattice should be partitioned into a set of space-filling polyhedra, which are centered about each of the nuclei, and that the field may be chosen to be centrally symmetrical within each of these polyhedra. Within each polyhedron, Hartree's equations may then be replaced by the equation

$$-\frac{\hbar^2}{2m}\Delta\psi + V(r)\psi = \epsilon\psi \tag{2}$$

where $V(r)$ is the approximate spherically symmetrical field. The solutions of (2) in spherical polar coordinates have the form

$$\psi = \frac{f_l(r,\epsilon)}{r}\Theta_m^l(\theta,\varphi) \tag{3}$$

where $f_l(r,\epsilon)$ is a radial function that satisfies the radial equation

$$-\frac{\hbar^2}{2m}\frac{d^2f}{dr^2} + \left[V(r) + \frac{\hbar^2}{2m}\frac{l(l+1)}{r^2}\right]f = \epsilon f \tag{4}$$

and $\Theta_m^l(\theta,\varphi)$ is a surface harmonic

$$\Theta_m^l = \sqrt{\frac{(l-m)!}{(l+m)!}\frac{2l+1}{4\pi}}P_l^m(\cos\theta)e^{im\varphi}. \tag{5}$$

Bloch functions may be constructed from functions of the type (3) by forming series of those which are associated with the same value of ϵ. The coefficients in these series may be determined from appropriate boundary conditions which we shall discuss below. This procedure forms the basis of the cellular method.

The manner in which cells are chosen depends upon the lattice for which computations are being made. For monatomic crystals in which all atoms are translationally equivalent, the most convenient cell is chosen by taking the polyhedron whose plane faces bisect orthogonally the lines joining an atom with its nearest neighbors. Figures 1 and 2 show cells of this type for monatomic face-centered and body-centered cubic lattices. The cell may be chosen in the same way when more than one atom is present in the unit cell if they are equivalent, as in diamond or in closed-packed hexagonal crystals. Figures 3 and 4 show this type of cell for the crystals that were just mentioned. If the atoms in the unit cell are not equivalent, as in sodium chloride, the cells may not be chosen

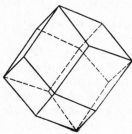

Fig. 1.—The cellular polyhedron for a monatomic face-centered cubic lattice.

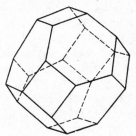

Fig. 2.—The cellular polyhedron for a monatomic body-centered cubic lattice.

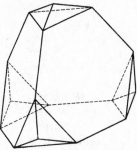

Fig. 3.—The cellular polyhedron for diamond.

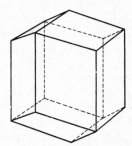

Fig. 4.—The cellular polyhedron for a close-packed hexagonal lattice.

on the basis of symmetry alone. Each case of this type can be handled in many ways.

As we mentioned above, within any cell, the Bloch function $\psi_\mathbf{k}$, associated with the energy ϵ and wave number $\mathbf{k}$, may be expressed[1] in terms of a series of the type

$$\psi_\mathbf{k}(\mathbf{r}) = \sum_{l,m} b_\mathbf{k}{}^{l,m} f_l(r, \epsilon(\mathbf{k})) \Theta_m^l(\theta, \varphi). \tag{6}$$

[1] Wigner and Seitz, *op. cit.*; J. C. Slater, *Phys. Rev.*, **45**, 794 (1934).

The important practical problem associated with this series is that of determining the b from the boundary conditions which the Bloch functions must satisfy. These conditions are the following: (a) ψ_k and its first derivative must be continuous at the boundary points between neighboring polyhedra, and (b) ψ_k must satisfy the relation

$$\psi_k(r + \tau) = \psi_k e^{2\pi i k \cdot r} \tag{7}$$

where τ is a translation vector of the lattice. It turns out that these conditions can be satisfied only for discrete values of ϵ for a given wavenumber vector k. The permissible solutions furnish us the desired relationship between ϵ and k. It may easily be seen that the boundary conditions need be satisfied only for points within a single unit cell, for the form of the function at any point outside this cell is connected with a value inside by Eq. (7). Let us suppose that condition a has been satisfied at all the interfaces between polyhedra in a given unit cell. By definition, the vectors that join opposite faces of the remaining surface are primitive translations of the lattice, since these faces constitute the boundary of the unit cell. Moreover, the points on the faces are the only ones in the unit cell to which the condition (7) can apply.

It has not been feasible in any of the work that has been carried through up to the present time to satisfy conditions a and b at all points of the surfaces of the polyhedra. Instead, all but a finite number of the terms in the series (6) are discarded, and boundary conditions are satisfied at just enough points to determine the coefficients of all these terms. The only justification for this procedure lies in the belief that the series (6) should converge rapidly for small values of k, since then the wave length of the Bloch function is large compared with the dimensions of the cell. Results that have been derived by using this method will be presented in the following chapters.

Shockley[1] made an extensive test of the cellular method in a case in which exact solutions are known, namely, in which $V(r)$ is constant. He found that when a small number of boundary points is used the approximation is satisfactory for zones which normally are occupied but that it usually is very bad for excited states.

Several improvements[2] on the cellular method have been proposed since Shockley's work; but only one of these, namely, the method of Herring and Hill, has been applied to practical problems. These workers assumed that the ψ_k functions at a given point on the zone may be expressed as a finite sum of free-electron functions of the type

[1] W. Shockley, *Phys. Rev.*, **52**, 866 (1937).
[2] J. C. Slater, *Phys. Rev.*, **51**, 846 (1937); G. Wannier, *Phys. Rev.*, **53**, 671 (1938); C. C. Herring and A. G. Hill, *Phys. Rev.* (to appear).

$$\psi_{\mathbf{k}} = \sum_{i=1}^{n} a_i{}^{\mathbf{k}} e^{2\pi i \mathbf{k}_i \cdot \mathbf{r}} \tag{8}$$

in which the constants $a_i{}^{\mathbf{k}}$ and $\mathbf{k}_i$ were chosen by use of group theory. The matrix components of the crystalline potential that connect approximate wave functions of the type (8) at the corresponding points in several zones were then computed, and the resulting matrix was diagonalized. In this way, approximate values of $\epsilon(\mathbf{k})$ for the boundary points were obtained. This method, which was applied by Herring and Hill to beryllium (*cf.* Sec. 81), evidently is a special case of the perturbation method described in Sec. 61, the functions (8) being the appropriate linear combinations in the zero-order approximation.

74. The Hartree Field.—In order to determine a self-consistent Hartree field within each of the polyhedra in the cellular approximation, it is necessary, first of all, to adopt a starting field or charge distribution from which wave functions may be computed. This field may be chosen in many ways. For example, when one is dealing with a monatomic solid, the ion-core field plus the field arising from a uniform distribution of valence electrons may be used. In any case, the starting potential V_I and the starting charge distribution ρ_I are related by the equation

$$V_I(\mathbf{r}_1) = \int \frac{\rho_I(\mathbf{r}_2)}{r_{12}} d\tau_2. \tag{1}$$

Bloch functions $\psi_{\mathbf{k}}$ may be constructed from the starting field by use of the method described in Sec. 73, which combines the solutions of the equation

$$-\frac{\hbar^2}{2m}\Delta\psi + V_I^s(\mathbf{r})\psi = \epsilon\psi \tag{2}$$

where V_I^s is the spherically symmetric part of (1) in a given polyhedron. A new electronic charge distribution, $e\sum_{\mathbf{k}}|\psi_{\mathbf{k}}(\mathbf{r})|^2$, where $\mathbf{k}$ is summed over all occupied levels, is determined by these $\psi_{\mathbf{k}}$, and a new Hartree field

$$V_{II}(\mathbf{r}_1) = e^2 \int \frac{\Sigma|\psi_{\mathbf{k}}(\mathbf{r}_2)|^2}{r_{12}} d\tau_2 + V_c(\mathbf{r}_1) \tag{3}$$

may be determined from this distribution. In (3), $V_c(\mathbf{r}_1)$ is the total potential from the rigid ion cores; that is,

$$V_c(\mathbf{r}_1) = \sum_{i,\alpha} v_{i,\alpha}(\mathbf{r}_1) \tag{4}$$

where $v_{i,\alpha}(\mathbf{r}_1)$ is the potential at $\mathbf{r}_1$ that arises from the core of the αth atom in the ith unit cell. The entire potential (3) may be written in the form of a lattice sum, such as the sum in Eq. (4), by expressing the first term of (3) in the form

$$\sum_{\mathbf{k}} e^2 \int \frac{|\psi_{\mathbf{k}}(\mathbf{r}_2)|^2}{r_{12}} d\tau_2 = \sum_{i,\alpha} \int \frac{\rho_{i,\alpha}(\mathbf{r}_2)}{r_{12}} d\tau_2$$

where $\rho_{i,\alpha}$ is the value of $\sum_{\mathbf{k}} e^2 |\psi_{\mathbf{k}}|^2$ within the αth polyhedron of the ith unit cell.

Usually, $V_{\mathrm{II}}(\mathbf{r})$ will not be spherically symmetrical within a polyhedron, since it may contain all surface harmonics that are compatible with the symmetry of the polyhedron. Hence, the closest agreement that may be expected, is that V_{I}^s and the spherically symmetric part of V_{II} should be identical. This will not be the case, unless by fortunate chance. Hence, it usually is necessary to choose a new field V_{III} as the starting field for another computation. There are no general rules for choosing V_{III} in such a way that the field V_{IV}, which is derived from its wave functions in the way in which V_{II} was derived from the solutions for V_{I}, will be closer to V_{III} than V_{II} was to V_{I}. The convergence is often swift in a monatomic lattice of equivalent atoms if V_{III} is taken as the mean of V_{I} and V_{II}, but this scheme does not work very well in solids that contain two or more different kinds of atom. The factors that govern the speed of convergence have not been investigated in any general way.

The final wave functions that are derived from a self-consistent Hartree field may differ appreciably from the solutions of Fock's equations, for exchange terms are neglected in Hartree's equations. Unfortunately, the exchange terms usually cannot be included merely by adding one-electron potential terms (*cf.* Chap. VI). There are special cases, however, in which they may be included very simply; we shall discuss these in the next section.

75. Exchange Terms*.—There are two cases in which the exchange terms have been handled rigorously, namely, the cases of perfectly bound electrons and perfectly free electrons. In the first case, the atoms are so far apart that the electronic wave functions of separate atoms do not overlap appreciably; in the second, the potential field in which the electrons move is so nearly constant that the one-electron functions have the form $e^{2\pi i \mathbf{k} \cdot \mathbf{r}}$. We shall discuss these two cases in detail.

a. Rigidly Bound Electrons (Narrow Bands).—Atoms and ions when they are far apart affect only the electrostatic field in their own vicinity. Hence, in this case, the atomic or Heitler-London approximation is the most accurate of the one-electron approximational methods, and the Bloch approximation is as accurate only when it is identical with the Heitler-

London method. We saw in Chap. VIII that the two schemes are equivalent when there are completely closed shells if determinantal eigenfunctions are used. Let us consider the way in which this equivalence appears in Fock's equations for the two systems. For simplicity, we shall deal with a monatomic lattice of atoms whose valence electrons form closed shells of the s^2 type. We shall let $\psi(\mathbf{r} - \mathbf{r}(n))$ represent the Heitler-London wave function of the electron that is centered about the nucleus at the position $\mathbf{r}(n)$. Fock's equation for $\psi(\mathbf{r} - \mathbf{r}(n))$ has the form

$$-\frac{\hbar^2}{2m}\Delta\psi(\mathbf{r}_1 - \mathbf{r}(n)) + \left\{ V(\mathbf{r}_1 - \mathbf{r}(n)) + \right.$$
$$\left. \int \frac{|\psi(\mathbf{r}_2 - \mathbf{r}(n))|^2}{r_{12}}d\tau_2 \right\}\psi(\mathbf{r} - \mathbf{r}(n)) = \epsilon\psi(\mathbf{r}_1 - \mathbf{r}(n)) \quad (1)$$

where $V(\mathbf{r}_1 - \mathbf{r}(n))$ is the ion-core field of the atom at $\mathbf{r}(n)$ and the integral is the coulomb potential of the other electron on this atom. Exchange integrals do not occur because, by assumption, electrons on the same atoms have opposite spin and those on different atoms do not overlap appreciably. Outside a given atom, the coulomb field of the electrons cancels the ion-core field; hence, there are no coulomb terms between different atoms. Since exchange terms are absent, Hartree's and Fock's equations are identical in this particular case.

Now, let us consider Hartree's and Fock's equations in the Bloch approximation. As long as the closed shells do not overlap, we should be able to write the Bloch functions in the form

$$\psi_\mathbf{k}(\mathbf{r}) = \frac{1}{\sqrt{N}}\sum_n e^{2\pi i\mathbf{k}\cdot\mathbf{r}(n)}\psi(\mathbf{r} - \mathbf{r}(n)) \quad (2)$$

where $\mathbf{k}$ is the wave-number vector, $\psi(\mathbf{r} - \mathbf{r}(n))$ is the normalized one-electron function that is centered about the atom at $\mathbf{r}(n)$, and N is the total number of atoms in the lattice. In a filled zone, each value of $\mathbf{k}$ is associated with a pair of electrons having opposite spin. In the following paragraphs of this section, summations over $2\mathbf{k}$ will imply summation over both types of states associated with the N values of $\mathbf{k}$ in a zone.

Hartree's equations for the $\psi_\mathbf{k}$ are

$$-\frac{\hbar^2}{2m}\Delta\psi_\mathbf{k}(\mathbf{r}_1) + \sum_{2\mathbf{k}'}{}' e^2 \int \frac{|\psi_{\mathbf{k}'}(\mathbf{r}_2)|^2 d\tau_2}{r_{12}} + \sum_n v_n(\mathbf{r}_1)\psi_\mathbf{k}(\mathbf{r}_1) = \epsilon\psi_\mathbf{k} \quad (3)$$

where $\displaystyle\sum_{2\mathbf{k}'}{}'$ extends over all pairs of electrons in the zone, except one of the pairs having wave number $\mathbf{k}$, and $v_n(\mathbf{r}_1)$ is the ion-core field of the atom at $\mathbf{r}(n)$. The potential

$$\int \frac{|\psi_k|^2}{r_{12}} d\tau_2, \tag{4}$$

arising from a single electron, is negligible at any point in the lattice as long as the crystal contains a large number of atoms. Hence, it is immaterial whether the sum in (3) extends over all electrons or all electrons except one and the primes may be deleted. Using (2), we have

$$\sum_{2k}|\psi_k(\mathbf{r})|^2 = \sum_{2k}\sum_{n,m} e^{2\pi i \mathbf{k} \cdot [\mathbf{r}(n) - \mathbf{r}(m)]} \psi(\mathbf{r} - \mathbf{r}(n)) \psi^*(\mathbf{r} - \mathbf{r}(m)). \tag{5}$$

The terms for which $\mathbf{r}(n)$ differs from $\mathbf{r}(m)$ vanish because the ψ do not overlap. Thus,

$$\sum_{2k}|\psi_k(\mathbf{r})|^2 = \sum_{n} 2|\psi(\mathbf{r} - \mathbf{r}(n))|^2 \tag{6}$$

where the factor 2 appears because of spin. Outside the mth atom, the potential arising from the term $2|\psi(\mathbf{r} - \mathbf{r}(m))|^2$ in (6) and the ion-core field $v_n(\mathbf{r})$ cancel one another. Hence, only the term in (6), arising from $|\psi(\mathbf{r} - \mathbf{r}(n))|^2$, need be considered in the vicinity of the nth atom, for the other terms are canceled by the ion-core terms. Thus, near the nth atom, (3) reduces to

$$-\frac{\hbar^2}{2m}\Delta\psi(\mathbf{r}_1 - \mathbf{r}(n)) + \left\{ 2e^2 \int \frac{|\psi(\mathbf{r}_2 - \mathbf{r}(n))|^2}{r_{12}} d\tau_2 + v_n(\mathbf{r}_1) \right\} \psi(\mathbf{r}_1 - \mathbf{r}(n)) =$$
$$\epsilon\psi(\mathbf{r}_1 - \mathbf{r}(n)). \tag{7}$$

This equation is not the same as Eq. (1), because of the factor 2 which appears in the coulomb integral. It is easy to trace this spurious screening term to the fact that the electrons are completely uncorrelated in the total wave function on which Hartree's equations are based (*cf.* Chap. VI). In the Hartree approximation of Bloch's scheme, the probability of an electron being at a given atom is determined only by the average charge distribution $2|\psi|^2$ of other electrons on the atom. Actually, other electrons tend to stay away from this atom when the given electron is there both because of the electron repulsion and because of exchange.

Let us consider next the Fock approximation. In this case, we have, in addition to the terms on the left-hand side of (3), the exchange terms

$$-\sum_{\substack{k' \\ \| \text{ spin}}} e^2 \left[\int \frac{\psi_{k'}^*(\mathbf{r}_2)\psi_k(\mathbf{r}_2)}{r_{12}} d\tau_2 \right] \psi_{k'}(\mathbf{r}_1) \tag{8}$$

where the sum extends only over electrons of one kind of spin. This sum need not be primed if the prime is dropped in (3), since the additional terms just cancel one another. Using (2) once again, we find that (8) is equal to

$$-\frac{e^2}{N^{\frac{3}{2}}}\sum_{\mathbf{k}'}\left\{\sum_n e^{2\pi i\mathbf{k}'\cdot\mathbf{r}(n)}\psi(\mathbf{r}_1 - \mathbf{r}(n))\right\}$$
$$\left\{\sum_m\int\frac{e^{2\pi i(\mathbf{k}-\mathbf{k}')\cdot\mathbf{r}(m)}|\psi(\mathbf{r}_2 - \mathbf{r}(m))|^2 d\tau_2}{r_{12}}\right\}. \quad (9)$$

The cross terms in the second factor of the sum have been dropped because the ψ on different atoms do not overlap. This equation, in turn, may be written in the form

$$-\frac{e^2}{N^{\frac{3}{2}}}\sum_{n,m}\sum_{\mathbf{k}'}e^{2\pi i\mathbf{k}'\cdot[\mathbf{r}(n)-\mathbf{r}(m)]}e^{2\pi i\mathbf{k}\cdot\mathbf{r}(m)}\psi(\mathbf{r}_1 - \mathbf{r}(n))\int\frac{|\psi(\mathbf{r}_2 - \mathbf{r}(m))|^2}{r_{12}}d\tau_2. \quad (10)$$

The sum $\sum_{\mathbf{k}'}e^{i\mathbf{k}'\cdot[\mathbf{r}(n)-\mathbf{r}(m)]}$ is equal to $N\delta_{n,m}$, however, for $\mathbf{k}'$ is summed over the points of a zone. Hence, (8) reduces to

$$-\frac{e^2}{\sqrt{N}}\sum_n\left\{e^{i\mathbf{k}\cdot\mathbf{r}(n)}\psi(\mathbf{r}_1 - \mathbf{r}(n))\int\frac{|\psi(\mathbf{r}_2 - \mathbf{r}(n))|^2}{r_{12}}d\tau_2\right\}. \quad (11)$$

Thus, Fock's equation for $\psi(\mathbf{r}_1 - \mathbf{r}(n))$ is

$$-\frac{\hbar^2}{2m}\Delta\psi(\mathbf{r}_1 - \mathbf{r}(n)) + \left\{e^2\int\frac{|\psi(\mathbf{r}_2 - \mathbf{r}(n))|^2}{r_{12}}d\tau_2 + v_n(\mathbf{r}_1)\right\}\psi(\mathbf{r}_1 - \mathbf{r}(n)) =$$
$$\epsilon\psi(\mathbf{r}_1 - \mathbf{r}(n)),$$

which is identical with (1). Thus, the exchange integrals remove a part of the spurious screening that occurs in Hartree's equations (7). This fact shows, however, that we cannot expect exchange to compensate for all the inadequacies of Hartree's scheme, even in the simple example discussed above. The exchange correlation affects only electrons with parallel spin and does not alter the spurious screening of electrons with antiparallel spin. The remaining defect may be corrected only by solving the many-body problem by a method that is more accurate than the one-electron approximation.

Let us consider an example in which the atoms do not have closed shells and the zones are not completely filled. We cannot expect the Heitler-London and Bloch schemes to lead to identical results in this case, but we can examine the relative merits of the two. We shall assume that the atoms have a single valence electron outside the closed shells and shall designate the ion-core field for this electron by $V(r)$. The equation for the Heitler-London function associated with the atom at $\mathbf{r}(n)$ is

$$-\frac{\hbar^2}{2m}\Delta\psi(\mathbf{r} - \mathbf{r}(n)) + V(\mathbf{r} - \mathbf{r}(n))\psi(\mathbf{r} - \mathbf{r}(n)) = \epsilon\psi(\mathbf{r} - \mathbf{r}(n)) \quad (12)$$

as long as the atoms do not overlap. Under these conditions, the total energy of the lattice, relative to a system of N ionized atoms is $N\epsilon$, and is independent of the assignment of spins to the electrons. This solution evidently is as accurate as the ion-core field.

In the Bloch approximation, the wave functions have the form (2) and the $\mathbf{k}$ may be assigned spins in many different ways. The $\mathbf{k}$ range over an entire zone if the spins are parallel, for example, but many other arrangements are possible. The total wave function is a single determinant in two cases, namely, the states of highest multiplicity, in which all spins are parallel, and the state of zero multiplicity, in which all spins are paired. In both these cases, Hartree's equations have the form

$$-\frac{\hbar^2}{2m}\Delta\psi(\mathbf{r}_1 - \mathbf{r}(n)) + \left\{ V(\mathbf{r}_1 - \mathbf{r}(n)) + \right.$$
$$\left. e^2 \int \frac{|\psi(\mathbf{r}_2 - \mathbf{r}(n))|^2}{r_{12}} d\tau_2 \right\}\psi(\mathbf{r}_1 - \mathbf{r}(n)) = \epsilon\psi(\mathbf{r}_1 - \mathbf{r}(n)), \quad (13)$$

which differs from (12) by spurious screening terms. The exchange terms of Fock's equations remove this term in the case in which all spins are parallel, for then the $\mathbf{k}$ range over an entire zone and the exchange term is identical with (8). On the other hand, if the spins are paired and only the lower half of a zone is filled, the exchange term for $\psi_{\mathbf{k}}$ is [*cf.* Eq. (9)]

$$-\frac{e^2}{N^{\frac{2}{3}}}\sum_{\mathbf{k}'}\left\{ \sum_n e^{2\pi i\mathbf{k}'\cdot\mathbf{r}(n)}\psi(\mathbf{r}_1 - \mathbf{r}(n)) \right\}\left\{ \sum_m \int \frac{e^{2\pi i(\mathbf{k} - \mathbf{k}')\cdot\mathbf{r}(m)}|\psi(\mathbf{r}_2 - \mathbf{r}(m))|^2 d\tau_2}{r_{12}} \right\}$$

where $\mathbf{k}'$ is summed over half a zone. This may be rewritten in the form

$$-\frac{e^2}{N^{\frac{2}{3}}}\sum_{n,m}\sum_{\mathbf{k}'} e^{2\pi i\mathbf{k}'\cdot[\mathbf{r}(n) - \mathbf{r}(m)]}e^{2\pi i\mathbf{k}\cdot\mathbf{r}(m)}\psi(\mathbf{r}_1 - \mathbf{r}(n)) \int \frac{|\psi(\mathbf{r}_2 - \mathbf{r}(m))|^2}{r_{12}} d\tau_2$$

[*cf.* Eq. (10)]. The summation

$$\sum_{\mathbf{k}'} e^{2\pi i\mathbf{k}'\cdot[\mathbf{r}(n) - \mathbf{r}(m)]}$$

is not equal to $N\delta_{n,m}$, for $\mathbf{k}'$ does not range over an entire zone. If we assume, however, that the atoms are so far apart that the summation is negligible when $\mathbf{r}(n) \neq \mathbf{r}(m)$, (13) reduces to $(N/2)\delta_{n,m}$. Hence, in this case, Fock's equation for ψ in the Bloch scheme is

$$-\frac{\hbar^2}{2m}\Delta\psi(\mathbf{r}_1 - \mathbf{r}(n)) + \left\{V + \frac{e^2}{2}\int\frac{|\psi(\mathbf{r}_2 - \mathbf{r}(n))|^2}{r_{12}}d\tau_2\right\}\psi(\mathbf{r}_1 - \mathbf{r}(n)) =$$
$$\epsilon\psi(\mathbf{r}_1 - \mathbf{r}(n)). \quad (14)$$

The exchange term cancels only half the spurious screening term. This fact shows that the Bloch scheme may be very inaccurate[1] for some states when the bands are very narrow. Thus, the energy of the lattice in the Bloch approximation is

$$N\frac{e^2}{4}\int\frac{|\psi(\mathbf{r}_1)|^2|\psi(\mathbf{r}_2)|^2}{r_{12}}d\tau_{12} \quad (15)$$

higher than the energy in the Heitler-London approximation when spins are paired. The integral (15) is equal to

$$\frac{3}{32}\frac{e^2}{a_h},$$

or about 2.9 ev, for hydrogenic $1s$ functions.

As the atoms are brought nearer and nearer, the terms in (13) for $\mathbf{r}(n) \neq \mathbf{r}(m)$ may be neglected no longer and the exchange terms reduce part of the screening effect of electrons on different atoms. This type of correlation effect does not occur in the Heitler-London scheme. Thus, a part of the advantage of the Heitler-London scheme over the Bloch scheme begins to disappear as atoms begin to overlap. It is for this reason, among others, that the Bloch scheme may be used in competition with the Heitler-London scheme in the computation of binding energies of actual solids.

b. Perfectly Free Electrons.—As a working model for discussing the case of perfectly free electrons, we shall consider a system of N electrons in a box that contains a uniform positive charge distribution of total magnitude Ne. The positive charge compensates for the over-all repulsion of the electrons and makes the system stable. The exchange terms for this system have not been treated in the Heitler-London approximation. Although this solution probably would be very poor kinematically because the metallic properties of the model are not apparent in the Heitler-London approximation (*cf.* Sec. 66, Chap. VIII), the cohesive energy probably would compare favorably with that derived on the basis of the Bloch approximation.

Since the positive-ion distribution is uniform in our model, the Bloch functions have the free wave form

---

[1] Caution must be used in applying Bloch functions to discussions of ferromagnetism; for, as we see here, the Bloch approximation gives a spurious ferromagnetism in a case in which states of all spin actually have the same energy.

$$\psi_{\mathbf{k}} = \frac{1}{\sqrt{V}} e^{2\pi i \mathbf{k} \cdot \mathbf{r}} \qquad (16)$$

where V is the volume of the crystal. Let us assume that all the lowest electronic levels are doubly occupied. The value, k_0, of k at the top of the filled region is given by[1]

$$k_0 = \left(\frac{3N}{8\pi V} \right)^{\frac{1}{3}}, \qquad (17)$$

and the exchange term in the equation for $\psi_{\mathbf{k}}$ is

$$-\frac{e^2}{V^{\frac{2}{3}}} \sum_{\mathbf{k}'} e^{2\pi i \mathbf{k}' \cdot \mathbf{r}_1} \int \frac{e^{2\pi i (\mathbf{k} - \mathbf{k}') \cdot \mathbf{r}_2}}{r_{12}} d\tau_2, \qquad (18)$$

which is identical with

$$-\frac{e^2}{V^{\frac{2}{3}}} e^{2\pi i \mathbf{k} \cdot \mathbf{r}_1} \sum_{\mathbf{k}'} \int \frac{e^{2\pi i (\mathbf{k}' - \mathbf{k}) \cdot (\mathbf{r}_1 - \mathbf{r}_2)}}{r_{12}} d\tau_2. \qquad (19)$$

This integral is independent of $\mathbf{r}_1$ and may be evaluated by direct methods.[2] The result is that (19) reduces to

$$-C_k \frac{e^{2\pi i \mathbf{k} \cdot \mathbf{r}_1}}{\sqrt{V}}$$

where

$$C_k = e^2 k_0 \left(2 + \frac{k_0^2 - k^2}{k_0 k} \log \left| \frac{k + k_0}{k - k_0} \right| \right)$$

$$= 0.306 \frac{e^2}{r_s} \left[2 + \frac{1}{\alpha} (1 - \alpha^2) \log \left| \frac{1 + \alpha}{1 - \alpha} \right| \right] \qquad (20)$$

in which $\alpha = k/k_0$ and r_s is defined by the equation

$$\frac{4\pi}{3} r_s^3 = \frac{V}{N}.$$

Hence, (19) is equal to a constant times $\psi_{\mathbf{k}}$, or

$$A \psi_{\mathbf{k}} = -C_k \psi_{\mathbf{k}} \qquad (21)$$

for perfectly free electrons, where A is the Dirac exchange operator. The function $C_k r_s / e^2$ is shown in Fig. 5.

The mean value of $-C_k$ for all electrons is equal to twice the exchange energy per electron. We shall evaluate this energy directly.[3] The total exchange energy is

---

[1] *Cf.* Eq. 20, Sec. 26; also, Sec. 49.

[2] P. A. M. Dirac, *Proc. Cambridge Phil. Soc.*, **26**, 376 (1930); J. Bardeen, *Phys. Rev.*, **49**, 653 (1936).

[3] Wigner and Seitz, *op. cit.*

$$-\frac{1}{V^2}\sum_{\mathbf{k},\mathbf{k'}}e^2\int\frac{e^{2\pi i(\mathbf{k'}-\mathbf{k})\cdot(\mathbf{r_1}-\mathbf{r_2})}}{r_{12}}d\tau_1 d\tau_2$$

$$\cong -\frac{e^2}{V^2}\int e^{2\pi i(\mathbf{k'}-\mathbf{k})\cdot(\mathbf{r_1}-\mathbf{r_2})}\rho_k\rho_{k'}d\mathbf{k'}\,d\mathbf{k}\,d\tau_1 d\tau_2 \qquad (22)$$

where $\rho_k = V$ is the density of points in $\mathbf{k}$ space and the summation has been replaced by an integration. This replacement is allowable for

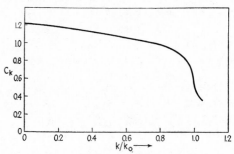

FIG. 5.—The negative of the exchange energy for perfectly free electrons as a function of k/k_0. The energy units are e^2/r_s.

ordinary-sized crystals since their states are very dense. Integrating over $\mathbf{k}$ and $\mathbf{k'}$, we find that (22) reduces to

$$-4\pi e^2\int\frac{(2\pi k_0 r_{12}\cos 2\pi k_0 r_{12} - \sin 2\pi k_0 r_{12})^2}{r_{12}^5}d\tau_1 d\tau_2 = -4\pi e^2 V k_0^4. \quad (23)$$

k_0 may be replaced by its value (17), and then the mean exchange energy per electron is found to be

$$-\frac{e^2}{4\pi^{\frac{1}{3}}}n_0^{\frac{1}{3}}3^{\frac{1}{3}} \qquad (24)$$

where n_0 is the number of electrons per unit volume. In terms of r_s, (24) becomes

$$-0.458\frac{e^2}{r_s}. \qquad (24a)$$

The implications of the exchange terms for perfectly free electrons were discussed in Sec. 49, Chap. VI. We saw there that the exchange terms have the effect of keeping electrons of antiparallel spin apart (*cf.* Fig. 2, Chap. VI). The absence of such correlational effects for electrons of antiparallel spin in the Fock-Bloch approximation constitutes a large source of error. The Heitler-London approximation will furnish some correlation between electrons of both kinds of spin by keeping them on separate atoms; but as we have seen in molecular problems, such as that of H_2, this method of introducing correlations is not very accurate. Thus, it seems reasonable to say that both one-electron schemes have comparable errors when the electronic interactions are appreciable. We shall discuss the correlation terms for the Bloch scheme in the next section.

76. Correlation Correction for Perfectly Free Electrons.—The total correlation correction for the free-electron model in which the positive charge distribution is uniform has been investigated most thoroughly by Wigner.[1] A part of this correction is the exchange energy

$$-0.458\,\frac{e^2}{r_s}, \tag{1}$$

which was derived in Sec. 75 (*cf.* Eq. 24*a*). We shall be interested here in the additional term that arises primarily from electrons of antiparallel spin.

The simplest case in which the total correlation may be estimated is that in which the electron density is so low that the electronic kinetic energy is negligible. This is the case in which r_s is very large. The electrons then will form the most stable lattice arrangement, which, according to the Madelung type computations, is a body-centered cubic arrangement. Its energy, relative to the energy of a perfectly uniform negative charge distribution, is

$$-0.746\,\frac{e^2}{r_s}. \tag{2}$$

Hence, the correlation correction to the Bloch-Fock scheme, which is the difference between (1) and (2), is

$$-0.288\,\frac{e^2}{r_s} \tag{3}$$

for low electron density. The expression that is valid for small values of r_s should approach this asymptotically.

The details of Wigner's calculations for high electron density are too involved for discussion in a book of this type. We shall present only a brief summary of his procedure and results.

In the starting approximation, in which the correlation term (3) is zero, the total electronic wave function may be taken in the form

$$\frac{1}{(N/2)!}\begin{vmatrix}\psi_1(x_1)\dots\dots\dots\psi_1(x_{\frac{N}{2}})\\ \cdot\qquad\qquad\cdot\\ \cdot\qquad\qquad\cdot\\ \cdot\qquad\qquad\cdot\\ \psi_{\frac{N}{2}}(x_1)\dots\dots\psi_{\frac{N}{2}}(x_{\frac{N}{2}})\end{vmatrix}\begin{vmatrix}\psi_1(y_1)\dots\dots\dots\psi_1(y_{\frac{N}{2}})\\ \cdot\qquad\qquad\cdot\\ \cdot\qquad\qquad\cdot\\ \cdot\qquad\qquad\cdot\\ \psi_{\frac{N}{2}}(y_1)\dots\dots\psi_{\frac{N}{2}}(y_{\frac{N}{2}})\end{vmatrix} \tag{4}$$

[1] E. Wigner. *Phys. Rev.*, **46**. 1002 (1934); *Trans. Faraday Soc.*, **34**, 678 (1938).

where the x refer to electrons of one spin, the y to electrons of opposite spin, the ψ are free-electron wave functions, and N is the total number of electrons. This function does not satisfy the Pauli principle, but it has the same energy as though it did. Exchange effects, which are the principal consequence of the Pauli principle in the one-electron approximation, are given correctly by (4) since all terms of given spin are contained in the same determinant.

Wigner considers the following modified form of (4)

$$\begin{vmatrix} \psi_1(x_1; y_1, \ldots, y_{\frac{N}{2}}) & \cdots & \psi_1(x_{\frac{N}{2}}; y_1, \ldots, y_{\frac{N}{2}}) \\ \cdot & & \cdot \\ \cdot & \cdot & \cdot \\ \cdot & & \cdot \\ \psi_{\frac{N}{2}}(x_1; y_1 \ldots y_{\frac{N}{2}}) & \cdots & \psi_{\frac{N}{2}}(x_{\frac{N}{2}}; y_1, \ldots y_{\frac{N}{2}}) \end{vmatrix} \begin{vmatrix} \psi_1(y_1) & \cdots & \psi_1(y_{\frac{N}{2}}) \\ \cdot & & \cdot \\ \cdot & \cdot & \cdot \\ \cdot & & \cdot \\ \psi_{\frac{N}{2}}(y_1) & \cdots & \psi_{\frac{N}{2}}(y_{\frac{N}{2}}) \end{vmatrix} \quad (5)$$

in order to obtain functions from which to construct a better total wave function. Here, the $\psi(y_i)$ are the wave functions for free electrons. The $\psi(x_i, y_1, \ldots, y_n)$ are to be determined by the condition that the mean energy of (5) should be a minimum. The ψ are then used in a new total wave function from which a new total energy may be computed. This new total wave function evidently will not be a rigorous solution of the complete Schrödinger equation, but it is a closer approximation than the function based on (4).

The correlation energy obtained by Wigner in this way is accurate only for high electron densities because of his approximate computational methods. He found that this result may be joined to (3) by the function

$$-\frac{0.288e^2}{r_s + 5.1a_h}, \quad (6)$$

which he estimates is accurate to within 20 per cent.

An expression similar to (20) in the preceding section that will give correlation energy as a function of k has not been developed. However, we may compute the correlation energy of the uppermost electron in the filled band. Suppose that a number of electrons are removed from the top of the band so that the total number is equal to N_e instead of N, the total number in a neutral lattice. The correlation energy will change as a result, and the new value may be derived by taking into account the change in density, that is, by changing r_s to the value $(N/N_e)^{\frac{1}{3}}r_s$. Hence, if the total correlation energy is expressed in the form

$$E_c = N_e g\left(\left(\frac{N}{N_e}\right)^{\frac{1}{3}} r_s\right) \quad (7)$$

where, according to (6), g is

$$g(r) = -e^2 \frac{0.288}{r + 5.1a_h}, \tag{8}$$

the correlation energy of the upper electrons is

$$\left(\frac{\partial E_c}{\partial N_e}\right)_{N_e=N} = g(r_s) - \frac{1}{3}g'(r_s)r_s. \tag{9}$$

This expression may be used in considering energy changes during a process in which an electron is removed or added to a system, such as during thermionic or photoelectric emission or when an electron jumps from the conduction band to a vacant inner-shell level during X-ray emission.

CHAPTER X

THE COHESIVE ENERGY

77. Introduction*.—The degree to which the computed energy of a system agrees with an accurate observed value is a measure of the accuracy of the wave functions that are used in the theoretical computations, because of the variational theorem. For this reason, computations of the cohesive energies of solids occupy an important position in the development of the theory of solids. The existing calculations deal with simple substances, such as the monovalent metals, the alkali halides and hydrides, and rare gas solids, all of which will be discussed in this chapter under three headings: metals, ionic crystals, and molecular crystals. There have been no accurate computations that deal with valence crystals, such as diamond.

Before beginning the detailed discussion of cases, we shall derive several useful equations. The cohesive energy of a solid is defined as the difference between the energy of the crystal in the normal, bound state, at absolute zero of temperature, and the energy of the isolated atoms or molecules of which it is composed. If the surface energy is neglected, the cohesive energy is proportional to the total number of atoms or molecules in the lattice and may be expressed conveniently in units of electron volts per molecule or in the thermochemist's unit of kilogram-calories per mol. This energy is equal to the actual heat of sublimation only when the substance evaporates into the atomic or molecular constituents to which the separated system is referred.

Let us derive an approximate expression for the cohesive energy in the general case in which there are ν atoms per molecule and m molecules per unit cell of the crystal. Then, if $E_A{}^\beta (\beta = 1, \cdots, \nu)$ is the energy of the βth neutral atom and Ψ is the complete electronic wave function for N unit cells of the crystal, the cohesive energy E_c, relative to a system of separated atoms, is

$$E_c = mN \sum_{\beta=1}^{\nu} E_A{}^\beta - \frac{1}{Nm} \int \Psi^* H \Psi d\tau(x_1, \cdots, z_n, \zeta_1, \cdots, \zeta_n) \quad (1)$$

where H is the complete Hamiltonian of the crystal and n is the total number of electrons. We shall develop this in the important case in which Ψ is expressed in terms of the solutions of Fock's or Hartree's equations.

345

If the one-electron functions for the βth atom are $\varphi_k{}^\beta(x,y,z)$, where k ranges over the n_β electrons of the atom, and if the total wave function of the atom is $\Psi_A{}^\beta(x_1, \cdots, z_n, \zeta_1, \cdots, \zeta_{n_\beta})$, the energy of the atom is

$$E_A{}^\beta = \sum_{k=1}^{n_\beta} \int \varphi^*{}_k{}^\beta \left(-\frac{\hbar^2}{2m}\Delta - \frac{Z_\beta e^2}{r_{\beta k}} \right) \varphi_k{}^\beta d\tau(x,y,z,\zeta) +$$

$$\frac{1}{2}\sum_{k,l} \int \Psi^*{}_A{}^\beta \frac{e^2}{r_{kl}} \Psi_A{}^\beta d\tau(x_1, \cdots, \zeta_{n_\beta}). \quad (2)$$

The notation that is used in this equation is the same as that which was introduced in Chap. VI. $-Z_\beta e^2/r_{\beta k}$ is the potential of an electron in the field of the nucleus, whose charge is $Z_\beta e$, and e^2/r_{kl} is the interaction potential of the kth and lth electrons. The last term in (2) may be expressed as a sum of coulomb and exchange integrals of the type

$$I_{kl}{}^\beta = e^2 \int \frac{|\varphi_k(\mathbf{r}_1)|^2|\varphi_l(\mathbf{r}_2)|^2}{r_{12}} d\tau_1' d\tau_2' \quad (3a)$$

and

$$K_{kl}{}^\beta = e^2 \int \frac{\varphi_k{}^*(\mathbf{r}_1)\varphi_l{}^*(\mathbf{r}_2)\varphi_k(\mathbf{r}_2)\varphi_k(\mathbf{r}_1)}{r_{12}} d\tau_1' d\tau_2' \quad (3b)$$

which have coefficients depending upon the multiplicity of Ψ_A. We shall not be concerned with the numerical values of these coefficients at present but shall express the two sums in the form

$$\frac{1}{2}\sum_{k,l}\alpha_{kl}{}^\beta I_{kl}{}^\beta \quad \text{and} \quad -\frac{1}{2}\sum_{k,l}\beta_{kl}{}^\beta K_{kl}{}^\beta. \quad (4)$$

The wave functions of those g_β electrons on the βth atom that belong to rigid closed shells are practically unchanged in passing from the free to the bound state. We shall designate the terms in (2) and (4) that involve only these g_β electrons by the symbol $\sum_g^\beta$. Similarly, the terms that involve these electrons and the remaining $n_\beta - g_\beta$ electrons on the atom will be designated by $\sum_{g,n-g}^\beta$, and the terms that involve only the $n_\beta - g_\beta$ electrons outside closed shells will be designated by $\sum_{n-g}^\beta$. The first set of terms $\sum_g^\beta$ is canceled by the similar term of Eq. (1) that appears in the expression for the total energy of the solid and may be dropped from consideration. The two other terms are

$$\sum_{g,n-g}^{\beta} = \sum_{k=1}^{g} \sum_{l=g_\beta}^{n_g} (\alpha_{kl}{}^\beta I_{kl}{}^\beta - \beta_{kl}{}^\beta K_{kl}{}^\beta), \tag{5}$$

$$\sum_{n-g}^{\beta} = \sum_{k=g}^{n_\nu} \int \varphi_k{}^\beta * \left(-\frac{h^2}{2m}\Delta - \frac{Z_\beta e^2}{r_{\beta k}} \right) \varphi_k{}^\beta d\tau(x,y,z) +$$

$$\frac{1}{2}\sum_{k,l=g}^{n_\nu} (\alpha_{kl}{}^\beta I_{kl}{}^\beta - \beta_{kl}{}^\beta K_{kl}{}^\beta). \tag{6}$$

The first of these expressions is the energy of the interaction between the valence and closed-shell electrons; the second is the total energy of the valence electrons, minus this interaction. In the two cases in which the one-electron functions are solutions of either Hartree's or Fock's equations, namely,

$$H^H \varphi_i{}^\beta = \epsilon^\beta(H)\varphi_i{}^\beta,$$
$$H^F \varphi_i{}^\beta = \epsilon^\beta(F)\varphi_i{}^\beta, \tag{7}$$

respectively, it may be verified readily that the sum of (5) and (6) reduces to

$$E_A{}^\beta(H) = \sum_{k=g}^{n_\nu} \epsilon_k{}^\beta(H) - \frac{1}{2}\sum_{k,l=g}^{n_\beta} (\alpha_{kl}{}^\beta I_{kl}{}^\beta - \beta_{kl}{}^\beta K_{kl}{}^\beta),$$

$$E_A{}^\beta(F) = \sum_{k=g}^{n_\nu} \epsilon_k{}^\beta(F) - \frac{1}{2}\sum_{k,l=g}^{n_\beta} (\alpha_{kl}{}^\beta I_{kl}{}^\beta - \beta_{kl}{}^\beta K_{kl}{}^\beta). \tag{8}$$

Let us now designate the one-electron wave functions of the valence electrons in the crystal by φ_i, for which i ranges over values from 1 to n', where n' is the total number of valence electrons. We shall assume that the total wave function Ψ has unit multiplicity. The total energy of the crystal then is (*cf.* Chap. VI)

$$E = \int \Psi^* H \Psi d\tau = \sum_{i=1}^{n} \int \varphi_i(\mathbf{r}_1)\left[-\frac{h^2}{2m}\Delta + V(\mathbf{r}_1) \right]\varphi_i(\mathbf{r}_1)d\tau_1 +$$

$$\frac{1}{2}\sum_{i,j=1}^{n} \int \frac{|\varphi_i(\mathbf{r}_1)|^2|\varphi_j(\mathbf{r}_2)|^2}{r_{12}}d\tau_{12} - \frac{1}{2}\sum_{i,j}^{n} e^2 \int \frac{\varphi_i^*(\mathbf{r}_1)\varphi_j^*(\mathbf{r}_2)\varphi_i(\mathbf{r}_2)\varphi_j(\mathbf{r}_1)}{r_{12}}d\tau_{12} +$$

$$\frac{1}{2}\sum_{\alpha,\beta} \frac{Z_\alpha Z_\beta e^2}{r_{\alpha\beta}} + \frac{1}{2}\sum_{\alpha,\beta} V_{\alpha\beta} + \sum_{\alpha,i} \epsilon_{\alpha i} \tag{9}$$

in which the self-energy of the rigid ion cores is neglected. Here, V is the coulomb ion-core field, $V_{\alpha\beta}$ is the exchange interaction between ion α

and ion β, and $\epsilon_{\alpha i}$ is the exchange interaction between φ_i and the αth ion. Just as in Eqs. (8), the first term in (9) may be expressed in terms of the energy parameters of Hartree's or Fock's equations and the exchange and coulomb integrals. The energy per molecule of the crystal may then be obtained by dividing (9) by Nm, and an approximate expression for the cohesive energy per molecule may be obtained by subtracting (9) from the expression for the total energy of the free atoms that is derived by adding terms of the type (8).

Thus, the cohesive energy may be expressed in terms of energy parameters and coulomb and exchange integrals for the atomic and crystalline systems in the Hartree or Fock approximations. Since the practical difficulties of evaluating these parameters and integrals are very large, progress has been made only in those cases in which simple approximational methods, such as those described in the last chapter, can be used. There is a tendency for the errors that are introduced by use of the one-electron approximation to compensate one another when this approximation is employed in both the atomic and crystalline states, for the computed energies are too high in both cases. Thus, the computed cohesive energy may be larger or smaller than the correct value, depending upon whether or not the correlational error in the atomic state is larger than that of the solid. In solids such as ionic and molecular crystals in which the Heitler-London approximation can be used for both the crystalline and the atomic states, the correlation error is nearly the same in both cases, and we may expect to obtain a good value of the cohesive energy. On the other hand, the correlation energy of the valence electron on a free alkali atom is much smaller than that of the valence electrons in the corresponding metal. Hence, the one-electron approximation cannot be expected to give a good value of the cohesive energy in this case.

A. METALS

78. The Alkali Metals.—The cohesive energies of the three alkali metals lithium, sodium, and potassium have been computed[1] to a similar degree of accuracy by using the approximate methods that were described in the last chapter. We shall discuss the three alkali metals at the same time.

a. The Ion-core Field.—The first step in computing the cohesive energy of any substance is to obtain an ion-core field that takes into account the

---

[1] Li: F. Seitz, *Phys. Rev.*, **47**, 400 (1935); J. Bardeen, *Jour. Chem. Phys.*, **6**, 367 (1938).

Na: E. Wigner and F. Seitz, *Phys. Rev.*, **43**, 804 (1933); **46**, 509 (1934). E. Wigner, *Phys. Rev.*, **46**, 1002 (1934). Bardeen, *op. cit.*

K: E. Gorin, *Physik. Z. Sowj.*, **9**, 328 (1936).

interaction between the closed-shell electrons and the valence electrons. In both lithium and sodium, it was found possible to construct a radial potential field $v_c(r)$ having the property that the eigenvalues of the equation

$$-\frac{\hbar^2}{2m}\Delta\psi + v_c(r)\psi = E\psi \tag{1}$$

closely reproduce the observed atomic-term values. From the standpoint of Fock's equations, it is possible to say that in these cases the exchange interaction between core and valence electrons may be replaced by an ordinary potential term. In lithium, for example, the best field that could be obtained by a trial-and-error method duplicated the atomic values to within a few tenths of a per cent. The field for sodium, which was derived by Prokofjew[1] for another purpose, leads to term values that agree with the observed ones to within about 1 per cent. Thus, the computed atomic 3s function has an energy of 0.381 Rydberg unit, whereas the observed value is 0.378. Gorin attempted to construct a similar field for potassium, but he found that this could not be done with sufficient accuracy. Presumably, in this case the exchange terms cannot be replaced even approximately by ordinary potential terms. As a result, Gorin used a Hartree ion-core field and evaluated the exchange integrals between the valence and core electrons by direct methods. The ionization potential that he obtained in this way is 0.2934 Rydberg unit, which should be compared with the measured value of 0.3190 Rydberg unit.

   b. *Application of the Cellular Approximation.*—All the alkali metals form body-centered cubic lattices for which the polyhedron that should be used in the cellular approximation is the truncated octahedron shown in Fig. 2, Chap. IX. It may be assumed, for simplicity, that these polyhedra can be replaced by spheres of equal volume. The error that is made in doing so can be shown to be negligible and will be discussed below. Since each of the spheres is electrostatically neutral, the coulomb potential in a given cell that arises from any other cell is zero. Hence, all that is necessary is to consider the coulomb field arising from the charge in a given cell in the sphere approximation. For this reason, we shall restrict the following discussion to a single sphere.

   When deriving the electronic wave function within a sphere, we may neglect the potential of the valence-electron distribution in the first approximation. This procedure is permissible because the electronic charge distribution turns out to be very nearly constant and its potential is a slowly varying function that may be included readily by perturbation methods in a later approximation.

---

[1] W. Prokofjew, *Z. Physik*, **58**, 255 (1929).

Thus, the first problem to be solved in determining electronic wave functions is that of finding solutions of Eq. (1) that satisfy boundary conditions implied by the Bloch form

$$\psi_{\mathbf{k}} = \chi_{\mathbf{k}} e^{2\pi i \mathbf{k} \cdot \mathbf{r}} \tag{2}$$

(*cf.* Chap. VIII). We may expect the lowest eigenfunction to be one for which $\mathbf{k}$ is zero and which has cubic symmetry relative to any nucleus. The lowest order surface harmonics that possess this symmetry are s and g functions of the type

$$s = 1$$

$$g = \frac{1}{r^4}\left[(x^2 y^2 + y^2 z^2 + z^2 x^2) - \frac{1}{3}(x^4 + y^4 + z^4) \right]. \tag{3}$$

Hence, ψ_0 should have approximately the form

$$\psi_0 = f_s(r) + g f_g(r) \tag{4}$$

where f_s and f_g are radial functions. The equations for continuity of the normal gradient of ψ at the points (100) and (111) on the sphere of radius r_s are

$$f_s'(r_s) - \tfrac{1}{3}f_g'(r_s) = 0,$$
$$f_s'(r_s) + \tfrac{2}{9}f_g'(r_s) = 0, \tag{5}$$

where r_s is the radius of the sphere and the prime indicates differentiation with respect to r. The solutions of (5) are either

$$\psi_0 = f_s(r) \qquad \text{with} \qquad f_s'(r_s) = 0 \tag{6}$$

or

$$\psi_0 = g f_g(r_s) \qquad \text{with} \qquad f_g'(r_s) = 0. \tag{7}$$

Since an s function should have lower energy than a g function that satisfies the same boundary conditions, we should use the first of these conditions for the lowest state. If the actual polyhedron were used instead of a sphere, each of the two equations (6) would involve different values of r_s, one of which would be the distance from the center of the polyhedron to a point on the surface in the (100) direction, and the other of which would be the distance to a point in the (111) direction. We should then obtain two solutions that involve both the s and g functions. One of these, however, would be predominantly an s function and the other predominantly a g function.

We may conclude that within the sphere ψ_0 is that s function which goes into the lowest atomic s function when the lattice is expanded and which satisfies the condition $\psi_s'(r_s) = 0$. A relative scale plot of this function for the value of r_s corresponding to the actual lattice constant is shown in Fig. 1 for sodium. The energy of ψ_0 as a function of r_s,

$\epsilon_0(r_s)$, is shown in Fig. 2 for the three alkali metals.   The full curve for potassium includes the exchange interaction between the valence and closed-shell electrons.   These energy curves resemble very closely the

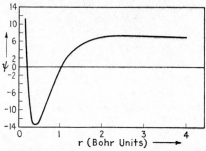

Fig. 1.—The lowest wave function of metallic sodium.   It should be noted that this function is practically constant for over ninety per cent of the atomic volume.

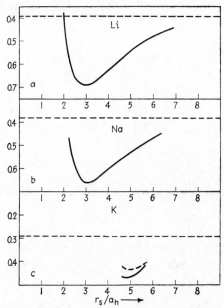

Fig. 2.—The $\epsilon(r_s)$ curves for *a* lithium, *b* sodium, and *c* potassium.   The energy scale is in Rydberg units (one Rydberg unit equals 13.54 ev).   The horizontal dotted lines represent the normal-state atomic energies.   The dashed curve for potassium represents the case in which the valence-electron closed-shell exchange energy is neglected.   Exchange is included in the full curve.

characteristic energy versus internuclear distance plots for diatomic molecules and show that the stability of metals is related to the fact that the spatial distribution of potential in a solid allows some of the electrons to have a lower energy than they have in the free atom. A part of this decrease in energy is connected with the decrease in kinetic energy that may be associated with increased smoothness of the wave functions, and part is connected with the fact that the center of gravity of the electronic charge in a given cell is nearer the nucleus.

Only two electrons are in the state going with $\mathbf{k} = 0$. In order to find the other wave functions in first approximation, it is necessary to solve (1) for those cases in which $\mathbf{k}$ is not zero. Since only half of the first zone is filled in the normal state of the alkali metals, it is not necessary[1] to find the exact form of the $\epsilon(\mathbf{k})$ curve near the boundary of this zone in order to compute the cohesive energy. With this in mind, we shall compute $\psi_{\mathbf{k}}$ and $\epsilon(\mathbf{k})$ by a method that is accurate near the center of the zone but not accurate near the boundary.

If we substitute (2) in Eq. (1), we find that $\chi_{\mathbf{k}}$ must satisfy the equation

$$-\frac{\hbar^2}{2m}\Delta\chi_{\mathbf{k}} + v_c\chi_{\mathbf{k}} - \frac{\hbar^2}{m}2\pi i\mathbf{k} \cdot \operatorname{grad} \chi_{\mathbf{k}} = \epsilon'(\mathbf{k})\chi_{\mathbf{k}} \tag{8}$$

where

$$\epsilon'(\mathbf{k}) = \epsilon_0(\mathbf{k}) - \frac{\hbar^2}{2m}\mathbf{k}^2. \tag{9}$$

The solution of (8), as given by the Rayleigh-Schrödinger perturbation method,[2] is

$$\epsilon'(\mathbf{k}) = \epsilon_0 - \frac{\hbar^2}{m}2\pi i\int \psi_0{}^*\mathbf{k} \cdot \operatorname{grad} \psi_0 d\tau + \frac{\hbar^4}{m^2}4\pi^2\sum_\nu \frac{|\int\psi_\nu{}^*\mathbf{k} \cdot \operatorname{grad} \psi_0 d\tau|^2}{\epsilon_0 - \epsilon_\nu} + \cdots, \tag{10}$$

$$\chi_{\mathbf{k}} = \psi_0 - \frac{\hbar^2}{m}2\pi i\sum_\nu \psi_\nu \frac{\int\psi_\nu\mathbf{k} \cdot \operatorname{grad} \psi_0 d\tau}{\epsilon_0 - \epsilon_\nu} + \cdots, \tag{11}$$

where the ψ_ν are the solutions of (8) for $\mathbf{k} = 0$ and ψ_0 is the lowest s function. The first integral in (10) is zero; the second term may be simplified in the following way. The only function ψ_ν for which $\int\psi_\nu\mathbf{k} \cdot \operatorname{grad} \psi_0 d\tau$ does not vanish is one having p symmetry, that is, one having any one of

---

[1] As we see from Fig. 17, p. 300, the closest point of approach to the zone boundary occurs in the (110) direction of wave-number space. Since the ratio of k_0 to the value of k at the zone is only 0.88 in this direction, it is possible that a small correction for deviations from the free-electron $\epsilon(\mathbf{k})$ curve should be made (cf. footnote 2, p. 366).

[2] Bardeen, op. cit., has developed an alternative method for solving Eq. (7) which involves the assumption that the solution may be expressed in the form

$$\chi_{\mathbf{k}} = f(r) + \mathbf{k} \cdot \mathbf{r}g(r)$$

where $f(r)$ and $g(r)$ are radial functions within each sphere. Evidently, $f(r)$ is ψ_0 and $g(r)$ is $f_p(r)$ in Eq. (18).

the forms

$$x\varphi_\nu(r), \qquad y\varphi_\nu(r), \qquad z\varphi_\nu(r), \tag{12}$$

where $\varphi_\nu(r)$ is a radial function.   In each of these three cases, the matrix components are, respectively,

$$k_x \int \frac{x^2}{r^2} \varphi_\nu(r) \psi_0'(r) d\tau, \qquad k_y \int \frac{y^2}{r^2} \varphi_\nu \psi_0' d\tau, \qquad k_z \int \frac{z^2}{r^2} \varphi_\nu \psi_0' d\tau. \tag{13}$$

Since the integrals in these terms are equal, Eqs. (9) and (10) may be written in the form

$$\epsilon'(\mathbf{k}) = \epsilon_0 + \frac{\hbar^4}{m^2} \mathbf{k}^2 \sum_\nu \frac{64\pi^4}{9} \frac{|\int \varphi_\nu \psi_0' r^2 dr|^2}{\epsilon_0 - \epsilon_\nu} \tag{14}$$

$$\chi_\mathbf{k} = \psi_0 - \frac{\hbar^2}{m} 2\pi i \sum_\nu \mathbf{k} \cdot \mathbf{r} \varphi_\nu \frac{4\pi}{3} \frac{\int \varphi_\nu \psi_0' r^2 dr}{\epsilon_0 - \epsilon_\nu}. \tag{15}$$

The p functions that are satisfactory solutions of (1) in the sphere approximation are those which satisfy the boundary condition $\varphi_\nu(r_s) = 0$. This fact may be proved by setting up equations similar to (5).   When these p functions have been computed, both $\epsilon'(\mathbf{k})$ and $\psi_\mathbf{k}$ may be determined by evaluating the integrals in (14) and (15).   It is not evident from what has been said that the higher order perturbation terms are negligible, but a practical examination of these for all three metals shows that they actually are so for values of $\mathbf{k}$ in the first zone that are not too near the zone boundary.

It is worth noting that, in the approximation in which Eq. (14) is valid, $\epsilon(\mathbf{k})$ may be written as

$$\epsilon(\mathbf{k}) \equiv \epsilon'(\mathbf{k}) + \frac{\hbar^2}{2m} \mathbf{k}^2 = \epsilon_0 + \frac{\hbar^2}{2m^*} \mathbf{k}^2 \tag{16}$$

where[1]

$$\frac{1}{m^*} = \frac{1}{m} \left(1 + \frac{\hbar^2}{m} \sum_\nu \frac{32\pi^2}{9} \frac{|\int \varphi_\nu \psi_0' r^2 dr|^2}{\epsilon_0 - \epsilon_\nu} \right). \tag{17}$$

Similarly, $\psi_\mathbf{k}$ may be expressed in the form

$$\psi_\mathbf{k} = e^{2\pi i \mathbf{k} \cdot \mathbf{r}} (f_s(r) + 2\pi i \mathbf{k} \cdot \mathbf{r} f_p(r)) \tag{18}$$

---

[1] We shall see in Chap. XVII that the terms in the coefficient of $1/m$ in Eq. (17) are related to the f factors of radiation theory.

where f_s is the lowest s function and

$$f_p(r) = -\frac{\hbar^2}{2m}\sum_\nu \varphi_\nu(r)\frac{4\pi}{3}\frac{\int \varphi_\nu \psi_0' r^2 dr}{\epsilon_0 - \epsilon_\nu}. \tag{19}$$

The radial integrals that appear in the preceding equations were determined for several values of r_s for each of the three metals. The values of the quantity m/m^* that were computed[1] from these [cf. Eq. (17)] are listed in Table LII. It turns out in the case of sodium that the

TABLE LII.—VALUES OF m/m^* FOR THE ALKALI METALS

(The values for the observed values of r_s are given in boldface type.)

r_s/a_h	m/m^*
Li	
3.00	0.584
3.21	**0.653**
3.32	0.684
Na	
3.80	1.079
3.96	**1.069**
4.12	1.059
K	
4.82	**1.72**
5.06	1.59
5.34	1.48
5.47	1.44

summation in (17) "accidentally" vanishes for values of r_s in the vicinity of the observed one, so that this ratio is practically unity. The ratio is less than unity in lithium and greater than unity in potassium. A small term that takes into account the variation with **k** of the interaction between the valence electrons and the closed-shell electrons has been included in the case of potassium.

The expression (16) for $\epsilon(\mathbf{k})$ is the same as the expression for the energy of free electrons that was used in the theory of metals discussed in Chap. IV, with the difference that ϵ_0 replaces $-W_a$. The quantity m^* is, as before, the effective electron mass. Since this mass is practically equal to the ordinary electronic mass in the case of sodium, this metal should behave more like an ideal metal than either lithium or potassium. The same point is indicated by the fact that ψ_0 for sodium (cf. Fig. 1) is almost constant in about 90 per cent of the volume of the sphere, a fact which shows that the wave functions are closely equal to $Ae^{2\pi i \mathbf{k}\cdot\mathbf{r}}$, where A is a constant.

As long as Eq. (16) is valid, the electrons in their normal state completely occupy a sphere centered about the origin of **k** space, just as in the

[1] The values for Na and Li were computed by Bardeen using the method discussed in footnote 2, p. 352.

Sommerfeld theory. The radius k_0 of the occupied sphere of points in **k** space is given by the relation

$$\frac{4\pi}{3}k_0^3 V = \frac{N}{2} \tag{20}$$

where V is the volume of the crystal, N is the total number of electrons, and the 2 in the denominator of the right-hand side arises from spin degeneracy. The mean value of

$$\frac{h^2}{2m^*}\mathbf{k}^2 \tag{21}$$

for these electrons will be called the mean Fermi energy in the following discussion, since it corresponds to the mean kinetic energy $\bar{\epsilon}$ that the electrons would possess if they were distributed according to Fermi-Dirac statistics. Actually, (21) is a combination of kinetic and potential energy so that the mean value of this quantity is not the mean value of $-\hbar^2\Delta/2m$. The mean Fermi energy is easily found to be

$$\epsilon_F = \frac{3}{5}\frac{h^2}{2m^*}k_0^2 = \frac{3}{10}\frac{h^2}{m^*}\left(\frac{3n_0}{8\pi}\right)^{\frac{2}{3}} \tag{22}$$

where $n_0 = N/V$ is the number of electrons per unit volume. This is identical with the expression for $\bar{\epsilon}$ that was derived in Sec. 26, Chap. IV. Since $n_0 = 1/(4\pi r_s^3/3)$, Eq. (22) is

$$\epsilon_F = \frac{3}{10}\frac{h^2}{m^*}\left(\frac{9}{32\pi^2}\right)^{\frac{2}{3}}\frac{1}{r_s^2} = \frac{1.105}{r_s^2}\frac{e^2}{a_h}\left(\frac{m}{m^*}\right). \tag{23}$$

Values of this energy are listed in Table LIII.

It is interesting to note that the quantity

$$-(\epsilon_0 + \epsilon_F + \epsilon_I), \tag{24}$$

where ϵ_I is the negative of the atomic energy, that is, the ionization energy, agrees closely with the cohesive energies in the cases of lithium and sodium. In the first case, (24) is 39 kg cal/mol, and the observed value is also 39; in the second case, (24) is 24.4 kg cal/mol, and the observed value is 26. These computed values are given for the observed values of r_s at which (24) actually does have a minimum. The agreement is not so close in the case of potassium, which we shall discuss separately below.

In taking (24) to be the cohesive energy per atom, it is effectively assumed that the field acting upon an electron is essentially that of the ion core in the polyhedron in which the electron is momentarily found. Thus, it is assumed that no more than one electron can be in a given cell at one time. The agreement between (24) and the observed cohesive energies in the cases of lithium and sodium suggests that on the average the electrons actually do avoid one another in this way.

The cohesive energy of potassium, computed from Eq. (24), is 6 kg cal/mol, whereas the observed value is 22.6. This discrepancy is surprising at first, for we might expect the properties of the alkali metals to vary continuously as we pass down the periodic chart from lithium to cesium. Gorin believes that the error in potassium is related to flaws in the Hartree field on which the computations are based, for this field does not reproduce the atomic energy levels with the same accuracy as the fields used for lithium and sodium (cf. part a of this section) even when exchange terms are included. Thus, the error in the lowest level of the atom is 0.735 ev without exchange and 0.347 ev with exchange.

Since the exchange terms are larger in the solid than in the free atom, because the center of gravity is nearer the nucleus in the solid, it seems

TABLE LIII.—VALUES OF $\epsilon_0 + \epsilon_I$ AND ϵ_F FOR THE ALKALI METALS
(The values of ϵ_I, the ionization energy, are theoretical values.)

r_s	$\epsilon_0 + \epsilon_I$	ϵ_F	$-(\epsilon_0 + \epsilon_I + \epsilon_F)$	Observed cohesive energy
Li($\epsilon_I = 5.365$ ev $= 123.4$ kg cal/mol)				
3.00	−84.6 kg cal/mol	44.7	39.9	
3.21	**−82.6**	**43.6**	**39.0**	**39 kg cal/mol**
3.32	−81.5	42.7	38.8	
Na($\epsilon_I = 5.159$ ev $= 118.7$ kg cal/mol)				
3.80	−75.3	51.4	23.9	
3.96	**−71.3**	**46.9**	**24.4**	**26**
4.12	−67.2	42.9	24.3	
K($\epsilon_I = 3.973$ ev $= 91.4$ kg cal/mol)				
4.82	. . . .	. . . .	. . . .	**22.6**
5.06	−41.0	42.7	−1.7	
5.34	−40.6	35.7	4.9	
5.47	−39.7	33.1	6.6	

reasonable to expect that the entire correlational effect between the valence and closed-shell electrons increases in passing from the free atom to the solid. Now the exchange and correlation interaction energies of the valence and closed-shell electrons in the free atom are, respectively, 0.388 ev and 0.347 ev. Gorin attempted to correct the absolute error of the lowest level in the solid by multiplying the closed-shell valence-electron exchange energy of the solid by a factor

$$\frac{(0.388 + 0.347)}{0.388} = 1.89.$$

The correction induced by this is listed in Table LIV and increases the binding energy, as computed by Eq. (24), to about 14.5 kg cal/mol for the

TABLE LIV.—CORRECTED VALUES OF $\epsilon_0 + \epsilon_I$ FOR POTASSIUM OBTAINED BY INCREASING CLOSED-SHELL VALENCE-ELECTRON EXCHANGE BY A FACTOR 1.89
($\epsilon_I = 4.333$ ev = 99.7 kg cal/mol)

	$\epsilon_0 + \epsilon_I$	ϵ_F	$-(\epsilon_0 + \epsilon_I + \epsilon_F)$	Observed cohesive energy
4.82	......	....	....	**22.6**
5.06	−51.5	42.5	9.0	
5.34	−48.1	34.5	13.6	
5.47	−46.2	31.9	14.3	

point $r_s = 5.4$ where the total energy is a minimum. The agreement between the observed and calculated energies is now comparable with that found above for lithium and sodium. The value of the lattice parameter at which the minimum occurs is much too large, however, a fact showing that the corrections of Table LIV are not adequate for small interatomic distances.

c. *The Influence of Coulomb Terms.*—We shall now proceed to correct the equations in part *b* by considering the effects of the coulomb term

$$\sum_k e^2 \int \frac{|\psi_k(\mathbf{r}_2)|^2 d\tau_2}{r_{12}}. \qquad (25)$$

We could evaluate this term, using the wave function (9), and determine a new set of functions to replace (9) by placing the result in Eq. (1). The new solutions then could be employed in a reiteration of this procedure and the process repeated until a self-consistent set of functions is found. Actually, the change in the wave functions is negligibly small at the end of the first step in this procedure. The charge distribution

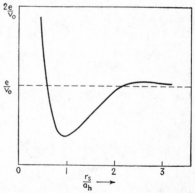

FIG. 3.—The charge distribution in the unit sphere of lithium. The ordinates are expressed in units of e/v_0, where v_0 is the atomic volume.

$$\sum_k |\psi_k|^2 = \sum_k |\chi_k|^2, \qquad (26)$$

obtained from the functions (10), is shown in Fig. 3 for lithium when $r_s = r_0$. It should be noted that this function is practically constant over the major part of the unit sphere. This condition is satisfied even better in sodium and not quite so well in potassium. Since the potential that arises from a constant charge distribution turns out to have a negligible effect on the wave functions (10), as we shall see, we may conclude that the effect will also be negligible when (26) is used. The electronic potential at a point r, arising from a spherical charge distribution of constant density $-e/v_0$ where v_0 is the volume of the sphere, is

$$V(r) = \frac{3}{2}\frac{e^2}{r_s} - \frac{1}{2}\frac{e^2r^2}{r_s^3} \qquad (r \leq r_s). \qquad (27)$$

The constant term does not affect the wave function, and so we need consider only the term in r^2. The correction to ψ_0 arising from this term is

$$-\frac{e^2}{2r_s^3}\sum_{\mathbf{k}}' \psi_{\mathbf{k}}\frac{\int\psi_{\mathbf{k}}^*r^2\psi_0 d\tau}{\epsilon_0 - \epsilon_{\mathbf{k}}} \qquad (28)$$

where the $\psi_{\mathbf{k}}$ obviously form the family of s functions that satisfy the condition $\psi'(r_s) = 0$. An upper limit to the value of the integrals $\int\psi_{\mathbf{k}}^*r^2\psi_0 d\tau$ may be obtained by evaluating the quantity

$$\sum_{\mathbf{k}}'\left|\int\psi_0^*r^2\psi_{\mathbf{k}}d\tau\right|^2 = \int\psi_0 r^4\psi_0 d\tau - \left|\int\psi_0 r^2\psi_0 d\tau\right|^2.$$

Since ψ_0 is practically constant, this is

$$(\tfrac{3}{7} - \tfrac{9}{25})r_s^4 = \tfrac{12}{175}r_s^4.$$

Thus, the maximum value of any of the integrals in (28) is $0.13e^2/r_s$. The difference between the first two s levels in the alkali metals is of the order of $e^2/2a_h$. Hence, the ratio of the coefficient of any $\psi_{\mathbf{k}}$ to that of ψ_0 is at most $0.26/r_s$, which varies from 0.08 to 0.05 for the three alkali metals. Actually, only the lowest of the $\psi_{\mathbf{k}}$ has a coefficient of this magnitude; the others play a less important role because the energy denominators increase and the value of the integral $\int\psi_{\mathbf{k}}^*r^2\psi_0 d\tau$ decreases with increasing **k**. In addition, it seems likely that the upper limit is too high, for it is determined by adding the absolute values. On the whole, then, we may conclude that the correction term to ψ_0 changes the function by only a few per cent. The same conclusion can be drawn for the other functions, for they are practically equal to $\psi_0 e^{2\pi i\mathbf{k}\cdot\mathbf{r}}$. The total error made by neglecting the change in wave functions can be shown to be less than 1 kg cal/mol, which is less than the computational error of the present work.

By way of summary, it may be said that the solutions of (1) that are given by (2) and (11) in the sphere approximation also are solutions of Hartree's equations for the lattice.

d. The Influence of Exchange Terms.—It was demonstrated in Sec. 75, Chap. IX, that the exchange operator A is diagonal in the special case in which the eigenfunctions are free-electron waves $e^{2\pi i \mathbf{k} \cdot \mathbf{r}}$. In this case, the solutions of Hartree's and Fock's equations are identical. Now, we saw in part *c* that the solutions of Hartree's equations are very nearly free-electron waves for all three alkali metals under consideration. Hence, the exchange operator should be almost diagonal in these cases, and we may assume that the solutions of Fock's equations are identical with (10). This fact is a very fortunate one, for unless it were true the problem of treating the alkali metals would involve many more difficulties.

The influence on the exchange energy of the small p term in (18) may be estimated in the following way. For simplicity, we shall assume that $\psi_\mathbf{k}$ has the form

$$\psi_\mathbf{k} = e^{2\pi i \mathbf{k} \cdot \mathbf{r}}(\alpha + i\gamma \mathbf{k} \cdot \mathbf{r}) \tag{29}$$

where both α and γ are constants. This is equivalent to assuming that f_s and f_p are constant. In this case, the quantity $A \cdot \psi_\mathbf{k}$, where A is the exchange operator, is

$$-e^2 \sum_{\mathbf{k}'} \int \frac{(\alpha - i\gamma \mathbf{k}' \cdot \mathbf{r}_2)(\alpha + i\gamma \mathbf{k}' \cdot \mathbf{r}_1)(\alpha + i\gamma \mathbf{k} \cdot \mathbf{r}_2)}{|\mathbf{r}_1 - \mathbf{r}_2|} e^{2\pi i[(\mathbf{k} - \mathbf{k}') \cdot \mathbf{r}_2 + \mathbf{k}' \cdot \mathbf{r}_1]} d\tau_2.$$

If we make the transformation $\mathbf{r}_3 = \mathbf{r}_2 - \mathbf{r}_1$, we may,[1] to terms in γ^2, change this to

$$-e^2(\alpha + i\gamma \mathbf{k} \cdot \mathbf{r}_1) e^{2\pi i \mathbf{k} \cdot \mathbf{r}_1} \sum_{\mathbf{k}'} \int \frac{(\alpha + i\gamma \mathbf{k}' \cdot \mathbf{r}_3)(\alpha - i\gamma \mathbf{k}' \cdot \mathbf{r}_3)}{|\mathbf{r}_3|} e^{2\pi i (\mathbf{k} - \mathbf{k}') \cdot \mathbf{r}_3} d\tau_3.$$

Hence, the functions (18) are eigenfunctions of A to terms in γ^2. We shall find that the γ^2 terms are very small when we discuss the value of the exchange energy in part *e*.

e. The Energy in the One-electron Approximation.—We may now evaluate the energy of the crystal in the one-electron approximation. This energy is

$$E = \int \Psi^* H \Psi d\tau$$

---

[1] This approximation is equivalent to assuming that the wave functions have the form

$$\psi_\mathbf{k} = \alpha e^{i\frac{\gamma}{\alpha} \mathbf{k} \cdot \mathbf{r}}.$$

where Ψ is the determinantal eigenfunction, formed of the functions (2), and

$$H = \sum_i \left(-\frac{\hbar^2}{2m}\Delta_i + V_i \right) + \frac{1}{2}\sum_{i,j}' \frac{e^2}{r_{ij}} + \frac{1}{2}\sum_{\alpha,\beta}' \frac{e^2}{r_{\alpha\beta}} \qquad (30)$$

is the total Hamiltonian of the lattice.   In Eq. (30),

$$V_i = \sum_\tau v_c(|\mathbf{r}_i - \mathbf{r}(n)|) \qquad (31)$$

where $v_c(|\mathbf{r}_i - \mathbf{r}(n)|)$ is the potential at $\mathbf{r}_i$ arising from the ion at the position $\mathbf{r}(n)$, and the last term is the interaction potential of the ions.   The latter has the same form as for point charges, for the ion cores are so far apart that they do not overlap appreciably.[1]

The mean value of the operator (30) is

$$2\sum_\mathbf{k} \int \psi_\mathbf{k}^* \left(-\frac{\hbar^2}{2m}\Delta + V \right)\psi_\mathbf{k} d\tau + 4\sum_{\mathbf{k},\mathbf{k}'}' \frac{e^2}{2}\int \frac{|\psi_\mathbf{k}(\mathbf{r}_1)|^2 |\psi_{\mathbf{k}'}(\mathbf{r}_2)|^2}{r_{12}}d\tau -$$
$$2\sum_{\mathbf{k},\mathbf{k}'} \frac{e^2}{2}\int \frac{\psi_\mathbf{k}^*(\mathbf{r}_1)\psi_{\mathbf{k}'}^*(\mathbf{r}_2)\psi_\mathbf{k}(\mathbf{r}_2)\psi_{\mathbf{k}'}(\mathbf{r}_1)}{r_{12}}d\tau + \frac{1}{2}\sum_{\alpha,\beta}' \frac{e^2}{r_{\alpha\beta}} \qquad (32)$$

where the sums over $\mathbf{k}$ and $\mathbf{k}'$ extend over the occupied values of $\mathbf{k}$, excluding spin, and the factors 2, 4, and 2 in the first three terms give the results of spin summation.   We may split the first integral into integrals over each of the N cells of the lattice.   Since the component integrals must be the same for every cell, they are equal to

$$N2\sum_\mathbf{k} \int_v \psi_\mathbf{k}^* \left(-\frac{\hbar^2}{2m}\Delta + V \right)\psi_\mathbf{k} d\tau, \qquad (33)$$

in which the integration extends over a single cell.   From (31), we may derive the relation

$$\sum_\mathbf{k} 2\int_{v_n} \psi_\mathbf{k}^* V\psi_\mathbf{k} d\tau = \sum_\mathbf{k} 2\int |\psi_\mathbf{k}|^2 v_c(|\mathbf{r}-\mathbf{r}(n)|)d\tau + \sum_{n'\neq n}\sum_\mathbf{k} 2\int |\psi_\mathbf{k}|^2 v_c(|\mathbf{r}-\mathbf{r}(n')|)d\tau$$

$$(34)$$

where the integration extends over the nth cell.   In the second term, $v_c(|\mathbf{r} - \mathbf{r}(n')|)$ may be replaced by $-e^2/|\mathbf{r} - \mathbf{r}(n')|$ since the field is coulombic outside the n'th cell.   Hence, if the fact that the distribution $\sum_\mathbf{k} 2|\psi_\mathbf{k}|^2$

---

[1] The contribution to the energy from the exchange and van der Waals terms is less than 0.2 kg cal/mol so that these terms are negligible for the cohesive energy. They are important, however, when the elastic constants are computed.

is spherically symmetric within a given cell is taken into account, (34) reduces to

$$\sum_{\mathbf{k}} 2 \int_{v_n} |\psi_{\mathbf{k}}(\mathbf{r})|^2 v_c(|\mathbf{r} - \mathbf{r}(n)|) d\tau - \sum_{n'}{}' \frac{e^2}{|\mathbf{r}(n) - \mathbf{r}(n')|}, \tag{35}$$

so that (33) may be replaced by

$$N \sum_{\mathbf{k}} 2 \int_{v_n} \psi_{\mathbf{k}}{}^* \left(-\frac{\hbar^2}{2m} \Delta + v_c(|\mathbf{r} - \mathbf{r}(n)|) \right) \psi_{\mathbf{k}} d\tau - \frac{1}{2} \sum_{\alpha,\beta}{}' \frac{e^2}{r_{\alpha\beta}}, \tag{36}$$

since

$$N \sum_{n'}{}' \frac{e^2}{|\mathbf{r}(n) - \mathbf{r}(n')|} = \frac{1}{2} \sum_{\alpha,\beta}{}' \frac{e^2}{r_{\alpha\beta}}. \tag{37}$$

The coulomb term in (32) may be split into two parts, namely, one for which $\mathbf{r}_1$ and $\mathbf{r}_2$ lie in the same cell, and another in which they lie in different cells. In the sphere approximation, the second term is equal to the mutual potential of a set of point charges, that is, to

$$\frac{1}{2} \sum_{\alpha,\beta}{}' \frac{e^2}{r_{\alpha\beta}}, \tag{38}$$

which cancels the last term in (36).

Upon combining these results, we find that (32) reduces to

$$2 \sum_{\mathbf{k}} \epsilon(\mathbf{k}) + 4N \sum_{\mathbf{k},\mathbf{k}'} \frac{e^2}{2} \int_\sigma \frac{|\psi_{\mathbf{k}}(\mathbf{r}_1)|^2 |\psi_{\mathbf{k}'}(\mathbf{r}_2)|^2}{r_{12}} d\tau_{12} -$$
$$2 \sum_{\mathbf{k},\mathbf{k}'} \frac{e^2}{2} \int \frac{\psi_{\mathbf{k}}{}^*(\mathbf{r}_1) \psi_{\mathbf{k}'}{}^*(\mathbf{r}_2) \psi_{\mathbf{k}}(\mathbf{r}_2) \psi_{\mathbf{k}'}(\mathbf{r}_1)}{r_{12}} d\tau_{12}. \tag{39}$$

Thus, the total energy differs from the sum of the energy parameters for all doubly occupied states by the self-potential of the charge distribution in each of the polyhedra and the exchange energy of all electrons. That the self-potential and exchange terms would enter in just this way might easily have been predicted on the basis of the discussions of parts c and d.

The self-energy of the charge distribution within a unit cell has been evaluated numerically for several values of r_s in each of the three metals under discussion; the results are listed in Table LV. It should be noted that the actual energy is very close to $0.6e^2/r_s$, which is the self-energy of a constant charge distribution, for the observed lattice constant in lithium and sodium. There is a considerable deviation, however, in the case of potassium, which probably is related to the errors in the potassium field that were discussed in part b.

TABLE LV.—COMPARISON OF COULOMB AND EXCHANGE ENERGIES FOR ACTUAL
ELECTRONS AND FOR FREE ELECTRONS AT PARTICULAR VALUES OF r_s
(In kg cal/mol)

	Coulomb		Exchange	
	Actual	$0.6e^2/r_s$	Actual	$-0.458e^2/r_s$
$Li(r_s = 3.21a_h)$	114.8	116.3	−90.2	−88.9
$Na(r_s = 3.96a_h)$	93.8	94.4	−72.0	−72.0
$K(r_s = 5.06a_h)$	82.2	74.3	−57.5	−56.6

The exchange terms are complicated by the fact that the integrals cannot be broken up into integrals over single polyhedra in any simple manner. They may be evaluated, however, by replacing the functions (18) with the simpler functions (29). This approximation actually is a good one in the alkali metals, for f_s is very nearly constant and f_p is small. The exchange energy has been evaluated by direct computation in this case and the result is

$$N\left(-0.458\frac{e^2}{r_s} - 1.05\eta\,\frac{e^2}{r_s} - 1.09\eta^2\frac{a_h^4 e^2}{r_s^5}\right) \qquad (40)$$

where η is the constant appearing in ψ_k when it is expressed in the form

$$\psi_k = \frac{e^{2\pi i k \cdot r}}{\sqrt{v}}\left(1 + i\frac{\sqrt{5}\eta}{r_s}k \cdot r\right). \qquad (41)$$

When η is zero, (41) is a free-electron wave function and the expression (40) reduces to

$$-0.458N\frac{e^2}{r_s},$$

which, as we have seen in Sec. 75 of the last chapter, is the exchange energy for free electrons. Numerical values of the quantity (40) appear in Table LV. The terms in η^2 actually are very small, a fact showing that the exchange operator is almost constant (cf. part d).

f. Justification of the Sphere Approximation.—Before discussing the cohesive energy, we shall justify the approximation in which the polyhedron surrounding each atom is replaced by a sphere. It will be shown that the expression for the total energy does not differ appreciably from that derived in part e, if the wave functions (18) are used at points inside the polyhedron instead of at those in the sphere.

We may dispose of the exchange terms at once by observing that the principal term of (40), namely, $-0.458e^2/r_s$, was derived by using functions which extend throughout the lattice. The terms in η may be influenced slightly by the sphere approximation, but they are so small that this variation cannot be important.

Next, let us discuss Eq. (32). V may be expressed rigorously in the form (31), so that we are interested in the sum

$$\sum_{\mathbf{k}} 2 \int_v \psi_{\mathbf{k}}^* \left(-\frac{\hbar^2}{2m}\Delta + v_c(|\mathbf{r} - \mathbf{r}(n)|) \right)\psi_{\mathbf{k}} d\tau \tag{42}$$

where the integration now extends over the polyhedron.

Although $\psi_{\mathbf{k}}$ satisfies Eq. (1), we cannot set the sum (42) equal to $2\Sigma\epsilon_{\mathbf{k}}$, for the normal gradient of $\psi_{\mathbf{k}}$ is not continuous at the boundary of the polyhedron. In other words, since $\Delta\psi_{\mathbf{k}}$ is singular at the boundary, it is necessary to determine the contribution to the integral that is associated with the singularity. It may be shown (see the papers on sodium, footnote 1, page 348) that the necessary addition to the energy is

$$-\frac{\hbar^2}{2m}\sum_{\mathbf{k}}\sum_{\tau} \int_{v_{\tau}} \psi_{\mathbf{k}}^* \operatorname{grad} \psi_{\mathbf{k}} \cdot d\boldsymbol{\sigma} \tag{43}$$

where the integrals extend over the surface of a polyhedron and the second sum extends over all cells in the lattice. This integral is less than 0.005 ev for ψ_0 and may be neglected. Since f_p in (13) is small near the boundary, the integral is also negligible in other cases.

The remaining terms in the expression for the total energy, namely,

$$N\sum_{\mathbf{k}}\sum_{n'\neq n} 2 \int_{v_n} |\psi_{\mathbf{k}}|^2 v_c(|\mathbf{r} - \mathbf{r}(n')|)d\tau +$$

$$\frac{e^2}{2}\int \frac{\left[2\sum_{\mathbf{k}}|\psi_{\mathbf{k}}(\mathbf{r}_1)|^2\right]\left[2\sum_{\mathbf{k'}}|\psi_{\mathbf{k'}}(\mathbf{r}_2)|^2\right]}{r_{12}} d\tau_1 d\tau_2 + \frac{1}{2}\sum_{\alpha,\beta}\frac{e^2}{r_{\alpha\beta}}, \tag{44}$$

may be broken into two expressions, namely, one for the self-energy of the charge in a polyhedron, and one for the mutual interaction energy of the set of polyhedra. In order to do so, all that is necessary is to write the second term of (44) in the form

$$N\sum_{n}\frac{e^2}{2}\int_{v_n} \frac{\left[2\sum_{\mathbf{k}}|\psi_{\mathbf{k}}(\mathbf{r}_1)|^2\right]\left[2\sum_{\mathbf{k'}}|\psi_{\mathbf{k'}}(\mathbf{r}_2)|^2\right]}{r_{12}} d\tau_1 d\tau_2 +$$

$$\sum_{n,n'}{}' \frac{e^2}{2}\int_{v_n}\int_{v_{n'}} \frac{\left[2\sum_{\mathbf{k}}|\psi_{\mathbf{k}}(\mathbf{r}_1)|^2\right]\left[2\sum_{\mathbf{k'}}{}'|\psi_{\mathbf{k'}}(\mathbf{r}_2)|^2\right]}{r_{12}} d\tau_1 d\tau_2. \tag{45}$$

The first term is the sum of the self-energies of the electronic charge distribution in each cell, and the second is the interaction energy of the electronic charges in different cells. The sum of the second term in (45) and the first and last terms in (44) is the total electrostatic interaction energy of the polyhedra. In the sphere approximation, this sum was assumed to be zero, and (44) was evaluated by computing the self-energy of the charge in the sphere.

The electronic charge density in each of the three alkali metals is very nearly equal to e/v throughout all parts of the polyhedron, except in the region S near the nucleus where the eigenfunctions have nodes. Since the distribution is spherically symmetrical in this region, the potential outside S is the same as though the polyhedron contained a positive point charge at its center and a uniform negative distribution of density e/v. Hence, except for the value of the self-energy of the electronic charge in S, (44) has the same value as for a lattice of positive point charges that contains a uniform distribution of negative charge. The self-energy of this lattice may be computed by the methods discussed in Chap. II and is

$$-0.8958\frac{e^2}{r_s}$$

per ion. The interaction energy of the positive and negative charge in a given polyhedron of this lattice, as determined by direct numerical calculation, is

$$-1.4939\frac{e^2}{r_s}.$$

If we designate by $\Delta\epsilon$ the difference between the self-energy of the actual electronic charge distribution in the region S and the self-energy of a constant electronic distribution in the same region, (44) is equal to

$$N\left(0.598\frac{e^2}{r_s} + \Delta\epsilon\right) \tag{46}$$

where N is the number of atoms. The result corresponding to (46) is, in the sphere approximation,

$$N\left(0.6\frac{e^2}{r_s} + \Delta\epsilon\right) \tag{47}$$

where $0.6e^2/r_s$ is the self-energy of a sphere of constant charge distribution. The difference, namely, $-N0.002e^2/r_s$, is never more than 0.2 kg cal/mol and may be neglected.

The close agreement between (46) and (47) is not a fortuitous coincidence but rests upon the fact that the field outside a polyhedron actually is practically zero, as assumed in the sphere approximation.

g. *Energy in the Fock Approximation.*—The complete expression for the cohesive energy of the alkali metals in the one-electron approximation is

$$N\left(\epsilon_0 + \frac{1.105}{r_s^2}\frac{e^2}{a_h} + \frac{0.598e^2}{r_s} + \Delta\epsilon - \frac{0.458e^2}{r_s} + \epsilon_I\right), \qquad (48)$$

TABLE LVI.—CONTRIBUTIONS TO THE COHESIVE ENERGY OF THE ALKALI METALS
(The values for K contain the corrected values of $\epsilon_0 + \epsilon_I$ of Table LIV.)

r_s	$\epsilon_0 + \epsilon_I + \epsilon_F$	Coulomb term	Exchange energy (40)	Cohesive energy in Fock approximation	Correlation energy	Final cohesive energy	Observed
				Li			
3.00	−39.9 kg cal/mol	124.6	−95.1	10.4 kg cal/mol	−22.3	32.7	
3.21	**−39.0**	**114.8**	**−90.2**	**14.4**	**−21.7**	**36.1**	**39.0**
3.32	−38.8	110.9	−87.1	14.0	−21.5	35.5	
				Na			
3.80	−23.9	98.3	−75.1	0.7	−20.3	21.0	
3.96	**−24.4**	**93.8**	**−72.0**	**2.6**	**−19.9**	**23.5**	**26.0**
4.12	−24.3	89.6	−69.2	3.9	−19.6	24.5	
				K			
4.82	....	....	......	....	......	....	**22.6**
5.06	− 9.0	81.7	−57.3	−15.4	−17.8	2.4	
5.34	−13.6	69.7	−54.1	− 2.0	−17.3	15.3	
5.47	−14.3	67.6	−52.5	− 0.8	−17.1	16.3	

which includes only the principal term in the expression (4) for the exchange energy. This result is listed in Table LVI in the column headed "Cohesive energy in Fock approximation." The minimum values of the cohesive energies and the corresponding values of the lattice constant appear in Table LVII. A striking feature of these results is

TABLE LVII.—COMPARISON OF OBSERVED AND CALCULATED VALUES OF THE COHESIVE ENERGY AND LATTICE CONSTANT IN THE FOCK APPROXIMATION AND FINAL APPROXIMATION

	Cube-edge distance, Å			Cohesive energy, kg cal/mol		
	Observed	Fock	Final	Observed	Fock	Final
Li	3.46	3.50	3.50	39	14.6	36.2
Na	4.25	4.56	4.51	26	4.1	24.5
K	5.20	5.86	5.82	23	− 0.7	16.5

the fact that the cohesive energy determined by the one-electron approximation is smaller than the observed value by about the amount that one would expect from consideration of the atomic and molecular problems for which correspondingly accurate solutions have been found. It seems reasonable to ascribe most of the error, which is about 1 ev per electron, to the neglect of correlations between electrons having opposite spin.[1]

Since the one-electron wave functions are nearly the same as those for free electrons, we may use Wigner's expression for the correlation energy, which was discussed in Sec. 76, namely,

$$-\frac{0.288}{r_s + 5.1a_h}e^2. \tag{49}$$

The final energies are given in the next to the last column of Table LVI and in Table LVII.

The calculations for lithium probably are the most significant since they are the most accurate. It is not easy to trace the source of the error of 3 kg cal/mol, but it probably arises from an error in the expression (49) for the correlation energy.[2] It is possible, however, that the effective ion-core field also contributes to this error, for it may not adequately represent the valence-electron and closed-shell interaction in the solid. The error in sodium probably has the same origin as that in lithium, whereas a large part of the error in potassium, which was discussed in part b, undoubtedly is connected with the inaccuracies in Hartree's closed-shell wave functions.

79. Metallic Hydrogen.—Although a metallic modification of hydrogen is unknown, Wigner and Huntington[3] have made a computation of the properties of this hypothetical substance in order to estimate the conditions under which it should be stable. This computation was carried through on the assumption that the metallic lattice would be

---

[1] Tables LVI and LVII contain a summary of values that the writer regards as the "best" results of the papers listed in footnote 1, p. 348.

[2] C. Herring has pointed out to the writer that two types of correction to the equation $\epsilon = h^2k^2/2m^*$ may be expected in the case of lithium. In the first place, there may be a slight downward curvature in the (110) direction because of the proximity of the zone boundary. This effect presumably will also occur in sodium and probably is not large. In addition, since the $\epsilon(\mathbf{k})$ curve for lithium is well below the free-electron curve near $\mathbf{k} = 0$ and we may expect it to rise in higher zones, there is probably a positive term of the order k^4 in a more accurate representation of $\epsilon(\mathbf{k})$. This term probably does not affect the Fermi energy appreciably but may be responsible for a decrease in level density near the edge of the filled region that is important for other properties, such as the conductivity and paramagnetic susceptibility (see Chaps. XV and XVI).

[3] E. WIGNER and H. B. HUNTINGTON, *Jour. Chem. Phys.*, **3**, 764 (1935).

body-centered, so that the computational procedure closely resembles that used for the alkali metals. The principal difference is that the ion core is a proton which rigorously has a coulomb field. This simplification makes it possible to evaluate the wave functions and energies analytically when the sphere approximation is used. The coulomb, exchange, and correlation energies were estimated by the same methods that were used for the alkali metals. Figure 4 shows the electronic energy per electron as a function of r_s. The binding energy, which involves a small correction for the zero-point vibrational energy, is found to be 10.6 kg cal/mol for a density 0.59. The corresponding values for the molecular forms are 52.4 kg cal/gram-atom and 0.087. The difference in energy shows clearly why the ordinary form is not metallic.

If it is assumed that the observed compressibility of the molecular form, namely, $3 \cdot 10^{-9}$ cm²/dyne, is constant for a large change in volume, it is found that the energy of the molecular form would be increased by only 0.92 kg cal/mol when the density is changed from 0.087 to 0.59, so that there would be no tendency to change to a metallic form. Actually, the compressibility decreases with decreasing volume. Even if it became large enough to make the change from the molecular to the metallic phase possible, however, the pressure required would be at least 400,000 atmospheres which is not attainable at present.

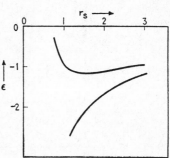

FIG. 4.—The lower curve is the $\epsilon_0(r_s)$ curve for metallic hydrogen. The upper curve is the total energy per electron. The origin of the energy scale is the energy of an ionized hydrogen atom, so that only the values of the upper curve relative to one Rydberg unit are of interest for cohesion. Abscissa is expressed in Rydberg units.

80. Monovalent Noble Metals.—Fuchs[1] treated the cohesion of copper along the lines developed in the preceding sections and found that the interaction between closed shells and the exchange and correlation interaction between valence and closed-shell electrons play a much more important role in this metal than in the alkali metals. The reasons for this may be found from an investigation of Hartree's wave functions for atomic copper, for these show that about $0.4e$ of the $10e$ charges of the newly completed $3d$ shell lie outside the sphere whose volume is equal to the volume of the unit cell. This means that the effect of binding on the d-shell electrons is nearly as important as the effect on the valence electrons and that the discrete atomic d levels are split into bands in the solid. Since it would be difficult to treat all eleven electrons by the

[1] K. FUCHS, *Proc. Roy. Soc.*, **151**, 585 (1935); **153**, 622 (1936); **157**, 444 (1936).

Bloch scheme in a computation of the cohesive energy, Fuchs assumed that the d shell is nearly rigid.

The closed-shell one-electron functions may be expressed in the form

$$\psi_{nl,m} = f_{nl}(r)\Theta_l^m(\theta,\varphi) \tag{1}$$

where $f_{nl}(r)$ is a radial function associated with the radial and orbita

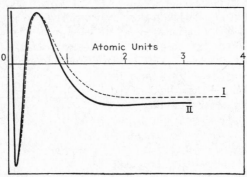

Fig. 5.—The lowest energy wave functions of metallic copper. The dotted curve corresponds to a case in which exchange is neglected and the full curve to one in which it is included. (*After Fuchs.*)

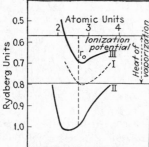

Fig. 6.—The $\epsilon_0(r_s)$ curves of metallic copper without exchange (curve I) and with exchange (curve II). Curve III represents the mean energy per electron after adding the Fermi energy. (*After Fuchs.*)

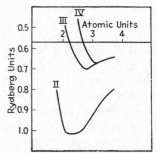

Fig. 7.—Curves II and III are the same as in Figure 6. Curve IV includes the ion-ion exchange repulsion. (*After Fuchs.*)

angular momentum quantum numbers n and l, respectively, and Θ_l^m is a surface harmonic. In a given closed shell, the z component of angular momentum quantum number m ranges over all integer values from $-$ to l, and each ψ occurs once with each of the two possible spin orientations. If it is assumed that the valence electrons in the metal are so well correlated that only one appears in a given cell at a time, Fock's equation for the valence-electron wave function $\varphi(\mathbf{r}_1)$ is

$$-\frac{\hbar^2}{2m}\Delta\varphi(\mathbf{r}_1) + v_c\varphi(\mathbf{r}_1) - e^2\sum_{n,l}\sum_{m=-l}^{l}\int\frac{\psi_{nl,m}{}^*(\mathbf{r}_2)\psi_{nl,m}(\mathbf{r}_1)\varphi(\mathbf{r}_2)}{r_{12}}d\tau_2 = \epsilon\varphi(\mathbf{r}_1).$$

$$(2)$$

Here, v_c is the coulomb potential of the ion core, and the summation extends over those ψ in the closed shells which have the same spin as φ.

Fuchs solved Eq. (2) for the radial function of the $4s$ type that satisfies the boundary conditions $\varphi'(r_s) = 0$. In doing so, he evaluated the coulomb and exchange terms by the use of the Hartree one-electron wave functions for atomic copper. The solutions that were obtained with and without the exchange terms are shown in Fig. 5 along with Hartree's $3d$ function. In Fig. 6, $\epsilon_0(r_s)$ is represented for both approximations. The exchange energy is 2.5 ev for the value of r_s corresponding to the observed lattice constant. Since this is a large fraction of the cohesive energy of 3.1 ev, we may conclude that the correlation interaction between closed-shell electrons and valence electrons cannot be neglected in an accurate computation. This term could be included roughly in the manner developed by Gorin for potassium, but the labor is not justified in the present case because of other approximations that are made.

The Fermi energy was not determined by use of the perturbation method described in Sec. 78. Instead, it was evaluated more roughly by means of a perturbation scheme in which the entire periodic potential field of the ion cores $V(\mathbf{r})$ is treated as a perturbation. The starting wave functions in this case are free waves, and the energy of the perturbed functions $\chi_{\mathbf{k}}e^{2\pi i\mathbf{k}\cdot\mathbf{r}}$ is

$$\epsilon'(\mathbf{k}) = \frac{\hbar^2}{2m}\mathbf{k}^2 + V_{000} + \frac{1}{v^2}\sum_{\mathbf{k}'}\frac{|\int e^{-2\pi i(\mathbf{k}-\mathbf{k}')\cdot\mathbf{r}}V(\mathbf{r})d\tau|^2}{\epsilon(\mathbf{k}) - \epsilon(\mathbf{k}')}$$

$$(3)$$

(cf. Sec. 61). Here V_{000} is the integral of $V(\mathbf{r})$ over a unit cell,

$$\mathbf{k}' = \mathbf{k} + \mathbf{K},$$

where $\mathbf{K}$ is any lattice vector in the reciprocal lattice, and v is the volume of the unit cell. Fuchs retained only that term of the second-order sum in (3) that belongs to the lowest zone and derived the expression

$$\frac{1}{r_s^2}(2.21 - 0.10075V_g^2r_s^2)$$

$$(4)$$

for the mean Fermi energy. This result is expressed in Rydberg units when r_s is given in Bohr units. V_g is the integral:

$$V_g = |\int e^{-4\pi i\mathbf{g}\cdot\mathbf{r}}V(\mathbf{r})d\tau|$$

where **g** is the vector joining the origin of **k** space to the nearest point of the first zone boundary for the face-centered lattice. Fuchs assumes that V_g, which is approximately equal to one-half the energy gap at the first zone boundary, is of the order of magnitude 2 ev. The second term in (4) is then practically negligible, and the Fermi energy is the same as for free electrons. This conclusion is not fully justified, for the same line of reasoning would lead to a Fermi energy less than the value for free electrons in the case of potassium, whereas Gorin's work shows that the opposite is true. We may conclude that the other terms in (3) usually are not negligible,[1] so that the approximation (4) is not reliable. The sum of $\epsilon_0(r_s)$ and the free-electron Fermi energy is shown in Fig. 6.

Fuchs assumed that the valence-electron self-energy is exactly balanced by the exchange and correlation energies, just as in the alkali metals. The error made in doing this undoubtedly is smaller than that introduced in estimating the Fermi energy from (4).

Finally, the very important ion-ion repulsion term was estimated by means of a modified Fermi-Thomas method. When this result is added to the previous results, the energy curve shown in Fig. 7 is obtained. The closed-shell interaction correction is of the order of 0.5 ev when is equal to r_0 and rises very rapidly as the lattice constant decreases. As we shall see in Sec. 82, the fact that the compressibility of the mono-valent noble metals is less than that of the alkali metals can be associated with this interaction term.

The cohesive energy and the computed lattice constant of copper are listed in Table LVIII. This energy was obtained by subtracting the

TABLE LVIII.—THE OBSERVED AND CALCULATED COHESIVE ENERGY AND LATTICE CONSTANT OF COPPER

	Cube edge, Å	Cohesive energy, kg cal/mol
Calculated	4.2	33
Observed	3.6	81

observed ionization energy of atomic copper from the energy curve of Fig. 7. The large error in the calculated value cannot be associated with an error in a single term in this case, as it could in the case of lithium and sodium. In spite of this fact, the computations do give an indication

[1] Except in the case of lithium, the actual free-electron band does not correspond to the lowest zone, so that there are terms in (3) with positive as well as negative energy denominators.

[2] In subsequent work on the elastic constants, which is discussed in Sec. 82, this interaction was also used.

of the relative importance of the various terms that influence the cohesive energy in the monovalent noble metals.

81. Metallic Beryllium.—Herring and Hill[1] have given a detailed treatment of metallic beryllium along the lines of the preceding sections. When compared with the monovalent metals, this case presents two additional complicating features, namely, the fact that the lattice is close-packed hexagonal instead of cubic, so that there are two atoms per unit cell, and the fact that two zones are nearly completely occupied, so that the free-electron approximations cannot be applied without careful investigation. The atomic cell for beryllium is shown in Fig. 4 of Chap. IX, and the first two zones are shown in Fig. 18 of Chap. VIII. Since the computations contain more details than it seems advisable to present

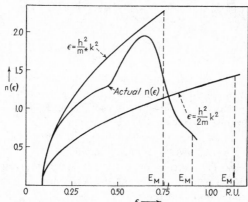

FIG. 8.—The $n(\epsilon)$ curve for beryllium. This is compared with the free-electron $n(\epsilon)$ curves for $m/m^* = 1$ and 0.62.

here, we shall survey only the general outline of their work and compare their results with experimental material.

To begin with, they obtained a self-consistent Hartree field for the valence electrons in the metal, using Hartree's atomic field for the $(1s)^2$ core. It should be mentioned at this point that all computations were carried out for values of r_s, the radius of the atomic sphere, both larger and smaller than the observed value, as well as for the observed value $(2.37a_h)$. Wave functions for a number of points in **k** space were then computed, the functions for the center points of the zone being determined by the method used for ψ_0 for the alkali metals, and the functions at the zone boundaries being determined by the use of the free-electron perturbation scheme discussed in Sec. 73. From the energies of these functions, a level-density curve was obtained and the mean Fermi energy computed.

[1] C. C. HERRING and A. G. HILL, *Phys. Rev.* (to appear).

A complicating feature of this process is the fact that the exchange interaction between valence and core electrons had to be included, as in the case of potassium. A comparison of the actual distribution curve with that for perfectly free electrons and that for free electrons having the effective mass m^* determined from the curvature of the $\epsilon(\mathbf{k})$ curve near $\mathbf{k} = 0$ is shown[1] in Fig. 8. The vertical lines represent the top of the occupied regions of levels in the three cases. It may be seen that the actual density function has a strong minimum near the top of the filled region—a fact that seems to occur generally among the alkaline earth metals. Values of m/m^* for several values of r_s are given in Table LIX. It may be noted that the ratio is less than unity as in lithium.

TABLE LIX.—VALUES OF m/m^* FOR THE VALENCE ELECTRONS OF BERYLLIUM DETERMINED FROM THE CURVATURE OF $\epsilon(\mathbf{k})$ NEAR THE BOTTOM OF THE FILLED BAND

r_s, a_h	m/m^*
2.07	0.422
2.37	**0.616**
2.67	0.697

Next, Herring and Hill attempted to make a more accurate estimate of exchange than would correspond to the use of the free-electron value. This proved to be very difficult, but they came to the conclusion that the exchange probably does not deviate by more than about 6 per cent from the free-electron value.

In lieu of a better alternative, they employed the free-electron correlation energy. The correlation energy is not greatly larger than the uncertainty in the exchange, so that this procedure probably does not introduce an important new error.

The results of the computation are listed in Table LX and are compared with computed quantities. The theoretical values are expressed

TABLE LX.—COMPARISON OF OBSERVED AND COMPUTED VALUES OF THE COHESIVE ENERGY, LATTICE PARAMETER, AND COMPRESSIBILITY OF BERYLLIUM

	Calculated	Observed
Cohesive energy (kg cal/mol).............	53 to 36	75
Equilibrium value of $r_s(a_h)$...............	2.23 to 2.57	2.37
$1/\beta \cdot 10^{-12}$ (cgs).........................	0.87 to 1.32	1.25

in terms of the limiting values obtained as a result of several methods of approximating the various quantities.

To explain the discrepancy in cohesive energy, Herring and Hill suggest the possibility that exchange and correlation energies for the

---

[1] Figure 8 is not the final level-density curve obtained by Herring and Hill, but resembles it closely.

electrons near the top of the filled region, where the free-electron approximation undoubtedly is worst, may be considerably larger than their computations indicate.   In connection with this, they note that the work function computed from their results by a method to be discussed in the next chapter is in very bad agreement with the best observed value.   This discrepancy would also be decreased if the exchange and correlation energies of the uppermost electrons were increased.

82. The Elastic Constants of Metals.—It is pointed out in footnote 1, page 94, that the elastic constants c_{ij} that enter in the relation

$$X_i = \sum_j c_{ij} x_j \qquad (i, j = 1, \cdots, 6),$$

between the six independent components X_i of the stress tensor and the six independent components x_j of the symmetric strain tensor are given by the equation

$$c_{ij} = \frac{\partial^2 E(x_1, \ldots, x_6)}{\partial x_i \partial x_j} \tag{1}$$

where E is the energy per unit volume of the crystal as a function of the strains.   Thus, the elastic constants may be computed if the energy of the crystal as a function of homogeneous atomic displacements is known.

a. Compressibility.—The simplest energy change to compute is that accompanying a uniform compression in all directions.   In this case,

$$\begin{aligned} x_1 = x_2 = x_3 &\equiv \xi, \\ x_4 = x_5 = x_6 &= 0, \end{aligned} \tag{2}$$

so that the change in energy per unit volume is

$$\delta E = \tfrac{1}{2}(c_{11} + c_{22} + c_{33})\xi^2 + (c_{12} + c_{23} + c_{31})\xi^2. \tag{3}$$

Since the relative change in volume $\delta V/V$ is 3ξ for small displacements, Eq. (3) may be placed in the form

$$\delta E = \frac{1}{2\beta}\left(\frac{\delta V}{V}\right)^2, \tag{4}$$

in which the *compressibility* β is related to the c by the equation

$$\frac{1}{\beta} = \frac{1}{9}[(c_{11} + c_{22} + c_{33}) + 2(c_{12} + c_{23} + c_{31})]. \tag{5}$$

In cubic crystals,

$$\begin{aligned} c_{11} &= c_{22} = c_{33}, \\ c_{12} &= c_{23} = c_{31}, \end{aligned}$$

so that

$$\frac{1}{\beta} = \frac{1}{3}(c_{11} + 2c_{12}),\qquad(6)$$

whereas, in hexagonal crystals,

$$c_{11} = c_{22},$$
$$c_{23} = c_{31},$$

whence

$$\frac{1}{\beta} = \frac{1}{9}(2c_{11} + c_{33} + 2c_{12} + 4c_{31}).\qquad(7)$$

Now, $\delta V/V$ is simply related to the relative change $\delta r_s/r_s$ in the radius of the atomic sphere by the equation

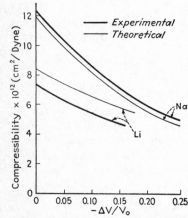

FIG. 9.—Comparison of the observed and calculated compressibilities of lithium and sodium as functions of the relative change in volume $-\Delta V/V_0$. (*After Bardeen.*)

$$\frac{\delta V}{V} = 3\frac{\delta r_s}{r_s}.$$

Hence, if $E(r_s)$ is the energy per unit volume of the crystal as a function of r_s,

$$\frac{1}{\beta} = \frac{r_s^2}{9}\frac{\partial^2 E}{\partial r_s^2};$$

or if $\epsilon(r_s)$ is the energy per atom,

$$\frac{1}{\beta} = \frac{1}{12\pi r_s}\frac{\partial^2 \epsilon(r_s)}{\partial r_s^2}.\qquad(8)$$

This quantity evidently may be computed from the $\epsilon(r_s)$ curves discussed in preceding sections. These values are given in Table LXII.

Bardeen[1] has used the computed $\epsilon(r_s)$ curves for lithium and sodium to compute the compressibilities for a range of values of r_s and has compared these with values obtained from Bridgman's room-temperature results[2] by extrapolation to absolute zero of temperature. The computed and observed curves are given in Fig. 9. The pressure required to produce the maximum change of volume in these cases is of the order of magnitude 40,000 kg/cm². Bardeen suggests that the disagreement in the case of lithium arises from neglect of the effect of the discontinuity of the $\epsilon(\mathbf{k})$ curve at the first zone boundary (*cf.* footnote 2, page 366).

b. Other Relations.—In order to compute all the elastic constants, it is necessary to determine the energy change of the crystal for deforma-

[1] BARDEEN, *op. cit.*, 372.
[2] P. W. BRIDGMAN, *Proc. Am. Acad. Sci.*, **72**, 207 (1938).

tions other than a homogeneous compression. The most extensive work of this kind has been carried out by Fuchs[1] for the alkali metals. He determined the energy change for the following two additional types of dilation:

1. *Equal contraction and expansion, respectively, along two cube edges, which leave the volume unchanged.*—This deformation may be expressed in terms of the fractional displacements along the three axes which will be designated by α_x, α_y, and α_z. If the z direction is that along which the distances are unchanged, we have

$$\alpha_z = x_3 = 0,$$
$$\alpha_x = x_1 = \xi,$$
$$\alpha_y = x_2 = -\xi,$$
$$x_4 = x_5 = x_6 = 0.$$

Thus, in this case,

$$\delta E = (c_{11} - c_{12})\xi^2,$$

and

$$c_{11} - c_{12} = \frac{1}{2} \frac{\partial^2 E}{\partial \xi^2}. \tag{9}$$

2. *Shearing strain in a plane parallel to two cube edges.*—In this case, the compressional strains are zero, and only the shear strains are finite. Thus the dilatation is described by the relations

$$x_1 = x_2 = x_3 = 0; \qquad x_4 = \xi; \qquad x_5 = x_6 = 0.$$

Hence,

$$\delta E = \tfrac{1}{2}c_{44}\xi^2.$$

It is evident that these deformations, unlike the deformation that determines the compressibility, distort the cells into noncubic forms, thus deforming the spheres of the sphere approximation into ellipsoids. Fuchs treated the changes in each of the contributions to the total energy that were discussed in the preceding sections in the following way:

i. Since the lowest wave function ψ_0 is practically constant near the boundary of the polyhedron, Fuchs assumed that it is not appreciably changed by a dilatation that does not alter the volume. Thus, ϵ_0 was regarded as a function of r_s alone.

ii. Fermi energy. Since the Fermi energy of a free-electron gas depends only upon the volume, this assumption was retained in treating the alkali metals.

[1] Fuchs, *ov. cit.*

iii. The coulomb energy. We saw in part *f*, Sec. 78, that the coulomb interaction of the electrons and ions is the same as the self-energy of a set of positive point charges in a uniform cloud of negative charge, except for regions very near to the nuclei. Since the form of the wave functions in these regions is not affected by distortions that do not change the volume, the change in coulomb energy for the distortions 1 and 2 is the same as the change in the electrostatic self-energy of the simple lattice. The methods for computing this energy were discussed in Chap. II for the case in which the atoms are at lattice positions. The changes for the distortions 1 and 2 are simply related to second derivatives of these expressions for the appropriate lattice.

iv. Exchange energy and correlation energy. It was assumed that these depend only on the volume, as for perfectly free electrons.

v. Noncoulomb ion-ion interaction. The interaction energy of the closed shells is very sensitive to the interatomic distance and, hence, affects the elastic constants appreciably even in cases, such as the alkali metals, in which its contribution to the cohesive energy is small. Fuchs determined the effect of this interaction in the alkali metals by use of a repulsive term of the Born-Mayer type

$$C_{12}be^{\frac{r_1+r_2-r}{\rho}}$$

which was discussed in Chap. II, and a van der Waals term of the type

$$\frac{A}{r^6},$$

which was discussed in Chap. VII. The methods used for determining C_{12} and A need not be discussed again here. In the case of copper, the closed-shell interaction was taken from the work discussed in Sec. 80. This interaction leads to a slightly expanded lattice.

To summarize, the elastic constants are determined by terms *iii* and *v*.

In the alkali metals, the coulomb contribution is two to three times larger than the contribution from ion-ion interaction, whereas the situation is reversed in the noble metals, as may be seen in Table LXI. A comparison of observed and computed constants is given in Table LXVII. It may be seen from Table LXI that the comparatively high rigidity of the noble metals arises from the closed-shell interactions.

It is interesting to examine the extent to which these constants satisfy the conditions[1] for isotropy and the Cauchy-Poisson relations, which are, respectively,

$$2c_{44} = c_{11} - c_{12}$$

[1] See footnote 1, p. 94 and footnote 1, p. 106.

TABLE LXI.—COMPOSITION OF THE ELASTIC CONSTANTS OF MONOVALENT METALS
(AFTER FUCHS)
(In units of 10^{11} dynes/cm^2)

	Coulomb contribution	Ion-ion interaction
Li		
$c_{11} - c_{12}$	0.339	−0.02
c_{44}	1.263	0.086
Na		
$c_{11} - c_{12}$	0.143	−0.02
c_{44}	0.532	0.048
K		
$c_{11} - c_{12}$	0.0644	0.02
c_{44}	0.240	0.020
Cu		
$c_{11} - c_{12}$	0.573	4.53
c_{44}	2.57	6.4

TABLE LXII.—COMPARISON OF OBSERVED AND CALCULATED VALUES OF THE ELASTIC
CONSTANTS OF MONOVALENT METALS
(In units of 10^{11} dynes/cm^2)

	$1/\beta$	$c_{11} - c_{12}$	c_{44}	c_{11}	c_{12}
Li					
Calculated....................	1.30	0.341	1.349	1.53	1.19
Na					
Calculated....................	0.88	0.141	0.580	0.97	0.83
Observed.....................	∼0.85	0.145	0.59	∼0.95	∼0.80
K					
Calculated....................	0.41	0.062	0.260	0.45	0.39
Cu					
Calculated....................	14.1	5.1	8.9	17.5	12.4
Observed.....................	13.9	5.1	8.2	18.6	13.5

and

$$c_{12} = c_{44}.$$

It is readily seen from the results of Table LXII that neither condition is closely satisfied and that the alkali metals[1] are very far from isotropic. As was pointed out in Sec. 19, this fact accounts for a large part of the anomalies in the specific-heat curves of these metals.

83. Cohesion of Alloys.—It was seen in Sec. 3 that alloys usually have a small heat of formation. There have been no extensive computations of these heats. Mott,[2] however, has attempted to estimate the difference in energy between completely ordered and completely disordered β brass, which is a body-centered metal containing equal numbers of copper and zinc atoms (cf. Sec. 3). He assumed that the additional valence electrons of zinc cluster mainly about the zinc ions and that the potential near a zinc ion is greater than that near a copper ion by an amount

$$\varphi(r) = \frac{e}{r}e^{-qr} \tag{1}$$

where q is a constant. Since this potential vanishes as e^{-r} at large distances, it follows that its use is equivalent to the assumption that the zinc atoms are neutral. The screening constant q was evaluated by comparing the observed resistivity of the alloy with that computed on the assumption that the difference between the potential of the two ions is given by (1) (cf. Sec. 130) and was found to be $2.7 \cdot 10^8$ cm^{-1}, or

$$\frac{1}{q} = 0.37 \text{ Å},$$

which is only about one-quarter of the radius of the zinc atom. When this value of q is used, the charges in the zinc and copper polyhedra of the body-centered lattice are $0.075e$ and $-0.075e$, respectively, which corresponds to an electrostatic interaction energy per atom of

$$-\frac{1.017}{a}(0.075)^2e^2 \tag{2}$$

where a is the cube-edge distance and 1.017 is the appropriate Madelung constant. This energy is 0.027 ev for β brass. Since the mean potential at an ion arising from neighbors would be zero in a perfectly disordered lattice, it follows that (2) is the electrostatic ordering energy. In addition, Mott estimated the decrease in exchange repulsive energy in going

---

[1] The experimental values for sodium were obtained by Quimby and Siegel (see Sec. 19).

[2] N. F. MOTT, *Proc. Phys. Soc.*, **49**, 258 (1937).

from the disordered to the ordered state and obtained a value of 0.013 ev per atom. Thus the total ordering energy is 0.04 ev per atom. The total change in energy in the transition from order to disorder as obtained by integrating the specific-heat curve of Fig. 43, Chap. I, is about 0.043 ev per atom, which is the same order of magnitude as the computed value. This computation suggests that the largest source of ordering energy is the Madelung term, as in ionic crystals.

84. Simplified Treatments of Cohesion.[1]—In addition to the preceding work, in which the computation of cohesive properties is based entirely on the Schrödinger equation, there have been several treatments of cohesion that start from other points. Among these treatments, the two most important are those which start from semiempirical equations of state and interrelate measured quantities and those which use the Fermi-Thomas statistical equation. We shall discuss these briefly.

a. The Semiempirical Method.—Perhaps the most extensive work of this kind is that of Grüneisen,[2] who assumed, following a suggestion of Mie, that the atoms in monatomic substances interact in pairs with a potential energy relation of the type

$$\epsilon(r) = -\frac{a}{r^m} + \frac{b}{r^n} \tag{1}$$

in which r is the interatomic distance and a, b, m, and n are positive constants, n being larger than m. This assumption is analogous to that of the Born theory of ionic crystals in which $m = 1$ and $a = e_1 e_2$. According to (1), the total energy of the crystal at absolute zero of temperature is

$$E = -\frac{a}{2} N \sum_i \frac{1}{r_i^m} + \frac{b}{2} N \sum_i \frac{1}{r_i^n} \tag{2}$$

in which the sum extends over all values of the distance r_i between a given atom and the others.

Three relations among the four parameters in (2) were determined by the condition that this expression give the observed values of the atomic volume, cohesive energy, and compressibility of the solid at absolute zero.

The temperature-dependent free energy of the lattice was obtained by adding to (2) the free-energy function corresponding to Debye's specific-heat law. When this was done, it was found that

---

[1] This type of work is extensively discussed in the book by J. C. Slater, *Introduction to Chemical Physics* (McGraw-Hill Book Company, Inc., New York, 1939).

[2] G. Grüneisen, see *Handbuch der Physik*, vol. X.

$$\frac{n+2}{6} = -\frac{d \log \Theta_D}{dV} \tag{3}$$

where Θ_D is the characteristic temperature and V is the molar volume. It may be shown that the right-hand side of (3) is equal to the quantity

$$\frac{-V(\partial V/\partial T)_p}{C_p(\partial V/\partial p)_T}. \tag{4}$$

Thus a fourth relation among the parameters was determined by the condition that $(n+2)/6$ be equal to the measured values of (4).

Using the resulting total free-energy function, Grüneisen was able to correlate a number of properties of metals and of monatomic insulators such as diamond. For example, Table LXIII gives a comparison of

TABLE LXIII.—COMPARISON OF OBSERVED VALUES OF EXPANSION COEFFICIENTS WITH THOSE COMPUTED BY THE USE OF GRÜNEISEN'S THEORY
(The values of α_l are given in cgs units.)

Temperature interval, °K	$\alpha_l \cdot 10^6$	
	Calculated	Observed
Diamond		
84.8–194.1	0.16	0.18
194.1–273.2	0.61	0.58
273.2–296.2	0.97	0.97
296.2–328	1.17	1.17
328 –351	1.37	1.45
Copper		
20.4– 80.5	4	3.8
82 –289	14.0	14.2
289 –523	17.4	17.2
523 –648	18.7	18.6
648 –773	19.5	19.6

observed mean values of the expansion coefficients of diamond and copper, taken for a range of temperature ΔT, and the mean values computed from Grüneisen's equation of state.

It is clear that the function (2) cannot be expected to give the proper elastic constants since it would lead to the Cauchy-Poisson relations, which are not usually satisfied in metals (*cf.* Sec. 82).

Modifications of Grüneisen's plan that are based on more accurate information of the cohesive forces in metals have been developed by a number of people, principally Rice[1] and Bardeen.[2] These methods have in common the property that they express the absolute zero energy of the entire metal as a sequence of terms which vary inversely as different powers of the volume, semitheoretical and empirical data being used to evaluate the constants. This type of modification has been pushed furthest by Bardeen, who used it to discuss the behavior of alkali metals at very high pressures and to correlate a number of Bridgman's measurements.

It was seen in Sec. 78 that the cohesive properties of the alkali metals are given closely by the quantity $-(\epsilon_0 + \epsilon_I + \epsilon_F)$, in which $\epsilon_0 + \epsilon_I$ is the energy of the electron of zero wave number relative to the energy of the free atom, in the sphere approximation, and ϵ_F is the Fermi energy. This result depends upon the fact that the exchange and correlation effects combine in such a way as to allow on the average only one electron in a given polyhedron at a given time. Now, by integrating Eq. (1), Sec. 78, with appropriate simplifying assumptions, it is possible to show that $\epsilon_0(r_s)$ has the approximate form

$$\epsilon_0(r_s) = \frac{a}{r_s^3} - \frac{3}{r_s} \tag{5}$$

where a is a constant that varies from solid to solid. In order to obtain a slightly more general result, Bardeen assumed that ϵ_0 actually can be expressed in the form

$$\epsilon_0(r_s) = \frac{a}{r_s^3} - \frac{c}{r_s} \tag{6}$$

to a higher degree of accuracy. We shall not consider a proof of (5) necessary, for (6) is a valid assumption if taken in the same spirit as Grüneisen's relation (1). Now, we saw in part f, Sec. 78, that the Fermi energy varies as

$$2.21\alpha \frac{e^2}{r_s^2} \tag{7}$$

in which $\alpha = m/m^*$. Since r_s is proportional to $v^{\frac{1}{3}}$, where v is the atomic volume, the total energy of the crystal in the sphere approximation may be expressed in the form

$$E(v) = A\left(\frac{v_0}{v}\right) + B\left(\frac{v_0}{v}\right)^{\frac{2}{3}} - C\left(\frac{v_0}{v}\right)^{\frac{1}{3}} \tag{8}$$

where A, B, and C are constants for a given metal and v_0 is the observed atomic volume.

[1] O. K. Rice, *Jour. Chem. Phys.*, **1**, 649 (1933).
[2] Bardeen. *op. cit.*, 372.

Bardeen determined the parameters A, B, and C by adjusting them so that (8) would give the empirical values of $E(v_0)$, v_0, and the compressibility. Thus, since E must have a minimum when $v = v_0$, we obtain

$$\frac{C}{3} = A + \frac{2}{3}B. \tag{9}$$

Moreover, if the value of E at absolute zero is E_0 and if the value of the compressibility is β_0, we find in addition

$$-E_0 = 2A + B,$$
$$\frac{3}{\beta_0} = 2A + \frac{2B}{3}. \tag{10}$$

Values of A, B, and C determined from these equations by the use of empirical data are given in Table LXIV for all the alkali metals. The

TABLE LXIV.—COMPARISON OF EMPIRICAL AND THEORETICAL VALUES OF THE
CONSTANTS IN BARDEEN'S EMPIRICAL EQUATION OF STATE
(The constants are expressed in units of 10^{-12} erg per atom.)

	Li	Na	K	Rb	Cs
A (empirical)...............	1.4	4.3	5.1	4.2	3.7
B (empirical)...............	8.4	1.2	− 2.1	− 0.6	0.0
B (theoretical)...............	4.5	3.0	2.3	1.7	1.5
C (empirical)...............	20.8	15.4	11.2	11.4	11.2
C (theoretical)...............	20.9	16.2	13.2	12.3	11.4

values of B and C are compared with the values obtained using the value $c = 3$, corresponding to Eq. (5), and the free-electron value of α in (7). It may be seen that (5) is a good approximation, whereas (7) is not good if α is assumed to be a constant. We have seen that α actually varies appreciably with v in lithium and potassium [see Eq. (17), Sec. 78; and Table LII]; hence, this discrepancy in the value of B is not surprising. Bardeen suggests that the true Fermi energy probably varies in the manner

$$B\left(\frac{v_0}{v}\right)^{\frac{2}{3}} + B'\left(\frac{v_0}{v}\right) \tag{11}$$

in which the second term takes into account the variation of α with v.

As a test of the validity of the relation (8), Bardeen compared the volume-pressure curves obtained from this equation with those obtained by extrapolating Bridgman's measurements to absolute zero of temperature. The theoretical volume at pressure p is determined by the equation

$$\left(\frac{\partial E}{\partial V}\right)_p = -p, \tag{12}$$

which leads to the relation

$$pv_0 = y^4(y - 1)[2A + \tfrac{2}{3}B + A(y - 1)] \tag{13}$$

in which

$$y = \left(\frac{v_0}{v}\right)^{\frac{1}{3}}.$$

The comparison is shown in Fig. 10. It may be observed that the agreement is fairly good in all the alkalies, the largest deviation occurring for rubidium. In addition, the experimental curve for cesium shows a break indicative of a phase change.

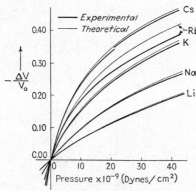

FIG. 10.—Comparison of the observed and calculated relative changes in the volume of the alkali metals as functions of pressure. The break in the experimental curve for cesium is discussed in the text. (*After Bardeen.*)

The effect of ion-ion exchange interaction is neglected in Eq. (8). This varies as

$$Ae^{-\frac{r}{\rho}} \tag{14}$$

and should contribute higher power terms to (8). One might expect these higher power terms to enter first for cesium, since it has the highest compressibility. Bardeen assumed that the expression (8) is valid for both a face-centered and a body-centered lattice with given values of the constants and that additional different terms should be added in the two cases in order to include the ion-ion interaction. He determined ρ in Eq. (14) from the results of Mayer and Bleick's computation of the exchange interaction between neon atoms, and he determined A by the methods used in the Born-Mayer theory (*cf.* Sec. 11). He found that a polymorphic change from a body-centered to a face-centered lattice should occur at about the same pressure as the observed change indicated in Fig. 10, which suggests that the change actually is of this type.

b. The Fermi-Thomas Method.—In the Fermi-Thomas[1] statistical treatment of the many-electron problem, it is assumed that the electrons are effectively free at each point, so that the mean kinetic energy of the electrons at a point r is related to the density $n(r)$ by the equation for perfectly free electrons, namely,

$$\epsilon_k(r) = \frac{3}{10}\frac{h^2}{m}\left[\frac{3n(r)}{8\pi}\right]^{\frac{2}{3}} \tag{15}$$

[cf. Eq. (20), Sec. 26.]. This assumption clearly is rigorous only if the change in potential is very small over the distance of the electronic wave length, a condition that is not satisfied near the nuclei of atoms. In their original treatments of neutral atoms, Fermi and Thomas assumed that the kinetic energy of the fastest electrons, namely,

$$\epsilon_m(r) = \tfrac{5}{3}\epsilon_k(r), \tag{16}$$

is equal to the negative of the potential energy $-e\varphi(r)$ at r. Thus, $n(r)$ and $\varphi(r)$ are related by the equation

$$e\varphi(r) = \frac{h^2}{2m}\left[\frac{3n(r)}{8\pi}\right]^{\frac{2}{3}},$$

or

$$n(r) = \frac{8\pi}{3}\left[\frac{2me}{h^2}\varphi(r)\right]^{\frac{3}{2}}. \tag{17}$$

Now, φ satisfies Poisson's equation

$$\Delta\varphi = 4\pi e n(r); \tag{18}$$

and if $n(r)$ is eliminated from (18) by means of Eq. (17),

$$\Delta\varphi = \frac{32\pi^2}{3}e\left(\frac{2me}{h^2}\right)^{\frac{3}{2}}\varphi^{\frac{3}{2}}. \tag{19}$$

This equation is solved for neutral atoms with the boundary conditions that φ be zero at infinity and vary as Ze/r near the origin. It yields reasonably good qualitative distribution functions for the electrons in heavy atoms.

Modifications of Eq. (19) that are valid for systems more general than neutral atoms have been developed by Dirac[2] and by Lenz and Jensen.[3] The Lenz-Jensen scheme, which is a variational one, is formally equivalent to the original Fermi-Thomas scheme, inasmuch as the Eulerian

[1] See, for example, L. BRILLOUIN, Die Quantenstatistik (Julius Springer, Berlin, 1930), for a survey of early work on the Fermi-Thomas theory.

[2] P. A. M. DIRAC, Proc. Cambridge Phil. Soc., 123, 714 (1929).

[3] H. JENSEN and W. LENZ, Z. Physik, 77, 713, 722 (1932).

equation of their variational principle reduces to Eq. (19) when applied to a neutral atom. Dirac's scheme, however, is more general, for it contains an additional term that decreases the energy at a point r by an amount equal to the exchange energy of one of a system of electrons in a region where the density is n, namely,

$$-0.916e^2\left[\frac{4\pi n(r)}{3}\right]^{\frac{1}{3}}. \tag{20}$$

Thus, Lenz and Jensen's method is equivalent to Hartree's when applied to perfectly free electrons, whereas Dirac's scheme is equivalent to Fock's.

Slater and Krutter[1] applied Dirac's method to a system of electrons that are in a lattice of point positive charges but did not obtain a minimum in the energy versus interatomic distance curve. This result is not surprising; for the correlation energy, which is neglected in this method, is a large fraction of the cohesive energy in the alkali metals, which correspond most closely to Slater and Krutter's model.

Gombas[2] has applied Lenz and Jensen's method to lattices that correspond to the alkali and alkaline earth metals. The lattices are even less stable in this approximation than in Dirac's; however, Gombas added a number of correction terms in order to compensate for the errors of the method. Thus, he added the free-electron exchange and correlation energies, which are sufficient to make the lattices stable. In addition, he added ion-ion interaction correction terms and valence-electron ion-core exchange terms. In this way, he has obtained energies that approximate the observed ones closely. The success of this procedure in the case of the alkali metals undoubtedly lies in the fact that the valence electrons are very nearly free so that the results obtained from the Fermi-Thomas method are nearly the same as those obtained from Hartree's method.

B. IONIC CRYSTALS

85. Sodium Chloride.—The most significant, purely quantum mechanical computations of the cohesive energies of ionic crystals are those which have been carried out on sodium chloride and on lithium hydride by Landshoff[3] and Hylleraas,[4] respectively. The first of these will be discussed in the present section. Landshoff based his work on a

---

[1] J. C. SLATER and H. KRUTTER, *Phys. Rev.*, **47**, 559 (1935).

[2] P. GOMBAS, *Z. Physik*, **95**, 687 (1936); **99**, 729 (1936); **100**, 599 (1936); **104**, 592 (1937); **108**, 509 (1938).

[3] R. LANDSHOFF, *Z. Physik*, **102**, 201 (1936); *Phys. Rev.*, **52**, 246 (1937). The writer is indebted to Landshoff for the values given in Table LXV.

[4] E. A. HYLLERAAS, *Z. Physik*, **63**, 771 (1930).

Heitler-London approximation in which he used the solutions of Fock's equations for Na^+ and Cl^- that were determined by Fock and Petrashen and by Hartree and Hartree, respectively. For simplicity, he did not attempt to compute the absolute energy of the lattice. Instead, he determined the energy of the crystal relative to the theoretical energy of the free ions as obtained from the one-electron functions. The advantage of this procedure lies in the fact that the internal energies of the ions do not appear in the final expression for the cohesive energy and do not need to be evaluated. Since the absolute accuracy of the solutions of Fock's equations for the ions has not been determined, Landshoff's results cannot furnish us with an estimate of the absolute accuracy of the Heitler-London approximation when applied to the solid. Cohesive energies computed by Landshoff's method might turn out to be larger than the observed value, as we have seen in Sec. 77.

It should be recalled that the Heitler-London and Bloch schemes are identical in cases such as the present one in which the Heitler-London scheme contains only closed shells. Thus, the Bloch scheme should lead to results of comparable accuracy.

The one-electron functions associated with neighboring ions are not orthogonal for the observed lattice spacing, for they overlap appreciably. Since it is convenient to use an orthogonal set of one-electron functions, Landshoff orthogonalized the free-ion functions in the following approximate manner: Let ψ_μ designate the free-ion wave functions that are centered about different nuclei. Landshoff showed that the following linear combinations of the ψ,

$$\chi_\mu = \psi_\mu(1 + \tfrac{3}{8}S_{\mu\mu}{}^2) - \tfrac{1}{2}\sum_{\eta}{}' S_{\eta\mu}\psi_\eta \tag{1}$$

where

$$\begin{aligned} S_{\nu\mu} &= \int \psi_\nu{}^* \psi_\mu d\tau, \\ S_{\mu\mu} &= 0, \\ S_{\mu\mu}{}^2 &= \sum_\eta S_{\eta\mu}{}^* S_{\mu\eta}, \end{aligned} \tag{2}$$

satisfy the conditions

$$\int \chi_\mu{}^* \chi_\nu d\tau = -\frac{3}{4}\sum_\eta S_{\mu\eta}S_{\eta\nu} + O(S^3),$$

$$\int |\chi_\mu|^2 d\tau = 1 + O(S^3), \tag{3}$$

in which $O(S^3)$ designates terms that contain the "overlap" integrals $S_{\mu\nu}$ to the third and higher powers. Thus, the χ are orthogonal to the approximation in which $-\frac{3}{4}\sum_{\lambda}S_{\mu\lambda}S_{\lambda\mu}$, and higher order terms are negligible.

Landshoff proceeded on the assumption that these terms are negligible and treated the φ as though they were orthogonal. It is not possible to tell from his results to what extent this assumption actually is justifiable.

Under these conditions, the mean total energy of the Hamiltonian operator,

$$H = -\frac{\hbar^2}{2m}\sum_i\Delta_i - \sum_{i,\alpha}\frac{e^2Z_\alpha}{r_{i\alpha}} + \frac{1}{2}\sum_{i,j}'\frac{e^2}{r_{ij}} + \frac{1}{2}\sum_{\alpha,\beta}'\frac{Z_\alpha Z_\beta}{r_{\alpha\beta}}, \qquad (4)$$

is

$$E = \sum_\mu 2\int \chi_\mu{}^*(\mathbf{r}_1)\left(-\frac{\hbar^2}{2m}\Delta - \sum_\alpha\frac{e^2Z_\alpha}{r_{1\alpha}}\right)\chi_\mu(\mathbf{r}_1)d\tau +$$

$$\sum_{\mu,\nu}' 2e^2\int\frac{|\chi_\nu(\mathbf{r}_1)|^2|\chi_\mu(\mathbf{r}_2)|^2}{r_{12}}d\tau_{12} -$$

$$\sum_{\mu,\nu}' e^2\int\frac{\chi_\mu{}^*(\mathbf{r}_1)\chi_\nu{}^*(\mathbf{r}_2)\chi_\mu(\mathbf{r}_2)\chi_\nu(\mathbf{r}_1)}{r_{12}}d\tau_{12} + \frac{1}{2}\sum_{\alpha,\beta}\frac{e^2Z_\alpha Z_\beta}{r_{\alpha\beta}}. \qquad (5)$$

The factor 2 enters in the first two terms because of spin.

If the χ are expressed in terms of the ψ by means of Eq. (1), Eq. (5) becomes

$$E = \sum_\mu 2(\mu|h|\mu) + \sum_{\mu,\nu}'[2(\mu\nu|g|\mu\nu) - (\mu\nu|g|\nu\mu)] + \sum_\mu S_{\mu\mu}{}^2\Big\{2(\mu|h|\mu) +$$

$$2\sum_\nu[2(\mu\nu|g|\mu\nu) - (\mu\nu|g|\nu\mu)]\Big\} - \sum_{\mu,\eta}S_{\mu\eta}\Big\{2(\eta|h|\mu) + \sum_\nu[2(\eta\nu|g|\mu\nu) -$$

$$(\eta\nu|g|\nu\mu)]\Big\} + \sum_{\alpha,\beta}'\frac{Z_\alpha Z_\beta e^2}{r_{\alpha\beta}} \qquad (6)$$

where

$$(\eta|h|\mu) = \int\psi_\eta{}^*\left(-\frac{\hbar^2}{2m}\Delta - \sum_\alpha\frac{e^2Z_\alpha}{r_{1\alpha}}\right)\psi_\mu d\tau, \\[2mm] (\eta\nu|g|\mu\lambda) = e^2\int\frac{\psi_\eta{}^*(\mathbf{r}_1)\psi_\nu{}^*(\mathbf{r}_2)\psi_\mu(\mathbf{r}_1)\psi_\lambda(\mathbf{r}_2)}{r_{12}}d\tau_{12}. \qquad (7)$$

All second-order terms of the type

$$\sum_{\eta,\lambda}S_{\mu\nu}S_{\nu\lambda}(\eta\lambda|g|\mu\nu)$$

have been discarded in deriving Eq. (6).

Expression (6) may be simplified considerably if the fact that the ψ are solutions of Fock's equations is used, for then the integrals involving Δ may be expressed in terms of the energy parameters of Fock's equations and coulomb and exchange integrals. Many of these terms cancel when the resulting equation is subtracted from the expression for the energy of the system of free ions. The final equation for the cohesive energy is

$$
E_c = \sum_\mu \left[\int |\psi_\mu(\mathbf{r}_1)|^2 \left(-\sum_{\alpha \neq \mu} \frac{Z_\alpha e^2}{r_{\alpha 1}} \right) d\tau_1 + \sum_\nu{}' 2(\mu\nu|q|\mu\nu) + \right.
$$
$$
\left. \frac{1}{2}\sum_{\alpha,\beta}{}' \frac{Z_\alpha Z_\beta e^2}{r_{\alpha\beta}} \right] + \sum_\mu \left\{ \sum_\nu{}' (\mu\nu|g|\nu\mu) + \right.
$$
$$
\left. \sum_\eta S_{\mu\eta} \left[\int \psi_\mu{}^* \psi_\mu \left(\sum_{\beta \neq \mu} - \frac{Z_\beta e^2}{r_{1\beta}} \right) d\tau_1 + 4 \sum_{\nu \neq \eta}{}' (\mu\nu|g|\mu\nu) \right] \right\}. \quad (8)
$$

The first, or coulomb, term is the electrostatic energy of a lattice of ions of which the electronic charge distribution is given by the functions $|\psi_\mu|^2$. This differs from the value $-1.748 Ne^2/r_0$, corresponding to the Madelung energy (cf. Chap. II), because neighboring ions overlap. Landshoff found that only the overlapping of neighboring Na^+ and Cl^- ions is important. There are two terms in the correction to the Madelung value, namely, a positive term I_1, which arises from the repulsion between the electrons, and a negative one $-I_2$, which arises from the attraction between the electrons and the nuclei. Although I_1 and I_2 turn out to be of the order of magnitude of several electron volts per ion pair at the observed value of r_0, the anion-cation distance, Landshoff found that they nearly cancel one another so that the total coulomb correction is small.

The exchange terms were evaluated in a straightforward manner. Only the exchange terms between nearest neighbors are important, but for these ions both the conventional exchange terms

$$
C = \sum_\mu \sum_\nu{}' (\mu\nu|g|\nu\mu)
$$

and the term, which we shall call B, that includes the factors $S_{\mu\nu}$, are very large. For example, C is -0.348 ev per ion pair, and B is 0.645 ev for $r_0 = 5a_h$. The two terms compensate for one another, however, and their sum is much smaller.

The coulomb and exchange terms appear in Table LXV along with the Madelung energy. The cohesive energy has a maximum of 179.3 kg cal/mol at $r = 5.34a_h$ which should be compared with the observed value of 183 ± 10 kg cal/mol at $r_0 = 5.4a_h$.

TABLE LXV.—CONTRIBUTIONS TO THE COHESIVE ENERGY OF SODIUM CHLORIDE
(r_0 is the nearest-neighbor distance.   In kg cal/mol)

r_0/a_h	Madelung energy	Coulomb correction	Exchange interaction	Cohesive energy	$I_{v.d.w.}$	Final cohesive energy
5.0	−217.7	0.3	41.1	176.3	−4.2	180.5
5.1	−213.5	0.1	35.5	177.9	−3.7	181.6
5.2	−209.4	−0.1	30.6	178.9	−3.3	182.2
5.3	−205.4	−0.2	26.4	179.2	−3.0	182.2
5.4	−201.6	−0.3	22.8	179.1	−2.7	181.8
6.0	−181.5	−0.3	9.5	172.3	−1.5	173.8

	Final values	
	Calculated	Observed
r_0/a_h	5.25	5.4
E_c	183	183

It is difficult to estimate the absolute accuracy of this one-electron approximation; for on the one hand the absolute accuracy of Fock's approximation for the free ions is not known, and on the other Landshoff does not give a numerical estimate of the magnitude of the neglected terms in E_c.   If Landshoff's approximations are valid, the agreement between his results and experiment indicates that the Heitler-London method that is based on solutions of Fock's equations for the free ions is fairly good.   As the one-electron functions become more accurate, the center of gravity of the electronic charge on an ion approaches the nucleus, and the wave functions of different ions overlap less.   This in turn decreases the repulsive terms and increases the computed value of E_c.   Since Landshoff's result leaves little room for improvement, the one-electron functions are probably very good.

In higher approximations, both van der Waals and polarization terms should be added to the preceding results.   Landshoff has evaluated the van der Waals term by means of the expressions developed in Chap. II and has obtained the following equation:

$$E_{v.d.w} = \frac{17.6}{(r_0/a_h)^6} \frac{e^2}{a_h}.$$

This adds about 3 kg cal to the result of Table LXV and decreases r_0 to $5.25a_h$.

86. Lithium Hydride.—Hylleraas[1] has treated lithium hydride in essentially the same way that Landshoff has treated sodium chloride. He employed one-electron wave functions of hydrogenic type with nuclear screening in order that the entire computation might be performed analytically. The principal objection to this procedure is that his approximate hydrogenic wave functions do not lead to very good binding energies for the free H^- and Li^+ states. The one-electron wave functions are

$$\psi = e^{-\left(Z - \frac{5}{16}\right)\frac{r}{a_h}}$$

(1)

where Z is 1 for hydrogen and 3 for lithium. The ionic energies derived from these functions are

	Observed, ev	Calculated, ev
H^-	-0.718	0.745
Li^+	75.28	73.72

In other words, H^- is not stable in the approximation in which the functions (1) are employed. This fault is reflected in the fact that the electronic distribution of H^- which is obtained from the hydrogenic functions is not very accurate. This error would not be very serious if it were not for the fact that Hylleraas found the H^--H^- interaction to be

TABLE LXVI.—CONTRIBUTIONS TO THE COHESIVE ENERGY OF LITHIUM HYDRIDE
(In kg cal/mol)

r_0/a_h	Madelung energy	H-Li interaction		H-H, coulomb plus exchange	Total cohesive energy
		Coulomb	Exchange		
3.84	-283.1	33.8	67.4	-34.0	215.9
4.16	-261.3	21.9	46.9	-26.0	218.5
4.56	-238.5	15.0	26.0	-14.0	211.5

	Final values	
	Calculated	Observed
r_0/a_h	4.08	3.84
E_c	218	218.5

[1] E. A. HYLLERAAS, *Z. Physik*, **63**, 771 (1930).

large and to favor binding.   The contribution from this interaction term probably would be much smaller if more accurate wave functions were used.

We shall not dwell on the details of Hylleraas' computations since they involve exactly the same approximations as those of the preceding case.   Although his results were given in analytical form we shall list numerical values.   Table LXVI contains the values of the quantities which were discussed in the last section.   The H-Li and H-H interaction terms are listed separately.

The principal contribution to the H-H interaction is the coulomb correction, for the exchange term is practically negligible.   As we mentioned above, this correction undoubtedly would be less if more accurate wave functions had been used.

87. The Elastic Constants of Ionic Crystals.—The methods discussed in Sec. 82 evidently could be applied to compute the elastic constants of sodium chloride and lithium hydride.   Actually, only the compressibility of sodium chloride has been evaluated.   It is given in Table LXVII.

TABLE LXVII.—THE RECIPROCAL OF THE COMPRESSIBILITY OF SODIUM CHLORIDE
(AFTER LANDSHOFF)
(In units of 10^{12} dynes/cm^2)

	Calculated	Observed
$1/\beta$	4.35	4.16

C. MOLECULAR CRYSTALS

88. Computations of Cohesive Energy.—The principal source of intermolecular cohesion in nonpolar molecular crystals and in many polar molecular crystals is the van der Waals force.   The quantum mechanical methods of computing this force were described in Chap. VII. We saw there that the first approximation term in the expression for the van der Waals energy of two molecules varies as A/r^6, where r is the interatomic distance, and that the next term varies as B/r^8.   This interaction term is not the only one, however, for just as in ionic crystals (*cf.* Chap. II) there are other sources of interaction energy.   Those which are most important in the simple cases to be considered here are the following: (*a*) The electrostatic interaction term which arises from the "static" charge distributions on the molecules.   Although the molecules considered here are neutral, they are not spherically symmetrical and, for this reason, have an electrostatic interaction.   (*b*) The repulsive term, which in the Born-Mayer theory varies as $be^{-\frac{r}{\rho}}$ where b and ρ are constants.   The force arising from this term varies more rapidly with distance than the van der Waals force, so that the van der Waals energy is larger than the repulsive energy at the observed intermolecular distance.

There is a continuous gradation between those molecules which rigidly retain the electronic structure of the free molecule in entering the solid and those which become as highly deformed as the constituents of valence crystals. For this reason, there is no sharp dividing line between valence and molecular types, as we have seen in Sec. 8. Since the cohesion of valence compounds is characterized by exchange energies that favor cohesion in the Heitler-London approximation, it follows that the repulsive term changes its sign to favor binding in the course of the transition from ideal molecular crystals to valence types. Only the more ideal molecular types, which are characterized by very low heats of sublimation, have been considered in any detail up to the present time.

a. London's Calculations.—London[1] first suggested that the electrostatic and repulsive energy terms are very small in comparison with van der Waals terms, and he computed the cohesive energies simply by evaluating the van der Waals term for the observed interatomic distance. As we shall see from the more accurate work described in parts *b* and *c*, this hypothesis sometimes is valid within the comparatively large error of computations of the van der Waals energy but is often very inaccurate. In addition, London treated diatomic and triatomic molecules as though they were spherically symmetrical, in order that he might use the equations that were derived in Sec. 58. We shall see in part *c* that this probably is only a fair approximation.

According to Eq. (19), Sec. 58, the van der Waals interaction energy ϵ_v of two molecules is, in first approximation,

$$\epsilon_v = -\frac{A}{r^6} \tag{1}$$

where

$$A = \tfrac{3}{4}h\nu_0\alpha^2, \tag{2}$$

in which α is the polarizability and ν_0 is a mean excitation frequency. The sum of terms of type (1) for a face-centered cubic lattice containing N molecules is

$$E = -N\frac{59A}{d^6} \tag{3}$$

where d is the cube-edge distance. In evaluating A, London assumed that ν_0 should be closely equal to the principal oscillator frequencies that appear in empirical equations for the refractive index of the gas of each kind of molecule (*cf.* Sec. 148, Chap. XVII). He used these frequencies

[1] F. London, *Z. physik. Chem.*, **11B**, 222 (1930).

in the cases in which they have been measured and spectroscopically determined series-limit frequencies in the other cases. In addition, he employed measured polarizabilities.

Computed values of E are given in Table LXVIII for a number of molecular crystals. Not all these have face-centered cubic crystals, but London assumed that the error made in assuming they have is small.

TABLE LXVIII.—THE VAN DER WAALS ENERGIES OF A NUMBER OF MOLECULAR CRYSTALS AS DETERMINED BY LONDON'S APPROXIMATE EQUATION

Molecule	Lattice	Calculated E, kg cal/mol		Measured E
		ν_0 from refraction	ν_0 from spect.	
Ne	f.c.c.	0.47	0.40	0.52
A	f.c.c.	2.08	1.83	1.77
N_2	h.c.p.	1.64	1.61	1.50
O_2	h.c.p.	1.69	1.48	1.74
CO			1.86	
CH_4	f.c.c.	2.42	2.47	2.40
NO		2.89	2.04	
HCl	f.c.c.		4.04	4.34
HBr	f.c.c.		4.53	4.79
HI			6.50	
Cl_2	Fig. 78, Chap. I		7.18	6.0

b. *More Accurate Calculations for the Rare Gases.*—It was mentioned in Sec. 58 that the repulsive-energy terms have been computed in the case of helium and neon. Using Mayer and Bleick's equation for the neon repulsive energy and assuming that this term is valid only for nearest neighbors, Deitz[1] found that the total repulsive energy is 0.35 kg cal/mol which, in absolute magnitude, is almost as large as London's value of the van der Waals energy, namely, 0.47 kg cal. This suggests that London's value of the van der Waals term is about half as large as the true value and that most of the agreement in Table LXVIII is fortuitous. The discrepancy presumably arises from the fact that Margenau's higher order term is neglected, for this may be as much as half as large as London's term.

c. *Carbon Dioxide.*—Sponer and Bruch-Willstäter[2] have computed the cohesive energy of solid carbon dioxide, taking into account all three of the terms that were mentioned in the introductory paragraphs. It may be recalled from the discussion of Sec. 7 that this solid, which is a typical molecular crystal, consists of a face-centered cubic arrangement

[1] V. DEITZ, *Jour. Franklin Inst.*, **219**, 459 (1935).
[2] H. SPONER and M. BRUCH-WILLSTÄTER, *Jour. Chem. Phys.*, **5**, 745 (1937).

of CO_2 molecules that is similar to the lattice of FeS_2 (Fig. 64, Chap. I). The cohesive energy is 8.24 kg cal/mol.

In first approximation, the molecule was treated as a spherically symmetric unit. The van der Waals energy that was obtained for the observed lattice distance by use of the London-Margenau equations of Sec. 58 and the observed polarizability and ionization energies is 6.0 kg cal/mol of which 3.6 arises from London's term and 2.4 from Margenau's. In the next approximation, the molecule was treated as though it were composed of two centers, which in practice are regarded as the effective centers of charge of the two oxygen ions. Since the electronic distributions on these ions are presumably distorted from spherical form, it was assumed that the positions of the centers lie between the carbon and oxygen ions. Using several reasonable values of these positions, Sponer and Bruch-Willstäter obtained values of the van der Waals energy ranging between 9.6 and 7.6 kg cal/mol.

The electrostatic energy was computed on the assumption that two excess electrons of the oxygen ions are localized at the centers mentioned in the preceding paragraph and that the carbon ion, which is midway between the centers, has a positive charge of $4e$. In this way, it was found that the electrostatic energy varies between -0.1 and -0.5 kg cal/mol.

The constants in the repulsive term were determined from measured values of the compressibility and expansion coefficient by the method described in Sec. 9, Chap. II. The resulting value of the repulsive energy is about -1.1 kg cal. A summary of the computed quantities is given in Table LXIX.

TABLE LXIX

	Contribution to Cohesive Energy, kg cal/mol
Van der Waals (one center)	6.0
Van der Waals (two centers)	Between 9.6 and 7.6
Electrostatic energy	Between -0.1 and -0.5
Repulsive term	-1.1
Total result (two center)	~ 7
Observed	8.24

CHAPTER XI

THE WORK FUNCTION AND THE SURFACE BARRIER

89. The Principles Involved in Computing the Work Function.— There is a close correlation between the work function of a clean metal surface and the volume properties of the metal. In general, the work function is high[1] if the cohesive energy is high, and vice versa. On the other hand, the work function may be appreciably affected if the surface is altered by oxidation or by the deposition of a fraction of an atomic layer of another metal. These facts indicate that the work function is determined both by the binding properties and by the surface structure. The correct relationship between these factors was first pointed out by Wigner and Bardeen.[2] We shall begin by discussing the principles involved in their work and then shall present detailed computations.

Let us consider a semiinfinite crystal that is bounded by a plane, as shown in Fig. 1. The distribution of ions in regions of the main body of the crystal far from the surface is not influenced by the presence of the surface and may be computed under the assumption that the lattice is perfectly periodic. On the other hand, the distribution in the cells near the surface should be different from the distribution in internal cells because the potential changes rapidly near the boundary. The type of difference depends upon the kind of crystal, the orientation of the surface,

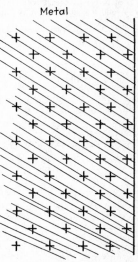

Fig. 1.—A semiinfinite crystal. The distribution of ions in internal regions is the same as that determined by X rays. The surface distribution, however, may be different.

and the kind of adsorbed atoms or ions; for this reason, it must be investigated separately in each case. Fortunately, this surface-sensitive region does not affect the bulk properties of the solid in any

---

[1] This correlation was first pointed out by A. Sommerfeld, *Naturwissenschaften,* **15**, 825 (1927); **16**, 374 (1928). See also J. Frenkel, *Z. Physik,* **49**, 31 (1928).

[2] E. WIGNER and J. BARDEEN, *Phys. Rev.,* **48**, 84 (1935); J. BARDEEN, *Phys. Rev.,* **49**, 653 (1936).

practical way, for its volume is of the order of $1/N$ times the volume of the crystal, if there are N^3 cells in the lattice, and N is of the order of 10^7 for a single crystal of ordinary size.

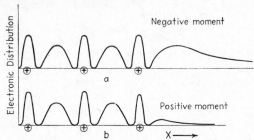

FIG. 2.—Schematic diagram showing cases in which the electronic distribution may lead to negative and positive dipole layers. In case a the electronic distribution extends beyond the surface a good deal; in case b the extension is small.

The only effect of the surface cells in which we need be interested at present is the way in which they influence the dipole moment of the

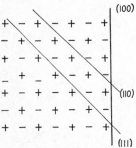

FIG. 3.—The intersection of different surface planes with a (010) plane. If the ions are not displaced relative to the positions in an ideal lattice, the (100) surface has zero dipole moment because there are alternate positive and negative charges. In the (110) surface the surface charges shown are positive, however, those surface charges in (010) planes above and below the one illustrated are negative so that there is no surface dipole in this case. On the other hand, the charges in a (111) surface are either all positive or all negative; hence, there is a dipole layer in this case.

surface. If the crystal is a monatomic cubic metal, such as sodium, the dipole moment of interior cells is zero. The dipole moment of cells near the surface usually is not zero, however, for the potential field in this region is not cubically symmetrical. Hence, these cells effectively give the surface a dipole moment (*cf.* Fig. 2). In polar crystals, such as sodium chloride, in which the unit cell may be chosen in such a way that it has a dipole moment, the surface moment depends both upon the way in which the surface cuts through the lattice and upon the distortion of the surface cells (*cf.* Fig. 3). In any case we shall designate the component of the dipole moment per unit area in the direction normal to the surface by P_n. The important property of this moment for our purposes is the fact that it raises the coulomb potential inside the lattice by a term $-4\pi P_n$.

Let us consider the influence of the surface on the work function from the standpoint of Koopmans' theorem (*cf.* Sec. 67, Chap. VIII). This theorem states that the energy required to remove an electron in the state ψ_k from the crystal is equal to the negative of the parameter $\epsilon(\mathbf{k})$ in Fock's equation.

$$-\frac{\hbar^2}{2m}\Delta\psi_{\mathbf{k}}(\mathbf{r}_1) + \left[V(\mathbf{r}_1) + e^2\int\frac{\rho(\mathbf{r}_2)}{r_{12}}d\tau_2 + A(\mathbf{r}_1)\right]\psi_{\mathbf{k}}(\mathbf{r}_1) = \epsilon(\mathbf{k})\psi_{\mathbf{k}}(\mathbf{r}_1). \quad (1)$$

Here, $V(\mathbf{r}_1)$ is the total ion-core potential, $\rho(\mathbf{r}_2)$ is the valence-electron distribution, and A is the exchange operator. The quantity

$$V(\mathbf{r}_1) + e^2\int\frac{\rho(\mathbf{r}_2)}{r_{12}}d\tau_2$$

is the Hartree potential of the crystal, which, at points inside the lattice, differs from the potential for a crystal in which the surface dipole is zero by the term $e4\pi P_n$. $A(\mathbf{r}_1)$ arises from the exchange correlation hole, which is confined to the vicinity of the electron, and is not affected by the surface as long as the electron is inside the crystal. $\psi_{\mathbf{k}}$ has the form $\chi_{\mathbf{k}}e^{2\pi i\mathbf{k}\cdot\mathbf{r}}$ inside the lattice if we employ the Bloch scheme; and although it is different in the cells near the surface, the total volume of these cells is so small that they may be neglected in computing integrals involving ψ that extend over the entire lattice.

Multiplying (1) by $\psi_{\mathbf{k}}^*$ and integrating, we find

$$\epsilon(\mathbf{k}) = -\frac{\hbar^2}{2m}\int\psi_{\mathbf{k}}^*\Delta\psi_{\mathbf{k}}d\tau_1 + \int|\psi_{\mathbf{k}}|^2\left[V'(\mathbf{r}_1) + e^2\int\frac{\rho(\mathbf{r}_2)}{r_{12}}d\tau_2\right]d\tau_1 +$$
$$\int\psi_{\mathbf{k}}^*A\psi_{\mathbf{k}}d\tau_1 + 4\pi eP_n. \quad (2)$$

Here,

$$V'(\mathbf{r}_1) + \int\frac{e^2\rho(\mathbf{r}_2)}{r_{12}}d\tau_2$$

is the Hartree potential for a lattice having no surface dipole. Thus, except for the term $4\pi eP_n$, the work function $-\epsilon(\mathbf{k})$ is determined by volume integrals. The introduction of correlation effects does not alter this conclusion, for correlation terms, like exchange terms, arise from a hole in the vicinity of the electron.

When an electron is removed from a metal, the remaining electrons concentrate in the interior of the solid in order to keep this region electrostatically neutral. At first sight, it appears as though this effect might invalidate the previous conclusions. This is not so, however, for the energy change accompanying the concentration process is equal to the difference between the electrostatic energy of a volume and a surface distribution of 1 electronic unit. This difference, which is of the order e^2/L, where L is the diameter of the crystal, is about 10^{-8} ev for ordinary specimens.

90. The Internal Contribution to the Work Function.—Wigner and Bardeen have evaluated the volume part of the expression (2) for the

uppermost electrons in the filled levels of the metals discussed in Chap. X
It may be assumed that the metals are uncharged without introducing
an error; for, in the first place, additional charge would accumulate at the
surface, leaving the interior of the metal neutral, and, in the second place,
this surface charge is never high enough in actual cases to alter the surface
dipole layer appreciably.   An ordinary specimen of metal has about 10
surface cells which contain about 10^6 esu of electronic charge.   If or
per cent of this charge were removed, the field near the metal would be
raised to about one million volts, which is as high as practical value
ordinarily go; however, the dipole layer would not be altered by more
than a few per cent.

According to the sphere approximation, which is reliable in the alkali
metals (cf. Chap. X), the Hartree potential within any cell may be
determined by the charge inside that cell, for the electronic and ionic
charges in other cells cancel one another.   Thus, the first two terms in
Eq. (2) of the preceding section, namely,

$$-\frac{\hbar^2}{2m}\int \psi_k{}^*\Delta\psi_k d\tau_1 + \int |\psi_k|^2\left[V'(\mathbf{r}_1) + e^2\int \frac{\rho(\mathbf{r}_2)}{r_{12}}d\tau_2 \right]d\tau_1,$$

are equal to

$$\frac{V}{v}\left\{ \int_v \psi_k{}^*\left(-\frac{\hbar^2}{2m}\Delta + v_c \right)\psi_k d\tau_1 + \frac{V}{v}e^2\int_v |\psi_k|^2\left(\int_v \frac{\rho_s(\mathbf{r}_2)}{r_{12}}d\tau_2 \right)d\tau_1 \right\} \quad (1$$

where the integrals extend over a single cell, V/v is the ratio of the volume
of the crystal to that of a cell, v_c is the ion-core field inside the cell, and
ρ_s is the electronic distribution in the cell.   The first term in this equation
is equal to

$$\epsilon_0 + \frac{h^2}{2m^*}k^2 \quad (2$$

where ϵ_0 is the energy parameter in the equation

$$-\frac{\hbar^2}{2m}\Delta\psi_0(\mathbf{r}) + v_c(\mathbf{r})\psi_0(\mathbf{r}) = \epsilon_0\psi(\mathbf{r})$$

(cf. Sec. 78, Chap. X) and $(h^2/2m^*)k^2$ is the additional energy of ψ_k.
If we replace the polyhedral cell by a sphere and assume that both $e\rho_s$
and $eV|\psi_k|^2/v$ are constant and equal to e/v, the second integral is simply
twice the self-energy of a spherical charge distribution, namely, $1.2e^2/r_s$,
where r_s is the radius of the sphere.

The exchange integral in Eq. (2) of the preceding section was evaluated in Sec. 75, Chap. IX, for plane waves and is equal to

$$-0.306\frac{e^2}{r_s}f\left(\frac{k}{k_0} \right)$$

where

$$f\left(\frac{k}{k_0}\right) = 2 + \frac{k_0}{k}\left[1 - \left(\frac{k}{k_0}\right)^2\right]\log\left|\frac{k_0 + k}{k_0 - k}\right|.$$

An explicit equation for the correlation energy of any given electron has not been derived; however, the correlation energy of the electrons at the top of the band is (*cf.* Sec. 76, Chap. IX)

$$g(r_s) - \tfrac{1}{3}g'(r_s)r_s$$

where

$$g(r_s) = -e^2\frac{0.288}{r_s + 5.1a_h}.$$

If we add these results, we find that the energy ϵ_w required to remove an uppermost electron is given by the equation

$$-\epsilon_w = \epsilon_0 + \frac{h^2}{2m^*}k_0^2 + 1.2\frac{e^2}{r_s} - 0.612\frac{e^2}{r_s} + g(r_s) - r_s\frac{g'(r_s)}{3} + 4\pi eP_n \quad (3)$$

where $4\pi eP_n$ is the surface dipole term.

Now, $(h^2/2m^*)k_0^2$ is equal to five-thirds of the mean Fermi energy ϵ_F of the crystal and may be replaced by this quantity. If we then replace ϵ_0 by the value obtained from the expression for the cohesive energy per atom ϵ_c, namely,

$$-\epsilon_c = Z\left[\epsilon_0 + \epsilon_F + \frac{0.6e^2}{r_s} - \frac{0.458e^2}{r_s} + g(r_s)\right] + I(Z), \quad (4)$$

where Z is the number of valence electrons per atom and $I(Z)$ is the ionization potential of the free atom, we find

$$\epsilon_w = \frac{1}{Z}\left[\epsilon_c + I(Z)\right] + \left[-\frac{2}{3}\epsilon_F - 0.6\frac{e^2}{r_s} + \frac{0.458e^2}{3r_s} + \frac{r_sg'(r_s)}{3} - 4\pi eP_n\right]. \quad (5)$$

Wigner and Bardeen derived this equation by another method, namely, by computing the energy of the crystal as a function of the number of electrons and ions, N_e and N_i, respectively. The work function ϵ_w is then the derivative of this energy with respect to N_e. The advantage of this procedure is that the work required to remove an ion or a neutral atom may be computed from the same expression.[1]

All the quantities in (5) except P_n have been computed for lithium, sodium, and potassium in Chap. X, so that it is possible to determine $\epsilon_w + 4\pi eP_n$ for these metals. The values are given in Table LXX.

---

[1] The expression for the energy required to remove an ion contains the surface dipole with opposite sign. Hence, the work required to remove a neutral atom does not depend on the dipole moment.

TABLE LXX

	r_s, a_h	$\epsilon_w + 4\pi e P_n$, ev	Experimental value
Li	3.28	2.19	2.28
Na	4.00	2.15	2.25
K	4.97	1.87	2.24

The close agreement between the observed values of ϵ_w and the computed values of $\epsilon_w + 4\pi e P_n$ for lithium and sodium suggests that the surface dipole moment is very small for clean metal surfaces. This conclusion is borne out by explicit computations of P_n that Bardeen has made for sodium.[1]

On the basis of the work discussed in Sec. 81 and Eq. (5), Herring and Hill have found the work function of beryllium to be negative by about 1.7 ev, if it is assumed that the dipole layer is zero. This result is in disagreement with the best observed value, which is about 4 ev. The discrepancy suggests either that the width of the occupied region of levels is much less than these investigators have found, or that beryllium usually has a tightly bound surface layer of electronegative atoms.

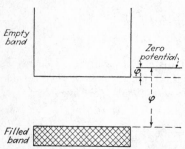

FIG. 4.—Schematic energy-level diagram of an insulator. The zero of potential is assumed to be slightly above the bottom of the empty band. The work function φ_1 for inserting an electron is smaller than that for removing an electron from the filled band φ by the energy difference of the filled and empty band.

91. The Work Function in Non-metallic Crystals.—All the fundamental principles used in the previous section in discussing the work function of metals can also be applied to nonmetals. In general, the energy required to remove an electron from the solid or to put one in depends both upon the volume characteristics of the substance and upon the surface dipole layer. As an example, let us consider the energy required to put an electron into a neutral sodium chloride crystal. This energy is less than the energy required to remove an electron from the uppermost levels of the filled band by an amount equal to the gap between the filled and unfilled bands (cf. Fig. 4).

We shall consider a somewhat idealized case in which the surface is a (100) plane and in which the ions near the surface retain the same relative positions as those in the interior. Under these conditions, the ordinary lattice potential along the dotted line in Fig. 5 is zero because

<hr />

[1] BARDEEN, op. cit.

points on this line are equidistant from positive and negative ions.   This statement is rigorously true inside the crystal only if we assume that the ions act as point charges.   We shall see in Chap. XIII that the incoming electron is distributed principally about the sodium ions, very much as in metallic sodium.   For this reason, we may assume that the energy of the electron is equal to the energy of a sodium ion in the Madelung field at the position of a sodium ion plus the additional energy by which this level is lowered by the development of band structure.   The Madelung potential at a sodium ion is $1.74e^2/r_0$, and the ionization potential of a sodium atom is $-0.19e^2/a_h$.   Since the sodium-sodium distance in the salt is about the same as in the metal, we shall assume that the depression due to band formation is also the same.   According to Fig. 2b, Chap. X, this is $0.11e^2/a_h$.   Hence, the work function is[1]

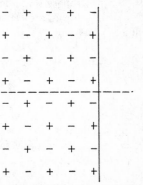

$$\varphi_1 \cong 1.74\frac{e^2}{r_0} - 0.19\frac{e^2}{a_h} - 0.11\frac{e^2}{a_h}$$

$$= 0.03\frac{e^2}{a_h} \cong 0.8 \text{ ev.}$$

This actually is the sum of the internal and surface contributions; however, there is no surface dipole layer in the present case because the surface is a (100) plane that contains equal numbers of positive and negative charges.   If we were to deal with another plane, such as a (111) plane,

Fig. 5.—The potential along the dotted line is zero because points on this line are equidistant from positive and negative charges.

or were to alter the interionic distances near the surface, we could compute the surface dipole term by computing the Madelung potential at a sodium ion.   The difference between this value and $1.74e^2/r_0$ would then be $-4\pi eP_n$.

It is doubtful whether the energy gained by an electron on entering an insulator is always as small as the value computed above for a typical alkali halide.   The photoelectric work function of insulators, such as cuprous oxide, that absorb in the visible is[2] of the order of 5 ev, a fact indicating that the width of the forbidden region is about 2 ev and the work function φ_1, discussed above, about 3 ev.

---

[1] A similar value has been obtained by N. F. Mott, *Trans. Faraday Soc.*, **34**, 500 (1938) using slightly different reasoning.

[2] See, for example, A. L. Hughes and L. A. DuBridge, *Photoelectric Phenomena* (McGraw-Hill Book Company, Inc., New York, 1932); R. Fleischmann, *Ann. Physik,* **5**, 73 (1930); R. J. Cashman (paper 179, program Washington Meeting, American Physical Society, 1940).

92. Thermionic Emission and the Temperature Dependence of the Work Function*.—In Sec. 30, Chap. IV, we derived the Richardson Dushman equation for the thermionic electron emission from unit area of a metal

$$I = A(1 - r)T^2 e^{-\frac{W}{kT}}. \tag{1}$$

Here, W is the work function, which was assumed to be a constant for the entire surface, A is a universal constant, 120 amp/cm²-deg², and r is the electronic reflection coefficient. In this section, we shall reexamine the relations that enter into Eq. (1) in the light of the previous work of this chapter.

Let us suppose that the metal is at temperature equilibrium with an external electron atmosphere. If we may assume that the electron cloud behaves as a perfect gas, which is a reasonable assumption as long as the density is small, the number of electrons that pass from the outside to the inside per unit area in unit time is

$$\frac{p(1 - r)}{(2\pi mkT)^{\frac{1}{2}}}, \tag{2}$$

where p is the external pressure, m is the electron mass, and k is Boltzmann's constant. This should be equal to the number that evaporates from a unit area of the surface since the system is at equilibrium. Hence the thermionic current is

$$I = \frac{ep(1 - r)}{(2\pi mkT)^{\frac{1}{2}}}. \tag{3}$$

The equilibrium pressure $p(T,V)$, which is a function of the temperature T and volume V of the crystal, may be related to the heat ΔH required to sublime 1 mol of electrons from the metal at constant pressure by the Clausius-Clapeyron equation

$$\left(\frac{\partial \log p}{\partial T}\right)_V = \frac{\Delta H}{RT^2}, \tag{4}$$

where R is the gas constant. If the specific heat of the electrons inside the metal is neglected,

$$\Delta H = N_A W(V,T) + \tfrac{5}{2}RT \tag{5}$$

where $W(V,T)$ is the work function of the metal when the volume is V and the temperature is T, and $5R/2$ is the molar heat at constant pressure of the electron gas. The integral of (4) is

$$\log p = \int \frac{\Delta H}{RT^2} dT + C \equiv -\frac{\Delta H}{RT} + \int \left(\frac{d\Delta H}{dT}\right)_V \frac{dT}{RT} + C'$$

$$= -\frac{\Delta H}{RT} + \frac{\Delta S}{R} \tag{6}$$

where

$$\Delta S = \int \left(\frac{d\Delta H}{dT}\right)_v \frac{dT}{T} + C'R$$

is the entropy change that accompanies evaporation of a mol of electrons. We may substitute ΔH from Eq. (5). According to the third law of thermodynamics,[1] the constant term in ΔS should be chosen in such a way that the entropy change associated with $W(V,T)$ is

$$\int_0^T \left[\frac{dW(V,T)}{dT}\right]_v \frac{dT}{T}$$

and the entropy change of the gas is

$$R(\tfrac{5}{2} \log T + \tfrac{5}{2} + j)$$

where

$$j = \log \frac{2(2\pi m)^{\frac{3}{2}}}{h^3} k^{\frac{5}{2}}$$

is the chemical constant. Thus, (6) may be written

$$\log p = -\frac{W(V,T)}{kT} + \frac{1}{k}\int_0^T \left[\frac{dW(V,T)}{dT}\right]_v \frac{dT}{T} + \frac{5}{2} \log T + j. \qquad (7)$$

Substituting this in (3), we obtain

$$I = A(1 - r)T^2 e^{-\frac{W(V,T)}{kT} + \frac{1}{k}\int_0^T \left(\frac{dW(V,T)}{dT}\right)_v \frac{dT}{T}} \qquad (8)$$

where A is the coefficient that occurs in Eq. (1).

It should be noted that Eq. (8) differs from Eq. (1) by the factor

$$e^{\frac{1}{k}\int_0^T \left(\frac{dW}{dT}\right)_v \frac{dT}{T}} \qquad (9)$$

which is unity for all temperatures only if $(dW/dT)_v = 0$. The appearance of this term indicates that the method used to derive Eq. (1) is faulty whenever the work function is temperature-dependent. Since the correction term (9) arises from the entropy of electrons in the solid, we see that the simple model used to derive (1) is in error because we neglected interactions between the electrons and the solid that cannot be described adequately by a simple potential barrier. Equation (8) could be derived on the basis of statistical mechanics, but it would be necessary to consider the entire solid in doing so.

Before investigating the importance of the temperature dependence of W, we shall introduce convenient definitions of the work function and

[1] See P. S. EPSTEIN, *Textbook of Thermodynamics* (John Wiley & Sons, Inc., New York, 1937).

of the thermionic coefficient that were developed by Becker and Brattain.
In ordinary thermionic experiments, the emitted current I is measured
over a certain temperature range, and the quantity $\log (I/T^2)$ is then
plotted as a function of $1/T$. The experimental work function and
thermionic coefficient, $W^*(T)$ and $A^*(T)$, are defined, respectively
as the slope of this curve and intercept of the tangent on the $\log (I/T^2$
axis (cf. Fig. 6); that is,

$$W^* = kT^2\left[\frac{d \log (I/T^2)}{dT}\right]_p,$$

$$\log A^* = \log \frac{I}{T^2} + T\left[\frac{d \log (I/T^2)}{dT}\right]_p$$

$$= \log \frac{I}{T^2} + \frac{W^*}{kT}. \tag{10}$$

These derivatives are taken at constant pressure because the specimen
is kept in a vacuum during the experiments.

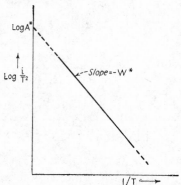

FIG. 6.—Diagrammatic representation of the definitions of A^* and W^*.

If Eq. (8) is substituted in these equations and if the relation

$$\left(\frac{\partial}{\partial T}\right)_p = \left(\frac{\partial}{\partial T}\right)_V + \left(\frac{\partial V}{\partial T}\right)\left(\frac{\partial}{\partial V}\right)_T$$

$$\equiv \left(\frac{\partial}{\partial T}\right)_V + \alpha V\left(\frac{\partial}{\partial V}\right)_T \tag{11}$$

is used, where $\alpha = (\partial V/\partial T)/V$ is the coefficient of volume expansion,
it is found that

[1] J. A. BECKER and W. H. BRATTAIN, Phys. Rev., 45, 694 (1934).

$$W^* = T^2\left[\left(\frac{\partial}{\partial T}\right)_V + \alpha V\left(\frac{\partial}{\partial V}\right)_T\right]\left[-\frac{W(V,T)}{T} + \int_0^T\left(\frac{\partial W}{\partial T}\right)_V\frac{dT}{T}\right]$$

$$= W(V,T) + \alpha T^2 V\left(\frac{\partial}{\partial V}\right)_T\left[-\frac{W(V,T)}{T} + \int_0^T\left(\frac{\partial W}{\partial T}\right)_V\frac{dT}{T}\right], \quad (12)$$

$$\log\frac{A^*}{(1-r)A} = \frac{1}{k}\int_0^T\left(\frac{\partial W}{\partial T}\right)_V\frac{dT}{T} + \frac{\alpha V}{k}\left(\frac{\partial}{\partial V}\right)_T\left[-\frac{W(V,T)}{T} + \int_0^T\left(\frac{\partial W}{\partial T}\right)_V\frac{dT}{T}\right]. \quad (13)$$

Wigner[1] has attempted to estimate some of the terms in (13), using the expression for the work function of sodium developed in Sec. 90. His treatment is not rigorous, but it shows that the terms on the right-hand side of (13) are not negligible even for simple metals and that the value of $\log\frac{A^*}{A(1-r)}$ is of the order unity, that is, lies between $+5$ and -5. $W(V,T)$ may be related to the work function $W(V,0)$ computed in the previous sections in the following way: $W(V,T)$ is defined as the energy required to remove an electron from the metal at temperature T. The total energy of the crystal when no electrons have been removed is

$$E = E_D\left(\frac{\Theta_D}{T}\right) + E_e$$

where E_D is the vibrational energy of the lattice, which we shall express in terms of Debye's characteristic temperature Θ_D, and E_e is the electronic energy, which is not temperature-dependent if the small electronic specific heat is neglected. If n electrons are removed at temperature T, the new energy E' is

$$E' = E_D + E_e + nW(V,T). \quad (14)$$

Suppose that n electrons are removed at absolute zero of temperature instead and that the metal is then raised to temperature T. The resulting energy, which is again E', is

$$E_D\left(\frac{\Theta_D + \Delta\Theta_n}{T}\right) + E_e + nW(V,0) \quad (15)$$

where $\Delta\Theta_n$ is the change in the characteristic temperature that results from the removal of n electrons. Equating (14) and (15), we obtain

$$W(V,T) = \frac{1}{n}\frac{\Delta\Theta_n}{T}E_D'\left(\frac{\Theta_D}{T}\right) + W(V,0). \quad (16)$$

[1] E. WIGNER, Phys. Rev., **49**, 696 (1936). See also K. F. HERZFELD, Phys. Rev., **35**, 248 (1930).

Wigner finds that the contribution to (13) from the last term[1] in Eq. (16) is about -3.6 for sodium. It turns out that this result is balanced somewhat by the contribution from the first term in (16). There is, however, no reason for expecting the two terms to cancel.

Direct measurements of the temperature coefficient of $W(V,T)$ in Eq. (5) have been carried out by Krüger and Stabenow[2] for molybdenum, tungsten, and tantalum. These workers measured the heat lost by a wire during thermionic emission. The temperature variation is within the experimental error in molybdenum and has the value $0.6 \cdot 10^{-4}$ ev/deg per electron in both tungsten and tantalum. If the integral in the expression (9) is evaluated with the use of this coefficient and with the assumptions that $(\partial W/\partial T)$ approaches zero near absolute zero, so that the contribution from the lower limit of integration may be neglected, and the temperature coefficients at constant volume and constant pressure are practically the same, it is found that $\log \left[\dfrac{A^*}{A(1-r)} \right] \sim -5.1$.

Hence, A^* would be about 0.66 amp/deg-cm^2 if r were zero. Unfortunately, these workers did not measure A^* on the specimens for which this work was carried out. The values of A^* measured by other workers are about one hundred times larger.

---

[1] Since the last term is independent of temperature, it is the work function discussed in the preceding sections of this chapter.

[2] F. Krüger and G. Stabenow, *Ann. Physik*, **22**, 713 (1935).

CHAPTER XII

THE EXCITED ELECTRONIC STATES OF SOLIDS

93. Introduction.—Both the Bloch and the Heitler-London approximations have been used to treat the excited states of solids. Although these methods have not been rigorously tested in particular cases, qualitative and semiquantitative arguments may be used to show that one approximation is more suitable than another in a given case. For example, we may expect that the Bloch approximation is more suitable[1] to use in a discussion of the excited states of metals because it alone leads[2] to the low-lying continuous, conducting levels that are characteristic of these solids (*cf.* Sec. 66). Similarly, we shall find that the Heitler-London approximation is more applicable to the lower levels of molecular and ionic crystals.

In this chapter, we shall discuss the general principles upon which computations of excited states are now based and shall also present some simple results. This discussion begins with a survey of the uses of the Bloch method and is followed by a similar survey of the Heitler-London scheme. Problems in which intermediate approximations are applicable will be discussed in later sections.

94. Excited States in the Band Scheme.—The band scheme is based upon a one-electron approximation in which the ψ have the form

$$\psi_{\mathbf{k}} = \chi_{\mathbf{k}} e^{2\pi i \mathbf{k} \cdot \mathbf{r}} \tag{1}$$

and satisfy Fock's equations

$$-\frac{\hbar^2}{2m}\Delta\psi_{\mathbf{k}} + (V + A)\psi_{\mathbf{k}} = \epsilon(\mathbf{k})\psi_{\mathbf{k}} \tag{2}$$

where V is the coulomb or Hartree potential and A is the exchange operator. The entire wave function of the solid may be constructed from determinants of wave functions of type (1). V and A are not appreciably altered if one of the $\psi_{\mathbf{k}}$ in the set is replaced by another, since the ψ

---

[1] Direct evidence for the qualitative correctness of the Bloch approximation is also obtained from a study of the soft X-ray emission spectra of metals (*cf.* Sec. 104).

[2] It should be pointed out that the Heitler-London scheme would also include the Bloch states if we considered atomic wave functions of the type associated with continuous spectra as well as the localized wave functions of the type associated with the discrete atomic levels. We shall explicitly avoid including the first type of wave function, however.

extend over the entire lattice and have small amplitude in any given cell. Hence, V and A may be chosen the same for both normal and excited states. It follows from Koopmans' theorem that $\epsilon(\mathbf{k}) - \epsilon(\mathbf{k}')$ is the energy required to excite the crystal from a given state to the one in which $\psi_\mathbf{k}$ is replaced by $\psi_{\mathbf{k}'}$. Thus, the possible excited levels may be obtained from the one-electron energy diagrams of the zone scheme. The highest occupied zone is not completely filled in a metal, whence the lowest states of the solid as a whole have a quasi-continuous system of energy levels. Since the conduction properties of metals require this type of continuum, the band approximation is naturally suited for a semiquantitative description of these solids. As we have seen in Chap. X, the process of improving the zone approximation for a metal does not simply effect a compromise between the Bloch and Heitler-London approximations but consists in treating correlations more accurately. This does not mean that some atomic properties are not retained in passing from the free atoms to the solid, for the functions $\chi_\mathbf{k}$ in Eq. (1) preserve many of the features of atomic wave functions.

The band scheme can be applied to insulators as well as to metals. In these cases, the highest occupied zone is completely filled in the normal state, so that the first excited level is a finite distance above the lowest level. We have seen in Sec. 64 that the lowest state of an ionic or molecular crystal is described with equal accuracy by either the Bloch or the Heitler-London scheme. We shall see in the next section, however, that the Heitler-London scheme leads to excited levels that are not contained in the Bloch approximation. For this reason, the zone scheme is not always adequate for a qualitative description of insulators.

95. Excited States in the Heitler-London Scheme.—Let us apply the Heitler-London scheme to sodium chloride, which is a typical insulator. We shall attempt to follow the behavior of the lowest atomic and ionic energy levels as the ions are brought together to form the normal lattice and are kept in crystalline arrangement during the process. The excited ionic levels will be neglected for the moment since they cannot be treated properly without including a discussion of continuous spectra.

At infinite separation, the ionic and atomic levels of sodium and chlorine are as illustrated in the right-hand side of Fig. 1 in which the halogen-ion level is given relative to that of the neutral atom and the level of neutral sodium is given relative to that of Na^+. Thus, the normal state of Cl is at -3.8 ev, and the lowest level of neutral sodium is at -5.2 ev. Other levels are neglected for the present. The minimum energy required to transfer an electron from a halogen ion to an alkali ion is -1.4 ev at infinite separation. Since this value is negative, the infinitely separated system is more stable as a set of neutral atoms than as a set of ions. This situation is gradually altered as the ions approach one another. If r_0 is the distance between nearest neighboring ions in the lattice, the electrostatic potential at the negative ions is $1.748e^2/r_0$ and that at the positive ions is the negative of this. Thus, the halogen-ion levels and alkali-ion levels are respectively raised and lowered in accordance with the equations

$$\left.\begin{aligned}
\epsilon(\text{Cl}^-) &= -3.8 - \frac{47.3}{(r_0/a_h)}, \\
\epsilon(\text{Na}^+) &= -5.2 + \frac{47.3}{(r_0/a_h)}.
\end{aligned}\right\} \tag{1}$$

in which the unit of energy is the electron volt. These equations are valid only as long as the ions do not overlap appreciably. When they

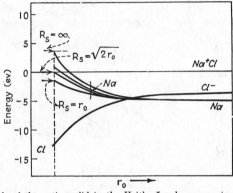

Fig. 1.—Levels of the entire solid in the Heitler-London approximation based on an ionic model. At large separations the state of neutral atoms is most stable because the ionization potential of metal atoms usually is larger than the electron affinity of the halogen atom. This situation is reversed as the atoms are brought together, because the Madelung energy favors the ionic state. Corresponding to each level of the metal atom there are an infinite number of levels of the entire solid, each of which is related to a particular value of R_s in Eq. (3).

do overlap, additional energy terms should be added to (1) in order to include the effects of exchange and correlation interactions. Since these terms are only about 10 per cent of the electrostatic terms, we shall neglect them temporarily. The energy $\Delta\epsilon$ required to transfer an electron from a halogen ion to a *distant* alkali metal ion then is the difference between the two terms in (1), namely,

$$\Delta\epsilon = -1.4 + \frac{94.6}{(r_0/a_h)} \ (\text{ev}). \tag{2}$$

This is 16.5 ev for the normal interionic distance of $5.29a_h$ in sodium chloride. There are, however, an infinite number of levels lying below this one, for an energy e^2/R less than (2) is required to transfer the excited electron from a halogen ion to an alkali ion that is at a finite distance R. That this statement is true may be seen from the fact that the Madelung potential at a given alkali ion is decreased by e^2/R if an electron is removed from a chlorine atom at distance R. Thus, the normal and excited levels

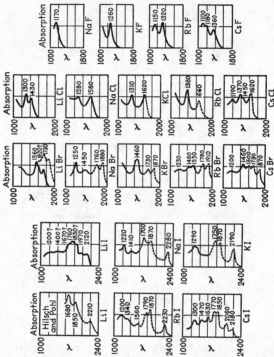

Fig. 2.—The ultraviolet absorption bands of the alkali halide crystals. The dotted portions of the curves represent measurements by Hilsch and Pohl. [*After Schneider and O'Bryan, Phys. Rev.,* **51**, 293 (1937).]

are disposed as in Fig. 1, in which the excited levels are separated from the ground state by an amount

$$\Delta\epsilon_s = -1.4 + \frac{94.6}{(r_0/a_h)} - \frac{27.08}{(R_s/a_h)} \text{ (ev).} \tag{3}$$

At the observed lattice distance, the first excited level, which is 11.3 ev above the ground state, corresponds to $R_s = 5.29a_h$, the distance between nearest neighboring ions, whereas the next level, which is 12.9 ev above the ground state, corresponds to $R_s = \sqrt{2}r_0$. Hilsch and Pohl[1] first pointed out that the difference in the quantum energies of the first two

[1] R. Hilsch and R. W. Pohl, *Z. Physik,* **59**, 812 (1930).

ultraviolet absorption bands of sodium chloride is very nearly equal to the difference between these computed excited levels. The first band has its peak at 1580 Å (*cf*. Fig. 2), and the second has its peak at 1280 Å; these values correspond to energies of 7.8 and 9.6 ev, respectively. The difference between these energies is 1.8 ev, which agrees closely with the difference of 1.6 ev for the calculated absorption energies. von Hippel[1] has attempted to refine this simple calculation of the first excited levels by including corrections for the perturbation of the neutral alkali and halogen atoms. His computed values and the observed values are compared in Table LXXI.

TABLE LXXI.—COMPARISON OF COMPUTED AND OBSERVED ENERGY DIFFERENCES BETWEEN THE GROUND STATE AND THE FIRST EXCITATION STATE OF THE ALKALI HALIDES (AFTER VON HIPPEL)

	Observed, ev	Calculated, ev
NaCl	7.8	8.3
KCl	7.6	8.0
RbCl	7.4	7.7
LiBr	6.7	8.1
NaBr	6.5	7.4
KBr	6.6	7.3
RbBr	6.4	7.1

The excited discrete levels, shown in Fig. 1, actually are highly degenerate, for the excited electron may be removed from any one of the N halogen ions of the crystal and may be carried to any one of its g_s neighboring alkali metal ions that are at distance R_s without altering $\Delta\epsilon_s$. Thus, the levels are Ng_s-fold degenerate. Since $\sum_s g_s$ is equal to N, the total number of alkali metal ions, we see that the total number of excited levels in the system is N^2.

In the simple model used above, the first excited level is sixfold degenerate, if spin is neglected, for each chlorine ion has six equidistant neighboring alkali ions. This degeneracy is partly accidental, for the six functions do not have the proper symmetry to have the same energy in a cubic crystal. Thus, the degenerate levels would split if interatomic interactions were taken into account. In first approximation, the new functions should be linear combinations of the six functions ψ_i that are localized on the separate alkali ions. The electronic distribution of the new functions should be spread over all six neighboring ions. The lowest state evidently is the symmetrical function that is formed by adding all

[1] A. VON HIPPEL, *Z. Physik*, **101**, 680 (1936).

six ψ_i, and is analogous to an atomic s function. Above this, there are
a triply degenerate set that is analogous to the three atomic p functions

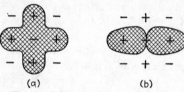

and a doubly degenerate level that
has no atomic analogue. Two of
the four possibilities for a two-dimensional case are shown in Fig. 3.

Let us now consider the continuum of levels corresponding to
ionization of the halogen ion. The
continuum remains essentially unchanged as long as the ions are far
apart; however, the free electrons
begin to be perturbed appreciably
when the ions occupy a considerable

(a)   (b)

Fig. 3.—Schematic diagram of the wave
functions of excited electrons in ionic
crystals. Function a is the analogue of an
atomic s function and distributes the
electronic charge equally about all of the
neighboring positive ions. Function b
is the analogue of a p function and has
opposite signs in the two "wings."

fraction of the volume of the crystal. Since the new levels should be
computed by determining the wave
functions of the free electrons in the
field of the entire crystal, they are
obviously the same as the excited
levels that are computed on the
basis of the Bloch approximation.
We know, however, that the Bloch
bands contract into the levels of
the excited states at infinite separation. Hence, we may conclude that
the levels of the continuum tend to
cross and combine with the excited
discrete levels of the Heitler-London
scheme, broadening these lines into
bands. The extent to which this
broadening actually takes place depends upon the interatomic distance. Only the series limit is
affected in a case in which the lattice
is highly extended, whereas in the
opposite extreme of a highly com-
pressed lattice the continuum overlaps even the lowest level, making
the solid a metal. In the intermediate case the lowest level is discrete, as shown schematically in Fig.
4. One important case is that cor-

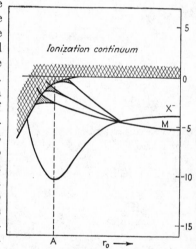

Fig. 4.—Schematic diagram showing
the behavior of the levels of an ionic crystal
as the atoms are brought together (see
also Fig. 1). At large distances the
neutral system is stable, whereas at
intermediate distances the ionic system is
stable. The bands corresponding to the
ionization continuum broaden and spread
and may actually overlap all of the discrete
levels. The broadening of the excited
nonconducting levels at A corresponds to
the formation of excitation bands, which
is discussed in the next section. The
energy units are electron volts for NaCl.

responding to the point A of Fig. 4 in which the continuum has overlapped all the excited levels except the lower ones.   In this case, the first excited levels should be determined approximately by means of the Heitler-London scheme, as von Hippel has done, whereas the higher states should be determined by the Bloch approximation.   Experimental evidence, which will be presented in the next chapter, indicates that this corresponds closely to the state of affairs in the simpler ionic crystals.   It should be noted that we have removed some of the degeneracy of the lowest excited levels in Fig. 4 before they merge with the continuum. The origin of this effect is discussed in the next section.

Frenkel,[1] who was one of the first to discuss correctly the possible relationships between normal and excited states in insulators, has called the lower excited levels of Fig. 4 that have not mingled with the continuum at *A* "excitation levels," since they are analogous to the excited states of atoms.   Similarly, he has called the higher levels that should be treated by the band approximation "ionization levels," since they are analogous to the ionized states of atoms.   The ionization states have already been discussed in Chap. VIII.   We know from this presentation that the crystal should become photoconducting when excited to these levels since the excited electron is then free to roam throughout the lattice.   On the other hand, the excited electron remains fixed relative to the atom from which it came in the excitation states.   Hence, we should not expect photoconductivity to accompany optical excitation to these states.   Photoconductivity actually does not seem to occur as a result of absorption in the first fundamental absorption bands of the alkali halides, a fact which indicates that the first excited levels in these solids correspond to excitation states.[2]

It should be mentioned in passing that Valasek[3] has presented good experimental evidence that the excited X-ray levels in salts such as sodium chloride and potassium chloride should be described by the atomic scheme, or by the exciton scheme which is discussed in the next section, rather than by the band scheme.

---

[1] J. Frenkel, *Phys. Rev.*, **37**, 17 (1931); **37**, 1276 (1931).

[2] The cases of the alkaline earth oxides and sulfides, such as zinc oxide and zinc sulfide, are still uncertain, for the structure of the absorption bands of these solids has not been thoroughly investigated.   A recent experimental investigation of this problem for the alkali halides has been carried out by L. P. Smith and J. N. Ferguson (see paper 177, program of Washington Meeting, American Physical Society, 1940). These observers find photoconductivity in the long wave length tail of the fundamental bands, but not in the interior.   According to the exciton viewpoint (see next section) this conductivity arises either from direct ionization of impurity atoms or lattice-defect atoms (Chap. XIII), or from secondary ionization of these atoms by excitons.   In view of results of this kind, however, it must be admitted that there is no conclusive experimental evidence that photoconductivity would not occur at least in the tail of the fundamental band of a pure perfect crystal.

[3] J. Valasek, *Phys. Rev.*, **47**, 896 (1935); **53**, 274 (1938).

Let us consider another example of a system to which the Heitler-London method may be applied, namely, that of an insulating crystal which contains a neutral impurity atom. As we shall see in the next chapter, this system corresponds to a semi-conductor, such as ZnO, in which there are interstitial zinc atoms. At present, we shall be interested in the case in which the ionization potential of the neutral atom is less than the energy of the first absorption band of the pure crystal and the interstitial atom occupies a position of zero potential at which it is symmetrically surrounded by positive and negative ions. These conditions are approximately fulfilled in zinc oxide. We may

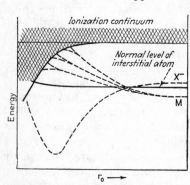

Fig. 5.—Behavior of the levels of an interstitial atom in an ionic crystal. The dotted lines correspond to the energy-level curves of the bulk material. In this case, the excitation and ionization energies of the bulk material are larger than those of the interstitial atom at the equilibrium position, which corresponds to the minimum of the lower curve. At small interatomic distances, the lowest level of the impurity atom may merge with the ionization continuum. This probably does not happen in actual cases.

expect, by analogy with the case of sodium chloride, that the energy required to transfer an electron from a negative to a positive ion is negative at infinite separation, for the electron affinities of negative ions are usually smaller than the ionization potential of metal atoms. This situation is altered as the atoms are brought together, for the positive ions are surrounded principally by negative ones, and vice versa. At infinite separation, the ionization continuum of an interstitial atom lies at the same position as that of the negative ions and must blend into the Bloch bands of the entire solid as the atoms are brought together, because these are the levels of a free electron in the lattice. The normal state of an interstitial atom behaves in the manner shown symbolically in Fig. 5. In this case, the level remains discrete until it merges with the spreading ionization continuum.

96. Excitation Waves.—As we have seen in the preceding section, there is reason to believe that there are nonconducting excited levels beneath the conducting states described by the Bloch approximation in insulators such as sodium chloride. We shall attempt to describe these excitation states more fully in the present section, using a simple model that was considered first by Frenkel[1] and by Peierls[2] and more recently by Slater and Shockley.[3]

[1] FRENKEL, *op. cit.; Physik. Z. Sowj,* **9,** 158 (1936).
[2] R. PEIERLS, *Ann. Physik,* **13,** 905 (1932).
[3] J. C. SLATER and W. SHOCKLEY, *Phys. Rev.,* **50,** 705 (1936).

Let us consider a system of identical atoms that possess only one electron and are ordered in a simple crystalline array. For simplicity, we shall neglect electron spin and shall assume that both the lowest and first excited levels are nondegenerate. It will be evident that the important conclusions which may be drawn for this simple system are valid for a similar system of atoms of any type in another lattice.

Let us assume that the atoms are located at the positions

$$\mathbf{r}(n) = n_1 \tau_1 + n_2 \tau_2 + n_3 \tau_3 \tag{1}$$

where the τ are the primitive translations of the lattice and the n range over all integer values. We shall let ψ_n and ψ'_n be the normal and excited wave functions of the electron at $\mathbf{r}(n)$, and we shall assume[1] that we are dealing with a case in which the wave functions on different atoms overlap so little that the ψ are practically the same as atomic functions. ψ_n and ψ'_n then are orthogonal to one another and to the wave functions of electrons on other atoms. A wave function for the lowest state of the entire system may be constructed by taking a determinant of the form

$$\Psi_0 = \frac{1}{\sqrt{N!}} \begin{vmatrix} \psi_1(\mathbf{r}_1) \dots \dots \dots \dots \dots \dots \dots \dots \dots \psi_1(\mathbf{r}_N) \\ \cdot \qquad\qquad\qquad\qquad\qquad\qquad \cdot \\ \cdot \qquad\qquad\qquad\qquad\qquad\qquad\qquad \cdot \\ \cdot \qquad\qquad\qquad\qquad\qquad\qquad\qquad\qquad \cdot \\ \psi_N(\mathbf{r}_1) \dots \dots \dots \dots \dots \dots \dots \dots \dots \psi_N(\mathbf{r}_N) \end{vmatrix} \tag{2}$$

where N is the total number of electrons and of atoms. The mean energy of this wave function is

$$E_0 = \int \Psi_0{}^* H \Psi_0 d\tau \tag{3}$$

where H is the Hamiltonian of the entire system. This integral may be expanded in terms of the eigenvalues of the ψ and the exchange and coulomb integrals between atoms.

Let us consider next wave functions for the case in which one atom is excited. The system of wave functions Ψ_n obtained by replacing ψ_n in (1) by ψ'_n are not the best excited wave functions, for the integrals

$$E_{mn} = \int \Psi_m{}^* H \Psi_n d\tau \qquad m \neq n \tag{4}$$

do not vanish. It is easy to show in fact that, under the orthogonality conditions on the ψ_n and ψ'_n that were assumed above, (4) is equal to

$$e^2 \int \frac{\psi'_m(\mathbf{r}_1)\psi_n(\mathbf{r}_2)\psi_m(\mathbf{r}_1)\psi'_n(\mathbf{r}_2)}{r_{12}} d\tau_{12} - e^2 \int \frac{\psi'_m(\mathbf{r}_1)\psi_n(\mathbf{r}_2)\psi_m(\mathbf{r}_2)\psi'_n(\mathbf{r}_1)}{r_{12}} d\tau_{12} \tag{5}$$

[1] It should be emphasized that the following approximation is accurate only when the atoms are not too close together. There may be no nonconducting excited states if the atoms are pushed together sufficiently (*cf.* Secs. 66 and 95).

when $m \neq n$. When $m = n$, (4) may be expanded in terms of the energy levels of the normal and excited atoms and the exchange and coulomb integrals between normal and excited atoms. We shall designate the difference between E_{nn} and E_0, which is the order of magnitude of the resonance energy of an isolated atom, by

$$\epsilon = E_{nn} - E_0. \tag{6}$$

We shall attempt to diagonalize the N-dimensional matrix formed by the E_{mn}. This process is equivalent to finding those linear combinations Ψ' of the Ψ_n which have the form

$$\Psi' = \sum_n a_n \Psi_n, \tag{7}$$

in which the a satisfy the equations

$$\sum_m a_m E_{mn} = E' a_n. \tag{8}$$

It follows from the symmetry of the crystal that E_{mn} depends only upon the difference between the integer sets m and n. This fact suggests that we should reduce the N equations (8) to the same form by making the substitution

$$a_m = a_k e^{2\pi i \mathbf{k} \cdot \mathbf{r}(m)} \tag{9}$$

where $\mathbf{k}$ is a vector in the reciprocal lattice of the crystal. Equation (8) then becomes

$$E'_{\mathbf{k}} = E_{nn} + \sum_l{}' E_{n,n+l} e^{2\pi i \mathbf{k} \cdot \mathbf{r}(l)} \tag{10}$$

where $E_{n,n+l} = E_{m+r,m+r+l}$, if l and r are arbitrary integer sets. The prime in this summation indicates that the term for $l = 0$, which appears outside the sum, is to be excluded. The normalized wave function associated with the wave number $\mathbf{k}$ may be found by substituting Eq. (9) in (7) and is

$$\Psi_{\mathbf{k}} = \frac{1}{\sqrt{N}} \sum_n e^{2\pi i \mathbf{k} \cdot \mathbf{r}(n)} \Psi_n. \tag{11}$$

The independent values of $\mathbf{k}$ range over a single zone if there is one atom per unit cell and the excited state is nondegenerate, whereas they range over αg zones if there are α atoms per unit cell and the degeneracy of the excited level is g.

In view of the assumption that overlapping is small, it is reasonable to assume that E_{mn} is zero for all except nearest neighbors. If I is the value of E_{mn} in this case, (9) may be written

$$E'_{\mathbf{k}} = E_{nn} + I\sum_{\tau} e^{2\pi i \mathbf{k}\cdot\tau} \tag{12}$$

where τ is to be summed over nearest neighbors. Equation (12) is simply

$$E'_{\mathbf{k}} = E_{nn} + I(\cos 2\pi k_x a + \cos 2\pi k_y a + \cos 2\pi k_z a) \tag{13}$$

in the case of a simple cubic lattice having lattice constant a. This equation, which is similar to the equations derived in Sec. 65 for the case of narrow conduction bands, shows that the excited levels form a band the width of which is of the order of magnitude I. Equations (10) and (11) are valid only as long as I is appreciably smaller than ϵ. Otherwise, more atomic states must be considered in diagonalizing the Hamiltonian matrix.

The excited atom is not localized in the states described by (11); instead it is distributed throughout the crystal. By constructing wave packets, it is easy to show that the excitation moves with the group velocity

$$\mathbf{v} = \frac{1}{h}\,\mathrm{grad}_{\mathbf{k}}\,E_{\mathbf{k}} \tag{14}$$

in the energy state $E_{\mathbf{k}}$ (*cf.* Sec. 68).

The current associated with a given wave function $\Psi_{\mathbf{k}}$ is the mean value integral

$$\mathbf{I}_{\mathbf{k}} = \frac{Neh}{2mi}\int{}' (\Psi_{\mathbf{k}}\,\mathrm{grad}_1\,\Psi_{\mathbf{k}}{}^* - \Psi_{\mathbf{k}}{}^*\,\mathrm{grad}_1\,\Psi_{\mathbf{k}})d\tau', \tag{15}$$

which may be reduced to

$$\frac{eh}{2mi}\sum_{mn}[e^{2\pi i \mathbf{k}\cdot[\mathbf{r}(m)-\mathbf{r}(n)]} - e^{2\pi i \mathbf{k}\cdot[\mathbf{r}(n)-\mathbf{r}(m)]}]\int{}' \Psi_m\,\mathrm{grad}_1\,\Psi_n d\tau'. \tag{16}$$

If Ψ_m and Ψ_n are expanded by substituting their determinantal form, it is found that $\int'\Psi_m\,\mathrm{grad}_1\,\Psi_n dr'$ vanishes for $m \neq n$, for one or more vanishing integrals of the type

$$\int\psi_n\psi'_m d\tau, \qquad \int\psi'_n\psi_m d\tau, \qquad \int\psi_n\psi_m d\tau, \qquad \int\psi'_m\psi'_n d\tau$$

appear in each term.[1] Hence, (15) is zero and the excitation waves carry no current. We may, if we choose, regard the excitation wave as though

---

[1] If the overlap integrals for immediate neighbors do not vanish, the current will not be strictly zero, but will be very small, corresponding to motion of an electron an atomic distance.

it were an uncharged particle, created by exciting the crystal, that may move about the lattice. This convenient concept was first introduced by Frenkel, who called the imaginary particle an "exciton."

The selection rules for optical transitions from the ground state Ψ_0 to the excitation state Ψ_k are determined by the integral

$$\int \Psi_0{}^* \left(\sum_i \text{grad}_i \, e^{-2\pi i \eta \cdot r_i}\right) \Psi_k d\tau(x_1, \cdots, z_n) \tag{17}$$

where i is summed over all N electrons and $\mathbf{n}$ is the wave-number vector for the light quantum.[1] Substituting from Equation (11), we find that (17) becomes

$$\sqrt{N} \sum_n e^{2\pi i \mathbf{k} \cdot \mathbf{r}(n)} \int \Psi_0 \, \text{grad} \, e^{-2\pi i \eta \cdot \mathbf{r}} \Psi_n d\tau(x_1, \cdots, z_n) \tag{18}$$

where $\mathbf{r}$ is the coordinate vector of any one of the electrons. If the determinantal form of the Ψ is used, the integral in this equation is reduced to

$$\frac{1}{N} \int \psi_n \, \text{grad} \, e^{-2\pi i \eta \cdot \mathbf{r}} \psi_n' d\tau(x,y,z), \tag{19}$$

which determines the selection rules for optical transitions in isolated atoms. Ordinarily, $\mathbf{n}$ is so small that $e^{-2\pi i \eta \cdot \mathbf{r}}$ does not vary appreciably over a single atom and may be replaced by $e^{-2\pi i \eta \cdot \mathbf{r}(n)}$. Thus, (17) becomes

$$\sum_n e^{2\pi i (\mathbf{k}-\eta) \cdot \mathbf{r}(n)} \frac{1}{\sqrt{N}} \int \psi_n \, \text{grad} \, \psi_n' d\tau = \sqrt{N} \left(\int \psi_n \, \text{grad} \, \psi_n' d\tau\right) \delta_{\mathbf{k}\eta}. \tag{20}$$

It may be concluded that the transition probability is zero unless the condition

$$\mathbf{k} = \mathbf{n} \tag{21}$$

is fulfilled and unless the excited state ψ_n' is one to which transitions from ψ_n are allowed. Since $\mathbf{n}$ is very small, (21) is equivalent to the condition that $\mathbf{k} = 0$.

If we consider, instead of the system described above, one such as sodium chloride, in which the chlorine ion has excited states when in the crystal, the practical problem of constructing the excited states is complicated by the fact that the constituent ions contain more than one electron. This should not affect the qualitative results of the preceding discussion, such as that the width of the excitation band increases with

[1] The optical properties associated with excitation bands are discussed in Chap. XVII.

increasing interionic interaction and that the excitation wave carries no current.

The lowest states of the electronegative ions in simple ionic crystals are S-like, since these ions have closed-shell configurations, whereas the first excited states have P-like symmetry, as we have seen in the last section. Since optical transitions between these types of level are allowed, the condition (21) determines the selection rules.

Wannier[1] has proposed a very simple semiquantitative method of looking at the excitation bands in insulators. If we remove an electron from the highest filled band of an insulator, we produce a positive charge that should be able to move about freely in an undistorted lattice. In the Bloch picture, the excited electron is independent of the positive hole and is also able to move about freely. Actually, the hole and the electron should attract one another with a force that is coulomb-like at large distances. Wannier has shown that the excitation bands are analogous to the discrete levels of a hydrogen atom in the sense that in these states the electron and hole revolve about one another in closed orbits. The different levels in a given band correspond to the different translational levels of an excited hydrogen atom. On the basis of this picture, Wannier has derived a set of simple approximate equations from which the wave functions and energy levels of the exciton may be determined.

[1] G. H. WANNIER, *Phys. Rev.*, **52**, 191 (1937).

CHAPTER XIII

THE ELECTRONIC STRUCTURE OF THE FIVE SOLID TYPES

97. Introduction.—The present chapter, in which we shall present a survey of the electronic constitution of the normal and excited states of the five solid types, is the central chapter of the book since all the preceding chapters are preparatory for it. A large part of this discussion is necessarily qualitative and probably will remain so until computational technique has been developed much further. Thus, we shall use the one-electron approximations freely in cases in which they do not lead to qualitatively incorrect results. In other cases, we shall employ the method of description in which the energy levels of the entire solid are used.

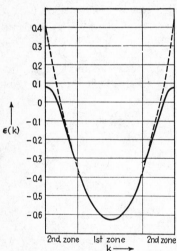

Fig. 1.—The $\epsilon(\mathbf{k})$ curve for the (110) direction of sodium (full line). The dotted line is the free-electron parabola. The energy scale is in Rydberg units. (*After Slater.*)

A. METALS

98. General Remarks.—Although the correlation terms that were discussed in Chaps. IX and X probably are important in a quantitative determination of any property of a metal, it is unlikely that they often affect the qualitative properties. The possible exceptions occur in connection with those low-temperature effects, such as superconductivity, which are not well understood at present. For this reason, we shall discuss the valence electrons of simple metals on the basis of the band approximation. Many of the qualitative properties of d-shell electrons can also be treated adequately in this way. The method is not entirely satisfactory, however, for many other properties of d-shell electrons can be explained better with the Heitler-London approximation. This fact shows that neither of the one-electron schemes is very good in this case and that the d shells should be treated as a whole. This more accurate procedure has been used only in a few cases, such as in the

420

spin-wave theory of ferromagnetism which is developed in Chap. XVI. It is usually assumed at present that the accurate solution would yield the same result as the one-electron schemes in those cases in which the latter appear to give a satisfactory description of affairs.

99. Simple Metals. *a. The Alkali Metals.*—We discussed most of the known facts concerning the electronic levels of the alkali metals in Chap. X. A few additional results follow.

The zone structure of sodium[1] has been investigated by several workers. All the results show that the gaps are very narrow and that the effective electron mass is close to unity. In view of Shockley's investigation of the empty lattice by the cellular method (*cf.* Sec. 73), we may

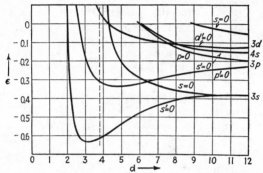

Fig. 2.—The dependence of the energy bands of sodium on interatomic distance *d*. It should be observed that the *s*- and *p*-level bands overlap strongly at the observed value of *d*. This behavior is characteristic of the simpler metals. The energy scale is in Rydberg units. (*After Slater.*)

say that the electrons in sodium are free, within the accuracy of this method. The zone structure determined by Slater is shown in Fig. 1, and the dependence of energy levels on interatomic distance is shown in Fig. 2. It should be noted that the *s*- and *p*-level bands overlap a great deal at the actual interatomic spacing. This overlapping of *s* and *p* levels is characteristic of all metals.

Bardeen[2] has pointed out that exchange terms have a very important effect on the density of electronic levels near the top of the filled region when the electrons are nearly free. In this case, the exchange energy is

$$\epsilon_{ex} = -0.306\left(2 + \frac{1 - \frac{k^2}{k_0^2}}{k/k_0}\log\frac{|k_0 + k|}{|k_0 - k|}\right)\frac{e^2}{r_s} \tag{1}$$

[1] E. WIGNER and F. SEITZ, *Phys. Rev.*, **43**, 804 (1933); **46**, 509 (1934). J. C. SLATER, *Phys. Rev.*, **45**, 794 (1934); *Rev. Modern Phys.*, **6**, 209 (1934).

[2] J. BARDEEN, *Phys. Rev.*, **50**, 1098 (1936).

so that the total dependence of electronic energy on k is

$$\epsilon(\mathbf{k}) = \epsilon_0 + \frac{h^2}{2m}k^2 + \epsilon_{ex}. \tag{2}$$

The number of levels having values of k in the range from k to $k + dk$ is

$$dn = 8\pi k^2 V dk = 8\pi k^2 V \frac{1}{d\epsilon/dk} d\epsilon. \tag{3}$$

We may see from Fig. 5 of Chap. IX, that $d\epsilon_{ex}/dk$ is infinite when k is equal to k_0. Thus, the density of levels is much smaller at $k = k_0$ than it would be if ϵ were simply a parabolic function of k. Since the electronic specific heat should be proportional to the density of levels in this region, it follows that the electronic specific heat of a free-electron gas, in which the exchange interaction is included, should be less than the value

$$k\frac{\pi^2}{2}\frac{kT}{\epsilon_0},$$

which was derived in Sec. 27. In fact, Bardeen has shown that the specific heat should vary as $-(\log T)/T$ at low temperatures when $\epsilon(\mathbf{k})$ has the form of Eq. (2). The low-temperature specific heat of the alkali metals has not been measured accurately enough to check this behavior. It is possible that correlation terms have an effect which may compensate for the effect of exchange.[1]

This infinity in the slope of the exchange energy is accompanied by a singularity in curvature. In fact, the mean value of the second derivative of (1) becomes infinite as $\log |k_0 - k|$ when $k \to k_0$, a fact implying that the electronic mass of the uppermost electrons in the Fermi band approaches zero at the absolute zero of temperature. This singularity is usually ignored in computations of such effects as conductivity and diamagnetism (see Chaps. XV and XVI) because of the undetermined influence of correlations. This procedure seems to be justified at room temperature by the fact that results obtained are usually in good agreement with experiment.

The zone structure of lithium[2] was investigated by Millman, using the cellular approximation. His results show that the effective electron mass is greater than unity, in this case, as we already have seen in Sec. 78. In all other respects, the zone scheme is like that of sodium.

The matrix elements that determine the transition probabilities for optical absorption are zero for perfectly free electrons and are undoubt-

[1] This possibility was pointed out by E. Wigner, *Trans. Faraday Soc.*, **34**, 678 (1938).

[2] J. MILLMAN, *Phys. Rev.*, **47**, 286 (1935).

edly small for the nearly free electrons in the alkali metals.   Were this not so, the alkali metals would probably be colored, since the first allowed transition of the type $\mathbf{k} \rightarrow \mathbf{k} + \mathbf{K}$ should occur at 1.5 ev in sodium and at about 2 ev in lithium, according to the zone diagrams for these metals.

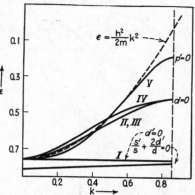

Fig. 3.—The $\epsilon(\mathbf{k})$ curves for the (110) direction of copper. The five branches I to V meet at $k = 0$, giving a five-fold degenerate point. Curve V corresponds to the *s-p* band in simple metals and is nearly the same as the free-electron curve. Actually all levels are mixtures of *s*, *p*, and *d* states. Curves I to IV and the lowest curve correspond to the *d* band. The energy scale is in Rydberg units. (*After Krutter.*)

   b. The Noble Metals Copper, Silver, and Gold.—The monovalent noble metals differ from the alkali metals in that they have newly completed *d* shells in the atomic configurations.   These *d* levels lie so close to the *s* levels that configurations such as $3d^{10}4s$ and $3d^9 5s^2$ in the free atoms are about $1\frac{1}{2}$ ev apart.   When the atoms are brought together, the *s* and *d* levels split into overlapping bands.   Naturally, the *d*-electron band is narrower than the *s-p* band because the *d* electrons are partly screened by the others.

   The first investigation of the *d* band was made by Krutter,[1] who applied the cellular method to copper.   When computing the *s-p* bands,

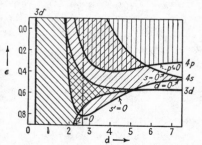

Fig. 4.—Dependence of energy bands of copper on interatomic distance *d*.   It should be observed that the *s*, *p*, and *d* bands overlap appreciably at the actual interatomic distance.   The energy scale is in Rydberg units.   (*After Krutter.*)

he assumed that the field within each cell is that of the free Cu^+ ion; when computing the *d* band, he used a field obtained from the $3d^9 4s$ configuration of Hartree's atomic wave functions for copper.   In view of the discussion of Chap. X, we may say that these simplifications are reasonable for semiquantitative work.   Figures 3 and 4 show the dependence of the bands upon interatomic distance, and the reduced-zone structure in the (110) direction at the observed lattice distance.   The second figure shows that the limit of the filled region is far above the uppermost

[1] H. M. Krutter, *Phys. Rev.*, **48**, 664 (1935).

level of the d band, so that this band is completely filled. It is doubtful whether the d band actually is as wide as Krutter's work indicates. Hartree neglected exchange and correlation terms in deriving the field used in computing these wave functions for copper, so that they extend farther from the nucleus and overlap more than they should. The five d-band $\epsilon(\mathbf{k})$ curves meet at $k = 0$ in Krutter's results. His approximate method of applying the cellular scheme is responsible for this degeneracy, for in a more accurate solution the level would split into a twofold and a threefold degenerate level.[1]

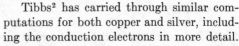

n(ϵ)

0.45 0.35 0.25 0.15 0.05
$\epsilon \longrightarrow$

Fig. 5.—The density of energy levels in calcium. The contributions from the first and second zones are indicated separately. The energy scale is in Rydberg units. (*After Manning and Krutter.*)

The reddish color of copper is attributed to an optically induced electronic transition from the d band to the s-p band. According to Krutter's results, the minimum difference between levels for which a transition is allowed is of the order of 3 ev in the (100) direction, which implies strong absorption in the blue region of the visible spectrum. Since silver is not so strongly colored as copper, we may conclude that the difference between the d and s bands is larger for this metal. The difference presumably decreases again in gold since it is colored.

Tibbs[2] has carried through similar computations for both copper and silver, including the conduction electrons in more detail.

c. Calcium.—The only extensive calculation on the zone structure of the alkaline earth metals, aside from that for beryllium, which was discussed in Sec. 81, is the work of Manning and Krutter[3] for calcium. This metal has a face-centered cubic lattice, and the methods and approximations used in obtaining the energy contours were similar to those used by Krutter for copper.

The alkaline earth metals should be insulators for large interatomic spacing since the atoms have closed-shell configurations in the normal states. The conductivity arises from overlapping of the s, p, and d bands at the observed interatomic distance. Manning and Krutter

---

[1] This fact may be derived from a group-theoretical treatment of crystalline wave functions. A cubic crystal cannot have wave functions for $\mathbf{k} = 0$ that are higher than threefold degenerate. Thus, the fivefold degenerate atomic d function splits into a twofold and a threefold degenerate level.

[2] S. R. Tibbs, *Proc. Camb. Phil. Soc.*, **34**, 89 (1938).

[3] M. F. Manning and H. M. Krutter, *Phys. Rev.*, **51**, 761 (1937).

found that this overlapping occurs not in the three principal crystallographic directions, but instead in the (021) direction.   Figure 5 shows the density of levels as a function of energy for the first band and part of the second.   According to this result, the amount of overlapping actually is very small, a fact which suggests that calcium, like beryllium, is very nearly an insulator.

100. Metals with Irregular Structures.—Mott and Jones[1] have made a detailed investigation of the zone structure of metals such as mercury, white tin, and bismuth that have unusual valencelike structures.   In all these metals, it is found that the edge of the filled system of energy levels is very close to a prominent zone boundary, that is, to a boundary that corresponds to a strong X-ray reflecting plane.   Since the energy gaps

probably are large at a boundary of this type (*cf.* Sec. 62), there is only a small amount of overlapping of the filled and unfilled zones in these metals.   Calcium, which was discussed in part *c* of the preceding section, is a simple case of this type.   Figure 6 shows the prominent zone boundary for the bismuth lattice which contains five electrons per atom.

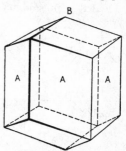

This observation that irregular metals possess nearly filled bands gives a very satisfactory phenomenological explanation of the fact that their properties lie between those of ideal metals and of valence types.   The gaps in an ideal valence crystal are wide enough to keep the occupied and the unoccupied zones apart, for these solids are insulators.   On the other hand,

Fig. 6.—The prominent zone boundary for bismuth. This contains $10n$ states so that it is nearly filled. The overlapping of the levels of this zone and the next is believed to occur at points *A*. (*After Jones.*)

the gaps are very narrow in ideal metals.   Since the gaps in irregular metals are intermediate between those of these two cases, we may expect that other properties should be intermediate.

The question of why the irregular metals choose those structures which have nearly filled bands rather than others in which the properties are more metallic can be answered accurately only by computing the lattice energy for a nonmetallic and a typically metallic structure, as has been done in the case of hydrogen (*cf.* Sec. 79).   Since such computations have not yet been carried out, we must be satisfied for the present with the chemist's type of answer, namely, that the constituent atoms of irregular metals bear a resemblence to hydrogen in that they prefer to form a structure in which the atoms are coordinated, as in valence crystals.

---

[1] N. F. Mott and H. Jones, *Theory of the Properties of Metals and Alloys*, pp. 162 *ff.*

101. Transition Metals.—Mott[1] and Slater[2] have used a zone model as the basis for qualitative discussions of the transition metals as a class. Two types of band are employed in this model, namely, a wide low-density valence-electron band, which arises from the s and p atomic states, and a narrow high-density d-electron band (*cf*. Fig. 7). This scheme of levels occurs in copper, as we have seen in Sec. 99. The essential difference between copper and the transition metals is that the d band is not completely filled in the latter.

An important property of the ferromagnetic transition metals that is readily explained by the zone theory is that their gyromagnetic ratio is nearly equal to 2. A mechanical moment may be induced in a ferromagnetic substance by magnetizing it, and the ratio of the magnetic moment, expressed in units of the Bohr magneton, to the angular momentum, expressed in units of $\hbar$, is called the gyromagnetic ratio.[3] If the magnetic moment arises purely from orbital motion, the ratio should be unity (*cf*. Chap. V); if it arises from electronic spin, the ratio should be 2; and if it arises from a combination of spin and orbital motion, it should lie between zero and 2. The fact that the value usually is nearly 2 (for example, the value for iron is 1.93) indicates that most of the orbital angular momentum that the d electrons possess in the free atom is "quenched" in passing from the gas to the solid and that principally the spin magnetic moment remains. The orbital angular momentum is negligible in the band scheme,[4] since the zones are symmetrically filled in such a way that the electrons may be paired in groups which move in opposite directions with equal velocities. Accurate measurements such as the one cited above for iron show that the orbital magnetic moment is not entirely quenched, a fact indicating that the band approximation is not entirely accurate.

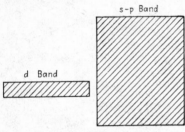

Fig. 7.—Schematic diagram of the relative positions and widths of the d band and s-p band (*cf*. Fig. 4 for copper). The narrow d band has room for ten electrons per atom, whereas the s-p band has room for only two.

The cellular method has been applied in detail to only one transition metal, namely, tungsten, which is discussed later, in part *b*. Slater

[1] N. F. Mott, *Proc. Phys. Soc.*, **47**, 571 (1935).

[2] J. C. Slater, *Phys. Rev.*, **49**, 537 (1936); *Jour. Applied Phys.*, **8**, 385 (1937).

[3] See E. C. Stoner, *Magnetism and Matter* (Methuen & Company, Ltd., London, 1934); see also S. J. Barnett, *Rev. Modern Phys.*, **7**, 129 (1935).

[4] This point has been carefully investigated by H. Brooks (paper to appear in *Phys. Rev.*).

however, has used the level system computed by Krutter for copper to discuss some of the details of the iron-group metal alloys, assuming that the relative positions of the d and the s-p bands do not change very much throughout the iron group (*cf*. Sec. 103).   The density of levels in the s-p and d bands of copper is shown in Fig. 8; the vertical lines indicate the extent to which these levels would be filled in transition metals with different numbers of electrons per atom.

a. The Iron-group Transition Metals.—We shall now discuss several properties of the iron-group transition metals on the basis of the band picture.

1. Cohesion.—One of the striking properties of the cohesive energies of the transition metals immediately preceding copper, silver, and gold is

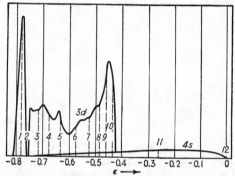

FIG. 8.—The density of levels in the iron-group series.   The vertical lines designate the limit to which the levels are filled in the elements having the number of electrons corresponding to the integers given.   Thus the d band is completely filled in copper (11 electrons) and is not quite filled in nickel (10 electrons).   This figure is based on Krutter's work on copper.   The abscissa is expressed in Rydberg units.   (*After Slater.*)

the fact that they usually are larger than the cohesive energies of the monovalent metals.   This fact is illustrated by the following sequences:

Ni	85 kg cal/mol	Cu	81 kg cal/mol
Pd	110 kg cal/mol	Ag	68 kg cal/mol
Pt	127 kg cal/mol	Au	92 kg cal/mol

Mott has proposed the following qualitative interpretation of this fact. The electrons in the s-p band are principally responsible for the cohesion of all these metals, since the d shells are nearly filled.   Fuchs has estimated the d-shell interaction in copper (*cf*. Sec. 80) and has found it to be of the order of 0.5 ev per atom.   If it is assumed that the electronic levels are very nearly the same in the transition-group metals and in the simple metals that immediately follow them (for example, in the sequence from iron to copper), it should be expected that the differences in binding properties arise from differences in the way in which the levels are filled.

Let us consider iron, cobalt, nickel, and copper. In the first three cases, the d band is not completely filled, and the s-p band presumably is filled to the same height as the d band. Since the density of d levels is very high, this fact means that the s-p band is filled to very nearly the same point in each case, if it is assumed that the relative positions of the two bands remain fixed. On the other hand, the d band of copper is completely filled, so that the additional electrons fill the s-p band to a point far above the top of the d band. Mott points out that the bottom of the s-p band, the top of the filled region, and the mean energy per atom in copper are related in the manner shown in Fig. 9, in which curve I is the bottom of the s-p band, curve II is the top of the filled region, curve III is the mean energy, and the zero of energy is referred to the state in which all atoms are infinitely separated. Since the electrons between curves II and

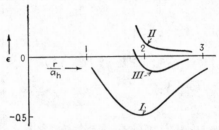

Fig. 9.—Schematic representation of the electron energies in copper. Curve I is the $\epsilon_0(r_s)$ curve; curve II is the top of the filled band; and curve III is the mean energy. Mott points out that curve III would be lowered if there were fewer electrons in the s-p band.

III have more energy than the average, it might be expected that curve III would be lowered if the electrons in the upper part of the filled region were either removed or placed in lower levels. These electrons are essentially removed in the transition metals, according to the zone model. Hence, it may be expected that, in these cases, curve II and curve III are lower than in copper. It is only fair to say that this argument can be used only in a qualitative way, since a shift in the filling of one-electron levels has associated with it changes in exchange and correlation energies that cannot be included in a simple energy diagram.

It has been suggested by other workers[1] that at least part of the binding of the transition metals is related to a lowering of the center of gravity of the occupied d levels in passing from the free atom to the solid. Recent work on tungsten, which is discussed below, indicates that this effect is probably the largest source of cohesive energy in the platinum series of transition metals. Whether or not it is important in the iron group remains to be seen.

[1] F. Seitz and R. P. Johnson, *Jour. Applied Phys.*, **8**, 84, 186, 246 (1937).

In addition, Pauling,[1] reasoning on the basis of the empirical knowledge of the properties of transition metals, has suggested that some of the d electron wave functions combine with the s-p functions to form a scheme of levels in which there is a larger number of valence electrons per atom than the one s-p function per atom suggested by Krutter's band scheme for copper.   In principle, Pauling's suggestion is equivalent to assuming that there are two types of d-electron band in the transition metals, namely, a wide band (see Fig. 10), which is similar to the s-p band but contains 2.6 of the 5 d functions per atom of given spin, and a narrow band, which contains the remaining 2.4 d functions.   The first class of function (*A* type), along with the s-p func-

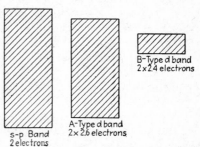

tions, is responsible for the large cohesive energy of the transition metals since its existence implies an increase in the number of binding electrons in the metal, whereas the second class of function (*B* type) is responsible for the magnetic properties in a manner that will be described under 2.   It was pointed out in Sec. 99, in connection with Krutter's work on copper, that the five d zones should break into two separate systems containing two and three zones, respectively (see footnote 1, page 424), and that Krutter's approximation does not give this splitting.

FIG. 10.—The energy bands in the iron-group metals according to Pauling. The s-p functions combine with some of the d functions to form two broad bands, namely, the left-hand band, which is designated as the s-p band, but which contains a mixture of d functions, and the A type of d band, which has room for 1.6 electrons per atom of given spin.   The B type of d-electron band is narrow and has room for 2.4 electrons per atom of given spin.

It is possible that Pauling's scheme of levels would be obtained from Krutter's if the band approximation were applied with a higher degree of accuracy.   We shall see in Sec. 104, however, that the experimentally determined levels do not seem to agree with Pauling's assumptions.

   2. *The filling of levels in the ferromagnetic elements.*—This topic was previously introduced in Sec. 27, Chap. IV, in which we used the band theory to explain the low-temperature specific heats of transition metals. It was postulated there that in the ferromagnetic elements the half of the d band associated with one kind of electron spin is completely filled and that the saturation magnetic intensity, expressed in Bohr magnetons per atom, is equal to the number Δn_d of unoccupied levels per atom in the other half of the d band.   The number of electrons per atom n_v in the s-p band may be computed from this hypothesis.   If m is the total number

[1] L. Pauling, *Phys. Rev.*, **54**, 899 (1938).

of d and s-p electrons in the atom, it follows that Δn_d, n_v, and m must satisfy the relation

$$10 - \Delta n_d + n_v = m$$

or

$$n_v = m + \Delta n_d - 10.$$

We find, using the measured saturation moments, that n_v is, respectively, 0.2, 0.7, and 0.6 for iron, cobalt, and nickel. The first value undoubtedly is too small if the cohesive energy of iron is to be explained in terms of the energy of s-p electrons, whereas the other two are reasonable. This low value suggests either that half the d band is not filled in this case or that the d band is lower in iron than in the metals following it and the cohesive energy is related to this lowering. The second possibility is not entirely unreasonable, for, as was mentioned above, the cohesive energy of tungsten (see part b of this section) seems to be related entirely to the behavior of the d band. The first possibility is readily explained on the basis of Pauling's suggestion; for in iron the B type band of Fig. 10 would be completely drained of the 2.4 electrons per atom having one type of spin and 0.2 electron per atom having the other type of spin would be removed, whereas in cobalt and nickel only a fraction of the electrons in half this band would be removed. We are not able to decide between the two alternatives on the basis of the present knowledge of energy levels, however, and in subsequent discussions we shall arbitrarily assume that the first is correct.

The Heitler-London scheme has been used with considerable success in discussing the spin-aligning forces of ferromagnetism, as will be seen in Chap. XVI. Since this approximation does not provide a satisfactory simple explanation of the low-temperature electronic heat of the transition metals, it cannot be used in place of the band theory for all purposes.

3. *The paramagnetic transition metals.*—We have already seen in Sec. 29 that the band theory cannot explain even semiquantitatively both the specific heat and the magnetic susceptibility of the paramagnetic transition metals. This failure lends additional support to the statements made above concerning the limitations of the band approximation when applied to d-shell electrons.

b. Tungsten.—The band structure of metallic tungsten, which has a body-centered lattice and whose atoms possess six valence electrons and a newly filled f shell outside a rare gas configuration, has been investigated by Manning and Chodorow.[1] These workers assumed that the f shell is unaffected by solid binding and obtained wave functions and energy levels for the remaining six electrons per atom. In first approxi-

---

[1] M. F. Manning and M. I. Chodorow, *Phys. Rev.*, **56**, 787 (1939).

mation, an effective field for these electrons was obtained from a charge distribution derived by renormalizing the parts of the free-atom wave functions lying within the atomic sphere of the lattice. Valence-electron wave functions and $\epsilon(\mathbf{k})$ contours were then computed with the use of this field. In second approximation, the charge density obtained from the results of the first approximation was used to compute a new field. This second approximation was very nearly self-consistent.

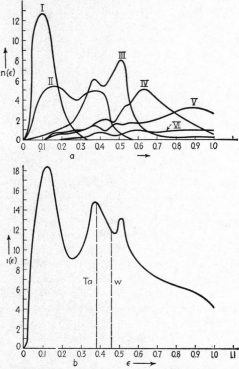

Fig. 11.—$n(\epsilon)$ curves for the five d zones and one s-p zone of tungsten. The energy scale is in Rydberg units. (*After Manning and Chodorow.*)

The $n(\epsilon)$ curves for the six lowest zones are shown in Fig. 11a. The set labeled with roman numerals I to V correspond to the five d bands, and curve VI corresponds to the s-p band. The total $n(\epsilon)$ curve for all six bands is given in Fig. 11b. The limit of the filled regions of tungsten and of the neighboring element tantalum are marked by vertical lines. According to these results, the number of electrons in the s-p band is of the order of magnitude 0.1 electron per atom in both these metals. Manning and Chodorow estimate that the center of the filled region is about 8 ev below the mean of the occupied levels of the free atom. These results imply that the large cohesive energy of tungsten, namely, 210 kg cal/mol, arises from the fact that practically all of the six valence electrons may occupy the low-energy portion of the d band. In addition, they imply that the contribution to cohesion from the s-p electrons, which are responsible for most of the cohesion in simple metals, is negligible in this case.

Using the computed $n(\epsilon)$ curves, Manning and Chodorow estimated the electronic heat of tungsten and tantalum by the use of the equations derived in Chap. IV. The observed and calculated values are given in Table LXXII.

TABLE LXXII.—A COMPARISON OF OBSERVED AND CALCULATED ELECTRONIC HEATS
OF TUNGSTEN AND TANTALUM
(In units of 10^{-4} cal/deg-mol)

	Theoretical	Experimental	
		Low temperature	High temperature
W	$4.8T$	. . . .	$5.1T$
Ta	$6.2T$	$27T$	$7T$

The low-temperature value for tantalum was obtained indirectly from conductivity measurements near absolute zero, and the high-temperature values were obtained after subtracting the $3R$ lattice vibrational heat and questionable $C_p - C_v$ corrections from the observed molar heats. A discussion of this work may be found in the original paper by Manning and Chodorow. It is difficult to say whether the discrepancy between low-temperature and high-temperature values implies error in the simple theory of electronic heats developed in Chap. IV or in the treatment of experimental results. In any case, the agreement between the theoretical results and the high-temperature values is excellent.

102. Simple Substitutional Alloys.—In the experimental survey of Chap. I, it was seen that the Hume-Rothery electron-atom ratio rule correlates the solid-phase portions of the phase diagrams of different substitutional alloy systems. This rule states that a given phase occurs for a fixed electron-atom ratio in a number of different alloy systems. As a result of an extensive investigation, Jones[1] has found that the edge of the filled region of levels lies close to a prominent zone boundary when the Hume-Rothery rule is satisfied. This observation allows us

---

[1] H. JONES, *Proc. Roy. Soc.*, **144**, 225 (1934); **147**, 396 (1934). MOTT and JONES, *op. cit.*, Chap. V.

to replace the Hume-Rothery rule by the statement that the stable alloy phases have nearly filled systems of zones.

As we mentioned in Sec. 100, it is not evident why a nearly filled zone system should be more stable than another, and this fact has not yet received a completely satisfactory explanation. Figure 12 shows

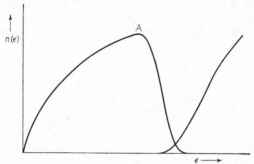

Fig. 12.—Schematic behavior of density of levels near a prominent zone boundary of a metal (*cf.* Fig. 5 for calcium and Fig. 8 of Chap. X for beryllium).

the behavior of the density of states per unit energy range near a zone boundary at which the gaps are large. The density of levels in the lower zone increases at first as the zone boundary is approached because the $\epsilon(\mathbf{k})$ curves bend over. After this rise, the density falls and approaches the axis sharply. It should rise sharply again in the higher zone in the manner illustrated. We may conclude that the two zones overlap in the substitutional alloys from the fact that these alloys are metallic conductors. Jones assumes that the maximum A of the lower zone occurs at that value of energy for which the contours in wave-number space just touch the zone boundary. This assumption has been justified by a detailed treatment of the $\epsilon(\mathbf{k})$ curves for a number of zones with the use of simplified models (*cf.* Sec. 65). He then postulates that the stability of a phase increases as the levels are filled to the point A and then decreases rapidly beyond this point, because the average energy of the additional energy is much larger than the mean energy of all electrons. If this assumption is true, the electron-atom ratio associated with A should be the value for which the phase is most stable.

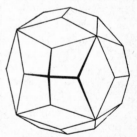

Fig. 13.—The prominent zone boundary for the γ brass phase. This zone contains room for 90 electrons per unit cube of the lattice. (*After Jones.*)

Four phases are commonly met in simple substitutional alloy systems. The α phase is face-centered cubic, the β phase is body-centered cubic, the γ phase has a more complex cubic structure, and the η phase is close-packed hexagonal.   The first zones for the α, β, and η phases were shown in Chap. VIII.   The prominent zone for the γ phase is shown in Fig. 13.   This zone contains 90 states for a cubic cell of 52 atoms, or 1,731 states per atom.   The Hume-Rothery electron-atom ratio is $\frac{21}{13}$, or about 1.615.   Jones has computed the ratio corresponding to the point A, Fig. 12, for each of these four phases, assuming that the energy contours are spheres.   His values are given in Table LXXIII along with Hume-Rothery's fractional estimates.   The two values differ slightly but agree equally well with the experimental values.

TABLE LXXIII

Phase	Hume-Rothery's fractional value	Jones' value
α	. . . .	1.362
β	$\frac{3}{2} = 1.5$	1.480
γ	$\frac{21}{13} = 1.615$	1.538
η	$\frac{7}{4} = 1.75$	1.7

103. Alloys Involving Transition Metals.—The properties of transition metals explained most readily by the band theory are (1) the quenching of ferromagnetism by the addition of nontransition metals that form solid solutions and (2) the dependence of the saturation magnetic moment of ferromagnetic alloys on atomic composition.   We shall discuss these two topics together.

If similar phases of the iron-group elements and copper and zinc have practically the same zone structure, we may expect that the magnetic properties of their alloys depend principally upon the extent to which zones are filled, that is, upon the electron-atom ratio.   We shall use the following two principles in correlating the saturation magnetic moments:

1. The number of valence electrons per atom in the s-p band of all ferromagnetic metals is about 0.7.   If, in addition, we were to accept Pauling's postulate, we should also assume that the A type d band of Fig. 10, which contains 2.6 electrons per atom, is filled or nearly filled and that the electrons are removed from or added to the B type band.

2. The saturation magnetization, expressed in Bohr units per atom, is approximately equal to the number of holes per atom in the d band. The word "approximately" is inserted because the saturation magnetization seems to be less in iron, as we have seen in Section 101.   We shall try to give additional insight into this point in the following paragraphs.

Before presenting a general survey of results, we shall consider two typical cases.   Suppose that some of the nickel atoms in a specimen of nickel are replaced by copper atoms. Since copper has one electron per atom more than nickel, we should expect that each copper atom which is added has the same effect as if one electron were added to the bands of pure nickel. These additional electrons enter the d band and should decrease the number of holes at the rate of 1 per copper atom. Since there is 0.6 hole per atom in pure nickel, we should expect the saturation magnetization to decrease linearly with the concentration of copper and to be zero when the atom fraction of copper is 0.6. Sadron's measurements show this to be the case (see Fig. 54, Chap. I). We should expect zinc atoms to have twice the effect of copper atoms since zinc has two valence electrons instead of one. This is also found to be true.

Let us consider next the effect of alloying nickel and cobalt. Nickel has 0.6 hole per atom in the d band and cobalt has approximately 1.7. According to the band model, the number of holes per atom in the alloy that contains an atomic fraction of x nickel atoms and $(1 - x)$ cobalt atoms should be

$$n_x = 0.6x + 1.7(1 - x), \quad (1)$$

so that the saturation magnetic moment should be n_x Bohr magnetons per atom. This rule is closely obeyed in the nickel-cobalt system, as we shall see below.

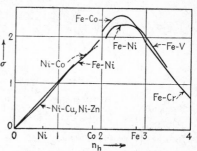

Fig. 14.—Relation between the saturation magnetization, expressed in magnetons per atom, and the number of holes per atom. (*After Slater.*)

Figure 14 shows[1] the relation between the saturation magnetization per atom and the number of holes per atom for a number of substitutional alloys of the iron-group elements. The number of holes per atom in the d band is computed by the use of equations of the type (1) on the assumption that there are 0.7 s-p electron per atom in all transition metals except nickel, which has 0.6. Nontransition elements are assumed to have a negative value, corresponding to 0.7 minus their valence (that is, -0.3 for copper and -1.3 for zinc). If α_a and α_b are the number of holes per atom in the pure metals A and B, the number per atom in the alloy that contains a fraction f_a of A and f_b of B is

$$n_h = \alpha_a f_a + \alpha_b f_b, \quad (2)$$

analogous to (1). The abscissa of Fig. 14 is the value of n_h computed from (2); the ordinate is the saturation magnetic moment per atom σ.

[1] J. C. Slater, *Jour. Applied Phys.*, **8**, 385 (1937).

A straight line corresponds to strict proportionality between these two quantities. It should be observed that this relationship is closely obeyed until the point at which n_h is about 2.2. The experimental curve then bends over smoothly and approaches the axis along what appears to be a straight line. The small value of the saturation magnetization of iron that was mentioned in Sec. 101 is in accordance with this bending. Pauling's assumption of the existence of A type and B type d bands is primarily based upon this fact, for in his picture the B type band is completely drained of electrons of one type of spin at the composition corresponding to the peak of the curves of Fig. 14, that is, when n_h is about 2.4. As one passes farther to the right, toward iron, manganese,

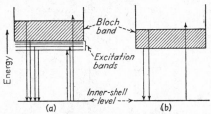

Fig. 15.—In case a there are excitation bands for the inner-shell electrons, and both continuous bands and discrete lines may be expected in emission. In case b the excitation bands have been absorbed into the continuum. The shaded region indicates the filled part of the bands. It should be observed that emission occurs from occupied valence levels, absorption to unoccupied ones.

and chromium, the remaining electrons are removed from the B type band and σ decreases linearly.

Mott and Jones[1] have used essentially the same principles to determine the number of holes in several paramagnetic transition metals. For example, the paramagnetism of palladium decreases when it is alloyed with gold. Since the paramagnetism vanishes when a fraction of 0.55 palladium atoms has been replaced by gold atoms, these workers conclude that palladium contains 0.55 hole per atom.

104. Level Densities from Soft X-ray Emission Spectra.—Experimental values of the level-density curves of the valence electrons in metals may be obtained from the soft X-ray emission and absorption spectra of metals.[2] These curves have particular value in deciding whether the excitation-band picture should be applied to metals as well as to insulators or whether the band approximation is accurate for qualitative work.

[1] Mott and Jones, *op. cit.*, pp. 199–200.
[2] See the survey article by H. W. B. Skinner, *Reports on Progress in Physics* V (1938), (Cambridge University Press, 1939). This possibility was first pointed out by W. V. Houston, *Phys. Rev.*, **38**, 1797 (1931).

Let us suppose that an electron is missing from an inner shell of an ion in a metal. We may describe these inner-shell levels in the Heitler-London approximation and may designate the wave function of the missing electron, which is localized about a single ion, by $\psi_f(\mathbf{r})$. If the Heitler-London or the excitation-band picture were valid for the excited states of this electron, there would be a set of discrete levels or a set of nonconducting excitation states beneath the ionization limit; the latter corresponds to the beginning of the Bloch levels (*cf.* Fig. 15). Thus, the emission and absorption spectra would consist of a continuum corresponding to transitions between the Bloch band and the lowest level and of discrete lines corresponding to transitions between the excitation levels and the lowest levels. On the other hand, if there are

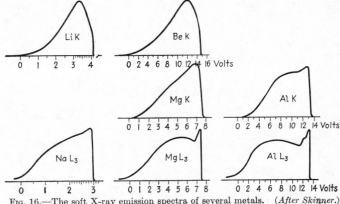

Fig. 16.—The soft X-ray emission spectra of several metals. (*After Skinner.*)

no excitation bands[1] because the ionization band has broadened enough to absorb them, only the continuum should be present. In the transition from one of these cases to the other, we may expect the intensity of the discrete lines to predominate over that of the continuum at first, then decrease, and finally disappear.

The observed soft X-ray emission spectra of lithium, sodium, beryllium, magnesium, and aluminum, as determined by O'Bryan and Skinner[2] and Farineau,[3] are given in Fig. 16. The lithium and beryllium bands arise from transitions to the $1s$ level (K band), and the two magnesium

---

[1] The absence of excitation levels would imply that the free electrons so completely screen the hole in the ion core that there is not enough potential for a discrete level. The quantitative description of this effect would require a very accurate treatment of the many-electron problem.

[2] H. M. O'Bryan and H. W. B. Skinner, *Phys. Rev.,* **45,** 370 (1934).

[3] J. Farineau, *Compt. rend.,* **203,** 540 (1936), **204,** 1108 (1937), **204,** 1242 (1937), **205,** 365 (1937); *Nature,* **140,** 508 (1937). See footnote 1, p. 440, also.

and aluminum bands arise, respectively, from transitions to the $1s$ level and $2p$ level (L_{III} band). In contrast with the absorption spectra of the alkali halides (*cf.* Sec. 95), these spectra show no indication of strong discrete lines, which implies that the excitation levels are not prominent. In order to determine the extent to which the low-energy tails of these bands are vestiges of the transitions from excitation states, we must examine the theory of emission more closely.

We shall assume[1] that the periodic wave functions may be expressed in the form

$$\psi_{\mathbf{k}} = \chi_0(\mathbf{r})e^{2\pi i \mathbf{k} \cdot \mathbf{r}}$$

where χ_0 is an s function in the vicinity of the nucleus. The intensity of the line emitted in the jump from $\psi_{\mathbf{k}}$ to ψ_f then is proportional to

$$|\textstyle\int \psi_{\mathbf{k}}{}^* \operatorname{grad} \psi_f d\tau|^2 \tag{1}$$

(*cf.* Sec. 43). For small values of $\mathbf{k}$ or $\mathbf{r}$, we may expand the exponent in powers of $\mathbf{k} \cdot \mathbf{r}$ and keep only the first two terms. Thus, (1) becomes

$$|\textstyle\int \chi_0(1 - 2\pi i \mathbf{k} \cdot \mathbf{r}) \operatorname{grad} \psi_f d\tau|^2. \tag{2}$$

If ψ_f is a p function, the integral of the term in (2) that involves $\mathbf{k}$ vanishes, leaving

$$|\textstyle\int \chi_0 \operatorname{grad} \psi_f d\tau|^2, \tag{3}$$

which is independent of $\mathbf{k}$. This result should be valid for fairly large values of $\mathbf{k}$ since ψ_f is localized exceedingly close to the nucleus. Thus, the intensity of the emission band as a function of energy should depend only upon the density of levels and, as a result, should vary as $\sqrt{\epsilon}$ near the low-energy side. On the other hand, if ψ_f is an s function, the term corresponding to (3) vanishes and the remaining term can be reduced to

$$\mathbf{k}^2 \left| \int \chi_0 x \frac{\partial \psi_f}{\partial x} \right|^2, \tag{4}$$

which varies as $\mathbf{k}^2$ or as ϵ near the low-energy side of the band. Since the density of levels varies as $\sqrt{\epsilon}$ in this energy region, it follows that the intensity of the band should vary as $\epsilon^{\frac{3}{2}}$.

Now, the L_{III} emission curves for sodium, aluminum, and magnesium correspond to the first of these two cases. It is clear from Fig. 16 that the intensity starts out much more slowly than $\sqrt{\epsilon}$. A comparison of

[1] Houston, *op. cit.*; H. Jones, N. F. Mott and H. W. B. Skinner, *Phys. Rev.*, **45**, 379 (1934).

the actual curve and the $\sqrt{\epsilon}$ curve that would fit the sodium curve[1] most closely is given in Fig. 17a. This suggests that there is a residue of the excitation bands; however, it is also possible, although not probable, that the measured tail is due to a background emission of impurity atoms. Since the K emission curves of Fig. 16 correspond to the second of the two cases, they should rise as $\epsilon^{\frac{3}{2}}$. Figure 17b gives a comparison of an $\epsilon^{\frac{3}{2}}$ curve and the observed curve for lithium. The high-energy cutoff of the theoretical curve is chosen so that the width is the same as that computed in Sec. 78. The discrepancy on the low-energy side may have the explanation suggested in Fig. 17a for sodium. The discrepancy on the high-energy side indicates that the actual density of levels does not

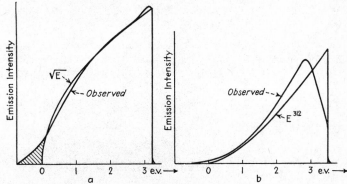

FIG. 17.—(a) Comparison of the actual emission band of sodium and that expected from the simple band theory. It is suggested that the shaded region represents the contribution from vestiges of the excitation bands. The energy scale is in electron volts. (b) Comparison of the actual emission band of lithium and that predicted by the band theory.

vary as $\sqrt{\epsilon}$ in lithium. The premature peak in the curve for lithium has not received a satisfactory explanation. It is possible that it is related to a rapid variation in exchange and correlation energy at the top of the filled region; however, if this is the case, it is not easy to see why sodium does not have a similar peak. If the difference between the curves for lithium and sodium is real, we may expect the electronic specific heat of sodium to be more normal than that of lithium, for its level density is more nearly like that for perfectly free electrons.

It is interesting to note that the beryllium curve of Fig. 16 behaves as though the levels of a single zone were almost completely occupied, a fact indicating that this metal is very nearly an insulator. In mag-

[1] In these comparisons, the effect of both exchange and correlation on the calculated band widths is neglected because the second, which tends to compensate for the broadening effect of the first (see footnote 1, p. 422) is not precisely known.

nesium and aluminum, however, the curves appear as though two zones overlapped extensively.

TABLE LXXIV.—OBSERVED AND CALCULATED WIDTHS OF THE SOFT X-RAY EMISSION
BANDS (AFTER SKINNER)

(The theoretical values are those computed in Chap. X.   Values in parentheses are
free-electron values)

	Observed, ev	Calculated, ev
Li	4.1 ± 0.3	3.4
Na	3.4 ± 0.2	3.2
Be	14.8 ± 0.5	(13.8)
Mg	7.6 ± 0.3	(7.2)
Al	13.2 ± 0.5	(12.0)

The widths of a number of the emission bands determined by O'Bryan and Skinner, and others, are given in Table LXXIV and are compared with the theoretical values in cases in which the latter have been computed.   The values

$$\frac{h^2}{2m}\left(\frac{3n_0}{8\pi}\right)^{\frac{2}{3}}$$

for perfectly free electrons are given in parentheses in the other cases.

The emission bands of metals containing filled or partly filled d bands have been investigated by a number of workers,[1] among whom are Bearden, Shaw, Beeman, Friedman, Saur, Gwinner, and Farineau. Farineau's curves for nickel, copper, and zinc are shown in Fig. 18.   Since the density of levels in the d band presumably is much higher than that in the s-p band, we may conclude that practically all of this structure arises from the d band.   One of the important features of these curves is the fact that there is a single peak in the cases of copper and nickel and not two, as we might expect from Krutter's work.   Krutter's curve is shown in the diagram for copper.   If this peak were to be associated with

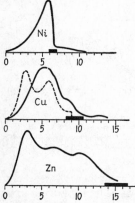

FIG. 18.—The emission curves for nickel, copper, and zinc. Presumably this emission arises mainly from the d band. The difference between copper and nickel is mainly due to the fact that copper has one more valence electron than nickel. Zinc, however, has an entirely different structure so that the band structure should be different.

[1] J. A. BEARDEN and C. H. SHAW, *Phys. Rev.*, **48**, 18 (1935); W. W. BEEMAN and H. FRIEDMAN, *Phys. Rev.*, **56**, 392 (1939); E. SAUR, *Z. Physik*, **103**, 421 (1936); E. GWINNER, *Z. Physik*, **108**, 523 (1938).

Pauling's B type of d band (see Fig. 10), we should expect that it would occur nearer the high-energy cutoff of the curves, at least in the case of nickel. Assuming that these density curves are trustworthy, we must conclude both that the d bands shift to some extent in going from copper to nickel and that the one-electron approximations discussed in previous sections are not very accurate when applied to d electrons. The curves for copper and zinc are considerably different; however, this fact is not surprising, for the crystalline symmetry is different in the two cases.

B. IONIC CRYSTALS

105. Plan of Treatment.—In the preceding chapter, we saw: (a) that the lowest state of ideal ionic crystals may be treated approximately by either the Heitler-London or the band scheme; (b) that the lower, non-conducting, excited states may be treated by the method of excitation waves, at least in the case of the alkali halides; and (c) that the higher excited states may be treated by the band scheme. In the following sections, we shall apply these approximations to several alkali halide crystals and alkaline earth oxide and sulfide crystals the experimental properties of which have been investigated with some degree of completeness. The first two sections apply to crystals having ideal, undistorted lattice structures, and the following section applies to ideal crystals having lattice distortions of a type that will be described in more detail in that section.

106. The Alkali Halides.—Zone-structure calculations have been made for two alkali halides, namely, lithium fluoride and sodium chloride. In addition, computations have been made for lithium hydride, which resembles the alkali halides closely since negative hydrogen ions behave like ions of a halogen. In an atomic picture, the eight valence electrons per unit cell of the alkali halides completely occupy the outer s and p shells of the negative ions. In the band scheme, the same electrons occupy four zones, one of which connects adiabatically with the ionic s level and the three others of which connect with the ionic p level. Figure 19 illustrates the manner in which the ionic levels broaden into the bands of the zone theory as the ions are brought together. The levels of the negative ions are depressed and those of the positive ions are raised because of the Madelung field. In addition, the levels break into bands when the ions begin to overlap. At the observed lattice distance, the s and p bands are separated from one another and from the higher unfilled band which connects with the lowest level of the metal ion. The ionic levels do not split into bands in the Heitler-London approximation but remain discrete, roughly following the center of gravity of the bands (*cf.* the discussion in Sec. 64 concerning the connec-

tion between the Heitler-London and Bloch energy parameters). If
the squares of the 2s wave function of atomic lithium and of the radial
part of the p wave function of ionic fluorine are plotted in such a way
that each distribution is centered about points which are separated by
the distance between the lithium and fluorine ions in the crystal, it is
found that the peaks of the distributions overlap and that the center of
gravity of the charge of an undistorted lithium atom would lie in the
shells of the surrounding halogen ions (*cf.* Fig. 1, Chap. II, for lithium
hydride).

In treating sodium chloride, Shockley[1] chose an effective chlorine
charge distribution by normalizing Hartree's chlorine ion wave functions

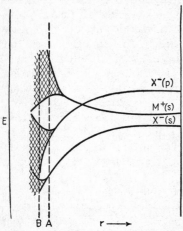

within a sphere the volume of which
is equal to that of the unit cell of
sodium chloride. He assumed that
only eight electrons are in this cell
at any one time, because of correlation effects, and computed the effective field for a given electron in the
sphere by taking the charge of the remaining seven into account. The
effective field for an electron near a
sodium ion was taken as the ion-core
field used in the computations on
metallic sodium. To these effective
ion fields, Shockley added the
Madelung field of the surrounding
ions. In addition, he subtracted
from the sodium ion-core field the
field arising from a single electron
that is spread uniformly over the six
surrounding halogen ions. The missing electron is assumed to be at the

Fig. 19.—Schematic representation
of the manner in which the ionic levels
of the constituents of ionic crystals
break into bands in the band approximation. At *A* the *s* and *p* bands of the
negative ion are separate, whereas at *B*
they overlap.

sodium ion. Boundary conditions were applied to the wave functions
in three different ways: (1) by neglecting the sodium wave functions
and treating the lattice as though it were composed of chlorine ions, (2)
by satisfying only sodium-chlorine boundary conditions, and (3) by
satisfying both sodium-chlorine and chlorine-chlorine boundary conditions. The equations employed in the first case were those derived by
Krutter for copper, and Shockley derived similar equations for the two
other cases. The $\epsilon(\mathbf{k})$ curves that were obtained from these equations are
shown in Figs. 20 *a,b*.

[1] W. Shockley, *Phys. Rev.*, **50**, 754 (1936).

More recently, Tibbs[1] has investigated the conduction band of sodium chloride by a similar method and has shown that the effective mass of the conduction electrons is close to unity.

Attempts to construct self-consistent fields for the lattice were made in the treatments of lithium fluoride and hydride.[2] This task is much more difficult in the case of diatomic crystals than in the case of mona-tomic ones, because the total charge in each polyhedron of the unit cell, as well as the relative distribution within a given polyhedron, must be the same in the initial and final solutions. In these ionic crystals, the lattice was divided into cubes of equal volume centered about each of the ions, and the cubes were replaced by equivalent spheres. The Li^+

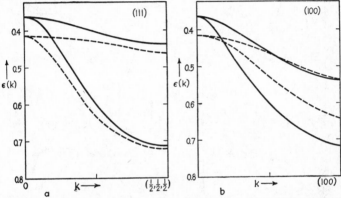

Fig. 20.—(a) $\epsilon(\mathbf{k})$ curve for the (111) direction of sodium chloride. Only the p-band curves are given. The full curve corresponds to the results obtained by neglecting the sod-ium wave functions; the dotted lines correspond to the most accurate procedure described in the text. (b) Same for the (100) direction. (After Shockley.)

ion-core field was taken from the work on metallic lithium, and the field of the $(1s)^2$ ion core of F^- was taken from Hartree's work. The remain-ing eight electrons per unit cell were treated by a self-consistent method. The charge distribution of the valence electrons was not determined by computing wave functions for all values of $\mathbf{k}$ and taking an average. Instead, it was assumed that the average of the four wave functions associated with $\mathbf{k} = 0$ is the same as the average distribution of all electrons. This approximation is justified by the fact that the mean charge distribution in the alkali metals is practically the same as the distribution for $\mathbf{k} = 0$. Boundary conditions were satisfied at different points of the polyhedra for several different values of $\mathbf{k}$, group theoretical

[1] S. R. Tibbs, Trans. Faraday Soc., **35**, 1471 (1939).
[2] D. H. Ewing and F. Seitz, Phys. Rev., **50**, 760 (1936).

methods being used to choose the appropriate combination of zonal harmonics in each case.

Figure 21 shows the agreement between initial and final charge distributions in the final trials for lithium fluoride, and Figs. 22 and 23 give plots of the s- and p-band functions for $\mathbf{k} = 0$ for both lithium fluoride and lithium hydride.

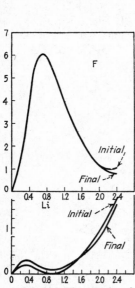

Fig. 21.—The initial and final charge distributions in the Li and F polyhedra in the last computation for the self-consistent field in lithium fluoride. The abscissae are in Bohr units. (*After Ewing and Seitz.*)

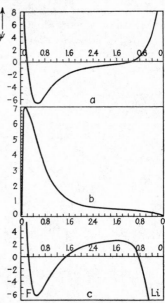

Fig. 22.—Wave functions for $\mathbf{k} = 0$ in lithium fluoride. a is the s-band function; b is the p-band function and c is the function for the first unoccupied band. It should be noted that a and b are distributed principally about the F^- ion, whereas c is distributed about both ions. The abscissae are in Bohr units.

The integral of the charge distribution inside the lithium sphere of lithium fluoride is $0.95e$ in the Hartree approximation. This result is unquestionably too large since neither exchange nor correlation effects were taken into account. A somewhat better value might have been obtained by including these terms in the way Shockley did, namely, by excluding one unit of electronic charge in determining the field inside the fluorine sphere. A rough estimate shows that this procedure would probably reduce the charge in the lithium sphere to about $0.5e$, which, in turn, would leave about 0.05 valence electron in the sphere the radius of which is equal to the classical lithium-ion radius. This estimate

furnishes a good justification for the Born-Mayer approximation in which
the charge in this sphere is assumed to be zero, for the correction to the
Madelung energy would be only
about 5 per cent, which is less than
the importance of repulsive terms.

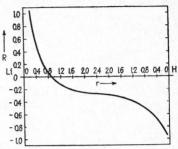

Figures 24 and 25 show the $\epsilon(\mathbf{k})$
curves for several important crystallographic directions.[1] The values
corresponding to the dots are the
actual computed cases, whereas the
full lines were obtained by interpola-
tion. The upper curve of the second
band for lithium fluoride is doubly
degenerate in the (100) and (111)
directions. This degeneracy also appears in Shockley's results.

FIG. 23.—The wave function for
$\mathbf{k} = 0$ in the filled band of lithium
hydride. The abscissae are in Bohr
units. (*After Ewing and Seitz.*)

One important feature of the zone scheme is the fact that the uppermost filled bands are several volts wide. Although these values probably
are too large, because Hartree fields were used in obtaining them, their
order of magnitude seems unquestionably to be correct.

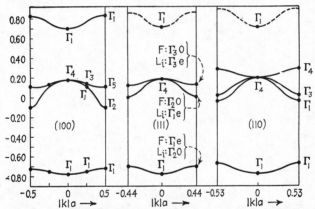

FIG. 24.—$\epsilon(\mathbf{k})$ curves of lithium fluoride for three prominent crystallographic directions.
The Greek letters are crystallographic term symbols; a is the interionic distance. The
lower two band systems are filled; the upper band is empty. The energy is expressed in
Rydberg units. (*After Ewing and Seitz.*)

In summarizing this discussion of the normal states of the alkali
halides, we may say that the charge distribution in the lattice is very

---

[1] It is interesting to note that the s curves increase with increasing $|\mathbf{k}|$, whereas the
p curves tend to decrease. This fact was pointed out in Sec. 65 on the basis of the
narrow-band approximation.

nearly the same as if the crystal were composed of free positive and negative ions. The lowest s band of the zone picture is very narrow so that the halogen s shell is not appreciably perturbed. The width of the p band is of the order of 1 ev, which indicates that neighboring ions overlap appreciably and that the exchange interaction is of the order of 1 ev. The magnitude of the band width also indicates that the effective mass of a free hole in the p band is comparable with the mass of an electron.

The wave functions for $\mathbf{k} = 0$ in the first unoccupied zone of lithium

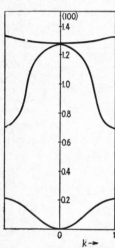

Fig. 25.—Same as Fig. 24 for the (100) direction in lithium hydride. There is only one filled band in this case so that the upper two bands are empty. These bands correspond to the occupied p bands of lithium fluoride (Fig. 24). The energy is expressed in Rydberg units.

fluoride are shown in Fig. 22. Here, the electronic charge is distributed more or less uniformly between the positive and negative ions. The closest vertical distance between the filled and unfilled bands of Fig. 24 is 7.5 ev for the end point of the zone in the (110) direction. This difference should be the energy required to induce photoconductivity in the pure crystal, and it should be greater than the first absorption energy. Actually it is less, since the fundamental peak of lithium fluoride occurs below 1000 Å, which makes the absorption peak greater than 12 ev. The Hartree approximation, can be blamed for this discrepancy, for the exchange terms would lower the filled band much more than the empty one. This tendency for the exchange energy to be smaller in magnitude for an excited electron than for a normal one is shown for perfectly free electrons in Fig. 5, Chap. IX.

We have seen in Sec. 95, Chap. XII, that the position of the nonconducting excited levels of the alkali halides can be estimated fairly closely by the use of an atomic model. According to this work, the first excitation band should lie about 12 ev above the lowest state in lithium fluoride, a value that agrees closely with the threshold absorption frequency for lithium fluoride. The complete structure of the first ultraviolet absorption bands of lithium fluoride and sodium chloride has not been measured. In a typical case such as that of sodium bromide, illustrated in Fig. 26, it seems natural to assume, in analogy with the absorption spectra for atoms, that the peaks A, B, C, and D correspond to transitions to excitation levels and that the ionization edge falls in the short wave length foot of the band, that is, at about 1200 Å. The corresponding points for lithium fluoride and sodium chloride (*cf.* Fig. 2 of Chap. XII) undoubtedly lie at energies greater than 14 ev and 12 ev, respectively.

107. Alkaline Earth Oxides and Sulfides.—Although the cellular method has not been applied to any of the alkaline earth salts, it is possible to draw several plausible conclusions about their zone structures by indirect reasoning. The interatomic exchange terms, which give rise to most of the repulsive forces in ionic crystals and arise mainly from the metal-ion negative-ion interactions, are about four times larger in the oxides and sulfides than in the halides. Since these exchange energies are closely related to the widths of the occupied bands (*cf.* Sec. 64), we may expect that the bands are much broader in the oxides and sulfides. Referring to Fig. 19, we may expect that the point B where

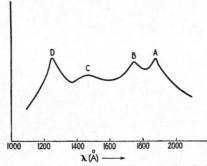

Fig. 26.—The structure of the first ultraviolet absorption band of sodium bromide.

the s and p bands are very close corresponds to the alkaline earth oxides and sulfides, if the point A corresponds to the alkali halides.[1]

It is possible to construct reasonable energy-level diagrams for some of the alkaline earth salts by use of energy-level data derived from the Born cycle and from spectroscopic measurements on the free ions. As examples, we shall take zinc oxide and zinc sulfide, which have properties that are typical of other members of this group of salts. The crystal structure of zinc oxide is the wurtzite lattice, which is also the high-temperature structure of zinc sulfide. The low-temperature form of zinc sulfide has the zincblende lattice which is similar to diamond.

We shall begin by considering the energy necessary to remove an electron from a free negative ion and place it on a free zinc ion. The

---

[1] This overlapping of s and p bands occurs in diamond, as is shown in Sec. 109. Hence we may expect it to be associated with some valence characteristics. H. M. James and V. A. Johnson (*Phys. Rev.*, **56**, 119 (1939)) have shown, in fact, that the charge distribution in zinc oxide is not perfectly ionic.

total electron affinities of O^{--} and S^{--} have been determined approximately by the Born cycle and are about -7 and -4 ev, respectively. Lozier[1] determined experimentally the affinity of neutral oxygen for one electron and found that it is 2.2 ± 0.2 ev, which shows that the negative affinity of oxygen for two electrons is due entirely to the second electron and that the energy necessary to remove one electron from O^{--} is about -9 ev. The energy level of O^{--} is plotted[2] relative to the normal state of O^{-} in the right-hand column of Fig. 27. Since O^{--} and S^{--} have about

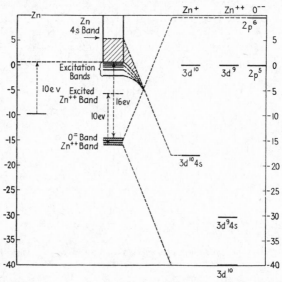

Fig. 27.—The ionic energy-level relations in zinc oxide. The levels of Zn^{+}, Zn^{++}, and O^{--} are given on the right. The behavior of these when a lattice is formed is shown by the dotted lines. The relative position of the levels in a neutral zinc atom is given on the left. (See footnote 9 on page 450).

the same classical radii as F^{-} and Cl^{-}, respectively, we may conclude that the electronic structure of the first pair of ions is very similar to that of the second pair.

Now, the affinities of the halogens decrease as one passes down the periodic chart. For this reason, we shall assume that the affinity of a sulfur atom for a single electron is about 1 ev less than that of an oxygen ion, which makes the energy of S^{--} about 5 ev relative to the energy of S^{-}. This is indicated in the right-hand column of Fig. 28. The energy levels of zinc ions have been measured spectroscopically and are given in the second and third columns from the right in Figs. 27 and 28. Since the

[1] W. W. Lozier, *Phys. Rev.*, **46**, 268 (1934).

[2] F. Seitz, *Jour. Chem. Phys.*, **6**, 454 (1938).

energy of Zn^+ relative to Zn^{++} is 17.9 ev, the energy required to remove an electron from a free oxygen ion and place it on a free zinc ion is -26.9 ev.   The same quantity is -22.9 for S^{--} and Zn^{++}.   Let us now arrange the ions in the lattices of ZnO and ZnS and gradually decrease the lattice constants from infinity.   The lattice potential at the position of the positive ions is negative, and the potential at the negative ions is positive, so that the levels of the negative ions are depressed and those of the positive ions are raised during this process.   This change is indicated by the dotted lines of Figs. 27 and 28, which show the total shift as computed from the Madelung potentials of the zinc oxide and zinc sulfide lattices.

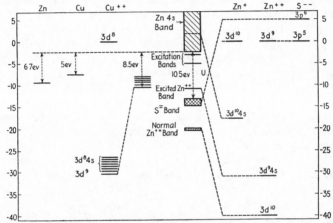

FIG. 28.—Same as Fig. 27 for zinc sulfide.   The positions of levels of neutral zinc and copper and of Cu^{++} are given on the left.   (See footnote 1 on page 450.)

At the actual interatomic distances, the discrete levels of the free ions broaden into bands characteristic of band structure, and excitation bands appear[1] below the first unoccupied zone, which is adiabatically connected with the lowest level of free Zn^+.   The energy required to remove an electron from a negative ion and place it on a zinc ion at infinity, as found from these diagrams, is about 16 ev for zinc oxide and 10.5 ev for zinc sulfide.   This transition evidently corresponds to ionization.   The energy required to carry the electron from a negative ion to a near-lying zinc ion is less, of course, as is indicated in the figures. The first absorption band of zinc sulfide has been measured roughly, and the experimental work seems to show that the first excitation band should be about 6 ev above the ground state.   The value of the absorption energy obtained from Fig. 28 is somewhat larger than this.   Figure

---

[1] It is possible, however, that the excitation levels may merge with the ionization levels.

27 indicates that the first absorption band of zinc oxide may correspond to internal excitation of the zinc ion, that is, to the transition

$$Zn^{++}(3d^{10}) \rightarrow Zn^{++}(3d^94s).$$

If this is true, the first absorption peak of zinc oxide should lie near 1200 Å. There does not seem to be any work on the ultraviolet absorption properties of this salt.

Diagrams[1] such as Figs. 27 and 28 may be constructed for any ionic crystal for which the necessary data are known. Some additional uses for these diagrams will be found in Sec. 113.

108. Equilibrium Atomic Arrangements for Excited States.—The orderly lattice arrangement of atoms in crystals that is determined by X-ray analysis is the equilibrium distribution for the lowest electronic state. There is no reason for expecting the same distribution to be stable for excited electronic states of insulators. In fact, there is reason to expect the opposite, for each excited electronic state of diatomic or polyatomic molecules has its own equilibrium atomic arrangement. Since the dependence of excited electronic levels on atomic arrangement has not been investigated in a detailed and quantitative way, we shall have to be contented with qualitative pictures.[2]

Let us consider an ideal ionic crystal, such as one of the alkali halides or one of the alkaline earth salts. If we neglect thermal effects, the atoms occupy lattice sites in the normal electronic state. Suppose that we now use electrons or light quanta to excite the crystal to a higher electronic level. An excited electron and a hole are then produced, and the two should move together if the excited state is not conducting. The crystal still has the equilibrium atomic arrangement of the lowest level immediately after electronic excitation because of the Franck-Condon principle. Now, as we saw in the last chapter, the excited levels occur in systems of quasi-continuous bands, each level of which corresponds to an exciton moving with a definite velocity. If the exciton is produced by optical absorption, it usually is moving slowly, because the selection rules forbid transitions in which the wave-number vector of the exciton lies very far from the center of the zone and because the group velocity grad_k $\epsilon(k)/h$ is zero when the wave number is zero. This selection rule is not valid, of course, if the excitation is induced by means of cathode rays or alpha particles, which have appreciable momentum; hence, the exciton may move more swiftly in these cases. If the exciton is regarded as an

---

[1] In both Figs. 27 and 28, the positions of levels of neutral atoms are represented on the left. The ionization energies of these atoms should be altered because of polarization effects such as those discussed in Sec. 112.

[2] Cf. A. VON HIPPEL, Z. Physik, **101**, 680 (1936). F. SEITZ, op. cit., p. 150; Trans. Faraday Soc., **35**, 74 (1939).

excited ion, it is easy to see that the lattice near it is under stress for the normal atomic arrangement, for an excited ion and a normal ion usually interact differently with their neighbors. These stresses would set the excited atom into oscillation about a new equilibrium position if the exciton were permanently at rest. If it is moving even slowly, however, the atoms near the exciton may not have time to move very far during the short time that the exciton is near. For example, the time required for an exciton that is moving at 10^6 cm/sec to traverse a distance of 10^{-8} cm is 10^{-14} sec, and the time required for an atom to make one oscillation is about ten times this. We may expect, however, that some lattice vibrations about the normal atomic positions are stimulated and that the exciton is slowed because it dissipates this vibrational energy. Thus, the exciton should eventually drop to the lowest excited energy state, and it should ordinarily be at rest in this state because $\mathrm{grad_k}\ \epsilon(\mathbf{k})$ is zero at the lowest point of the exciton band in simple crystals (*cf.* Sec. 96 of the previous chapter). The atom on which the exciton comes to rest should be set into violent vibrational motion because of the stresses mentioned above. This kind of motion is strongly damped since the atoms are strongly coupled. Thus, the localized vibrational energy should be dissipated during a time of the order of 10^{-13} sec, which is the atomic oscillation period, by the production of elastic waves that radiate from the vibrating atom.

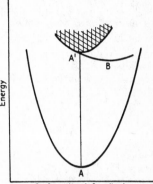

FIG. 29.—Schematic representation of the possible positions of normal and excited states of an insulator as functions of atomic coordinates. The lowest curve corresponds to the normal state and its minimum determines the normal atomic arrangement. The cross-hatched region corresponds to the levels of moving excitons. Their minima are at position *A* because the lattice cannot come to complete equilibrium when the excitons move. The discrete level that separates from this band represents the state of an exciton at rest. Its minimum is not at *A*, because the interactions between normal and excited atoms are different.

The possible disposition of the normal and excited levels of the crystal is shown symbolically in Fig. 29. The abscissa represents the configurational coordinates of the lattice, that is, the interatomic distances, and the ordinate is the energy of the crystal. The lowest level corresponds to the normal electronic state so that its minimum *A* corresponds to the normal atomic arrangement of the lattice. The second curve represents the lowest excited state, in which the exciton is at rest. The minimum *B* corresponds to the values of the configurational coordinates for which the excited atom and its neighbors are at equilibrium. The quasi-continuous band of levels represents the excited states in which the exciton is in motion. These levels have their minima at the same point as the normal state because the stresses are not localized when the exciton is in motion. During excitation, the system jumps from A to A'; as soon as the exciton comes to rest, the state of the system "slides" toward B with the emission of elastic waves.

There are two conceivable arrangements of the normal and excited levels. In the first (Fig. 30a), the excited state has its minimum B within the minimum of the lower curve. In the second case (Fig. 30b), the minimum E is outside the lower curve. The system slides to B in the first instance and may then jump to the point C on the lower curve by emitting a light quantum. It slides to D in the second case

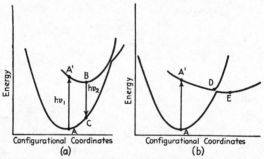

Fig. 30.—Possible arrangements of the normal level and the level of an exciton at rest. In a the minimum of the excited level is inside the lower curve so that fluorescent emission of frequency ν_2 may follow absorption of ν_1. In case b the minimum of the excited level is outside that of the normal level. Hence, fluorescence cannot occur.

and may then slide either to A or to E. All the energy is dissipated immediately in the form of lattice waves in the first case. The system may rest at E indefinitely in the second, storing a part of the excitation energy. At temperatures above absolute zero, the system oscillates about E and should eventually pass over D and down to A. The crystal may be fluorescent in the case of Fig. 30a, since some of the absorbed energy may be radiated as light; however, it is not fluorescent in the second case. Fluorescence has never been observed[1] unambiguously to accompany absorption in the fundamental bands of ideal crystals, which indicates that the second case occurs commonly.

C. VALENCE CRYSTALS

109. The Zone Structure of Diamond.—In spite of the importance of the subject, practically no quantitative work has been done on valence

[1] See *ibid.*

binding in crystals.   The reason for this lack, as we mentioned in Sec. 97, is that the simplest atoms entering into valence crystals have so many valence electrons which are appreciably affected by the crystalline binding that the computations are more complicated than for simple metals or salts.   The existing work consists of a semiquantitative investigation of the zone structure of diamond and a qualitative discussion of the most appropriate form of the Heitler-London functions for the atoms in valence crystals with a tetrahedral arrangement of nearest neighbors. Carbon and silicon are the principal atoms to which these considerations apply.

The fact that the lowest state of atomic carbon is degenerate and that diamond is an insulator shows that the energy levels of the entire solid vary very much as the atoms of the lattice are brought together.   The lowest level is highly degenerate at infinite separation, and it must broaden into a dense band just as the lowest levels of metal atoms do.   Since the crystal actually is an insulator, we must conclude that a single level separates from this dense array and is the lowest level at the observed interatomic distance (*cf.* Fig. 31).   A similar situation occurs in ionic crystals, for the highly degenerate state corresponding to free neutral atoms is more stable than the state of free ions at infinite separation; in this case, however, the two states cross before appreciable splitting occurs, since the Madelung energy favors an ionic state.

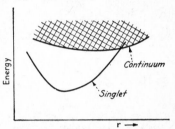

Fig. 31.—The dependence of the levels of diamond on interatomic distance (schematic). At large distances the lowest state is highly degenerate; however, a nondegenerate discrete level separates at smaller distances.

For this reason, the situation is much easier to understand in ionic crystals.   A simple picture of the same type has not been developed for diamond, although the separation of the singlet level can be shown to occur in the band approximation, as will be seen below.   In fact, the separation occurs for such a simple carbon field that we may expect the effect is determined primarily by the crystal structure, that is, is connected with the way in which wave functions are diffracted by the diamond lattice.

   a.   The Band Approximation.—Kimball,[1] who is responsible for a semiquantitative zone treatment of diamond, satisfied boundary conditions at the center points of the four faces of the atomic polyhedron, which is shown in Fig. 3, Chap. IX.   There are two polyhedra of this type in the unit cell.   The boundary conditions, which were taken as the continuity of ψ and its normal gradient at these points, require that ψ be

[1] G. E. KIMBALL, *Jour. Chem. Phys.*, **3**, 560 (1935).

expanded in terms of four surface harmonics. Kimball took these to be one s function and three p functions, the field that was used to obtain the radial parts being the one derived by Torrance for the $2s$ and $2p$ functions of atomic carbon. The energy bands are shown[1] in Fig. 32. The atomic $2s$ level splits into two zones since there are two atoms per unit cell, whereas the atomic $2p$ function splits into six zones. These two systems of zones overlap at $r = 2.7a_h$ and then split into two new systems which contain four zones each. Two of the lower set of zones have zero width in Kimball's approximation, but they probably would

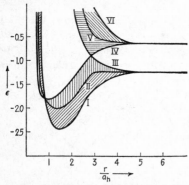

FIG. 32.—The band structure of diamond. It should be noted that the s-p bands overlap and break into two separate systems. From the standpoint of the entire solid this corresponds to the behavior of Fig. 31. (*After Kimball.*)

have finite width if more boundary points and surface harmonics had been used. Since the lower zone system is just exactly filled by the eight electrons per unit cell, the crystal is an insulator at the observed lattice distance in Kimball's approximation. There is no isolated low-lying group of four zones at large distances, however, so that the crystal should be a metal when r is greater than $2.7a_h$. From the standpoint of the entire crystal, this means that a single level separates from a continuum when r is $2.7a_h$.

Kimball found that his initial and final charge distributions were not the same, which shows that his starting field was inaccurate. That he still obtained the separation of bands which is needed to make diamond an insulator suggests, as we mentioned above, that this separation is determined primarily by the crystal structure. Thus, it is likely that carbon would be a metal if the atoms were placed together in one of the simple close-packed lattices.

The difference between the filled and unfilled bands of Fig. 32 is 7 ev, which corresponds to an absorption peak near 1700 Å. It is probable, however, that excitation levels lie between these bands. The energy-level splitting is so large that the atomic-perturbation method of Sec. 96 cannot be used as an argument in favor of these levels. Instead, they should be treated by a more general method, such as that suggested by Wannier.

b. The Heitler-London Approximation.—The properties of tetravalent carbon in saturated hydrocarbon compounds suggest to the chemist that

[1] A similar figure has been derived by F. Hund, *Physik. Z.*, **36**, 888 (1935), on the basis of simpler reasoning.

carbon has tetrahedral directional properties. For this reason, he assigns a tetrahedral bond structure to carbon and assumes that this atom prefers to join with other carbon atoms or with hydrogen atoms along these bonds. The structure of diamond supports this viewpoint since each carbon atom in it is surrounded tetrahedrally by four other carbon atoms.

Pauling[1] and Slater[2] have independently established a set of principles that may be used to understand this tetrahedral character. Let us consider molecular hydrogen for a moment. As we have seen in Sec. 56, the stability of this molecule arises from the following two facts: (1) The field between two protons is stronger than the field of one. (2) Two electrons may share this region and minimize their repulsive energy by correlating their motion so that they are not there at the same time. When the binding is largest, the atomic distributions are distorted in such a way that the wave functions extend along the line of centers, where the field is largest. On the basis of results such as this, Pauling and Slater suggest that the observed atomic arrangements in valence compounds are those for which the Heitler-London functions have the largest peaks along the line of centers. In applying this principle to carbon, they assume that the carbon bonds are so strong that the $2s$ and the $2p$ electrons should be treated on an equal footing. Kimball's results support this assumption, for in his model the s levels split into bands that are as wide as the p bands and the two types of state become thoroughly mixed. Pauling found by direct computation that the four orthogonal functions which are linear combinations of one s function and three p functions and which have maximum directional localization

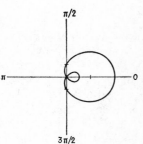

Fig. 33.—Polar plot of the Pauling bond function (1) for carbon.

extend toward the four corners of a tetrahedron. The equation of any one of these may be placed in the form

$$f(r)(\tfrac{1}{2} + \tfrac{3}{2}\cos\theta) \tag{1}$$

where $f(r)$ is a radial function and θ is the polar angle measured from the directional axis. The angular part of (1) is shown in Fig. 33.

It is evident that the Slater-Pauling principle cannot be rigorous, for carbon also forms a stable crystal in which the coordination is not tetrahedral, namely, graphite.

[1] L. Pauling, *Jour. Am. Chem. Soc.*, **53**, 1367 *ff.* (1931).
[2] J. C. Slater, *Phys. Rev.*, **37**, 481 (1931).

The electronic distribution that can be derived from Kimball's model undoubtedly shows directional properties similar to those of Pauling's tetrahedral functions.   Since neither scheme has been used to make a quantitative computation of the cohesive energy of diamond, it is not possible to say which would give a better binding energy.

The same principle may be applied in discussing other simple valence bonds such as those that occur between silicon atoms in solid silicon or between silicon and oxygen in silica.   The second case, in which each silicon atom is surrounded by four oxygen atoms and each oxygen atom by two silicon atoms, is complicated somewhat by the fact that oxygen, in place of having two electrons available for binding, actually lacks two electrons from a complete p shell.   As we have seen in the previous discussion on solids, these holes may be treated like positively charged electrons.   Hence, we may say that the silicon-oxygen bond in silica occurs between a directed valence electron of silicon and a directed hole of oxygen.   We should expect these bonds to be polar because holes behave like positive charges.

D. SEMI-CONDUCTORS

110. General Principles.—There are two types of semi-conductor, namely, those which contain impurities and those which do not.   Many polar salts, such as zinc oxide, belong to both classes.   We shall be principally interested in pure semi-conductors, since their experimental properties have been studied more systematically than those of impure semi-conductors.

The principles determining the electronic conductivity of pure salts that have been given appropriate heat-treatment were first recognized and developed by Schottky and Wagner.[1]   They pointed out, for example, that the electronic conductivity of pure zinc oxide can be associated with interstitial zinc atoms the presence of which may be understood in terms of ordinary principles of statistical mechanics.   We shall present their work in a form that is modified in keeping with the treatment of statistical mechanics used in this book.

There are two main divisions of pure semi-conductors,[2] namely, those which conduct by free electrons and those which conduct by holes. These two types may be distinguished experimentally by the sign of the Hall coefficient; for the first class has the normal sign, that is, the same sign as the alkali metals and bismuth, and the second has opposite sign. It is easy to construct a system of electronic energy levels that explains

[1] W. Schottky and C. Wagner, *Z. physik. Chem.*, **11B**, 163 (1930).   C. Wagner, *Z. physik. Chem.*, **22B**, 181 (1933).

[2] See, for example, the review by B. Gudden, *Ergebnisse exakt. Natur.*, **13**, 223 (1934); also Chap. I of this book

qualitatively the difference between these two types. Consider, for example, a typical insulator having the system of filled and unfilled levels shown in Fig. 34a. If we add foreign atoms or distort the atoms in another way, we may expect to introduce new electronic levels in the forbidden regions. The electronic charge associated with these states is localized about the distortion or impurity atom.[1] The detailed properties of the additional levels vary from case to case and should be discussed separately in each one. If all the discrete levels are occupied, as in Fig. 34b, the electrons near the conducting bands may be thermally excited to this band, thereby making the crystal an electronic conductor. We discussed the properties of this type of semi-conductor in Chap. IV

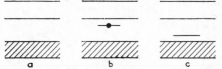

FIG. 34.—(a) The filled and unfilled levels in an insulator. (b) The discrete impurity level is occupied by an electron. This substance may be an electronic semi-conductor if the electron may be thermally excited to the empty band. (c) The level is unoccupied. This substance may be a hole semi-conductor with an 'anomalous' Hall coefficient, if electrons may be thermally excited from the filled band to this level leaving free holes in the lower band.

and found that the low-temperature conductivity σ should vary according to the equation

$$\sigma = n_b^{\frac{1}{2}} \frac{4\sqrt{2}}{3} \frac{e^2 l_0}{h^{\frac{3}{2}}} (2\pi m^* kT)^{\frac{1}{4}} e^{-\frac{\Delta\epsilon}{2kT}} \tag{1}$$

where n_b is the number of bound states per unit volume, l_0 is the mean free path, m^* is the effective mass of the free electrons, and $\Delta\epsilon$ is the activation energy for freeing the electrons (*cf.* Fig. 35). The sign of the Hall coefficient is normal in this case. If the discrete electronic levels are unoccupied (Fig. 34c), the electrons in the filled region may be thermally excited to the lowest unoccupied level, leaving free holes in the band. These holes should behave like free positively charged electrons and should give the solid a Hall coefficient whose sign is opposite to that of the preceding case. Equation (1) should also be valid in this case if the constants n_b, l_0, m^*, and $\Delta\epsilon$ are reinterpreted in terms of free and bound holes. We shall discuss the electronic levels of several solids in more detail below.

[1] The behavior of models of this type has been discussed by A. H. Wilson, *Proc. Roy. Soc.*, **133**, 458 (1931); **134**, 277 (1931); R. H. Fowler, *Proc. Roy. Soc.*, **140**, 505 (1933); **141**, 56 (1933).

At temperature T, the equilibrium state of an insulator is determined by the condition that its free energy

$$A = E - TS$$

be a minimum.   This quantity is simply E at zero temperature so that the equilibrium condition presumably demands that the ideal crystalline arrangement be most stable at this temperature, since E probably is then a minimum.   This arrangement need not be the most stable above the absolute zero; for if the entropy associated with a distorted arrangement, such as that caused by placing some atoms in interstitial positions or by removing normal atoms, is large enough, the distorted state is more stable than the normal one.   Let us consider a simplified system consisting of a monatomic lattice of N atoms.   We shall let ϵ be the energy that is necessary to remove one atom from a typical lattice site and to place it at the surface in a normal position.   It will be assumed that this energy is independent of the number n of atoms removed for small values of n. If we neglect any changes in the vibrational energy that may occur as a result of this transposition, the total entropy change is determined by the number of possible ways in which the n vacancies may be distributed among the N sites.   This number evidently is $N!/n!(N-n)!$ so that the entropy S is

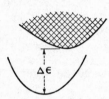

Fig. 35.—Schematic representation of the excited and normal levels in a semi-conductor containing an impurity atom. The minimum of the lower curve corresponds to the equilibrium atomic arrangement when the discrete level of the foreign atom is occupied. The minimum of the upper continuum corresponds to the equilibrium arrangement when the electron is ionized. The two minima are different because the ionized foreign atom does not interact with the lattice in the same way as the un-ionized one. For this reason the energy $\Delta\epsilon$ for thermal ionization is usually less than the energy for optical ionization, for the second process must obey the Franck-Condon principle and corresponds to a vertical jump in this diagram.

$$S = k \log \frac{N!}{n!(N-n)!} \simeq -kn \log \frac{n}{N}. \quad (2)$$

Thus, the free energy

$$A = n\epsilon + kTn \log \frac{n}{N} \quad (3)$$

is a minimum when

$$\frac{n}{N} = e^{-\frac{\epsilon}{kT}}.$$

This result shows that we should expect some deviations from the ideal crystalline state at any finite temperature.   If ϵ is 1 ev and T is 1000°K, which are values that can reasonably occur, we find

$$\frac{n}{N} \sim 10^{-4.3}.$$

The principles used in this computation may be applied to other cases, most important of which for the theory of semi-conductors are the polar crystals having the composition M_mX_n, where M is a metal atom and X is an electronegative atom. The four independent types of deviation from ideal arrangement that may occur in these solids are as follows: There may be vacancies (*a*) in the metal lattice or (*b*) in the electronegative lattice; and there may be (*c*) interstitial metal atoms, or (*d*) interstitial electronegative atoms. These four types of *lattice defect* may occur in any one of the various possible combinations. We shall discuss a few actual cases in the following sections.

111. The Alkali Halide Semi-conductors.—Hilsch, Pohl,[1] and their numerous collaborators have made extensive investigations of semi-conducting alkali halide crystals that are produced by heating these solids in alkali metal vapor until they become colored. Figure 36 shows

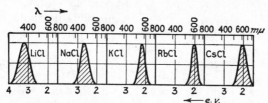

Fig. 36.—*F*-center absorption bands at room temperature in various alkali halides. The wave-length scale is in units of 10^{-7} cm. (*After Pohl.*)

the spectral dependence of the new absorption band in several cases. Identical discoloration may be produced by X-ray or cathode-ray bombardment. The discoloration is not so durable in these cases, however, for it may be removed by relatively mild heating that does not affect the discoloration produced by heating the crystals in alkali vapor. From the intensity of the absorption bands, it is possible to determine the number of absorption centers that are responsible for the discoloration. This number depends upon the method used to prepare the colored crystal and varies from 10^{15} to 10^{19} per cubic centimeter in the specimens ordinarily used in experiments. Pohl has named these absorbing centers "*F* centers," (*Farbzentren*). We shall use the same term.

The electronic conductivity of the alkali halides containing *F* centers becomes appreciable above 200°C and is superimposed upon the ionic conductivity, which is also appreciable. The two may be separated by measuring either the conductivities of separate clear and colored specimens or the conductivity before and after the *F* centers have been removed. This removal can be accomplished by placing the colored

[1] See the survey article by R. W. Pohl, *Physik. Z.*, **39**, 36 (1938).

crystal in a constant field at the temperatures at which the conductivities are measured. The F centers then migrate to the anode and disappear. During this procedure, the conductivity drops, as is shown in Fig. 37. Electronic conductivity may be induced at low temperatures by illuminating the crystal with light in the absorption bands. We shall discuss this photoconductivity in detail in Chap. XV.

The ionic conductivity of the alkali halides usually is caused by the migration of both positive and negative ions, as may be determined by transport measurements, such as those discussed in Chap. I. Frenkel[1] first pointed out that the migrating positive and negative ions probably do not move by squeezing past one another, as they would in an ideal

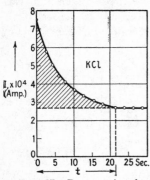

lattice, since the activation energy[2] that would be required for this process is much larger than the activation energies determined by measurements of the temperature dependence of conductivity. He estimated that the activation energy in an ideal lattice would be about the same as the cohesive energy. This is about 7 ev in sodium chloride, whereas the observed value[3] of the activation energy is only 1.9 ev. For this reason, Frenkel postulated that alkali halide crystals normally contain vacancies in both the positive- and negative-ion sites. We may conclude that these vacancies are present in equal numbers in uncolored crystals, for otherwise these crystals would be charged. They presumably have the same thermodynamical origin as the vacancies in the simple monatomic lattice discussed at the end of the previous

Fig. 37.—Decrease in the current of colored potassium chloride as the F centers are removed by conduction. The initial conductivity is due to ions and electrons, the final conductivity to ions alone. The temperature was 580°C; the field intensity was 300 volts per cm. (*After Pohl.*)

section and can be discussed in terms of the theory used there.

It is possible to give two reasonable pictures of the discoloration of alkali halides by X rays. In both pictures, it is assumed that the primary action of the X rays is to free an electron from one of the inner shells of an atom of the lattice and that the discoloration center is associated with the absorption properties of this electron when it subsequently becomes trapped in the lattice. The most apparent trapping positions are the vacancies in the negative-ion lattice and the positive ions. The

[1] J. Frenkel, *Z. Physik*, **35**, 652 (1926).

[2] The activation energy is the least energy required to interchange two atoms. This is discussed in detail in the next chapter.

[3] W. Lehfeldt, *Z. Physik*, **85**, 717 (1933).

vacancies should have an affinity because the Madelung potential is positive at these positions, whereas the positive ions should have an affinity because an electron should be able to polarize the surrounding lattice and produce a stable discrete level lying below the conduction bands discussed in Sec. 106 of this chapter.   Evidence obtained from investigations of photoconductivity seems to support the first interpretation of the trapping position and rule out the second.   If any alkali atom could trap an electron and produce an F center, the mean free path for trapping of free electrons should be independent of the density of F centers and of vacancies in the negative-ion lattice, since the number of these is far less than the number of alkali metal ions.   The experimental work on the photoconductivity of crystals containing F centers shows that the density of trapping points is far less than the density of alkali metal ions and depends upon the density of F centers.   Hence, it may be concluded that F centers are electrons trapped in vacant halogen-ion sites.[1]   There has been no experimental evidence to show that electrons are ever trapped by the positive ions.

When an alkali halide crystal is heated in alkali metal vapor, some of the atoms of the vapor presumably become absorbed on the surface and lose their electrons.   These electrons then wander into the crystal and occupy vacant halogen positions, producing F centers.   The ions left behind may then diffuse into the lattice, decreasing the number of positive-ion vacancies and keeping the volume of the crystal unchanged. Let us suppose that the crystal is at temperature T and that it normally contains n vacancies in the positive- and negative-ion lattice.   In addition, let us suppose that it is placed in a container of which the volume V is much larger than that of the crystal and which contains N_A neutral alkali metal atoms in vapor form.   If n_F atoms are absorbed into the crystal, the number of vacancies in the positive- and negative-atom sites is decreased from n to $n - n_F$.   The mixing entropy associated with the vacancies is

$$-2k(n - n_F) \log \frac{n - n_F}{N} \tag{1}$$

where N is the number of ions in the lattice.   In addition, the n_F electrons that occupy the halogen sites have the entropy

$$-kn_F \log \frac{n_F}{N}, \tag{2}$$

since they may occupy any of the N sites.   The vapor has the entropy

[1] This interpretation of F centers is due to J. H. de Boer, *Rec. trav. chim. Pays-Bas*, **56**, 301 (1937), and has been developed by R. W. Gurney and N. F. Mott, *Trans. Faraday Soc.*, **34**, 506 (1938).

$$-k(N_A - n_F) \log \frac{(N_A - n_F)}{C} \tag{3}$$

where

$$C \cong \frac{V}{h^3}(2\pi\mu kT)^{\frac{3}{2}}$$

in which μ is the atomic mass. Thus, the total free energy as a function of n_F and n is

$$A(n_F, n) = n_F\epsilon + (n - n_F)\epsilon' + kT\left[2(n - n_F) \log \frac{n - n_F}{N} + n_F \log \frac{n_F}{N} + \right.$$
$$\left. (N_A - n_F) \log \frac{(N_A - n_F)}{C} \right] \tag{4}$$

where ϵ is the energy required to dissociate an atom into an F center and

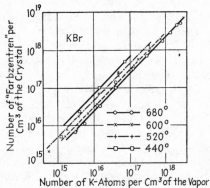

Fig. 38.—Relation between the density of F centers in a potassium bromide crystal and the density of alkali metal atoms in the vapor for different temperatures. (*After Pohl.*)

an ion and ϵ' is the energy required to form normal lattice defects. Then A is a minimum when

$$\frac{n_F}{(N_A/C)} = Ne^{\frac{\epsilon}{kT}}. \tag{5}$$

According to this result, the ratio of the concentration of F centers to the concentration of atoms in the vapor should be constant. Figure 38 shows that this relation is obeyed in KBr over a wide range of concentrations.[1] In addition, the temperature dependence is the same as that predicted by (5). Rögener finds experimentally that $\epsilon = -0.25$ ev for KBr and -0.10 ev for KCl.

[1] Pohl, *op. cit.*

If the equilibrium density of F centers is established at one temperature and alkali metal pressure and the crystal is then cooled, the excess F centers should coagulate into colloidal globules of alkali metal unless the cooling takes place so rapidly that the higher density is frozen in.

Following a plan of Gurney and Mott,[1] we may obtain a rough idea of the energy levels of an F center by using classical methods. The Madelung potential at a vacant negative-ion site is $-Ae^2/r_0$ as long as the surrounding ions have perfect crystalline order. Here, A is the Madelung constant and r_0 is the nearest-neighbor distance. This poten-

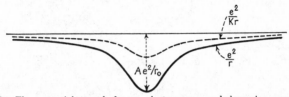

FIG. 39.—The potential trough for an electron near a halogen-ion vacancy. The full curve represents the potential when polarization is neglected, the dotted curve the potential when it is not.

tial is of the order of 9 ev for sodium chloride. At large distances from the vacant site, the total potential is

$$-\frac{e^2}{r} + V_M \tag{6}$$

where V_M is the periodic Madelung potential of a normal lattice and e^2/r is the potential arising from the vacancy. The mean value of V_M is very close to zero for an electron placed in an alkali halide crystal (*cf.* Sec. 91), and so we are justified in dropping it from (6) in a good approximation. The remaining potential then varies smoothly between the value $-Ae^2/r_0$ at the vacant site and the value $-e^2/r$ at large distances, in the manner shown in Fig. 39. If there were no electron in the vacant site, the surrounding ions would be displaced from their equilibrium positions for the normal lattice. We shall assume, however, that they are nearly in their normal positions when an electron is present, for this electron should have nearly the same electrostatic effect on the neighboring ions as a halogen ion. When the electron is at large distances from the site, it is partly screened from the excess positive charge by the polarization charge that it induces in the crystal. Hence, the potential at large distances should vary roughly as e^2/n^2r instead of as e^2/r, where n is the refractive index. Thus, the potential well in which the electron is trapped should have the form of the second curve, Fig. 39. This trough has an infinite number of discrete levels, which end in a series

[1] R. W. GURNEY and N. F. MOTT, *Proc. Phys. Soc.* (sup.), **49**, 32 (1937).

limit at zero energy, since the field is coulomb-like at large distances. The lowest of the discrete states is an s state, and the next two are s and p states. The optical transition from the lowest s state to the lowest p state should have the greatest probability and should correspond to the absorption band of the colored crystal. This transition should not lead directly to photoconductivity, for the excited electron is bound to the vacant site just as the normal electron is. The excited state lies nearer the continuum, however, so that the probability for a thermal transition to the continuum is higher from it than from the lowest level. In order to account for the observed photoconductivity, we must assume that the electron actually becomes free by thermal excitation. The probability for this excitation should decrease with decreasing temperature and vanish at absolute zero. This disappearance of photoconductivity near the absolute zero is actually observed.[1] We shall discuss the effect further in Sec. 134.

F centers are not the only color centers that may be introduced into alkali halide crystals.[2] Thus, Pohl and his coworkers have found that a stoichiometric excess of halogen atoms may be produced by heating the crystal in halogen vapor. Most of the coloration lies in the near ultraviolet region of the spectrum in this case. Although these centers have not been investigated so fully as F centers, it seems probable that they are neutral halogen atoms in halogen sites. The F center absorption band may be destroyed and a new band may be introduced in the far ultraviolet by heating a crystal containing F centers in hydrogen vapor. In this case, it is believed that the hydrogen atoms diffuse to the F centers and form H^- ions at these positions. The far ultraviolet absorption band of these "U centers" presumably corresponds to the first excitation frequency of the internally absorbed hydrogen ions.

112. Zinc Oxide.—Zinc oxide that has been formed at low temperatures is a pure white substance having no appreciable electronic conductivity. After being heated to a high temperature, it develops a brownish hue and is a good electronic conductor at room temperature. A preliminary discussion of these properties was given in Sec. 37, which deals with the free-electron theory of semi-conductors. We saw there that the room-temperature conductivity obeys Eq. (1), Sec. 110 of this chapter. For the purposes of this discussion, this equation may be placed in the form

$$\sigma = A e^{-\frac{\epsilon'}{kT}} \tag{1}$$

---

[1] *Ibid.*

[2] Recently H. Pick has investigated the optical and electrical properties of colored halides containing divalent strontium halides, *Ann. Physik*, **35**, 73 (1939).

where ϵ' and A are not strongly temperature-dependent in the range over which conductivity is measured.   It was also seen in Sec. 37 that these constants are influenced by the pressure of oxygen in which the specimen is heated.   Since A is related to the mean free path l_0 and the density of centers n_b by the equation

$$A = 0.024 l_0 n_b^{\frac{3}{2}} T^{\frac{1}{4}} \text{ ohm}^{-1} \text{ cm}^{-1}$$

and since Hall-effect measurements show that l_0 is practically constant for a given specimen, we may conclude that the variation of A with oxygen pressure implies a variation in n_b. This variation has been investigated by Wagner and Baumbach;[1] we shall now discuss their results.

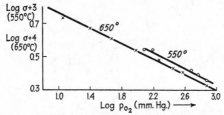

FIG. 40.—Dependence of the conductivity of zinc oxide on vapor pressure of oxygen
(After Baumbach and Wagner.)

Figure 40 shows the variation of high-temperature conductivity with oxygen vapor pressure for a specimen that has been heated at two temperatures.   Practically all the n_b bound electrons are free at the temperatures at which these measurements are made, for $\Delta\epsilon$ then is appreciably less than kT. Hence, the curves of Fig. 40 give directly the dependence of n_b on oxygen pressure p_O.   They show that

$$n_b = c p_O^{-\frac{1}{n}} \qquad (2)$$

where $n \simeq 4.2$.

All the experimental results may be explained satisfactorily if we assume that the heated zinc oxide loses oxygen atoms from the surface and leaves excess zinc atoms, which become ionized and diffuse into the interstices of the lattice.   Wagner has ruled out the alternative possibility that vacancies are produced in the oxygen lattice and that the conduction electrons are those which might normally occupy these vacancies by showing that the negative-ion transport number is very small compared with the positive-ion transport number. The small observed positive ionic current is carried either by ionized interstitial

[1] H. H. v. BAUMBACH and C. WAGNER, *Z. Physik Chem.*, **22B**, 199 (1933).

zinc atoms or by the normal zinc ions.   There is ample room for interstitial zinc atoms in the zinc oxide lattice because it has the porous wurtzite structure.

The displacement of equilibrium with changing oxygen pressure may be treated in the following way.   We shall let ϵ'' be the energy required to produce a singly charged interstitial zinc ion, a free electron in the lattice, and one atom of gaseous oxygen, which is attached to another oxygen atom to form an O_2 molecule.   We shall let n be the number of interstitial Zn^+ ions and N_{O_2} the total number of O_2 molecules in the gas, where N_{O_2} is much larger than n.   The total free energy then is

$$A(n,N_{O_2}) = n\epsilon'' + kT\left[n \log \frac{n}{N} + n \log \frac{n}{C} + \left(N_{O_2} + \frac{n}{2}\right) \log \frac{\left(N_{O_2} + \dfrac{n}{2}\right)}{C}\right] \tag{3}$$

where the first entropy term is that of the interstitial ions, N being the total number of interstitial sites, the second term is the entropy of the free-electron gas, and the third term is the entropy of the O_2 molecules. The equilibrium value of n is

$$nN_{O_2}^{\frac{1}{2}} = Be^{-\frac{\epsilon''}{kT}} \tag{4}$$

where B is a constant.   Hence, n should vary as $p_{O_2}^{-\frac{1}{2}}$ according to this simple theory.   The result is in reasonable agreement with the observed variation of $p_{O_2}^{-\frac{1}{4.3}}$.

In the first approximation, we might treat the energy levels of the interstitial zinc atoms as though they were free atoms in a homogeneous, polarizable medium.   The principal effect of the polarizability[1] is to decrease the distance between the ground state and the continuum, as we have seen in the previous section.   Suppose that we had a hydrogen atom in a medium of refractive index n.   Then, the potential between the electron and proton would be $-e^2/n^2r$, where r is the radial distance between the two particles.   The presence of n in the potential energy changes the Rydberg constant to R/n^4 where R is the normal value for a free atom.   The refractive index of zinc oxide is about 2, so that we should expect the ionization energy to be lowered by a large factor, of the order of magnitude 10.   The same qualitative result should apply to zinc, which has an ionization potential of 9.36 ev, and should lower the ionization energy of the interstitial atom to about 1 ev.   The observed values of ϵ' in Eq. (1), however, are even lower than this value.   For example, ϵ' is commonly less than 0.01 ev in a specimen that has been

[1] N. F. Mott and R. W. Gurney, *Proc. Phys. Soc.* (sup.), **49**, 32 (1937).

heated for a long while in a vacuum. Moreover, Fritsch (*cf.* Sec. 37) has found that ϵ' in Eq. (1) varies with the pressure of oxygen and has shown that ϵ' increases as the density of interstitial atoms decreases. This effect indicates that the interstitial zinc atoms interact with one another and in some way decrease the distance between the bound and free levels. The density of interstitial atoms is of the order of 10^{18}, according to Hall-effect measurements, so that this interaction is conceivable only if the radius of the interstitial atom is ten times larger than the radius of a normal zinc atom. Now, the radius of a hydrogen atom in a medium of index n would be n^2 times larger than the radius of a normal atom. Thus, it is possible that the electrons in the interstitial atoms move in very large orbits because the surrounding medium is highly polarizable.

113. Cuprous Oxide and Other Substances That Involve Transition Metals.—Cuprous oxide is a very useful semi-conductor, but its highly intricate properties are only partly understood. The most reliable evidence seems to show that it has hole conductivity, which indicates that it has either a deficiency of metal atoms or an excess of oxygen atoms. Wagner and his coworkers[1] have evidence to show that the copper ion is much more mobile than oxygen, and they conclude from this that the conducting oxide probably contains vacancies in the copper-ion lattice. These are shown schematically in Fig. 41. Since the deficient copper ion carries away an electron, the lattice should contain one

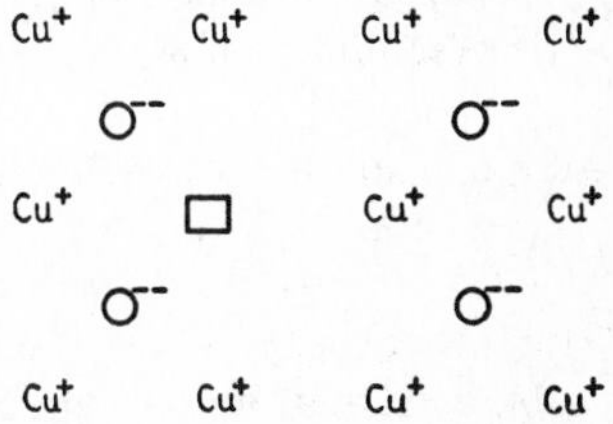

Fig. 41.—Schematic representation of copper vacancies in cuprous oxide. The vacancy leaves the lattice with a deficiency of one electron.

electron hole for each vacancy. This hole may normally reside either on a copper ion, turning a Cu^+ ion into a Cu^{++} ion, or on an oxygen ion, turning an O^{--} ion into an O^- ion. Wagner suggests that the first picture is more probable since copper is commonly bivalent. According to this view, the conductivity of cuprous oxide should result from the motion of the hole from one copper ion to another. The most stable position for the hole should be near the vacancy since there is an excess negative charge at that position.

De Boer and Verwey[2] have attempted to make a systematic classification of other semi-conductors containing metals with partly filled $3d$ shells. They computed the energy of electrons on atoms and ions near vacant sites, using atomic-model methods similar to those that we have

[1] C. Wagner, *Physik. Z.*, **36**, 721 (1935).

[2] J. H. de Boer and E. J. W. Verwey, *Proc. Phys. Soc.* (sup.), **49**, 59 (1937).

used in Secs. 95 and 107. We shall discuss their results for three cases, namely, nickel oxide (NiO), cuprous iodide, and potassium iodide. Although the last case does not involve transition metals, they consider it for comparative purposes.

In nickel oxide, which has sodium chloride structure and an oxygen excess, some metal-ion sites are vacant. De Boer and Verwey conclude that the electrons removed from the lattice with the positive ions are taken from two nickel ions alongside the vacancy, leaving two Ni^{+++} ions. These holes may be thermally freed making the crystal conducting.

Cuprous iodide has the zincblende structure and is a halogen excess semi-conductor. The workers conclude that in this case there are neutral iodine atoms at the iodine sites alongside the metal-ion vacancies. This case should be contrasted with that of cuprous oxide, discussed above, in which the hole is believed to reside on the positive ions. De Boer and Verwey estimate that in copper iodide the hole would be 1 ev less stable at a copper-ion site than at an iodine site.

Potassium iodide is also an excess halogen semi-conductor. The computational evidence indicates that there are positive-ion vacancies and neutral halogen atoms nearby, just as in copper iodide.

De Boer and Verwey also point out that the zone approximation is much less accurate for d-shell electrons than for electrons in s-p levels. It is probably true that the lowest level of the entire solid in a salt containing an odd number of d electrons per unit cell is separated from the higher levels by a large gap, even though the lowest levels should be quasi-continuous in the Bloch approximation. A case in point seems to be CoO which has the sodium chloride lattice with one cobalt ion per unit cell. Since this ion has seven d electrons, the salt should be a metallic conductor, according to the zone theory. Actually, it is not, a fact which shows that the ordinary rules for predicting metallic character cannot always be applied to d-electron groups.

E. MOLECULAR CRYSTALS

114. Survey.—There has been no explicit work on the electronic energy levels of molecular crystals. Apparently, it is safe to assume that the lower excited states may be treated by the methods of excitation waves to a high degree of accuracy. The widths of the excitation bands should be small, for the intermolecular forces are small. One consequence of this fact is that the spacing between the lower excitation bands should be nearly the same as the spacing between the electronic levels of the free molecules. In addition, there should be additional bands below the ionization continuum that correspond to the transfer of an electron from a molecule to one of its neighbors. The principles which determine all the levels should be enough like those which have been used successfully in connection with ionic crystals to require no further comment here. This subject probably could be developed considerably if experimental investigations of the absorption spectra of molecular crystals were carried out in the near ultraviolet and Schumann regions.

115. The Transition between the Solid Types.—In Chap. I, we attempted to show the interrelation between the solid types by means of Fig. 82. We may now discuss this diagram again, using our knowledge of the electronic states. The ideal metals, which are on the left, possess broad, incompletely filled bands when described by the zone approximation. They cannot be described adequately by the Heitler-London approximation, since the lowest energy levels of the entire solid are quasi-continuous.

As we move to the right in Fig. 82, the energy bands separate into filled and unfilled sets. This transition takes place gradually, being well advanced in metals such as calcium, bismuth, and graphite and complete in diamond and possibly boron. The separation of bands also occurs as we move from ideal substitutional alloys to ionic crystals. In this case, substances such as Mg_3Sb_2 occupy the intermediate positions that correspond to bismuth, and so forth, in the monatomic case. Ideal valence and ionic crystals may be described in terms of both the Heitler-London and the Bloch scheme. Neither is completely satisfactory when used alone, however, even for qualitative work, and the two approximations must be combined to form a complete picture. The atomic functions must be greatly perturbed in constructing the best Heitler-London functions for the lowest state of these solids. This is indicated by the fact that the best functions in valence crystals have directional properties and the best functions in ionic crystals are closely like the functions of free ions. In the energy-level diagram of the entire solid, a singlet separates from the quasi-continuous levels of metals as we move from left to right.

Passing still farther to the right, we come to molecular solids, which usually are described more satisfactorily by the Heitler-London approximation than by the Bloch approximation, since the Heitler-London functions are very nearly the same as those of the free molecular units. The lowest level of the entire solid is a discrete singlet, and the higher levels that lie below the ionization limit presumably are grouped into narrow excitation bands.

Semi-conductors are insulating crystals that have additional electronic states because they contain lattice defects or foreign atoms.

CHAPTER XIV

THE DYNAMICS OF NUCLEAR MOTION. PHASE CHANGES

116. The Adiabatic Approximation*.—We have treated the nuclear coordinates as parameters in practically all the preceding discussion because we were interested primarily in the stationary electronic states. We shall now examine the extent to which this procedure may be justified and shall discuss a scheme for treating electronic and nuclear motion together. This scheme was employed by Born and Oppenheimer[1] in connection with the stationary states of molecules and has been used subsequently in similar problems.[2]

The complete Hamiltonian operator, $\mathfrak{H}_c$, for a crystal is

$$\mathfrak{H}_c = -\sum_i \frac{\hbar^2}{2m}\Delta_i - \sum_a \frac{\hbar^2}{M_a}\Delta_a + \frac{1}{2}\sum_{i,j}' \frac{e^2}{r_{ij}} +$$
$$V_{ei}(x_1, \cdots, z_n, \xi_1, \cdots, \zeta_f) + V_{ii}(\xi_1, \cdots, \zeta_f)$$
$$= H - \sum_a \frac{\hbar^2}{2M_a}\Delta_a. \tag{1}$$

The indices i, j, $\ldots$, n extend over all electrons, and the indices a, b, $\ldots$, f extend over all ions. M_a is the mass of the ath ion, V_{ei} is the electron-ion interaction potential, V_{ii} is the interaction potential of the rigid ions, and H is the electronic Hamiltonian of Chap. VI, in which the nuclear kinetic-energy terms were neglected. In Chap. VI, V_{ei} was taken to have the form

$$\sum_i V(\mathbf{r}_i).$$

Since the ions were regarded as being fixed, $V(\mathbf{r}_i)$ had the periodicity of the lattice. We must now include the dependence of V on the nuclear coordinates, since they will also be treated as dynamical variables.

---

[1] M. BORN and J. R. OPPENHEIMER, *Ann. Physik*, **84**, 457 (1927).

[2] H. PELZER and E. WIGNER, *Z. physik. Chem.*, **15B**, 445 (1932); E. WIGNER, *Z. physik. Chem.*, **19B**, 203 (1932); H. EYRING, *Jour. Chem. Phys.*, **3**, 107 *ff.* (1935) (*cf.* review article in *Chem. Rev.*, **17**, 65 (1935); L. FARKAS and E. WIGNER, *Trans. Faraday Soc.*, **32**, 708 (1936). A critique of this method as applied to problems in chemical reactions is given by E. Wigner, *Trans. Faraday Soc.*, **34**, 29 (1938); see also J. O. Hirschfelder and E. P. Wigner, *Jour. Chem. Phys.*, **7**, 616 (1939).

Although the exact characteristic functions Φ of $\mathfrak{H}_c$ usually are intricate functions of the x and the ξ, we shall attempt to use approximate solutions of the form

$$\Phi_{r\alpha}(x_1, \cdots, z_n, \xi_1, \cdots, \zeta_f) = \Psi_r(x_1, \cdots, z_n, \xi_1, \cdots, \zeta_f)\chi_{r\alpha}(\xi_1, \cdots, \zeta_f, t), \quad (2)$$

where Ψ_r is an electronic function of the type used in previous chapters, in which the ξ were regarded as parameters, and $\chi_{r\alpha}(\xi_1, \cdots, \zeta_f)$ is a function of the nuclear coordinates. This approximation is commonly called the adiabatic approximation, because at each instant the electronic distribution is taken to be the same as though the nuclear coordinates were at rest at the positions they have at this instant. This assumption obviously can be true only if the electrons move much more rapidly than the ions. We shall see presently that the accuracy of the approximation depends on the fact that ionic masses are great relative to the electron mass.

If the function (2) is substituted in the equation

$$\mathfrak{H}_c\Phi = -\frac{\hbar}{i}\frac{\partial\Phi}{\partial t} \quad (3)$$

and if it is recalled that

$$H\Psi_r(x_1, \cdots, z_n, \xi_1, \cdots, \zeta_f) = E_r(\xi_1, \cdots, \zeta_f)\Psi_r, \quad (4)$$

it is found that

$$-\sum_a \frac{\hbar^2}{2M_a}\chi_{r\alpha}\Delta_a\Psi_r - \sum_a \frac{\hbar^2}{M_a}\operatorname{grad}_a\Psi_r\cdot\operatorname{grad}_a\chi_{r\alpha} - \sum_a \frac{\hbar^2}{2M_a}\Psi_r\Delta_a\chi_{r\alpha} +$$
$$E_r(\xi_1, \cdots, \zeta_f)\Psi_r\chi_{r\alpha} = -\frac{\hbar}{i}\Psi_r\frac{\partial\chi_{r\alpha}}{\partial t}. \quad (5)$$

Multiplying this by Ψ_r^* and integrating over the electronic coordinates, we obtain

$$\left[-\sum_a \frac{\hbar^2}{2M_a}\int\Psi_r^*\Delta_a\Psi_r d\tau(x_1, \cdots, z_n)\right]\chi_{r\alpha} -$$
$$\sum_a \left[\frac{\hbar^2}{M_a}\left(\int\Psi_r^*\operatorname{grad}_a\Psi_r d\tau\right)\cdot\operatorname{grad}_a\chi_{r\alpha}\right] - \sum_a \frac{\hbar^2}{2M_a}\Delta_a\chi_{r\alpha} +$$
$$E_r(\xi_1, \cdots, \zeta_f)\chi_{r\alpha} = -\frac{\hbar}{i}\frac{\partial\chi_{r\alpha}}{\partial t}. \quad (6)$$

It will now be shown that the first term may ordinarily be neglected for stationary-state problems. In order to do so, two extreme cases will be considered, namely, that in which the electrons are perfectly free and that in which they are completely bound.

The one-electron wave functions are of the form $e^{2\pi i \mathbf{k} \cdot \mathbf{r}}$ in the first case and hence are practically independent of nuclear coordinates. Thus, the first two terms in (6) are vanishingly small.

In the opposite case, we may, for simplicity, regard the wave function as though it were composed of one-electron functions of the type

$$f_a(x_i - \xi_a, y_i - \eta_a, z_i - \zeta_a). \tag{7}$$

Under this condition, we have the relation

$$\sum_a \operatorname{grad}_a \Psi = -\sum_i \operatorname{grad}_i \Psi; \tag{8}$$

hence, the first term in (6) is simply

$$\frac{m}{M} \sum_i \int \Psi_r{}^* \left(-\frac{\hbar^2}{2m}\Delta_i \right) \Psi_r d\tau \tag{9}$$

if it is assumed that there is only one type of ion. Since the quantity (9) is equal to m/M times the mean kinetic energy of the electrons, it is normally negligible because the ratio m/M is at most 1/1,840.

The second term also may be dropped in stationary-state problems, for Ψ can then be chosen as a real function. Hence,

$$\int \Psi^* \operatorname{grad}_a \Psi = \frac{1}{2} \operatorname{grad}_a \int |\Psi|^2 d\tau = \operatorname{grad}_a 1 = 0.$$

The final equation for $\chi_{r\alpha}$ is

$$\sum_a - \frac{\hbar^2}{2M_a}\Delta_a\chi_{r\alpha} + E_r(\xi_1, \cdots, \zeta_f)\chi_{r\alpha} = -\frac{\hbar}{i}\frac{\partial \chi_{r\alpha}}{\partial t}, \tag{10}$$

which has the form of a Schrödinger equation in which E_r is the effective nuclear potential function. This equation has stationary-state solutions of the form

$$\chi_{r\alpha}(\xi_1, \cdots, \zeta_f, t) = \lambda_{r\alpha}(\xi_1, \cdots, \zeta_f)e^{-\frac{i\mathfrak{E}_{r\alpha}t}{\hbar}} \tag{11}$$

where $\lambda_{r\alpha}$ satisfies the equation

$$-\sum_a \frac{\hbar^2}{2M_a}\Delta_a\lambda_{r\alpha} + E_r(\xi_1, \cdots, \zeta_r)\lambda_{r\alpha} = \mathfrak{E}_{r\alpha}\lambda_{r\alpha} \tag{12}$$

and $\mathfrak{E}_{r\alpha}$ is the constant total energy of the system. It is evident that $\mathfrak{E}_{r\alpha}$ is the mean value of $\mathfrak{H}_c$ in the approximation in which the first two terms of (6) are negligible.

Although the mean value of $\mathfrak{H}_c$ for the function (2) is accurate to within terms of the order of m/M times the electronic kinetic energy if Ψ and χ satisfy Eqs. (4), (11), and (12), it does not follow that all mean values are equally accurate,[1] for we know from the variational theorem that the energy is stationary for small variations in the wave functions. As we have seen in Sec. 39 the accuracy of the mean values of other quantities is usually of the order of magnitude $\sqrt{m/M}$.

117. A Qualitative Survey of the Theory of Phases.—Equation (12) of the preceding section, which is the equation for the stationary states of nuclear motion, is usually very difficult to solve accurately because $E_r(\xi_1, \cdots, \zeta_f)$ usually is an extremely intricate nonseparable function of the nuclear variables. This may be realized from the fact that this equation should yield a description of all types of phases of matter from solids to gases.

Let us consider the behavior of the function $E_0(\xi_1, \cdots, \zeta_f)$ that is associated with the lowest electronic wave function. This is the effective potential field in which the ions usually move. E_0 approaches a constant value corresponding to the normal energy of the constituent atoms when the atoms are separated by more than 10^{-6} cm, (*cf.* Fig. 1). As the nuclei are brought together, E_0 usually decreases to a certain minimum value and then increases again as the nuclei are crowded more closely. The depth of

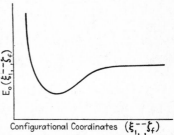

Fig. 1.—Schematic behavior of the electronic energy of the lowest state of the entire crystal as the interatomic distance is varied. For large separations E_0 approaches a constant, whereas it decreases and then increases again as the atoms are brought closer together.

this absolute minimum relative to the value of E_0 for large separations is a measure of the cohesive energy of the solid. In addition to this absolute minimum there may be secondary minima corresponding to atomic configurations that may be metastable at very low temperatures. A part of the purpose of the next section is to examine the relative stability of the minima of this type that correspond to crystalline arrangements.

If the coordinates of any atom or group of atoms are varied slightly when the system is at an equilibrium point, we may expect E_0 to increase. The change in E_0 is not the same for all directions of variation but depends upon the crystalline binding. Now, if we rearrange the atoms in any way that does not alter the crystal structure, the initial and final energies are the same. Since E_0 increases for changes near the equilibrium values, it follows that this function passes through a maximum

[1] The accuracy of the adiabatic approximation in special cases is discussed by H. Pelzer and E. Wigner and by E. Wigner (*cf.* footnote 2, p. 470).

during the rearrangement from one equilibrium distribution to another. The minimum value of this maximum for all possible rearrangement paths, that is, the height of the "saddle point" of E_0 in the potential barrier that separates the two minima, is called the "activation energy" for the rearrangement (*cf.* Fig. 2). Paths leading through this saddle point are the ones ordinarily followed when rearrangements take place thermally.[1]

We may now describe the stationary states of nuclear motion qualitatively, using these concepts. In the very lowest state, which is described by the wave function λ_{00}, the system is localized near the absolute minimum of $E_0(\xi_1, \cdots, \zeta_f)$. The actual distribution of nuclei is given by the function $|\lambda_{00}|^2$, which will be described more fully in the next section. The energy parameter $\mathfrak{E}_{00}$, associated with the lowest state, is slightly greater than the minimum of E_0, the difference being the zero-point energy of the atoms. Since the probability distribution function decreases very rapidly in the regions where E_0 is greater than $\mathfrak{E}_{00}$, it follows that the individual atoms are statistically localized near their equilibrium positions as long as the energy per atom is less than the saddle point of the barrier surrounding the equilibrium position. There is a chance that a large part of the zero-point energy of the system may become localized in one atom, allowing it to move away; however, the likelihood of a large fluctuation usually is very small for the normal state as we shall see in the next section.

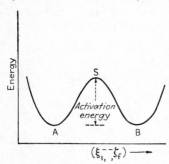

Fig. 2.—Schematic diagram showing the behavior of the energy of the system when the atoms are rearranged without changing the structure. A and B correspond to the minimum energy arrangements. During the rearrangement from A to B the energy increases, passes through a maximum, and decreases. The lowest maximum occurs at the saddle point S of the potential hill separating A and B. The height of S above A and B is the activation energy.

As we go to higher energy states, the probability of finding the entire system at regions away from the equilibrium position increases because there is a larger range of configuration space in which $\mathfrak{E}_{0\alpha} - E_0$ is positive. When the mean energy per atom becomes comparable with the height of the saddle point for a given rearrangement, this rearrangement may take place spontaneously with an appreciable probability. The very lowest states in which rearrangements occur appreciably are those in which the system is still crystalline and in which a small fraction of the atoms are diffusing about, whereas the higher states correspond to

---

[1] *Cf.* the survey article by H. Eyring, *Chem. Rev.*, **17**, 65 (1935).

liquid phases in which there is no lattice structure and in which atoms correlate their positions only with those of their nearer neighbors.[1] Thus the gradation of stationary states from the crystalline phase to the liquid phase is perfectly continuous. The phenomenon of melting, which ordinarily occurs abruptly with the absorption of heat, does not imply any discontinuity in the allowable energy states but is a process in which a range of possible states is jumped over for reasons that are described in Sec. 121. The states skipped during melting are those associated with glasses and supercooled liquids.

The states for which the total energy is greater than that of the system of free molecules correspond to the gaseous phase. In this phase each molecule has enough energy on the average to surmount the barriers holding the atoms of the solid or liquid phase together.

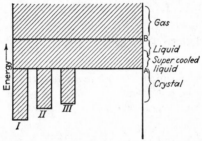

Fig. 3.—Schematic diagram showing the types of energy states of a system of atoms. I, II, and III represent relative minima of the $E_0(\xi_1, \cdots, \zeta_n)$ function that correspond to different allotropic phases. The energy states below A correspond to crystalline phases in which the atoms are vibrating. Long-distance order vanishes at A and the range of glasses, supercooled liquids, and liquids lies between A and B. The levels near A are usually skipped during melting. The gaseous phase lies above B.

When two or more different crystalline arrangements of the ξ correspond to relative minima of $E_0(\xi_1, \cdots, \zeta_f)$, only the lowest is thermodynamically stable at the absolute zero of temperature. The time required to bring about thermodynamical equilibrium may be very long, however, if the system gets caught in one of the higher minima at low temperature. For this reason, several different phases of a substance may be stable in a practical sense.

We may summarize this qualitative discussion by means of the energy-level diagram shown in Fig. 3. There usually are several types of stable states, labeled I, II, III, etc., which correspond to the different polymorphic forms. Above each of these is a range of energy, terminated by the line A, in which the system possesses lattice symmetry. The

---

[1] Evidence for this type of atomic correlation is given by X-ray analysis of liquids. See, for example, A. H. Compton and S. K. Allison, *X-rays in Theory and Experiment* (D. Van Nostrand Company, Inc., New York, 1935).

region above A, which is the domain of glasses or supercooled liquids, gradually blends into the less viscous liquid state. The states corresponding to energies near A usually are thermodynamically unstable at all temperatures. The line B, which marks the point at which the entire system is in the gaseous phase, depends upon the volume in which the system is kept. B actually may fall below A, in which case the solid sublimes before melting.

118. Low-energy States.—As stated in the last section, we shall assume that the relative minima of $E_0(\xi_1, \cdots, \zeta_f)$ occur for lattice arrangements of the nuclear coordinates. Let us consider a minimum of this type and describe the lattice in terms of the notation of Sec. 22, Chap. III. It will be assumed that the corner points of the unit cells are specified by the vectors

$$\mathbf{r}(p) = p_1\tau_1 + p_2\tau_2 + p_3\tau_3 \tag{1}$$

where the p_i are integers and the τ_i are the primitive translations. In place of the variables ξ_a, η_a, and ζ_a, we may introduce the variables

$$\mathbf{r}_\alpha(p_1,p_2,p_3) = \mathbf{r}(p_1,p_2,p_3) + \varrho_\alpha \tag{2}$$

where the $\mathbf{r}_\alpha$ are the position vectors, relative to the origin of coordinates of the n atoms in the cell specified by p_1, p_2, p_3 and ϱ_α is the position of these atoms relative to the corner point. In addition, we may introduce the variables

$$x_\alpha{}^i(p_1,p_2,p_3), \qquad (i = 1, 2, 3), \tag{3}$$

for the coordinates of the displacement of the αth atom from its equilibrium position.

If $E_0(\xi_1, \cdots, \zeta_f)$ is expanded in terms of the $x_\alpha{}^i(p)$, we need retain only the quadratic term in the first approximation. The problem of finding the stationary states then reduces to the normal coordinate problem that was discussed in Sec. 22, Chap. III. We know, from the results derived there, that the quadratic terms in E_0 may be reduced to the sum of squares by making the normal coordinate substitution

$$x_\alpha{}^i(p) = \frac{1}{\sqrt{NM_\alpha}} \sum_{t,\sigma} a_t(\mathfrak{d}) \xi^i{}_{\alpha,t}(\mathfrak{d}) e^{2\pi i\sigma \cdot \mathbf{r}_\alpha(p)} \tag{4}$$

where $a_t(\mathfrak{d})$ is the complex amplitude of the tth normal mode of wave number $\mathfrak{d}$, and $\xi^i{}_{\alpha,t}(\mathfrak{d})$ is the complex direction vector of the displacement of the αth atom. If the $a_t(\mathfrak{d})$ are replaced by the real amplitudes

and

$$\begin{aligned}\alpha_t(\mathfrak{d}) &= \frac{a_t(\mathfrak{d}) + a_t{}^*(\mathfrak{d})}{\sqrt{2}} \\ \alpha_t(-\mathfrak{d}) &= \frac{a_t(\mathfrak{d}) - a_t{}^*(\mathfrak{d})}{\sqrt{2}i},\end{aligned}\Bigg\} \tag{5}$$

the normal form of E_0^q is

$$E_0^q = \tfrac{1}{2}\sum_{t,\sigma} 4\pi^2 \nu_t^2(\mathbf{d})\alpha_t^2(\mathbf{d}) \tag{6}$$

where $\mathbf{d}$ is summed over an entire zone.   There are $3nN$ $\alpha_t(\mathbf{d})$ in all, for $\mathbf{d}$ has N independent values and t has $3n$ values.

Since the kinetic energy T has the form

$$T = \tfrac{1}{2}\sum_{t,\sigma} \dot{\alpha}_t^2(\mathbf{d}) \tag{7}$$

when expressed in terms of the α, the Hamiltonian function of the system is

$$H = \tfrac{1}{2}\sum_{t,\sigma}[p_t^2(\mathbf{d}) + 4\pi\nu_t^2(\mathbf{d})\alpha_t^2(\mathbf{d})] \tag{8}$$

where $p_t(\mathbf{d})$ is the momentum variable conjugate to $\alpha_t(\mathbf{d})$.   The corresponding quantum operator is

$$H = \frac{1}{2}\sum_{t,\sigma}\left[-\hbar^2\frac{\partial^2}{\partial\alpha_t^2(\mathbf{d})} + 4\pi^2\nu_t^2(\mathbf{d})\alpha_t^2(\mathbf{d}) \right], \tag{9}$$

which may be separated into operators for each normal coordinate. Hence, the stationary-state wave function has the form

$$\Lambda(\cdots, \alpha_t(\mathbf{d}), \cdots) = \prod^{t,\sigma}\lambda_{n(t,\sigma)}(\alpha_t(\mathbf{d})) \tag{10}$$

where $\lambda_n(\alpha_t(\mathbf{d}))$ satisfies the Schrödinger equation for a simple harmonic oscillator, namely,

$$-\frac{\hbar^2}{2}\frac{\partial^2\lambda_n}{\partial\alpha_t^2(\mathbf{d})} + 2\pi^2\nu_t^2(\mathbf{d})\alpha_t^2(\mathbf{d})\lambda_n = \epsilon_n\lambda_n. \tag{11}$$

The total energy in the state (10) relative to the minimum of E_0 is

$$E(\cdots, n_t(\mathbf{d}), \cdots) = \sum_{t,\sigma}\epsilon_{n(t,\sigma)} = \sum_{t,\sigma}[n(t,\mathbf{d}) + \tfrac{1}{2}]h\nu_t(\mathbf{d}). \tag{12}$$

Let us examine the properties of the lowest state.   The normalized eigenfunctions of the wave equation (11) are

$$\lambda_0(\alpha_t(\mathbf{d})) = \left(\frac{4\pi\nu}{h}\right)^{\frac{1}{4}} e^{-\frac{1}{2}\frac{4\pi^2\nu_t^2(\sigma)\alpha_t^2(\sigma)}{h\nu_t(\sigma)}} \tag{13}$$

so that

$$\Lambda_0 = Ce^{-\sum_{t,\sigma}\frac{1}{2}\frac{4\pi^2\nu_t^2(\sigma)\alpha_t^2(\sigma)}{h\nu_t(\sigma)}}. \tag{14}$$

As an approximation, we may replace the exponent by

$$-\frac{1}{h\bar{\nu}}\left[\sum_{t,\sigma}\frac{1}{2}4\pi^2\nu_t^2(\sigma)\alpha_t^2(\sigma)\right] = -\frac{E_0^q}{h\bar{\nu}} \tag{15}$$

where E_0^q is the potential energy and $\bar{\nu}$ is a mean oscillator frequency, which is of the order of magnitude $k\Theta_D/h$, Θ_D being the characteristic temperature of the substance. The probability of finding the system in the volume element $\prod^{t,\sigma} d\alpha_t(\sigma)$ is

$$\left[\prod^{t,\sigma}\left(\frac{4\pi\nu_t(\sigma)}{h}\right)^{\frac{1}{2}} d\alpha_t(\sigma)\right] e^{-2\frac{E_0^q}{h\bar{\nu}}}, \tag{16}$$

and the probability of finding E_0^q lying in the range from E to $E + dE$ is

$$C'e^{-\frac{2E}{h\bar{\nu}}}E^{\frac{3N}{2}-1}dE \tag{17}$$

where C' is a constant. This function has a very steep maximum of width $h\bar{\nu}$ at the value of E_0 satisfying the equation

$$E = \tfrac{3}{4}Nh\bar{\nu}.$$

The fluctuation in the position and energy of a single atom may be estimated by expressing Eq. (16) in terms of the atomic coordinates. If we keep all atoms except one fixed at their equilibrium positions, E depends on the displacement variable x of this atom as the function kx^2, where k is related to $\bar{\nu}$ in order of magnitude by the equation

$$k \cong 4\pi^2\bar{\nu}^2M$$

in which M is the atomic mass. Hence, the probability depends on x^2 through a factor of the form

$$e^{-\frac{4\pi^2\bar{\nu}M}{h}x^2}.$$

The half width of this distribution, namely, $(1/2\pi)\sqrt{h/\bar{\nu}M}$, is 10^{-9} cm for $\bar{\nu} \cong 10^{14}$ sec^{-1} and $M = 2 \times 10^{-24}$, the mass of hydrogen, a fact showing that the range of fluctuation ordinarily is small compared with interatomic distances.

119. Polymorphism.—When $E_0(\xi_1, \cdots, \zeta_f)$ has a relative minimum for two or more crystallographic phases, the thermodynamically stable one at the absolute zero of temperature is that having the lowest energy. Another arrangement may be more stable, however, at high temperatures. We may obtain a simple interpretation of this fact in the following way.

According to Boltzmann's theorem,[1] the relative probability of finding the αth modification in the energy state E_α at temperature T is

$$G(E_\alpha)e^{-\frac{E_\alpha}{kT}} = e^{-\frac{E_\alpha - TS(E_\alpha)}{kT}} \qquad (1)$$

where $G(E_\alpha)$ is the degeneracy of the energy level E_α and $S = k \log G(E_\alpha)$ is the entropy associated with this level. The function (1) has an extremely steep maximum at the value of E_α satisfying the equation

$$\frac{dE_\alpha}{dS_\alpha} = T, \qquad (2)$$

as may be proved by setting the derivative of (1) equal to zero. The sharpness of this peak may be appreciated from the fact that E_α is of the order of calories for an ordinary-sized crystal, whereas the fluctuations in E_α are of the order of kT, which is about 10^{-20} cal at ordinary temperatures. The condition (2) allows us to specify the equilibrium state of a given modification at any temperature very simply, for this state corresponds to the point on the $E(S)$ curve at which the slope is T (cf. Fig. 4). It should be observed, in passing, that the condition that the function (1) be a maximum is that the function $A(E) = E - TS(E)$ be a minimum. Since A is the thermodynamical free energy, this condition is identical with the thermodynamical condition for determining the stable state. The numerical value of A at any tempera-

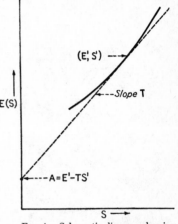

FIG. 4.—Schematic diagram showing the relationship between E, S, T, and A. The full line is the $E(S)$ curve as determined, for example, by solving the Schrödinger equation. The equilibrium state at temperature T is the state corresponding to the point (E', S') where the slope of $E(S)$ is T. The intercept of this tangent with the energy axis is the free energy. The specific heat at temperature T is related to the second derivative by the equation

$$C_V = \frac{\left(\frac{dE}{dS}\right)}{\left(\frac{d^2E}{dS^2}\right)}.$$

ture is determined from the quantities in Fig. 4 by extrapolating the tangent line to the energy axis.

The relationship between E and S may be derived very easily when the lattice frequencies are all the same. In this case, the number of quanta available for distribution among the $3N$ degrees of freedom of the crystal is

[1] See the footnotes on statistical mechanics in Chaps. III and IV.

$$n = \frac{E - E'}{h\nu} \qquad (3)$$

where E' is the energy of the lowest state and ν is the vibrational frequency of the modes. The degeneracy G of this state is the total number

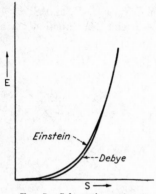

Fig. 5.—Schematic representation of the $E(S)$ curves for the lattice vibrations of a crystal. The Einstein curve rises more rapidly than the Debye curve because there are fewer ways of dividing E into quanta if the Einstein frequency distribution is used.

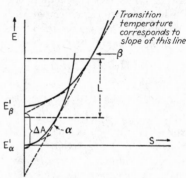

Fig. 6.—$E(S)$ curves for two crystalline phases. Both curves resemble those of Fig. 5; however, the curve rises more rapidly in the crystal having the higher vibrational frequency (the α phase in this case). Since the curves cross, they have a common tangent and a phase change will occur under equilibrium conditions at the temperature corresponding to the slope of the common tangent. The latent heat is the energy difference at the points of tangency, etc.

of ways in which these quanta may be distributed among the modes, namely,

$$G = \frac{(3N + n - 1)!}{(3N - 1)!\, n!}. \qquad (4)$$

Hence, by the use of Stirling's approximation, we find

$$S = k \log G \cong k[(3N + n) \log (3N + n) - 3N \log 3N - n \log n]. \qquad (5)$$

S and dE/dS are zero when $n = 0$ and S increases monatomically with E (*cf.* Fig. 5). The $E(S)$ curve rises more slowly for small values of S if the Debye distribution of frequencies is used. We may determine the value of E and S at any temperature T by employing the value of n that minimizes the free energy $E - TS$. Using E and S as determined by Eqs. (3) and (5), we find

$$n = 3N \frac{1}{e^{\frac{h\nu}{kT}} - 1}.$$

When E and S are computed from this, it is found that

$$A = E - TS = E' - 3NkT \log \frac{e^{\frac{h\nu}{kT}}}{e^{\frac{h\nu}{kT}} - 1}. \tag{6}$$

Let us consider two different crystallographic modifications α and β of a substance and determine the transition temperature for a case in which Eq. (5) may be used for both phases. The fundamental frequencies of the two phases will be designated by ν_α and ν_β, respectively. If it is assumed that E_β' is greater than E_α', so that the α phase is most stable at low temperatures, and that ν_α is greater than ν_β, the two $S(E)$ curves have the form shown in Fig. 6. Since $S_\alpha(E - E_\alpha')$ rises more slowly than $S_\beta(E - E_\beta')$, the two curves cross and have a common tangent line.

The ratio of the probabilities P_α and P_β of finding the system in either the first or the second phase at temperature T is

$$\frac{P_\alpha}{P_\beta} = e^{-\frac{[E_\alpha(T) - TS_\alpha(T)] - [E_\beta(T) - TS_\beta(T)]}{kT}}$$

$$= e^{-\frac{A_\alpha(T) - A_\beta(T)}{kT}} \tag{7}$$

where A_α and A_β are the free energies of the two phases. The ratio (6) is either very great or very small except for a narrow temperature range in which A_α and A_β differ by a factor of order of magnitude kT. Thus, the transition temperature T' is given by the thermodynamical equation

$$A_\alpha(T') = A_\beta(T'). \tag{8}$$

It may be seen from the construction of Fig. 6 that this condition is satisfied at the temperature corresponding to the slope of the tangent line of the two $E(S)$ curves. The α phase is stable below this temperature, and the β phase is stable above; moreover, the heat of the transition is equal to the difference L between the energies of the two tangent points.

According to Eq. (6), the free energies of the two phases at temperature T are

$$\left. \begin{array}{l} A_\alpha(T) = E_\alpha' - 3NkT \log \dfrac{e^{\frac{h\nu_\alpha}{kT}}}{e^{\frac{h\nu_\alpha}{kT}} - 1} \\[20pt] A_\beta(T) = E_\beta' - 3NkT \log \dfrac{e^{\frac{h\nu_\beta}{kT}}}{e^{\frac{h\nu_\beta}{kT}} - 1}. \end{array} \right\} \tag{9}$$

The difference of these free energies is

$$\Delta A_{\alpha\beta} = E_{\alpha}' - E_{\beta}' - 3NkT \left[\log e^{\frac{h(\nu_{\alpha} - \nu_{\beta})}{kT}} - \log \frac{e^{\frac{h\nu_{\alpha}}{kT}} - 1}{e^{\frac{h\nu_{\beta}}{kT}} - 1} \right], \quad (10)$$

which is zero at the temperature defined by the equation

$$e^{-\frac{E_{\alpha}' - E_{\beta}'}{3NkT'}} = \frac{e^{-\frac{h\nu_{\alpha}}{kT'}} - 1}{e^{-\frac{h\nu_{\beta}}{kT'}} - 1}. \quad (11)$$

The necessary and sufficient condition that must be satisfied if this equation is to have a root is that ν_{α} should be greater than ν_{β} if E_{β}' is greater than E_{α}'.

The condition replacing (11) when all the $3N$ frequencies are different is

$$e^{-\frac{E_{\alpha}' - E_{\beta}'}{kT}} = \frac{\prod^{\nu_{\alpha}} (e^{-\frac{h\nu_{\alpha}}{kT'}} - 1)}{\prod^{\nu_{\beta}} (e^{-\frac{h\nu_{\beta}}{kT'}} - 1)}, \quad (12)$$

as may be seen by using the expression

$$E_{\alpha}' + kT \log \prod^{\nu_{\alpha}} (1 - e^{-\frac{h\nu_{\alpha}}{kT}}) \quad (13)$$

for the free energy of a system of oscillators.[1] Equation (12) is difficult to solve directly even when there is a simple relation between frequency and the wave number. In practical work, it actually is simpler to compute numerically the free energies of the phases and to find the temperature at which these functions are equal. The specific heats of none of the phases for which transitions have been investigated thoroughly obey either the Einstein or the Debye law, however, so that there is no need for discussing these computations in detail. Instead, we shall discuss several actual cases. It should be mentioned at this point that the credit for the first intensive investigations of the thermal

---

[1] This expression may be derived from the partition functions discussed in Sec. 18, Chap. III, by the use of the relation

$$-kT \log f = A$$

connecting the free energy of a system and its total partition function. [See, for example, R. H. Fowler, *Statistical Mechanics* (Cambridge University Press, 1936).]

effects associated with allotropy is due to Nernst,[1] who used this subject as the cornerstone in establishing his heat theorem.

a. Tin.—The transition between black and white tin has been studied fairly completely and has practical interest because it is responsible for tin disease which may impair the protective coating of tinned metals. The low-temperature, or black, form has the diamond structure; the high-temperature form has a complex tetragonal lattice. The thermodynamical transition point was determined most accurately by Cohen and van Eijk[2] who measured the temperature at which the emf of an electrolytic cell in which the two electrodes are made of the different phases vanishes. This temperature is 292°K. The specific heats of

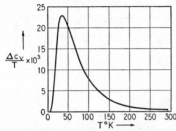

FIG. 7.—The $\Delta c_V/T$ curve for gray and white tin. The specific heat of white tin is the larger and accounts for the phase change. The ordinates are cal/deg².

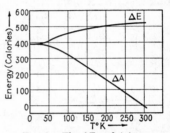

FIG. 8.—The ΔE and ΔA curves for gray and white tin. Below the transition temperature, A is smaller for the gray modification. The point at which ΔA becomes zero is the transition temperature.

both phases were measured by Lange[3] to within a few degrees of absolute zero; the difference of these specific heats divided by T is shown in Fig. 7. The specific-heat curves do not obey the Debye law closely, but they do approach $3R$ at high temperatures, showing that the oscillator model is probably accurate. The transition heat ΔE of the phase change was measured by Brönsted and was found to be 535 cal at the transition temperature and 399 cal at absolute zero. The complete transition-heat curve is shown in Fig. 8. The corresponding free-energy curve ΔA, which may be determined by computing $\Delta E(T)$ and $\Delta S(T)$ from the empirical data under the condition that $S(T)$ vanish at absolute zero, is shown in the same figure. The same curve may be obtained from $\Delta E(T)$ alone by solving the Gibbs-Helmholtz equation

[1] W. NERNST, *The New Heat Theorem* (E. P. Dutton & Company, Inc., New York, 1926).

[2] E. COHEN and C. VAN EIJK, *Z. physik. Chem.*, **30A**, 601 (1899).

[3] F. LANGE, *Z. physik. Chem.*, **110A**, 360 (1924).

[4] J. N. BRÖNSTED, *Z. physik. Chem.*, **65**, 744 (1909).

$$\Delta A = \Delta E - T\frac{\partial \Delta A}{\partial T}$$

under the third-law condition that $\partial \Delta A/\partial T$ be zero at absolute zero. ΔA crosses the axis at 295°, showing good agreement with Cohen and van Eijk's directly measured value.

b. Sulfur.—The transition of sulfur from rhombic to monoclinic form at 368.5°K was investigated by Nernst,[1] Brönsted,[2] and a number of other workers. It is worth mentioning that this transition is the first recorded case of allotropy.[3] The specific-heat curves deviate considerably from the Debye form, and their difference is shown in Fig. 9. The transition temperature as computed from the point at which ΔA vanishes is 370°K.

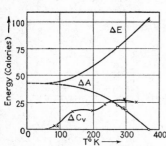

Fig. 9.—The ΔE, ΔA and ΔC_V curves for the two phases of sulfur.

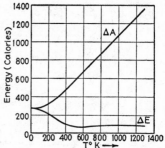

Fig. 10.—The ΔA and ΔE curves for diamond and graphite. The latter has the lower free energy at all temperatures at ordinary pressures.

c. Carbon.—Detailed measurements of the transition heat of the diamond-graphite phase change by Roth and Wallasch[4] show that graphite has the lower energy at ordinary temperatures and pressures. The difference, however, is only 160 ± 30 cal at room temperature. The lower curve of Fig. 10 shows the temperature dependence of this difference. Since the characteristic temperature of diamond is higher than that of graphite at ordinary pressures, it follows from the preceding discussion that graphite is more stable than diamond at all temperatures. The free-energy difference curve of Fig. 10 supports this conclusion, for it rises away from the value at absolute zero.

Simon[5] has made a thermodynamical estimate of the pressure dependence of the free-energy curve and has concluded that pressures in the

[1] Nernst, *op. cit.*

[2] J. N. Brönsted, *Z. physik. Chem.*, **55**, 371 (1906).

[3] Mitscherlich, *Ann. Physik*, **88**, 328 (1852).

[4] Roth and Wallasch, *Ber. deut. Chem. Ges.*, **46**, 896 (1913).

[5] F. Simon, *Handbuch der Physik*, Vol. X, p. 376 (Julius Springer, Berlin, 1926).

neighborhood of fifty thousand atmospheres would be needed to reverse the equilibrium at the high temperatures at which the rate of change is appreciable.

An outstanding exception to the rule that the amplitude of the zero-point oscillation is small compared with the interatomic distance seems to occur in one of the condensed phases of helium. The phase diagram of helium at low temperatures is shown[1] in Fig. 11. In the immediate vicinity of absolute zero, this substance forms a true solid if the pressure is above 25 atmospheres; however, at lower pressures, it forms two liquid phases, which are known as liquid helium I and liquid

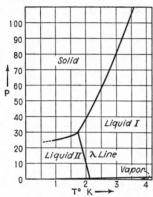

Fig. 11.—The low-temperature phase diagram of helium. There is no solid phase below 25 atmospheres. The ordinate is expressed in atmospheres.

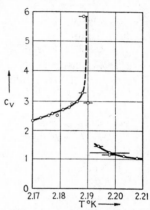

Fig. 12.—The anomaly in the specific heat of liquid helium at the λ-point. The ordinate is expressed in cal/gram-degree. (*After Keesom and Keesom.*)

helium II, and no ordinary type of solid. The density of the liquid phases is about 0.70 relative to that of the high pressure solid phase. The transition point between helium I and helium II, which is known as the λ point, is distinguished by several striking effects. Thus, it is found that the specific-heat curve[2] has the discontinuity shown in Fig. 12 and that the fluidity[3] and thermal conductivity increase very much in passing below the λ point. In addition, it is found[4] that liquid helium II exhibits the mechanical analogue of the thermoelectrical effects observed

[1] W. H. Keesom and K. Clusius, *Proc. Acad. Sci. Amsterdam*, **35**, 320 (1932); W. H. Keesom and H. P. Keesom, *Proc. Acad. Sci. Amsterdam*, **35**, 736 (1932).

[2] W. H. Keesom and H. P. Keesom, *Physica*, **2**, 557 (1935).

[3] W. H. Keesom and H. P. Keesom, *Physica*, **2**, 557 (1935); B. V. Rollin, *Physica*, **3**, 266 (1936); J. F. Allen, R. Peierls, and M. Z. Uddin, *Nature*, **140**, 475 (1937).

[4] J. O. Wilhelm, A. D. Misener, and A. R. Clark, *Proc. Roy. Soc.*, **151**, 342 (1935); E. F. Burton, *Nature*, **135**, 265 (1935), **142**, 72 (1938).

in metals, for mechanical flow is induced as a result of temperature gradients.

Guided by the similarity of the specific-heat curve shown in Fig. 12 and the specific-heat curve observed during the transition between the ordered and disordered state in alloys such as β brass in which the components are present in equal numbers (*cf.* Fig. 43, Chap. I), Fröhlich[1] suggested that the two liquid helium phases represent ordered and disordered phases of a crystal. In particular, he suggested that the phases have the diamond structure (Fig. 4, Chap. I), which may be regarded as a body-centered cubic lattice in which half the atoms are replaced by vacancies, and that the disordering process consists in the interchange of atoms and vacancies. Thus, according to this picture, liquid helium I would correspond to the phase in which there is no long-distance order and liquid helium II would correspond to the partly ordered phase. There are, however, the following two objections to Fröhlich's model: (1) It should be expected that the liquid helium II phase would become more and more solidlike as the temperature is lowered and ordering increases, whereas it is actually found that the viscosity seems to become smaller and smaller. (2) London[2] showed, on the basis of the Slater-Kirkwood expression for the interaction energy of two helium atoms, that the diamond type of lattice is unstable relative to the interchange of vacancies and atoms, so that Fröhlich's ordering process is unlikely.

As a result of this work, London suggested that the amplitude of the zero-point oscillations in these liquid phases is so large that the atoms should be treated as though free in the same sense that the electrons in a metal are free. Since the helium atoms obey Einstein-Bose statistics instead of Fermi-Dirac statistics, London suggested that a qualitative insight into the properties of the two liquid helium phases might be obtained by treating them as a degenerate Bose-Einstein gas. The thermal and mechanical properties of a gas of this type may be obtained by methods analogous to those used in the Sommerfeld theory of metals, the function

$$f(\epsilon) = \frac{1}{Ae^{\frac{\epsilon}{kT}} - 1}$$

replacing the Fermi-Dirac function. It is found that the specific-heat curve of this gas has the singularity shown in Fig. 13 which begins when the particles start to condense in the lowest energy state. London suggested that this singularity corresponds to the λ point of liquid helium

[1] H. Fröhlich, *Physica*, **4**, 639 (1937).

[2] F. London, *Nature*, **141**, 643 (1938), **142**, 612 (1938); *Phys. Rev.*, **54**, 947 (1938).

and that the differences between the curves of Figs. 12 and 13 arise because the helium atoms interact and thus are not perfectly free. The temperature T_0 at which the singularity occurs in Fig. 13 would be 3.14°K for a perfect gas of helium atoms having the observed density of the liquids. This actually is fairly close to the observed λ point at 2.19°K. London has also shown that many of the unusual thermal and mechanical properties of liquid helium II can be given a qualitative explanation on the basis of his simple model.

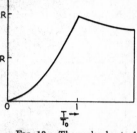

Fig. 13.—The molar heat of a Bose-Einstein gas near the degeneracy temperature T_0. The temperature T_0 is equal to

$$\frac{h^2}{2\pi Mk}\left(\frac{3n_0}{2.612}\right)^{2\!/\!3}$$

where M is the atomic mass and n_0 is the number of atoms per unit volume.

120. The Effect of Electronic Excitation on Phase Changes.—The Gibbs-Helmholtz equation, namely,

$$A = E + T\frac{\partial A}{\partial T}, \tag{1}$$

may be integrated and placed in the form

$$A = -T\int_0^T \frac{E}{T'^2}dT', \tag{2}$$

which makes it possible to compute the free energy when the function $E(T)$ is known. Since

$$E(T') = E(0) + \int_0^{T'} C(T'')dT'' \tag{3}$$

where $C(T)$ is the molar heat,

$$A = E(0) + T\int_0^T \frac{1}{T'^2}\int_0^{T'} C(T'')dT''. \tag{4}$$

Thus, A may be computed from the molar heat. We may conclude from this equation that A is affected appreciably by a given part of a system only when this part contributes to the specific heat. Since the electrons do not contribute appreciably to the specific heat in simple metals and insulators, as is evidenced by their obedience to Dulong and Petit's law above the characteristic temperature, we may conclude that phase changes in these solids are not influenced by the electronic excitation which ordinarily occurs. This is not true of substances containing transition-element atoms, however, for the electronic specific heat usually is comparable with the $3R$ value at sufficiently high temperatures.

An important case in which the electronic excitation probably plays a role is afforded by iron. As we have seen in Sec. 2 the α or body-centered phase is stable at all temperatures in the range from 0°K to the melting point at 1803°K except for a region extending from 1174° to 1674°K in which the face-centered γ phase is stable. It is possible to understand this behavior by using the information given by the specific-heat curves of Fig. 17, Chap. I. We may see from this figure that the γ phase has the lower characteristic temperature; hence, we may conclude from the discussion of the preceding section that it would be the stable phase at high temperatures if the free energy were determined by the lattice vibrations alone. Actually, the electronic specific heat of the α phase is larger than that of the γ phase at temperatures above 580°K, and this difference tends to compensate for the "advantage" the γ phase receives from the larger value of the vibrational specific heat below 200°C.

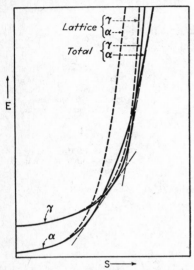

The situation that probably exists is indicated schematically in Fig. 14 in which the dotted lines represent the contribution to the $E(S)$ curves of the two phases from lattice vibrations. If these were the actual curves, the γ phase would be stable at high temperatures. The electronic specific heat alters the curves and leads to the full lines. Since the electronic specific heat of the α phase is the larger at high temperatures, the $E(S)$ curve of this phase is altered most. We may conclude that the $E(S)$ curve for the γ phase crosses that for α twice in the manner shown, two tangent lines being thus produced.

Fig. 14.—Probable behavior of the $E(S)$ curves of the α and γ phases of iron. The α curve is initially lower, but rises more rapidly because the characteristic temperature of the α phase is larger. At higher values of S, the larger electronic heat of the α phase reverses the curves.

If the electronic specific heats of α and γ iron were accurately measured,[1] it would be possible by the use of Eq. (4) to test the preceding picture by direct computations of the contributions to A.

Since cobalt has a similar reappearance of the low-temperature phase at high temperatures, we may conclude that Fig. 14 also applies to it.

121. Melting.—We shall not attempt to give a survey of the present status of the theory of liquids, for to do so would carry us too far afield, but we shall mention briefly the process of melting.[2]

[1] An analysis of the entire specific-heat curve of nickel that is oased on the use of the low-temperature value has been given by E. C. Stoner, *Phil. Mag.*, **22**, 81 (1936).

[2] Survey articles on the theory of liquids are as follows: K. F. Herzfeld, *Jour.*

The quadratic approximation for the function $E_0(\xi_1, \cdots, \zeta_f)$, which was discussed in Sec. 118, is not valid when the amplitude of atomic vibrations becomes comparable with the interatomic distances, because the potential well about any atom flattens in the directions in which there are saddle points (*cf.* Fig. 15), such as the saddle points corresponding to the interchange of two atoms. This flattening should cause the density of levels $G(E)$ to increase[1] more rapidly than for the quadratic approximation, so that the entropy may increase in the manner shown in Fig. 16. If the $S(E)$ curve has an inflection point, as in Fig. 16, the

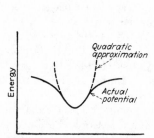

FIG. 15.—Schematic representation of the behavior of the atomic potential well. This deviates from the simple parabolic well assumed in the ordinary theory of specific heats, when the atomic displacements become large.

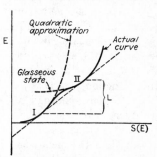

FIG. 16.—Schematic representation of the $E(S)$ curve for an actual solid (full line). The effect of anharmonic forces is to increase the entropy of higher energy states.

system jumps from the state I to the state II at a temperature T given by the slope of the line joining these two points. Below this temperature, the system is crystalline; above it, the system presumably is in the liquid state since this is the phase in which atomic rearrangements take place fairly freely, as is evidenced by the ability of liquids to flow. If ϵ is the height of the activation hill for the interchange of atoms, the number of atoms having enough energy to pass over this hill should vary with temperature in the manner

$$Ae^{-\frac{\epsilon}{kT}} \tag{1}$$

where A is nearly constant. Hence, the flowability, that is, the reciprocal of the viscosity, should vary with temperature as the quantity (1). The

Applied Phys., **8**, 319 (1937); H. Eyring and J. Hirschfelder, *Jour. Phys. Chem.*, **41**, 249 (1937); R. H. Ewell, *Jour. Applied Phys.*, **9**, 252 (1938); J. E. Lennard-Jones, *Physica*, **4**, 941 (1937); N. F. Mott and R. W. Gurney, *Reports on Physics Progress*, Vol. V (Cambridge University Press, 1939).

[1] A part of this increase may be regarded as a mixing entropy (*cf. ibid.*).

energy L (cf. Fig. 16), by which the system jumps during the transition, is the latent heat of fusion. It should be clear from the figure that this heat is connected with the change in entropy ΔS by the equation

$$\Delta S = \frac{L}{T}.$$

The states immediately below II, which are skipped during melting, describe supercooled liquids, whose flowability decreases with decreasing temperature and which become glasses at low temperatures. The $E(S)$ curve of such a glass may depart from the full curve of Fig. 15 (cf. the dotted section) when rearrangements no longer take place. Thus, the entropy may decrease rapidly, although the atomic arrangement is not crystalline and the energy is not so low as for a perfect crystal.

Mott[1] has used a simple model of the liquid state to compute a partition function, which he has employed successfully in relating some properties of the liquid and solid phases. He assumed that the liquid state is dynamically similar to a solid, inasmuch as the individual atoms are vibrating, and he assumed that the mean liquid vibrational frequency ν_l is lower than the mean solid frequency ν_s because the liquid is less rigid than the solid. He neglected the contribution to the partition function from the interchange of atoms so that the relationship between the solid and liquid phase in this model is essentially the same as that we have found for allotropic modifications. Thus, the partition functions for the liquid and solid phases are

$$\left.\begin{aligned} f_l &= e^{-\frac{E_l}{kT}}\left(\frac{e^{\frac{h\nu_l}{kT}}}{e^{\frac{h\nu_l}{kT}}-1}\right)^{3N}, \\ f_s &= e^{-\frac{E_s}{kT}}\left(\frac{e^{\frac{h\nu_s}{kT}}}{e^{\frac{h\nu_s}{kT}}-1}\right)^{3N}, \end{aligned}\right\} \qquad (2)$$

where E_s is the lowest state of the crystal, E_l is the lowest state of the liquid, that is, the energy of the supercooled liquid at absolute zero, and N is the number of atoms. The condition for equilibrium at the melting temperature T_M is

$$\left(\frac{\nu_s}{\nu_l}\right)^3 = e^{\frac{E_l-E_s}{RT_M}}, \qquad (3)$$

in which it is assumed that kT_M is much greater than $h\nu_s$. $E_l - E_s$ is very nearly equal to L, the latent heat of fusion, whence

[1] N. F. MOTT, *Proc. Roy. Soc.*, **146**, 465 (1934).

$$\left(\frac{\nu_s}{\nu_l}\right)^3 = e^{\frac{L}{RT_M}}, \tag{4}$$

which is Mott's relation.

Mott has tested this relation for metals by comparing the ratio ν_s/ν_l computed from (4) with the ratio derived from conductivity measurements. We shall discuss the connection between the electrical conductivity σ and the vibrational frequency of the lattice in the next chapter and shall find that

$$\frac{\sigma_l}{\sigma_s} = \frac{\nu_l^2}{\nu_s^2} \tag{5}$$

in simple cases. In other words, we have

$$\frac{\sigma_s}{\sigma_l} = e^{\frac{2L}{3RT_M}}. \tag{6}$$

The extent to which this equation is satisfied may be seen from Table LXXV. It is valid for the simpler metals, but it fails for the metals with

TABLE LXXV.—COMPARISON OF OBSERVED VALUES OF σ_s/σ_l WITH THOSE COMPUTED FROM MOTT'S RELATION (6) BY THE USE OF OBSERVED VALUES OF L AND T_M

	σ_s/σ_l	
	Observed	Calculated
Li	1.68	1.84
Na	1.45	1.77
K	1.55	1.75
Rb	1.61	1.76
Cs	1.66	1.75
Cu	2.07	1.97
Ag	1.9	2.0
Au	2.28	2.22
Al	1.64	2.0
Cd	2.0	2.3
Pb	2.07	1.87
Sn	2.1	3
Tl	2.0	2.3
Zn	2.09	2.3
Hg	~4	2.23
Bi	0.43	5.0
Ga	0.58	4.5
Sb	0.67	5.6

unusual structures.   This failure is explained in part by the fact that the relation (5) is not obeyed by the transition metals.

Equation (6) should apply to allotropic forms of a metal at the transition temperature in cases in which this temperature is much larger than the characteristic temperature of the two solids.   There do not seem to be enough experimental data available to test the relation in any cases of this kind.

Herzfeld, Mayer, and Kane[1] have computed the free energies of the rare gas solids, relative to the free energies of the gaseous constituents, as functions of the lattice constants and have found that these solids would undergo a discontinuous expansion if they were superheated.   In terms of a diagram of the type of Fig. 12, their results indicate that the first inflection point in the $E(S)$ curve of one of these solids occurs, not because the individual atomic potential wells are anharmonic, but because the curvature of the well decreases as the crystal expands.   At the present time, it is not possible to say whether or not this effect makes an alteration of the qualitative picture of melting, presented above, necessary in all solids.

In the Debye approximation, the molar free energy of the solid may be written in the form

$$A(r) = E_0(r) + \tfrac{3}{2}N_A h\nu_m + A_D(\nu_m, T, r) \tag{7}$$

where $E_0(r)$ is the molar electronic energy, $\tfrac{3}{2}N_A h\nu_m$ is the zero-point vibrational energy, $A_D(\nu_m, T, r)$ is the free energy obtained from a Debye function, namely,

$$A = -T \int_0^T \frac{dT}{T^2} E_D(\nu_m, T), \tag{8}$$

and ν_m is the maximum frequency of the lattice.   The energy $E_0(r)$ was expressed in terms of a van der Waals interaction term of the form

$$-\frac{C}{r^6} \tag{9}$$

and a repulsive term of the form

$$CBe^{-\frac{r}{\rho}}. \tag{10}$$

Thus,

$$E_0(r) = C\left\{ -\frac{14.5}{r^6} + 12Be^{-\frac{r}{\rho}}\left[1 + \frac{1}{2}e^{-\frac{(\sqrt{2}-1)r}{\rho}} + \cdots \right]\right\} \tag{11}$$

[1] K. F. Herzfeld and Maria G. Mayer, *Phys. Rev.*, **46**, 995 (1934); Brother Gabriel Kane, *Jour. Chem. Phys.*, **7**, 603 (1939).

in which the factor 14.5 is the coefficient in the sum of terms (9) for a face-centered lattice and the coefficient of $12B$ is the sum of terms (10). In each case, ρ was given several values. For example, the values 0.345 Å and 0.209 Å were used in all cases. The first is the empirical value of Born and Mayer (see Chap. II), and the second is the value found by Bleick for neon. Corresponding values of C and B were then determined with the condition that (11) plus the zero-point energy should give the observed lattice constant and cohesive energy at low temperatures. The process of determining these constants actually involves a reiteration procedure; for ν_m, which occurs in the zero-point energy, was determined from the theoretical expressions for the elastic constants, which in turn involve C and B.

Figure 17 shows the computed and observed values of the internal pressure $-\partial A/\partial V$ of krypton for several temperatures near the observed melting temperature, namely, 116.0°K. The two sets of curves correspond to the values 0.345 and 0.209, respectively. The ordinary equilibrium volume is determined by the condition that $\partial A/\partial V$ vanish. The interesting feature of these curves is that those for temperatures above 108° and 91°K, respectively, are positive everywhere, which indicates that the body-centered crystal becomes unstable. This behavior is tied up with the fact that the elastic constants, and hence ν_m, decrease

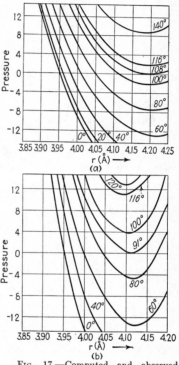

FIG. 17.—Computed and observed values of the internal pressure of solid krypton for several temperatures. *a* corresponds to the value 0.345Å of the repulsive parameter and *b* to the value 0.209Å. The minima that occur at high temperatures imply that the crystals become unstable. (*After Kane.*)

with increasing lattice constant and thus raise A_D. Thus, the free energy presumably could be decreased by rearranging the atoms into another phase, such as the liquid, and at the same time increasing the atomic volume. Since the solid phase becomes unstable even relative to the gaseous phase, it is natural to suggest that melting is forced by the disruption of the crystal.

122. Atomic Diffusion*.—Atomic diffusion has been studied with fair experimental accuracy[1] in a number of metals. The processes that have been treated are the diffusion of constituents in substitutional and interstitial alloys and the self-diffusion in those monatomic metals having radioactive isotopes.[2] Although the diffusion of atoms in insulating crystals has not been investigated so thoroughly as diffusion in metals, the general facts probably are very similar in the two cases.

It is usually assumed that diffusion in solids obeys the space-time diffusion equation

$$\frac{dc}{dt} = \text{div} D \text{ grad } c \tag{1}$$

where c is the concentration of the diffusing atoms and D is the diffusion coefficient, which usually depends on c. In order to translate D into an atomic constant in a simple case, let us assume that the distance between atomic planes in the direction of diffusion is δ and that the probability that an atom moves from one plane to the next in unit time is d. If n_1 is the number of diffusing atoms per unit area and n_2 is the number in the neighboring plane, the number per unit area that passes from plane 1 to plane 2 in unit time is

$$\frac{dN}{dt} = d(n_1 - n_2). \tag{2}$$

The quantity $n_1 - n_2$ is δ^2 times the concentration gradient in the direction normal to the plane, however, so that Eq. (2) is

$$\frac{dN}{dt} = \delta^2 d\mathbf{r}_1 \cdot \text{grad } c \tag{3}$$

where $\mathbf{r}_1$ is a unit vector normal to the plane. Equation (3) reduces to (1) if we set

$$D = d\delta^2. \tag{4}$$

This method of reasoning can be used to convert D into an atomic constant d, in more complex cases than the one treated here.

It is found experimentally that for fixed concentration D depends upon temperature in the manner

$$D = Ae^{-\frac{\epsilon}{kT}} \tag{5}$$

[1] See the review articles by R. F. Mehl, *Jour. Applied Phys.*, **8**, 174 (1937); R. M. Barrer, *Proc. Phys. Soc.* **52**, 58 (1940).

[2] The radioactive indicator method was first used by G. von Hevesy, W. Seith, and A. Keil (*cf. Z. physik. Chem.*, **37**, 528 (1931); *Z. Physik*, **79**, 197 (1932)).

where both A and ϵ are practically constant. Using equations of the type (4), we may translate A into an atomic constant. Thus, in the case in which (4) is valid, we may write

$$A = a\delta^2$$

where a is the "jumping frequency" for a given atom when kT becomes large compared with ϵ. It is also found experimentally[1] that these frequencies usually are of the order of 10^{13}. Figure 18 shows the dependence of log D on $1/T$ for the diffusion of gold in lead and illustrates a typical case in which Eq. (5) is valid. A number of values[2] of ϵ that are determined from the slopes of curves of this type are given in Table LXXVI.

There are three conceivable simple mechanisms for diffusion of atoms A in a solid AB, namely, the following:

1. Atoms A and B may interchange places, squeezing by one another in the normal lattice.

2. The A atoms may diffuse individually through interstices.

3. The diffusion may take place with the help of vacancies, the atoms moving only into vacant adjacent sites. It may be postulated that in this case the crystal with vacancies is thermodynamically more stable than one without vacancies.

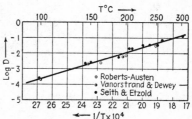

FIG. 18.—Temperature dependence of the diffusion coefficient of gold in lead. (*After Mehl.*)

The first mechanism has the disadvantage that if it were valid one might expect the activation energy for the process in which two atoms squeeze by one another to be very high, of the order of the cohesive energy, whereas the values in Table LXXVI are uniformly less than the cohesive energies. For this reason, this process is ordinarily ruled out.

---

[1] This fact was pointed out by S. Dushman and I. Langmuir, *Phys. Rev.*, **20**, 113 (1922), who suggested that an approximate value of a should be obtained from the relation $a = \epsilon/h$. Since ϵ is of the order of magnitude 10^{-14} erg, the values of a obtained in this way are of the order of magnitude 10^{13}.

[2] Most of these values are taken from footnote 1, p. 494. The values for self-diffusion in copper, gold, zinc and bismuth have the following origin.

Cu: J. STEIGMAN, W. SHOCKLEY, and F. C. NIX, *Phys. Rev.*, **56**, 13 (1939).

Au: SAGRUBSKIJ, *Physik. Z. Sowj.*, **12**, 118 (1937); MCKAY, *Trans. Faraday Soc.*, **34**, 845 (1938).

Zn: F. BANKS, H. DAY and P. MILLER (see program Washington Meeting, Am. Phys. Soc., 1940).

Bi: W. SEITH and A. KEIL, *Z. Elektrochem.*, **39**, 538 (1933).

If it does occur, it is easy to understand why a is of the order of magnitude 10^{13}, for this is the magnitude of atomic vibrational frequencies.

The second process should require a smaller activation energy than the first only in crystals that have interstitial sites sufficiently large to accommodate the atoms A. Since the interstices of metals forming substitutional alloys usually are much smaller than the atomic size, it seems probable that this mechanism occurs only in interstitial alloys, such as iron carbide and nitride, and in semi-conductors, such as zinc oxide, that have interstitial atoms (*cf.* Sec. 112). We saw, in the case of zinc oxide, that the fraction of interstitial atoms f is governed by an equation of the type

$$f = n_{O_2}^{-\frac{1}{4}} B' e^{-\frac{\epsilon''}{kT}} \qquad (6)$$

where n_{O_2} is the density of oxygen in the surrounding vapor and B' is a constant. If the probability that one of these atoms jumps is

$$\tilde{\nu}_i = \tilde{\nu}^i e^{-\frac{\epsilon'}{kT}}, \qquad (7)$$

TABLE LXXVI.—ACTIVATION ENERGIES FOR METALLIC DIFFUSION

Solvent	Solute	Activation energy, ev
Cu	Cu	2.5
	Zn (9.58%)	1.8
	Zn (29.08%)	1.8
	Sn (10%)	1.7
Zn	Zn	0.75
Pb	Pb	1.2
	Sn	1.04
	Tl	0.91
	Bi	0.81
	Cd	0.78
	Ag	0.66
	Au	0.57
Ag	Au	2.5
Au	Ag	3.0
Bi	Bi	⊥1.3
		∥6.1
Fe	C	1.6
	N	1.5

the atomic diffusion coefficient d should be

$$d = \tilde{\nu}^i n_{O_2}{}^{-\frac{1}{2}} B' e^{-\frac{\epsilon' + \epsilon''}{kT}}.$$ (8)

In this case, both a and ϵ are composite quantities.

One should expect a much smaller value of the activation energy for the third process than for the two others in the case of substitutional alloys.[1] Since the number of vacancies should be less than the total number of atoms in this case, d should be a composite quantity as in Eq. (8). According to the discussion in Sec. 110, the fraction of vacancies f is given by the equation

$$f = e^{-\frac{\epsilon''}{kT}}$$

where ϵ'' is the energy required to remove an atom from an interior site to the surface. Hence,

$$d = \tilde{\nu}_\infty^v e^{-\frac{\epsilon' + \epsilon''}{kT}}$$ (9)

where $\tilde{\nu}_\infty^v e^{-\frac{\epsilon'}{kT}}$ is the probability per unit time that an atom and a vacancy on neighboring sites change places.

It is reasonable to suppose that the third mechanism occurs in substitutional alloys, but this supposition has not been conclusively demonstrated.

We may develop an equation for the jump frequency, using elementary principles of the theory of reaction rates. We shall treat the problem generally enough so that the results are applicable both to diffusing interstitial atoms and to diffusing vacancies although we shall refer to the diffusing particle as an interstitial atom.

As the atom moves from one equilibrium position to another, the energy of the system rises through a maximum at the saddle point of the barrier between minima. According to the theory of reaction rates, we may regard the jumping process as an act in which the system is thermally excited to the saddle point S through which it then passes. The probability per unit time $\tilde{\nu}$ that this process occurs for a given atom is then equal to the rate at which atoms pass through S divided by the total number of interstitial atoms. For simplicity, we shall assume that the energy of the system depends only upon the three positional coordinates of the jumping atom; moreover, we shall assume that the potential is nearly constant for a short distance along the direction of flow

---

[1] Theoretical work of H. Huntington, in progress, indicates that the activation energy for interstitial diffusion in copper is about three times larger than that for the third process.

through the saddle point. The rate at which the atoms pass through S then is equal to the number of atoms per unit length of S times their mean velocity. As long as only a small fraction of the diffusing atoms are at the saddle point at any one time, the number of atoms n_s per unit length of S is equal to the total number of interstitial atoms n times the ratio of the partition function per unit length of S to the partition function of the interstitial atom at its equilibrium position.

We shall assume that the interstitial atom in an equilibrium position is at rest, for in this book this assumption is usually made in computing the number of interstitial atoms. It is incorrect, since the interstitial atom actually oscillates about an equilibrium position. The error made in this way, however, cancels in taking the product of n and the probability of finding an atom at S.

In addition, we shall assume that the forces acting on an atom in the saddle point are also harmonic in the two directions orthogonal to the direction of flow. If ν_s is the vibrational frequency, the partition function for these 2 degrees of freedom then is

$$f = \left(\frac{1}{1 - e^{-\frac{h\nu_s}{kT}}} \right)^2. \tag{10}$$

We shall be interested primarily in the case in which $h\nu_s$ is much smaller than kT. Then, f is

$$\left(\frac{kT}{h\nu_s} \right)^2. \tag{11}$$

The partition function per unit length in the direction of flow is equal to the partition function for a one-dimensional gas at a point where the potential energy is ϵ_s, where ϵ_s is the height of S above the equilibrium position. This function is

$$\frac{(2\pi MkT)^{\frac{1}{2}}}{h} e^{-\frac{\epsilon_s}{kT}}. \tag{12}$$

Hence, the complete partition function per unit length of the saddle point is

$$f_s = \frac{1}{\alpha} \left(\frac{kT}{h\nu_s} \right)^2 \frac{(2\pi MkT)^{\frac{1}{2}}}{h} e^{-\frac{\epsilon_s}{kT}}, \tag{13}$$

where α is the number of saddle points of height ϵ_s about a given equilibrium position. This number, which should depend upon crystal symmetry as well as upon the type of ions in the lattice, could be as large as forty-eight for a cubic crystal but usually is near unity.

The mean velocity with which the atoms pass through the saddle point is

$$\bar{v} = \frac{\int_0^\infty e^{-\frac{Mv^2}{2kT}} v\, dv}{\int_{-\infty}^\infty e^{-\frac{Mv^2}{2kT}} dv}$$

$$= \left(\frac{kT}{2\pi M}\right)^{\frac{1}{2}}. \tag{14}$$

Hence, the jump frequency $\tilde{\nu}$ is

$$\tilde{\nu} = f_s \bar{v} = \frac{1}{h^3} \frac{(kT)^3}{\alpha v_s^2} e^{-\frac{\epsilon_s}{kT}}. \tag{15}$$

This equation has the form

$$\tilde{\nu} = \nu_\infty e^{-\frac{\epsilon_s}{kT}}$$

where

$$\nu_\infty = \frac{(kT)^3}{\alpha h^3 v_s^2}.$$

According to this result, the diffusion coefficient D has the form

$$D = d\delta^2 = \delta^2 \frac{n}{N} \frac{1}{\alpha h^3} \frac{(kT)^3}{v_s^2} e^{-\frac{\epsilon_s}{kT}}, \tag{16}$$

where n is the number of interstitial atoms or vacancies per unit volume and N is the total number of atoms per unit volume. This equation will be used in Sec. 132.

123. The Phase Boundaries of Alloys.—By applying the principle of minimum free energy, it is possible to derive the equations that determine the phase boundaries of alloys. Let us consider two binary alloy phases α and β of a pair of monatomic metals A and B. The necessary condition that the two phases be in thermodynamical equilibrium evidently is that the free energy of the entire system remain stationary if atoms are taken from one alloy to the other.

We shall assume that there are N_a A atoms and N_b B atoms in the entire system and that the total number N remains fixed when the atoms are taken from one alloy to the other. In addition, it will be assumed that there are n_a A atoms and n_b B atoms in the α phase. Thus, there are $(N_a - n_a)$ A atoms and $(N_b - n_b)$ B atoms in the β phase. The composition of the phases may then be specified by means of the fractions x_α and x_β of A atoms, which are, respectively,

$$x_\alpha = \frac{n_a}{n_a + n_b} \quad \text{and} \quad x_\beta = \frac{N_a - n_a}{N - n_a - n_b}. \tag{1}$$

Now, if $A_\alpha(x)$ is the free energy of a specimen of the α phase that contains N atoms of both types and has composition x and if $A_\beta(x)$ is the same quantity for the β phase, the total free energy of the system is

$$A = \frac{n_a + n_b}{N} A_\alpha \left(\frac{n_a}{n_a + n_b} \right) + \frac{N - n_a - n_b}{N} A_\beta \left(\frac{N_a - n_a}{N - (n_a + n_b)} \right). \quad (2)$$

Thus, the conditions for equilibrium, which are

$$\frac{\partial A}{\partial n_a} = 0, \qquad \frac{\partial A}{\partial n_b} = 0, \quad (3)$$

lead to the equations

$$A_\alpha - A_\beta + (1 - x_\alpha) A_\alpha'(x_\alpha) - (1 - x_\beta) A_\beta'(x_\beta) = 0,$$
$$A_\alpha - A_\beta - x_\alpha A_\alpha'(x_\alpha) + x_\beta A_\beta'(x_\beta) = 0, \quad (4)$$

in which $A' = \partial A / \partial x$. By subtracting these, we may derive the equation

$$A_\alpha'(x_\alpha) = A_\beta'(x_\beta), \quad (5)$$

which, when substituted in either equation, gives the additional relation

$$A_\alpha' = \frac{A_\alpha - A_\beta}{x_\alpha - x_\beta}. \quad (6)$$

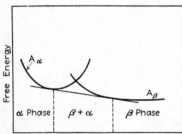

Fig. 19.—Schematic representation of the condition for determining the phase boundaries of alloys.

Equations (5) and (6) state that the boundaries of the α and β phases are determined by the points at which the slopes of the two free-energy curves are equal and have a common tangent (cf. Fig. 19). To the left of the point of tangency on the A_α curve, the α phase alone is stable, whereas to the right of the corresponding point on the A_β curve the β phase is stable.

Jones[1] has applied the relations (5) and (6) to the boundaries of the α and β brass types of phase of substitutional alloys that were discussed in Sec. 3. These phases are, respectively, face-centered and body-centered cubic and, as was seen in Sec. 102, are stable for a range of electron-atom ratios near the values for which the zones of the two lattices are filled to the points of highest level density. Jones assumed that all energies except the Fermi energy are practically the same for disordered specimens of the two phases so that the filling of the one-electron levels alone determines the relative energies at absolute zero of temperature. The $n(\epsilon)$ curves for the face-centered and body-centered lattices of the brass (Cu-Zn) system that are determined by the approximate methods

[1] H. Jones, *Proc. Phys. Soc.*, **49**, 243 (1937).

discussed in Sec. 65 are shown in Figs. 20a and b.   The energy difference per atom of the two phases is shown in Fig. 20c as a function of the electron-atom ratio n_e.   It should be observed that the $n(\epsilon)$ curves are identical to about 6 ev, which corresponds to an electron-atom ratio of about 0.95.   The $n(\epsilon)$ curve for the face-centered lattice rises to a peak in the energy range just above this, so that this phase has a lower Fermi energy.   The β phase then has its peak, and the relative energy curve changes sign.   It is clear that when the composition corresponds to the intercept of the relative energy curve with the n_e axis, that is, when the electron-atom ratio n_e is 1.44, the system would be most stable if it consisted of a quantity of α phase having a lower value of n and a quantity of β phase having a higher value.   On the other hand, if n_e is near 1.2, it would be necessary to raise the energy of the α phase a great deal in order to form a small quantity of β phase. Thus, we should expect the α phase to be stable at this point.   The actual values of n_e at the phase boundary points, as determined from Eqs. (5) and (6) for the absolute zero of temperature by replacing the free energy by the energies computed from Fig. 20a, are

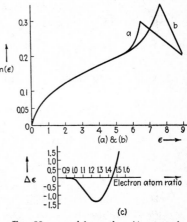

$$n_{e,\alpha} = 1.409,$$
$$n_{e,\beta} = 1.447.$$

FIG. 20.—a and b are the $n(\epsilon)$ curves for the face-centered and body-centered structures, respectively.   Curve c is the relative energy $\Delta\epsilon$ of the two phases as a function of the electron-atom ratio.   The energy scales are in electron volts.   (*After Jones.*)

As may be seen from Fig. 20b, these values lie very close to the point where the relative energy curve intercepts the axis.

Jones extended this work to higher temperature ranges by adding mixing entropy terms such as those considered in the theory of order and disorder.   If a given disordered phase has n_α A atoms and n_β B atoms, its mixing entropy is

$$S = k \log \frac{(n_a + n_b)!}{n_a! n_b!}, \tag{7}$$

in which $(n_a + n_b)!/n_a! n_b!$ is the total number of ways of rearranging the A and B atoms among the $n_a + n_b$ sites.   When this is expanded by means of Stirling's approximation, it becomes

$$S = -Nk[x \log x + (1 - x) \log (1 - x)] \tag{8}$$

where $N = n_a + n_b$ and x is defined by Eq. (1). Thus, if $E(x)$ is the energy of the N atoms as a function of x,

$$A(x) = E(x) + NkT[x \log x - (1 - x) \log (1 - x)]. \qquad (9)$$

Using free-energy functions of this type and $E(x)$ curves obtained from the data of Fig. 20, Jones computed the phase boundaries of the α and β phases of the Cu-Zn and Cu-Al systems as functions of temperature. The observed and calculated curves are shown in Fig. 21. They agree as closely as one might expect in view of the simplifying assumptions made in this work.

Jones has applied similar computations to the liquidus and solidus curves of substitutional alloys, which are briefly described in Sec. 3,

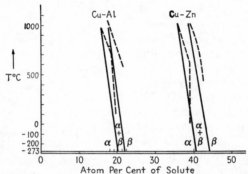

Fig. 21.—A comparison of the observed and calculated phase boundaries of the α and β phases of Cu-Zn and Cu-Al. The full curves are the theoretical ones; the broken lines are experimental. (*After Jones.*)

and has shown that the dependence of these curves upon composition may be adequately explained if the solid phases are assigned free-energy functions of type (9).

124. Order and Disorder in Alloys. *a. Experimental Discussion.*[1]— X-ray diffraction studies of substitutional alloys show in many cases that each type of atom is localized at a definite site in the unit cell, just as the constituents of ionic crystals are localized at definite positions. As the temperature is raised in these cases, the degree of order may decrease, even though the crystalline arrangement is maintained. This decrease is made evident by the fact that it is no longer possible to tell precisely which kind of atom occupies a given site. The order may decrease continuously, as in β brass, or it may undergo an abrupt change, as in Cu_3Au. These cases and others are discussed in Sec. 3, Chap. I. Alloys in which the order changes abruptly usually have an abrupt change in

---

[1] See the previous discussion in Sec. 3.

heat content at the transition temperature whereas those in which the change is continuous do not, although there may be a discontinuity in specific heat at the point at which all sites become equivalent.[1] The two types of phase change, characterized respectively by Cu_3Au and $CuZn$, are said to be of the first and second kinds.

Although two-component alloys have been investigated most widely, these are not the only substances in which order and disorder occur. Ketelaar[2] has found, for example, that silver and copper mercuric iodides (Ag_2HgI_4 and Cu_2HgI_4) show complex order and disorder changes which resemble closely those found in Cu_3Au. The low-temperature modifications have the tetragonal structure shown in Fig. 22 in which mercury atoms occupy the eight corners of a nearly cubic cell. The iodine atoms are distributed tetrahedrally about four of the eight corners, forming a face-centered lattice, and the silver or copper atoms are arranged at the centers of the four vertical faces. The iodine atoms do not change their relative positions as the temperature is raised. The metal atoms, however, make an abrupt change, becoming uniformly distributed over the eight corners and six face centers of the cube. It is evident that two of these fourteen sites must be vacant on the average. Hence, the disordering process involves silver atoms, mercury atoms, and vacancies in the ratio 2:1:1. The behavior of the ionic conductivity of Ag_2HgI_4 is shown in Fig. 23.

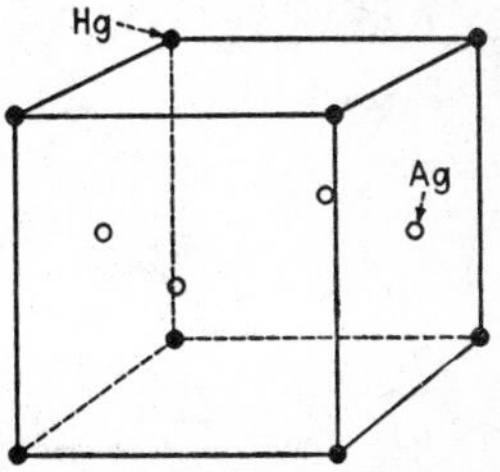

FIG. 22.—The positions of mercury and silver atoms in the ordered, low-temperature phase of Ag_2HgI_4. The iodine atoms, which are not shown, are distributed tetrahedrally about the Hg atoms. In the high-temperature phase the Hg atoms, the Ag atoms, and the vacancies at the centers of the top and bottom faces of the cube become mixed. The low-temperature form is slightly tetragonal; the high-temperature form is cubic.

b. Qualitative Principles.—We shall not devote space to a detailed treatment of the more advanced theories of order and disorder since discussions of these may be found in other writings;[3] however, we shall give a brief discussion of the principles involved and of the simpler theories.

It is clear that the disordered alloy has a higher entropy than the ordered one. If we neglect any difference in the vibrational entropy of the ordered and disordered state, we may estimate the maximum change in mixing entropy that accompanies disordering by computing the

---

[1] The discontinuous behavior of the elastic constants of Cu_3Au in the vicinity of the ordering temperature has been investigated by S. Siegel, *Phys. Rev.*, **57**, 537(1940).

[2] J. A. A. KETELAAR, *Z. physik. Chem.*, **26B**, 327 (1934), **30B**, 53 (1935); *Z. Krist.*, **87**, 436 (1934).

[3] See the survey article by F. C. Nix and W. Shockley, *Rev. Modern Phys.*, **10**, 1 (1938).

number of arrangements associated with the completely ordered and disordered states. In the simplest system, namely, that in which there are two types of atom present in equal numbers N, the number of arrangements associated with the ordered state is unity because all atoms of a given type are equivalent. The number of arrangements in the completely disordered state is the number of ways of distributing N atoms among $2N$ sites, namely $(2N)!/(N!)^2$. Hence, the maximum increase in mixing entropy associated with disordering is

$$k \log \frac{(2N)!}{(N!)^2} \cong 2Nk \log 2. \tag{1}$$

A similar calculation may be made for any system.

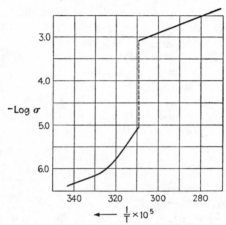

Fig. 23.—The conductivity of Ag_2HgI_4 near the transition temperature. σ is expressed in ohm^{-1} cm^{-1}. (*After Ketelaar.*)

The energy of the crystal presumably increases as we pass from the ordered to the disordered state, for otherwise the ordered state would not be stable at low temperatures. Hence, the energy versus entropy curve should rise with increasing disorder. Figure 24 shows two possible ways in which this curve may behave. In the first case, the $E(S)$ curve has an inflection point so that a tangent line may be drawn to two parts of the curve. Thus, the entropy, and hence the order, should show an abrupt change at the temperature equal to the slope of this tangent line, and there should be a latent heat, just as in melting. In the second case, the $E(S)$ curve has positive curvature so that there is no discontinuity in order.

The actual behavior of a solid is not necessarily determined by the $E(S)$ curve for disordering alone. It is possible that the vibrational frequencies of the crystal may decrease in passing from the ordered to the disordered state; the disordered phase then has a higher vibrational entropy than the ordered one at corresponding temperatures. Thus, the ordered and disordered phases may behave like different allotropic phases, and the transition may occur abruptly even though the $E(S)$ curve for disordering alone would predict a gradual decrease of order. The theories of order and disorder that are discussed in the article referred to in footnote 3, page 503, do not yet interpret the experimental material in a quantitative way. The reason for this lack is, of course, that the complete entropy and energy changes which accompany disordering are very difficult to compute, just as are the changes of these quantities

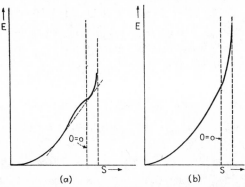

FIG. 24.—Two possible behaviors of the $E(S)$ curves for order-disorder changes. In case (a) there is an inflection point, whence long-distance order may appear and disappear abruptly with a latent heat. In the second case there is no inflection point and the transition is continuous. The curvature changes at the entropy corresponding to zero long-distance order ($O = 0$), so that there is a discontinuity in specific heat. The right-hand vertical line corresponds to the entropy for zero short-distance order in each case.

during melting. The results of this work, however, leave little doubt that the qualitative principles are now understood.

    c. Definitions of Order.—There are two interesting types of order, namely, long-distance order, which measures the extent to which the positions of atoms in different cells of the lattice are correlated, and short-distance order, which measures the extent to which the positions of neighboring atoms are correlated. The first type of order is responsible for the Bragg reflection of X rays by lattices; the second is responsible for the diffraction rings of liquids and glasses. Following Bethe,[1] we may define these two types of order mathematically in the following way.

    When long-distance order is discussed, the lattice may be divided into as many types of site as are occupied by different atoms in the completely ordered state. Thus, there are two types of site in β brass and in

[1] H. BETHE, *Proc. Roy. Soc.*, **150**, 552 (1935).

Cu_3Au, and there are three types in Ag_2HgI_4.   In crystals in which two types of site are present in equal numbers, the long-distance order O is defined as the difference between the probability that an atom will occupy its own kind of site and the probability that the other kind of atom will occupy this site.   Thus,

$$O = P_a(A) - P_a(B) \tag{2}$$

where $P_a(A)$ is the probability that an A atom occupies its own site and $P_a(B)$ is the probability that a B atom does.   The four probabilities $P_a(A)$, $P_a(B)$, $P_b(B)$, $P_b(A)$ obviously satisfy the equations

$$P_a(A) + P_b(A) = 1, \qquad P_a(A) + P_a(B) = 1,$$
$$P_a(B) + P_b(B) = 1 \qquad P_b(A) + P_b(B) = 1, \tag{3}$$

which shows that there is only one independent P.   In the case of Cu_3Au in which there are three times as many A atoms as B atoms the long-distance order may be defined by the equation

$$O = P_a(A) - 3P_a(B). \tag{4}$$

The interrelations between the P are

$$P_a(A) + P_a(B) = 1, \qquad P_b(B) + P_b(A) = 1,$$
$$P_b(B) + 3P_a(B) = 1, \qquad P_a(A) + \tfrac{1}{3}P_b(A) = 1, \tag{5}$$

so that again there is only one independent P.   It is clear that the order parameters defined by Eqs. (2) and (4) are unity in the state of highest long-distance order and are zero when there is no long-distance order. This convenient fact is the principal reason for selecting these combinations of the P, for any one of them could serve as a measure of long-distance order.

Long-range order is not so easy to define in systems such as Ag_2HgI_4 that have three kinds of site, for there is then more than one independent P.   Let us consider Ag_2HgI_4 as an example, designating the nine probabilities by

$$P_{Ag}(Ag), P_{Ag}(Hg), P_{Ag}(V), \cdots, P_V(V)$$

where the subscripts refer to sites and V is the symbol for a vacancy. These probabilities are interrelated by the following six equations:

$$\left.\begin{aligned} P_{Ag}(Ag) + P_{Ag}(Hg) + P_{Ag}(V) &= 1. \\ P_{Hg}(Ag) + P_{Hg}(Hg) + P_{Hg}(V) &= 1. \\ P_V(Ag) + P_V(Hg) + P_V(V) &= 1. \end{aligned}\right\} \tag{6}$$

$$\left.\begin{aligned} P_{Ag}(Ag) + \tfrac{1}{2}P_{Hg}(Ag) + \tfrac{1}{2}P_V(Ag) &= 1. \\ 2P_{Ag}(Hg) + P_{Hg}(Hg) + P_V(Hg) &= 1. \\ 2P_{Ag}(V) + P_{Hg}(V) + P_V(V) &= 1. \end{aligned}\right\} \tag{7}$$

Only five of these equations are independent so that it is necessary to know four of the P before the average distribution of atoms in the unit cell can be given. Thus, it is not possible to express the degree of long-distance order in terms of a single parameter as it was in the preceding cases.

Short-distance order is also easy to define in two-component alloys. In the completely ordered state, a given kind of atom has a definite arrangement of atoms in the neighboring sites. We may specify the short-distance order in any state by giving the difference σ between the fraction of atoms in a shell surrounding a given atom that have the same arrangement as in the perfectly ordered state and the fraction of atoms that have not. The size of the cell may be varied to suit the case at hand. The quantity σ evidently is equal to unity in the completely ordered state and to zero in the completely random state.

The concept of long-distance order was introduced into the theory of order and disorder first because it is measured directly by ordinary X-ray diffraction data. Bethe[1] pointed out, however, that short-range order actually is a more fundamental quantity since the interatomic energy is determined primarily by it.

d. Elementary Theories of Order and Disorder.—The earliest theory of order and disorder was developed by Gorsky[2] and applied to two-component systems of the type AB. However, an equivalent theory developed later by Bragg and Williams[3] undoubtedly is responsible for the more recent interest in the subject. In these earlier theories, it was assumed that the long-distance order O existing at temperature T is determined by the energy V required to take an atom from an ordered position to a disordered one. This assumption is expressed by the equation

$$O = O(V,T). \tag{8}$$

It was also assumed that V is a function of the long-distance order so that there is a second relation

$$V = V(O,T). \tag{9}$$

The relations (8) and (9) are sufficient to determine O as a function of T alone.

Gorsky derived explicit forms for Eqs. (8) and (9) in the case of an alloy of composition AB. Since there are thermal fluctuations, there is a finite probability that each atom will leave its position, diffuse

[1] *Ibid.*

[2] W. Gorsky, *Z. Physik,* **50,** 64 (1928).

[3] W. L. Bragg and E. J. Williams, *Proc. Roy. Soc.,* **145,** 699 (1934); **151,** 540 (1935).

through the lattice, and fall into a vacant site. Gorsky assumed that the probabilities per unit time of A and B atoms leaving their own sites are equal as are the probabilities that they will leave improper sites. We shall designate these probabilities by l_p and l_i, respectively. In addition, he assumed that the probabilities that the free atoms will fall into any vacant proper sites are equal, as are the probabilities that they will fall into improper sites. We shall designate these by f_p and f_i, respectively. If N is the total number of atoms of a given kind, n is the number of vacant sites, and α is the fraction of atoms on proper sites, the equilibrium equations for proper and improper sites are

$$N\alpha l_p = n^2 f_p,$$
$$N(1 - \alpha)l_i = n^2 f_i. \tag{10}$$

Solving these equations and using the relation

$$O = 2\alpha - 1, \tag{11}$$

we obtain

$$O = \frac{1 - (f_i l_p / f_p l_i)}{1 + (f_i l_p / f_p l_i)}. \tag{12}$$

Gorsky assumed that f_p and f_i are not strongly temperature-dependent and that they are nearly equal so that their ratio is practically unity. Since l_p and l_i are temperature-dependent, he assumed that their ratio is

$$\frac{l_p}{l_i} = \frac{e^{-\frac{\epsilon_p}{kT}}}{e^{-\frac{\epsilon_i}{kT}}}. \tag{13}$$

Using these relations, we find

$$O = \tanh \frac{\epsilon_p - \epsilon_i}{2kT}. \tag{14}$$

Evidently, $\epsilon_p - \epsilon_i$ is proportional to V, the energy required to remove an atom from an ordered site to a disordered one. Hence, (14) is

$$O = \tanh \frac{\beta V}{kT} \tag{15}$$

where β is a proportionality factor. This equation has the form (8).

Were V independent of O, O would decrease slowly with increasing temperature and would approach zero when T becomes infinite. It is clear that V must depend upon order, however, for there is no difference between proper and improper sites in the completely disordered state. Hence, Gorsky assumed that V varies linearly with O in the manner

$$V = V_0 O \tag{16}$$

where V_0 is a constant. This equation corresponds to (9).

The solution of Eqs. (15) and (16) is given by the equation

$$O = \tanh \frac{\beta V_0 O}{kT}. \tag{17}$$

This has two roots, namely, the root $O = 0$, which is independent of temperature, and a root which is unity when T is zero and zero when

$$T_c = \frac{\beta V_0}{2k} \tag{18}$$

and varies continuously in between (*cf.* Fig. 25). The relation described by the second root agrees qualitatively with that observed in β brass.

One obvious objection to the details of Gorsky's treatment is the fact that he assumes a questionable diffusion process in deriving Eq. (15).

Bragg and Williams[1] modified the principles used in Gorsky's treatment and extended the field of application of the method. In earlier work, they introduced rate processes in order to derive equations equivalent to (15). Williams[2] subsequently showed that this procedure is not necessary and that the equations may be derived on the basis of statistical mechanics. We shall discuss their work from the later standpoint.

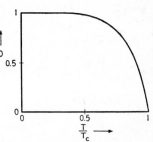

Fig. 25.—The $O(T)$ curve obtained from Gorsky's theory of order and disorder.

Let us consider a case in which there are n_a A atoms and a sites and n_b B atoms and b sites. Generalizing Eqs. (2) and (4), we may define the long-range order by the equation

$$O = P_a(A) - \frac{n_a}{n_b} P_a(B) = P_a(A)\left(1 + \frac{n_a}{n_b}\right) - \frac{n_a}{n_b}$$

$$= p\left(1 + \frac{x}{1-x}\right) - \frac{x}{1-x} = \frac{p-x}{1-x} \tag{19}$$

where, for simplicity, we have replaced $P_a(A)$ by the symbol p, and $\dfrac{n_a}{n_a + n_b}$, the fraction of a sites, by x. Let us now compute the entropy associated with a given value of order. The $xN'p$ A atoms in the xN' a positions ($N' = n_a + n_b$) may be distributed in

$$n_1 = \frac{(xN')!}{(xN' - xN'p)!(xN'p)!} \tag{20}$$

[1] Bragg and Williams, *op. cit.*

[2] E. J. Williams, *Proc. Roy. Soc.*, **152**, 231 (1935). See also R. H. Fowler, *op. cit.*

independent ways, whereas the $(xN' - xN'p)$ A atoms in the $(N' - xN')$ b sites may be distributed in

$$n_2 = \frac{(N' - xN')!}{(N' - 2xN' + xN'p)!(xN' - xN'p)!} \tag{21}$$

ways. Since the total number of arrangements is the product of n_1 and n_2, the entropy[1] is

$$
\begin{aligned}
S &= k \log n_1 n_2 \\
&= C - N'k[x(1 - p) \log (1 - p) + xp \log p + \\
&\quad (1 - 2x + xp) \log (1 - 2x + xp) + (x - xp) \log (x - xp)] \tag{22}
\end{aligned}
$$

where C is independent of p.

The energy of the disordered state relative to the ordered one is

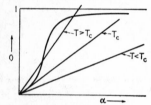

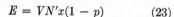

$$E = VN'x(1 - p) \tag{23}$$

where $N'x(1 - p)$ is the number of atoms that have been moved from ordered to disordered positions. The value of p for which the free energy $E - TS$ computed from (22) and (23) is a minimum satisfies the equation

FIG. 26.—$O(\alpha)$ curve for $x = \frac{1}{4}$. As the temperature is raised from absolute zero the $\alpha = V/kT$ lines become tangent at the origin before the other intercept has reached the origin.

$$\log \frac{(1 - p)^2 x}{p(1 - 2x + xp)} = -\frac{V}{kT}. \tag{24}$$

If we replace p by O, using Eq. (19) we obtain an equation connecting O and V, namely,

$$O = 1 - \frac{[4x(1 - x)(e^\alpha - 1) + 1]^{\frac{1}{2}} - 1}{2x(1 - x)(e^\alpha - 1)} \tag{25}$$

where $\alpha = V/kT$. This equation reduces to Gorsky's equation (15) when $x = \frac{1}{2}$, a fact showing that β should be $\frac{1}{4}$. The expansion of Eq. (25) in the neighborhood of $\alpha = 0$ is

$$O(\alpha) = x(1 - x)\alpha + \frac{1}{2}x(1 - x)(1 - 2x)^2\alpha^2 + \cdots. \tag{26}$$

Hence, S usually starts out with a finite slope and positive curvature, the exceptional case being $x = \frac{1}{2}$. If it is assumed that the relation (9) is

$$V = V_0 O,$$

as Gorsky did, the transition is of the first kind in all cases, except that in which $x = \frac{1}{2}$. This follows from the fact, illustrated in Fig. 26, that the

---

[1] In this computation, only the mixing entropy is considered. Actually, the change in vibrational entropy should be included as well.

$V(O)$ line becomes tangent to the $O(V)$ line at the origin before the other intercept approaches the origin when the $O(V)$ curve has positive curvature.

e. Bethe's Treatment of the Problem.—Bethe provided a new and important approach to the problem of order and disorder by pointing out that the ordering energy and entropy are determined primarily by the short-range order since neutral atoms interact with short-range forces. Thus, the equation of state may be determined by considering short-range order, and the long-distance order may be obtained as a by-product. In addition to recognizing this principle, Bethe developed an approximate method for computing the partition function of the system. Discussions of this and of subsequent theoretical work may be found in the review article listed in footnote 3, page 503.

125. Free Rotation in Crystals. *a. Experimental Survey.*—The specific-heat curves of molecular solids frequently show peaks resembling those observed during order-disorder transitions in alloys. Two interpretations of these peaks have been given in the theoretical development of the subject, namely, the hypothesis due to Pauling[1] that the peaks accompany the onset of free molecular rotation and the hypothesis due to Frenkel[2] that the molecules undergo only torsional oscillations both above and below the transition temperature and that they have less relative orientation above the transition than below. At least in the case of ammonium chloride, which has been investigated very thoroughly by Lawson,[3] the evidence seems to be in favor of Frenkel's hypothesis, as we shall see below. The observed cases may be classified as follows:

1. Nonpolar molecular crystals, such as CH_4, N_2, O_2.
2. Ammonium salts.
3. Polar molecular crystals.

We shall discuss these categorically.

1. *Nonpolar cases.*—Although carbon atoms of solid methane form a face-centered lattice below 89°K, there is no direct evidence concerning the position of the hydrogen atoms. The specific-heat curve possesses the changes, shown in Fig. 80, Chap. I, near 20°K, but these are not accompanied by the appearance of a latent heat. As the temperature is raised through this transition region, the molar volume increases abruptly from 30.57 to 36.65 cm³. There is no other obvious change in crystal structure during the transition. It is assumed, however, that the hydrogen atoms are localized below the transition temperature and are not localized above it.

[1] L. PAULING, *Phys. Rev.*, **36**, 430 (1930).
[2] J. FRENKEL, *Acta Physicochemica*, **3**, 23 (1935).
[3] A. W. LAWSON, *Phys. Rev.*, **57**, 417 (1940).

Solid oxygen and nitrogen seem to possess similar transitions at 23.7° and 35.4°K, respectively. The experimental work indicates that there are large hysteresis effects associated with the transition in these cases, so that the results are not so definite as for methane.

2. *Ammonium salts.*—A number of ionic crystals that contain the NH_4 radical, such as the ammonium halides, ammonium sulfate, and ammonium nitrate, have specific-heat curves that show anomalies similar to those observed in methane. The curve for ammonium chloride is shown in Fig. 69 of Chap. I. As a result of a very careful set of experiments, Lawson has shown in this case that the specific heat at fixed volume does not exhibit nearly so high a peak as the specific heat at constant pressure and that C_V is $9R$ above the transition temperature, corresponding to torsional oscillations of the molecules rather than free rotation. Thus his results support Frenkel's hypothesis rather than Pauling's in this case. It seems likely that Frenkel's picture is also valid in the other ammonium salts and probably in solids of polar molecules, but it does not appear safe to draw conclusions concerning other cases.

3. *Polar molecular crystals.*—Many crystals that are composed of polar molecules, such as solid hydrogen chloride, hydrogen iodide, and hydrogen sulfide, behave in a way similar to the substances already mentioned and yet show important differences. For example, hydrogen chloride forms a cubic crystal above 98.8°K in which the chlorine nuclei are localized in a face-centered cubic lattice. Since this fact clearly means that the molecules are not parallel, we may safely assume that they are more randomly oriented. The lattice changes abruptly to a tetragonal face-centered form at 98.8°K with the appearance of a latent heat. We may conclude that the molecules have higher relative orientation in the low-temperature form of the substance. Apparently, the intermolecular forces and lattice frequencies are sufficiently different in the two states that the crystal behaves as though it were undergoing an allotropic phase change.

Hydrogen bromide and iodide behave more nearly like nonpolar crystals since they do not exhibit a latent heat during the transition; however, their specific-heat curves have very large discontinuities.

b. *Pauling's Theory and Fowler's Extension.*—Pauling's hypothesis was treated in a semiquantitative manner by Fowler. Since this work resembles that on order and disorder discussed in the previous section, we shall discuss it briefly. Pauling assumed that the potential energy of a molecule in a lattice depends upon its angular orientation relative to the crystallographic axes. Let us consider a lattice of nonpolar diatomic molecules and specify the position of a molecule relative to the orientation for minimum energy by a polar angle θ and an azimuthal angle ϕ. When ϕ is fixed, we may expect that the energy varies with

θ in the manner shown in Fig. 27, the direction $\theta = \pi$ being equivalent to $\theta = 0$. If V_0 is large enough, the lowest energy levels of the molecule E_i correspond to states of oscillation about the equilibrium orientation so that the molecules should not rotate at very low temperatures. They should rotate, however, at sufficiently high temperatures. Pauling realized that the height of the potential-energy curve depends upon the amount of rotation of the other molecules, for otherwise the specific-heat curve would not be discontinuous.

It is clear that the principles embodied in Pauling's picture of the onset of free rotation are the same as those used to explain the order-disorder transitions in alloys. This was first pointed out by Fowler[1] who applied the equivalent of a Bragg-Williams approximation to Pauling's theory in the case of a lattice of polar diatomic molecules. He assumed that the aligning potential may be expressed in the form

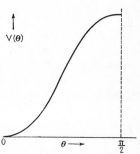

FIG. 27.—The variation of $V(\theta)$ as a function of θ for a nonpolar diatomic molecule. The equilibrium position is $\theta = 0$.

$$V = -V_0 \cos \theta \qquad (1)$$

and that there are enough negative energy levels to justify the use of classical mechanics when one is computing the partition function near the transition temperature. The partition function then is

$$f(T) = \frac{1}{h^2} \int e^{\frac{-\frac{1}{2I}\left(p_\theta{}^2 + \frac{p_\varphi{}^2}{\sin^2 \theta}\right) + V_0 \cos \theta}{kT}} dp_\theta dp_\varphi \sin \theta d\theta d\varphi$$
$$= \frac{2I}{\hbar^2} \frac{k^2 T^2}{V_0} \sinh \frac{V_0}{kT}. \qquad (2)$$

Thus the specific heat is

$$C_V = Nk \left[2 - \left(\frac{V_0}{kT \sinh \dfrac{V_0}{kT}} \right)^2 \right]. \qquad (3)$$

Now, V_0 should depend upon temperature because it is affected by the prevailing degree of rotation, which is a temperature-dependent quantity. In a treatment of the present problem that is closely patterned after the Bragg and Williams treatment of order and disorder, the degree of rotation R would be defined in a physically reasonable way; it would

[1] R. H. FOWLER, *Proc. Roy. Soc.*, **149**, 1 (1935); see also *Statistical Mechanics* (Cambridge University Press, 1936).

then be assumed that V_0 depends upon this variable in some explicit manner. This equation and Eq. (2) would then be analogous to Eqs. (8) and (9) of the preceding section and would lead to a relation between the degree of rotation and temperature. In this scheme, different types of transition could be treated by varying the relation between V_0 and the degree of rotation. Instead, Fowler used a fixed relation between V_0 and R and treated different transition-types by taking different definitions of R. We shall discuss two of his cases.

In the first case, he defined the nonrotating molecules as those satisfying the relation

$$\frac{1}{2I}\left(p_\theta{}^2 + \frac{p_\varphi{}^2}{\sin^2\theta}\right) < \beta V_0 \tag{4}$$

where β is an adjustable parameter. The molecules specified by Eq. (4) evidently have kinetic energy less than βW for any angular orientation. The fraction R of molecules that are rotating is then given by the equation

$$R = \frac{f_0}{f} \tag{5}$$

where f is the partition function (2) and

$$
\begin{aligned}
f_0 &= \frac{2\pi}{h^2}\int_0^\pi e^{\frac{V_0\cos\theta}{kT}}\int_{(p_\theta{}^2+p_\varphi{}^2/\sin^2\theta)>2I\beta V_0} e^{-\frac{p_\theta{}^2+p_\varphi{}^2/\sin^2\theta}{2IkT}}\,dp_\theta dp_\varphi \\
&= \frac{2IkT}{\hbar^2}\frac{kT}{V_0}e^{-\frac{\beta V_0}{kT}}\sinh\frac{V_0}{kT}.
\end{aligned}
\tag{6}
$$

Thus,

$$R = e^{-\frac{\beta V_0}{kT}}. \tag{7}$$

For simplicity, the dependence of V_0 on R was taken as

$$V_0 = V_0'(1 - R) \tag{8}$$

where V_0' is a constant. Equations (7) and (8) determine the relation between R and temperature. It is easily seen that R is unity at

$$T_c = \frac{\beta V_0}{k}$$

and that

$$R = 1 - \frac{2T}{T_c}\left(1 - \frac{T}{T_c}\right)$$

near the transition point. The specific heat below T_c is

$$C_V = Nk\left(2 - \frac{x^2}{\sinh^2 x}\right) + N_0 V_0\frac{dR}{dT}\left(\frac{\cosh x\,\sinh x - x}{\sinh^2 x}\right)$$

where

$$x = \frac{V_0(1 - R)}{kT},$$

and is Nk above T_c. Moreover, the specific heat is continuous at T_c in this case, although it has the maximum shown in Fig. 28 below T_c.

Fowler was able to alter the definition of order in such a way as to obtain a more abrupt change. If it is assumed that the nonrotating

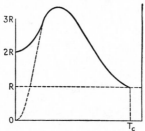

Fig. 28.—The type of molar heat curve obtained from Eq. (3). Above T_c the value is R, as for diatomic molecules. The full line corresponds to the classical case treated by Fowler. The broken line represents the effect of quantum mechanical modification.

molecules satisfy the relation

$$\frac{1}{2I}\left(p_\theta{}^2 + \frac{p_\varphi{}^2}{\sin^2 \theta}\right) < \frac{V_0}{2}(1 + \cos \theta)$$

instead of the relation (4), it is found that

$$1 - R = \tanh \frac{V_0}{2kT},$$

which is analogous to Gorsky's equation of the preceding section. When coupled with Eq. (7), this relation leads to a discontinuous specific-heat curve, as in the case of alloys. It is possible to obtain transitions of the first kind in a similar way.

Note: The topics of nucleation and rates of simple phase changes in solids, which we have omitted for reasons of space, should properly be included in this chapter. A discussion of these topics that is in accord with the presentation of the preceding sections has been given by R. Becker, *Ann. Physik*, **32**, 128 (1938). Becker shows that many of the facts concerning the rates of simple phase changes may be explained semiquantitatively on the assumption that the energy of the surfaces of misfit between the new and the old phase is such that only relatively large nuclei are stable.

CHAPTER XV

THEORY OF CONDUCTIVITY

In the present chapter, we shall be interested in three types of conductivity, namely, metallic conductivity, ionic conductivity, and photoconductivity. The first of these was discussed in Chap. IV on the basis of the free-electron gas model and will be redeveloped in the first part of this chapter following a method that was first used by Houston and Bloch. The two other topics will be discussed in subsequent parts of the chapter.

A. METALLIC CONDUCTIVITY

126. Summary of Older Equations.—The Lorentz-Sommerfeld theory of metallic conduction is based upon Boltzmann's equation of state

$$\mathbf{v} \cdot \operatorname{grad}_r f + \boldsymbol{\alpha} \cdot \operatorname{grad}_v f = b - a \tag{1}$$

where $\mathbf{v}$ is the electronic velocity, $\boldsymbol{\alpha}$ is the electronic acceleration,

$$\operatorname{grad}_r = \mathbf{i}\frac{\partial}{\partial x} + \mathbf{j}\frac{\partial}{\partial y} + \mathbf{k}\frac{\partial}{\partial z},$$

$$\operatorname{grad}_v = \mathbf{i}\frac{\partial}{\partial v_x} + \mathbf{j}\frac{\partial}{\partial v_y} + \mathbf{k}\frac{\partial}{\partial v_z},$$

f is the statistical distribution function, which gives the number of particles per unit volume having velocity v_x, v_y, v_z, and a and b are collision terms. This equation was derived in Sec. 31 and was solved on the assumption that f has the form

$$f = f_0 + v_x \chi(v) \tag{1a}$$

where f_0 is the distribution function in the absence of a field and χ is an undetermined function that is small compared with f_0. In addition, the quantity $b - a$, which gives the difference between the numbers of particles entering and leaving a unit volume of phase space because of collisions, was computed on the assumption that the electrons make elastic collisions with the ions of the lattice. It was found to have the value

$$b - a = -\frac{v_x v \chi(v)}{l_0} \tag{2}$$

where l_0 is the mean free path, which is assumed to be independent of velocity.   If f_0 is the Maxwell-Boltzmann distribution function

$$Ae^{-\frac{\epsilon}{kT}}$$

where A is a constant, the conductivity σ_M is

$$\sigma_M = \frac{4}{3} \frac{n_f l_0 e^2}{\sqrt{2\pi m^* kT}} \tag{3}$$

where n_f is the number of free electrons per unit volume, etc. [*cf.* Eq. (5), Sec. 36].   Since the mean velocity $\bar{v}_M$ in Maxwell-Boltzmann statistics is

$$4\sqrt{\frac{kT}{2\pi m^*}}, \tag{4}$$

Eq. (3) may be placed in the form

$$\sigma_M = \frac{n_f l_0 e^2 \bar{v}_M}{3kT}. \tag{5}$$

On the other hand, if f_0 is the Fermi-Dirac distribution function

$$A' \frac{1}{e^{\frac{\epsilon - \epsilon_0'}{kT}} + 1},$$

σ_F is

$$\sigma_F = \frac{n_0 l(\epsilon_0') e^2}{m^* v(\epsilon_0')} \tag{6}$$

[*cf.* Eq. (13), Sec. 32], where $v(\epsilon_0')$ is the velocity of the electrons at the top of the filled band.

Equation (3) gives the proper order of magnitude for the conductivity at room temperature if l_0 is taken as the interatomic distance and if n_f is the total number of electrons per unit volume.   The temperature dependence is wrong, however, for the observed conductivity varies as $1/T$ near room temperature.

Equation (6) is incorrect if l_0 is given the same value as in the preceding case because $v(\epsilon_0')$ is between ten and one hundred times larger than $\sqrt{kT/m^*}$, if m^* is the electronic mass.   Moreover, the temperature dependence is also wrong.   In order to justify the use of this equation, which is more reasonable than (3) since electrons actually obey the Pauli principle, it is necessary to assume that l is temperature-dependent and at room temperature is between ten and one hundred times larger than the interatomic distance.   It must also be assumed that l approaches infinity at low temperatures in order to explain the observed increase in conductivity with decreasing temperature (*cf.* Fig. 3).   This type of temperature dependence would imply, however, that the assumptions going into the derivation of Eq. (2) are also in error.

Houston[1] and Bloch[2] reopened the problem of metallic conductivity by investigating the way in which electrons interact with a crystal lattice on the basis of quantum mechanics. We shall discuss this work and subsequent refinements in the next section. It will be seen that inelastic electron-lattice collisions are of primary importance in determining the resistance; however, the amount of energy given to the lattice by the electrons is small, so that Sommerfeld's equation (6) is not badly in error.

127. The Collisions between Electrons and Lattice Vibrations in Monovalent Metals*.—Houston[1] first pointed out that the mean free path of an electron in a perfect nonoscillating lattice should be infinite. This is very easy to see in the Bloch scheme; for then the one-electron functions have the form

$$\psi_k = \chi_k e^{2\pi i k \cdot r},$$

and the velocity of the electron in a given state is

$$v = \text{grad}_k \, \epsilon(k)/h.$$

In the absence of any perturbation, an electron should continue in this state indefinitely.

An ordinary metal does not satisfy these ideal conditions for two reasons: (1) Its lattice is undergoing thermal oscillations, and (2) it usually contains imperfections, such as impurities and lattice defects. Both these effects may scatter electrons and thus make the mean free path finite.

The temperature oscillations should decrease with decreasing temperature and become very small at absolute zero. This fact provides a satisfactory qualitative explanation of the great rise in conductivity at low temperatures. The imperfections, on the other hand, should not be affected appreciably by decreasing temperature and should account for the residual resistance at low temperatures. Moreover, since the imperfections should depend upon the previous history of a specimen, we should expect the residual resistance to vary from specimen to specimen, as is observed.

From a wave standpoint, we may say that the atoms of a perfect lattice scatter electrons coherently, that is, in a manner that resembles the Laue diffraction of X rays. Hence, before an electron can be scattered in a perfect lattice it must occupy a level at the boundary of a zone, and the level to which it can jump must be vacant. These conditions are not ordinarily satisfied by an appreciable fraction of the conduction electrons. We shall see below that the scattering due to thermal vibrations may be regarded as the analogous coherent scattering by a lattice

---

[1] W. V. Houston, *Z. Physik*, **48**, 449 (1928); *Phys. Rev.*, **34**, 279 (1929).

[2] F. Bloch, *Z. Physik*, **52**, 555 (1928); **59**, 208 (1930).

which is periodically deformed by a vibrational wave, the distorted crystal behaving like a grating with a grating constant equal to the wave length of the lattice wave. Since this type of scattering is also limited by Laue conditions, a given vibrational mode can deflect a given electron only through definite angles. It is generally assumed that the temperature-independent scattering which gives rise to residual resistance is essentially incoherent, that is, that the scattering centers are arranged so haphazardly that they may be treated as though independent of one another. We shall discuss this in more detail in Sec. 130.

We shall devote the rest of this section to a quantitative discussion of the scattering of electrons by lattice vibrations. All quantitative treatments of this topic have been simplified by means of the assumption that the electronic energy depends only upon $k = \sqrt{k_x^2 + k_y^2 + k_z^2}$, the scalar wave number. This condition is closely satisfied in the monovalent metals, and thus the discussion of the present section should apply most nearly to them.

The differences between the various treatments of the problem of electron scattering lie in the different assumptions that have been made regarding the interaction between the electrons and the lattice. Let us consider a simple monatomic metal containing N atoms and designate the equilibrium positions of the atoms by the variables

$$\mathbf{r}(p) = p_1\boldsymbol{\tau}_1 + p_2\boldsymbol{\tau}_2 + p_3\boldsymbol{\tau}_3 \tag{1}$$

where the p are integers and the $\boldsymbol{\tau}$ are primitive translations. We shall designate the displacement of this atom from its equilibrium position by $\mathbf{R}(p)$. In the quadratic approximation, $\mathbf{R}$ may be expressed in the form

$$\mathbf{R}(p) = \frac{1}{\sqrt{N}} \sum_{t,\sigma} \frac{a_t(\boldsymbol{\sigma})}{\sqrt{M}} \boldsymbol{\xi}_t(\boldsymbol{\sigma}) e^{2\pi i \sigma \cdot \mathbf{r}(p)} \tag{2}$$

[*cf.* Eq. (4), Sec. 118], where the a are the complex amplitudes and the $\boldsymbol{\xi}$ are the unit polarization vectors of the normal modes, the $\boldsymbol{\sigma}$ are the wave-number vectors, which extend over the N values in a single zone, and M is the atomic mass. Since each atom is a center of symmetry in our simple lattice and $\mathbf{R}(p)$ is real, the $\boldsymbol{\xi}$ are real vectors and

$$a_t(\boldsymbol{\sigma}) = a_t(-\boldsymbol{\sigma}).$$

As we have seen in Sec. 22 it is convenient to define real amplitudes $\alpha_t(\boldsymbol{\sigma})$ in terms of the a by means of the equations

$$\left. \begin{aligned} \alpha_t(\boldsymbol{\sigma}) &= \frac{a_t(\boldsymbol{\sigma}) + a_t{}^*(\boldsymbol{\sigma})}{\sqrt{2}}, \\ \alpha_t(-\boldsymbol{\sigma}) &= \frac{a_t(\boldsymbol{\sigma}) - a_t{}^*(\boldsymbol{\sigma})}{\sqrt{2}i}. \end{aligned} \right\} \tag{3}$$

a. The Perturbing Potential.—In the earliest work on the quantum theory of conductivity by Bloch,[1] Brillouin[2] and Bethe,[3] it was assumed, for convenience, that as the ions move the electronic charge is deformed in such a way that the potential of an electron at the point $\mathbf{r}'$ in the deformed lattice is the same as that at the point $\mathbf{r}$ in the undeformed one. Here $\mathbf{r}$ and $\mathbf{r}'$ are connected by the equation

$$\mathbf{r}' = \mathbf{r} + \mathbf{R}(\mathbf{r}) \tag{4}$$

where $\mathbf{R}(\mathbf{r})$ is obtained from Eq. (2) by replacing $\mathbf{r}(p)$ by $\mathbf{r}$. Thus,

$$V_d(\mathbf{r} + \mathbf{R}(\mathbf{r})) = V_n(\mathbf{r}) \tag{5}$$

in which V_n is the potential in the undeformed lattice and V_d is the perturbed potential. In first approximation the perturbing potential $\delta V(\mathbf{r})$ is

$$\delta V = V_d(\mathbf{r}) - V_n(\mathbf{r}) = -\mathbf{R} \cdot \operatorname{grad} V_n(\mathbf{r}). \tag{6}$$

Nordheim[4] objected to this assumption because he believed that the important part of the field is that near the ions which moves almost unchanged as the nuclei oscillate. For this reason, he suggested that the perturbing potential should be obtained by treating the lattice as though it were a system of rigid oscillating atoms. In this case,

$$V(\mathbf{r}) = \sum_p v(\mathbf{r} - [\mathbf{r}(p) + \mathbf{R}(p)])$$

where v is the potential of an atom. Thus, the perturbing potential then is

$$\delta V(\mathbf{r}) = -\sum_p \mathbf{R}(p) \cdot \operatorname{grad} v(\mathbf{r} - \mathbf{R}(p)). \tag{7}$$

The most satisfactory discussion of the potential has been given by Bardeen[5] who obtained it by a self-consistent field method. His result, which should be valid for the monovalent metals, is more nearly like Bloch's than like Nordheim's, the reason being that the volume of space near the ions in which the atomic potential $v(\mathbf{r})$ is large is so small that the "rigid" part of the field actually can be neglected. We shall not discuss the derivation of Bardeen's results in full mathematical detail but refer the reader to the original paper. His final equation for the matrix components $\delta V_{\mathbf{k},\mathbf{k}'}$ of the perturbing potential connecting

[1] Bloch, *op. cit.*
[2] L. Brillouin, *Quantenstatistik* (Julius Springer, Berlin, 1931).
[3] H. Bethe, *Handbuch der Physik* XXIV/2 (1933).
[4] L. Nordheim, *Ann. Physik*, **9**, 607 (1931).
[5] J. Bardeen, *Phys. Rev.*, **52**, 688 (1937).

the states of electronic wave number $\mathbf{k}$ and $\mathbf{k}'$ is

$$\delta V_{\mathbf{k},\mathbf{k}'} = \sum_{\natural,\sigma,\alpha} D_t(\mathfrak{d} + \mathbf{K}_\alpha)\alpha_t(\mathfrak{d}) \cos \gamma_t(\mathfrak{d}) \delta(\mathbf{k}', \mathbf{k} + \mathfrak{d} + \mathbf{K}_\alpha) \tag{8}$$

in which the $\mathbf{K}_\alpha$ are principal vectors in the reciprocal lattice, $D_t(\mathfrak{d} + \mathbf{K}_\alpha)$ is a function of $|\mathfrak{d} + \mathbf{K}_\alpha|$ alone, and $\gamma_t(\mathfrak{d})$ is the angle between $\mathfrak{d}$ and $\xi_t(\mathfrak{d})$.

We shall now discuss the manner in which the resistivity is related to the matrix components of the perturbing potential in the general one-electron case as well as that in which Eq. (8) is valid.

b. The Selection Rules for Electronic Collisions.—It was seen in Sec. 43 that the probability $P_{\alpha\beta}$ that a perturbed system will change from a state of energy E_α to a state of energy E_β in time t is

$$P_{\alpha\beta} = \left(\frac{t}{\hbar}\right)^2 |V_{\alpha\beta}|^2 \frac{\sin^2 x}{x^2} \tag{9}$$

where

$$x = \frac{(E_\alpha - E_\beta)t}{2\hbar} \tag{10}$$

and $V_{\alpha\beta}$ is the matrix component of the perturbing potential connecting the two states:

$$V_{\alpha\beta} = \int \Phi_\beta{}^* V \Phi_\alpha d\tau. \tag{11}$$

For sufficiently long times, we may replace $\dfrac{\sin^2 x}{x^2}$ by

$$\frac{2\pi\hbar}{t}\delta(E_\alpha - E_\beta)$$

so that

$$P_{\alpha\beta} = \frac{2\pi t}{\hbar}|V_{\alpha\beta}|^2\delta(E_\alpha - E_\beta). \tag{12}$$

We shall apply these results to the problem of conductivity, regarding the entire crystal as a single system. The unperturbed wave functions of this system have the form [*cf.* Eqs. (2) and (11), Sec. 116]

$$\Phi_{r,\alpha}(x_1, \cdots, z_n, \xi_1, \cdots, \zeta_f) = \Psi_r(x_1, \cdots, z_n)\Lambda_{r\alpha}(\xi_1, \cdots, \zeta_f)$$

where Ψ_r is the electronic wave function and $\Lambda_{r\alpha}$ is the nuclear-coordinate wave function. In the present case, in which the harmonic approximation is employed, the wave functions $\Lambda_{r\alpha}$ are

$$\Lambda(\alpha) = \prod^{\sigma,t}\lambda_{n_t(\sigma)}(\alpha_t(\mathfrak{d})). \tag{13}$$

The α appearing in this equation are defined by Eqs. (3), and the λ satisfy the harmonic-oscillator equation

$$-\frac{\hbar^2}{2}\frac{\partial^2\lambda_n}{\partial\alpha_t^2(\mathfrak{o})} + 2\pi^2\nu_t^2(\mathfrak{o})\alpha_t^2(\mathfrak{o})\lambda_n = \epsilon_{n_t}(\mathfrak{o})\lambda_{n_t}(\mathfrak{o}) \tag{14}$$

where

$$\epsilon_{n_t(\sigma)} = [n_t(\mathfrak{o}) + \tfrac{1}{2}]h\nu_t(\mathfrak{o}). \tag{14a}$$

It is assumed, of course, that the electronic wave function is a determinant of Bloch functions. Hence, the entire state of the system may be specified by the electronic wave numbers $\mathbf{k}$, the electronic-spin quantum numbers, and the vibrational quantum numbers of the lattice. Since Bardeen's perturbation potential discussed in part a is the sum of identical one-electron terms that are independent of spin, those matrix components connecting states in which spin quantum numbers differ, or in which more than one wave-number vector is different, vanish. The nonvanishing components connect states for which the changing wave number satisfies the condition[1]

$$\mathbf{k}' = \mathbf{k} + \mathfrak{o} + \mathbf{K}_\alpha \tag{15}$$

where $\mathbf{k}$ is its initial value and $\mathbf{k}'$ its final value [*cf.* Eq. (8)]. We must now find the matrix components of the quantity in the right-hand side of Eq. (8) for the nuclear-coordinate wave functions. This term involves the nuclear coordinates $\alpha_t(\mathfrak{o})$ through the function

$$\alpha_t(\mathfrak{o})\,\cos\gamma_t(\mathfrak{o}) \tag{16}$$

which appears as a coefficient of $D_t(\mathfrak{o} + \mathbf{K}_\alpha)$ in Eq. (8). The matrix components of (16) vanish for all states except those in which $n_t(\mathfrak{o})$ differs by an integer, because $\Lambda(\alpha)$ is a product of one-dimensional harmonic-oscillator functions. The nonvanishing components of (16) are

$$[\alpha_t(\mathfrak{o})\,\cos\gamma_t(\mathfrak{o})]_{n,n'} = \sqrt{\frac{n+1}{2}\frac{\hbar}{2\pi\nu_t(\mathfrak{o})}}\,\delta(n',\,n+1)\,\cos\gamma_t(\mathfrak{o}) +$$
$$\sqrt{\frac{n}{2}\frac{\hbar}{2\pi\nu_t(\mathfrak{o})}}\,\delta(n',\,n-1)\,\cos\gamma_t(\mathfrak{o}). \tag{17}$$

We may summarize these results by saying that an electron may change its quantum number from $\mathbf{k}$ to $\mathbf{k}'$ in a single collision, where $\mathbf{k}$ and $\mathbf{k}'$ satisfy Eq. (15). This change must also satisfy the Pauli principle; that is, the state $\mathbf{k}'$ must be unoccupied. At the same time, one and only one of the three modes of vibration of given $\mathfrak{o}$ may change its vibrational quantum number by unity. This quantum number may

---

[1] This relationship is essentially Laue's equation for the diffraction of a wave of wave number $\mathbf{k}$ by a lattice having lattice constant $1/|\sigma|$.

decrease only if the initial value is 1 or greater than 1, that is, if the mode is initially in an excited state. Since energy must be conserved during this collision according to Eq. (12), $\mathbf{k}'$ and $\mathbf{\delta}$ must satisfy one of the relations

$$\begin{aligned}\epsilon(\mathbf{k}') &= \epsilon(\mathbf{k} \pm (\mathbf{\delta} + \mathbf{K}_\alpha)) + h\nu_t(\mathbf{\delta}),\\ \epsilon(\mathbf{k}') &= \epsilon(\mathbf{k} \pm (\mathbf{\delta} + \mathbf{K}_\alpha)) - h\nu_t(\mathbf{\delta})\end{aligned} \tag{18}$$

It should be mentioned here that Peierls[1] was the first person to point out that values of $\mathbf{K}_\alpha$ different from zero should be considered in the preceding equations. As we shall see below, these additional terms make an appreciable contribution to the resistivity.

The matrix component for a given change of state of the system is obtained by choosing the components of (8) for one of the three vibrational modes of given $\mathbf{\delta}$ and by using Eq. (17). When this matrix component is substituted in Eq. (12), we obtain the probability for the process

$$\begin{aligned}\mathbf{k} &\to \mathbf{k}',\\ n_t(\mathbf{\delta}) &\to n_t(\mathbf{\delta}) \pm 1,\end{aligned} \tag{19}$$

all other quantum numbers remaining fixed. The total probability that an electron of given $\mathbf{k}$ is scattered in given time when the vibrational system is initially in the state specified by a given set of vibrational quantum numbers is obtained by summing the probability for an individual process over all values of $\mathbf{\delta}$, $\mathbf{K}_\alpha$, and t with the different alternatives in sign.

For the purposes of the following discussion, we shall write the square of the matrix element of the perturbing potential connecting the electronic states $\mathbf{k}$ and $\mathbf{k} + \mathbf{\delta} + \mathbf{K}_\alpha$ and the vibrational states $n_t(\mathbf{\delta})$ and $n_t(\mathbf{\delta}) + 1$ or $n_t(\mathbf{\delta}) - 1$ in the form

$$A_t(\mathbf{k}, \mathbf{\delta} + \mathbf{K}_\alpha) \cdot \begin{cases} n_t(\mathbf{\delta}) + 1\\ n_t(\mathbf{\delta}) \end{cases} \tag{20}$$

where

$$A_t(\mathbf{k}, \mathbf{\delta} + \mathbf{K}_\alpha) = |D_t(\mathbf{\delta} + \mathbf{K}_\alpha)|^2 \frac{\hbar}{4\pi\nu_t(\mathbf{\delta})} \cos^2 \gamma_t(\mathbf{\delta}). \tag{21}$$

c. *The Computation of $b - a$ for Temperatures above the Characteristic Temperature.*—According to Eq. (9) and the preceding results, the total probability $P(\mathbf{k}, \mathbf{\delta} + \mathbf{K}_\alpha)$ that an electron in state $\mathbf{k}$ shall make a transition to another state $\mathbf{k} + \mathbf{\delta} + \mathbf{K}_\alpha$ is

$$P(\mathbf{k}, \mathbf{\delta} + \mathbf{K}_\alpha) = \sum_t A(\mathbf{k}, \mathbf{\delta} + \mathbf{K}_\alpha)\{[n_t(\mathbf{\delta}) + 1]\omega(\epsilon(\mathbf{k}') - \epsilon(\mathbf{k}) - h\nu_t(\mathbf{\delta})) +$$

$$n_t(\mathbf{\delta})\omega(\epsilon(\mathbf{k}') - \epsilon(\mathbf{k}) + h\nu_t(\mathbf{\delta}))\} \tag{22}$$

[1] R. Peierls, *Ann. Physik*, **12**, 154 (1932).

where

$$\omega(\epsilon) = \frac{4 \sin^2 \dfrac{\epsilon t}{2\hbar}}{\epsilon^2} \tag{23}$$

and

$$\mathbf{k}' = \mathbf{k} + \mathfrak{d} + \mathbf{K}_\alpha.$$

It is assumed at this point that the state $\mathbf{k}'$ is unoccupied. We may conveniently note that $h\nu_t(\mathfrak{d})$ is small compared with $\epsilon(\mathbf{k})$ or $\epsilon(\mathbf{k}')$ for the electrons near the top of the filled region since $k\Theta_D$, which corresponds to the maximum value of $h\nu_t(\mathfrak{d})$, is less than 0.05 ev for most metals, whereas $\epsilon(k_0)$ is at least 1 ev for all metals. This means that the electronic energy is very nearly conserved during the collisions discussed in this part of the present section. Hence, as a practical approximation, we shall replace the ω in (22) by $\omega(\epsilon(\mathbf{k}') - \epsilon(\mathbf{k}))$. As will be seen below, this approximation is justifiable as long as T is appreciably larger than Θ_D. With this simplification, $P(\mathbf{k}, \mathfrak{d} + \mathbf{K}_\alpha)$ becomes

$$P(\mathbf{k}, \mathfrak{d} + \mathbf{K}_\alpha) = \sum_t A_t(\mathbf{k}, \mathfrak{d} + \mathbf{K}_\alpha)[2n_t(\mathfrak{d}) + 1]\omega(\epsilon(\mathbf{k}') - \epsilon(\mathbf{k}))$$

$$\equiv B(\mathbf{k}, \mathbf{k} + \mathfrak{d} + \mathbf{K}_\alpha)\omega(\epsilon(\mathbf{k}') - \epsilon(\mathbf{k})). \tag{22a}$$

In a practical problem in which we know only that a metal is at temperature T, we are not able to give the $n_t(\mathfrak{d})$ in this equation particular integer values. Instead, we can know only the average values, which we shall assume are given by the equation

$$n_t(\mathfrak{d}) = \frac{1}{e^{\frac{h\nu_t(\sigma)}{kT}} - 1}$$

(*cf.* Sec. 18).

It was remarked above that Eq. (22a) gives the transition probability only when the state $\mathbf{k}'$ is unoccupied. If $f(\mathbf{k}')$ is the probability that this state is occupied, the probability that it is not occupied is $[1 - f(\mathbf{k}')]$. Under equilibrium conditions in the absence of an external field, we may assume that f has the value

$$f_0(\mathbf{k}) = \frac{1}{e^{\frac{\epsilon(\mathbf{k}) - \epsilon_0'}{kT}} + 1}, \tag{24}$$

corresponding to Fermi-Dirac statistics.

We may now compute the collision terms in Boltzmann's equation for statistical equilibrium. The total number of electrons per unit volume leaving a unit volume in $\mathbf{k}$ space per unit time because of collisions is

$$a = \frac{d}{dt} \int f(\mathbf{k}) B(\mathbf{k,k'}) \omega(\epsilon(\mathbf{k}) - \epsilon(\mathbf{k'}))[1 - f(\mathbf{k'})]\rho(\mathbf{k'})d\tau(\mathbf{k'})$$

where $\rho(\mathbf{k'})$ is the density of levels at the point $\mathbf{k'}$ and the integration extends over those points in wave-number space for which the selection rules are satisfied. Similarly, the number of electrons per unit volume entering the unit volume of $\mathbf{k}$ space because of collisions is

$$b = \frac{d}{dt} \int f(\mathbf{k'}) B(\mathbf{k,k'}) \omega(\epsilon(\mathbf{k}) - \epsilon(\mathbf{k'}))[1 - f(\mathbf{k})]\rho(\mathbf{k'})d\tau(\mathbf{k'}).$$

Hence, $b - a$ is

$$b - a = \frac{d}{dt} \int B(\mathbf{k,k'}) \omega(\epsilon(\mathbf{k'}) - \epsilon(\mathbf{k})) \{f(\mathbf{k'})[1 - f(\mathbf{k})] -$$
$$f(\mathbf{k})[1 - f(\mathbf{k'})]\} \rho(\mathbf{k'})d\tau(\mathbf{k'}). \quad (25)$$

This evidently vanishes when $f(\mathbf{k})$ has the form of Eq. (24). In the case in which there is an electrical field, we shall assume that f has the form

$$f(\mathbf{k}) = f_0(\mathbf{k}) + k_x g(\mathbf{k}) \quad (26)$$

where $g(\mathbf{k})$ is a small function that depends only upon $\epsilon(\mathbf{k})$. This assumption evidently is equivalent to that of Eq. (1a), Sec. 126. If (26) is substituted in Eq. (25) and only first-order terms are kept, it is found that

$$b - a = \frac{d}{dt} \int B(\mathbf{k,k'}) \omega(\epsilon(\mathbf{k}) - \epsilon(\mathbf{k'}))[k_x'g(\mathbf{k'}) - k_x g(\mathbf{k})]\rho(\mathbf{k'})d\tau(\mathbf{k'}). \quad (27)$$

We shall now integrate this under the assumption that

$$\epsilon(\mathbf{k}) = \frac{h^2 \mathbf{k}^2}{2m}$$

and that, when $|\mathbf{k}| = |\mathbf{k'}|$, $B(\mathbf{k,k'})$ depends only upon $|\mathbf{k}|$ and the angle θ between $\mathbf{k}$ and $\mathbf{k'}$. Then, $\rho(\mathbf{k'})$ is a constant equal to $2V$, and

$$d\tau(\mathbf{k'}) = k'^2 dk' \sin\theta d\theta d\varphi$$
$$= \left(\frac{dk'}{d\epsilon'}\right) k'^2 d\epsilon(k') \sin\theta d\theta d\varphi$$

where θ and φ are the polar angles of the vector $\mathbf{k'}$ measured relative to $\mathbf{k}$. Making use of the relations

$$\int_{-a}^{a} F(\epsilon)\omega(\epsilon)d\epsilon \cong \frac{2\pi t}{\hbar}F(0), \quad (28)$$

for $at >> \hbar \pi$ and

$$\int_0^{2\pi} (k_x - k_x')d\varphi = 2\pi k_x(1 - \cos \theta),$$

we find

$$b - a = -\left(\frac{dk}{d\epsilon}\right)Vk^2k_xg(\epsilon)\frac{8\pi^2}{h}\int_0^{\pi} B(\mathbf{k},\mathbf{k}')(1 - \cos \theta) \sin \theta d\theta$$

where the integration extends over θ.

Comparing this equation with the corresponding equation for Sommerfeld's theory [cf. Eq. (2), Sec. 126], namely,

$$b - a = -\frac{v_x v\chi(v)}{l},$$

we may conclude by analogy that the mean free path l is

$$\frac{1}{l(k)} = 16\pi^3\left(\frac{dk}{d\epsilon}\right)^2 Vk^2 \int_0^{\pi} B(\mathbf{k},\mathbf{k}')(1 - \cos \theta) \sin \theta d\theta \qquad (29)$$

since $v_x\chi(v)$ is replaced by $k_xg(\epsilon)$ in the present problem and since

$$v = \frac{1}{\hbar}\frac{d\epsilon}{dk}.$$

Before this result can be substituted in Eq. (6) of the preceding section for the conductivity, it must be shown that k_xg and $v_x\chi$ have the same form. In order to do so, we must solve Boltzmann's equation

$$\left(\frac{\partial f}{\partial t}\right)_{\text{drift}} = b - a.$$

In the Lorentz-Sommerfeld case, the solution of this is (cf. Sec. 31)

$$v_x\chi = e\mathsf{E}_x l\frac{v_x}{v}\frac{\partial f_0}{\partial v_x}. \qquad (30)$$

Now, in the present case,

$$\left(\frac{\partial f}{\partial t}\right)_{\text{drift}} = -\frac{e\mathsf{E}_x}{\hbar}\frac{\partial f}{\partial \epsilon}\frac{\partial \epsilon}{\partial k_x} - \frac{1}{\hbar}\frac{\partial \epsilon}{\partial k_x}\frac{\partial f}{\partial x}. \qquad (31)$$

We may assume that f is independent of x and may retain only the first part of (26) in the remaining term in (31). We obtain

$$\left(\frac{\partial f}{\partial t}\right)_{\text{drift}} = -\frac{e\mathsf{E}_x}{\hbar}\frac{\partial f_0}{\partial \epsilon}\frac{\partial \epsilon}{\partial k_x}. \qquad (32)$$

Thus, the equation to be solved is

$$e\mathsf{E}_x v_x\frac{\partial f_0}{\partial \epsilon} = \frac{g}{l}\frac{k_x}{\hbar}\left(\frac{d\epsilon}{dk}\right)$$

where l is given by Eq. (29).   The solution of this is

$$k_x g = e\mathsf{E}_x l \frac{v_x}{v} \frac{\partial f_0}{\partial \epsilon},$$ (33)

which is identical with Eq. (30); hence, we may use the value of $l(k)$ given by Eq. (29) in all the equations of the Lorentz-Sommerfeld theory. Thus, the conductivity is

$$\sigma = \frac{e^2 n_0 l(k_0)}{m v(k_0)}.$$ (34)

It is sometimes convenient to write Eq. (34) in the form

$$\rho = \frac{1}{\sigma} = \frac{m v(k_0)}{e^2 n_0} \frac{1}{l(k_0)}$$ (35)

where ρ is the resistivity. When coupled with Eq. (29), this form shows clearly the way in which the resistivity depends upon the matrix components of the perturbing potential.

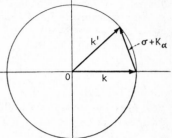

FIG. 1.—The relationship between **k**, **k′** and $\sigma + \mathbf{K}_\alpha$.   The center of the circle is at the origin of **k** space and the radius of the circle is $|\mathbf{k}| = |\mathbf{k}'|$.   The values over which $\sigma + \mathbf{K}_\alpha$ is integrated are given by the chords that connect **k** and **k′**.

d. The Numerical Computation of the High-temperature Conductivity.— We shall now outline the way in which Bardeen computed $1/l$.   According to the equations of parts *b* and *c*

$$B(\mathbf{k},\mathbf{k}') = \sum_t |D_t(\sigma + \mathbf{K}_\alpha)|^2 \frac{\hbar}{4\pi \nu_t(\sigma)} \cos^2 \gamma_t(\sigma) [2 n_t(\sigma) + 1].$$ (36)

This must be substituted in Eq. (29), and the result must be integrated over θ.   Before this can be done, it is necessary to investigate the dependence of σ and $\mathbf{K}_\alpha$ upon θ.   The relations between **k, k′,** σ, and $\mathbf{K}_\alpha$ are given by the equations

$$\left.\begin{aligned} \mathbf{k}' - \mathbf{k} &= \sigma + \mathbf{K}_\alpha, \\ \mathbf{k}'^2 = \mathbf{k}^2 &= |\mathbf{k} + \sigma + \mathbf{K}_\alpha|^2, \end{aligned}\right\}$$ (37)

which show that the allowed values of $\sigma + \mathbf{K}_\alpha$ are the chords of a sphere of radius $|\mathbf{k}|$ that pass through the point **k** (*cf.* Fig. 1).   Moreover, each allowed value of $\sigma + \mathbf{K}_\alpha$ satisfies this relation only once.   The relation between $|\sigma + \mathbf{k}_\alpha|$ and θ, the angle between **k** and **k′**, may be found by use of elementary geometry and is

$$|\sigma + \mathbf{K}_\alpha| = 2k \sin \frac{\theta}{2}.$$ (38)

Hence, all the terms of $B(\mathbf{k}, \mathbf{k}')$ except $\dfrac{\cos^2 \gamma_t(\mathfrak{o})}{4\pi \nu_t(\mathfrak{o})}$ and $[2n_t(\mathfrak{o}) + 1]$ depend upon θ alone. A part of the complication arising from the additional terms may be removed by assuming that

$$\nu_t(\mathfrak{o}) = c\sigma$$

where c is independent of t and $\mathfrak{o}$. This relation is not rigorous since the velocity of vibrational waves usually depends upon wave number even in an isotropic solid. With this assumption, we may make use of the relation

$$\Sigma \cos^2 \gamma_t(\mathfrak{o}) = 1,$$

which is valid because the three directions of polarization of lattice waves are orthogonal. We are then left only with the complication that part of the coefficients in the terms of B depend upon $|\mathfrak{o} + \mathbf{K}_\alpha|$ and part depend upon $|\mathfrak{o}|$. As long as $\mathfrak{o} + \mathbf{K}_\alpha$ lies in the first zone, $\mathbf{K}_\alpha$ is zero, so that these terms depend only upon $|\mathfrak{o}|$. Bardeen has pointed out that $|\mathfrak{o}|$ is very near to its maximum in the monovalent metals when $\mathbf{K}_\alpha$ is not zero. This fact can be made evident by observing that in monovalent metals the circle of radius $2k_0$ in $\mathfrak{o} + \mathbf{K}_\alpha$ space, which determines the allowed values of $\mathfrak{o} + \mathbf{K}_\alpha$, usually is not close to points $\mathbf{K}_\alpha$ other than the origin. Hence, we may replace $\nu_t(\mathfrak{o})$ by $c\sigma_m$ in those terms of $B(\mathbf{k}, \mathbf{k}')$ for which $\mathbf{K}_\alpha \neq 0$. This assumption evidently decreases the theoretical resistivity to some extent.

As a further simplification, it may be assumed that the first zone of wave-number space is a sphere of radius

$$\sigma_m = 2^{\frac{1}{3}} k_0.$$

The vector $|\mathfrak{o} + \mathbf{K}_\alpha|$ extends outside this sphere whenever

$$\sin \frac{\theta}{2} > \frac{2^{\frac{1}{3}}}{2} = \frac{1}{2^{\frac{2}{3}}}$$

or whenever $\theta > 79°$.

Finally, it will be assumed that the temperature is so high that

$$n_t = \frac{1}{e^{\frac{h\nu_t}{kT}} - 1}$$

may be replaced by $kT/h\nu_t$.

With these assumptions, we have

$$B(\mathbf{k}, \mathbf{k}') = \frac{2kT}{9c^2 NM} G(u)^2, \tag{39a}$$

when $\sin \theta/2 < 2^{-\frac{3}{4}}$, and

$$B(\mathbf{k},\mathbf{k}') = \frac{2kT}{9c^2NM}\frac{4k_0^2u^2}{\sigma_m^2}G(u)^2, \tag{39b}$$

when $\sin \theta/2 > 2^{-\frac{3}{4}}$, where

$$u = \sin \frac{\theta}{2}$$

and $G(u)$ is a somewhat intricate function that may be derived in a straightforward manner from the coefficients $D_t(\mathbf{d} + \mathbf{K}_\alpha)$ in Eq. (8).

It should be observed that the value of $B(\mathbf{k},\mathbf{k}')$ for which $\sin \theta/2$ is less than $2^{-\frac{3}{4}}$ joins continuously with that for which $\sin \theta/2$ is greater than $2^{-\frac{3}{4}}$, since

$$\frac{4k_0^2 2^{-\frac{3}{4}}}{\sigma_m^2} = 1.$$

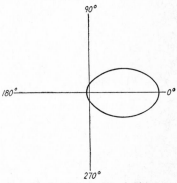

The function $G(u)^2$, which determines the angular distribution of scattering, decreases from a relative value of 1 to a value of about 0.1 in the range extending from $\theta = 0$ to $\theta = \pi$, as is shown in Fig. 2. Hence, collisions in which the electron is scattered in the forward direction are most probable.

FIG. 2.—A relative plot of $G(u)^2$ as a function of the polar angle θ.

Substituting the foregoing value of $B(\mathbf{k},\mathbf{k}')$ in Eq. (29), we obtain

$$\frac{1}{l(k)} = \frac{32\pi^3kT}{9c^2NM}\left(\frac{dk}{d\epsilon}\right)_{k=k_0}^2 k_0^2 \cdot 2^{-\frac{3}{4}}C^2 \tag{40}$$

where C^2 is defined by the equation

$$C^2 = 2^{\frac{3}{4}}\left[\int_0^{2^{-\frac{3}{4}}} G(u)^2u^3du + \int_{2^{-\frac{3}{4}}}^1 G(u)^2\frac{4k_0^2u^2}{\sigma_m^2}u^3du\right]. \tag{41}$$

Hence, if we use the relation $\Theta_D = hc\sigma_m/k$ the conductivity is

$$\sigma = \frac{4e^2k_0}{h^3\pi}\left(\frac{d\epsilon}{dk}\right)_{k=k_0}^2\left(\frac{Mk\Theta_D^2}{T}\right)\frac{1}{C^2}. \tag{42}$$

It should be observed that the effect of the relatively large value of $G(u)$ in the direction of forward scattering is partly compensated by the coefficients of this function in the integrands of (41). Actually the second integral represents about 40 per cent of C^2. A list of computed and observed values of the conductivity of a number of monovalent metals is given in Table LXXVII. The important simplifying assumptions made in deriving these values are that the electrons are perfectly free and that the lattice potential at the surface of the equivalent sphere (see Chap. X) is equal to $\epsilon(0)$. These assumptions are most closely satisfied by sodium, for which the agreement between observed and calculated values is best. The theoretical values usually are larger

TABLE LXXVII.—COMPARISON OF OBSERVED AND CALCULATED VALUES OF THE CONDUCTIVITY OF SEVERAL MONOVALENT METALS AT 0°C

[These values are taken from the review article by J. Bardeen, *Jour. Applied Phys.*, **11**, 88 (1940). In 10^4 ohm-cm]

Metal	Observed	Calculated
Li	11.8	28
Na	23.4	23
K	16.4	20
Rb	8.6	33
Ca	5.3	22
Cu	64	174
Ag	66	143
Au	49	142

than the measured ones, a fact showing that the computed values of $1/l$ should be larger. Bardeen estimates that about 10 or 15 per cent of the difference is due to the fact that ν is replaced by ν_m in the terms of B for which $\mathbf{K}_\alpha$ is not zero.

e. Other Computations.—Other workers have obtained results comparable with Bardeen's on the basis of somewhat different assumptions. We mentioned, for example, Bloch's assumption of deformable ions and Nordheim's assumption of rigid ions in part *a*. The first of these leads to an equation similar to (42) in which the constant C is given by an expression different from (41) which involves the electronic potential in the undeformed lattice. Peterson and Nordheim[1] have used the potential function for sodium, determined by the methods discussed in Chap. X, to compute Bloch's C and have found that this value leads to a conductivity about three times smaller than the experimental value given in Table LXXVII. This fact indicates that the actual fluctuations in potential are less than those given by the deformable-atom picture, so that scattering is less. The rigid-ion picture is not very well founded, as we saw in part *a*, and has not actually been used as the basis for a quantitative computation.

Peterson and Nordheim have proposed another method for determining the electron scattering in metals that is simpler, although less accurate

[1] E. L. PETERSON and L. W. NORDHEIM, *Phys. Rev.*, **51**, 355 (1937).

than Bardeen's.   They assume that the electronic wave functions have
the form

$$\chi_0 e^{2\pi i \mathbf{k} \cdot \mathbf{r}},$$

where $|\chi_0|^2$ varies inversely as the change in atomic volume, when the
lattice is perturbed by a vibrational wave of wave number σ.   They
then expand the perturbed wave function in terms of the unperturbed
functions and compute the matrix components of the perturbing potential from the coefficients.   The value of the square of these components is

$$(\delta V)^2{}_{\mathbf{k}, \mathbf{k}+\sigma} = \frac{\pi^2 h^4 \sigma^6}{4 m^2} |a_t(\sigma)|^2,$$

which may be used to compute $1/l$ in a way similar to that discussed
in the preceding section.   Peterson and Nordheim neglect the terms for
which $\mathbf{K}_\alpha \neq 0$, and obtain

$$\frac{C}{\epsilon_0'} = 0.84$$

for all metals.   Although this result agrees within about 10 per cent with
values of the same quantity computed by Bardeen, Bardeen points out
that the neglect of terms for which $\mathbf{K}_\alpha \neq 0$ is a serious omission, for if
they were included, the value of C/ϵ_0' would be increased by a factor of
the order 2.

Mott and Jones[1] have developed another simplified method for treating the resistivity at high temperatures.   They compute the probability
that an electron is scattered in a single polyhedron on the assumption
that the fluctuations of potential within a given polyhedron may be
handled as though independent of the fluctuations in other cells.   The
total scattering probability is then determined by adding the contributions from each cell.   This approximation is equivalent to assuming
that the atoms have individual oscillation frequencies, as in the Einstein
theory of specific heats, and is roughly valid as long as T is appreciably
larger than the characteristic temperature.   Since the scattering depends
upon the square of the atomic amplitude, which varies as $\sqrt{T}$, the
ordinary linear temperature dependence of resistance is obtained very
simply in this theory.

f. Low Temperature.—The first extensive investigation of the low-
temperature conductivity was carried through by Bloch.[2]   His work
follows closely the procedure presented above for high temperature,

[1] N. F. MOTT and H. JONES, *The Theory of the Properties of Metals and Alloys*
(Oxford University Press, New York, 1936).

[2] F. BLOCH, *Z. Physik*, **59**, 208 (1930).

although he made the additional simplifying assumptions that the scattering is isotropic and that the term for which $\mathbf{K}_\alpha$ is not zero may be neglected. On the whole, his computation, which will not be presented here, is more intricate because the assumption that

$$\omega(\epsilon(\mathbf{k}') - \epsilon(\mathbf{k}) \pm h\nu) = \omega(\epsilon(\mathbf{k}') - \epsilon(\mathbf{k}))$$

may not be made at low temperatures. Bardeen has corrected Bloch's results to conform to the use of his own interaction potential. This means, in principle, that he repeated the computation of parts a to c on the assumption that T is less than Θ_D. The results show that the ratio of the high-temperature conductivity σ_2 to the low-temperature conductivity σ_1 is

$$\frac{\sigma_2}{\sigma_1} = 497.6 \left(\frac{\epsilon_0'}{C}\right)^2 \left(\frac{T_1}{\Theta_D}\right)^4 \frac{T_1}{T_2} \tag{43}$$

where $T_1 << \Theta_D << T_2$. In other words, the results predict that the low-temperature conductivity should vary as T^{-5}. This temperature dependence was also found by Bloch who derived the relation

$$\frac{\sigma_2}{\sigma_1} = 497.6 \left(\frac{T_1}{\Theta_D}\right)^4 \frac{T_1}{T_2} \tag{44}$$

in place of (43).

The physical origin of this T^{-5} law may be understood in the following way. If we schematize the collision process by saying that the electrons make collisions with the quanta of lattice vibrations, the mean free path should contain a factor $1/T^3$ because the density of quanta varies as T^3 when T is well below the characteristic temperature. In addition, the collisions become less effective as the temperature decreases, for only the lattice waves of smaller wave number are excited. In fact, the mean wave number $\bar{\sigma}$ is of the order of magnitude kT/hc at temperature T, where c is the acoustical velocity. Consider an electron that is traveling in the direction of the field and has wave number $\mathbf{k}$. After a collision, its wave number is $\mathbf{k} + \sigma$, where σ is the wave number of the quantum with which it has collided. Since σ ranges over a sphere, the component of momentum in the direction of the field is not changed on the average by a factor of the order of magnitude $\bar{\sigma}/k$; instead, the change is of the order of magnitude $\bar{\sigma}^2/k^2$. Thus, the number of collisions required to stop the electron is of the order of magnitude $k^2/\bar{\sigma}^2$, which varies as $1/T^2$, whence the effective mean free path for stopping varies as

$$\frac{1}{(T^3 \cdot T^2)} = \frac{1}{T^5}.$$

If the atoms scattered the electrons independently at low temperatures, as Mott and Jones in their simplified theory assumed that they do at high temperatures, the density of quanta would decrease as $e^{-\frac{h\nu}{kT}}$ with decreasing temperature, where ν is a constant, and the resistance would decrease much more rapidly than T^5.

In this connection, Peierls[1] has raised the objection to the low-temperature theory that thermal equilibrium is assumed without adequate proof. He points out that in the Bloch-Bardeen type of treatment the electrons make only small-angle collisions at low temperatures, at least in the monovalent metals, in which the top of the filled region is not near a zone boundary, and that the number of low-frequency quanta excited may not correspond to equilibrium. This objection has not yet been fully cleared.

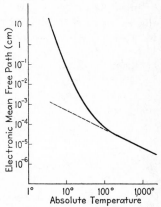

Fig. 3.—The electronic mean free path in silver as determined by equating the expression (6) of Sec. 126 for the theoretical conductivity to Grüneisen's empirical function for silver.

Extensive experimental work of Grüneisen[2] shows that the T^{-5} law is closely obeyed at low temperatures. It is not possible to distinguish between Eqs. (43) and (44), however, because the characteristic temperature cannot be fixed closely enough. Grüneisen has also found empirically that the reciprocal of the conductivity of many simple metals is given closely over a wide temperature range by the function

$$\frac{1}{\sigma} = AG(T) = AT^5 \int_0^{\frac{\Theta}{T}} \frac{x^5 dx}{(e^x - 1)(1 - e^{-x})} \tag{45}$$

if the constants A and Θ are properly chosen (cf. Fig. 3). Usually, Θ is close to the characteristic temperature of the substance. At high temperatures, this function approaches the value $AT/4$; at low temperatures, it approaches the value $124.4AT^5$. Thus, according to this empirical relation, the ratio of the high-temperature conductivity to the low-temperature conductivity is the same as Bloch's relation (44).

It is probably well to bear in mind that the basis of the T^{-5} law is intimately connected with the validity of Debye's T^3 law for specific heats. Since the apparent experimental verification of the latter at temperatures above 10°K is open to the criticisms discussed in Secs. 20

[1] R. PEIERLS, *Ann. Physik*, **4**, 121 (1930).
[2] E. GRÜNEISEN, *Ann. Physik*, **16**, 530 (1933).

and 23, the experimental test of the T^{-5} law is not so significant as it seems at first sight.

g. Critique of the High-temperature Treatment of Conductivity.—Kretschmann[1] has pointed out that, in the preceding treatments of the theory of conductivity, the electronic states are described as though their energy were accurately defined to well within the limits of the changes in energy occurring during transitions. Since these changes are of the order of magnitude $h\nu_m$, where ν_m is the maximum Debye frequency, he suggests this description can be accurate only if the perturbing effect of the lattice vibrations is much less than $h\nu_m$. The effect of the lattice is measured by the mean time between collisions, namely, $\tau = l/v$ where l is the mean free path and v is the mean velocity. At room temperature, $l \sim 10^{-5}$ cm in a good conductor and $v \sim 10^7$ cm/sec so that $\tau \sim 10^{-12}$ sec. Hence, the condition that should be satisfied, if the perturbing effects are small compared with $h\nu_m$, is that

$$\frac{h}{\tau} << h\nu_m$$

or

$$\nu_m >> \frac{1}{\tau} \sim 10^{12}. \tag{46}$$

Actually, the lattice frequencies are also of the order of magnitude 10^{12} sec^{-1}.

Peierls[2] has suggested in this connection that Kretschmann's criticism would be accurate only if the delta-function approximation of Eq. (12) had to be employed from the start. Actually, we have been able to use the form (9) of the perturbation equation until reaching Eq. (29) because the matrix component $|V_{\alpha\beta}|^2$ in (9) is a slowly varying function of the variable ϵ. The condition under which the relation (29) is valid is that the integrand should vary slowly within the range of ϵ in which $\omega(\epsilon)$ has its peak. Since the g in (27), which are expressed in terms of the Fermi-Dirac distribution function, vary within a range kT near the edge of the filled region, it follows that we must have

$$\frac{h}{kT} < t < \tau \tag{47}$$

instead of (46). This is slightly less restrictive than Kretschmann's condition, although room temperature is a borderline temperature even for good conductors.

[1] E. Kretschmann, *Z. Physik*, **87**, 518 (1934); **88**, 786 (1934).

[2] R. Peierls, *Z. Physik*, **88**, 786 (1934); *Helvetica Phys. Acta*, **7** (Sup.), 24 (1934).

h. The Effect of Electronic Coupling.—Houston[1] has suggested that the effects of electrostatic coupling between electrons which are disregarded in using wave functions constructed of one-electron functions may appreciably alter the quantitative values of the computed conductivity. He has pointed out, for example, that because of this coupling collisions may occur in which two or more electrons change their wave-number vectors at the same time. The wave-number vectors of individual electrons need not obey Eq. (15) in such collisions since this equation is replaced by one concerning the behavior of the total wave-number vectors of all electrons. In view of the excellent results Bardeen obtained for the theoretical conductivity in the case of sodium, it seems unlikely that Houston's conclusions are important in the case of the simpler metals at ordinary temperatures; however, they are possibly important at low temperatures for establishing thermal equilibrium.

128. Other Simple Metals.—We should not expect the equations developed in the preceding section to apply quantitatively to divalent metals in which the distribution of levels near the top of the filled band is different from that for perfectly free electrons as Manning's results show (Sec. 99). In agreement with this, it is observed that the resistivity of these metals is about four times larger than the resistivity of the monovalent metals preceding them in the periodic chart, even though the former have twice as many electrons and nearly the same lattice parameters and characteristic temperatures as the latter. The increase in resistivity may be understood qualitatively from the fact that the effective number of free electrons, that is, the number in the energy range of width kT near the top of the band, is smaller in the divalent metals than in the monovalent metals, for these electrons alone transport the current. No quantitative computations of the resistivity have been carried out.

Jones[2] has pointed out that the principle that accounts for the comparatively high resistivity of the divalent metals should also apply to those metals, such as bismuth, which lie between ideal metallic and valence types, since they also have nearly filled zones.

129. The Temperature-dependent Resistivity of the Transition-element Metals.—The resistivity of transition-element metals, such as iron, cobalt, and nickel, usually is higher than that of the simpler metals, such as copper, following them in the periodic chart and having nearly the same lattice parameters. We have seen in Chap. XIII that the transition metals differ from the simpler metals by having unfilled d levels with the same energy as the lowest unoccupied s-p levels. Since the s-p levels in transition-element metals are very nearly the same as those

---

[1] W. V. Houston, *Phys. Rev.*, **55**, 1255 (1939).

[2] H. Jones, *Proc. Roy. Soc.*, **147**, 396 (1934). See also Mott and Jones, *op. cit.*

in the simple metals which follow, the resistivity arising from transitions of the conduction electrons between *s-p* levels should be nearly alike in the two cases. Mott[1] was the first to suggest that the *s-p* electrons are principally responsible for the current in transition metals and that the additional transitions from *s-p* levels to the unfilled *d* levels accounts for most of the additional resistivity. As proof of the first of these suggestions, it is usually pointed out that the conductivity of any single band of electrons may be placed in the form

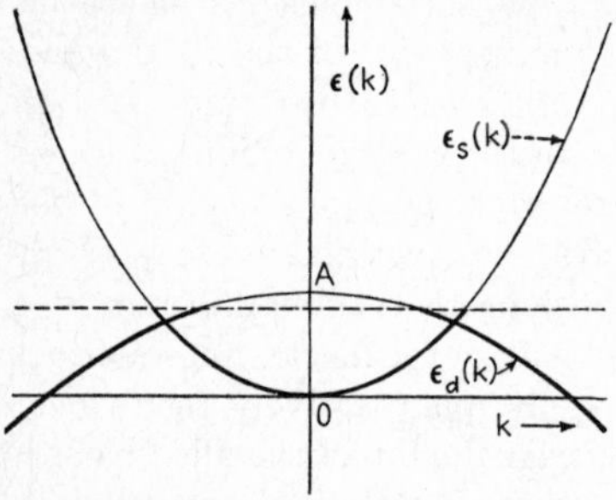

FIG. 4.—The relative positions of the *s* and *d* levels in the transition metals [*cf.* Eqs. (1) and (2)]. The dotted line represents the top of the filled region.

$$\sigma = \frac{n_0 e^2}{m^*} \tau \qquad (1)$$

[*cf.* Eq. (6), Sec. 126] where

$$\tau = l/v(\epsilon_0')$$

is the collision time and m^* is the effective mass of the electrons in the band. Since m^* for *d*-shell electrons is much larger than the electronic mass, it is assumed that (1) is small in comparison with the conductivity of the *s-p* electrons. Mott developed a simple mathematical theory of the scattering of *s-p* electrons arising from transitions to the *d* band; however, Wilson[2] has since given a more extensive treatment, which we shall discuss here.

Wilson assumed that the energy states in both the *s-p* bands and the *d* bands may be treated with the Bloch approximation and that the energy in each band is a quadratic function of the wave number in the reduced-zone scheme. The $\epsilon(\mathbf{k})$ curves for the two overlapping bands then appear as in Fig. 4. We shall select the zero of energy so that the energy $\epsilon_s(\mathbf{k})$ of the *s-p* electrons is

$$\epsilon_s(\mathbf{k}) = \frac{h^2}{2m_s} \mathbf{k}^2 \qquad (2)$$

and the energy $\epsilon_d(\mathbf{k})$ of the *d* levels is

$$\epsilon_d(\mathbf{k}) = A - \frac{h^2}{2m_d} \mathbf{k}^2 \qquad (3)$$

where m_s and m_d are the effective electron masses in the two bands.

In the one-electron approximation, the selection rules for transitions from the *s-p* band to the *d* band should be the same as those derived in

[1] N. F. MOTT, *Proc. Phys. Soc.*, **47**, 571 (1935); *Proc. Roy. Soc.*, **153**, 699 (1936), **156**, 368 (1936).

[2] A. H. WILSON, *Proc. Roy. Soc.*, **167**, 580 (1938).

Sec. 127 for transitions between levels of the s-p band, namely,

$$\mathbf{k'} - \mathbf{k} = \mathbf{d} + \mathbf{K}_\alpha, \tag{4}$$

$$\epsilon(\mathbf{k'}) = \epsilon(\mathbf{k}) \pm h\nu_t(\mathbf{d}), \tag{5}$$

where $\mathbf{k}$ is the wave number of the initial state, $\mathbf{k'}$ is that of the final state, $\mathbf{d}$ is the wave number of the vibrational mode causing the transition, and $\mathbf{K}_\alpha$ is a principal vector in the inverse lattice.

We may treat separately the contributions to the resistivity from the collisions in which the s-p electrons jump to s-p levels and those in which they jump to d levels, since we are dealing with a one-electron approximation. The first contribution was discussed in Sec. 127 for the monovalent metals. The result for the present case should differ from the result found there only in the fact that k_0 should be replaced by a value appropriate for the s-p levels in the transition metals. Since there is 0.6 free electron per atom in nickel, for example, we have

$$k_0 = \left(\frac{3}{8\pi}0.6n_0\right)^{\frac{1}{3}} \tag{6}$$

where n_0 is the number of atoms per unit volume. If this is substituted in the equations of Sec. 127, it is found that the resistivity increases by a factor $(0.6)^{-\frac{1}{3}}$ compared with the resistivity for a monovalent metal.

Wilson developed an expression for the additional resistivity that is valid for both high and low temperatures, using simplifying assumptions which will now be outlined. Let us assume that the lattice frequencies are distributed according to the Debye theory and that the longitudinal and transverse modes have the same value of $\nu(\mathbf{d})$. The probability for a transition in time t to a vacant state in the s-p band from one in the d band then is

$$\alpha_d(\mathbf{k},\mathbf{k'})\frac{\hbar}{2\pi\nu(\mathbf{d})} \cdot \begin{cases} [n(\mathbf{d}) + 1]\omega(\epsilon_d(\mathbf{k'}) - \epsilon_s(\mathbf{k}) - h\nu(\mathbf{d})) & (7a) \\ n(\mathbf{d})\omega(\epsilon_d(\mathbf{k'}) - \epsilon_s(\mathbf{k}) + h\nu(\mathbf{d})) & (7b) \end{cases}$$

where α_d is a function of $\mathbf{k}$ and $\mathbf{k'}$ that is equivalent to $2\pi\nu_t(\mathbf{d})/\hbar$ times the quantity $A_t(\mathbf{k},\mathbf{k'})$ appearing in Eq. (20), Sec. 127. Case $(7a)$ corresponds to a collision in which the vibrational mode of wave number $\mathbf{d}$ gains a quantum, and $(7b)$ to one in which it loses a quantum. The probabilities for the reverse transitions are

$$\alpha_d(\mathbf{k},\mathbf{k'})\frac{\hbar}{2\pi\nu(\mathbf{d})} \cdot \begin{cases} n(\mathbf{d})\omega(\epsilon_d(\mathbf{k'}) - \epsilon_s(\mathbf{k}) - h\nu(\mathbf{d})) \\ [n(\mathbf{d}) + 1]\omega(\epsilon_d(\mathbf{k'}) - \epsilon_s(\mathbf{k}) + h\nu(\mathbf{d})) \end{cases}. \tag{8}$$

As we remarked in part e, Sec. 127, the approximation

$$\omega(\epsilon_d(\mathbf{k'}) - \epsilon_s(\mathbf{k}) \pm h\nu(\mathbf{d})) \cong \omega(\epsilon_d(\mathbf{k'}) - \epsilon_s(\mathbf{k}))$$

may be employed at high temperatures but not at low temperatures. If we designate the electronic distribution functions for the levels of the s and d bands by f_s and f_d, we obtain the following expression for the contribution to the collision terms from the s-p to d transitions:

$$b - a = \frac{d}{dt}\int \alpha_d \frac{\hbar}{2\pi\nu(\mathfrak{d})}\Bigg(\{[n(\mathfrak{d}) + 1]f_s(1 - f_d) - n(\mathfrak{d})f_d(1 - f_s)\}$$
$$\omega(\epsilon_d(\mathbf{k'}) - \epsilon_s(\mathbf{k}) - h\nu(\mathfrak{d})) + \{n(\mathfrak{d})f_s(1 - f_d) - [n(\mathfrak{d}) + 1]f_d(1 - f_s)\}$$
$$\omega(\epsilon_d(\mathbf{k'}) - \epsilon_s(\mathbf{k}) + h\nu(\mathfrak{d}))\Bigg)\rho(\mathbf{k'})d\tau(\mathbf{k'}) \quad (9)$$

where the integration extends over the values of $\mathbf{k'}$ satisfying Eq. (4). Wilson integrated this under the following assumptions.

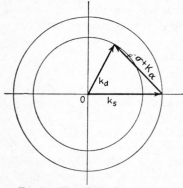

FIG. 5.—The relationship between k_{0s}, k_{0d} and $\sigma + \mathbf{K}_\alpha$ in the case of the transition metal. The outer circle is the boundary of the filled region in the s-p band, whereas the inner circle is the boundary in the d band, (*cf.* Fig. 3). These circles usually should not coincide. The vector $\sigma + \mathbf{K}_\alpha$ joins $\mathbf{k}_d$ and $\mathbf{k}_s$, so that its minimum value is

$$|\mathbf{k}_{0s}| - |\mathbf{k}_{0d}|.$$

a. α_d is a constant. This is equivalent to assuming isotropic scattering.

b. f_s and f_d may be placed in the form

$$\left.\begin{aligned} f_s &= f_{0,s} + k_x g_s(\epsilon_s(\mathbf{k})), \\ f_d &= f_{0,d} + k_x g_d(\epsilon_d(\mathbf{k})), \end{aligned}\right\} \quad (10)$$

where $f_{0,s}$ and $f_{0,d}$ are the Fermi-Dirac distribution functions for the field free problem and the g are comparatively small functions.

c. The functions g in (10) have the form

$$g(\epsilon) = C(\epsilon)\frac{\partial f}{\partial \epsilon}.$$

This functional form was also assumed in the cases discussed in Sec. 127.

d. The values of $n(\mathfrak{d})$ are given by the equation

$$n(\mathfrak{d}) = \frac{1}{e^{\frac{h\nu(\sigma)}{kT}} - 1}.$$

As in Sec. 127, it is convenient to integrate over the values of the vector $\mathfrak{d} + \mathbf{K}_\alpha$ instead of the values of $\mathbf{k'}$; there is, however, an important difference between the present and preceding cases. In the preceding case, the values of $\mathfrak{d} + \mathbf{K}_\alpha$ ranged over a sphere of radius k_0 that passed through the origin (*cf.* Fig. 1). Hence, the range of $|\mathfrak{d} + \mathbf{K}_\alpha|$ extended from zero to $2k_0$. In the present case, however, the values k_{0s} and k_{0d}

of the wave number of the electrons at the limit of the filled regions in the s and d zones should usually be different, so that the vectors $\mathbf{d} + \mathbf{K}_\alpha$ should range over a sphere of radius k_{0d} that is centered at a point k_{0s} (*cf.* Fig. 5). This sphere evidently will not pass through the origin, unless by accident. Since the lowest value of $|\mathbf{d} + \mathbf{K}_\alpha|$ in the integration is $|k_{0s} - k_{0d}|$, the vibrational modes of longest wave length usually do not play a role in scattering electrons between the s-p and d bands. Only these modes are active at sufficiently low temperatures, however, whence the resistance arising from the s-p- to d-band transitions should drop to zero at low temperatures much more rapidly than the resistance arising from s-p- to s-p-band jumps.

Wilson's result for the resistivity arising from the s-p- to d-band transitions at temperature T is

$$\rho = \frac{2m_s m_d}{n_0 e^2 \epsilon_0'^{\frac{1}{2}}} P_{sd} \left(\frac{T}{\Theta_D}\right)^3 \int_{\frac{\Theta'}{T}}^{\frac{\Theta_D}{T}} \frac{z^3 dz}{(e^z - 1)(1 - e^{-z})} \tag{11}$$

where Θ_D is the characteristic temperature,

$$\epsilon_0' = \frac{h^2}{2m_s} k_{0s}^2,$$

$$P_{sd} = \left(\frac{3}{4\pi}\right)^{\frac{1}{3}} \frac{3\pi^3 \epsilon_0'^2}{\sqrt{2} m_s^{\frac{1}{2}} Mak\Theta_D},$$

and $k\Theta'$ is the energy of the lowest vibrational frequency that scatters electrons between the s-p and d bands, that is,

$$k\Theta' = h|k_{0s} - k_{0d}|c$$

where c is the velocity of the elastic waves. The a in the denominator of P_{sd} is the lattice parameter.

At high temperatures, Eq. (11) approaches the value

$$\rho \cong \frac{m_s m_d}{n_0 e^2 \epsilon_0'^{\frac{1}{2}}} \left(\frac{T}{\Theta_D}\right) P_{sd} \left(1 - \frac{\Theta'^2}{\Theta_D^2}\right),$$

which has the linear temperature-dependence characteristic of the transitions within the s-p band. At low temperatures, however, Eq. (11) approaches zero as

$$e^{-\frac{\Theta'}{T}}.$$

The only extensive measurement of low-temperature resistance seems to be for platinum, which does not show the anomaly one would expect if an appreciable part of its resistivity were described by Eq. (11).

De Haas and de Boer,[1] who made the measurements, found that the resistivity ρ can be fitted by the function

$$\rho = CG\left(\frac{\Theta_D}{T}\right) + 1.5 \cdot 10^{-6}\rho_0 T^2 \qquad (12)$$

where ρ_0 is the room-temperature resistivity, and G is a Grüneisen function [cf. Eq. (45), Sec. 127]. An interpretation of the term in T^2 is given below. There are several reasonable explanations of the fact that a part of the resistance does not obey Eq. (11). (a) It is possible that Θ' is accidentally very small for platinum and that the experiments are not accurate enough to distinguish between a contribution to the resistivity of the form $G(\Theta/T)$ and one of the form of Eq. (11) with $\Theta' = 0$. (b) It is possible that the d-shell electrons should not be treated by the Bloch theory, so that the selection rule (4) is not applicable in the present case. (c) It is possible that the s-p- to d-band transitions actually are negligibly small and that the resistance may be explained by extension of the theory of Sec. 127. (d) It is possible that the $\epsilon(\mathbf{k})$ relation for the d electrons is so different from the free-electron relation (3) that there are many directions in $\mathbf{k}$ space for which $\mathbf{k}_{0s}$ and $\mathbf{k}_{0d}$ are equal. In this case, the very long lattice waves would always play a role, and the d-shell resistivity would not decrease so rapidly as an exponential function with decreasing temperature. This problem can be settled only on the basis of more extensive work.

Baber[2] has interpreted the term in Eq. (12) that varies as T^2 in terms of an enhancement of the transitions between s-p levels due to the presence of the holes in the d band. If the holes were rigidly fixed, they would behave like impurity atoms and would give rise to a temperature-independent scattering. If the band approximation may be used, however, the holes are also able to move and should also be scattered. Since they must obey the Pauli principle during these transitions and since the way in which the levels are occupied is temperature-dependent, the resistivity turns out to be temperature-dependent. Baber assumed that the interaction potential of an electron and a hole has the form[3]

$$V(r) = \frac{e^2}{r}e^{-qr}$$

and showed that the observed T^2 term in platinum may be derived by the use of reasonable numerical values of q.

---

[1] W. J. DE HAAS and J. H. DE BOER, *Physica*, **1**, 609 (1934).
[2] W. G. BABER, *Proc. Roy. Soc.*, **158**, 383 (1937).
[3] The reason for using this potential is discussed in the next section.

130. Residual Resistance.   The Resistivity of Alloys*.—Nordheim[1] was the first to point out that the residual resistance of nonsuperconducting metals probably is due to the scattering of electrons by lattice imperfections such as impurity atoms and flaws.   This qualitative explanation agrees very well with the fact that the residual resistance of a specimen depends upon its previous history.   If Q_i is the cross section for scattering by a given kind of imperfection, such as an impurity atom, and n_i is the density of imperfections, the mean free path l_i for scattering by these imperfections is given by the equation

$$\frac{1}{l_i} = n_i Q_i. \tag{1}$$

Thus, according to Eq. (6), Sec. 126, the residual resistance ρ_i due to this type of imperfection is

$$\rho_i = \frac{mv(\epsilon_0')}{n_0 e^2} n_i Q_i, \tag{2}$$

which is temperature-independent.   If Q_i is about $10^{-17} \mathrm{cm}^2$, which is a customary atomic cross section, and if n is 10^{18} cm^{-3}, which is the concentration of impurities in a reasonably pure specimen of metal, l_i is of the order 0.1 cm.   The mean free path for the scattering due to lattice vibrations, which was discussed in the previous sections, approaches this value at temperatures near 15°K in a good conductor such as silver.

In a sense, a disordered alloy may be viewed as a metal in which the impurity content is very high.   Hence, if Nordheim's picture is correct, it should be possible to compute the contribution to the resistance of alloys from the disordered atoms by a method similar to that used above in estimating the residual resistance.   We shall discuss this resistance on the basis of a procedure developed by Nordheim.

It is known from the discussion of Sec. 127 that the quantities determining the resistance are the squares of the matrix components of potential connecting electronic states.   We shall assume for simplicity that the potential V may be written as the sum of potential terms arising from each atom:

$$V = \sum_p v_p(\mathbf{r} - \mathbf{r}(p)) \tag{3}$$

where $v_p(\mathbf{r} - \mathbf{r}(p))$ is the potential of the atom at the position $\mathbf{r}(p)$. We shall also assume that v_p is zero outside the pth cell.   The matrix components of V then are

$$V_{\mathbf{k},\mathbf{k}'} = \sum_p \int \psi_{\mathbf{k}}{}^* v_p \psi_{\mathbf{k}'} d\tau. \tag{4}$$

[1] Nordheim, *op. cit.*

The integrand in the pth term in this series is finite only in the pth cell; moreover, the origin of each integrand may be shifted in such a way that it falls at $\mathbf{r}(p)$. Thus, the series may be written in the form

$$V_{\mathbf{k},\mathbf{k}'} = \sum_p e^{2\pi i(\mathbf{k}'-\mathbf{k})\cdot\mathbf{r}(p)} f_{p,\mathbf{k}\mathbf{k}'} \tag{5}$$

where

$$f_{p,\mathbf{k}\mathbf{k}'} = \int \psi_{\mathbf{k}}{}^*(\mathbf{r}) v_p(\mathbf{r}) \psi_{\mathbf{k}'}(\mathbf{r}) d\tau, \tag{6}$$

in which the integration extends over the cell centered at $\mathbf{r} = 0$. We shall retain the index p because v_p varies from cell to cell if there is more than one kind of ion present. The square of the absolute value of (5) is

$$|V_{\mathbf{k},\mathbf{k}'}|^2 = \sum_{q,p} e^{2\pi i(\mathbf{k}'-\mathbf{k})\cdot[\mathbf{r}(q)-\mathbf{r}(p)]} f^*_{p,\mathbf{k}\mathbf{k}'} f_{q,\mathbf{k}\mathbf{k}'}. \tag{7}$$

Let us suppose that there are s kinds of atom in the alloy and that the fraction of the ith kind is p_i. Then, the mean value of $f_{p,\mathbf{k}\mathbf{k}'}$ is

$$\overline{f_{\mathbf{k}\mathbf{k}'}} = \sum_{i=1}^{s} p_i f_{i,\mathbf{k}\mathbf{k}'} \tag{8}$$

where $f_{i,\mathbf{k}\mathbf{k}'}$ is the value of the integral (6) for the ith kind of atom. If we use this mean value of $f_{p,\mathbf{k}\mathbf{k}'}$ in place of the $f_{p,\mathbf{k}\mathbf{k}'}$ in Eq. (5), the sum vanishes since

$$\sum_{p,q} e^{2\pi i(\mathbf{k}'-\mathbf{k})\cdot[\mathbf{r}(q)-\mathbf{r}(p)]}$$

is zero, if $\mathbf{k}' - \mathbf{k}$ is not a principal vector in the reciprocal lattice. This result is not surprising, for the case in which $f_{p,\mathbf{k}\mathbf{k}'}$ is independent of p is that of a perfect lattice, in which electrons are scattered only as a result of Bragg reflection or thermal oscillations. With this in mind, Nordheim assumed that the contribution to the effective squared matrix component from the pth atom is the difference of $|f_{p,\mathbf{k}\mathbf{k}'}|^2$ and the square of the mean value $|\overline{f_{\mathbf{k}\mathbf{k}'}}|^2$. Although this difference is negative for an atom for which $|f_{p,\mathbf{k}\mathbf{k}'}|^2$ is less than average, the square of the total effective matrix component, namely,

$$|V_{\mathbf{k},\mathbf{k}'}|^2_{\text{eff}} = n(\overline{|f_{p,\mathbf{k}\mathbf{k}'}|^2} - |\overline{f_{\mathbf{k}\mathbf{k}'}}|^2), \tag{9}$$

is positive. Here,

$$\overline{|f_{p,\mathbf{k}\mathbf{k}'}|^2} = \sum_i p_i |f_{i,\mathbf{k}\mathbf{k}'}|^2, \tag{10}$$

and n is the total number of atoms. The use of Eq. (9) is equivalent to assuming that the part of the atomic scattering that is greater or less than the average is incoherent. This evidently can be true only if the alloy has no secondary long-distance order, such as can occur in β brass.

Let us consider a case in which the alloy contains two kinds of atom, A and B, and designate the fraction of A atoms by x. For simplicity we shall consider a unit volume. The value of the quantity (9) then is

$$nx(1 - x)(f_{a,\mathbf{kk}'} - f_{b,\mathbf{kk}'})^2. \tag{11}$$

This function, which is the analogue of the function $B(\mathbf{k}, \mathbf{\delta} + \mathbf{K}_\alpha)$ appearing in the theory of lattice vibrational scattering, leads to the equation

$$\frac{1}{l_x} = 16\pi^3 \left(\frac{dk}{d\epsilon}\right)_0^2 k_0^2 nx(1 - x) \int_0^\pi (f_{a,\mathbf{kk}'} - f_{b,\mathbf{kk}'})^2 (1 - \cos\theta) \sin\theta\, d\theta \tag{12}$$

if we assume that (11) depends only upon θ. If, in addition, we assume that the scattering is isotropic and that $(dk/d\epsilon)_0 = m/h^2 k_0$, we obtain

$$\frac{1}{l_x} = \frac{32\pi^3 m^2}{h^4} nx(1 - x)(f_{a,\mathbf{kk}'} - f_{b,\mathbf{kk}'})^2.$$

We shall place this equation in the form

$$\frac{1}{l_x} = nx(1 - x)Q' \tag{13}$$

where the quantity

$$Q' = \frac{32\pi^3 m^2}{h^4}(f_{a,\mathbf{kk}'} - f_{b,\mathbf{kk}'})^2$$

is the atomic cross section.

The resistivity ρ_x', that is associated with this type of scattering, then is

$$\rho_x' = \frac{mv(\epsilon_0')}{e^2} Q'x(1 - x) = \frac{h}{2e^2}\left(\frac{3n}{\pi}\right)^{\frac{1}{3}} Q'x(1 - x). \tag{14}$$

Thus, Nordheim's theory predicts that the temperature-independent part of the resistivity of a disordered solid solution should vary with concentration as $x(x - 1)$. This prediction has been checked in the silver-gold system. It is found that the experimental values of the additional resistivity can be fitted closely by Eq. (14) with

$$Q' = 0.635 \cdot 10^{-16} \text{ cm}^2.$$

Computed and observed values of ρ_x' are listed in Table LXXVIII.

As we mentioned previously, Eq. (14) is not valid in a range of concentration in which the alloy has an ordered phase for then more of the scattering is coherent than is assumed in using Eq. (9). Suppose, for example, that the lattice sites of the two-component alloy can be divided into two sets, namely, a sites, which are occupied by A atoms alone in the perfectly ordered alloy, and b sites, which are occupied by B atoms alone in the same phase. The a sites and the b sites may then be regarded as independent lattices when the temperature-independent resistivity is computed. We know from our previous result that the incoherent scattering is zero when each kind of site is occupied by only one type of atom. Hence, the Nordheim type of resistivity of the perfectly ordered lattice is zero. In a partly ordered state in which there are some B atoms in a sites and some A atoms in b sites and in which there still is a difference between a and b sites, the scattering by the atoms in the two lattices may be computed separately by the use of Eq. (9), the mean values $\overline{f_{a,kk'}}$ and $\overline{f_{b,kk'}}$ evidently being the appropriate matrix components to use in each case. When the long-distance order vanishes, the a and b sites become identical and Eq. (14) is again valid.

TABLE LXXVIII

x	0.01	0.025	0.316	0.629
ρ'_x (obs) $\times 10^6$	0.35	0.86	7.3	8.2
ρ'_x (calc) $\times 10^6$	0.35	0.88	7.6	8.2

The topic of the resistivity in disordered and ordered alloys has been considered in a high degree of detail on the basis of the Bragg-Williams theory by Muto,[1] who found that the resistivity arising from disordering should depend on the long-distance order parameter O in accordance with the equation

$$\rho_0 = A + BO + CO^2$$

in which A, B, and C are temperature-dependent. Muto has shown that this relation is in reasonable agreement with experiment in several interesting cases.

Mott[2] has used an extended form of Nordheim's theory to discuss the resistivity of substitutional alloys of copper (cf. Figs. 45 and 49, Chap. I). Let us suppose that an atom having $Z + 1$ electrons outside closed shells is placed in a monovalent metal. In the immediate vicinity of the foreign atom and outside its closed shells, we may expect the potential to be larger than that at the corresponding position near the monovalent atom by an amount

$$\frac{Ze^2}{r}$$

---

[1] T. Muto, Sci. Papers, Inst. Phys. Chem. Res., **30**, 99 (1936); **31**, 153 (1937).

[2] N. F. Mott, Proc. Cambridge Phil. Soc., **32**, 281, 1936.

where r is the distance from the nucleus. The valence electrons will swarm around the more highly charged ion preferentially, however, so that we may expect the difference in potential to vanish for large values of r. Mott assumed that the actual difference varies as

$$\frac{Ze^2}{r}e^{-qr} \tag{15}$$

where q is a constant, $1/q$ being the mean radius of the swarm of valence electrons. The function (15) evidently is an explicit form for the part of the atomic potential that gives rise to incoherent scattering in Nordheim's theory. Using this function and assuming that the electrons are nearly free, Mott found that the increase in resistance per atom per cent of the foreign atom is

$$\rho_0 = \frac{v}{100}\left(\frac{Ze^2}{mv^2}\right)^2\left[\log\left(1 + \frac{1}{y}\right) - \frac{1}{(1+y)}\right] \tag{16}$$

where

$$y = \frac{q^2h^2}{16\pi^2mv^2}$$

and v is the electron velocity. This result explains qualitatively the curves shown in Fig. 49, Chap. I. If the foreign atom is a nontransition atom and if the atomic radii are nearly the same, as is true for the atoms that form good solid solutions when mixed, we should expect q to be a constant so that ρ_0 should increase as Z^2. It may be seen that the two curves in Fig. 49 are very nearly parabolic on the positive side of the origin. The value of $1/q$ in this case is of the order of magnitude 0.3 Å. The points obtained by adding transition metals to copper and silver also lie very nearly on parabolas, suggesting that the potential (15) may be used for negative values of Z, which correspond to the number of holes in the transition-element atoms relative to the monovalent atoms.

131. Superconductivity.—Superconductivity, which was discussed very briefly in Chap. I, has developed theoretically in two directions. (1) There has been a phenomenological development in which the observed properties of superconductors are discussed in terms of the functions of thermodynamics and of Maxwell's equations. (2) There has been a very rudimentary treatment in terms of the electron theory of solids.

The first development, which is surveyed extensively in a tract by London,[1] leads to the conclusion that, in addition to having a high electrical conductivity, a superconductor is a medium in which the

---

[1] F. LONDON, *Une conception nouvelle de la supra-conductabilité, actualités scientifiques et industrielles* (Hermann et Cie., Paris, 1937).

magnetic flux is zero, a fact implying a very large diamagnetism. Moreover, the entropy associated with the superconducting state is abnormally low compared with the entropy of a metal in its normal state. Although the unique magnetic characteristics of superconductors are not so spectacular as the electrical properties, their theoretical significance is no less important. The reader is referred to London's article for a detailed discussion of this type of work.

The development along the lines of the electron theory of solids is due principally to Slater[1] and is still in only the most qualitative stage. Slater has suggested that the levels of the entire metal (see Secs. 66 and 98) become discrete, or at least possess extremely low density, at the bottom of the spectrum. In essence, the levels in this region are to be regarded as the residues of the excitation states that occur in the atomic approximation when the atoms of the metal are widely separated. Because of the strong perturbations, however, they cannot be described in terms of ordinary exciton theory. Instead, Slater would regard them as possessing extremely intricate wave functions corresponding to a blend of many exciton states, so that the electrons in a comparatively large region of the metal are intimately correlated. Thus, in a schematic way, Slater regards the metal, when in one of these lower states, as an aggregate of very large molecular units extending over one hundred or more atomic distances and having discrete levels that are very finely spaced, but not so finely spaced as they would be if the units were as large as the entire specimen of metal. Being large compared with ordinary molecules, these units should have a large diamagnetism, and Slater suggests that their properties are basically those of a superconductor; hence, he would regard the low-lying low-density levels of the entire metal as the superconducting states. The system can occupy these levels only at temperatures near absolute zero since they have very low statistical weights and are favored only by their low energy. For the reasons discussed in Chap. XIV, we may expect the system to jump to states of higher energy and higher entropy at temperatures above absolute zero, the transition taking the form of one of the three possible types of phase change. The most probable high-entropy states are, of course, those described in the band approximation, in which the metal has normal properties. Presumably, these states differ from the low-entropy ones principally in the fact that the pseudomolecular structure responsible for superconductivity has melted; that is, the electronic motions are no longer correlated over many atomic distances, but only over a few, as in the approximations developed in preceding chapters. The precise form of the change from the superconducting to the ordinary state depends upon the density of levels of the entire solid in the region

[1] J. C. Slater, *Phys. Rev.*, **51**, 195 (1937); **52**, 214 (1937).

between the two types of level, and the theory has not developed to a point where this may be predicted. Experiment shows that the phase change is of the second kind, in which there is a discontinuity in specific heat, but no latent heat.

The difficulties that stand in the way of a quantitative test of any electronic theory of superconductivity obviously arise from the difficulties of handling wave functions for the entire solid in a degree of approximation sufficiently high to include states of the type discussed by Slater. There seems little likelihood that these difficulties will be surmounted in the immediate future.

B. IONIC CONDUCTIVITY

132. General Principles.—It is believed at present that the ionic conductivity of solids is closely connected with the type of lattice imperfections that occur in pure semi-conductors. This idea was first suggested by Frenkel[1] and has been substantiated by subsequent work, the most thorough investigation of the possible types of lattice defect having been made by Schottky and Wagner.[2] We shall begin by discussing their work.

Let us consider a crystal of composition MX, such as a monovalent metal halide or an alkaline-earth oxide or sulfide. If the crystal is entirely perfect in its equilibrium state, a volume ionic conductivity is found only if positive or negative ions leave their normal sites and wander in the lattice because of the influence of the field. In this case, the crystal lattice would develop imperfections as an effect of the field. It is evident, however, that a very large field would be required to dislodge an ion from its normal position, for the potential energy of an ion varies by an amount of the order of magnitude of 1 volt in an interatomic distance. Thus, a field of millions of volts per centimeter would be required to induce a current. Hence, it is necessary to assume that the ions carrying the volume ionic current are wandering before the field is applied and that the field simply disturbs the statistical distribution of motion.

There are two ways in which ions may move through the crystal (*cf.* Fig. 6). (1) They may move through interstitial positions which are unoccupied in the perfect crystal. (2) They may move by jumping into vacant sites. In the second case, it is common to say that the vacancies move through the lattice and carry the current. Specific examples of crystals in which these types of conductivity occur are discussed below.

[1] J. Frenkel, *Z. Physik*, **35**, 652 (1926).
[2] C. Wagner and W. Schottky, *Z. physik. Chem.*, **B11**, 163 (1930).   C. Wagner, *Z. physik. Chem.*, Bodenstein Fest., 177 (1931); **B22**, 181 *ff.* (1933).

In order to derive an equation for the conductivity of the interstitial ions or the vacancies, we shall employ the same model of the flow process that was used in the derivation of the equation for the jump frequency of diffusing atoms in Sec. 122. According to this work, the probability per unit time that the interstitial atom or the vacancy will jump in one of the α directions, in which there is a saddle point of height ϵ_s, is

$$\bar{\nu} = \frac{1}{\alpha} \frac{(kT)^3}{h^3 \nu_s^2} e^{-\frac{\epsilon_s}{kT}} \tag{1}$$

in the absence of an electrostatic field. Here, ν_s is the vibration frequency in the two directions in the saddle point that are normal to the direction of flow. We shall treat a simple model in which there are six saddle points of energy ϵ_s, which lie along the six axial directions relative to

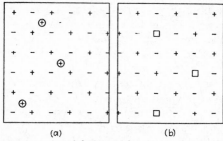

(a)   (b)

Fig. 6.—Schematic representation of the two modes of ionic motion. In (a) the interstitial ions $\oplus$ diffuse through the interstitial sites. In (b) the ions move via the vacancies $\square$.

the equilibrium point. Let us now assume that there is an electrostatic field of intensity E in the x direction. The saddle point lying in the direction of the field relative to a given equilibrium position is lowered by an amount $Ee\delta/2$ where $\delta/2$ is the distance between the equilibrium position and the saddle point and e is the charge on the ion or vacancy. Hence, the jump frequency for this barrier is changed to

$$\bar{\nu}_E^+ = \frac{1}{\alpha} \frac{(kT)^3}{h^3 \nu_s^2} e^{-\frac{\epsilon_s}{kT}} e^{\frac{Ee\delta}{2kT}}. \tag{2}$$

The saddle point in the opposite direction is raised by the same amount so that the jump frequency in the backward direction is

$$\bar{\nu}_E^- = \frac{1}{\alpha} \frac{(kT)^3}{h^3 \nu_s^2} e^{-\frac{\epsilon_s}{kT}} e^{-\frac{Ee\delta}{2kT}}. \tag{3}$$

Thus, the excess probability for jumping in the field direction is

$$\bar{\nu}_p = \bar{\nu}_E^+ - \bar{\nu}_E^- = \bar{\nu}\, 2 \sinh \frac{Ee\delta}{2kT}. \tag{4}$$

We shall be interested in fields so weak that $Ee\delta << kT$, in which case

$$\tilde{\nu}_p = \tilde{\nu}\frac{Ee\delta}{kT}. \tag{5}$$

Since the electrical polarization associated with each favorable jump is $e\delta$, the current i per unit area is

$$i = n\tilde{\nu}\frac{Ee^2\delta^2}{kT} \tag{6}$$

where n is the number of interstitial ions or vacancies per unit volume. Thus, the contribution σ to the conductivity from this flow is

$$\sigma = n\tilde{\nu}\frac{e^2\delta^2}{kT}. \tag{7}$$

The conductivity and ionic mobility are related by the equation

$$\sigma = ne\mu \tag{8}$$

whence

$$\mu = \tilde{\nu}\frac{e\delta^2}{kT}. \tag{9}$$

Equation (7) may be compared with the similar equation derived by Lorentz on the basis of a free-particle model, namely,

$$\sigma = n_f\frac{e^2\bar{v}l}{3kT} \tag{10}$$

[cf. Eq. (5), Sec. 126.]   Here, $\bar{v} = 4\sqrt{kT/2\pi M}$, l is the mean free path, and n_f is the density of free ions.   Equations (7) and (10) are formally equivalent if we make the correspondence

$$l \sim \delta, \qquad \frac{n_f\bar{v}}{3} \sim n\tilde{\nu}\delta.$$

If more than one kind of interstitial ion or vacancy is present in the lattice, the total ionic conductivity may be obtained by adding together the contributions of type (7) from each kind of ion.

The temperature dependence of n in Eq. (7) is determined by the particular way in which the lattice defects occur.   We have discussed two types of defect in Sec. 110, namely, those which occur in the alkali halides and those which occur in zinc oxide and zinc sulfide.   In the first case, it is believed that the metal-ion lattice and the halogen-ion lattice have equal numbers of vacancies so that the number of vacancies of a given kind is

$$n = Ne^{-\frac{\epsilon''}{2kT}} \tag{11}$$

when the crystal is in thermal equilibrium. Here, ϵ'' is the energy required to take an alkali-metal ion and a halogen ion from the interior to the surface of the crystal, and N is the total number of ions of a given kind. In the second case, it is believed that oxygen and sulfur atoms, respectively, evaporate and the excess metal atoms diffuse into the interstices of the crystal. The number of interstitial atoms in this case is

$$n = N_{O_2}^{-\frac{1}{2}} B' e^{-\frac{\epsilon''}{kT}} \tag{12}$$

where ϵ'' is the energy required to produce an interstitial zinc atom and an oxygen atom, the latter being bound to another to form a molecule in the vapor phase. These two cases do not exhaust the possible types. Schottky and Wagner have pointed out that there are in all the following three independent types.

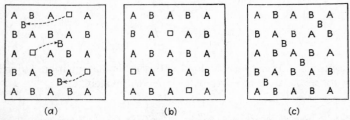

(a)   (b)   (c)

Fig. 7.—The three types of lattice defects. In (a) some of the B atoms have moved to interstitial places leaving vacancies. (b) B atoms have evaporated leaving vacancies in the lattice. There are no interstitial atoms. In the alkali halides there are equal numbers of A and B vacancies. An excess of one type or the other may be obtained, however, by heating the crystal in an appropriate vapor. (c) A fraction of the A atoms have evaporated from the surface leaving an excess of B atoms which diffuse into interstitial positions.

I (*cf.* Fig. 7a). There are interstitial M or X ions (or atoms) and there are in each of these cases an equal number of vacant M or X sites, respectively. If $n\epsilon$ is the energy required to remove n ions of a given kind to form n vacancies and n interstitial ions and if $-2kTn \log (n/N)$ is the entropy gained in doing so, the equilibrium value of n/N is

$$\frac{n}{N} = e^{-\frac{\epsilon''}{2kT}} \tag{13}$$

where N is the total number of ions of the given kind.

IIa (*cf.* Fig. 7b). Some M or X atoms evaporate, leaving an equal number of vacancies in M or X sites. Since the evaporating atoms must be neutral, there is an excess or a deficit of electrons in these two cases. These electrons or holes should reside near the vacancies in the lowest energy state. This case is similar to I, the difference being that there are no interstitial atoms in the lattice in the present case.

II*b*. We may classify separately the case in which the positive and negative ions leave in equal numbers. This ordinarily occurs in the alkali halides, for which the equilibrium value of n is given by an equation of the type (11).

III (*cf.* Fig. 7*c*). A fraction of one of the constituents may evaporate, leaving an excess of the other in interstitial sites. Zinc oxide and zinc sulfide, which were mentioned above, belong to this class.

We might expect that the deviations of type I and III, which involve interstitial atoms, should occur primarily in lattices that have large interstitial spaces, such as the zincblende and wurtzite structures, which have low coordination numbers.

Jost[1] has investigated the relative probability of case I and case II*b* for crystals having the sodium chloride structure, in which the interstitial spaces are comparatively small. If the values of n and ϵ'' in the two cases are distinguished by subscripts I and II, we have from Eqs. (11) and (13)

$$\frac{n_\mathrm{I}}{n_\mathrm{II}} = \frac{e^{-\frac{\epsilon_\mathrm{I}''}{2kT}}}{e^{-\frac{\epsilon_\mathrm{II}''}{2kT}}} = e^{-\frac{\epsilon_\mathrm{I}''-\epsilon_\mathrm{II}''}{2kT}} \tag{14}$$

Let us compute the difference $\epsilon_\mathrm{I}'' - \epsilon_\mathrm{II}''$ on the assumption that the interionic distances near a vacancy or near an interstitial ion are the same as for a perfect crystal. For simplicity, it may be assumed that the repulsive potential between ions varies as b/r^n where $n \sim 9$. The energy of an ion in a normal site then is (Sec. 11)

$$\epsilon = -1.746\frac{e^2}{a_0}\left(1 - \frac{1}{n}\right), \tag{15}$$

and the energy necessary to remove both a positive and a negative ion completely from the lattice is twice the negative of this. If, however, the ions are brought only to the surface, the energy should be -2ϵ minus the energy required to remove a positive and a negative ion from the surface, which is just the heat of sublimation per molecule. Hence, in the present approximation,

$$\epsilon_\mathrm{II}'' = -\epsilon. \tag{16}$$

The electrostatic energy of an interstitial ion in an undistorted sodium chloride lattice is zero because the relationship between the distribution of positive and negative ions is a symmetrical one. The distance between

[1] W. Jost, *Jour. Chem. Phys.*, **1**, 466 (1933); *Z. physik. Chem.*, **A169**, 129 (1934); *Physik. Z.*, **36**, 757 (1935). See also *Diffusion und chemische Reaktiones in festen Stoffen* (J. Steinkopf, Leipzig, 1937).

the center of the interstitial-ion site and the centers of the neighboring
ions is $a\sqrt{3}/2$, whence the repulsive energy for the interstitial atom is

$$\epsilon_r \sim 1.74 \frac{e^2}{na}\left(\frac{2}{\sqrt{3}}\right)^n \frac{2}{3}$$

where the factor $\frac{2}{3}$ enters because there are only four neighboring halogen
ions instead of six, as for an ordinary ion.   Hence,

$$\epsilon_I'' = -\epsilon + \epsilon_r,$$

and

$$\epsilon_I'' - \epsilon_{II}'' = \epsilon_r. \tag{17}$$

This difference is of the order of magnitude 1 ev for the alkali halides so
that

$$\frac{n_I}{n_{II}} \sim e^{-\frac{5,700}{T}}.$$

Thus, case IIb is much more probable than case I under the assumptions
made above.

It is clear, however, that these assumptions must be seriously in
error, for the magnitude of $\epsilon_{II}/2$ as given by Eq. (16) is about 3.7 ev for
sodium chloride.   According to Eq. (7), the activation energy for ionic
conductivity should be at least as large as this, whereas the observed
value[1] is only 1.90 ev.   A reasonable explanation of this discrepancy
is that the atoms or ions around vacancies and interstitial atoms become
displaced from their normal positions in the undeformed lattice and thus
lower the energy of the lattice.   The source of the additional energy is
not hard to find.   If an ion of charge e is removed from a site, the region
about the vacancy is left with an excess charge $-e$.   This charge should
polarize the surrounding lattice and the energy of the equilibrium state
should be lower than that of the undeformed lattice by the polarization
energy; moreover an interstitial ion should polarize the lattice in a
similar way.   If the charges occupied a spherical domain of radius r and
if the medium were continuous and had a static-field dielectric constant κ_s,
the polarization energy would be

$$\epsilon_p = -\frac{e^2}{2r}\left(1 - \frac{1}{\kappa_s}\right). \tag{18}$$

The energy actually should be computed with the use of a more detailed
atomic picture; however, we shall use Eq. (18) for an order-of-magnitude
estimate.   For vacancies, it will be assumed that r is equal to the mean
ionic radius, that is, to $a/2$, and for interstitial ions it will be assumed

[1] W. LEHFELDT, Z. Physik, 85, 717 (1933).

that r is equal to half the distance between the center of an interstitial site and the nearest ion site, namely, $a\sqrt{3}/4$. The energy required to produce an interstitial ion in case I then is

$$\epsilon_I'' = -\epsilon + \frac{e^2}{a}\left[\frac{1.74}{n}\left(\frac{2}{\sqrt{3}}\right)^n \frac{2}{3} - \left(1 - \frac{1}{\kappa_s}\right)\frac{2}{\sqrt{3}} - \left(1 - \frac{1}{\kappa_s}\right)\right], \quad (19)$$

and the energy required to produce a vacancy in each lattice is

$$\epsilon_{II}'' = -\epsilon - \frac{2e^2}{a}\left(1 - \frac{1}{\kappa_s}\right). \quad (20)$$

When the static dielectric constants of the alkali halides, which are of the order of magnitude 6, are used in these equations, the value of the additional negative terms in (19) are comparable with ϵ and reduce ϵ_{II}'' to values that are much more nearly in accord with the experimental activation energies. The difference between Eqs. (19) and (20) is

$$\epsilon_I'' - \epsilon_{II}'' \cong \frac{1.74e^2}{2a}\left[0.54 - 0.18\left(1 - \frac{1}{\kappa_s}\right)\right]. \quad (21)$$

The negative term reduces the difference to a value somewhat below that of Eq. (17) but does not reduce it enough to change the previous conclusion that the ratio (14) is very small. Jost and Nehlep[1] have made a more accurate estimate of $\epsilon_I'' - \epsilon_{II}''$ by taking into account the actual displacements of nearest ions. They find the value $0.40(1.74e^2/2a)$ for the extreme case in which κ_s is infinite. Since the difference should be larger than this in actual cases, we may conclude that deviations of type I do not occur in the crystals having sodium chloride structure for which only the electrostatic and repulsive terms of the Born theory are important. This includes practically all the alkali halides and probably the oxides and sulfides of beryllium, magnesium, and calcium.[2]

We saw in Chap. II that the van der Waals energy plays an important role in the halides of metals such as silver and thallium which have newly filled d shells. Jost and Nehlep have investigated the cohesive energy of interstitial metal ions in crystals of this type and have found that the contribution to $\epsilon_I'' - \epsilon_{II}''$ from the van der Waals term may reasonably reverse the sign of this difference. Thus, they find that in silver bromide, which has the sodium chloride structure, the correction to $\epsilon_I'' - \epsilon_{II}''$ should be

[1] W. Jost and G. Nehlep, Z. physik. Chem., **B32**, 1 (1936).

[2] A careful analysis of the contributions to the activation energy for electrolytic conductivity in sodium chloride has also been given by N. F. Mott and M. J. Littleton, Trans. Faraday Soc., **34**, 485 (1938).

$$-1.0\left(\frac{1.74e^2}{2a}\right),$$

which is sufficient to reverse the sign of (21). Although this result is not yet conclusive, it does show that case I is a possibility in some salts having sodium chloride structure.

Let us now discuss the ionic conductivity of the alkali halides on the assumption that they belong to class IIa. The number of vacancies of each kind then is $Ne^{-\frac{\epsilon''}{kT}}$, and the equation for the total ionic conductivity may be placed in the form

$$\sigma = N\frac{e^2\delta^2}{kT}e^{-\frac{\epsilon''}{2kT}}(\tilde{\nu}_+ + \tilde{\nu}_-) \tag{22}$$

where $\tilde{\nu}_+$ and $\tilde{\nu}_-$ are the jump frequencies for the two types of vacancy. The distance δ between neighboring like ions is the same in the two cases. In sodium chloride, the vacancies undoubtedly diffuse in the twelve (110) directions instead of in the six (100) directions as assumed in deriving (22). This fact does not impair the use of Eqs. (1) and (22) for an order-of-magnitude estimate of ν_s. According to Eq. (22) the ratio of the transport numbers of the two ions is determined by the ratio of the two terms in parenthesis. Tubandt's measurements on the transport numbers of the ions in NaCl show that the sodium-ion vacancies carry about 92 per cent of the current at 580°C. For this reason, we shall neglect $\tilde{\nu}_-$ in Eq. (22) for this substance. The equation may then be placed in the form

$$\sigma = Ae^{-\frac{\epsilon}{kT}} \tag{23}$$

where

$$\left.\begin{aligned}A &= N\frac{e^2\delta^2}{kT}\frac{1}{\alpha}\frac{(kT)^3}{h^3\nu_s^2}, \\ \epsilon &= \frac{\epsilon''}{2} + \epsilon_s.\end{aligned}\right\} \tag{24}$$

Lehfeldt's measured value of A (*cf.* Fig. 66, Chap. I) is 10^6 ohm^{-1} cm^{-1}. If we assume that

$$\delta = 2.8 \times 10^{-8} \text{ cm},$$
$$kT = 1.0 \times 10^{-13},$$
$$N = 2.24 \times 10^{22},$$

and solve for $1/\nu_s^2$, we find

$$\nu_s^2 \sim \frac{17 \times 10^{22}}{\alpha}, \tag{25}$$

or $\nu_s \sim 10^{11}$. This value, which is surprisingly low, implies that the saddle point is very flat in the direction at right angles to the direction of flow. It is possible, however, that the assumption made in deriving Eq. (7), namely, that the potential energy is determined by the position of the ion or vacancy alone, is far from correct and that many ions should be taken into account in computing the jump frequency of a vacancy. It is also possible, as has been suggested by Jost,[1] that ϵ is not temperature-independent but contains a linear term of the type aT. In this event, as in the case of thermionic emission, the measured value A^* of the intercept of the logarithmic plot would be a composite quantity of the form

$$A^* = Ae^{-\frac{a}{k}}$$

in which A is the computed constant (24). Jost has shown that the exponential factor in this equation may reasonably be of the order of magnitude 1,000.

At lower temperatures, the measured conductivities deviate from Eq. (23) in a way that depends upon the previous history of the crystal (*cf.* Fig. 66). This change may be described by saying that ϵ decreases with decreasing temperature. It is possible that the rate at which the equilibrium value of n is attained becomes so slow at temperatures below a temperature T' that the value of n for this temperature is retained. In this case, the temperature dependence of σ would be determined by ϵ_s alone. The slope of the low-temperature part of Lehfeldt's log σ versus $1/T$ plots for the alkali halides is usually about one-third that of the high-temperature part. Hence, if the preceding interpretation of this change is correct, this value shows that

$$\epsilon_s \sim \frac{1}{3}\left(\frac{\epsilon''}{2} + \epsilon_s\right),$$

or

$$\frac{\epsilon''}{2} \sim \frac{2}{3}\epsilon, \tag{26}$$

where ϵ is the high-temperature activation energy.

An alternative interpretation of the difference in slope of the low- and high-temperature portions of the curve is that the low-temperature conductivity arises from a very small number of ions that are situated at surfaces of internal cracks of the crystal and are more free to move than the ions in ordinary sites. It does not seem to be possible to decide between these alternatives at the present time.

---

[1] W. Jost, *Z. physik. Chem.*, **A169**, 129 (1934).

If Eq. (26) may be used, the ratio r of the number of vacancies to the total number of ions may be computed from the equation

$$r = e^{-\frac{\epsilon''}{2kT}} \cong e^{-\frac{2}{3}\frac{\epsilon}{kT}}.$$

Using the value $\epsilon = 1.90$ for sodium chloride, we find that $r \sim 10^{-6}$ at the melting point.

As we have seen above, it is not possible to say definitely in which category silver chloride and silver bromide may be placed, although Jost and Nehlep's computations indicate that they belong to class I. If the ratio of vacancies to normal atoms were large enough, it would be possible to distinguish between case I and case IIb by a comparison of the measured density and the density computed from X-ray data by assuming that there are no vacancies. In case I, the two should agree since there is one interstitial ion for each vacancy. In case IIb, however, the computed density should be larger than the measured one by a factor $(1 - r)^{-1}$ where r is the ratio of the number of vacancies to the number of normal ions. If we examine Fig. 66, Chap. I, we may see that the high-temperature slope is about twice the low-temperature value for silver chloride and silver bromide. Hence, in place of Eq. (26) we have

$$\epsilon'' = \epsilon. \tag{27}$$

For silver bromide, the value of r at the melting point computed[1] in this way is

$$r = e^{-\frac{\epsilon}{2kT}} = 10^{-2.8}. \tag{28}$$

The experimental accuracy of density and lattice constant measurements is not large enough[2] to detect the difference in densities that would arise from this value of r in case IIb. Wagner and Koch,[3] however, have used an ingenious indirect method, involving conductivity measurements, to determine the number of lattice defects in AgBr at various temperatures. Their method is based on the fact that the number of holes in AgBr may be increased by adding fixed quantities of $PbBr_2$. The lead salt forms a perfect solid solution when present in small concentrations. Since the lead ion is divalent, each ion added replaces two silver ions; hence, one vacancy is produced in the silver-ion lattice for each dissolved lead ion. The conductivity per vacancy may be determined from an investigation of the increase in conductivity as lead is

[1] F. Seitz, *Phys. Rev.*, **54**, 1111 (1938); **56**, 1063 (1939).

[2] Density measurements have been made by C. Wagner and J. Beyer, *Z. physik. Chem.*, **B32**, 113 (1936). See *ibid.* for the reason why the conclusions drawn by these workers are not trustworthy.

[3] E. Koch and C. Wagner, *Z. physik. Chem.*, **B38**, 295 (1938).

added, and the result may then be used to determine the number of vacancies in the pure crystal. Table LXXIX gives a comparison of the fraction of vacancies determined in this way with the fraction determined from Eq. (27). The computed values are off by a factor 5, a discrepancy that is not surprising in view of the simplifying assumptions used in the theoretical computation.

TABLE LXXIX.—COMPARISON OF THE FRACTION OF VACANT SITES IN SILVER BROMIDE AT VARIOUS TEMPERATURES AS DETERMINED BY EQ. (27) AND BY THE METHOD OF KOCH AND WAGNER

T, °C	r (theor.) $\cdot 10^3$	r (exp.) $\cdot 10^3$
300	0.71	4.0
250	0.36	1.8
210	0.18	0.76

If we compare Eq. (7) with Eq. (16) of Sec. 122 for the diffusion coefficient, namely,

$$D = \frac{n}{N}\bar{\nu}\delta^2,$$

we obtain the important relation

$$\sigma = N\frac{e^2}{kT}D. \tag{29}$$

This relation has been checked by von Hevesy and Seith[1] for PbI_2, in which the conductivity of Pb^{++} is known. They determined the diffusion coefficient of radioactive lead in single crystals of PbI_2 and compared this value with that computed from Tubandt's measured values of σ by means of relation (29). The two agree, within experimental error, in the temperature range from 255° to 290°C.

Von Hevesy and Seith have also measured the rate of diffusion of radioactive lead in $PbCl_2$. The positive-ion transport number is immeasurably small in this case, but by use of Eq. (29) they obtained the equation

$$\sigma(\text{ohm}^{-1}\,\text{cm}^{-1}) = 9.78 \cdot 10^{-4}e^{-\frac{4,180}{T}} + 1.15 \cdot 10^5e^{-\frac{15,000}{T}}$$

for the total conductivity. The first term is the chlorine-ion conductivity which was measured directly; the second term is the positive-ion conductivity which was computed by means of (29). The types of lattice defect that occur in $PbCl_2$ have not been determined, and so it is not

[1] G. VON HEVESY and W. SEITH, *Z. Physik*, **56**, 790 (1929); **57**, 869 (1929).

yet possible to give an interpretation of the large difference in the coefficients in this equation.

C. PHOTOCONDUCTIVITY

133. The Mean Free Path of Free Electrons in Ionic Crystals.—The semiempirical computations of the mean free path of free electrons in semi-conductors that were discussed in Part B of Chap. IV indicate that at room temperature the path is of the order of an interatomic distance, and is usually less than the mean free path of an electron in a metal. This result is not surprising, for the metal lattice is nearly at complete equilibrium when the conduction electrons move through it, whereas the ionic crystal is under stress.

We shall consider the simplified model of a crystal that is practically isotropic and shall discuss the scattering by the vibrational modes of frequency ν. If the kinetic energy ϵ of the free electron is greater than $h\nu$, the electron may lose energy to the lattice as well as gain it. It follows from the discussion of Sec. 127 that the probabilities that these processes will occur in unit time are of the form

$$\left.\begin{array}{c} A_\nu(\epsilon)(n_\nu + 1), \\ A_\nu(\epsilon)n_\nu, \end{array}\right\} \tag{1}$$

respectively, where in the isotropic case $A_\nu(\epsilon)$ is dependent only upon the electronic energy, and n_ν is the mean vibrational quantum number per oscillator, that is,

$$n_\nu = \frac{1}{e^{\frac{h\nu}{kT}} - 1}. \tag{2}$$

Thus, the total probability per unit time that the electron will be scattered by a lattice vibration of frequency ν is

$$P_\nu = A_\nu(\epsilon)(2n_\nu + 1) \qquad (\epsilon > h\nu). \tag{3}$$

In the case in which the electronic energy is less than $h\nu$, the electron cannot lose energy to the lattice so that

$$P_\nu = A_\nu(\epsilon)n_\nu \qquad (\epsilon < h\nu). \tag{4}$$

Thus, an electron having energy greater than $h\nu$ is scattered by the waves of frequency ν even at absolute zero of temperature, whereas one having energy less than ν is not. The total probability of scattering may be obtained by summing expressions of the type (3) and (4) over all lattice frequencies. We may expect that the terms for which $\nu < \epsilon/h$ will lead to a finite mean free path even at absolute zero.

We shall not discuss the computation of the function $A(\epsilon)$ in complete detail[1] but shall give a simplified discussion, due to Seeger and Teller,[2] of the case in which $n_\nu = 0$. This result may then be used to estimate the mean free path.

If it is assumed that a free electron in an ionic crystal is scattered isotropically by inelastic collisions in which it loses energy and that the average energy lost per collision is $h\nu_m$ where ν_m is the vibrational frequency of the optically active mode, the mean free path l and the energy loss per unit distance dW/dS should be related by the equation

$$h\nu_m = l\frac{dW}{dS}. \tag{5}$$

We shall use this equation to compute l after computing dW/dS by the method employed by Seeger and Teller. The equation evidently is valid only if the collisions are nearly isotropic. This condition is satisfied for electrons having energies not too large in comparison with $h\nu_m$ since the changes in velocity resulting from individual collisions then are comparable with the initial velocity. It is not satisfied, however, when the electronic energy is very large compared with $h\nu_m$, for then the collisions are predominantly through small angles, as in the case of electrons in metals.

An electron that is passing through an ionic crystal ordinarily moves so quickly, even when it has thermal energy, that the ions do not have time to come to complete equilibrium under the force of the electron. This fact is, of course, the basis of the Franck-Condon principle. For this reason, we shall assume that only that part of the polarization of the crystal arising from the electronic displacement is induced by the moving electron. The dielectric constant associated with this polarization is simply n^2, where n is the refractive index extrapolated for infinite wave length. The electrostatic field strength at a distance r from the electron then is

$$E = -\frac{e}{n^2 r^2}\mathbf{r}_1 \tag{6}$$

where $\mathbf{r}_1$ is a unit vector in the radial direction. Seeger and Teller use

[1] A discussion based entirely upon quantum mechanical perturbation methods may be found in the following papers: H. Fröhlich, *Proc. Roy. Soc.*, **160**, 230 (1937); H. Fröhlich and N. F. Mott, *Proc. Roy. Soc.*, **171**, 496 (1939). This discussion is more complete than the one in the present volume in the sense that temperature dependence is included; however, as Seeger and Teller point out, it is unreliable in the range of energy in which the mean free path is least because the mean time between collisions is of the order of magnitude 10^{-15} sec (see part *g*, Sec. 127). See also *Phys. Rev.*, **56**, 349 (1939).

[2] R. J. SEEGER and E. TELLER, *Phys. Rev.*, **54**, 515 (1938). See also *Phys. Rev.*, **56**, 352 (1939).

a slightly different constant for the coefficient of e/r^2, but the simple value $1/n^2$ is at least as accurate as theirs, in the writer's opinion.

Let us suppose that an electron moving with velocity v passes at a distance b from an ion having charge Ze. The force f normal to the direction of the electron velocity at time t is

$$f = \frac{1}{n^2} \frac{Ze^2}{b^2 + v^2t^2} \frac{b}{\sqrt{b^2 + v^2t^2}}, \tag{7}$$

in which the origin of t is chosen as the time of closest approach. The total impulse p transferred to the ion by the electron is the time integral of (7), namely,

$$p = \frac{2}{n^2} \frac{Ze^2}{vb}, \tag{8}$$

which is equivalent to the energy

$$\bar{\epsilon} = \frac{p^2}{2M} = \frac{2}{n^4} \frac{Z^2e^2}{Mb^2v^2}$$

where M is the ionic mass. Thus, the amount of energy that the electron loses to those ions lying within a cylindrical shell of radius b and thickness db in traveling unit distance is

$$\delta\left(\frac{dW}{dS}\right) = \bar{\epsilon} \frac{\pi b\, db}{d^3}, \tag{9}$$

in which d is the distance between like ions. If the lattice is of the type, such as sodium chloride, that is symmetrical in two types of ion, the total differential energy loss then is the sum of terms of type (9) for both types of ion. Setting

$$\frac{1}{\mu} = \frac{1}{M_1} + \frac{1}{M_2}$$

where M_1 and M_2 are the masses of the two kinds of ion, we may write this sum in the form

$$\delta\left(\frac{dW}{dS}\right) = \frac{2}{n^4} \frac{Z^2e^4}{\mu b^2v^2} \frac{\pi b\, db}{d^3}. \tag{10}$$

It may seem from Eq. (10) that $\delta(dW/dS)$ decreases as $1/b$, which shows that the *total* effect of distant ions is much greater than that of the ions through which the electron passes. Hence, we may integrate Eq. (10) with reasonable accuracy by treating the solid as though continuous. Thus,

$$\frac{dW}{dS} \cong \int \frac{2\pi Z^2 e^4}{n^4 \mu b^2 v^2} \frac{db}{b}, \tag{11}$$

in which the limits are yet to be specified.

We shall assume that the time τ during which the distance between the electron and ion is of the order of magnitude b may be taken as $2b/v$. In order that Eq. (11) shall be valid, this must be short compared

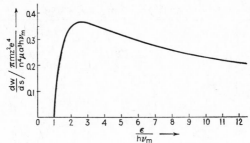

Fig. 8.—The semiclassical excitation function.

with the oscillational time of the optically active frequency, for otherwise the work done by the force averages to zero. Hence, we must have

$$\frac{2b}{v} < \frac{1}{2\pi \nu_m},$$

which gives as the upper limit of integration

$$b_{\max} = \frac{v}{4\pi \nu_m}.$$

As a lower limit, Seeger and Teller take the de Broglie wave length of the electron divided by 2π:

$$b_{\min} = \frac{\lambda}{2\pi} = \frac{h}{2\pi m v}.$$

Using these limits in Eq. (11), we find

$$\frac{dW}{dS} = \frac{2\pi Z^2 e^4}{n^4 \mu a^3 v^2} \log\left(\frac{mv^2/2}{h\nu_m}\right) \tag{12}$$

$$\equiv \frac{\pi m Z^2 e^4}{n^4 \mu a^3} \frac{1}{\epsilon} \log\frac{\epsilon}{h\nu_m},$$

where $\epsilon = mv^2/2$. This function, which is shown in Fig. 8, has the maximum value

$$\left(\frac{dW}{dS}\right)_{\max} = 0.37 \frac{\pi m Z^2 e^4}{n^4 \mu a^3 h\nu_m} \tag{13}$$

when $\epsilon = 2.7 h\nu_m$.

Combining Eqs. (1) and (10), we obtain

$$l = \frac{h\nu_m}{dW/dS} = \frac{n^4\mu a^3 h\nu_m}{\pi m Z^2 e^4} \frac{\epsilon}{\log \epsilon/h\nu_m}. \tag{14}$$

The minimum values of l, corresponding to $\epsilon = 2.7h\nu_m$, are listed in Table LXXX for several alkali halides.

TABLE LXXX.—THE MEAN FREE PATHS OF ELECTRONS IN IONIC CRYSTALS

	n^2	$l_{min} \cdot 10^8$ cm	$E_B \cdot 10^{-5}$, volts/cm	
			Calculated	Observed, room temperature
LiF	1.93	4.74	15.4	31
NaCl	2.19	5.0	4.7	15
KCl	2.19	4.0	4.8	10
KI	2.40	6.4	2.0	5.7
RbBr	2.40	7.4	1.6	6.3
RbI	2.85	12.9	0.81	4.9

These results obviously should apply only at the absolute zero of temperature and then only very approximately in the low-energy range where $\epsilon \sim h\nu_m$, for it has been assumed that the crystal possesses a single lattice vibrational frequency whereas actual crystals possess a continuous range of frequencies extending from zero to ν_m. Thus the actual dW/dS curve for absolute zero should not drop sharply to the axis at $\epsilon = h\nu_m$ but should continue smoothly to the origin.

Von Hippel[1] has suggested that dielectric breakdown in insulators occurs when the electrostatic field becomes so strong that on the average a free electron in the lattice can gain more energy from the field between collisions than it loses as a result of collisions. If E_B is the breakdown field, von Hippel's condition is

$$E_B l_{min} = h\nu_m. \tag{15}$$

By use of Eq. (1), this may be transformed to the form

$$E_B = \left(\frac{dW}{dS}\right)_{max}. \tag{16}$$

[1] A. VON HIPPEL, Jour. Applied Phys., 8, 815 (1937). Discussions of this and other theories of dielectric breakdown may be found in the following papers: H. Fröhlich, Proc. Roy. Soc., 160, 230 (1937), 172, 94 (1939). W. Franz, Z. Physik, 113, 607 (1939); Seeger and Teller, op. cit.; R. C. Buehl and A. von Hippel, Phys. Rev., 56, 941 (1939).

The results[1] that are obtained are shown in Table LXXX and are compared with von Hippel's measured values. The computed ones all are smaller than the observed ones. Seeger and Teller suggest that a part of this discrepancy may be due to the use of the actual electronic mass m in Eq. (11) instead of the effective mass for the interior of the crystal. Actually, it seems unlikely that the semiclassical computation of the mean free path is sufficiently trustworthy to merit accurate comparison with experimental results even if (15) is the correct condition for breakdown.

According to the discussion in the earlier part of this section, we should expect the mean free path to decrease with increasing temperature because the n_v in Eqs. (3) and (4) increase with increasing temperature. Thus, the breakdown strength of crystals should increase with increasing temperature if Eq. (15) is valid. Buehl and von Hippel[2] have observed that the dielectric strength of the alkali halides is decreased by cooling from room temperature to liquid-air temperature; however, the observed decrease is much more rapid than is to be expected from Eq. (15). It seems likely, at the present time, that the simple theory of dielectric breakdown needs important revision.[3]

Reasoning from his inability to detect a measurable Hall effect in photoconducting specimens of sodium chloride, potassium chloride, and potassium bromide crystals, Evans[4] has concluded that the mean free path in these crystals at room temperature is less than $4.5 \cdot 10^{-8}$ cm. The Hall effect is easily observable in zinc oxide and zinc sulfide (Sec. 37), and corresponds to mean free paths of 10^{-7} cm. The origin of the difference in properties of these two types of salt remains to be investigated.

134. Photoconductivity in Colored Alkali Halide Crystals.[5]—In this section and the next, we shall discuss the present status of the theory of photoconductivity. The most extensive interpretive work has been done on the photoconductivity of alkali halide crystals that contain F centers. As we have seen in Sec. 111, the properties of these crystals may be explained most reasonably by assuming that they contain more halogen-ion vacancies than alkali-metal-ion vacancies and that the

---

[1] These values differ somewhat from Seeger and Teller's because of differences in the form of Eq. (6).

[2] BUEHL and VON HIPPEL, *op. cit.*

[3] In an attempt to explain the observed temperature dependence, H. B. Sampson and the writer have pointed out that the primary excitation process of the free electrons is the production of excitons, rather than secondary electrons as is ordinarily assumed. Unless the excitons are dissociated by either the field or temperature, breakdown cannot occur. (To be published shortly.)

[4] J. EVANS, *Phys. Rev.*, **57**, 47 (1940).

[5] See the articles by R. W. Pohl surveying the experimental work, *Proc. Phys. Soc.*, **49** (1937); *Physik. Z.*, **39**, 36 (1938).

excess halogen vacancies are occupied by electrons. In addition to the alkali halides, we shall discuss photoconductivity in zinc sulfide and silver halide crystals.

Let us begin by considering the photoconductivity of colored sodium chloride. The other colored alkali halides behave in an essentially

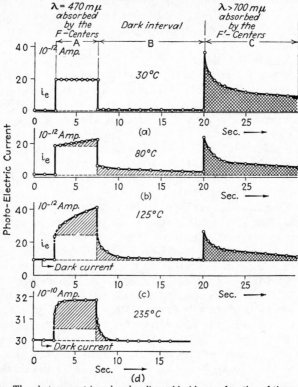

Fig. 9.—The photocurrent in colored sodium chloride as a function of time at different temperatures. During the interval A the crystal was irradiated with light in the F-center absorption band. During B the crystal was in the dark, and during C it was irradiated with infrared light. In (*a*) the crystal was at 30°C and only a primary current is observed during A; that is, the current varies abruptly when the light is turned on or off. Further current may be induced by infrared radiation *after* illumination with light in the F band. In (*b*) the temperature was 80°C and the current continued to rise after illumination began and did not drop to zero when the light was turned off. This secondary current is larger in cases (*c*) and (*d*). (*After Hilsch and Pohl.*)

similar way. At temperatures below 30°C, the photoconductivity begins abruptly when the crystal is exposed to light in the F-center absorption band, remains constant during a constant exposure that does not endure for too long a time, and drops abruptly to zero when the light is cut off (*cf.* Fig. 9*a*). A photocurrent that behaves in this way is said to be a *primary current.* Its properties may be explained[1] in terms of the following simple assumptions:

a. A fraction η of the absorbed light quanta free electrons, which wander about the lattice with thermal velocities.

b. In the presence of an electrostatic field of intensity E, the electrons drift in the direction of the field with mean velocity $\mu\mathsf{E}$ where μ is the mobility, which is temperature-dependent.

c. The electrons that do not reach the electrodes eventually become trapped. If λ is the mean distance traveled before trapping takes place, the mean distance ω an electron drifts in the direction of the field is

$$\omega = \mu\mathsf{E}\frac{\lambda}{\bar{v}} \tag{1}$$

where $\bar{v}$ is the mean velocity of random motion. We shall call ω the displacement distance.

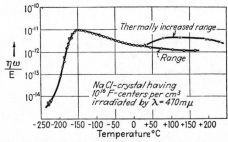

Fig. 10.—The function $\eta\omega/\mathsf{E}$ as a function of temperature for sodium chloride with *F*-centers. It should be noted that the photocurrent drops sharply below $-150°C$. The units of the ordinate scale are meter^2/volt. (*After Pohl.*)

If I is the intensity of the absorbed radiation per unit distance between electrodes, the measured current i should be

$$i = \eta\frac{I}{h\nu}e\omega, \tag{2}$$

according to these assumptions, where $h\nu$ is the energy of the absorbed light quanta. Since ω is proportional to E and η is presumably independent of field intensity, the quantity

$$\frac{ih\nu}{\mathsf{E}eI} = \eta\frac{\omega}{\mathsf{E}} \tag{3}$$

should be independent of E. This actually is found to be the case in the primary-current range. Figure 10 shows the temperature depend-

---

[1] *Ibid.*

ence of the measured value of (3) for sodium chloride.[1] It should be observed that $\eta\omega/\mathsf{E}$ varies relatively slowly from 25° down to $-150°C$ and then drops very rapidly, indicating that either η or ω decreases very rapidly near the absolute zero of temperature. A similar drop in photosensitivity has also been observed in potassium chloride.

Mott[2] has given an explanation of this decrease in terms of a decrease in the quantum yield η. If an F center corresponds to the lowest stable state of an electron and a vacancy in the halogen-ion lattice, the electrostatic field in which the electron moves should vary as $-e^2/n^2r^2$ at large distances from the vacancy. Hence, the electron should have more than one discrete level beneath the ionization continuum. By analogy with a hydrogen atom, or an alkali-metal atom, the lowest state of the system should be an s-like state and the strongest absorption band should correspond to the transition from this state to the lowest, discrete, p-like state, which Mott postulated is the F absorption band. Since the electron is bound in the p state, the crystal should not become photoconducting unless the electrons are thermally excited to the ionization continuum. Hence, there should be no photoconductivity at the absolute zero of temperature. If A is the relative probability that the electron jumps from the p state to the ground state in unit time without becoming free and if B is the relative probability that it becomes free in the same time, the quantum yield should be

$$\eta = \frac{B}{A + B} = \frac{1}{1 + (A/B)}. \qquad (4)$$

Mott assumes that A has the value 10^8 sec^{-1} and that B has the form

$$B = \nu e^{-\frac{\epsilon}{kT}} \qquad (5)$$

where ν is of the order of magnitude of an atomic oscillational frequency 10^{13} sec $^{-1}$ and ϵ is the energy required to ionize an electron in the excited state. Under these conditions,

$$\eta = \frac{1}{1 + 10^{-5}e^{\frac{\epsilon}{kT}}}. \qquad (6)$$

The condition, which must be satisfied if this is to begin dropping to zero at $100°K$, is that ϵ should be of the order of magnitude 0.01 ev.

Indirect evidence shows that the trapping centers are other F centers. As the crystal is illuminated[3] with light in the F absorption band in the

---

[1] G. GLASER and W. LEHFELDT, *Nachr. Kgl. Ges. Wiss. Göttingen*, **2**, 91 (1936).

[2] N. F. MOTT, *Proc. Phys. Soc.*, **50**, 196 (1938).

[3] Z. GYULAI, *Z. Physik*, **33**, 251 (1925); R. HILSCH and R. W. POHL, *Z. Physik*, **68**, 721 (1931).

region of temperature in which the primary photocurrent is observed, the F band gradually disappears and a new band appears on the long wave-length side of the F band. This new band, which is called the F' absorption band, ordinarily overlaps partly with the F band (cf. Fig. 11). Evidently, the F' band corresponds to the absorption of light by the centers that are formed by the trapped electrons. Measurements[1] on the displacement distance per unit field strength, ω/E, show that this quantity is inversely proportional to the concentration of F centers (cf. Fig. 12). This result suggests that the F centers act as trapping points for the free electrons and that an F' center consists of a

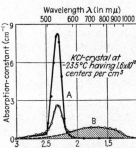

FIG. 11.—The F and F' bands of potassium chloride. A is the F band as it occurs before illumination with light in this band. After illumination, the intensity of the F band decreases and the F' band B occurs. The unit of wave length is 10^{-7} cm. The unit of ordinate scale is cm^{-1}. (*After Pohl.*)

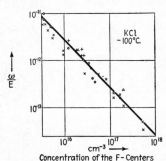

FIG. 12.—Plot of ω/E as a function of concentration of F centers. Since ω/E is the displacement distance per unit field intensity it follows that the mean free path for trapping is dependent upon the concentration of F centers. The ordinates are expressed in units of meter^2/volt. (*After Pohl.*)

vacancy plus two electrons. If this interpretation is correct, two F centers should be destroyed for each F' center formed. Pohl and his collaborators have found evidence showing that this condition actually is satisfied.

It is also found that the crystal containing F' centers becomes photoconducting when it is illuminated with light in the F' band. This shows that the electrons in F' centers may be freed by optical excitation, just as may those in an F band.

From the curve of Fig. 12, we may estimate the ratio of the mean free path l for scattering of free electrons and the cross section σ for capture of a free electron by an F center. Let us assume that η is unity at $-100°$C for potassium chloride. According to Eq. (1),

[1] G. GLASER, *Nachr. Kgl. Ges. Wiss. Göttingen*, **3**, 31 (1937).

$$\frac{\omega}{\mathsf{E}} = \mu \frac{\lambda}{\bar{v}}. \tag{7}$$

The mobility μ is related to the conductivity σ by the equation

$$\sigma = ne\mu \tag{8}$$

where n is the number of free electrons per unit volume. Since

$$\sigma = \frac{ne^2\bar{v}l}{3kT}, \tag{9}$$

according to classical theory [*cf.* Eq. (5) Sec. 126], we have

$$\mu = \frac{1}{3}\frac{e\bar{v}l}{kT}. \tag{10}$$

The mean distance λ for capture and the capture cross section Q are related to the density δ of capturing centers by the equation

$$\frac{1}{\lambda Q} = \delta. \tag{11}$$

Using Eqs. (7), (10), and (11), we obtain

$$\frac{\omega}{\mathsf{E}} = \frac{1}{3}\frac{e}{kT\delta}\frac{l}{Q}$$

or

$$\frac{l}{Q} = 3\frac{\omega}{\mathsf{E}}\delta\frac{kT}{e}. \tag{12}$$

If we employ the experimental values of ω/E and δ that are given in Fig. 12 and assume $Q \sim 10^{-16}$ cm^2, we find

$$l \sim 5 \cdot 10^{-9} \text{ cm.}$$

Although this estimated value is about one-tenth as large as the value computed in Sec. 133, it does not seem safe to draw any conclusions about the validity of the preceding equations from the discrepancy.

When a sodium chloride crystal that contains F' centers, which have been formed by illuminating the colored crystal in the F band, is heated above room temperature, the F' centers disappear[1] and are replaced by F centers. Hence, the trapped electrons may be thermally released and should be able to move farther than the displacement distance ω at high temperatures. This expectation is supported by the appearance of a secondary photocurrent above room temperature. It is found[2] that the

---

[1] GYULAI, *op. cit.;* HILSCH and POHL, *op. cit.*

[2] B. GUDDEN and R. W. POHL, *Z. Physik,* **31,** 651 (1925); W. THIELE, *Ann. Physik,* **25,** 561 (1936).

photocurrent continues to rise after illumination begins and approaches a saturation value. The rate at which this value is attained is greater at higher temperatures (*cf.* Fig. 9). In addition, the current does not disappear completely when the light is removed. Instead, it decreases abruptly by an amount equal to the initial rise and then gradually dies out. In Fig. 10, the double sets of points above 25°C correspond, respectively, to the initial and saturation photocurrents.

A complete mathematical treatment of the problem of secondary currents has not yet been developed. It seems likely, however, that the following three qualitative principles determine the behavior of the secondary current in all photoconductors, as well as in colored sodium chloride.

1. Electrons that are trapped after being released by light may be freed thermally at sufficiently high temperatures and may continue to drift toward the anode. This contribution i_s to the secondary current should not rise to a saturation value instantly if the electrons are trapped for a measurable time τ. Suppose that the electrons are optically freed at a rate n per unit time, so that the number in the crystal at the end of time t is nt. After being initially freed, the electrons move a distance ω and become trapped, giving rise to the primary current. If the probability that one of the nt trapped electrons is released in unit time is $1/\tau$, the total number released per unit time is nt/τ. If it is assumed that they move a distance ω and are again trapped, the secondary current as a function of time is

$$i_s = \frac{nt}{\tau}e\omega. \tag{13}$$

Deviations from this linear rise occur as soon as the electrons begin arriving at the anode at a rate comparable with the rate at which they are optically freed. The greatest possible value of i_s for uniform illumination between electrodes that are a distance d apart is $nde/2$ since each freed electron moves a mean distance $d/2$.

2. Additional electrons may enter the crystal from the cathode and move through the crystal to the anode. This flow from the cathode is induced by the space charge fields set up in the crystal by the displacement of the electrons in the primary current and in i_s. Either field emission from the cathode or an inherent dark electronic conductivity may serve to introduce the electrons into the crystal. We shall call this contribution to the current i_{II}. Since the space charge field should vary with time, immediately after illumination, i_{II} should also be a function of time. It may be difficult to separate i_{II} from i_s for this reason. The limiting value of the total photocurrent for long periods of illumination is determined by the dependence of i_{II} on the field strength near the

cathode; for if a sufficiently large space charge accumulates, the optically freed electrons will not be able to reach the cathode.

3. The migrating electrons may become permanently trapped to form anew the centers that were originally ionized by light quanta. In the alkali halides, this means that F centers may be formed by the recombination of freed electrons and halogen-ion vacancies. In a true equilibrium state, the rate at which this process occurs is equal to the rate at which the centers are ionized by light.

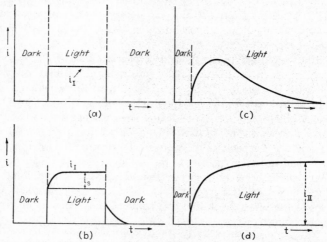

Fig. 13.—Schematic representation of the behavior of primary and secondary photocurrents. In (a) there is only a primary current i_I which corresponds to electrons actually released by light. These electrons ultimately become trapped. In (b), which is at a higher temperature, there is an additional current i_s corresponding to the flow of the thermally released trapped electrons. (c) corresponds to a case in which charge cannot pass from the cathode into the crystal. Polarization eventually reduces the current to zero in spite of continuous illumination. In (d) a current i_{II} flows from the cathode in the equilibrium state. All cases intermediate between (c) and (d) are possible.

Several different possible cases are illustrated schematically in Figs. 13a to d. In the first, there is only the primary current which rises and falls abruptly with changes in illumination. This current could not exist indefinitely if no charges entered the crystal to neutralize space charge. In Fig. 13b, we have the primary current and the secondary current i_s, which we have assumed reaches its saturation value. If there are no electrons flowing from the cathode, the primary current and the secondary current i_s eventually fall to zero because of polarization (Fig. 13c). The cathode current i_{II} prevents this drop and makes the final current finite. If i_{II} is large, the total current need not have a maximum (Fig. 13d), whereas if i_{II} is small, the current may rise to a peak and then fall asymptotically to a finite value.

Only the primary current flows in colored sodium chloride below 25°C. Since there is no direct evidence that electrons enter the crystal from the cathode between this temperature and 230°C, it is possible that only the primary current and i_s coexist in this range. Above 230°C, the electrons bound in F centers are freed thermally, and the current i_{II} undoubtedly occurs. The three components of current have not been separated experimentally, however.

Hilsch and Pohl[1] have treated a particular case of the general problem, namely, the case in which τ is so short that the primary current and i_s are inseparable and the rate at which electrons are trapped by ionized centers (effect 3) is negligible. They find that the total steady-state electronic current i is related to the effective primary current i_p by the equation

$$i = \frac{1}{1 - \gamma} i_p$$

where γ is the fraction of the dark conductivity before illumination that is due to electrons. If σ_e and σ_i are the electronic and ionic dark conductivities,

$$\gamma = \frac{\sigma_e}{\sigma_i + \sigma_e}.$$

They did not consider the transition current before the steady state is reached.

135. Photoconductivity of Zinc Sulfide and of the Silver Halides.[2]— Gudden and Pohl[3] have also made measurements on the photoconductivity of natural single crystals and artificially prepared powders of zinc sulfide that can be interpreted along the lines discussed in the preceding section. Although the impurity content of the single crystals is not discussed, we shall assume that the composition of crystals and powders is similar, since the photoconducting properties of both are nearly alike. The photoconducting powders are usually prepared by heating pure zinc sulfide either alone or in the presence of small quantities of salts of other metals, such as copper, manganese, or silver. It is believed (cf. Sec. 112) that small quantities of neutral metal atoms enter interstitial positions in the lattice as a result of the heating process and provide centers that may be ionized by the conductivity-inducing radiation in the near ultraviolet. The position of the spectral sensitivity curve is dependent upon the kinds of interstitial atom present, but it usually has its maximum near 3650Å and has a small tail in the blue region of the visible spectrum. Many of the photoconducting zinc sulfides, such as the pure heated

[1] R. HILSCH and R. W. POHL, *Z. Physik*, **108**, 55 (1937).

[2] Review of experiments: F. C. NIX, *Rev. Modern Phys.*, **4**, 723 (1932).

[3] B. GUDDEN and R. W. POHL, *Z. Physik*, **2**, 181, 361 (1920); **3**, 98 (1920); **4**, 206 (1921); **5**, 176 (1921); **6**, 248 (1921); **17**, 331 (1923); *Physik. Z.*, **23**, 417 (1922).

material and those activated by means of copper, silver, or manganese, luminesce brightly when excited with radiation lying in the region in which photoconductivity occurs. We shall discuss the correlation of these two effects below.

It is found that the photocurrents in the zinc sulfides are primary for electrostatic fields below 6,000 volts/cm. The primary current saturates, however, in sufficiently thin crystals; for example, Fig. 14 shows the saturation obtained for a crystal about 1 mm thick. When the maximum current is reached, one electron arrives at the anode for each light quantum that is absorbed, which indicates that the displacement distance ω is greater than the distance between electrodes. If, in

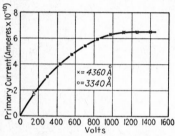

Fig. 14.—The saturation of the primary photocurrent with voltage in zinc sulfide. Saturation occurs when each electron flows to the cathode. (*After Hilsch and Pohl.*)

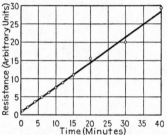

Fig. 15.—The resistance of a specimen of zinc sulfide as a function of time. The specimen had previously been irradiated with ultraviolet light. It was then placed in liquid air and irradiated with infrared radiation, starting at time zero. (*After Reimann.*)

Eq. (12) of the preceding section, we set ω equal to the thickness of the crystal for the value of E at which the current is half the saturation value and if we set l equal to 10^{-8} cm and Q equal to 10^{-16} cm^2, we find

$$\delta \sim 3 \cdot 10^{13}.$$

This implies that the density of trapping centers is very low compared with the density of interstitial atoms, which usually is about 10^{18} cm^{-3}. Moreover, since the shape of the saturation current is independent of light intensity, we cannot conclude that the trapping centers are the ionized centers. No satisfactory explanation of the trapping has yet been given.[1]

[1] In recent experiments based on a study of the decay of luminescence, R. P. Johnson, *Jour. Optical Soc.*, **29**, 387 (1939), has shown that there are at least two types of trapping center. One of these binds the electrons more tightly than the others and probably is the trapping center that is important for room-temperature photoconductivity. The other probably would also be important at lower temperatures.

The trapped electrons may be temporarily released either by heating the crystal to sufficiently high temperatures or by illuminating it with infrared light at any temperature. This freeing is made evident by the appearance of conductivity. Hence, if a crystal that has previously been illuminated with ultraviolet light is continuously illuminated with infrared light or is kept sufficiently warm, the electrons that are continuously being freed from the trapping centers should eventually recombine with the ionized interstitial atoms and the conductivity should gradually decrease. If we assume that the rate at which n ionized electrons and n ionized atoms recombine is proportional to the product of the number of each, we obtain the equation[1]

$$\frac{dn}{dt} = -\alpha n^2, \tag{1}$$

which has the integral

$$\frac{1}{n} = \alpha t + \frac{1}{n_0} \tag{2}$$

where n_0 is the number of photoelectrons at time $t = 0$. Under these conditions, the resistivity of the crystal that is illuminated with infrared light should increase linearly with time. A relationship of this type has been observed by Reimann[2] (*cf.* Fig. 15) on a specimen of zinc sulfide that was kept at liquid-air temperature and was continuously illuminated with infrared light after an initial excitation with ultraviolet light.

According to Eqs. (1) and (2), the rate at which electrons and interstitial ions recombine is given by the equation

$$\frac{dn}{dt} = -\frac{\alpha n_0^2}{(n_0 \alpha t + 1)^2}. \tag{3}$$

If light quanta of frequency ν were emitted during this recombination, the intensity $I(t)$ at time t would be

$$I(t) = h\nu \frac{dn}{dt} = -\frac{h\nu \alpha n_0^2}{(n_0 \alpha t + 1)^2} = \frac{I(0)}{(\beta t + 1)^2} \tag{4}$$

where

$$\beta = n_0 \alpha.$$

It is found experimentally that the luminescence of zinc sulfide decays very nearly in accordance with this equation at times not too near the initial time of excitation. This fact provides a possible explanation of the

---

[1] This discussion is valid only when there is a single trapping center. It has been generalized by Johnson, *op. cit.*

[2] A. L. REIMANN, *Nature*, **140**, 501 (1937).

luminescence of the salts. A more definite correlation between lumines-cence and recombination would be provided by simultaneous measurements of the decay of conductivity and of luminescence when a previously excited crystal is illuminated with infrared radiation or is warmed.

Currents higher than the saturation values of Fig. 13 may be obtained at room temperature by continuously illuminating the crystal with ultraviolet light while maintaining high electrostatic field intensity. Figure 16 shows the way in which the total charge removed from the crystal, as a function of time, deviates from linearity at different field

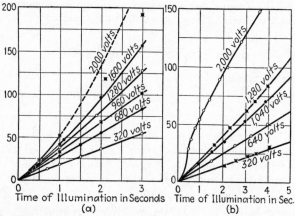

Fig. 16.—The total charge conducted by a specimen of zinc sulfide that was con-tinually illuminated with ultraviolet light as a function of time. At high field intensities the current rises above the saturation value of the primary current that is shown in Fig. 14. The scale of ordinates is such that 100 units equal 7.2×10^{-11} coulomb. (*After Hilsch and Pohl.*)

strengths. These secondary currents, like those in the colored halides, have not been studied quantitatively.

Photoconductivity that is qualitatively similar to that observed in zinc sulfide has been found in many other natural and artificial crystals. Most notable among these are selenium and the silver halides. The first of these is used in the photovoltaic cell and has been studied extensively for practical work. The results, however, are not very useful for interpre-tive work.

The photoconductivity of the similar salts, AgCl and AgBr, has been investigated by Hilsch and Pohl,[1] Toy and Harrison,[2] and Lehfeldt,[3] the work of the last of these investigators being the most extensive. The spectral sensitivity curves for stimulating conductivity extend

[1] R. Hilsch and R. W. Pohl, *Z. Physik*, **64**, 606 (1930).

[2] F. C. Toy and G. B. Harrison, *Proc. Roy. Soc.*, **127**, 613 (1930).

[3] Lehfeldt. *Nachr. Kgl. Ges. Wiss. Göttingen*, **1**, 170 (1935).

throughout the visible and into the near ultraviolet region of the spectrum, have peaks in the blue, and lie close to, if not actually in, the tail of the fundamental absorption band. Lehfeldt has found that in the range from room temperature to liquid-air temperature the quantum yield of photoelectrons is close to unity for all absorbed light.

It has often been suggested that electrons are freed with quantum yield unity throughout the fundamental absorption region, and that photoconductivity is observed in the tail of this region and not in its center only, because the reflection coefficient becomes large in the center. According to present view concerning the excited states of perfect insulating crystals (Chap. XII), the observed photoconductivity arises from impurities or lattice defects, at least in the case of the alkali halides.

As we saw in Sec. 132, Jost and Nehlep have given theoretical evidence that the lattice defects in silver halides are interstitial silver ions. Thus, it is possible that the source of photoelectrons is either the interstitial silver ion, if the ion carries an electron with it, or the negative ions near the vacancy, if the interstitial ion does not take an electron. Mott[1] has shown, however, that it is not unreasonable to suppose that photoelectrons are produced in these salts by thermal decomposition of excitons formed by absorption in the fundamental band.

Note: The theory of contact rectification of the type occurring in galena and copper oxide rectifiers has not been discussed in this chapter, in which it would naturally belong. This subject has passed through a gradual development and the most recent treatment, which seems to correlate most of the known facts, is that given by N. F. Mott, *Proc. Roy. Soc.*, **171**, 27, 281 (1939).

[1] N. F. Mott, *Proc. Roy. Soc.*, **167**, 384 (1938).

CHAPTER XVI

THE MAGNETIC PROPERTIES OF SOLIDS

136. Introduction.—It was seen in Chap. I that there are three main classes of solids as far as magnetic properties are concerned, namely, diamagnetic, paramagnetic, and ferromagnetic substances. Practically all simple insulators and about half the simple metals are diamagnetic, whereas all other insulators and metals, except for a few ferromagnetic substances, are paramagnetic. The ferromagnetic materials become paramagnetic when heated to sufficiently high temperatures, a fact showing that paramagnetism and ferromagnetism are intimately connected.

Diamagnetism is related to changes in the orbital motion of electrons that occur when atomic systems are placed in a magnetic field. It may be recalled that the current induced in a closed electrical circuit by a magnetic field is always in such a direction as to tend to keep the total flux unchanged. Thus, the circuit has, in effect, a negative susceptibility. This effect is retained even in systems of charges that must be treated by quantum mechanics and is responsible for diamagnetism. Paramagnetism, on the other hand, is related to the tendency of a permanent magnet to align itself in a magnetic field in such a way that its dipole moment is parallel to the field. In atomic systems, the permanent moment is the magnetic moment associated with electron spin in the simplest cases, but it may also be the permanent moment of an unfilled atomic shell that arises from a combination of spin and orbital motion. If a system is more stable when the atomic dipoles are parallel, the system is ferromagnetic at low temperatures. Ferromagnetism disappears at high temperatures for a reason similar to that for which solids melt, namely, because the nonferromagnetic state is more disordered and has a higher entropy than the ferromagnetic one. The moment-aligning forces in ferromagnetic substances are not the magnetic forces between dipoles but have electrostatic origin, as we shall see in Sec. 143.

137. The Hamiltonian Operator in a Magnetic Field.—According to the results of Sec. 42, the Hamiltonian operator for any system of electrons in an external electromagnetic field is

$$H = \sum_i \frac{1}{2m}\left(\mathbf{p}_i + \frac{e}{c}\mathbf{A}_i\right)^2 + V - \sum_i e\,\varphi_i + \sum_i \frac{e}{mc}\mathbf{d}_i \cdot \text{curl } \mathbf{A}_i \quad (1)$$

576

where $\mathbf{p}_i$ is the momentum operator for the ith electron, φ and $\mathbf{A}$ are the scalar and vector potentials of the external field, V is the internal electrostatic potential of the electronic system, and $-e\mathbf{\delta}_i/mc$ is the spin magnetic moment of the ith electron.[1] If the external field is uniform and of intensity $\mathbf{H}$, we may choose $\mathbf{A}$ as

$$\mathbf{A} = \tfrac{1}{2}\mathbf{H} \times \mathbf{r}. \tag{2}$$

For convenience, we shall take $\mathbf{H}$ to lie in the z direction. Under these conditions,

$$H = \sum_i \left[-\frac{\hbar^2}{2m}\Delta_i + \frac{\mathsf{H}_z e}{2mc}\frac{\hbar}{i}\left(x_i\frac{\partial}{\partial y_i} - y_i\frac{\partial}{\partial x_i} \right) + \frac{\mathsf{H}_z^2 e^2}{8mc^2}(x_i^2 + y_i^2) \right] + $$
$$V + \sum_i \frac{e}{mc}\mathsf{H}_z\sigma_{z_i}. \tag{3}$$

The operator

$$\sum_i \frac{\hbar}{i}\left(x_i\frac{\partial}{\partial y_i} - y_i\frac{\partial}{\partial x_i} \right) = \sum_i m_{z_i} \tag{4}$$

is the z component of the total angular momentum operator (*cf.* Sec. 40). If the two terms containing H_z to the first power are combined, they reduce to

$$\mathsf{H}_z\frac{e}{2mc}\sum_i (m_{z_i} + 2\sigma_{z_i}), \tag{5}$$

in which the negative of the coefficient of H_z is the z component of the total magnetic moment arising from both orbital motion and spin. This term accounts for the weak-field Zeeman effect in free atoms since its matrix components are usually much larger than those of the quadratic term

$$\frac{\mathsf{H}_z^2 e^2}{8mc^2}\sum_i (x_i^2 + y_i^2). \tag{6}$$

We may discuss the contributions to the total energy from the terms (5) and (6) separately for the inner-shell electrons and for the valence electrons. Some of the properties of the first type of electron, which may be treated like the electrons in free atoms or ions, will be presented in this section. These results can be applied to all the electrons in those ionic crystals whose constituent ions behave as if they were nearly free.

---

[1] The interaction between the field and the magnetic moments of the electrons, not considered in Sec. 42, is also included in this Hamiltonian.

We shall devote other sections to the valence electrons of other substances that require separate consideration.

Although the z component of the total angular momentum operator is a constant of motion for a free atom, the z component of magnetic moment usually is not, because of the factor 2 that appears as a coefficient of σ_{z_i} in (5). There is one important exceptional case, however, namely, that of Russell-Saunders coupling,[1] in which the spin-orbit interaction terms are small. This case occurs commonly among the atoms on the left-hand side of the periodic chart. We shall list the operators that are constants of motion in this case. The conventional form of the eigenvalues of each operator are also given.

a. The square of the total angular momentum

$$\left[\sum_i (\mathbf{m}_i + \mathbf{d}_i)\right]^2; \qquad \hbar^2 J(J+1). \tag{7}$$

b. The square of the orbital angular momentum

$$\left[\sum_i \mathbf{m}_i\right]^2; \qquad \hbar^2 L(L+1). \tag{8}$$

c. The square of the spin angular momentum

$$\left[\sum_i \mathbf{d}_i\right]^2; \qquad \hbar^2 S(S+1). \tag{9}$$

d. The z component of the total angular momentum

$$\sum_i (m_{z_i} + \sigma_{z_i}); \qquad \hbar J_z. \tag{10}$$

e. The z component of the total orbital angular momentum

$$\sum_i m_{z_i}; \qquad \hbar M_z. \tag{11}$$

f. The z component of the total spin angular momentum

$$\sum_i \sigma_{z_i}; \qquad \hbar S_z. \tag{12}$$

In the preceding equations, the quantum numbers L and M_z are allowed only integer values, whereas the quantum numbers J, S, and J_z are integers in atoms having an even number of electrons and are half integers in atoms having an odd number of electrons. The levels of

---

[1] *Cf.* E. U. Condon and G. H. Shortley, *The Theory of Atomic Spectra* (Cambridge University Press, 1935); G. Herzberg, *Atomic Spectra* (Prentice-Hall, Inc., New York, 1937).

given J are $(2J + 1)$-fold degenerate and the different degenerate states may be specified by the $2J + 1$ values of J_z that range from J to $-J$ in integer steps. In an ideal case of Russell-Saunders coupling, the levels group themselves into widely separated sets, called multiplets, which are specified by given values of L and S (*cf.* Fig. 1). The different levels in each multiplet are in turn specified by values of J that range by integer steps from the value $L + S$ to the value $|L - S|$. The separation of levels in the same multiplet, which is determined by a small interaction between spin and orbital motion that does not appear in Eq. (1), is given by the simple equation

$$E_{J+1} - E_J = \alpha(J + 1)$$

where α is a constant for a given multiplet.

In terms of these quantum numbers, the eigenvalues of the magnetic term (5) are

$$\mathsf{H}_z\beta\left[1 + \frac{J(J + 1) + S(S + 1) - L(L + 1)}{2J(J + 1)}\right]J_z \qquad (13)$$

where

$$\beta = \frac{e\hbar}{2mc}$$

is the Bohr magneton. We shall be interested in this result principally for the discussion of the magnetic effects of inner shells of the atoms in solids. Since J, L, and S are all zero for completely closed shells, the contribution to (5) arising from these shells is zero; however, they do contribute to the term (6). On the other hand, the quantum numbers usually are not zero for the unfilled inner shells of the transition-element atoms or of the rare earth atoms. Hence, the term (5) is important in these cases. It is easy to see that (6) is unimportant whenever the expression (13) is not zero, for if the coefficient of $\mathsf{H}_z\beta$ in (13) is of the order of unity and if the mean value of $(x_i^2 + y_i^2)$ in (7) is of the order of magnitude a_h^2, the ratio of (6) to (5) is

$$\mathsf{H}_z\frac{\hbar^3}{6ce^3m^2},$$

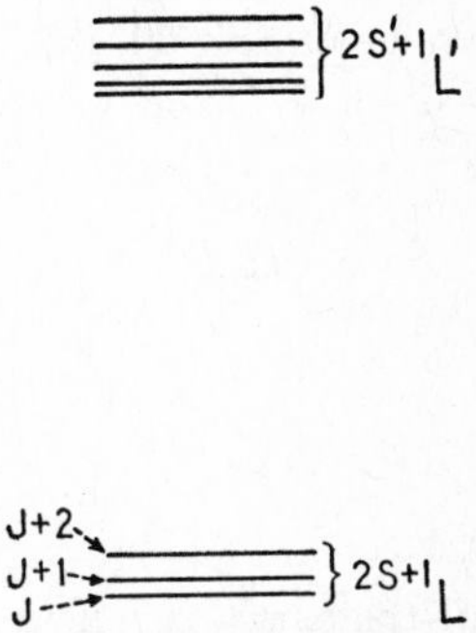

FIG. 1.—The distribution of multiplets in Russell-Saunders coupling. The distance between the multiplets, that is, between the groups of levels of given L and S, are large compared with the inner multiplet separations.

which is of the order of magnitude $\mathsf{H}_z \cdot 10^{-10}$ in cgs units. This is completely negligible for ordinary magnetic field strengths.

We shall now examine the connection between the energy states and the magnetic susceptibility of an atomic system. The susceptibility χ is, by definition, the ratio of the magnetic polarization per unit volume M and the magnetic field H; that is,

$$\mathsf{M} = \chi \cdot \mathsf{H}. \tag{14}$$

In the general case in which M and H are not in the same direction, χ is a tensor. We shall restrict the present discussion to the case in which χ is a constant. If the external magnetic field is changed by an increment $\Delta\mathsf{H}$, the change in energy of a system that is in the energy state E' and has a magnetic polarization per unit volume M is

$$\frac{\Delta E'}{V} = -\mathsf{M} \cdot \Delta\mathsf{H} = -\chi\mathsf{H} \cdot \Delta\mathsf{H}. \tag{15}$$

Hence, if H' is the scalar value of the magnetic field,

$$\chi = -\frac{1}{V\mathsf{H}'}\frac{\partial E'}{\partial \mathsf{H}'}. \tag{16}$$

If the system is at the absolute zero of temperature, E' is the lowest energy state E_0. Thus, in this case,

$$\chi = -\frac{1}{V\mathsf{H}'}\frac{\partial E_0}{\partial \mathsf{H}'}. \tag{17}$$

On the other hand, if the system is at a finite temperature T, the mean value of $\partial E'/\partial \mathsf{H}'$ is

$$\frac{\sum_i (\partial E_i/\partial \mathsf{H}')e^{-\frac{E_i}{kT}}}{\sum_i e^{-\frac{E_i}{kT}}} = -kT\frac{\partial \log f}{\partial \mathsf{H}'} \tag{18}$$

where

$$f = \sum_i e^{-\frac{E_i}{kT}} \tag{19}$$

is the partition function of the system. Hence,

$$\chi = \frac{kT}{V\mathsf{H}'}\frac{\partial}{\partial \mathsf{H}'} \log f. \tag{20}$$

Since $-kT \log f$ is also the free energy A of the system, (20) may be written in the form

$$\chi = -\frac{1}{\mathsf{H}'V}\frac{\partial A}{\partial \mathsf{H}'}, \tag{20a}$$

which is a generalization of (17).

Let us apply Eq. (20) to a simple system consisting of N independent atoms. We shall assume that the inner-multiplet spacing is so large compared with kT that all atoms are in the lowest $2J + 1$ states of a multiplet, which are degenerate in the absence of a field. This model applies to an ionic or molecular crystal in which some ions or atoms have incompleted inner shells so perfectly screened that they are the same as in a free atom. If we use Eq. (13) for the splitting of the lowest level, the partition function is

$$f = \left[\sum_{M_z = -J}^{J} e^{-\frac{H_z \beta g(J,L,S) M_z}{kT}} \right]^N \tag{21}$$

where

$$g = \left[1 + \frac{J(J + 1) + S(S + 1) - L(L + 1)}{2J(J + 1)} \right] \tag{22}$$

is the Landé g factor. Summing the series, we find

$$f = \left[\frac{\sinh (J + \frac{1}{2})\alpha}{\sinh \alpha/2} \right]^N \tag{23}$$

where

$$\alpha = \frac{H_z \beta g}{kT}. \tag{24}$$

Thus, the susceptibility is

$$\chi = \frac{N}{V} \frac{\beta g J}{H_z} B_J(\alpha), \tag{25}$$

in which

$$B_J(\alpha) = \frac{(J + \frac{1}{2}) \coth (J + \frac{1}{2})\alpha - \frac{1}{2} \coth \alpha/2}{J} \tag{26}$$

is the Brillouin function.[1] Values of B_J for several values of J are shown in Fig. 2. Since $B_J(\alpha)$ approaches unity for large values of α, the limiting value of M is $N\beta g J/V$ when βH_z is much larger than kT, which corresponds to complete alignment of the magnetic moment parallel to the magnetic field. For small values of α, B_J varies linearly with α so that, when βH_z is much smaller than kT, the magnetic polarization is

$$M(H) = \frac{N}{V} \frac{\beta^2 g^2 J(J + 1)}{3kT} H, \tag{27}$$

and the susceptibility is

$$\chi = \frac{N}{V} \frac{\beta^2 g^2 J(J + 1)}{3kT}. \tag{28}$$

---

[1] L. Brillouin, *Jour. phys.*, **8**, 74 (1927).

Equation (28) was first derived by Langevin[1] for the case of classical mechanics, in which $\beta g \sqrt{J(J+1)}$ is replaced by the permanent atomic magnetic moment.

If the atom has a closed-shell structure, so that J is zero, (28) vanishes and the susceptibility should be determined by the quadratic energy term (6). Since closed shells are spherically symmetrical, the matrix component of $\sum_i (x_i^2 + y_i^2)$ for the lowest state is

$$\tfrac{2}{3} n \overline{r^2} \tag{29}$$

where n is the total number of electrons in the atom and $\overline{r^2}$ is the mean

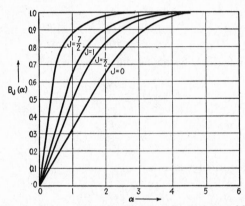

Fig. 2.—Values of $B_j(\alpha)$ for the values $J = 0, \tfrac{1}{2}, 1,$ and $\tfrac{7}{2}$.

value of r^2 for any electron. Thus, the magnetic energy per unit volume of a group of N atoms of this type is

$$\frac{\mathsf{H}_z^2 e^2}{12mc^2} \frac{Nn}{V} \overline{r^2}, \tag{30}$$

and their susceptibility is

$$-\frac{Nn}{V} \frac{e^2}{6mc^2} \overline{r^2}, \tag{31}$$

which corresponds to diamagnetism.

We have seen in the previous chapters that simple ionic crystals behave as though they were composed of spherically symmetric closed-shell ions of the constituent atoms. If there are N_α/V ions of the αth type per unit volume of this crystal, the diamagnetic susceptibility is

[1] P. Langevin, *Jour. phys.*, **4**, 678 (1905).

$$\chi = -\frac{e^2}{6mc^2}\sum \frac{N_\alpha n_\alpha \overline{r_\alpha^2}}{V} \tag{32}$$

where n_α and $\overline{r_\alpha^2}$ are, respectively, the number of electrons on the αth ion and the mean value of r^2 for this ion. Equation (32) may also be used to compute the diamagnetic contribution to the susceptibility from the closed shells of any solid. A discussion of the methods of computing the diamagnetic susceptibilities of various closed-shell ions and tables of numerical values of these susceptibilities may be found in Chap. VIII of Van Vleck's book.[1]

138. The Orbital Diamagnetism of Free Electrons*.

—A theorem of classical mechanics[2] states that a system of charges that are confined in a fixed volume but are otherwise free has zero magnetic susceptibility. If the system is not confined, each constituent charge is induced to move in a helical path and the total magnetic flux is decreased. The charges striking the wall, however, have their paths changed in such a way that their magnetic field nullifies the field of the rest. This result may be understood on the basis of the following formal argument. In classical mechanics, a magnetic field alters the direction of motion of a charge but does not change its speed. Hence, the distribution of energy states and, as a result, the partition function of a system of charges that is at equilibrium are unaltered by the magnetic field. According to Eq. (20) of the preceding section this means that the susceptibility is zero. The fact that the charges are confined assures us that the system is in equilibrium.

Landau[3] first pointed out that this theorem is not valid in quantum mechanics because the distribution of energy levels is altered by a magnetic field in the new scheme. This may be demonstrated as follows. The Schrödinger equation for a free electron in a magnetic field, as derived from the Hamiltonian (3) of the preceding section, is

$$-\frac{\hbar^2}{2m}\Delta\psi + \frac{H_z e}{2mc}\frac{\hbar}{i}\left(x\frac{\partial}{\partial y} - y\frac{\partial}{\partial x}\right)\psi + \frac{H_z^2 e^2}{8mc^2}(x^2 + y^2)\psi = \epsilon\psi, \tag{1}$$

in which it is assumed that the field is in the z direction. The spin term is neglected in the present section. If we make the transformation

$$\psi = \varphi(x,y,z)e^{i\frac{H_z e}{2c}\frac{xy}{\hbar}}, \tag{2}$$

Eq. (1) reduces to

$$-\frac{\hbar^2}{2m}\Delta\varphi + \frac{H_z e}{mc}\frac{\hbar}{i}x\frac{\partial\varphi}{\partial y} + \frac{H_z^2 e^2}{2mc^2}x^2\varphi = \epsilon\varphi, \tag{3}$$

[1] J. H. VAN VLECK, *The Theory of Electric and Magnetic Susceptibilities* (Oxford University Press, New York, 1932).

[2] *Cf. ibid.*

[3] L. LANDAU, *Z. Physik*, **64**, 629 (1930).

which does not explicitly contain y or z. This equation may be further simplified by means of the substitution

$$\varphi(x,y,z) = \lambda(x)e^{2\pi i(k_y y + k_z z)}, \tag{4}$$

for it then becomes

$$-\frac{\hbar^2}{2m}\frac{d^2\lambda}{dx^2} + \left[\frac{1}{2m}\left(\hbar k_y + \mathsf{H}_z\frac{e}{c}x\right)^2 + \frac{\hbar^2 k_z^2}{2m}\right]\lambda = \epsilon\lambda, \tag{5}$$

which is identical with the equation for a simple one-dimensional oscillator that is centered about the position

$$x' = -\frac{ch}{e\mathsf{H}_z}k_y \tag{6}$$

and has the natural frequency

$$\nu = \frac{1}{2\pi}\frac{\mathsf{H}_z e}{mc}. \tag{7}$$

Hence, the allowed values of ϵ are

$$\epsilon = \frac{\hbar^2 k_z^2}{2m} + \frac{\mathsf{H}_z he}{2\pi mc}\left(n + \frac{1}{2}\right), \tag{8}$$

where n is restricted to integer values.

The form of the total wave function, namely,

$$\psi_{n,k_y,k_z} = e^{2\pi i\left(\frac{\mathsf{H}_z e}{2ch}xy + k_y y\right)}\lambda(x)e^{2\pi i k_z z}, \tag{9}$$

shows that the motion in the z direction is the same as for a free particle having a component of momentum hk_z along this axis. Thus if L_z is the length of the container in the z direction, the number of states that have a fixed value of n and k_y and values of k_z lying in any range Δk_z is

$$N_{\Delta z} = \Delta k_z L_z. \tag{10}$$

The contribution to the total energy from the other two degrees of freedom is (8) minus $\hbar^2 k_z^2/2m$, or

$$\epsilon' = \frac{\mathsf{H}_z he}{2\pi mc}\left(n + \frac{1}{2}\right). \tag{11}$$

Each of the discrete levels of this two-dimensional system is highly degenerate. This degeneracy may be estimated in the following way. The parameter k_y in Eq. (8) is analogous to a y component of wave number so that the number of values of k_y in an allowed range Δk_y is

$$N_{\Delta k_y} = \Delta k_y L_y$$

by analogy with Eq. (10). The allowed range of k_y, however, is not unlimited as is that of k_z, for the point x' defined by Eq. (6), which is the center of gravity of the electron position in the x direction, should lie within the container. Thus, if the width of the box in the x direction is L_x, the total number of allowed values of k_y is

$$N_n = \frac{e\mathsf{H}_z}{ch}L_xL_y, \tag{12}$$

which is also the number of states of given ϵ'.

The distance $\Delta\epsilon'$ between successive values of the two-dimensional energy parameter is $\mathsf{H}_z he/2\pi mc$, according to Eq. (11). Hence, the

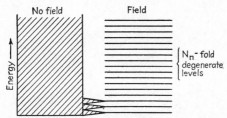

FIG. 3.—Schematic representation of the coalescence of levels of the two-dimensional system in a magnetic field. In effect, bundles of N_n levels of the continuum for perfectly free electrons combine to form discrete levels. The center of gravity of the bundle remains unchanged.

density of the states of the two-dimensional system is

$$\frac{N_n}{\Delta\epsilon'} = 2\pi m\frac{L_xL_y}{h^2}. \tag{13}$$

This is independent of H_z and is the same as the density of levels for a free particle in two dimensions. Thus, the magnetic field does not change the average density of levels although it does alter the detailed distribution.

We may obtain a qualitative picture of Landau's diamagnetism for a system obeying classical statistics from a discussion of the energy levels of the two-dimensional system. The quasi-continuous energy levels of the field-free system become discrete in the presence of the field. In effect, groups of N_n levels coalesce to form each member of the discrete set of levels (11) as is shown in Fig. 3. The individual groups coalesce about their center of gravity; that is, the quasi-continuous set from 0 to $\mathsf{H}_z he/2\pi mc$ coalesce to $\mathsf{H}_z he/4\pi mc$, and so forth. Now, in the absence of a magnetic field the lower levels of any group are preferentially filled if Boltzmann statistics are used. Hence, the mean energy of electrons in a group is less than the energy at the center point. Since all the particles have the energy of the single level into which the group coalesces in the presence of a magnetic field, the mean energy of the electrons is raised. Thus, the system is diamagnetic. This effect is much smaller if the particles obey Fermi-Dirac statistics and if the system is degenerate, for then only the few electrons at the top of the filled region have their mean energy altered by the field. There is a diamagnetic term, however, even in this case, as will be shown below.

We shall now compute the value of the susceptibility of the three-dimensional model for both classical and Fermi-Dirac statistics. In the classical case, the partition function for a single particle is

$$
\left.
\begin{aligned}
f &= \int_{-\infty}^{\infty} L_z dk_z \left\{ \sum_{n=0}^{\infty} \frac{e \mathsf{H}_z L_x L_y}{ch} e^{-\frac{\frac{h^2 k_z{}^2}{2m} + \mathsf{H}_z \beta (2n+1)}{kT}} \right\} \\
&= \frac{V e \mathsf{H}_z}{ch} \sqrt{\frac{2\pi m k T}{h^2}} \sum_{n=0}^{\infty} e^{-\frac{\mathsf{H}_z \beta (2n+1)}{kT}} \\
&= \frac{V e \mathsf{H}_z}{ch^2} \sqrt{2\pi m k T} \frac{1}{2 \sinh (\mathsf{H}_z \beta / kT)}
\end{aligned}
\right\}
\tag{14}
$$

where $V = L_x L_y L_z$. For normal field intensities, the susceptibility determined from this partition function by means of Eq. (20) of the preceding section is

$$
\chi = -\frac{1}{3} \frac{N}{V} \frac{\beta^2}{kT}
\tag{15}
$$

where N is the total number of particles.

For Fermi-Dirac statistics, the partition function f of a particle is given by the equation[1]

$$
\log f = 2 \sum_{n=0}^{\infty} \int_{-\infty}^{\infty} \frac{e \mathsf{H}_z V}{ch} dk_z \log \left\{ 1 + e^{\alpha - \frac{\frac{h^2 k_z{}^2}{2m} + \mathsf{H}_z \beta (2n+1)}{kT}} \right\}.
$$

Since α is very large for a degenerate gas, we may replace the logarithm in the integrand by

$$
\alpha - \frac{\left[\frac{h^2 k_z^2}{2m} + \mathsf{H}_z \beta (2n+1) \right]}{kT}
$$

and evaluate the summations only for those values of n and k_z for which this quantity is positive. This procedure is equivalent to assuming that the major part of the contribution to diamagnetism arises from the

[1] See, for example, R. H. Fowler, *Statistical Mechanics* (Cambridge University Press, 1937). The quantity α is equal to ϵ_0'/kT, as in Sec. 26.

electrons in the completely filled part of the Fermi distribution. The integration over k_z then takes place between the limits

$$\pm\frac{1}{h}(2mkT)^{\frac{1}{2}}\left[\alpha - \frac{H_z\beta(2n+1)}{kT}\right]^{\frac{1}{2}} \tag{16}$$

and leads to the result

$$\sum_n \frac{2eH_zV}{ch^2}\frac{4}{3}(2mkT)^{\frac{1}{2}}\left[\alpha - \frac{H_z\beta(2n+1)}{kT}\right]^{\frac{3}{2}}. \tag{17}$$

The summation over n extends from zero to $kT\alpha/2H_z\beta$ and may be obtained by use of the approximate equation

$$\sum_0^N F(n) \cong \int_{-\frac{1}{2}}^{N+\frac{1}{2}} F(n)dn - \frac{1}{24}\left[F'(n)\right]_{-\frac{1}{2}}^{N+\frac{1}{2}}. \tag{18}$$

The result is

$$\left[N\frac{h^2}{5m}\left(\frac{3N}{8\pi V}\right)^{\frac{2}{3}} - \frac{2\pi mV\beta^2}{3h^2}\left(\frac{3N}{\pi V}\right)^{\frac{1}{3}}H_z^2\right]\frac{1}{kT}, \tag{19}$$

in which the two terms arise, respectively, from the two terms of (18). The first term is independent of H_z, so that the second is entirely responsible for the magnetism of the system. The susceptibility determined from it is

$$\chi = -\frac{4\pi m\beta^2}{3h^2}\left(\frac{3N}{\pi V}\right)^{\frac{1}{3}}. \tag{20}$$

It is easy to see from the form of Eq. (17) that the summation over n would vanish in the approximation of Eq. (18) if it were carried out before the integration rather than after it. This shows that the diamagnetic term (20) is a three-dimensional effect and is related to a redistribution of electrons near the top of the filled band among different levels. The reason for the redistribution is illustrated schematically in Fig. 4, which represents a plane in wave-number space that is normal to the z axis. The circle is the limit of the occupied region. In the absence of a field, the allowable values of wave number are uniformly distributed, whereas, in the presence of a field H_z, the group of levels lying in the cylindrical shell parallel to the z axis associated with a range

$$\Delta\epsilon' = \frac{H_zhe}{2\pi mc}$$

coalesce to form levels going with a single value of n. Some of the levels near the boundary of the circle that were previously occupied now lie outside the circle. Others that previously were unoccupied now lie inside. Although the electrons move from the first set of levels to the second, the mean free energy remains larger than the value in the absence of a field by the second term in the brackets of Eq. (19).

Peierls[1] has pointed out that Eq. (18) is valid only when the condition

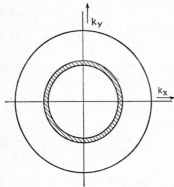

$$|F(m) - F(m - 1)| << F(m) \quad (21)$$

is satisfied for all values of m, which implies that $H_z\beta << kT$.[2] This condition is not satisfied for sufficiently low temperatures and high field strengths. Thus, $kT/H_z\beta$ is of the order unity for $T \sim 10°K$ and $H_z \sim 10$ kilogauss. Under these conditions, the susceptibility must be computed by using different approximational methods.

FIG. 4.—Schematic representation of the behavior of the electronic levels in the three-dimensional case. The diagram represents a cross section of wave-number space normal to the z axis, the field being in the z direction. The outer circle is the limit of the filled region. In the presence of a field the group of levels contained in the shaded cylindrical tube parallel to the z axis coalesce to form levels associated with a single value of n. Some of the levels of given k_z are raised and an equal number are lowered. This is unimportant for the electrons well inside the filled region, since the mean energy is unchanged. On the other hand, the effect is important for the electrons near the surface of the filled region and their mean energy is raised.

Suppose that we have a two-dimensional system of free particles at absolute zero of temperature and that the energy levels, in the presence of a field, are determined by Eq. (11), namely,

$$\epsilon' = H_z\beta(2n + 1).$$

When the degeneracy $eH_zL_xL_y/ch$ becomes greater than N, the total number of particles, all particles occupy the state for which $n = 0$ and the total energy is

$$E = NH_z\beta, \quad (22)$$

which corresponds to a constant magnetic moment and zero susceptibility. As H_z is lowered, E decreases until the degeneracy becomes less than N, whereupon some of the electrons begin filling levels for which n is unity. The energy then increases with decreasing field intensity, so that the system becomes paramagnetic when

$$H_z = \frac{chN}{eL_xL_y}.$$

[1] R. PEIERLS, *Z. Physik*, **80**, 763 (1933); **81**, 186 (1933).

[2] Otherwise it is possible that $\alpha - \dfrac{H_z\beta(2n + 1)}{kT}$ may be accidentally zero for $n = n'$ and very large for $n = n' + 1$.

It is easy to see that the general expression for the total energy is

$$E(n) = g\beta H_z^2 \{1 + 3 + \cdots + [2(n - 1) + 1]\} +$$
$$(N - ng H_z)H_z\beta(2n + 1) \Big\}$$
$$= NH_z\beta(2n + 1) - g H_z^2\beta n(n + 1) \qquad (23)$$

when

$$\frac{N}{n + 1} < g H_z < \frac{N}{n} \qquad (24)$$

where

$$g = \frac{e}{ch}L_x L_y.$$

The magnetic moment for fields in the range (24) is

$$M = -\frac{\partial E(n)}{\partial H_z} = -N\beta(2n + 1) + 2g H_z\beta n(n + 1),$$

which is equal to $N\beta$ when $g H_z = N/n$ and is $-N\beta$ when

$$g H_z = \frac{N}{(n + 1)}.$$

Thus, the moment abruptly changes sign at frequent intervals as the

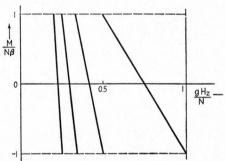

Fig. 5.—Behavior of the magnetic intensity as a function of field intensity at absolute temperature in the two-dimensional case.

field is lowered, the discontinuities occurring at points for which $g H_z$ is equal to N/n. This behavior is shown in Fig. 5.

The three-dimensional spectrum is not discrete, so that discontinuous changes in sign do not occur. Peierls[1] has shown, however, that oscillations in sign still occur. By a direct extension of the preceding work, he obtained the equation

$$-\frac{M}{V} = \frac{e^{\frac{3}{2}}}{\pi^2\hbar^{\frac{3}{2}}c^{\frac{3}{2}}}\frac{\epsilon_0'^{\frac{3}{2}}}{\beta^{\frac{1}{2}}}\sigma\left(\frac{\beta H_z}{kT}, \frac{\epsilon_0'}{kT}\right) \qquad (25)$$

[1] R. Peierls, *Z. Physik*, **81**, 186 (1933).

for the magnetic moment per unit volume.   Values of σ, determined by direct computation, are shown in Fig. 6.   Variations similar to this have been observed in bismuth and will be discussed in the next section.

139. The Orbital Diamagnetism of Quasi-bound Electrons*.—Peierls[1] has extended the theory of the diamagnetism of valence electrons to include the case in which the electrons are nearly bound.   It turns out in this case that there are three contributions to the susceptibility, namely, one that is identical with the susceptibility of atomic electrons, given by Eq. (31), Sec. 137, another that is a generalization of Eq. (20) of the previous section for perfectly free electrons, and a third that has no analogue in either the free or the bound models.   Although

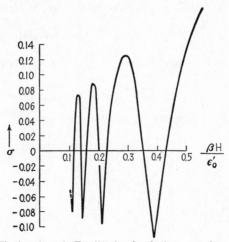

FIG. 6.—The function σ in Eq. (25) for the absolute zero of temperature.

Peierls's treatment of the problem, which is presented here, is valid only for nearly-bound electrons, Wilson[2] has shown that the second term also occurs for nearly free electrons and is a measure of the susceptibility accompanying the type of free-electron motion found in metals. As we shall see, this term may be used to explain the unusually large diamagnetism of bismuth and the γ phases.

Consider a set of weakly interacting atoms that are centered at the lattice positions $\mathbf{r}(n)$ given by the equation

$$\mathbf{r}(n) = n_1\boldsymbol{\tau}_1 + n_2\boldsymbol{\tau}_2 + n_3\boldsymbol{\tau}_3, \tag{1}$$

in which the $\boldsymbol{\tau}$ are primitive translations for a simple cubic lattice having lattice constant d.   If the interaction forces are neglected, the Schröd-

[1] *Ibid.*

[2] A. H. WILSON, *The Theory of Metals* (Cambridge University Press, 1936).

inger equation for the electron on the nth atom is

$$-\frac{\hbar^2}{2m}\Delta\varphi_n(\mathbf{r} - \mathbf{r}(n)) + \left\{V(\mathbf{r} - \mathbf{r}(n)) - \frac{ie\hbar}{2mc}\mathsf{H}_z\left(x\frac{\partial}{\partial y} - y\frac{\partial}{\partial x}\right) + \right.$$
$$\left. \frac{e^2\mathsf{H}_z^2}{8mc^2}(x^2 + y^2)\right\}\varphi_n = \epsilon\varphi_n, \quad (2)$$

in which φ_n and V are the atomic wave function and potential, respectively. This equation is different for each atom because the zero of the vector potential has been chosen to be the origin of coordinates. We may reduce the equations to a standard form by the transformation

$$\varphi_n = e^{-2\pi i\alpha\mathbf{r}(n)\cdot(\mathbf{r}\times\mathsf{H})}\psi(\mathbf{r} - \mathbf{r}(n)) \quad (3)$$

where $\alpha = e/2hc$. The equation satisfied by ψ, namely,

$$-\frac{\hbar^2}{2m}\Delta\psi(\mathbf{r} - \mathbf{r}(n)) + \left\{V(\mathbf{r} - \mathbf{r}(n)) - \frac{ie\hbar}{2mc}\mathsf{H}_z\left((x - n_xd)\frac{\partial}{\partial y} - \right.\right.$$
$$\left.\left. (y - n_yd)\frac{\partial}{\partial x}\right) + \frac{e^2\mathsf{H}_z^2}{8mc^2}[(x - n_xd)^2 + (y - n_yd)^2]\right\}\psi = \epsilon\psi, \quad (4)$$

is the same as the equation for $\varphi_0(\mathbf{r} - \mathbf{r}(n))$. Hence,

$$\varphi_n = e^{-2\pi i\alpha\mathbf{r}(n)\cdot(\mathbf{r}\times\mathsf{H})}\varphi_0(\mathbf{r} - \mathbf{r}(n)). \quad (5)$$

This result shows that the energy $\epsilon(\mathsf{H}_z)$ of each of the unperturbed functions is the same, namely,

$$\epsilon(\mathsf{H}_z) = \epsilon_0 + \chi_a\frac{\mathsf{H}_z^2}{2}, \quad (6)$$

where ϵ_0 is the energy in the absence of a field and χ_a is the atomic diamagnetic susceptibility:

$$\chi_a = \frac{e^2}{8mc^2}\int|\varphi|^2(x^2 + y^2)d\tau \quad (7)$$

[cf. Eq. (30), Sec. 137]. This is the first of the three contributions to the total susceptibility that were mentioned in the opening paragraph.

Equation (5) may be placed in another convenient form by employing the displacement operator $e^{-a\frac{\partial}{\partial x}}$, which has the property[1]

$$e^{-a\frac{\partial}{\partial x}}f(x) = f(x - a). \quad (8)$$

---

[1] This property of the operator $e^{-a\frac{\partial}{\partial x}}$ may easily be demonstrated by expanding it as a power series in terms of the exponent, and comparing the result of the operation of the operator on a function with Taylor's series.

Using this operator, we may write φ_n in the form

$$\varphi_0(\mathbf{r} - \mathbf{r}(n)) = e^{-\frac{i}{\hbar}\mathbf{r}(n)\cdot\mathbf{p}}\varphi_0(\mathbf{r}) \tag{9}$$

where

$$\mathbf{p} = \frac{\hbar}{i}\,\text{grad}.$$

Thus, Eq. (5) becomes

$$\varphi_n = e^{-2\pi i \mathbf{r}(n)\cdot\mathbf{P}}\varphi_0, \tag{10}$$

in which

$$\mathbf{P} = \left(\frac{\mathbf{p}}{\hbar} + \alpha \mathbf{r} \times \mathbf{H}\right). \tag{11}$$

We shall now consider interatomic perturbation terms on the assumption that the Hamiltonian operator for an electron near the nth atom is

$$H' = -\frac{\hbar^2}{2m}\Delta + \sum_m V(\mathbf{r} - [\mathbf{r}(n) - \mathbf{r}(m)]) - \frac{ie\hbar}{2mc}\mathsf{H}_z\left(x\frac{\partial}{\partial y} - y\frac{\partial}{\partial x}\right) + \frac{e^2\mathsf{H}_z^2}{8mc^2}(x^2 + y^2). \tag{12}$$

This is the same as the Hamiltonian operator in Eq. (2) except for the term

$$\sum_m' V(\mathbf{r} - [\mathbf{r}(n) - \mathbf{r}(m)]). \tag{13}$$

If we assume that the perturbed eigenfunction ψ may be expressed in the form

$$\psi = \sum_m a_m\varphi_m, \tag{14}$$

the eigenvalue equations for the a_m are

$$\epsilon' a_m = \sum_l u_{ml}a_l \tag{15}$$

where ϵ' is the eigenvalue of the perturbed system and

$$u_{ml} = \int\varphi_m{}^*H'\varphi_l d\tau = \int\varphi_0{}^*e^{2\pi i \mathbf{r}(n)\cdot\mathbf{P}}H'e^{-2\pi i \mathbf{r}(l)\cdot\mathbf{P}}\varphi_0 d\tau. \tag{16}$$

Peierls proved that the off-diagonal matrix elements (16) are matrix components of the operator

$$E = \epsilon(\mathsf{H}) + \sum_l A(l)e^{2\pi i \mathbf{r}(l)\cdot\mathbf{K}} \tag{17}$$

where

$$A(l) = \int \varphi_l^* \left\{ \sum_q{}' V(\mathbf{r} - \mathbf{r}(q)) \right\} \varphi_0 d\tau \tag{18}$$

and

$$\mathbf{K} = \mathbf{P} - \frac{e}{hc}(\mathbf{r}_0 \times \mathbf{H}), \tag{19}$$

in which $\mathbf{r}_0$ is by definition the matrix satisfying the relation

$$\mathbf{r}_0 \cdot \varphi_n = \mathbf{r}(n)\varphi_n.$$

In the presence of a magnetic field, $A(l)$ involves the field intensities through the function φ_n^*. This fact may be made explicit by writing φ_n^* in the form (10) with the help of Eq. (11). In the absence of a magnetic field,

$$\mathbf{K} = \frac{1}{2\pi i} \, \text{grad}, \tag{20}$$

and the eigenvalues of $\mathbf{K}$ are the wave-number vectors $\mathbf{k}$. Hence, in this case the eigenvalues of E are

$$\epsilon(\mathbf{k}) = \epsilon_0 + \sum_l{}' A_0(l) e^{2\pi i \mathbf{k} \cdot \mathbf{r}(l)} \tag{21}$$

where A_0 is the value of A when $\mathbf{H} = 0$. This result was previously derived in Sec. 65 by more direct means.

Using Eq. (17) Peierls computed the total partition function for the perturbed system by a method that will not be discussed here. The result is

$$\log f = \int \left\{ g\left(\epsilon(\mathbf{k}) + \frac{(\chi_a + \epsilon_1)\mathsf{H}_z^2}{2} \right) + \right.$$
$$\left. \frac{\mathsf{H}_z^2 \alpha^2}{24\pi^2} \left[\frac{\partial^2 \epsilon}{\partial k_x^2} \cdot \frac{\partial^2 \epsilon}{\partial k_y^2} - \left(\frac{\partial^2 \epsilon}{\partial k_x \partial k_y} \right)^2 \right] g''(\epsilon_0'(\mathbf{k})) \right\} d\tau \tag{22}$$

where

$$\left. \begin{aligned} g(\epsilon) &= \log (1 + e^{\frac{\epsilon_0' - \epsilon}{kT}}), \\ \epsilon_1(\mathbf{k}) &= -\frac{\alpha^2}{\mathsf{H}_z^2} \sum_l \left(e^{2\pi i \mathbf{k} \cdot \mathbf{r}(l)} \int \varphi_0(\mathbf{r} - \mathbf{r}(l)) \cdot [\mathbf{r}(l) \cdot (\mathbf{r} \times \mathbf{H})]^2 \right. \\ & \qquad\qquad \left. \left\{ \sum_n{}' V(\mathbf{r} - \mathbf{r}(n)) \right\} \varphi_0 d\tau \right), \end{aligned} \right\} \tag{23}$$

and the integration extends over all wave-number space. It should be observed that ϵ_1 is independent of H_z. If the first integrand in (22) is expanded to terms in H_z^2, it leads to the result

$$\int \left[g(\epsilon) + \frac{H_z^2}{2}(\chi_a + \epsilon_1)g'(\epsilon) \right] d\tau(\mathbf{k}), \tag{24}$$

in which the integral involving the first term is the partition function in the absence of a field. Since χ_a is a constant, the remaining terms may be placed in the form

$$\frac{H_z^2}{2} \left[-\frac{N\chi_a}{kT} + \int \epsilon_1 g'(\epsilon) d\tau(\mathbf{k}) \right]. \tag{25}$$

Thus, if we include the second term in the integrand of (22), we see that there are three contributions to the susceptibility, namely,

$$\chi_1 = -\frac{N}{V}\chi_a, \tag{26}$$

$$\chi_2 = \frac{kT}{V}\int \epsilon_1 g'(\epsilon(\mathbf{k})) d\tau, \tag{27}$$

$$\chi_3 = \frac{kT}{12\pi^2}\frac{\alpha^2}{V}\int\int \left[\frac{\partial^2\epsilon}{\partial k_x^2}\frac{\partial^2\epsilon}{\partial k_y^2} - \left(\frac{\partial^2\epsilon}{\partial k_x \partial k_y} \right)^2 \right] g''(\epsilon) d\tau(\mathbf{k}). \tag{28}$$

χ_a should be the same order of magnitude as the atomic susceptibility of gases, namely,

$$\frac{e^2}{6mc^2}\overline{r^2},$$

and should lead to a diamagnetic susceptibility of the order of 10^{-7}. A simple estimate of χ_2 shows that it is related to χ_1 in order of magnitude by the equation

$$\chi_2 \sim \frac{m}{m^*}\chi_1$$

where m is the true electronic mass and m^* is the effective mass of the electrons in the filled region. Thus, χ_2 would be negligible in the limiting case of very narrow bands.[1]

The factor $g''(\epsilon)$ in the integrand of (28) is

$$\frac{1}{kT}\frac{\partial}{\partial\epsilon}\left(\frac{1}{e^{\frac{\epsilon - \epsilon_0'}{kT}} + 1} \right) = \frac{1}{kT}\frac{\partial f}{\partial\epsilon}$$

where f is the Fermi function. This derivative has a sharp peak at the point $\epsilon = \epsilon_0'$ and satisfies the relation [see Eq. (29), Sec. 26]

$$\int \alpha(\epsilon)\frac{\partial f}{\partial\epsilon}d\epsilon = -\alpha(\epsilon_0').$$

[1] Wilson (*op. cit.*) has shown that χ_2 is zero for perfectly free electrons.

Hence, χ_3 may be placed in the form

$$\chi_3 = -\frac{1}{12\pi^2}\frac{\alpha^2}{V}\int_S\left(\frac{\partial^2\epsilon}{\partial k_x^2}\frac{\partial^2\epsilon}{\partial k_y^2} - \frac{\partial^2\epsilon}{\partial k_x\partial k_y}\right)\frac{2V}{|\mathrm{grad}_k\ \epsilon|}dS \qquad (29)$$

where the integral extends over the surface of the filled region in wave-number space. This expression reduces to Landau's equation (15) of the previous section when the electrons are perfectly free, that is, when ϵ is equal to $h^2k^2/2m$.

The quantity in parentheses in Eq. (29) becomes large whenever the curvature of $\epsilon(\mathbf{k})$ is large at the edge of the filled region. Since this may occur near the boundary of a zone according to the zone theory, χ_3 should be largest for metals such as the alkaline earths and bismuth that have nearly filled zones. Jones[1] has postulated that the five valence electrons per atom in bismuth extend just beyond a prominent zone that has room

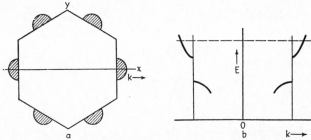

Fig. 7.—*a*, cross section of the prominent zone boundary of bismuth (see Fig. 6, Chap. XIII) normal to the preferred direction. The shaded regions are those in which it is believed the filled region extends into the outer zone. *b* represents schematically the behavior of the $\epsilon(\mathbf{k})$ curve for the line through the zone boundary shown in *a*. The dotted line in *b* is the top of the filled region. The curvature of the $\epsilon(\mathbf{k})$ curve is believed to be very large in the upper branch, so that the effective electron mass is small.

for five electrons per atom in several directions. In addition, he postulated that the behavior of the $\epsilon(\mathbf{k})$ curve near the zone is as in Fig. 7, so that the curvature is very great for the higher zone. The high diamagnetic susceptibility of this metal and of the alloys such as γ brass that have similarly filled zones may be understood in terms of this picture, for the integrand of Eq. (29) is large and positive for a part of the range of integration in all these substances.

In order to develop a semiquantitative theory of the diamagnetism of bismuth, Jones assumed that the energy contours are ellipsoids of revolution which are centered about the center points of the vertical plane faces of the zone in the six regions of Fig. 7 in which the filled region extends into the outer zones. Thus, if the z axis is parallel to the prin-

[1] H. JONES, *Proc. Roy. Soc.*, **144** 225 (1934); **147**, 396 (1934).

cipal axis of the crystal, the energy contours in any one of the six regions have the form

$$\epsilon(\mathbf{k}) = \frac{h^2}{2m}(\alpha_1 k_x^2 + \alpha_1 k_y^2 + \alpha_3 k_z^2) \tag{30}$$

where the origin is at the center point of the corresponding face. The constants α_1 and α_3 are, respectively, the ratios of the electron mass to the effective mass in the directions normal to the principal axis and along the principal axis. It follows from the symmetry of the prominent zone of bismuth that the ellipsoidal contours on opposite faces may be joined, six sets of completely ellipsoidal contours being thus produced.

If (30) is substituted in Eq. (29), it is found that the volume susceptibility in the z direction, expressed in cgs units, is

$$\chi = -0.122\sqrt{\epsilon_0'}\sqrt{\frac{\alpha_1^2}{\alpha_3}}10^{-6} \tag{31}$$

where ϵ_0' is the value of (30) at the top of the filled region expressed in ev. Thus, χ is largest in the direction in which α is smallest.

Now, the number of states per atom n_a within the ellipsoidal contour associated with ϵ_0' is

$$n_a = v_0 \frac{8\pi}{3}\left(\frac{2m\epsilon_0'}{h^2}\right)^{\frac{3}{2}}\frac{1}{\sqrt{\alpha_1\alpha_1\alpha_3}}, \tag{32}$$

in which v_0 is the atomic volume. This result may be derived by computing the volume of the ellipsoid and using the fact that there are $2V$ states per unit volume of $\mathbf{k}$ space, if V is the volume of the crystal. Jones evaluated n_a by noting that the temperature coefficient of resistance of the tin-bismuth alloy system changes from positive to negative as tin is added to bismuth, the value zero occurring when about 0.13 atom per cent of tin is present. Each bismuth atom that is replaced by a tin atom presumably carries with it one of the electrons from the overlapping region, for the valence of tin is one unit less than that of bismuth. Thus, if it is assumed that the point at which the temperature coefficient of resistance changes sign is the same as that at which the boundary of the filled region extends just to the first zone, it follows that the number of electrons per atom outside the inner zone in pure bismuth is 0.0013, or there is 0.0002 electron per atom in each of the six shaded regions of Fig. 7. From this, ϵ_0' may be computed by the use of Eq. (32), and $\sqrt{\epsilon_0'}$ in Eq. (31) may then be given a value.

Jones finds that the observed room-temperature values of $\chi_\perp$ and $\chi_\parallel$, which are listed in Table VI, Chap. I, may be obtained from Eq. (34) by assuming

$$\alpha_1 \sim 40 \quad \text{and} \quad \alpha_3 \sim 1$$

The assumption that the α in the plane normal to the principal axis are equal is justified by the fact that the susceptibility is the same in all directions normal to the principal axis.

The most extensive theoretical treatment of the low-temperature susceptibility of bismuth has been given by Blackman,[1] whose work is based on an extension of Peierls's theory. As we pointed out in the last section, it is observed experimentally that the susceptibility fluctuates

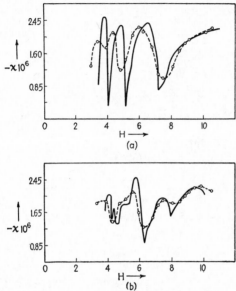

Fig. 8.—Comparison of the observed and calculated values of the low-temperature magnetic susceptibilities of bismuth in the plane normal to the principal axis. Curves *a* correspond to values along the *x* direction of Fig. 7, and curves *b* correspond to the *y* direction. The measured values, which are for 1.86°K, are represented by circles on the dotted curves.

with the field intensity in the neighborhood of absolute zero of temperature. For example, Figs. 8*a* and 8*b* show the variations[2] of $-\chi$ with field intensities at 1.86°K. The measurements *a* are for fields parallel to the *x* axis of Fig. 7, and the measurements *b* are for fields in the *y* direction. Thus, the effect is not the same in all directions normal to the principal axis. Fluctuations are not observed when the field is parallel to the principal axis. It may be mentioned that the curves of Fig. 8 for 1.86°K are closely similar to curves for 14.2°K in the region in which the abscissae

[1] M. BLACKMAN, *Proc. Roy. Soc.*, **166**, 1 (1938).

[2] W. J. DE HAAS and P. M. VAN ALPHEN, *Leiden Comm.*, 212 (1931); D. SHOENBERG and M. Z. UDDIN, *Proc. Roy. Soc.*, **156**, 687 (1936).

overlap. This suggests that the curves of Fig. 8 may safely be compared with theoretical curves computed for the absolute zero.

The fact that the susceptibility is not symmetrical about the principal axis in the low-temperature range indicates that the energy contours for the electrons which are responsible for the fluctuations cannot have the symmetry of the function (30), for this function is invariant under rotations about the z axis. Thus, Blackman was led to assume that the $\epsilon(\mathbf{k})$ function for these electrons is

$$\epsilon(\mathbf{k}) = \frac{h^2}{2m}(\alpha_1 k_x^2 + \alpha_2 k_y^2 + \alpha_3 k_z^2) \tag{33}$$

where α_1, α_2, and α_3 are different constants. At first sight, this appears to contradict Jones's results. Blackman points out, however, that the

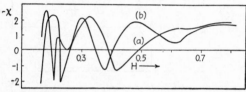

Fig. 9.—The susceptibility at absolute zero of temperature computed by assuming that there is only the type of electron corresponding to Eq. (33). The designations a and b have the same significance as in Fig. 8. The theoretical curves of Fig. 8 were derived by adding to the curves of this figure the susceptibility arising from the type of electron considered by Jones.

number of electrons responsible for the fluctuations probably is much smaller than the number determining the high-temperature susceptibility so that the two sets may occupy completely different parts of wavenumber space and make independent contributions to the susceptibility. In this connection, it should be emphasized that the zone shown in Fig. 7 and in Fig. 6, Chap. XIII, is not the first Brillouin zone but the fifth. Hence, it is very probable that there are regions in the outer zone other than those indicated in Fig. 7 in which the effective mass is very high. In support of this is the fact that the sign of χ fluctuates in the theoretical curves derived by Peierls (*cf.* Fig. 6) and Blackman (*cf.* Fig. 9) on the assumption of one type of electron, whereas the experimental susceptibility is always negative.

The susceptibility computed by Blackman for the absolute zero of temperature using (33) is shown in Fig. 9, which corresponds to the particular ratio $\alpha_1/\alpha_2 = 9.8$. The abscissa is the variable H/H_0, where H is the magnetic field intensity and H_0 is

$$H_0 = \frac{\epsilon_0'}{\beta(\alpha_1\alpha_2)^{\frac{1}{2}}}$$

in which ϵ_0' is the Fermi energy relative to the zero of (33). Blackman added a constant to his theoretical susceptibility in order to include Jones's type of diamagnetism and adjusted the parameters ϵ_0', α_1, α_2, and α_3 in order to fit the observed curves most reasonably. The results, which correspond to the parameters

$$\alpha_1 = 9.8, \qquad \alpha_2 = 1.0, \qquad \alpha_3 = 1.1 \cdot 10^3,$$
$$\epsilon_0' = 0.019 \text{ ev},$$

are shown in Fig. 8. The number of free electrons per atom in the part of the band responsible for the fluctuating susceptibility as computed from ϵ_0' is $1.2 \cdot 10^{-5}$, which is about 1 per cent of the number Jones found were responsible for the room-temperature susceptibility. As may be seen from Eq. (31), the large value of α_3 accounts for the absence of an observable fluctuating susceptibility in the z direction.

140. The Spin Paramagnetism of Valence Electrons*.—The origin of the paramagnetic behavior of many simple metals was first explained by Pauli in the elementary way described in Sec. 29. The value of the susceptibility obtained from this theory is

$$\chi = \frac{2\beta^2 g_s(\epsilon_0')}{V} \tag{1}$$

where β is the Bohr magneton, $g_s(\epsilon_0')$ is the density of energy levels of one spin at the top of the filled band, and V is the volume of the metal. Although this explanation, which involves the assumption that the two systems of energy states associated with opposite spins become displaced relative to one another in a field, is believed to be correct in principle, the simple computation requires modification for the following reasons:

a. The density function $g_s(\epsilon)$ is not necessarily the same as for free electrons.

b. The total energy of the solid cannot be expressed only as a function of one-electron energy terms, but also involves two-electron terms. Of these, the exchange and correlation energies are dependent upon the number of electrons having each kind of spin and affect the susceptibility in a way that cannot be included in the expression (1).

c. The orbital diamagnetism of inner closed shells and valence electrons, which was discussed in the previous sections, is comparable with the spin paramagnetism. These diamagnetic terms are so important in metals having newly filled d shells or nearly filled bands that they determine the sign of the susceptibility.

Let us assume that we have an electronic distribution, in which the first $(N + p)/2$ levels of electrons whose magnetic moments are parallel to the field are filled and in which the first $(N - p)/2$ levels of opposite spin are also filled. We shall designate the energy of the nth level from the bottom in the absence of a field by $\epsilon(n)$. When there is a field, the total energy in the Bloch-Hartree approximation is

$$E(p) = \int_0^{\frac{(N-p)}{2}} \epsilon_s(n)dn + \int_0^{\frac{(N+p)}{2}} \epsilon_s(n)dn - p\beta \mathsf{H}_z. \tag{2}$$

If this is expanded in terms of p, it is necessary to retain only the first-order terms, since we shall be interested in the case in which p/N is small. The result is

$$E(p) \cong E(0) + \frac{p^2}{4}\left(\frac{d\epsilon}{dp}\right)_{p=\frac{N}{2}} - p\beta \mathsf{H}_z. \tag{3}$$

We shall elevate and add to (3) the electronic-interaction terms for the case of free electrons.

It may be recalled that the exchange interaction energy arises from the interaction between electrons of parallel spin. When electrons of both spins are present in equal numbers, the exchange energy per electron is (*cf.* Sec. 75)

$$\epsilon_e = -\frac{3^{\frac{1}{3}}e^2}{2\pi^{\frac{1}{3}}}n_0^{\frac{1}{3}} \tag{4}$$

where n_0 is the total electron density. This may be expressed in terms of the radius r_s of the equivalent sphere and is then

$$-0.458\frac{e^2}{r_s}. \tag{5}$$

If there are $(N + p)/2$ electrons of one spin and $(N - p)/2$ of opposite spin, the total exchange energy is

$$E_e(p) = -\left[\frac{(N-p)}{2}\left(\frac{N-p}{N}\right)^{\frac{1}{3}} + \frac{(N+p)}{2}\left(\frac{N+p}{N}\right)^{\frac{1}{3}}\right]\frac{0.458e^2}{r_s}, \tag{6}$$

which becomes

$$E_e(p) = -N\frac{0.458e^2}{r_s} - \frac{2}{9}\frac{p^2}{N}\frac{0.458e^2}{r_s} \tag{7}$$

when expanded in terms of p.

We shall assume that the correlation energy arises only from electrons of opposite spin, for reasons discussed in Sec. 76. The correlation energy as a function of p then is

$$\begin{aligned}
E_c(p) &= -e^2\left\{\frac{(N-p)}{2}f\left(\left(\frac{N}{N+p}\right)^{\frac{1}{3}}r_s\right) + \frac{(N+p)}{2}f\left(\left(\frac{N}{N-p}\right)^{\frac{1}{3}}r_s\right)\right\} \\
&\cong -e^2\left\{Nf(r_s) + f'(r_s)\frac{5r_s}{9}\frac{p^2}{N} + \frac{1}{18}f''(r_s)r_s^2\frac{p^2}{N}\right\}
\end{aligned} \tag{8}$$

where

$$f = \frac{0.288}{r_s + 5.1a_h}.$$

The term in f' usually is at least ten times as large as the one in f''.

Combining Eqs. (3), (7), and (8), we obtain for the total energy

$$E_t(p) = E_t(0) + \alpha\frac{p^2}{N} - p\beta H_z \qquad (9)$$

where

$$\alpha = \frac{N}{4}\left(\frac{d\epsilon}{dp}\right)_{p=\frac{N}{2}} - \frac{2}{9}\frac{0.458}{r_s}e^2 - \frac{5}{9}f'(r_s)r_s e^2 - \frac{1}{18}f''(r_s)r_s^2 e^2. \qquad (10)$$

The expression (9) is a minimum for the value of p satisfying the equation

$$2\alpha p = N\beta H_z$$

or

$$p = N\frac{\beta H_z}{2\alpha}. \qquad (11)$$

If this is substituted in Eq. (9), we obtain

$$E_t(H_z) = E_t(0) - N\frac{\beta^2 H_z^2}{4\alpha}. \qquad (12)$$

Hence, the susceptibility is

$$\chi = n_0\frac{\beta^2}{2\alpha}. \qquad (13)$$

This reduces to Eq. (1) when the exchange and correlation terms are neglected, since

$$g_s(\epsilon') = \left(\frac{dp}{d\epsilon}\right)_{p=\frac{N}{2}}.$$

The terms in α are given in Table LXXXI for sodium and lithium. Although the first term is largest the others[1] are not negligible. Values of the total susceptibility given by Eq. (13) appear in the same table. In addition, values of the free-electron diamagnetism and the inner-shell diamagnetism are listed. The former were obtained by the use of Eq. (29), page 595, and the relation $\epsilon = h^2 k^2/2m^*$, the computed values of m^* being employed. The contribution to free-electron diamagnetism from exchange was not included since the corresponding term from correlations cannot be computed. The comparatively small ion-core terms were obtained from Van Vleck's book.

[1] Further details of this computation are to be published in the *Phys. Rev.*

The observed and calculated values of the susceptibility agree closely in the case of sodium, but the computed value for lithium is somewhat higher than the highest observed one. The agreement in the first case supports the general conclusion that the electrons in sodium are very nearly perfectly free, whereas the disagreement in the second case indicates that the relation $\epsilon = h^2\mathbf{k}^2/2m^*$ is not exact. The most reasonable source of the discrepancy would seem to be a comparatively small term in $\mathbf{k}^4$, which causes a lowering of the density of levels at the edge of the filled region and a corresponding decrease in paramagnetism.

TABLE LXXXI.—CONTRIBUTIONS TO α FROM THE TERMS IN EQ. (10) FOR LITHIUM AND SODIUM AND THE VALUE OF THE SUSCEPTIBILITIES GIVEN BY EQ. (13)
(The terms in the first row are expressed in electron volts; those in the second row in terms of 10^6 times the cgs unit of volume susceptibility.)

	$N(d\epsilon/dp)/4 = \epsilon_0'/3$	Exchange	Correlation	Total
Li	1.02	−0.86	0.19	0.35
Na	1.12	−0.70	0.19	0.61

	$\dfrac{n_0\beta^2}{2\alpha}$	Ion core	Diamagnetic	Total	Observed
Li	3.54	−0.05	−0.17	3.32	1.4–2.0
Na	1.11	−0.18	−0.23	0.70	0.63

It is interesting to note that the exchange energy is made more negative by increasing p [cf. Eq. (6)]. This shows that the exchange interaction of free electrons favors the alignment of spins. This tendency toward ferromagnetism ordinarily is more than compensated by the fact that both the Fermi energy and the correlation energy are raised when p increases. The change from paramagnetism to ferromagnetism can occur only when α becomes negative, for then Eq. (11) leads to a maximum rather than a minimum. Bloch[1] pointed out that the exchange term becomes larger than the Fermi term for sufficiently large values of r_s, since the first decreases as $1/r_s$ and the other as $1/r_s^2$. If we neglect the correlation terms, it follows from Eq. (10) that this occurs when

$$r_s > 6.03\frac{m}{m^*} a_h.$$

The limiting value of r_s for perfectly free electrons is about $6.0a_h$, which is larger than the value for any alkali metal. This fact and the fact

[1] F. BLOCH, Z. Physik, 57, 545 (1929).

that the correlation term does not favor spin alignment make it safe to say[1] that the observed ferromagnetism of transition metals should not be associated with nearly free electrons.

141. Paramagnetic Salts.—The theory of the properties of the para- magnetic rare earth and iron group element salts has been highly devel- oped in those cases in which the paramagnetic ions are so widely separated that the bands associated with their electrons are narrow and the atomic approximation is valid. We shall not discuss the details of this rather specialized topic here but refer the reader to other sources.[2]

142. Macroscopic and Local Field Corrections.—Suppose that the currents in an electromagnet are adjusted in such a way that the field at a given point in free space is H'. If the space is then occupied by a magnetic specimen, the orienting field that acts upon an atomic magnetic dipole is no longer H' because of the fields arising from the rest of the material. One part of the difference $H - H'$, namely, the demagnetiza- tion field, may be handled by classical methods. This contribution corresponds to the field of the effective surface distribution of magnetic charge that is induced on the specimen and is determined by the geomet- rical shape of the specimen. It usually varies from point to point, even when H' is constant; however, it is constant when the specimen has one of several possible shapes. In these cases, the correction takes the form

$$\Delta H_d = D M \tag{1}$$

where M is the intensity of magnetization and D is the *demagnetization constant*. D is -4π for a thin pillbox whose axis is parallel to the field and is $-4\pi/3$ for a sphere. Values for other cases have been listed by Stoner.[3] The correction (1) is negligible in substances having a small susceptibility since the fractional error made in neglecting it is of the order of χ.

Under certain conditions, it is convenient to discuss another type of correction field. Suppose, for example, that we are dealing with a dense paramagnetic gas of molecules having a molecular susceptibility χ_a. The magnetic moment per unit volume in this case is not simply

[1] *Cf.* E. Wigner, *Phys. Rev.*, **46**, 1002 (1934); *Trans. Faraday Soc.*, **34**, 678 (1938).

[2] See the following books and articles: Van Vleck, *op. cit.;* W. G. Penney and R. Schlapp, *Phys. Rev.*, **41**, 194 (1932), **42**, 666 (1932); W. G. Penney, *Phys. Rev.*, **43**, 485 (1935); A. Frank, *Phys. Rev.*, **39**, 119 (1932), **48**, 765 (1935); J. H. Van Vleck, *Phys. Rev.*, **41**, 208 (1932); O. M. Jordahl, *Phys. Rev.*, **45**, 87 (1934); F. H. Spedding, *Jour. Chem. Phys.*, **5**, 316 (1937); A. Siegert, *Physica*, **3**, 85 (1936), **4**, 138 (1937). A survey of the topic has recently been given by J. H. Van Vleck, *Reports of The Strassbourg Conference* (1939); to be published in *Ann. Inst. Henri Poincaré.*

[3] E. C. Stoner, *Magnetism and Atomic Structure* (E. P. Dutton & Co., Inc., New York, 1934).

$$\frac{N}{V}\chi_a \mathsf{H},\tag{2}$$

where H is the demagnetically corrected field, but is different because of the interaction between the molecular magnets. If H_l is the average local magnetic field acting on a given molecule, the mean magnetic moment per molecule is $\chi_a \mathsf{H}_l$ and the intensity of magnetization is

$$\mathsf{M} = n_0 \chi_a \mathsf{H}_l\tag{3}$$

where $n_0 = N/V$. In order to use this, we must know the relationship between H and H_l. Lorentz[1] was the first person to derive a relationship of this kind. He obtained the equation

$$\mathsf{H}_l = \mathsf{H} + \frac{4\pi}{3}\mathsf{M}\tag{4}$$

on the basis of the following assumptions:

a. The arrangement of molecules is either isotropic or cubic.

b. The relative orientation of magnetic moments is statistically the same for both near and distant molecules.

Assumption *b* is analogous to the assumption made in the Bragg-Williams theory of order-disorder (Sec. 124). If Eq. (4) is placed in Eq. (3), we find

$$\mathsf{M} = \frac{n_0\chi_a}{1 - (4\pi/3)n_0\chi_a}\mathsf{H}.\tag{5}$$

Thus, the susceptibility per unit volume is

$$\chi = n_0\frac{\chi_a}{1 - (4\pi/3)n_0\chi_a}.\tag{6}$$

It should be noted that, from the standpoint of electronic approximations, this and the following discussions of the local field implicitly assume that the magnetic units may be described in an atomic or molecular approximation. Thus, these discussions have significance only when the bands are narrow.

Lorentz's treatment is not completely satisfactory for the same reason that the Bragg-Williams theory of order is not satisfactory, namely, it does not take into account the fact that near neighbors are aligned more often than distant neighbors. We shall discuss two attempts that have been made to improve the theory.

Onsager[2] modified Lorentz's method of computing χ in the following way. For mathematical purposes, Lorentz had circumscribed an imaginary sphere about a given molecule and derived the term $4\pi\mathsf{M}/3$ in

---

[1] H. A. Lorentz, *The Theory of Electrons* (Teubner, Leipzig, 1906).

[2] L. Onsager, *Jour. Am. Chem. Soc.*, **58**, 1486 (1936).

Eq. (4) by considering the contributions to the field from the molecules inside and outside this sphere. In doing this, he assumed that the polarization is uniform. Onsager assumed that the sphere has physical reality as the volume within which the molecule is contained. In addition, he made the following two assumptions:

a. The polarization field in the magnetic medium outside the sphere is not uniform but is the same as the field about a hollow spherical cavity that contains a point dipole of magnitude **m**. One part of the polarization field M_H is induced by the constant applied field **H** and is fixed; the rest arises from the dipole and varies when the dipole changes its direction.

b. The field H_l inside the cavity, exclusive of the dipole field of the molecule, is the orienting field that acts upon a molecule. H_l is the sum of the external field **H** (with depolarization correction) and the field arising from the polarization outside the sphere. The second part of H_l is analogous to the $4\pi M/3$ term in the Lorentz equation (4).

In the case in which **m** is constant H_l may be determined as a simple solution of Laplace's equation and is

$$H_l = H + \frac{\mu - 1}{2\mu + 1}H + \frac{2(\mu - 1)}{(2\mu + 1)a^3}\mathbf{m} \tag{7}$$

where μ is the permeability and a is the radius of the spherical cavity. The second term, which arises from the polarization of the external medium induced by the field **H,** may be transformed to the form

$$\frac{4\pi\chi}{8\pi\chi + 3} \tag{8}$$

where χ is the macroscopic susceptibility $(1 - \mu)/4\pi$. This should be compared with the corresponding Lorentz term

$$\frac{4\pi\chi}{3}. \tag{9}$$

The third term in (7) is the reaction field of the dipole and is always parallel to **m**. It does not exert an orienting force so that it is unimportant when **m** is constant and may be dropped. Thus,

$$H_l = \left(1 + \frac{4\pi\chi}{8\pi\chi + 3}\right)H. \tag{10}$$

It should be noted that this approaches a limiting value $3H/2$ when the magnetic susceptibility becomes large, for the field then attempts to avoid the cavity. The same effect does not occur in Lorentz's approximation.

If this result is inserted into Eq. (3), the following implicit equation for χ is obtained:

$$\chi = \left(1 + \frac{4\pi\chi}{8\pi\chi + 3}\right)n_0\chi_a. \tag{11}$$

This equation has the solution

$$\chi = \frac{3}{16\pi}\left[-1 + 4\pi n_0\chi_a + \left(1 + \frac{8\pi}{3}n_0\chi_a + 16\pi^2 n_0^2\chi_a^2\right)^{\frac{1}{2}}\right]. \tag{12}$$

Equations (6) and (12) are identical with terms in $(n_0\chi_a)^2$ so that they do not give appreciably different results when $n_0\chi_a$ is much smaller than unity; however, they behave very differently in the region where $n_0\chi_a$ is near unity. Lorentz's expression (6) approaches infinity as $n_0 4\pi\chi_a/3$ approaches unity, whereas Onsager's expression remains finite, being equal to $3(1 + \sqrt{3})/8\pi$. This difference is very important when the atomic susceptibility satisfies the Curie law

$$\chi_a = \frac{C}{T}$$

for then χ_a becomes large at low temperatures. If Eq. (5) were correct, M would be finite even in the absence of a field when $n_0 4\pi\chi_a/3$ becomes unity. Since this effect implies ferromagnetism, Lorentz's theory implies that all substances obeying the Curie law should become ferromagnetic at sufficiently low temperatures. Onsager's equation, on the other hand, does not imply ferromagnetism since

$$\chi \sim \tfrac{3}{2}N\chi_a$$

for large values of χ_a. Simple calculations based on Lorentz's result show that the ferromagnetic Curie point should lie in the neighborhood of $0.1°K$ for most of the paramagnetic salts. Although several of these salts show ferromagnetic effects near this temperature, Van Vleck[1] believes that this ferromagnetism should be ascribed to the exchange coupling discussed in the next section. Hence, experimental evidence seems to support a modification of Lorentz's theory such as Onsager's.

Van Vleck has derived another relation that is valid at high temperatures and has a more rigorous foundation than either Lorentz's or Onsager's results. He included magnetic dipole-dipole interaction terms in the Hamiltonian function for a paramagnetic crystal of the type mentioned in the preceding section and computed the effect of these upon the partition function, using series expansion methods. If it is assumed that the atomic susceptibility χ_a satisfies the equation

$$\chi_a = \frac{\tau}{3T} \tag{13}$$

[1] J. H. Van Vleck, *Jour. Chem. Phys.*, **5**, 320 (1937).

where

$$\tau = \frac{g^2\beta^2 J(J+1)}{k},$$

Van Vleck's equation for χ may be written in the form

$$\chi = \frac{\tau}{3T}\left[1 + \frac{4\pi\tau}{3k}\frac{1}{T} + \left(\frac{4\pi\tau}{9k}\right)^2 \frac{1}{T^2} - \frac{\delta}{T^2} + \cdots \right] \tag{14}$$

where

$$\delta = \frac{Q\tau^2}{9}\left(1 + \frac{3}{8}\frac{1}{J(J+1)} \right). \tag{15}$$

Here, Q is an integer of the order of magnitude 10 that depends upon the crystal structure. If Onsager's expression for χ is expanded and a value of Q appropriate to his model is placed in Eq. (14), the two expressions are identical with the terms shown in (14). Lorentz's expression, on the other hand, does not give the same value for the terms that vary as $1/T^2$. Thus, Onsager's result is more accurate when T/τ is larger than unity.

Since Eq. (14) cannot be used at low temperatures, it is not possible to check the validity of Onsager's relation in this region by direct computation of the partition function.

Van Vleck has extended Eq. (14) for the case in which the levels of the paramagnetic ions are split by crystalline fields. This work will not be discussed further here.

One of the most direct supports of the Onsager-Van Vleck theory of local fields arises from its application to polar liquids and molecular solids.[1] These substances contain molecules having permanent dipole moments, so that the preceding theory can be taken over with little modification for a discussion of their electrical properties. Since the relative magnitudes of electrical polarizabilities are of the order of one thousand times larger than those of magnetic polarizabilities at corresponding temperatures, the temperature at which the form of the local field is important is much higher in the electrical case. If Lorentz's equation were valid, these substances should show the electrical analogue of ferromagnetism in cases in which intermolecular aligning forces other than the dipole-dipole force are relatively small. Actually, this effect is not observed when it would be expected. For example, it can be estimated that the electrical Curie point should occur at about 260°K in the case of HCl, whereas no anomalies are observed until 100°K, at which point molecular reorientation stops (*cf.* Sec. 125).

143. Ferromagnetism.—The theory of ferromagnetism has developed in stages starting from two different points, namely, from the atomic

[1] *Ibid.*, p. 556.

approximation and from the band approximation. The treatment that starts from the atomic approximation has value primarily for understanding the spin aligning forces in ferromagnetic media, whereas the band treatment, which was discussed in Sec. 101, has qualitative value for discussing the relation between the conduction electrons and the d-shell electrons. The first three parts of this section will be devoted to the atomic approximation and the fourth to the connection between this and the band approximation.

There is a close analogy between the atomic theory of ferromagnetism and the theory of order and disorder in alloys that was presented in Chap. XIV. In fact, the theory of ferromagnetism, which was developed first, was used as a pattern for the other. It will be seen that the magnetized state, like the ordered state of alloys, has a lower entropy than other possible states, so that it can occur only when it is favored by a low energy.

a. The Weiss Theory.—It was seen in the previous section that a relationship between the external magnetic field H and the local field H_l, of the type derived by Lorentz, namely,

$$H_l = H + \frac{4\pi}{3}M,\tag{1}$$

where M is the intensity of magnetization, can imply ferromagnetism if the atomic susceptibility χ_a becomes very large in a temperature range. Under this condition, the susceptibility χ, which is related to χ_a by the equation

$$\chi = n_0\frac{\chi_a}{1 - 4\pi n_0\chi_a/3}\tag{2}$$

where n_0 is the density of particles, becomes infinite when $4\pi n_0\chi_a/3$ is unity so that the magnetic moment per unit volume may be finite, even in the absence of a field. χ_a satisfies this condition at sufficiently low temperatures if it obeys Curie's law

$$\chi_a = \frac{A}{T}.\tag{3}$$

Hence, Eqs. (1) and (3) are sufficient for ferromagnetism. Even if the Lorentz equation were accurate, however, which it is not for the reasons discussed in the last section, it could not explain the ferromagnetism of iron, cobalt, and nickel, for the reasonable values of A are too small. Thus, if we use the relation

$$A = \frac{\beta^2 g^2 J(J+1)}{3k}\tag{4}$$

discussed in connection with the theory of paramagnetism and give J and g ordinary atomic values, we should expect ferromagnetism only below 1°K.

Weiss[1] arbitrarily dismissed these difficulties and assumed that in ferromagnetic material

$$\mathsf{H}_l = \mathsf{H} + \alpha\mathsf{M} \tag{5}$$

where α is a large constant of the order of magnitude 10^4. In addition, he assumed that the scalar value of atomic moment $\mathbf{m}$ is related to H_l by Eq. (25), Sec. 137, namely,

$$m(\mathsf{H}) = \beta g J B_J\!\left(\frac{\mathsf{H}_l\beta g}{kT}\right), \tag{6}$$

which is the generalization of Eq. (3) for strong fields. Equations (5) and (6) lead to the following implicit equation for M:

$$\mathsf{M}(\mathsf{H}) = n_0\beta g J B_J\!\left(\frac{\beta g(\mathsf{H} + \alpha\mathsf{M})}{kT}\right). \tag{7}$$

In his original work, Weiss actually used the classical analogue of the

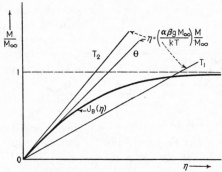

Fig. 10.—Schematic representation of a method of visualizing the roots of Eqs. (8a) and (8b). The lines correspond to (8b) for several temperatures, where $T_2 > T_1 > T_0$. The single curve represents (8a). It is assumed that H is zero, although the additive constant in (8b) usually cannot be shown on this scale anyway.

function B_J, which may be derived by allowing βg to approach zero and J to approach infinity in such a way that $\beta g J$ remains finite.

Equation (7) is equivalent to the two simultaneous equations

$$\frac{\mathsf{M}}{\mathsf{M}_\infty} = B_J(\eta), \tag{8a}$$

$$\eta = \frac{\beta g\mathsf{H}}{kT} + \left(\frac{\alpha\beta g\mathsf{M}_\infty}{kT}\right)\frac{\mathsf{M}}{\mathsf{M}_\infty}, \tag{8b}$$

[1] P. Weiss, *Jour. phys.*, **6**, 667 (1907). *Cf.* Stoner, *op. cit.*

where M_∞ is the saturation value of M. The roots of these equations, which may be pictured by the graphical method shown in Fig. 10, are shown in Fig. 11. The decrease of M/M_∞ with increasing temperature follows a continuous curve, so that the melting of ferromagnetism is a transition of the second kind in Weiss's theory. The Curie temperature, at which M vanishes, is

$$\Theta = \frac{\alpha\beta g M_\infty}{k}\frac{J+1}{3}. \tag{9}$$

Since M_∞ is of the order of magnitude of 1,000 gauss for the common

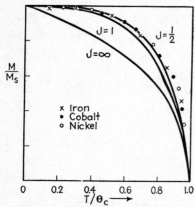

ferromagnetic metals, the value of Θ given by (9) is of the order of 0.1α. Thus, α must be of the order of magnitude 10^4, if Weiss's theory is to be adequate.

In Fig. 11, the observed values of M/M_∞ for iron, cobalt, and nickel as functions of T/Θ_c are compared with the computed functions for several values of J. It may be observed that the value $\frac{1}{2}$ fits the experimental work best, a fact suggesting that the magnetism arises almost entirely from spin. This is also supported by the fact that the gyromagnetic ratio is almost 2 (cf. Sec. 101).

FIG. 11.—Comparison of the observed saturation magnetization curves of iron, cobalt, and nickel and the computed curves for several values of J. The theoretical curve for $J = \frac{1}{2}$ fits the measured ones best.

The susceptibility above the Curie temperature may be found from the equation analogous to Eq. (2) of Lorentz's theory, namely,

$$\chi = n_0\frac{\chi_a}{1 - \alpha n_0\chi_a}. \tag{10}$$

If the value of $n_0\chi_a$ that may be derived from Eq. (6), namely,

$$n_0\chi_a = \frac{1}{\alpha}\frac{\Theta}{T}, \tag{11}$$

is substituted in Eq. (10), it is found that

$$\chi = \frac{\Theta}{\alpha(T - \Theta)}, \tag{12}$$

which is known as the Curie-Weiss law. The susceptibilities in the paramagnetic range have been investigated by Weiss[1] and coworkers. Figure 28, Chap. I, shows $1/\chi$ as a function of temperature for iron and nickel. If Eq. (12) were precisely valid, these curves should be straight lines; however, they are only approximately linear. In addition, the curve for iron shows a discontinuity because of the intrusion of the γ phase. Weiss and Foex have pointed out[2] that the curves for cobalt and nickel may be closely approximated by a series of straight lines in separate temperature regions. For this reason, it is suggested that the ferromagnetic metals have several magnetic allotropic phases above the Curie temperature and that a separate Curie-Weiss law is valid over the temperature range in which each phase is stable. A more reasonable interpretation is that the Curie-Weiss law is only a rough approximation to a more accurate equation. This is substantiated by more recent theoretical work which is discussed below.

It is possible to treat magnetocaloric effects on the basis of the Weiss theory.[3] Weiss postulated that the energy of magnetization E_m is related to the intensity of magnetization M by the equation

$$E_m = -\int_0^\mathsf{M} \mathsf{H}_l \cdot d\mathsf{M} \tag{13}$$

where H_l is the local field. It is implicitly assumed in this equation that the hypothetical local field H_l is an actual magnetic field. If we substitute the relation (5) in (13), we find

$$E_m = -\tfrac{1}{2}\alpha\mathsf{M}^2 \tag{14}$$

in the absence of an external field. Thus, the specific heat of magnetization is

$$c_m = -\frac{1}{2}\frac{\alpha}{\rho}\frac{d\mathsf{M}^2}{dT} \tag{15}$$

where ρ is the density. Since M varies most rapidly just below the Curie temperature and is zero above, the specific heat would rise to a peak at the Curie temperature and would then drop discontinuously to zero if Eq. (15) were valid. Although the areas under the experimental specific-heat curves are of the same order of magnitude as that of the theoretical curve, the forms of the two usually differ, inasmuch as a magnetic specific heat is observed above the Curie temperature. This fact is shown in Figs. 17 and 29 (Chap. I). The effect is largest in iron but is not negligible in the case of cobalt or nickel. It may be recalled that a similar discrepancy occurs between the specific heat predicted on

[1] P. Weiss, *Jour. phys.*, **5**, 129 (1924).

[2] *Ibid.*; G. Foex, *Ann. phys.*, **16**, 304 (1921).

[3] *Cf.* Stoner, *op. cit.*

the basis of the Bragg-Williams theory of order and disorder, and the observed specific heats. This discrepancy was removed in a qualitative way by taking into account short-distance order. We may conclude that a part of the error in the Weiss theory is related to the fact that it does not take into account the correlation of the magnetic moments of near-by atoms.

b. Heisenberg's Theory.—Heisenberg[1] first showed that the Weiss local field may be given a direct and simple explanation in the language of quantum theory. The principles involved in his work, which is based upon a Heitler-London approximation, may be demonstrated by the following simple problem.

Suppose that we have two atoms A and B that have one electron each[2] and are separated by a distance r_{ab}. We shall designate the atomic wave functions by ψ_a and ψ_b and the energies of the free atoms by ϵ. In addition, we shall assume that these states have no orbital angular momentum, so that all of the magnetic moment arises from spin. The possible antisymmetric wave functions of the complete system then are (*cf.* Secs. 48 and 56)

$$\Psi_I = [\psi_a(1)\psi_b(2) + \psi_a(2)\psi_b(1)][\eta_1(1)\eta_2(-1) - \eta_1(-1)\eta_2(1)],$$

$$\Psi_{II} = [\psi_a(1)\psi_b(2) - \psi_a(2)\psi_b(1)] \begin{cases} [\eta_1(1)\eta_2(1)] \\ [\eta_1(1)\eta_2(-1) + \eta_1(-1)\eta_2(1)], \\ [\eta_1(-1)\eta_2(-1)] \end{cases} \tag{16}$$

in an obvious notation. The first of these is the wave function of the singlet level, which has no spin moment, and the other three are the triplet functions, for which the spin quantum number S is unity. The second set of states evidently is the analogue of the set of ferromagnetic states of solids. We may assume that the interaction potential for the two atoms is

$$V_{ab} = \frac{e^2}{r_{ab}} + \frac{e^2}{r_{12}} - \frac{e^2}{r_{1b}} - \frac{e^2}{r_{2a}} \tag{17}$$

where r_{12} is the distance between the electrons, and r_{1b} and r_{2a} are the distances between a given nucleus and the electron on the other atom. The energies of the two types of state (16) are, respectively,

$$\begin{aligned} E_I &= E_c + J_e, \\ E_{II} &= E_c - J_e, \end{aligned} \tag{18}$$

where

$$E_c = 2\epsilon + \int |\psi_a(1)|^2 V_{ab} |\psi_b(2)|^2 d\tau_{12} \tag{19}$$

---

[1] W. Heisenberg, *Z. Physik,* **49,** 619 (1928).

[2] A treatment of this problem for the case in which each atom has more than one electron has been given by J. H. Van Vleck, *Jour. Chem. Phys.,* **6,** 105 (1938).

is the sum of the atomic energy and the coulomb interaction energy and

$$J_e = \int \psi_a{}^*(1)\psi_b{}^*(2) V_{ab}\psi_a(2)\psi_b(1)d\tau_{12} \tag{20}$$

is the exchange integral. It should be noted that V_{ab} in the integrals in Eqs. (19) and (20) could be replaced by e^2/r_{12} if the functions ψ_a and ψ_b were orthogonal. From Eq. (18), we see that the magnetic states Ψ_{II} are energetically stable relative to Ψ_I only if J is positive.

Equations (18) may be placed in a form that is significant for the theory of ferromagnetism. The square of the total spin operator

$$\Sigma^2 = (\mathbf{d}_1 + \mathbf{d}_2)^2 \tag{21}$$

is a constant of motion in each of the states Ψ_I and Ψ_{II}, the eigenvalues having the form $\hbar^2 S(S+1)$, where S is 0 and 1, respectively. If (21) is expanded, it becomes

$$\mathbf{d}_1^2 + \mathbf{d}_2^2 + 2\mathbf{d}_1 \cdot \mathbf{d}_2. \tag{22}$$

Since the individual spin angular momenta $\mathbf{d}_1^2$ and $\mathbf{d}_2^2$ are also constants of motion that have the eigenvalue $3\hbar^2/4$, it follows that $\mathbf{d}_1 \cdot \mathbf{d}_2$ is also a constant of motion and has the eigenvalue $-3\hbar^2/4$ when S is zero and $\hbar^2/4$ when S is unity. Employing the operator (21), we may place Eqs. (18) in the operator form

$$E = E_c + J_e\left(1 - \frac{\Sigma^2}{\hbar^2}\right) \tag{23}$$

or, using (22), in the form

$$E = E_c - \frac{J_e}{2} - J_e\frac{2\mathbf{d}_1 \cdot \mathbf{d}_2}{\hbar^2}. \tag{24}$$

If we now use the fact that the electronic magnetic moment $\mathbf{\mu}$ is $-2\beta\mathbf{d}/\hbar$, the spin-dependent part of (24) may be written as

$$-\frac{J_e}{2\beta^2}\mathbf{\mu}_1 \cdot \mathbf{\mu}_2. \tag{25}$$

Thus, apart from the dependence of J_e on interatomic spacing, the energy is determined by the relative orientations of the electron spins. It should be emphasized that this interaction energy is fundamentally electrostatic. Spin enters primarily as a consequence of the Pauli principle.

Bethe[1] has made a simple qualitative analysis of the conditions under which J is most likely to have a given sign. Let us suppose that the functions ψ_a and ψ_b have no nodes in the region where they overlap appreciably, so that the product $\psi_a(1)\psi_b(1)$ may be assumed to be positive

---

[1] H. BETHE, *Handbuch der Physik*, XXIV/2.

everywhere.   This condition is always satisfied if ψ_a and ψ_b are s functions that have nodes close to the nucleus but may also be satisfied in other cases if the nodal surfaces do not lie near the mid-point of the line connecting the centers of the two atoms.   Under this condition, the essentially positive terms

$$\frac{e^2}{r_{12}} + \frac{e^2}{r_{ab}} \tag{26}$$

favor ferromagnetism, whereas the negative terms

$$-\frac{e^2}{r_{b1}} - \frac{e^2}{r_{a2}} \tag{27}$$

do not.   The variable term e^2/r_{12} in (26) is larger when the product $\psi_a\psi_b$ is very large in a small volume of space than when the product is small in a large volume.   Moreover, the terms (27) are smallest when the overlapping region is as far from the nuclei as possible.   Hence, J is

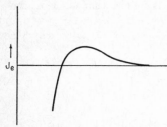

FIG. 12.—Behavior of J_e as a function of interatomic distance r.

most likely to be positive if (a) the distance r_{ab} is fairly large compared with the orbital radii and (b) the wave functions are comparatively small near the nuclei.   In both these cases, the product $\psi_a\psi_b$ is small at the nuclei and large near the mid-point between the atoms. Condition (b) is most fully satisfied when the orbital angular quantum number l is high since the wave functions start as r^l.   Hence, we should expect J to be positive for the interaction between unclosed shells of d or f electrons when the interatomic distance is large compared with the atomic radius. These conditions actually are satisfied by pairs of atoms in the metals of the iron group and rare earth type, in which the interatomic distances are determined primarily by the s-p valence electrons.   We shall see below that this qualitative argument can be applied to these metals, since the interaction between the d shells may be expressed as a sum of interactions between pairs of atoms.

It follows from the principles of the preceding discussion that the sign of J_e should depend upon the ratio x of the orbital radius and the interatomic distance in the manner shown in Fig. 12.   If x is close to unity, J_e should be negative; if it is large, J_e may be positive.   It is only fair to mention that Bethe's argument does not tell the entire story, for the diatomic molecules O_2 and NO, which have permanent magnetic moments, do not satisfy his conditions very well.

It may be shown[1] that in the Heitler London approximation the total energy of any number of electrons can be placed in the form

$$E = C - \frac{1}{2\hbar^2} \sum_{i,j}{}' J_{ij} 2\mathbf{d}_i \cdot \mathbf{d}_j, \tag{28}$$

which generalizes (24). Here C is a constant, J_{ij} is the exchange integral for the ith and jth electronic wave functions, and $\mathbf{d}_i$ and $\mathbf{d}_j$ are the spin operators of the ith and jth electrons. We shall be interested in the case in which each atom has one electron and in which J_e is appreciable only for nearest neighbors. Equation (12) then becomes

$$E = C - \frac{J_e}{\hbar^2} \sum_{\substack{\text{nearest} \\ \text{pairs}}} 2\mathbf{d}_i \cdot \mathbf{d}_j \tag{29}$$

where J_e is the exchange integral for neighboring atoms. The more general case in which there are several electrons per atom has also been considered, but we shall not treat it here since it does not lead to qualitatively different results.

Let us now discuss the number of states associated with different values of the z component of total magnetic moment. If there are N electrons, the largest value of the magnetic moment is βN, which occurs when all the spins are parallel and which can happen in only one way. The value $M\beta$, which occurs when there are $(N + M)/2$ moments parallel to the z axis and $(N - M)/2$ moments antiparallel to it, can happen in

$$n(M) = \frac{N!}{\left(\dfrac{N + M}{2}\right)! \left(\dfrac{N - M}{2}\right)!} \tag{30}$$

different ways. Thus, the state of maximum magnetic moment, which is energetically most stable when J_e is positive, has very low degeneracy, whereas the states of lower moment have larger values. Hence, we should expect the state of highest magnetization to occur only at low temperatures. The actual state at temperature T can be computed from the partition function; however, this computation is not easy to carry through directly because the $n(M)$ states (30) have different energies. Heisenberg assumed that the distribution of levels of given M may be approximated by the function

$$f_M(E) = \frac{n(M)}{(2\pi)^{\frac{1}{2}} \Delta_M} e^{-\frac{(E - \overline{E_M})^2}{\Delta_M{}^2}} \tag{31}$$

[1] *Cf.* J. H. Van Vleck, *The Theory of Electric and Magnetic Susceptibilities* (Oxford University Press, New York, 1932).

where $\overline{E_M}$ is the mean value of the energy levels of given M and Δ_M^2 is the mean square deviation from this mean. These quantities were computed by an approximate method that is discussed in Van Vleck's book. With this assumption, Heisenberg computed the partition function in a straightforward way,[1] after adding an energy term

$$2\beta \sum_i \mathfrak{d}_i \cdot \mathsf{H}$$

in order to include the effect of an external magnetic field. The magnetic equations obtained from this partition function are

$$\frac{\mathsf{M}}{\mathsf{M}_\infty} = \tanh \eta, \tag{32a}$$

in which

$$\eta = \frac{\beta \mathsf{H}}{kT} + \frac{1}{2}\left(\gamma - \frac{\gamma^2}{2}\right)\frac{\mathsf{M}}{\mathsf{M}_\infty} + \frac{\gamma^2}{4z}\left(\frac{\mathsf{M}}{\mathsf{M}_\infty}\right)^3 \tag{32b}$$

where z is the number of nearest neighboring atoms and

$$\gamma = \frac{zJ_e}{kT}.$$

Equations (32a) and (32b) are nearly the same as Weiss's equations (8a) and (8b) since $B_{\frac{1}{2}}(\eta)$ is equal to tanh η. The only difference lies in the term in η containing $(\mathsf{M}/\mathsf{M}_\infty)^3$. If this is dropped and the correspondence

$$\frac{\alpha\beta g}{kT} \sim \frac{1}{2}\left(\gamma - \frac{\gamma^2}{2}\right) \tag{33}$$

is made, the two systems of equations are identical. The field parameter α defined by (33) is a constant only at high temperatures, in which case

$$\alpha \cong \frac{zJ_e}{2\beta g}. \tag{34}$$

The Curie temperature is not related to this value of α by Eq. (9) but is given instead by the equation

$$\theta = \frac{2J}{k(1 - \sqrt{1 - 8/z})}, \tag{35}$$

which is real only if z is at least 8 and is positive only if J is positive. Thus, Heisenberg's treatment leads to conventional ferromagnetic behavior only for the more close-packed lattices. This means not that

---

[1] Another method of evaluating the partition function has been used by J. H. Van Vleck, *Phys. Rev.*, **49**, 232 (1936). This does not lead to qualitatively different results.

the crystal is not ferromagnetic at sufficiently low temperatures when z is less than 8 but only that the phase change is not of the second kind.[1]

Treatments of ferromagnetism using Heisenberg's model but employing distribution functions other than the Gaussian function (31) have been presented by several investigators.[2] By properly choosing this function, the difficulties associated with the imaginary behavior of (35) may be avoided.

The susceptibility above the Curie temperature does not conform to the Curie-Weiss law except when T is much greater than Θ. We shall not discuss the result since Heisenberg's model unquestionably is too simple to be applied quantitatively to actual ferromagnetic materials. It is important to know, however, that observed deviations from the law are not at variance with theory.

c. The Spin-Wave Treatment.—Heisenberg's treatment of ferromagnetism has the following weaknesses.

1. It is based upon a simple Heitler-London description in which the periodicity of the lattice is not taken into account.

2. An arbitrary approximation [*cf.* Eq. (31)] is used to obtain the distribution of levels. Since the thermal properties are strongly dependent upon this distribution, a more accurate description should be used for quantitative work.

Of the methods that have been employed to improve upon Heisenberg's work, we shall discuss that developed by Bloch[3] and extended more recently by Slater,[4] since it is the most fruitful. Although this treatment casts a new light upon the problem of ferromagnetism, its results are not radically different from those of Heisenberg's theory. For this reason, the older work can still be used for qualitative purposes.

It may be recalled that the Heitler-London approximation may be used to discuss the normal and lower excited states of insulators. When this is done, the lowest level is nondegenerate and the excited levels are very highly degenerate. Thus, if there are N atoms and the first excited one-electron state is g-fold degenerate, the first excited level is Ng-fold degenerate. A more accurate set of wave functions can be obtained by computing the matrix elements of the Hamiltonian connecting these Ng states and by diagonalizing the result. This problem, which was solved in Sec. 96 for a simple case in which the interatomic energy is

[1] This peculiar behavior of the Heisenberg model arises from the fact that the use of the Gaussian distribution is equivalent to assuming levels of arbitrarily low energy. Thus the $E(S)$ curve (Sec. 117) approaches the energy axis asymptotically with infinite slope, rather than with zero slope as it should.

[2] See, for example, F. Bitter, *Phys. Rev.*, **57**, 569 (1940).

[3] F. BLOCH, *Z. Physik*, **61**, 206 (1931).

[4] J. C. SLATER, *Phys. Rev.*, **52**, 198 (1937).

small, leads to the following results. The first excited wave functions Ψ_k are given by the equation

$$\Psi_k = a_k \sum_n e^{2\pi i k \cdot r(n)} \Psi_n \qquad (36)$$

where Ψ_n is the determinantal eigenfunction that is formed from the lowest wave function Ψ_0 by replacing the normal wave function ψ_n for the nth atom by the excited wave function ψ_n', and n is summed over all atoms. The Ψ_k, which evidently have wave characteristics, are called excitation waves. The energy associated with Ψ_k is

$$E_k = E_n + I \sum_\rho e^{2\pi i k \cdot \rho} \qquad (37)$$

where ϱ ranges over the vectors joining an atom with its nearest neighbors, E_n is the energy of the Ψ_n, and I is composed of integrals involving pairs of neighboring atoms.

Bloch constructed a set of magnetic wave functions that bear the same relation to Heisenberg's atomic functions that the excitation waves do to the Ψ_n in (36). Let us consider a system of N atoms, each of which has one valence electron. We shall assume that the one-electron wave functions $\psi(r - r(n)) = \psi_n$ are like atomic functions. For the basic nondegenerate wave function of the complete system, Bloch chose the state Φ_0 in which all electron spins are parallel. The energy E_0 of this state is

$$E_0 = N(\epsilon_0 + C - \tfrac{1}{2}J_e z) \qquad (38)$$

where ϵ_0 is the energy of a free atom, NC is the coulomb interaction energy of the system, J_e is the Heisenberg exchange integral (2) involving the ψ_n for pairs of neighboring atoms, and z is the number of nearest neighbors. The states Φ_n analogous to the Ψ_n in Eq. (1) are determinants of functions that differ from Φ_0 in that the spin of the electron on the nth atom has been reversed. These N functions have the same energy and have a z component of magnetic moment equal to $(N - 2)\beta$. The *spin waves* Φ_k, analogous to the excitation waves Ψ_k, are

$$\Phi_k = a_k \sum_n e^{2\pi i k \cdot r(n)} \Phi_n \qquad (39)$$

and have the energy

$$E_k{}^s = E_0 + 2J_e \sum_\rho (1 - e^{2\pi i k \cdot \rho}) \qquad (40)$$

where ϱ is summed over nearest neighbors as in Eq. (37). This function is shown schematically in Fig. 13.

It can be shown that, as long as the number of spin waves is small compared with N, the energy of the crystal in a state in which there are f spin waves of wave number $\mathbf{k}_1, \mathbf{k}_2, \ldots, \mathbf{k}_f$ is

$$E(\mathbf{k}_1, \cdots, \mathbf{k}_f) = E_0 + \sum_{\nu=1}^{f} \epsilon_s(\mathbf{k}_\nu) \tag{41}$$

where

$$\epsilon_s(\mathbf{k}_\nu) = E_{\mathbf{k}}{}^s - E_0. \tag{42}$$

Thus, the spin waves behave like elementary particles that are so nearly independent of one another that their energies are additive. It is evident that the z component of magnetic moment in the state having f spin waves is $\beta(N - 2f)$.

Under the restrictions for which Eq. (41) is valid, the necessary condition for ferromagnetism is that $E(\mathbf{k}) - E_0$ be positive, that is, that J_e be positive, for then Φ_0 is the lowest state of the system. This condition is identical with Heisenberg's.

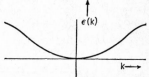

Fig. 13.—The schematic representation of the energy of Bloch's spin wave as a function of k in the ferromagnetic case. The curve is inverted in the nonferromagnetic one.

Bloch used Eq. (41) to compute the partition function for the system of electrons. There is no difficulty in determining the distribution of levels in this case, since one state is associated with each value of $\mathbf{k}$ in wave-number space. This partition function is

$$f = \sum_{f=0}^{\frac{N}{2}} e^{-\frac{E_0 - \beta H(N - 2f)}{kT}} \sum_{\mathbf{k}_1, \ldots, \mathbf{k}_f} \prod^{\nu=1, \cdots, f} e^{-\frac{\epsilon_s(\mathbf{k}_\nu)}{kT}}, \tag{43}$$

in which $-\beta H(N - 2f)$ is the field interaction term. At low temperatures, when only the lowest levels are excited, $\epsilon_s(\mathbf{k})$ may be replaced by the value

$$\epsilon_s(\mathbf{k}) \cong J_e \sum_{\rho} (\varrho \cdot \mathbf{k})^2. \tag{44}$$

Using this approximation, Bloch found that the magnetization M satisfies the equation

$$\frac{\mathsf{M}}{\mathsf{M}_\infty} = 1 - \alpha \frac{2.612}{2\pi^2} \left(\frac{kT}{J_e} \right)^{\frac{3}{2}} \tag{45}$$

where α depends upon the lattice and has the value $\frac{1}{4}$ for a face-centered lattice and $\frac{1}{2}$ for a body-centered lattice. This result may be placed in the form

$$\frac{\mathsf{M}}{\mathsf{M}_\infty} = 1 - \left(\frac{T}{\Theta}\right)^{\frac{3}{2}} \tag{46}$$

where Θ is the approximate Curie temperature, which for face-centered and body-centered lattices has the values

$$\Theta_{\text{f.c.}} = 9.7\frac{J_e}{k},$$

$$\Theta_{\text{b.c.}} = 6.1\frac{J_e}{k},$$

respectively.

Weiss[1] has made a careful experimental test of Bloch's $T^{\frac{3}{2}}$ law for iron and nickel. He found that a T^2 law holds above 70°K, but that the $T^{\frac{3}{2}}$ law applies in the range from 70° to 20°K. This verification of Bloch's result seems to be support for a spin-wave type of theory of ferromagnetism, although it must be admitted that Bloch's model, on which (46) was derived, is probably much too simple (see part d).

Slater[2] extended Bloch's method of determining the magnetic wave functions by carrying the perturbation procedure several steps further. The important differences between the two procedures are as follows:

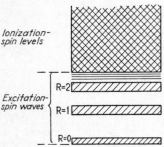

Ionization-spin levels

Excitation-spin waves

R=2

R=1

R=0

FIG. 14.—The energies of Slater's excitation-spin waves. The zero line represents the energy of the system when spins are parallel. The band $R = 0$ corresponds to Bloch's spin-wave curve (*cf.* Fig. 13) for which the electron having reversed spin remains on the same atom as the hole it leaves in the levels of opposite spin. The other discrete curves correspond to cases in which the electron having reversed spin is removed to an atom at distance R. The continuum represents the energy states of the electron and hole when they become completely free of one another.

a. Slater added to Bloch's spin waves (39) the wave functions $\Phi_{\mathbf{k},\mathbf{R}}$ that are defined by the equation

$$\Phi_{\mathbf{k},\mathbf{R}} = \sum_n e^{2\pi i \mathbf{k}\cdot\mathbf{r}(n)}\Phi_{n,\mathbf{R}} \tag{47}$$

where $\Phi_{n,\mathbf{R}}$ is constructed from Φ_0 by taking an electron from the atom at $\mathbf{r}(n)$ to the atom at $\mathbf{r}(n) + \mathbf{R}$ and reversing its spin. The functions (39) evidently are the special set for which $\mathbf{R}$ is zero. These excitation-spin waves have energies that can be represented schematically for each value of $\mathbf{R}$ by discrete curves in a one-dimensional diagram (*cf.* Fig. 14). The curve for $\mathbf{R} = 0$ corresponds to the curve of Fig. 13.

b. The matrix components of the Hamiltonian were computed for the system of excitation-spin waves (47), and the perturbing effect on

[1] P. WEISS, *Compt. Rend.*, **198**, 1893 (1934).
[2] *Ibid.*

the Bloch states of the states for which $\mathbf{R}$ is not zero was estimated from these components.

c. Slater did not employ ordinary atomic functions but used instead an orthogonalized system χ_n that was obtained from Bloch type functions $\chi_{\mathbf{k}}e^{2\pi i \mathbf{k}\cdot\mathbf{r}}$ of the ionization band approximation (cf. Chap. VIII) by means of the equation

$$\chi_n = \sum_{\mathbf{k}}\chi_{\mathbf{k}}(\mathbf{r})e^{2\pi i \mathbf{k}\cdot[\mathbf{r}-\mathbf{r}(n)]}. \tag{48}$$

This procedure has two important consequences: (1) All terms in the exchange integral (20) except those arising from e^2/r_{12} vanish because of the orthogonality conditions. (2) Some of the quantities in the expression for the perturbed function can be expressed in terms of characteristic quantities of the band approximation.

Slater's result for the energy of Bloch's spin waves, in the higher approximation, is

$$\epsilon_s(\mathbf{k}) = A\sum_{\rho}(1 - e^{2\pi i \mathbf{k}\cdot\rho}). \tag{49}$$

Here,

$$A = J_s - \frac{2W^2}{I_1} \tag{50}$$

where J_s is the exchange integral for the χ_n, namely,

$$J_s = \int \chi_n{}^*(1)\chi_{n+1}{}^*(2)\frac{e^2}{r_{12}}\chi_n(2)\chi_{n+1}(1)d\tau_{12}, \tag{51}$$

W is the width of the ionization band, and I_1 is essentially the difference between E_0 and the center of energy of the excitation-spin waves for $\mathbf{R}$ greater than zero (cf. Fig. 14). In the case in which there are f spin waves, the total energy $E(\mathbf{k}_1, \ldots, \mathbf{k}_f)$ may be obtained by substituting (49) into Eq. (41). When the atoms are widely separated, J_s is positive, W is very small, and A is then positive. On the other hand, W becomes very large when the atoms are close together, so that we may expect A to change its sign. Thus, in this approximation we should expect ferromagnetism only for widely separated atoms just as in the Heisenberg theory.

The behavior of the lowest spin-wave energy curve as the ionization band widens is shown in Fig. 15a and b. The levels of the ionization band occur at the series limit of the discrete curves of the excitation-spin-wave system and are indicated by the striped region. In the first case, the ionization band is narrow and the energy curve of the lowest spin wave is above E_0. In the second, the ionization band is so wide that it depresses the spin-wave curve below E_0.

d. Critique of the Theories of Ferromagnetism.—The Heisenberg and
spin-wave treatments of ferromagnetism are incomplete for metals
because the valence and *d*-shell electrons are not considered simultaneously. Actual ferromagnetic metals are made of atoms whose
total electronic states possess a mixture of *s-p* and *d*-electron characteristics. The *s-p* part of the atomic wave function is altered in such
a way that the solid has metallic properties, such as high conductivity,
and the *d* part produces ferromagnetism. Since the *g*-factor is ∼2 and
the observed atomic saturation moments are not integer multiples of
the Bohr magneton, we know that there is not an integer amount of

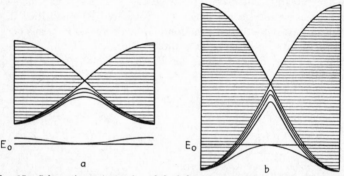

Fig. 15.—Schematic representation of the behavior of the spin-excitation levels as the
ionization band widens. In case *a* the ionization band is narrow and the system is ferromagnetic, whereas in case *b* the spin-wave curve is inverted so that the system is not
ferromagnetic. In both of these diagrams the ordinate is energy and the abscissa is the
difference of the wave-number vectors of the electron before and after excitation, which is
the wave-number vector of the exciton. Hence the lowest discrete curve is analogous to
that of Fig. 13. The shaded region represents the ionization-spin states, in which both
the electron and hole are free. This has zero width at the midpoint, which corresponds
to zero difference in wave number, because the $\epsilon(\mathbf{k})$ curve is the same for electrons of either
spin. (*After Slater.*)

d property per atom. Thus, the two characteristics are intimately
mixed, and in any complete theory of ferromagnetism they should be
discussed simultaneously.

The band approximation, which is based on Bloch type one-electron
functions, does an excellent job of picturing the balance between *s-p-* and
d-electron character. Since it is found that the *d* and *s-p* bands in
transition metals overlap, the relative number of electrons in each band
is determined by the condition that the energy be stationary under the
process of moving electrons from one band to the other, which means
essentially that the bands are filled to the same level. Thus, there is no
reason for expecting an integer number of *d* electrons per atom, for the
positions of the bands are determined by many factors. As we have
seen in Sec. 101, this picture can be used to correlate a large number

of the properties of transition metals that are related to d-electron character.

In order to explain ferromagnetism on the band theory, it unfortunately is necessary to assume arbitrarily that there are more electrons in the d band having one type of spin than the other. The excess in metals such as nickel and cobalt is so large that half the d band is completely filled, and the excess is slightly smaller in iron. If the band theory were accurate enough in the case of narrow bands to furnish a trustworthy explanation of this preponderance of electrons of one type of spin, the Heisenberg-Bloch-Slater type of treatment would be superfluous for most descriptive work. It is true that the exchange energy for Bloch functions, $\chi_k e^{2\pi i \mathbf{k} \cdot \mathbf{r}}$, favors ferromagnetism, but it is possible to show that for the narrow bands the correlation correction is just large enough to compensate for this effect in first approximation (*cf.* Sec. 75).

At first sight, it might seem possible to use the band scheme to determine the distribution of s-p and d electrons and to use the spin-wave scheme to handle the d electrons. This procedure cannot be carried out in a simple way, for wave functions that are more complicated than (47) would have to be employed, since there is not an integer number of d electrons per atom.

Thus, there does not seem to be a single, tractable, approximational scheme that can be used to develop satisfactory equations for all the properties of ferromagnetic metals. At present, we must use the spin-wave and band schemes in the separate domains in which they are individually most satisfactory.

It should be added that the Heisenberg and the spin-wave approximations are suited to discussions of ferromagnetism in ionic solids, such as magnetite and the rare earth salts, in which there is an integer number of magnetic electrons per atom.

144. Additional Application to Alloys.—We saw in Sec. 101 that many of the properties of ferromagnetic metals and alloys may be correlated on the basis of the band scheme. It is also possible to correlate other properties by the use of the Heisenberg type of theory in a way that will now be discussed.

Dehlinger[1] has attempted to construct semiquantitative exchange integral curves $J_e(r)$ of the type shown in Fig. 12 for the transition-metal atoms by the use of empirical information. Since the close-packed phases of both nickel and cobalt are ferromagnetic, it may be concluded that J_e is positive at the observed nearest-neighbor distance and that the complete curves have the forms shown in Fig. 16e in which the vertical dotted line represents the nearest-neighbor distance for the

---

[1] U. Dehlinger, *Z. Metallkunde,* **28,** 116 (1936); **28,** 194 (1936); **92,** 388 (1937).

close-packed structures. The face-centered, or γ, phase of iron is not ferromagnetic; however, the body-centered, or α, phase, in which the interatomic distance is slightly larger, is ferromagnetic. Hence, the $J_e(r)$ curve for iron (Fig. 16d) is negative at the dotted line and crosses the axis at larger values of r.

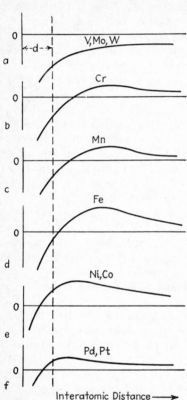

FIG. 16.—Hypothetical $J_e(r)$ curves for the magnetically important transition-metal atoms. Evidence discussed in the text indicates that the peak for nickel actually occurs to the left of the equilibrium spacing. (*After Dehlinger.*)

Dehlinger attempted to classify the $J_e(r)$ curves for the nonferromagnetic transition metals by studying their paramagnetism. He concluded that the corresponding J_e curve is nearly zero at the nearest-neighbor distance if the metals are strongly paramagnetic and if the paramagnetism increases with decreasing temperature. On the other hand, the crossing point is far away if the paramagnetism is weak or temperature-independent. It is possible, in the second case, that $J_e(r)$ is negative everywhere. Using considerations of this type, he arrived at the other curves of Fig. 16. It may be seen that in the cases of palladium and platinum he has concluded that the exchange integral is positive at the actual interatomic distances although the magnitude is small relative to that for the truly ferromagnetic metals.

Among the properties of ferromagnetic metals that are nicely explained in a qualitative way by Dehlinger's picture is the fact that their expansion coefficients change near the Curie point. Consider the case of iron, for example. It follows from the fact that the observed interatomic distance is on the left-hand side of the peaks of the $J_e(r)$ curves in this case that the additional interatomic force arising from exchange when the magnetic moments of two atoms are parallel instead of antiparallel tends to push them apart. The total force for the entire solid arising from this source is a maximum when all spins are parallel and decreases as the magnetization decreases, such as when the substance is heated, for then an increasing number of atoms have opposite spin. Since the maximum change in magnetization occurs in the neighborhood of the Curie point, it may be expected that the greatest decrease in the interatomic repulsive force arising from ferromagnetism occurs in this region of temperature. This decrease, however, should compensate for at least a part of the internal pressure that causes the solid to expand when heated. Hence, it may be expected that the expansion coefficient would decrease near the Curie point. In some cases, such as in invar steel, which is an iron-nickel-carbon alloy, the two effects almost compensate for a range of temperature, and the expansion coefficient is nearly zero.

Shockley[1] has pointed out that the expansion coefficient of nickel *increases* near its Curie point, showing that in this case the actual interatomic spacing is to the *right* of the peak of the $J_e(r)$ curve, and is not as is shown in Fig. 16d. A similar conclusion has been drawn by Bozorth[2] from the fact that the Curie point of the iron-nickel system passes through a maximum as nickel is added to iron. It presumably is safe to conclude that the atomic spacing in cobalt, which lies between iron and nickel, corresponds to a point near the peak of the $J_e(r)$ curve.

TABLE LXXXII.—THE SIGNS OF THE EXCHANGE INTEGRALS FOR NEAREST NEIGHBORING ATOMS AND FOR FARTHER NEIGHBORS IN CLOSE-PACKED PHASES OF A NUMBER OF BINARY ALLOYS

	Pt	Pd	Ni	Co	Fe	Mn	Cr	Mo	W
Pt	++								
Pd	++	++							
Ni	....	++	++						
Co	++	++	++	++					
Fe	++	....	++	++	−+				
Mn	....	++	++	−+	−+	−+			
Cr	++	....	−+	−+	−+	....	−−		
Mo	....	....	−−	−−	−−	....	....	−−	
W	....	....	−−	−−	−−	....	....	−−	−−

Dehlinger extended this type of semiempirical work to substitutional alloys and predicted the rudimentary properties of the $J_e(r)$ curves for a number of unlike atoms. The results of this investigation, which is discussed in more detail below, are listed in Table LXXXII. The first of the two signs in a given square represents the sign of $J_e(r)$ for nearest neighbors in the binary alloy formed of the atoms associated with the row and column in which the square is situated. The other sign is the sign of $J_e(r)$ for all farther neighbors. Thus, according to this diagram, J_e is

[1] W. SHOCKLEY, *Tech. Pub. Bell Tel. System*, **18**, 645 (1939).
[2] R. M. BOZORTH, *Tech. Pub. Bell Tel. System*, **19**, 1 (1940).

positive for iron and nickel atoms that are separated by the nearest-neighbor distances of the iron-nickel alloys and is also positive for all larger distances. Similarly, J_e is negative for first neighbors in nickel and chromium atoms and is positive for others in the nickel-chromium alloy system. The data in the squares lying in the principal diagonal apply to the interaction of pairs of atoms of the same kind and express nothing that is not contained in the curves of Fig. 16. If the conclusions contained in this table are correct, we should expect an alloy to be ferromagnetic only if it contains platinum, palladium, nickel, cobalt, iron, manganese, or chromium.

The method of deriving this information may be demonstrated by giving several examples. It is found that the saturation moment of nickel is raised when nickel atoms are replaced by iron. Thus, it may be concluded that the exchange integral is positive, for otherwise the magnetic moment of iron would set itself antiparallel to that of the nickel atoms and the magnetization would decrease. By assumption, the $J_e(r)$ curve has the form of Fig. 12; hence, $J_e(r)$ must be positive for larger distances. If small amounts of tungsten or chromium are added to nickel, the saturation moment is decreased, a fact indicating that $J_e(r)$ is negative at least for the nearest neighboring nickel-tungsten and nickel-chromium atoms in the corresponding alloys. These two cases differ, however, inasmuch as the Curie temperature increases rapidly with increasing tungsten content in the nickel-tungsten system and remains practically constant in the nickel-chromium system. The reason the Curie temperature does not fall is not difficult to understand. In both these cases, the nickel atoms immediately surrounding tungsten or chromium atoms have parallel moments at absolute zero of temperature. In order to reverse its moment, one of these nickel atoms must do enough work to overcome not only the nickel-nickel exchange interaction but also the nickel-tungsten or nickel-chromium interaction. Hence, if the interaction for antiparallel moments is more than the nickel-nickel interaction for parallel moments, the Curie temperature should rise when tungsten is added, as is observed. It may be concluded from the behavior of the Curie temperature that the nickel-chromium exchange energy is less negative than the nickel-tungsten energy. Dehlinger also concludes from the differences of the two cases that the nickel-tungsten exchange energy is negative for both nearest and more distant neighbors, whereas the chromium-nickel interaction is negative for first neighbors and positive for others.

Dehlinger has used the results of this scheme to correlate a number of interesting and important properties of ternary ferromagnetic alloys such as the Heusler alloys.

145. Ferromagnetic Anisotropy.—A semiquantitative theory of the magnetic anisotropy of cubic ferromagnetic substances has been developed by Van Vleck using an atomic model.[1] This anisotropy, which is made evident by the fact that there are easy and hard directions of magnetization in cubic metals (see Sec. 2), cannot be explained on the basis of exchange coupling between the spins of electrons on different atoms if there is one or less than one magnetic electron per atom, as in the case of nickel, for it may be shown[2] that this type of interaction always leads to isotropic expressions for the energy as a function of magnetization direction. Van Vleck suggested that the anisotropy is due to a coupling between spin and orbital angular momentum not unlike that which gives rise to the inner multiplet splitting in Russell-Saunders coupling. This coupling would not lead to anisotropy if the electronic distribution in the d shells were isotropic, as in an S state of a perfectly free ion; however, the d-shell wave functions are appreciably distorted because of crystalline binding, as we have seen in Sec. 99, which means that the d-shell distribution is anisotropic. Since this anisotropy is fixed relative to the crystal axes, the electronic spin becomes conscious of its orientation relative to the crystal through the coupling with the orbital motion.

Van Vleck assumed that the Hamiltonian for a ferromagnetic solid contains magnetic terms of the type

$$-\sum_{i,j}J_{ij}\mathbf{d}_i \cdot \mathbf{d}_j + A\sum_i \mathbf{m}_i \cdot \mathbf{d}_j + \tfrac{1}{2}\sum_{i,j}f_{ij}(\mathbf{r}_{ij}, \mathbf{m}_i \cdot \mathbf{m}_j, \mathbf{m}_i \cdot \mathbf{r}_{ij}, \mathbf{m}_j \cdot \mathbf{r}_{ij}), \quad (1)$$

in which $\mathbf{d}_i$ and $\mathbf{m}_i$ are the spin and orbital angular momentum operators of the electrons on the ith atom, $\mathbf{r}_{ij}$ is the radius vector connecting the ith and jth atoms, J_{ij} is the exchange integral for the two atoms, A is the spin-orbit coupling constant, f_{ij} is a polynomial expression in the arguments indicated, and the sums extend over all atoms. The first term evidently is the Heisenberg exchange term, which is responsible for ferromagnetism. The second term describes the coupling between spin and orbital motion, whereas the third term leads to an anisotropic electronic distribution. Van Vleck showed that the observed magnitude for nickel of the constant K_1 in Eq. (2), Sec. 2, may be explained by use of the energy terms (1) with theoretically reasonable values of A and f_{ij}.

The saturation magnetic moment in iron and cobalt is larger than one Bohr magneton per atom, and in these cases, it is possible to explain the magnetic anisotropy by using only the first term of (1); however, it

[1] J. H. Van Vleck, *Phys. Rev.*, **52**, 1178 (1937).

[2] R. Becker, *Z. Physik*, **62**, 253 (1930); see also *ibid.* reference 36.

is likely that the higher terms also are important in determining the details of the anisotropy in these cases as well.

Brooks[1] has recently shown that this topic may also be approached on the basis of the band approximation.

The problem of magnetostriction is closely connected with the problem of anisotropy; however, we shall not discuss it here.[2]

[1] H. BROOKS, *Phys. Rev.*, **57**, 570 (1940).
[2] See J. H. Van Vleck, *Phys. Rev.*, **52**, 1178 (1937).

CHAPTER XVII

THE OPTICAL PROPERTIES OF SOLIDS

146. Introduction.—The classical theory of the optical properties of solids is based upon Maxwell's equations for an uncharged polarizable medium, namely,

$$\text{div } (\mathsf{E} + 4\pi\mathsf{P}) = 0, \qquad \text{div } (\mathsf{H} + 4\pi\mathsf{M}) = 0,$$

$$\text{curl } \mathsf{E} = -\frac{1}{c}\frac{\partial(\mathsf{H} + 4\pi\mathsf{M})}{\partial t}, \qquad \text{curl } \mathsf{H} = \frac{1}{c}\frac{\partial\mathsf{E}}{\partial t} + \frac{4\pi}{c}\frac{\partial\mathsf{P}}{\partial t} + \frac{4\pi\mathsf{J}}{c}, \quad (1)$$

where P and M are the electric and magnetic polarization intensities and J is the current per unit area. We shall be interested only in the case in which M is small enough to be dropped. In practically all applications of these equations, it is found possible to assume that P and J are related to E by the equations

$$\begin{aligned}\mathsf{P} &= \boldsymbol{\alpha} \cdot \mathsf{E}, \\ \mathsf{J} &= \boldsymbol{\sigma} \cdot \mathsf{E},\end{aligned} \qquad (2)$$

where $\boldsymbol{\alpha}$ and $\boldsymbol{\sigma}$ are the polarizability and conductivity tensors of the system. Maxwell's theory does not give an explanation of the dependence of $\boldsymbol{\alpha}$ and $\boldsymbol{\sigma}$ upon frequency; the derivation of these relationships is the purpose of the atomic theory of solids.

Maxwell's theory of radiation is subject to direct experimental test whenever $\boldsymbol{\alpha}$ and $\boldsymbol{\sigma}$ may be measured without performing an optical experiment. Unfortunately, this includes only the long wave-length region of the spectrum that is employed in radio work. For shorter waves ($\lambda \ll 1$ cm), the results of Maxwell's theory must be employed to determine the constants, in lieu of values determined by use of atomic theory. Although this may seem to be only an experimental difficulty, it should be realized that the size of the electrical circuit which can resonate to radiation having frequency 10^{14} cycles per second is of the order of atomic dimensions. For this reason, it is necessary to have intimate knowledge of the theory of atomic systems before the results of optical experiments can be interpreted in a way that throws light upon the behavior of the charges in solids.

The classical theory of $\boldsymbol{\alpha}$ and $\boldsymbol{\sigma}$ was developed farthest by Lorentz[1] although important contributions have been made by other workers.[2]

---

[1] H. A. LORENTZ, *Theory of Electrons* (Teubner, Leipzig, 1906).

[2] P. DRUDE (see *ibid.*); C. ZENER, *Nature*, **132**, 968 (1933); R. DE L. KRONIG, *Nature*, **133**, 211 (1934).

This work is still very useful since some of its most significant results have not been modified. The quantum treatment of the optical properties is in principle just one of the fields of application of the theory of radiation developed in Chap. V. From a purely formal standpoint, it should only be necessary to apply Dirac's theory in order to determine the optical behavior of any solid. This formal procedure actually has not been followed very closely, however, and the subject has developed unevenly. Individual contributions have been made in order to obtain reasonably correct equations in a simple way rather than to obtain a self-consistent description of all properties. The reason for this procedure is, of course, that the rigorous theory is difficult to apply.

147. Classical Theory.—We shall discuss the solutions of Maxwell's equations (1) for an isotropic or cubic medium in which the electrical polarizability and the conductivity are constants instead of tensors. The equations then are

$$\text{div } \mathbf{E} = 0, \qquad \text{div } \mathbf{H} = 0,$$
$$\text{curl } \mathbf{E} = -\frac{1}{c}\frac{\partial \mathbf{H}}{\partial t}, \qquad \text{curl } \mathbf{H} = \frac{\epsilon}{c}\frac{\partial \mathbf{E}}{\partial t} + \frac{4\pi\sigma\mathbf{E}}{c}, \tag{1}$$

where

$$\epsilon = 1 + 4\pi\alpha \tag{2}$$

is the dielectric constant. There is one important point that should be kept in mind for future reference. The quantity

$$\alpha\frac{\partial \mathbf{E}}{\partial t}, \tag{3}$$

which appears in the fourth of Eqs. (1) when ϵ is replaced by use of (2), has the nature of a current—the polarization current. Maxwell believed that this current could be distinguished from the current $\mathbf{J}$ by the fact that the latter arises from the motion of obvious charge, such as that in conductors, whereas the former arises from hidden charge. In adopting an atomic viewpoint, we are no longer able to distinguish between the two types unambiguously; hence, we must be careful not to include the same current twice.

Since we are interested in periodic solutions of Eqs. (1), we shall employ complex values of $\mathbf{E}$ and $\mathbf{H}$ of the form

$$\mathbf{E}(x,y,z,t) = \mathbf{E}'(x,y,z)e^{2\pi i\nu t},$$
$$\mathbf{H}(x,y,z,t) = \mathbf{H}'(x,y,z)e^{2\pi i\nu t}.$$

Only the real parts of these functions will be regarded as physically interesting. The phases of the true current $\mathbf{J}$ and the polarization current $\alpha\partial\mathbf{E}/\partial t$ evidently differ by 90 deg when α and σ are real. For this reason, we may, if we choose, eliminate the term in $\mathbf{J}$ in the fourth of Maxwell's equations by replacing α with the complex polarizability α_c,

$$\alpha_c = \alpha + \frac{\sigma}{2\pi i \nu}, \tag{4}$$

or we may eliminate the term $\alpha \partial \mathbf{E}/\partial t$ by replacing σ by the complex conductivity

$$\sigma_c = \sigma + 2\pi i \nu \alpha. \tag{5}$$

The current defined by the equation

$$\mathbf{J} = \sigma \mathbf{E}$$

where σ is real is always in phase with the electrostatic field; hence, it constantly takes energy from the field. The mean power per unit volume that is lost in this way is

$$P = \overline{\mathbf{J} \cdot \mathbf{E}} = \sigma \overline{\mathbf{E}^2} \tag{6}$$

where the line indicates the average value. This relation may be used to show that the absorption is proportional to σ. Now the absorption coefficient η is defined by the equation

$$\frac{\partial \overline{W}}{\partial x} = -\eta \overline{W}$$

where $\overline{W}$ is the mean energy density and $\partial \overline{W}/\partial x$ is the decrease due to absorption alone. Since

$$\overline{W} = \frac{\overline{\mathbf{E}^2}}{4\pi}$$

and

$$P_l = \frac{\overline{W}}{t} = -c \frac{\partial \overline{W}}{\partial x} \tag{7}$$

for a unidirectional wave, it follows that

$$\eta = \frac{4\pi\sigma}{c}.$$

The polarization current, on the other hand, is 90 deg out of phase with $\mathbf{E}$ and does not remove energy from the field.

The plane-wave solutions of Eqs. (1) in which we are interested may be taken in the form

$$\begin{aligned}
\mathbf{E} &= \mathbf{E}_0 e^{2\pi i \nu \left(t - \frac{N}{c} \mathbf{\eta}_0 \cdot \mathbf{r} \right)} \\
\mathbf{H} &= \mathbf{H}_0 e^{2\pi i \nu \left(t - \frac{N}{c} \mathbf{\eta}_0 \cdot \mathbf{r} \right)}
\end{aligned} \tag{8}$$

where E_0 and H_0 are constant vectors, n_0 is a unit vector in the direction of propagation of the wave, and N is the complex index of refraction, which may be written

$$N = n - ik, \tag{9}$$

in which n is the ordinary index of refraction and k is the extinction coefficient.

The imaginary part of N evidently measures the damping of the wave. The equations connecting the quantities in (8), which may be derived by substituting for E and H in (1), are

$$n_0 \cdot E_0 = 0, \qquad n_0 \cdot H_0 = 0,$$
$$N n_0 \times E_0 = H_0, \qquad N n_0 \times H_0 = -\left(\epsilon + \frac{4\pi\sigma}{2\pi i\nu}\right)E_0. \tag{10}$$

These relations show that E_0 and H_0 are orthogonal to n_0 and to one another. For simplicity, we shall take n_0, E_0, and H_0 to lie in the x, y, and z directions, respectively. The last two equations then become

$$N E_y = H_z, \qquad N H_z = \left(\epsilon + \frac{2\sigma}{i\nu}\right)E_y. \tag{11}$$

Hence,

$$N^2 = \left(\epsilon + \frac{2\sigma}{i\nu}\right), \tag{12}$$

or

$$\left.\begin{aligned} n^2 - k^2 &= \epsilon, \\ nk &= \frac{\sigma}{\nu}. \end{aligned}\right\} \tag{13}$$

The phase angle between E and H is $\arctan k/n$, which may be expressed in terms of ϵ and σ by solving (13).

If σ is zero and ϵ is positive, it follows that

$$k = 0, \qquad n = \sqrt{\epsilon}.$$

Under these conditions, the wave is undamped, and E and H are in phase. This case evidently corresponds to the propagation of light through a perfectly transparent medium having index of refraction n. On the other hand, if ϵ is negative,

$$n = 0, \qquad k = \sqrt{-\epsilon},$$

and the wave is damped in the direction of propagation. This damping is not accompanied by absorption, however, for P in Eq. (7) vanishes if

σ is zero. These conditions evidently can be satisfied only if the medium is totally reflecting, which is shown explicitly by the extension of the theory of light that takes into account the behavior of a plane wave that is incident upon a plane surface of a medium.[1] According to this work, the reflection coefficient for normal incidence is

$$R = \frac{(n - 1)^2 + k^2}{(n + 1)^2 + k^2},\tag{14}$$

which is unity when n is zero.

If σ is not zero, the solutions of Eqs. (13) are

$$\left.\begin{aligned}n^2 &= \frac{\epsilon \pm \sqrt{\epsilon^2 + 4(\sigma/\nu)^2}}{2}, \\ k^2 &= \frac{-\epsilon \pm \sqrt{\epsilon^2 + 4(\sigma/\nu)^2}}{2}.\end{aligned}\right\}\tag{13a}$$

Thus, the medium is neither perfectly transparent nor perfectly reflecting. We shall discuss two cases of this type that are of particular interest.

a. Lorentz Treatment of Absorption and Dispersion in Insulators.—As early as 1880, Lorentz[2] showed that it is possible by use of a simple atomic model to account for the dispersive behavior of insulators near an absorption line. He postulated that insulating materials contain electrons that are bound to equilibrium positions by Hooke's law forces. We shall assume that these forces are isotropic and that the electrons are subject to a damping force proportional to the velocity. The equations of motion of an electron that is subject to a periodic electrostatic field directed along the y axis then is

$$m\frac{d^2y}{dt^2} + 2\pi m\gamma\frac{dy}{dt} + \kappa y = -e\mathsf{E}_0 e^{2\pi i\nu t}\tag{15}$$

where y is the displacement of the electron along the y axis, $2\pi m\gamma$ is the damping constant, κ is Hooke's constant, and E_0 is the amplitude of the electrostatic field. The solution of this equation is

$$\begin{aligned}y &= -\frac{e\mathsf{E}_0 e^{2\pi i\nu t}}{4\pi^2 m[(\nu_0^2 - \nu^2) + i\gamma\nu]} \\ &= -\frac{e}{4\pi^2 m}\frac{\mathsf{E}_0 e^{2\pi i\nu t}e^{-i\varphi}}{\sqrt{(\nu_0^2 - \nu^2)^2 + \gamma^2\nu^2}}\end{aligned}\tag{16}$$

where

$$\nu_0 = \sqrt{\frac{\kappa}{4\pi^2 m}}$$

---

[1] See, for example, P. Drude, *The Theory of Optics* (Longmans, Green & Company, New York, 1902).

[2] LORENTZ, *op. cit.*

is the natural frequency of the oscillator and

$$\varphi = \arctan \frac{\gamma \nu}{(\nu_0^2 - \nu^2)} \tag{17}$$

is the phase angle between the electronic motion and the field intensity. The current per unit area associated with this motion is

$$\mathbf{J} = -n_0 e \dot{y} = \frac{n_0 e^2}{4\pi^2 m} \frac{2\pi \nu e^{i\left(-\varphi + \frac{\pi}{2}\right)}}{\sqrt{(\nu_0^2 - \nu^2)^2 + \gamma^2 \nu^2}} \mathbf{E}_0 e^{2\pi i \nu t} \tag{18}$$

where n_0 is the number of oscillators per unit volume. Since the complex conductivity σ_c is the coefficient of $\mathbf{E}_0 e^{2\pi i \nu t}$ in this equation, we have

$$\sigma = \frac{n_0 e^2}{4\pi^2 m} \frac{2\pi \nu}{\sqrt{(\nu_0^2 - \nu^2)^2 + \gamma^2 \nu^2}} \sin \varphi = \frac{n_0 e^2}{4\pi^2 m} \frac{2\pi \gamma \nu^2}{(\nu_0^2 - \nu^2)^2 + \gamma^2 \nu^2}, \tag{19}$$

$$\alpha = \frac{n_0 e^2}{4\pi^2 m} \frac{1}{\sqrt{(\nu_0^2 - \nu^2)^2 + \gamma^2 \nu^2}} \cos \varphi = \frac{n_0 e^2}{4\pi^2 m} \frac{(\nu_0^2 - \nu^2)}{(\nu_0^2 - \nu^2)^2 + \gamma^2 \nu^2}. \tag{20}$$

It may be seen from Eq. (17) that φ is π or zero, depending upon the sign of $\nu - \nu_0$, whenever

$$|\nu - \nu_0| >> \gamma.$$

The phase angle varies between these limits in the manner shown in Fig. 1 as ν passes through the value ν_0, the width of the transition region being

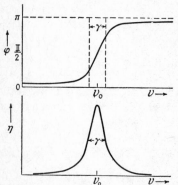

of the order of magnitude γ. Since σ is appreciably different from zero only in this transition region, this is the only region in which light may be absorbed. The absorption coefficient $4\pi\sigma/c$ is plotted in Fig. 1 and has

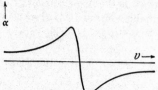

Fig. 1.—The phase angle φ and the absorption coefficient η of the assembly of oscillators as functions of frequency.

Fig. 2.—The polarizability of the system of oscillators.

a peak of half width γ, centered about ν_0.

The polarizability α, given by Eq. (20), changes sign as we pass through the absorption maximum because $\cos \varphi$ changes from positive to negative values. Its absolute value increases as $1/|\nu_0 - \nu|$, as ν

approaches the absorption region, but does not go to infinity because of the term in γ in the denominator of (20). The characteristic behavior of α is shown in Fig. 2. The dielectric constant ϵ, which is of immediate interest for investigating the optical properties, is

$$\epsilon = 1 + 4\pi\alpha = 1 + \frac{n_0 e^2}{m\pi} \frac{(\nu_0^2 - \nu^2)}{(\nu_0^2 - \nu^2)^2 + \gamma^2 \nu^2}. \tag{21}$$

Using these results and Eqs. (13), we may discuss the optical properties of the system of oscillators. In the region on the long wave-length side of ν_0 where $\nu_0 - \nu$ is much greater than γ, ϵ is positive and greater than unity, and σ is negligible. Hence, nk is zero, and $n^2 - k^2$ is positive. We may conclude that

$$k = 0, \qquad n^2 = \epsilon. \tag{22}$$

Thus, the system is transparent and has a refractive index greater than unity. This behavior is characteristic of most ionic and molecular crystals in the visible region of the spectrum. If we assume that n_0 is of the order of 10^{22} cm^{-3}, which is a normal atomic density, and that $\nu_0^2 - \nu^2$ is of the order of 10^{30} sec^{-2}, we find $\epsilon \sim 1.7$, or $n \sim 1.3$, which shows that this oscillator model can yield the correct magnitude for the optical quantities.

As we enter the absorption region, σ no longer is zero. ϵ is initially positive and passes through a maximum at

$$\nu = \nu_0 - \frac{\gamma}{2}.$$

It is readily found that

$$\sigma \sim 2\pi\nu\alpha$$

in the region of this maximum. Equations (13) then become

$$n^2 - k^2 = 1 + 4\pi\alpha, \qquad nk = 2\pi\alpha. \tag{23}$$

The solutions of (23) show that n is larger than $\sqrt{\epsilon}$ in this part of the absorption region and that k is of the order of magnitude $\sqrt{\alpha}$ for large values of α and of the order of magnitude α for small values. It follows from Eq. (14) that the reflection coefficient approaches unity when α becomes large.

At the center of the absorption line, α is zero, whereas σ has the maximum value

$$\sigma_m = \frac{n_0 e^2}{2\pi m \gamma}.$$

The values of n and k at this frequency are determined by the equations

$$n^2 = \frac{1 + \sqrt{1 + 4\sigma_m^2/\nu_0^2}}{2}, \qquad k^2 = \frac{-1 + \sqrt{1 + 4\sigma_m^2/\nu_0^2}}{2}.$$

As we pass to the other side of the absorption line, ϵ decreases and reaches a minimum value at

$$\nu = \nu_0 + \frac{\gamma}{2}.$$

Both n and k decrease as this occurs. If α reaches the value $-1/4\pi$, so that ϵ is zero, n and k are equal to

$$\sqrt{\frac{\sigma_m}{2\nu_0}}.$$

n is less than k in the region where ϵ is negative.

When $\nu - \nu_0$ is much greater than γ, σ is again zero and the medium no longer is absorbing. Whether it is perfectly reflecting or transparent here depends upon the sign of ϵ. Since α approaches zero as $-1/(\nu - \nu_0)$, ϵ is certainly positive for sufficiently high frequencies. The medium is transparent in this region although it is optically less dense than a vacuum. ϵ may be negative, however, in the non-absorbing region near the absorption line. When this happens, n is zero, k is finite, and the medium is totally reflecting (*cf.* Fig. 3). We shall see later that the optical properties of an ideal metal are similar to those of the system of oscillators on the high-frequency side of the center of the absorption line.

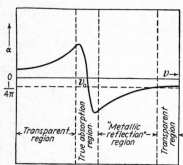

Fig. 3.—The polarizability and the four optical regions associated with an absorption line.

According to Eq. (19), the shape of an absorption line is determined by the function

$$\frac{\nu^2}{(\nu_0^2 - \nu^2)^2 + \nu^2\gamma^2}.$$

Since this is appreciable only in the region where ν_0 and ν are nearly equal, we may write

$$(\nu_0^2 - \nu^2)^2 = (\nu_0 - \nu)^2(\nu_0 + \nu)^2 = 4\nu^2(\nu_0 - \nu)^2.$$

We then obtain the relation

$$\frac{\nu^2}{(\nu_0^2 - \nu^2)^2 + \nu^2\gamma^2} \cong \frac{1}{4}\frac{1}{(\nu_0 - \nu)^2 + (\gamma/2)^2},\tag{24}$$

which is the same as the form (7) of Sec. 45 determined by quantum theory.

It is not possible to make more than an order-of-magnitude estimate of the breadth γ of the absorption line by means of classical theory. A lower limit is determined by radiation damping. According to classical theory,[1] an oscillating charge radiates energy at the rate

$$\frac{2}{3c^3}\overline{\dot{p}^2}$$

where p is the dipole moment, which is ey in the present case. For strictly periodic motion, this may be replaced by

$$\frac{2}{3}\frac{e^2}{c^3}\overline{\dddot{y}\dot{y}},$$

which is equivalent to assuming a damping force

$$\frac{2}{3}\frac{e^2}{c^3}\dddot{y} = -\frac{8\pi^2\nu^2}{3}\frac{e^2}{c^3}\dot{y}.$$

Thus, the damping frequency is

$$\gamma = \frac{4\pi\nu^2 e^2}{3mc^3}.$$

As we remarked in Sec. 45, this is of the order of magnitude 10^8 sec^{-1} for optical frequencies and is usually masked by the damping due to other sources, such as collisions.

Before leaving this topic, we should mention that the local field[2] correction has been neglected in deriving the equations for the optical properties of the assembly of oscillators. This correction may be included by use of Lorentz's theory when α is not too large. The macroscopic polarizability is related to the polarizability α_a of a single oscillator by the equation

$$\alpha = \frac{n_0\alpha_a}{1 - \dfrac{4\pi}{3}n_0\alpha_a}.\tag{25}$$

---

[1] See, for example, M. Abraham and R. Becker, *The Classical Theory of Electricity and Magnetism* (Blackie & Son, Ltd., London, 1932).

[2] See Sec. 142.

When the medium does not absorb,

$$n^2 = \epsilon = 1 + 4\pi\alpha,$$

or

$$\alpha = \frac{n^2 - 1}{4\pi}.$$

Thus, Eq. (25) may be transformed to

$$\frac{n^2 - 1}{n^2 + 2} = \frac{4\pi}{3}n_0\alpha_a. \tag{26}$$

In the absorbing case, we may write Eq. (12) in the form

$$N^2 - 1 = 4\pi\alpha_c$$

where α_c is the complex polarizability. Equation (25) is then generalized to

$$\alpha_c = \frac{n_0\alpha_{a,c}}{1 - \dfrac{4\pi}{3}n_0\alpha_{a,c}} \tag{27}$$

where $\alpha_{a,c}$ is the complex polarizability of an oscillator. When these two equations are combined, the relation replacing (26) is

$$\frac{N^2 - 1}{N^2 + 2} = \frac{4\pi}{3}n_0\alpha_{a,c}. \tag{28}$$

We shall discuss the application of these equations to particular cases below.

b. The Drude-Zener Treatment of Perfectly Free Electrons.—The classical treatment of the optical properties of metals that is based on the assumption of perfectly free electrons was developed by Drude,[1] Zener,[2] and Kronig.[3] The equation of motion for a free electron is

$$m\frac{d^2y}{dt^2} + 2\pi m\gamma\frac{dy}{dt} = -e\mathsf{E}_0 e^{2\pi i\nu t}, \tag{29}$$

which is identical with (15) except for the fact that κ is zero. Thus, we may anticipate that a system of free electrons behaves like a system of oscillators of frequency zero. In the present case, the damping term arises from the resistance of the metal, as we shall see more definitely below. The stationary solutions of Eq. (29) are

$$y = -\frac{e}{4\pi^2 m}\frac{\mathsf{E}_0 e^{2\pi i\nu t}}{-\nu^2 + i\gamma\nu},$$

---

[1] P. DRUDE, *op. cit.*
[2] C. ZENER, *op. cit.*
[3] R. DE L. KRONIG, *op. cit.*

from which we obtain

$$\sigma = n_0 \frac{e^2}{4\pi^2 m} \frac{2\pi\gamma}{\nu^2 + \gamma^2}, \tag{30}$$

$$\alpha = -n_0 \frac{e^2}{4\pi^2 m} \frac{1}{\nu^2 + \gamma^2}. \tag{31}$$

When ν is zero, the first of these equations is

$$\sigma_0 = n_0 \frac{e^2}{2\pi m \gamma}. \tag{32}$$

Comparing this with Eq. (6), Sec. 126, we see that

$$2\pi\gamma = \frac{v}{l} = \frac{1}{\tau}$$

where τ is the mean time between collisions, which is of the order of magnitude 10^{-13} sec at room temperature.

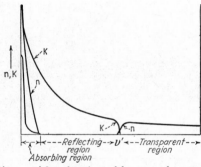

FIG. 4.—The quantities n and k as functions of frequency for a system of free electrons.

Since Eqs. (30) and (31) are identical with Eqs. (19) and (20) when ν_0 is zero, it follows that the optical properties of a system of free electrons should correspond closely to those of an insulator on the short wavelength side of the center of an absorption band. Thus, there should be an absorption region extending from zero frequency to $\nu \sim \gamma$, which in case n_0 is large enough should be followed by a nonabsorbing region in which ϵ is negative (*cf.* Fig. 4). Here, n is zero, and k is equal to $\sqrt{-\epsilon}$. Eventually, ϵ should become positive, since $|\alpha|$ decreases as $1/\nu^2$ with increasing frequency, and the system should become transparent. Hence, the system should be highly reflecting until α is $-1/4\pi$ and should then behave like an ordinary transparent insulator. The frequency ν' at which this transition occurs is so much larger than γ for ordinary densities of electrons that we may obtain it from Eq. (31) by setting γ

equal to zero. Thus,

$$\nu' = \sqrt{\frac{n_0 e^2}{\pi m}}. \tag{33}$$

We shall now discuss the three optical regions.

1. $\nu \ll \gamma$.—The optical relations corresponding to the absorbing region in which ν is much smaller than γ are

$$\left. \begin{aligned} n^2 &\cong \frac{n_0 e^2}{2\pi m} \frac{1}{\gamma^2} [-1 + \sqrt{1 + (\gamma/\nu)^2}] \cong \frac{n_0 e^2}{2\pi m \gamma \nu}, \\ k^2 &\cong \frac{n_0 e^2}{2\pi m} \frac{1}{\gamma^2} [1 + \sqrt{1 + (\gamma/\nu)^2}] \cong \frac{n_0 e^2}{2\pi m \gamma \nu}. \end{aligned} \right\} \tag{34}$$

These may be simplified by use of Eq. (32), for they then reduce to

$$n = k = \left(\frac{\sigma_0}{\nu}\right)^{\frac{1}{2}}. \tag{35}$$

If these values are substituted in Eq. (14) for R, we find

$$R = 1 - 2\left(\frac{\nu}{\sigma}\right)^{\frac{1}{2}}. \tag{36}$$

This relation has been tested experimentally in the far infrared region by Hagen and Rubens.[1] According to (36), the value of $(1 - R)\sqrt{\sigma}$ should be equal to $36.5/\sqrt{\lambda}$ if λ is measured in microns. Table LXXXIII shows the agreement between the observed and calculated values of this quantity for constantan.

TABLE LXXXIII.—VALUES FOR CONSTANTAN OF $(1 - R)\sqrt{\sigma}$

λ, microns	Observed	Calculated
4	19.4	18.25
8	13.0	12.90
12	11.0	10.54
25.5	7.36	7.23

2. $\nu \sim \nu'$.—In the region near ν', at which $4\pi\alpha$ becomes unity, ν is about one hundred times larger than γ so that σ/ν in Eqs. (13) is negligible in first approximation. These equations then are

$$\left. \begin{aligned} n^2 - k^2 &= 1 - \frac{n_0 e^2}{\pi m} \frac{1}{\nu^2}, \\ nk &= 0. \end{aligned} \right\} \tag{37}$$

[1] E. HAGEN and H. RUBENS, *Ann. Physik*, **14**, 936 (1904).

As we mentioned previously, n is zero and k is $\sqrt{-\epsilon}$ when ν is less than ν', and k is zero and n is $\sqrt{\epsilon}$ when ν is greater than ν'. Wood[1] has found that the alkali metals, in which the valence electrons are nearly free, satisfy these relations closely. Table LXXXIV contains the observed wave lengths at which the transition from the reflecting to the transmitting state occurs. These were determined by observations on thin films of the metals. The values calculated from Eq. (33) by the use of the true electronic mass and electron density are tabulated for comparison. In the cases of lithium and sodium, the values for the theoretical effective masses are also given.

TABLE LXXXIV

	Observed, Å	Calculated, Å	
		m	m^*
Li	1550	1500	1830
Na	2100	2090	2020
K	3150	2920	
Rb	3400	3220	
Cs	3800	3630	

3. $\nu' >> \nu >> \gamma$.—The theoretical and experimental results do not seem to agree very closely in the visible and near ultraviolet region where

$$\nu' >> \nu >> \gamma.$$

The values of σ and ϵ, given by Eqs. (30) and (31), in this region are

$$\sigma = \frac{n_0 e^2 \gamma}{2\pi m \nu^2} = \left(\frac{n_0 e^2}{2\pi m \nu}\right)^2 \frac{1}{\sigma_0}, \tag{38}$$

$$\epsilon = 1 - \frac{n_0 e^2}{\pi m \nu^2}. \tag{39}$$

Försterling and Freedericksz[2] obtained measurements of n and k for a number of metals, in the region from 1μ to 15μ, from which values of σ and ϵ may be determined by means of Eqs. (13). The observed values of ϵ usually agree with the theoretical ones determined from (39) to within about 10 per cent whereas the values of σ disagree by a comparatively large factor. It is found that the frequency dependence of the observed values is the same as that predicted by Eq. (38); however, the

[1] R. W. WOOD, Phys. Rev., 44, 353 (1933). See also R. W. WOOD and C. LUKENS, Phys. Rev., 54, 332 (1938).

[2] K. FÖRSTERLING and V. FREEDERICKSZ, Ann. Physik, 40, 201 (1913).

value of σ_0 that is required to give the proper magnitude is much smaller than the actual static conductivity. The two values of σ_0 are listed in Table LXXXV. In addition, the value of the effective electron mass that gives the best fit between the observed and computed ϵ curves is given in the cases of copper, silver, and gold.

TABLE LXXXV

	Ag	Au	Cu	Pt	Ir
Value of σ_0 for best fit..............	1.4	2.5	1.0	0.12	0.13
Actual value.....................	5.7	4.2	5.3	0.85	1.7
m^*/m..........................	1.07	1.13	2.56		

It should be pointed out[1] that the optical properties in this spectral region are determined by a thin surface layer of the metal. The *penetration distance* δ, in which the light intensity drops to $1/e$th of its initial value, is

$$\delta = \frac{\lambda}{4\pi k}$$

where λ is the wave length. This is about 200Å at 1μ for silver, since k is 5.62. It is possible that the conductivity σ_0 of a sheath of this thickness is considerably lower than that of the bulk material because of surface contaminations.

148. Quantum Formulation of the Optical Properties*.—We shall now develop the equations for the optical properties of solids in three idealized cases, namely: (a) the case of a system of isolated atoms, (b) a case in which the excited state of the system may be described by exciton waves, and (c) a system in which the electronic wave functions are determinants of Bloch one-electron functions. The results for the first two cases evidently may be applied to insulators such as molecular and ionic crystals, whereas the results for the third should apply to metals. The effect of nuclear motion will be neglected for the present.

a. A System of Isolated Atoms.—When the atoms or molecules in a solid are very loosely bound, its optical properties may be obtained from the equations that were derived for free atoms in Sec. 43. We saw there that the effective atomic polarizability for dispersive scattering of quanta of frequency ν is the tensor

$$\alpha_a = \sum_k{}' \frac{\mathbf{M}_{0k}\mathbf{M}_{k0}}{h} \frac{2\nu_{k0}}{\nu_{k0}^2 - \nu^2} \tag{1}$$

[1] This suggestion apparently was made first by A. H. Wilson, *The Theory of Metals* (Cambridge University Press, 1936).

[*cf.* Eq. (33), Sec. 43]. Here, k is summed over all excited states,

$$\mathbf{M}_{0k} = -\int \Psi_0^*(\Sigma e\mathbf{r}_i)\Psi_k d\tau \tag{2}$$

is the matrix component of the atomic dipole moment, and

$$\nu_{k0} = \frac{E_k - E_0}{h}. \tag{3}$$

Similarly, the probability $P_j(t)$ that an atom jumps from a state Ψ_0 to a state Ψ_j, if the energy density as a function of frequency is ρ_ν, is

$$P_j(t,\nu) = \frac{8\pi^3}{h^2}|\mathbf{M}_{0j}\cdot\mathbf{n}|^2\rho_\nu t\delta\left(\frac{E_i - E_0}{h} - \nu\right) \tag{4}$$

in the δ-function approximation of Sec. 43, where $\mathbf{n}$ is the direction of polarization of the radiation.   It should be possible to derive ϵ and σ from Eqs. (1) and (4).

Since it is convenient to treat cases in which ϵ and σ are constants instead of tensors, we shall usually consider examples in which (1) and (4) are independent of the direction.   In the present case, this is true if the atoms are in S states (*cf.* Sec. 43).

The individual terms of Eq. (1) are very similar to those derived in the last section for the polarizability of an oscillator in a nonabsorbing region, namely,

$$\alpha_{k0} = \frac{e^2}{4\pi^2 m}\frac{1}{\nu_{k0}^2 - \nu^2} \tag{5}$$

[*cf.* Eq. (20) in the case in which γ is negligible].   In fact, Eq. (1) may be written as

$$\boldsymbol{\alpha}_a = \sum_k \mathbf{f}_{k0}\alpha_{k0} \tag{6}$$

where

$$\mathbf{f}_{k0} = \frac{8\pi^2 m}{he^2}\nu_{k0}\mathbf{M}_{0k}\mathbf{M}_{k0} \tag{7}$$

is the oscillator strength of the transition from Ψ_0 to Ψ_k which evidently is a tensor quantity.   A theorem of the theory of atomic spectra states[1] that for free atoms the sum of all the $\mathbf{f}_{k0}$ connecting two levels is a multiple of the unit tensor.   For example, in the case in which the ground state is a $^1S_0$ state, $\mathbf{f}_{k0}$ is nonvanishing only for a $^1P_1$ state when the Russell-Saunders coupling scheme is valid.   If the three degenerate substates

[1] See, for example, E. U. Condon and G. H. Shortley, *The Theory of Atomic Spectra* (University Press, Cambridge, 1935); S. A. Korff and G. Breit, *Rev. Modern Phys.*, **4**, 471 (1932); G. Breit, *Rev. Modern Phys.*, **4**, 504 (1932).

are chosen to be eigenfunctions of the three components of orbital angular momentum, the sum of $\mathbf{f}_{k0}$ for the three states is

$$\sum_{k=1}^{3} \mathbf{f}_{k0} = \frac{8\pi^2 m}{3h} \nu_{k0} \sum_{k=1}^{3} \left[\left| \left(\sum_{i=1}^{n} x_i \right)_{k0} \right|^2 + \left| \left(\sum_{i=1}^{n} y_i \right)_{k0} \right|^2 + \left| \left(\sum_{i=1}^{n} z_i \right)_{k0} \right|^2 \right] \mathbf{I} \quad (8)$$

where $\mathbf{I}$ is the unit tensor. In the same case, the sum of the time derivatives of (4) for the transitions from the lowest state to the three degenerate P states is

$$\frac{dP_{l0}}{dt} = \frac{e^2 \pi}{mh} \frac{f_{l0}}{\nu_{l0}} \rho_\nu \delta(\nu_{l0} - \nu), \quad (9)$$

in which ν_{l0} is the frequency ν_{k0} and f_{l0} is the coefficient of $\mathbf{I}$ in Eq. (8).

The scalar coefficient of $\mathbf{I}$ in the equation corresponding to (8) for any two levels l and m is usually designated by f_{lm} and is called the f factor or the line strength for the two levels. The reason for the second designation is that f_{lm} generally occurs in the equation analogous to (9) and thus measures the relative line intensity. It follows that the f_{lm} may be determined from absorption and emission measurements.

Another theorem of quantum mechanics[1] states that

$$\sum_{l} f_{lm} = n, \quad (10)$$

in which l is summed over all levels and n is the total number of electrons in the atom. Since the ν_{k0} in Eq. (7) are positive for the lowest state in the atom, it follows that the f factors for this level are all positive. Hence, in this case, (10) states that the sum of the f factors for *absorption* from the lowest state is equal to n. The same conclusion cannot be drawn for the absorption f factors of an excited atom, for there may be additional factors connecting this and lower levels. Since the emission factors are negative, it follows that the sum of the absorption line strengths of an excited atom usually is greater than n. This fact is of importance in discussing the absorption f factors in the cases of monovalent atoms, such as the lighter alkali metals, in which the valence-electron wave functions can be derived from an effective ion-core field, as we have seen in Sec. 78. In this case, there is a theorem analogous to (10) for the valence electron, namely,

$$\sum_{l} f'_{lm} = 1, \quad (11)$$

[1] E. Wigner, *Physik. Z.*, **32**, 450 (1932); H. A. Kramers, C. C. Jonker, and T. Koopmans, *Z. Physik*, **80**, 178 (1932).

where f'_{lm} refers to the levels of the ion-core field.   In lithium, the lowest state is a 2s function, and there are no lower p functions; hence, the f_{lm} for the 2s state are all positive, and the sum of the absorption strengths of the valence electron should be unity.   In sodium, on the other hand, the lowest state is 3s.   Thus, in this case there is a fictitious 2p state lying below the 3s for which there is a negative f factor.   Hence, the sum of the absorption line strengths should be larger than unity.   The same is true in any of the heavier alkali metal atoms whose levels may be obtained from an effective ion-core field.

In atomic lithium,[1] the strengths for the valence-electron transitions have the values given in Table LXXXVI.

TABLE LXXXVI

Transition from 2s to	f
2p	0.7500
3p	0.0055
4p	0.0047
5p	0.0025
Continuum	0.24
Sum	1.01

Thus, 75 per cent of the oscillator strength is associated with the first excited level, and the sum of the f is close to unity.

According to Eq. (26) of the preceding section, the index of refraction in a nonabsorbing region should be related to the atomic susceptibility (6) by the equation

$$\frac{n^2 - 1}{n^2 + 2} = \frac{n_0}{3} \sum_l \frac{e^2}{\pi m} \frac{f_{l0}}{\nu_{l0}^2 - \nu^2} \tag{12}$$

where l is summed over all excited levels.   The f values of the rare gas atoms have been determined by expressing the observed index of refraction of the gas in a series of the type (12).   It is found[2] in this way that the total f value associated with the transition from the lowest level to the levels of the 1s2p configuration is 1.12 for helium, which has two electrons.   In the other rare gases, which have six p electrons in the outer shell, the corresponding numbers are as follows

Ne	2.37
Ar	4.58
Kr	4.90
Xe	5.61.

---

[1] B. TRUMPY, *Z. Physik*, **61**, 54 (1929).

[2] See, for example, K. L. WOLFF and F. K. HERZFELD, *Handbuch der Physik*, Vol. XX.

The absorption coefficient in the absorbing region may be determined from Eq. (9). According to this, the energy loss per unit volume at a point where the energy density is ρ_ν is[1]

$$\frac{dW}{dt} = \sum_l n_0 h \nu_l \frac{dP_l}{dt} = \frac{n_0 e^2 \pi}{m} \rho_\nu \sum_l f_l \delta(\nu_l - \nu)$$

where n_0 is the number of atoms per unit volume. Comparing this with Eq. (6) of the preceding section, we find

$$\sigma = \frac{n_0 e^2}{4m} \sum_l f_l \delta(\nu_l - \nu). \tag{13}$$

Hence, the absorption coefficient in this delta-function approximation is

$$\eta(\nu) = \frac{n_0 e^2 \pi}{mc} \sum_l f_l \delta(\nu_l - \nu) \tag{14}$$

according to which the absorption peaks should be infinitely narrow.

Combining Eqs. (6) and (13), we obtain for the complex atomic polarizability

$$\alpha_c = \alpha_a + \frac{\sigma_a}{2\pi \nu i} = \frac{e^2}{4\pi^2 m} \sum_l \left[\frac{f_l}{\nu_l^2 - \nu^2} - \frac{i\pi f_l \delta(\nu - \nu_l)}{2\nu} \right]. \tag{15}$$

It was found in Sec. 45 that the shape of an absorption line is given by the function

$$\frac{1}{(\nu_l - \nu)^2 + \Gamma^2} \tag{16}$$

instead of by a delta function. Hence, we should replace the delta function in Eqs. (13) and (15) by the function (16) multiplied by an appropriate normalization constant. Since the integral of (16) over all frequencies is π/Γ if ν_l is greater than Γ, the delta function should be replaced by

$$\frac{\Gamma/\pi}{(\nu_l - \nu)^2 + \Gamma^2}$$

or, if we use the approximation of Eq. (24) of the preceding section, by

$$\frac{2\gamma \nu^2/\pi}{(\nu_0^2 - \nu^2)^2 + \nu^2 \gamma^2} \tag{17}$$

---

[1] Since the atoms are in their lowest state, we may drop the subscript zero in f, P and ν.

where $\gamma = 2\Gamma$. At the same time, the coefficient $1/(\nu_i^2 - \nu^2)$ of the real part of (15) should be replaced by

$$\frac{(\nu_i^2 - \nu^2)}{(\nu_i^2 - \nu^2)^2 + \gamma^2\nu^2}. \tag{18}$$

Thus, (15) becomes

$$\left.\begin{aligned}
\alpha_c &= \frac{e^2}{4\pi^2 m}\sum_l f_l\frac{(\nu_i^2 - \nu^2) - i\gamma\nu}{(\nu_i^2 - \nu^2)^2 + \gamma^2\nu^2}\\
&= \frac{e^2}{4\pi^2 m}\sum_l\frac{f_l}{(\nu_i^2 - \nu^2) + i\gamma\nu},
\end{aligned}\right\} \tag{19}$$

which is analogous to the result that is obtainable from the classical equation (16) of the preceding section.

b. The Case of Excitation Waves.—In treating the absorption and dispersion in a solid whose normal and lowest excited states are described by excitation waves, we must use the equations that were derived in Sec. 96 for extended atomic systems. We shall discuss a simple model in which there is only one electron per atom. According to the discussion of Sec. 96, the normal-state wave function then is $1/\sqrt{N!}$ times a determinant of normalized one-electron functions of the atomic type $\psi(\mathbf{r} - \mathbf{r}(n))$ that are centered about each of the atomic positions

$$\mathbf{r}(n) = n_1\boldsymbol{\tau}_1 + n_2\boldsymbol{\tau}_2 + n_3\boldsymbol{\tau}_3. \tag{20}$$

The normalized excited states have the form

$$\Psi_{\mathbf{k},i} = \frac{1}{\sqrt{N}}\sum_n e^{2\pi i\mathbf{k}\cdot\mathbf{r}(n)}\Psi_{n,i} \tag{21}$$

where $\Psi_{n,i}$ is the wave function derived from Ψ_0 by replacing $\psi(\mathbf{r} - \mathbf{r}(n))$ by the excited atomic function $\psi_i(\mathbf{r} - \mathbf{r}(n))$, and $\mathbf{k}$ ranges over the points of a single zone. A band of levels of energy

$$E_i(\mathbf{k}) = E_i + I_i\sum_\rho e^{2\pi i\mathbf{k}\cdot\rho} \tag{22}$$

is associated with each value of i where E_i is the unperturbed energy of the state $\Psi_{n,i}$ and I_i is an integral involving neighboring excited and normal atoms. We shall discuss a case in which the lowest atomic function is an S state.

It was found in Sec. 96 that

$$\int \Psi_0 \text{ grad } \Psi_{\mathbf{k},i}d\tau = \frac{1}{\sqrt{N}}\left(\int \psi \text{ grad } \psi_i d\tau\right)\delta_{\mathbf{k},0} \tag{23}$$

where ψ is the normal atomic function and ψ_i is the excited function for the same atom [*cf.* Eqs. (17) to (20), Sec. 96]. We may conclude from this that the only allowed transitions take place between the excitation bands corresponding to atomic states between which transitions are allowed. The appearance of the factor $\delta_{k,0}$ in (23) implies that the excited states of the entire crystal must have the same wave number as the normal state in the reduced-zone sense, which is a generalization of the principle of conservation of momentum.

Now, the conductivity associated with these transitions is[1]

$$\sigma(\nu) = \frac{e^2 h}{24\pi^2 \nu m^2 V} \sum_{k,i} \left| N \int \Psi_{k,i} \text{ grad } \Psi_0 d\tau \right|^2 \delta(\nu_{0k,i} - \nu). \tag{24}$$

When the relation (23) is used, this becomes

$$\sigma(\nu) = \frac{n_0 e^2 h}{24\pi^2 \nu m^2} \sum_i \left| \int \psi \text{ grad } \psi_i d\tau \right|^2 \delta(\nu_i - \nu) \tag{25}$$

where

$$\nu_i = \frac{E_i - E_0}{h},$$

in which E_i and E_0 are, respectively, the unperturbed energies of $\Psi_{n,i}$ and Ψ_0. Equation (25) is identical with the expression for the conductivity of a system possessing n_0 isolated atoms per unit volume. Thus, it may be transformed into Eq. (13) by use of Eq. (8) and the equation

$$\left| \int \psi \text{ grad } \psi_i d\tau \right| = \frac{2\pi m \nu_i}{\hbar} \left| \int \psi \mathbf{r} \psi_i d\tau \right| \tag{26}$$

[*cf.* Eq. (24), Sec. 43].

[1] A detailed development of Eqs. (24) and (27) is omitted for brevity. These equations may be derived by the use of the semiclassical method by dividing the crystal into sections smaller than the wave length of light and larger than atomic dimensions and treating these sections both as specimens of the bulk solid and as molecular units to which the methods of Sec. 43 are applicable. Perturbed wave functions may be computed for the case in which a radiation field is present; and, from these, the power loss P due to transitions and the mean value $\mathbf{J}$ of the current operator (see Sec. 44) may be evaluated. The conductivity σ and the polarizability α are related to these quantities by the equations

$$P = \sigma \mathbf{E}^2 \quad \text{and} \quad \bar{\mathbf{J}} = \alpha \cdot \mathbf{E}.$$

It follows from these remarks that Eqs. (24) and (27) are valid only when the wave length of light is long compared with atomic dimensions. The second term in Eq. (27) arises from the part of Eq. (8), p. 223, involving the vector potential.

If Eq. (23) is substituted in the equation for the polarizability,[1] namely,

$$\alpha = \left[\frac{e^2 h}{24\pi^4 m^2 \nu^2 V} \sum_{k,i} \frac{\nu_{k,i0}}{\nu^2_{k,i0} - \nu^2} \left(N \int \Psi_{k,i} \text{ grad } \Psi_0 d\tau \right)^2 - \frac{e^2}{4\pi^2 m \nu^2} n_0 \right], \quad (27)$$

and Eq. (26) is used to simplify the result, it is found that

$$\alpha(\nu) = \left(\frac{2n_0}{3h\nu^2} \sum_i \nu_i^3 \frac{|\mathbf{M}_i|^2}{\nu_i^2 - \nu^2} - \frac{n_0 e^2}{4\pi^2 m \nu^2} \right) \quad (28)$$

where

$$|\mathbf{M}_i|^2 = \frac{e\hbar}{2\pi \nu_i m} \sum_{i=1}^{3} \left| \int \psi \text{ grad } \psi_i d\tau \right|^2 \quad (29)$$

is the dipole matrix component.

We may reduce this still further by use of the sum rule

$$\sum_l f_l \equiv \sum_i \frac{8\pi^2 m}{3he^2} \nu_i |\mathbf{M}_i|^2 = 1;$$

for if we subtract the equation

$$\frac{n_0 e^2}{4\pi^2 m \nu^2} \left(\sum_l f_l - 1 \right) = 0$$

from Eq. (28), we obtain

$$\alpha = \frac{n_0 e^2}{4\pi^2 m} \sum_l \frac{f_l}{\nu_i^2 - \nu^2} \quad (30)$$

which is equivalent to the expression for the polarizability of a system of isolated atoms.

We may conclude from the results of this part of the present section that the optical properties of a system that is described by means of excitation waves are the same as those of a system of free atoms. The absorption spectrum consists of discrete lines not because the energy levels do not form wide bands, but because the wave number of the exciton must be zero.

c. The Case of Bloch Wave Functions.—When the system may be described in the Bloch approximation, the lowest singlet wave function Ψ_0 is a determinant of functions

$$\psi_k = \chi_k e^{2\pi i k \cdot r}, \quad (31)$$

[1] See previous footnote.

in which each of $N/2$ ψ_k of lowest energy appear once with each spin. The excited singlet states are determinantal functions that are derived from Ψ_0 by replacing one or more of the ψ_k by excited functions $\psi_{k'}$. Since the matrix components in which we are interested are integrals of a one-electron operator, the interesting excited states differ from Ψ_0 by one Bloch function. If the state derived by replacing ψ_k with $\psi_{k'}$ is designated by $\Psi_{kk'}$, it follows from the discussion of Sec. 71 that

$$N \int \Psi_{kk'}{}^* \operatorname{grad}_1 \Psi_0 d\tau = (\int \chi_{k'}{}^* \operatorname{grad} \chi_k d\tau) \delta_{k',k+K}. \tag{32}$$

Thus, as was pointed out in Sec. 71, the allowed optical transitions correspond to "vertical" jumps in the reduced-zone scheme.

It should be noted that in the present case the electronic absorption spectra of the entire solid consist of broad bands instead of discrete lines, in contrast with the case of excitons. The reason for this is that in the transition from Ψ_0 to $\Psi_{kk'}$ the excited electron and the hole it leaves behind move independently of one another. Thus, the only restriction on electronic wave number is that the sum of the wave number k' of the excited electron and the wave number $-k$ of the hole be equal to a principal vector K in the reciprocal lattice, which may be satisfied for any value of k'.

The polarizability, which is given by Eq. (27), is

$$\alpha = \left[\frac{e^2 h}{24\pi^4 m^2 \nu^2 V} \sum_{k,K} \frac{\nu_{k,k+K}}{\nu^2{}_{k,k+K} - \nu^2} |\int \chi_{k+K}{}^* \operatorname{grad} \chi_k d\tau|^2 - \frac{e^2}{4\pi^2 m\nu^2} n_0 \right] \tag{33}$$

in the present case. The variable k extends over the occupied levels of the lowest state, once for each spin, and K is summed over all values of the principal wave-number vectors.

The sum rule for Bloch's one-electron functions[1] is

$$\frac{h}{6\pi^2 m} \sum_{K \neq 0} \frac{|\int \chi_{k+K}{}^* \operatorname{grad} \chi_k d\tau|^2}{\nu_{k,k+K}} + \frac{m}{3h^2} \Delta_k \epsilon(k) = 1. \tag{34}$$

If this equation is multiplied by $e^2/4\pi^2 m\nu^2$, summed over all values of k, and is combined with (33), it is found that

$$\alpha = \left[\frac{e^2 h}{24\pi^4 m^2 V} \sum_{k,K} \frac{|\int \chi_{k+K}{}^* \operatorname{grad} \chi_k d\tau|^2}{\nu_{k,k+K}(\nu^2{}_{k,k+K} - \nu^2)} - \sum_k \frac{e^2}{12\pi^2 h^2 \nu^2} \Delta_k \epsilon_k \right]. \tag{35}$$

The first part of this equation may be placed in the form

$$\frac{e^2}{4\pi^2 m V} \sum_{k,K} \frac{f_{k,K}}{\nu^2{}_{k,k+K} - \nu^2} \tag{36}$$

[1] See, for example, Wilson, *op. cit.*

where

$$f_{\mathbf{k},\mathbf{K}} = \frac{h}{6\pi^2 m} \frac{|\int \chi_{\mathbf{k}+\mathbf{K}}^* \text{ grad } \chi_{\mathbf{k}} d\tau|^2}{\nu_{\mathbf{k},\mathbf{k}+\mathbf{K}}}. \tag{37}$$

Equation (36) is the counterpart of the polarizability of insulators since it is associated with transitions between the lowest zone and others. It approaches (30) in the limiting case of narrow bands in which the $\chi_{\mathbf{k}}$ become atomic functions. The last term in (35) may be transformed to

$$-\frac{n_0 e^2}{4\pi^2 m^* \nu^2} \tag{38}$$

by making use of the relation

$$\frac{1}{m^*} = \frac{1}{3h^2}\Delta_{\mathbf{k}}\epsilon(\mathbf{k}).$$

Comparing Eq. (38) with Eq. (31) of the preceding section, namely,

$$\alpha = -\frac{n_0 e^2}{4\pi^2 m} \frac{1}{\nu^2 + \gamma^2},$$

we see that they are identical in the case in which the resistance damping γ is neglected and m^* is equal to m. Hence, (38) corresponds to the polarizability of free electrons.

The expression (24) for the conductivity becomes

$$\sigma(\nu) = \frac{e^2 h}{24\pi^2 \nu m^2 V} \sum_{\mathbf{k},\mathbf{K}} \left| \int \chi_{\mathbf{k}+\mathbf{K}}^* \text{ grad } \chi_{\mathbf{k}} d\tau \right|^2 \delta(\nu_{\mathbf{k},\mathbf{k}+\mathbf{K}} - \nu), \tag{39}$$

which may be used to discuss the absorption associated with transitions between bands. This expression is not valid at zero frequency, for it does not allow for the fact that free electrons may be continuously accelerated in a static electrostatic field. In the present approximation in which damping is neglected, σ should have an infinite peak when ν is zero. This term is absent because the perturbation scheme used in Sec. 43 was not applied properly in the aperiodic case of zero frequency. We need not discuss this case here, since it was treated extensively in Chap. XV.

149. Application to Metals.—We shall now discuss the application of the preceding theoretical results to metals. If the theory were applied accurately to all cases, we should be able to test it by comparing observed and computed values of n and k for a wide range of frequencies. This actually can be done only for several of the alkali metals and then only semiquantitatively. In other cases, we must be satisfied with a rough comparison of the peaks and minima of the observed absorption curves with those which should be expected from estimated levels of the band approximation.

a. The Alkali Metals.—It was seen in Chap. X that the valence electrons in alkali metals are very nearly free, for the occupied electronic levels can be expressed in the free-electron form

$$\epsilon(\mathbf{k}) = \frac{\hbar^2}{2m^*}\mathbf{k}^2$$

where m^* is a constant. The computed values of m/m^* for lithium, sodium, and potassium are

$$
\begin{array}{ll}
\text{Li} & 0.65 \\
\text{Na} & 1.07 \\
\text{K} & \sim 1.6 \; .
\end{array}
$$

Using the sum rule

$$1 - \sum_{\mathbf{K}} f_{\mathbf{k},\mathbf{K}} \equiv 1 - \sum_{\mathbf{K}} \frac{h}{2\pi^2 m} \frac{|\int \chi_{\mathbf{k}+\mathbf{K}}{}^* \operatorname{grad} \chi_{\mathbf{k}} d\tau|^2}{\nu_{\mathbf{k},\mathbf{k}+\mathbf{K}}} = \frac{m}{m^*}$$

[*cf.* Eq. (34) of the preceding section], we obtain a relation between m^* and the f factors that determine the absorption probability for transitions from one band to another. In sodium, m/m^* is very nearly unity so that $\sum_{\mathbf{K}} f_{\mathbf{k},\mathbf{K}}$ is very small. This means that the oscillator strengths for transitions from the lowest band to higher bands are small and that we should

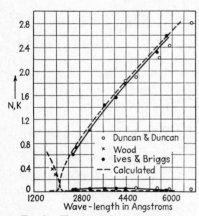

Fig. 5.—The quantities n, k, and nk, as functions of wave length, for sodium. In this case the agreement between observation and the simple theory is excellent. (*After Ives and Briggs.*)

expect the Zener-Kronig theory for perfectly free electrons to apply closely for this metal. This actually turns out to be the case. Experimental and theoretical[1] curves, which are shown in Fig. 5, agree closely down to 1850Å except for the fact that the observed value of nk, although very small in the reflecting region, is several times larger than the value computed from resistance damping by using the observed static resistivity. Since nk is proportional to the absorption coefficient, this fact implies that the absorption is higher than we should expect from the free-electron theory. It is possible that the discrepancy has the same explanation as that proposed for the cases discussed under 3, part (*b*),

[1] H. E. Ives and H. B. Briggs, *Jour. Optical Soc. Am.*, **27**, 181 (1937).

Sec. 147, namely, that the surface conductivity is less than the volume conductivity. It is also possible, however, that the volume absorption due to interband transitions begins in the visible region of the spectrum, for the energy-level diagram of sodium discussed in Sec. 99 (*cf.* Fig. 1) indicates that the lowest transition should occur at about 2 ev. In contradiction to this explanation is the fact that the peaks for the volume photoelectric effect lie far in the ultraviolet for the lighter alkali metals.

The theoretical values of m/m^* are appreciably different from unity for lithium and potassium. Hence, we not only should expect the absorption coefficient to be greater for these metals but should also expect the frequency at which the dielectric constants become zero to be displaced relative to the value for perfectly free electrons. There do not seem to be available measurements on n and k for lithium; however, Ives and Briggs[1] have made very accurate observations on potassium. Figure 6 shows the observed values of n and

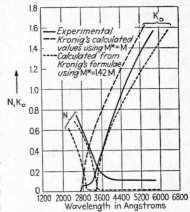

FIG. 6.—The quantities n, k, and nk for potassium. In this case the simple theory and experiment agree only if $m^* = 1.42m$ (see text). (*After Ives and Briggs.*)

k, which are compared with the theoretical curves obtained by use of $m^* = m$ and $m^* = 1.42m$. The second set of curves agrees with the experimental values much more closely than the first. Unfortunately, the corresponding value of m/m^* is less than unity rather than greater than unity, a fact suggesting that Gorin's estimate of m^* is not very accurate. As in the case of sodium, the observed value of nk is much larger than the theoretical one, although it is not possible to say whether or not the increase is due to internal absorption.

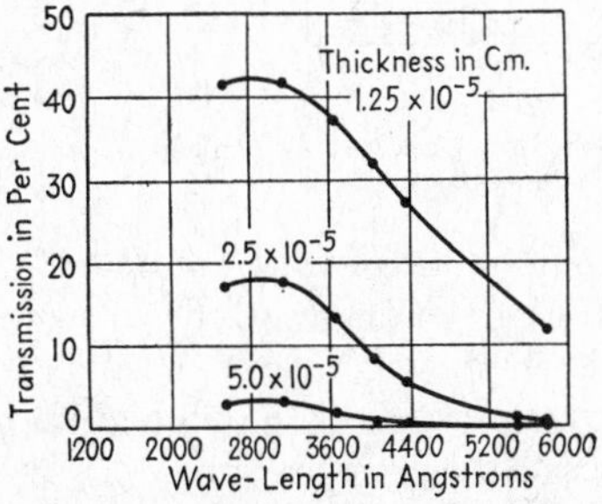

FIG. 7.—The transmission of several layers of cesium. The fall in transmission below 2800Å presumably implies nonvanishing interband transition probabilities. (*After Ives and Briggs.*)

We should mention in passing that Ives and Briggs[2] have also examined the optical properties of cesium and rubidium. Transmission

[1] H. E. IVES and H. B. BRIGGS, *Jour. Optical Soc. Am.*, **26**, 238 (1936).

[2] H. E. IVES and H. B. BRIGGS, *Jour. Optical Soc. Am.*, **27**, 395 (1937).

curves obtained by these workers for three different layers of cesium are shown in Fig. 7. The rise in transmission on the long wave-length side of the figure undoubtedly is related to the change in reflectivity; however, the fall on the short wave-length side is presumably related to interband absorption.

b. Copper, Silver, and Gold.—The extent to which the optical properties of a metal specimen are sensitive to its previous history is shown by

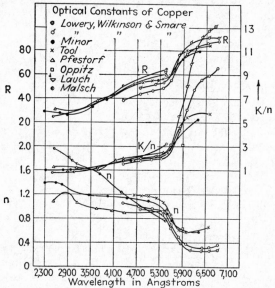

Fig. 8.—The quantities n, k/n, and R in per cent for copper as determined by various workers. (*After Nathanson.*)

the curves of Fig. 8, which contains a compilation of values of n and k/n for copper measured by several observers.[1] The shapes of the measured curves are the same, but the absolute values vary considerably from case to case. That the differences between the results for different cases are related to the treatment the surfaces of the specimens have received seems to be established beyond doubt. Lowery, Wilkinson, and Smare[2] have shown in the case of copper, for example, that k is increased and n is decreased when the metal surface is polished mechanically. Since nk increases during the polishing, it follows that the resistivity of the layer in which the light is reflected is increased. This effect can

---

[1] Taken from the review by J. B. Nathanson, *Jour. Optical Soc. Am.*, **28**, 300 (1938).

[2] H. Lowery, H. Wilkinson, and D. L. Smare, *Phil. Mag.*, **22**, 769 (1936).

be understood if the surface layer is made less perfectly crystalline as a result of polishing, for then the electronic mean free path is decreased. It is generally postulated,[1] at present, that the polished surface possesses a polycrystalline layer.

Typical *nk* curves for copper, silver, and gold are shown[2] in Fig. 9. The product *nk* is about ten times larger for these metals than for the alkalies, as measured by Ives and Briggs, a fact showing that the absorption is much larger in the former than in the latter. The peaks that occur near 2500Å and 4500Å in the case of copper are observed in almost all specimens, whereas the large rise that appears on the long wave-length side of 5700Å is very sensitive to surface treatment. For this reason, it is supposed that short wave-length peaks are related to volume absorp-

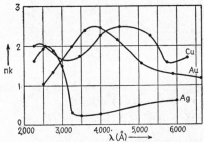

Fig. 9.—Typical *nk* curves for copper, silver, and gold (see text).   (*After Minor and Meier.*)

tion which would occur in an ideal specimen, whereas the peak in the red is associated with the ordinary resistivity. Mott and Jones[3] suggest that the peak near 4500Å is due to transitions from the filled *d* band to the vacant *s-p* levels and that the peak at 2500Å is due to transitions from the occupied *s-p* levels to a higher valence-electron band. This interpretation, rather than the inverse one, is supported by the following two facts: (1) Silver, which has *s-p* bands similar to those of copper but which has different *d* bands, also has a peak at 2500Å. (2) The peak at 4500Å shifts toward the blue as zinc is added to copper, and the *s-p* band is filled higher.[4] Since gold and silver have similar valence-electron structures, it might also be expected that gold would have a peak at 2500Å if the preceding interpretation is correct; however, this peak apparently does not occur.

[1] See, for example, L. H. Germer, *Phys. Rev.*, **50**, 659 (1936).

[2] R. S. Minor, *Ann. Physik*, **10**, 581 (1903); W. Meier, *Ann. Physik*, **31**, 1017 (1910).

[3] N. F. Mott and H. Jones, *The Theory of the Properties of Metals and Alloys* (Oxford University Press, New York, 1936).

[4] H. Lowery, H. Wilkinson, and D. L. Smare, *Proc. Phys. Soc.*, **49**, 345 (1937).

It seems natural to suppose that the large peak that is observed near 3700Å in gold has the same origin as the peak in the visible region for copper, for the atomic $d^{10}s$ and d^9s^2 configurations lie very close together in both cases. In any event, the presence of these peaks accounts for the characteristic colors of these metals, whereas the absence of one in silver explains the normal metallic color of this metal.

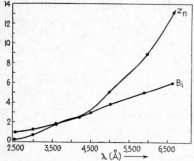

FIG. 10.—nk curves for zinc and bismuth. (*After Minor and Meier.*)

Kronig has applied the free-electron theory of Sec. 147 to Freedericksz's infrared optical measurements on copper, silver, and gold. The agreement was discussed under 3, part b, Sec. 147.

c. Divalent Metals.—Since the divalent metals have nearly filled overlapping bands, the absorption regions that correspond to transitions between the bands of these metals should lie nearer the red end of the spectrum than the corresponding absorption regions in the monovalent metals. The nk curve[1] for zinc, which is shown in Fig. 10, seems to show that these regions actually extend into the infrared. This conclusion is not entirely safe, however, for it is also possible that the resistivity of the specimen of zinc on which the measurements of Fig. 10 were made is high.

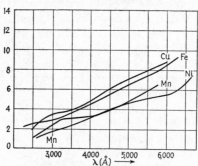

FIG. 11.—nk curves for several transition metals. (*After Minor and Meier.*)

d. Other Simple Metals.—Among the metals of higher valence with nearly filled bands, such as bismuth, antimony, white tin, and so forth, there apparently are available measurements only for bismuth. The nk curve for this metal, which is shown in Fig. 10, has a large rise in the

[1] See MINOR, *op. cit.;* MEIER. *op. cit.*

red, a fact suggesting that there is a peak in the infrared, as in the case of zinc.

It should be added that reflectivity measurements[1] indicate that aluminum is highly reflecting farther in the ultraviolet than most other metals. A reasonable explanation of this fact is given by the $n(\epsilon)$ curve for aluminum shown in Fig. 16, Chap. XIII, which indicates that all the valence electrons of aluminum are very nearly free. If we assume that they are free, the transition frequency ν' given by Eq. (33), Sec. 147, is near 800Å. Even if only one were free, however, the reflecting region would extend to about 1400Å.

e. Transition Metals. The nk curves for a number of transition metals are shown in Fig. 11. All these metals absorb strongly in the visible and near infrared, as might be expected from the fact that the unfilled s-p and d bands overlap.

150. Ionic Crystals.—The structure of the ultraviolet absorption bands of the alkali halides, which have been measured semiquantitatively by Hilsch and Pohl[2] and Schneider and O'Bryan,[3] are shown in Fig. 2, Sec. 95. At low temperatures, the regions of absorption consist of a number of narrow bands, each of which may be related to a transition between the lowest state and the state of wave-number zero in one of the excitation bands associated with the excited states of the halogen ion. According to the results of part *b*, Sec. 148, the absorption bands would consist of sharp lines if the transitions were purely electronic. As we have pointed out in Sec. 45, the observed width arises from the fact that lattice vibrations are stimulated during electron excitation.

The refractive indices of some of the alkali halides in the transmitting visible and ultraviolet regions of the spectrum as determined by Gyulai[4] are shown in Fig. 12. It may be observed that these curves exhibit the

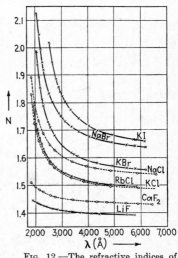

Fig. 12.—The refractive indices of a number of alkali halide crystals. (*After Gyulai.*)

[1] See, for example, the compilation of data in Landolt-Bornstein.

[2] R. HILSCH and R. W. POHL, *Z. Physik,* **44,** 421 (1927); **48,** 384 (1928); **64,** 606 (1930).

[3] E. G. SCHNEIDER and H. M. O'BRYAN, *Phys. Rev.,* **51,** 293 (1937).

[4] Z. GYULAI, *Z. Physik,* **46,** 80 (1928).

sharp rises which are to be expected on the long wave-length side of an absorption maximum.

Mayer[1] has attempted to correlate the measured absorption spectra and refractive indices of sodium chloride, potassium chloride, and potassium iodide by means of the theoretical results of Secs. 147 and 148. We have seen there that the atomic conductivity σ_a is related to the line strengths by the equation

$$\sigma_a(\nu) = \frac{e^2}{4m} \sum_l f_l \delta(\nu_l - \nu) \tag{1}$$

whereas the atomic polarizability in transparent regions is

$$\alpha_a = \frac{e^2}{4\pi^2 m} \sum_l \frac{f_l}{\nu_l^2 - \nu^2}. \tag{2}$$

The second quantity may be expressed in terms of the first by means of the equation

$$\alpha_a = \frac{1}{\pi^2} \int_0^\infty \frac{\sigma_a(\nu')}{\nu'^2 - \nu^2} d\nu', \tag{3}$$

which allows us to compute α_a from measured values of the absorption. The actual relationship between the index of refraction of a system of atoms and α_a is complicated by the local field correction. We have seen in Sec. 147 that in a transparent region

$$n^2(\nu) - 1 = \frac{4\pi}{3}\alpha(\nu) = \frac{4\pi n_0}{3} \frac{\alpha_a(\nu)}{1 - \frac{4\pi}{3}n_0\alpha_a} \tag{4}$$

if the Lorentz local field relations are employed [*cf.* Eq. (25)]. As we approach a single absorption frequency ν_0, α_a becomes

$$\alpha_a \sim \frac{e^2}{4\pi^2 m} \frac{f_0}{\nu_0^2 - \nu^2}$$

so that $\alpha(\nu)$ becomes

$$\alpha(\nu) \cong \frac{n_0 e^2}{4\pi^2 m} \frac{f_0}{\left(\nu_0^2 - \frac{e^2 n_0}{\pi m}f_0\right) - \nu^2}. \tag{5}$$

Hence, the effective absorption peak in the composite system occurs at

$$\nu_0' = \left(\nu_0^2 - \frac{e^2 n_0}{\pi m}f_0\right)^{\frac{1}{2}}. \tag{6}$$

[1] J. Mayer, *Jour. Chem. Phys.*, **1**, 270 (1933).

This displacement is not negligible in an ordinary solid, for $\sqrt{e^2 n_0/\pi m}$ may be as large as 10^{15} sec^{-1} in an ordinary crystal. Since the measured absorption peaks occur at these displaced positions, it is not allowable to compute α_a from (3) by assuming that σ_a is proportional to the observed absorption. Instead, Mayer assumed that

$$\alpha = \frac{1}{\pi^2} \int_0^\infty \frac{\sigma(\nu')}{\nu'^2 - \nu^2} d\nu' \tag{7}$$

where $\sigma(\nu')$ is proportional to the observed absorption coefficient. Since only the relative absorption curves $\eta(\nu)$ are measured, we may place this equation in the form

$$n^2(\nu) - 1 = C \int_0^\infty \frac{\eta(\nu')}{\nu'^2 - \nu^2} d\nu' \tag{8}$$

where C is a constant that is determined by comparing the observed and calculated values of n at one frequency. Mayer fitted the observed Schumann-region $\eta(\nu)$ curves analytically with a system of parabolic segments and one narrow rectangular peak. The height of this peak h was taken as an adjustable parameter which was determined along with C by fitting observed and calculated values of n. The resulting $\eta(\nu')$ functions were then used to integrate Eq. (8) analytically. In the final determination of C and h, a small correction to the observed dispersion was made for the contribution arising from the absorption peaks of the alkali ions in the soft X-ray region. The two observed values of $n(\nu)$ used to fix these parameters were obtained from measurements in the visible and in the far ultraviolet regions, respectively. Table LXXXVII gives a comparison of the observed and calculated values of $(n^2 - 1)$ at an intermediate frequency.

TABLE LXXXVII

	NaCl	KCl	KI
$(n^2 - 1)$(obs.)............	1.720 (2312 Å)	1.245 (2312 Å)	1.914 (3130 Å)
$(n^2 - 1)$(calc.)............	1.708	1.258	1.910
f........................	3.25	3.24	4.00

If we assume that these Schumann-region bands are associated with the electrons of the halogen ions, we may obtain the total optical strength f per ion from the equation

$$f = \frac{4m}{n_0 e^2} \int C\eta(\nu') d\nu' \tag{9}$$

where n_0 is the number of ions per cubic centimeter and $C\eta(\nu)$ is the final curve obtained by fitting (8). This equation is a consequence of Eqs. (5), (7), and (8). Calculated values of f are given in Table LXXXVII. It is interesting to note that the values for NaCl and KCl are closely alike, indicating that the optical strengths of the halogen ions are nearly the same in different compounds.

Mayer has used the functions $C\eta(\nu)$, determined by the preceding method, to evaluate the constants that appear in London's and Margenau's expressions for the van der Waals interaction of halogen ions. As we have mentioned in Sec. 12, this procedure leads to larger values of the interaction energy than are obtained by treating the halogens as though they had the same properties as neighboring rare gas atoms.

The fact that the optical strength of the halogen ions seems to be constant may be compared with the principle of additivity of refractivities, which has been evolved[1] from a study of the experimental refractive indices of ionic crystals. The molar refractivity of a crystal is defined by the equation

$$R = \frac{n^2 - 1}{n^2 + 2} V_M$$

where n is the refractive index and V_M is the molecular volume. It is evident from Eq. (26), Sec. 147, that R is a universal constant times the polarizability per molecule when the Lorentz equation for the local field is valid. Values of R, corresponding to the extrapolation of n to infinite wave length, are usually designated by R_∞. It is found experimentally that the values of R_∞ for a set of four simple ionic crystals AX, AY, BX, BY satisfy closely the additivity relations

$$R_{\infty,AX} - R_{\infty,AY} = R_{\infty,BX} - R_{\infty,BY},$$
$$R_{\infty,AX} - R_{\infty,BX} = R_{\infty,AY} - R_{\infty,BY}.$$

For example, the refractivities of several alkali halides satisfy the relations

$$R_{KCl} - R_{KBr} = 3.13,$$
$$R_{RbCl} - R_{RbBr} = 3.23,$$
$$R_{CsCl} - R_{CsBr} = 3.21. \tag{10}$$

This result suggests that we may speak with some significance of the refractivity of individual ions in the simpler ionic crystals. Using Eq. (6), Sec. 148, we obtain

$$R_\infty = \frac{N_A e^2}{3\pi m} \sum_l \frac{f_l}{\nu_l^2}.$$

[1] K. Fajans and G. Joos, *Z. Physik*, **23**, 1 (1923).

Hence, if the absorption frequencies ν_l^2 lie sufficiently close to one another,

$$R_\infty = \frac{N_A e^2}{3\pi m \overline{\nu^2}} f \tag{11}$$

where f is the total optical strength per ion and $\overline{\nu^2}$ is the mean value of the $1/\nu_l^2$. Since the position of the Schumann-region absorption bands in those halides having the same halogen ion are nearly the same, it follows that the additivity of the refractivities implies that the f factors for separate ions are additive.

By choosing the value 0.50 for the refractivity of the sodium ion, for reasons which we shall not discuss here, Fajans and Joos[1] have obtained the ion refractivities given in Table LXXXVIII from observed differences

TABLE LXXXVIII.—THE REFRACTIVITIES OF IONS (AFTER FAJANS AND JOOS)

Ion	R	Ion	R	Ion	R
F^-	2.5	Li^+	0.2	Be^{++}	0.1
Cl^-	9.00	Na^+	0.5	Mg^{++}	0.3
Br^-	12.67	K^+	2.23	Ca^{++}	1.3
I^-	19.24	Cs	6.24	Ba^{++}	4.3

of the type (10). When the value 9.00 for Cl^- is substituted in Eq. (11) along with the value $\bar{\nu} = 2.4 \cdot 10^{15}$ sec $^{-1}$ for the approximate center of gravity of the absorption bands of the alkali chlorides, it leads to

$$f_{Cl} = 3.4,$$

which is to be compared with the value 3.25 derived by Mayer (*cf.* Table LXXXVII). The corresponding values for I^- agree to about the same degree of accuracy.

151. Semi-conductors.—It was pointed out in Sec. 6 that there are two types of semi-conductor, namely, monatomic crystals such as silicon and selenium that contain impurities, and ionic crystals that either are impure or contain a stoichiometric excess of one constituent. Most prominent among the semi-conductors of the second kind are alkali halides with F centers, phosphorescent zinc sulfide, and similar alkaline-earth oxides and sulfides. The impurity or stoichiometric-excess atoms in all these semi-conductors have their own characteristic absorption bands that lie in the visible or near ultraviolet part of the spectrum. In the case of natural semi-conductors, such as silicon and the natural sulfides, which may have as much as 1 per cent of impurity, this absorption band may show up as an appreciable peak in the nk curve determined from reflection, even though it may overlap the fundamental absorption

[1] *Ibid.*

band of the substance. For example, Fig. 13 shows the nk curves for silicon and natural stibnite, MoS_2. The peaks that occur in the near ultraviolet are probably due to the impurities whose thermally or optically freed electrons make these substances semi-conductors. In most artificial semi-conductors, on the other hand, the number of impurity atoms is comparatively low so that they do not give rise to nk peaks of this magnitude. The absorption may be detected, however, by transmission measurements if the bands do not overlap the fundamental region.

The theory of dispersion has been applied to the F-center bands of the alkali halides by Smakula,[2] in order to determine the density of

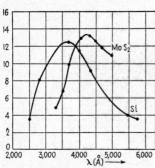

FIG. 13.—nk curves for silicon and natural MoS₂.

centers. It may be recalled that these bands probably arise from the excitation of electrons in vacant halogen sites in crystals containing an excess of alkali metal atoms, the absorption transition being analogous to the $1s$-$2p$ transition in atomic hydrogen (cf. Secs. 110 and 111). Since the electrons are coupled to the lattice, the observed absorption bands are much wider than the lines of free atoms at room temperature. The extinction coefficient of the bulk material is zero in the vicinity of the F-center bands, since these bands usually are far from the fundamental absorption region; moreover, the refractive index of the pure salt is usually constant in the vicinity of the F bands. We shall designate this constant by n'. For these reasons, Smakula assumed that the polarizability of the bulk material is given simply by the quantity

$$\frac{3}{4\pi}\frac{n'^2 - 1}{n'^2 + 2}.\tag{1}$$

In addition, he assumed that the complex polarizability of the electrons in the F centers may be represented by the corresponding polarizability function for a single absorption line, namely,

$$\frac{n_0 e^2}{4\pi^2 m}\frac{f}{\nu_F^2 - \nu^2 + i\nu_d'\nu}.\tag{2}$$

Here, n_0 is the density of F centers, ν_F is the frequency at the center of the absorption band, ν_d' is the damping frequency which is of the order of magnitude of 10^{14} sec^{-1}, and f is the oscillator strength of the transi-

[1] MINOR, op. cit.; MEIER, op. cit.

[2] A. SMAKULA, Z. Physik, **59**, 603 (1930).

tion. Since only one absorption peak is observed, we may expect that practically all the optical strength of the stoichiometric-excess electrons is centered in this band and that f should be near to unity. Using Eq. (28), Sec. 147, for the relation between the complex index of refraction and the polarizability, we obtain

$$\frac{(n - ik)^2 - 1}{(n - ik)^2 + 2} = \frac{n'^2 - 1}{n'^2 + 2} + \frac{n_0 f e^2}{3\pi m} \frac{1}{(\nu_F^2 - \nu^2) + i\nu'_d \nu}. \tag{3}$$

If we now set

$$n = n' + \Delta n \qquad \text{and} \qquad \nu = \nu_F + \Delta\nu$$

where Δn is a small quantity, we obtain for the real and imaginary parts of (3)

$$\Delta n = -\frac{n_0 f e^2}{18\pi m} \frac{(n'^2 + 2)^2}{n'} \frac{\Delta\nu(2\nu_F + \Delta\nu)}{\Delta\nu^2(2\nu_F + \Delta\nu)^2 + \nu'^2{}_d(\nu_F + \Delta\nu)^2}, \tag{3a}$$

$$k = \frac{n_0 f e^2}{18\pi m} \frac{(n'^2 + 2)^2}{n'} \frac{\nu'_d(\nu_F + \Delta\nu)}{\Delta\nu^2(2\nu_F + \Delta\nu)^2 + \nu'^2{}_d(\nu_F + \Delta\nu)^2}. \tag{3b}$$

The observed k curves, which are measured directly by the extinction of transmitted radiation, may be fitted closely by a function of the form (3b), as is shown in Fig. 14.

ν'_d may be eliminated from (3b) by expressing this quantity in terms of the value $\Delta\nu_{\frac{1}{2}}$ of $\Delta\nu$ for which k is half its maximum value k_m. When the resulting equation is solved for $n_0 f$, it is found that

$$n_0 f = \frac{9 k_m n' m \nu_F}{\pi(n' + 2)^2 e^2} \cdot \frac{\Delta\nu_{\frac{1}{2}}(2\nu_F + \Delta\nu_{\frac{1}{2}})}{(\nu_F + \Delta\nu_{\frac{1}{2}})^2}. \tag{4}$$

In making practical use of this equation, it is convenient to express the frequencies in electron volts and to express k in terms of the constant $\alpha(\nu)$ appearing in the equation

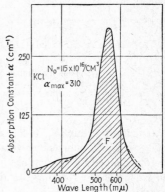

Fig. 14.—A comparison of the observed α curve for the F-centers of KCl and the curve of the form (3b) that fits it most closely. (*After Kleinschrod.*)

$$\frac{I}{I_0} = e^{-\alpha(\nu)d} \tag{5}$$

which expresses the decrease in intensity of light after it passes through a crystal of thickness d. The relation between k and α is

$$k = \frac{\alpha\lambda}{4\pi}. \tag{6}$$

When this is done, Eq. (4) becomes

$$n_0 f = 1.31 \cdot 10^{17} \frac{n'}{(n'^2 + 2)^2} \alpha_m \Delta \nu_{\frac{1}{2}} \frac{2\nu_F + \Delta \nu_{\frac{1}{2}}}{\nu_F + \Delta \nu_{\frac{1}{2}}} \tag{7}$$

where α_m, the value of α at the center of the absorption band, is expressed in inverse centimeters and the frequencies are expressed in electron volts.

If ν_F is much greater than $\nu_{\frac{1}{2}}$, as at low temperatures, we may simplify this equation to

$$n_0 f = 1.31 \cdot 10^{17} \frac{n'}{(n'^2 + 2)^2} \alpha_m W \tag{8}$$

where W is the width of the k curve at half maximum in electron volts.

Equation (7) has been used by Hilsch and Pohl[1] and their collaborators to determine the value of $n_0 f$ for F centers and for impurity

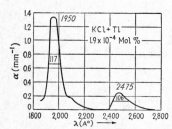

FIG. 15.—The impurity-induced α curve of a potassium chloride crystal containing thallium. (*After Koch.*)

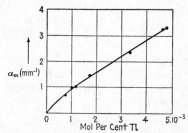

FIG. 16.—The curve obtained by plotting the optically determined values of α_m in crystals of KCl containing thallium as a function of the chemically determined number of impurity atoms.

atoms of other kinds. It has also been found possible to measure n_0 by direct chemical means in several of these crystals so that values of f may be determined by combining the results. Thus, in the case of F centers in KCl, Kleinschrod[2] has found that f is 0.81. The fact that this is not exactly unity indicates that the F-center electrons lose some optical strength because of interaction with the closed shells of the atoms present.

Koch[3] has combined optical and chemical measurements in a similar way to determine the f factors for the absorption bands of impurity thallium atoms in alkali halide crystals. It is found that small quantities of thallium halides may be dissolved in the alkali halides and that the resulting mixed crystals exhibit narrow absorption bands[4] in the ultra-

[1] See the survey by R. W. Pohl, *Physik. Z.*, **39**, 36 (1938).

[2] F. G. KLEINSCHROD, *Ann. Physik*, **27**, 97 (1936).

[3] W. KOCH, *Z. Physik*, **57**, 638 (1930).

[4] An interpretation of these peaks has been given by the writer, *Jour. Chem. Phys.*, **6**, 150 (1938).

violet below 3000 A.   A typical absorption curve is shown in **Fig. 15**.
In each case, there are two large peaks that seem to be closely related
to the energy levels of free monovalent thallium ions.   Figure 16 shows
the α_m versus n_0 curve obtained by combining optical and chemical
measurements.   From the slope of these, Koch obtains $f \sim 0.1$ for the
long wave-length band and $f \sim 0.6$ for the short wave-length band.

152. The Infrared Spectra of Ionic Crystals.—All polar compounds
possess infrared absorption bands that are associated with the stimulation
of oscillational motion of the atoms or ions.   Although the interatomic
forces in ionic crystals are comparable with electronic forces in atoms, the
vibrational spectra lie in the infrared because ionic masses are of the
order of magnitude 10^4 times larger than the electronic mass.   This fact
is made evident by the relation between the frequency ν and mass m
of an oscillator having force constant κ, namely,

$$\nu = \frac{1}{2\pi}\sqrt{\frac{\kappa}{m}}. \tag{1}$$

In order to discuss the optical effects associated with the lattice
vibrations, it is first necessary to obtain the expression for the dipole
moment of the lattice as a function of ionic displacements.   If we
arbitrarily define the dipole moment as zero when all the ions are at their
equilibrium positions $\mathbf{r}_\alpha(n)$, the dipole moment when the ions are at
positions $\mathbf{R}_\alpha(n)$ relative to the equilibrium positions is

$$\mathbf{M} = \sum_{\alpha,n} e_\alpha \mathbf{R}_\alpha(n). \tag{2}$$

Here, α extends over the ions in the unit cell, n extends over the cells
in the lattice, and e_α is the charge on the αth ion.   The variables $\mathbf{R}_\alpha(n)$
may be expressed in terms of normal coordinates of the form

$$\mathbf{R}_\alpha(n) = \sum_{s=1}^{3n}\sum_{\sigma} a_s(\sigma)\frac{\xi_{\alpha,s}(\sigma)}{\sqrt{NM_\alpha}}e^{2\pi i\sigma\cdot\mathbf{r}_\alpha(n)} \tag{3}$$

when the potential energy is a quadratic function of displacements (*cf.*
Sec. 22).   Here, $a_s(\sigma)$ is the amplitude of the normal mode of wave
number σ in which the αth atom is polarized in the direction $\xi_{\alpha,s}(\sigma)$, and
N is the total number of unit cells in the lattice.   We shall employ the
reduced-zone scheme so that σ extends over a single zone and s takes
values from 1 to $3n$, where n is the number of atoms in the unit cell.
When Eq. (3) is substituted in (2), it is found that

$$\mathbf{M} = \sqrt{N}\sum_\alpha e_\alpha \sum_s a_s(\sigma)\xi_{\alpha,s}(\sigma)\delta_{\sigma,0}. \tag{4}$$

Thus, only the modes of vibration associated with zero wave number in the reduced-zone scheme contribute to the dipole moment. The reason for this is that the contributions to **M** from different cells cancel one another in the other cases, since they have different phases.

Let us consider a cubic crystal such as sodium chloride that has two oppositely charged ions per unit cell. In this case, three of the normal modes associated with zero wave number are purely translational and, for this reason, do not contribute to the dipole moment. Hence, the dipole moment is determined by the remaining three modes, which correspond to the maximum frequency ν_m and represent oscillations in which the positive and negative ions move in opposite directions. Thus, the crystal is equivalent to a system of $3N$ diatomic oscillators of frequency ν_M. Since the polarizability of an oscillator is the same in quantum and in classical mechanics, it follows from the results of Sec. 147 that

$$\alpha_c = \frac{3n_0 e_i^2}{4\pi^2 \mu} \frac{1}{(\nu_m^2 - \nu^2) + i\gamma\nu} \tag{5}$$

where n_0 is the number of molecules per unit volume, μ is the reduced mass of the ions, e_i is the ionic charge, and γ is the damping frequency. As in the case considered in Sec. 147, this complex polarizability implies an absorption line of half-width γ at ν_m. Since an atomic mass rather than the electronic mass appears in the denominator of this equation, the polarizability arising from ionic oscillators is of the order of magnitude 10^{-4} times as large as the polarizability that would arise from an equal density of electronic oscillators of comparable frequency. For this reason, the index of refraction in the transparent visible and ultraviolet regions of most ionic crystals is determined almost entirely by the electronic absorption bands in the far ultraviolet.

In an ideal harmonic approximation, the damping frequency γ would be determined entirely by radiation damping and would have the value

$$\gamma = \frac{4\pi\nu_m^2}{3} \frac{e^2}{\mu c^3}$$

which is of the order of magnitude of 1 sec^{-1}, or about 10^{-14} ev for ordinary ionic crystals. The observed widths actually are far greater than this. For example, Fig. 17 shows observed[1] transmission curves for several specimens of sodium chloride at room temperature. It may be seen that the width of the peak is of the order of $5 \cdot 10^{12}$ sec^{-1}, or about 0.01 ev. It should also be observed that the transmission curves show more structure than should be expected from a single absorption

[1] R. B. Barnes, R. R. Brattain, and F. Seitz, *Phys. Rev.*, **48**, 582 (1935).

line.   Even more complicated structure has been observed under higher
dispersion in magnesium oxide, which has the same lattice structure as
sodium chloride.

A qualitative interpretation of this large damping and the accompanying structure was first given by Born and Blackman[1] on the basis of
classical mechanics.   They related the structure to cubic terms in the
expression for the potential energy of the ions that couple the optically
active modes of vibration to other modes.   Their work was later extended

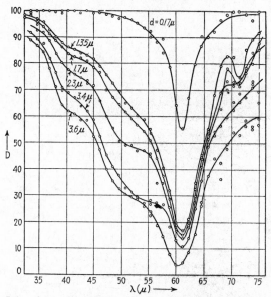

FIG. 17.—Infrared transition curves of several specimens of sodium chloride. The
abscissa is the wave length in microns. The numbers accompanying the curves are the
crystal thicknesses. (*After Barnes and Brattain.*)

by the use of quantum mechanics.[2]   Although this work provides the
machinery for a more complete theoretical investigation of the topic, a
thorough experimental treatment of the transmission properties of a
simple crystal over a range of temperatures is lacking at present.   For
this reason, it is not possible to say that the structure may be completely
interpreted in terms of anharmonic potential terms.   We shall present
briefly the principles employed in this theory.

[1] M. BORN and M. BLACKMAN, *Z. Physik*, **82**, 551 (1933); M. BLACKMAN, *Z.
Physik*, **86**, 421 (1933).
[2] BARNES, BRATTAIN, and SEITZ, *op. cit.*

In the quadratic approximation, the vibrational wave functions of the crystal have the form (*cf.* Sec. 118)

$$\Lambda_n(\cdots , \alpha_t(\eth), \cdots) = \prod^{t,\sigma} \lambda_{n_t(\sigma)}(\alpha_t(\eth)) \tag{6}$$

where the $\lambda_{n_t(\sigma)}$ are harmonic oscillator wave functions and the $n_t(\eth)$ are integers. The energy of this state has the same form as for an assembly of oscillators, namely,

$$E_n = \sum_{t,\sigma} \left(n_t(\eth) + \frac{1}{2} \right) h\nu_t(\eth). \tag{7}$$

During absorption, the system changes its state from Λ_n to the state Λ_{n+1} in which all quantum numbers are the same except for that of one of the optically active modes, which increases by unity. The energy difference between Λ_n and Λ_{n+1} is clearly $h\nu_m$.

The cubic perturbing potential has the form

$$V_c = \sum_{i,j,k} c_{ijk} \alpha_{t(i)}(\eth(i)) \alpha_{t(j)}(\eth(j)) \alpha_{t(k)}(\eth(k)) \tag{8}$$

where the c_{ijk} are constants. The limitations on the combinations of α that can occur in this series, which may be obtained from group theory, will not be discussed here.

In the perturbed scheme, the new wave functions Λ_n' have the form

$$\Lambda_n' = \Lambda_n + \sum_{n'} a_{n,n'} \Lambda_{n'} \tag{9}$$

where the $a_{n,n'}$ are given by the equations

$$a_{n,n'} = \frac{\int \Lambda_{n'} V_c \Lambda_n d\tau}{E_n - E_{n'}}, \tag{10}$$

in which the integration extends over the coordinate space of the variables $\alpha_t(\eth)$. Since V_c is the sum of cubic terms and the $\Lambda_{n'}$ are products of one-dimensional oscillator functions, it follows that each state Λ_n is now coupled with states in which three quantum numbers differ from those of Λ_n by one unit. Thus, if the system is in the state Λ_n', it may make an optical transition not only to Λ_{n+1}' but to any other state in which a wave function Λ_{n+1} appears in the sum in Eq. (9). For this reason, the optical strength of the absorption process is distributed throughout many states, and the absorption band is broader than in the quadratic approximation. We may expect the width of this absorption band to increase with increasing temperature because the amplitudes $\alpha_t(\eth)$ in (8) increase with increasing temperature. This effect has been observed qualitatively.

153. Special Topics.—There are a number of interesting topics concerning the optical properties of solids that limitations on space do not permit us to discuss in detail.   For the benefit of readers who are interested, we shall outline several of these topics briefly and give the principal references.

a. The Photoelectric Effect in Metals.—In the interior of a metal, the only allowed optical transitions take place between bands in accordance with the selection rules discussed in the previous section, namely, that the transition must be vertical in the reduced-zone scheme.   Tamm and Schubin[1] have pointed out that additional absorption may take place near the surface since the wave functions are not periodic in this region and the selection rules employed in Sec. 149 are not valid.   Although the second type of absorption is relatively unimportant in a discussion of the optical properties of metals, since only about one quantum in five hundred is absorbed in this way in passing through the surface, it is extremely important for the photoelectric effect, for electrons that are excited near the surface are in an excellent position to get out of the metal. The first detailed treatment of the surface photoelectric effect was carried through by Mitchell[2] and has been extended by several workers.[3]   We shall discuss a treatment given by Hill that is closely patterned after Mitchell's work and has been applied to the case of the alkali metals.

Hill assumed that the electronic potential is a constant $-W_a$ inside the metal and that the electrons are restrained from pouring out by a barrier at the surface.   In the detailed computations, he considered two types of barrier, namely, a square barrier for which the potential jumps abruptly from $-W_a$ to zero, and an image-force barrier of the form

$$V(x) = \begin{cases} -\dfrac{e^2}{4x + e^2/W_a} & x \geq 0 \\ -W_a & x \leq 0 \end{cases}$$

(see Fig. 9, Chap. IV).   Since the internal optical absorption is zero in this model, because the electrons are free, it can be used only for a discussion of the surface effect.   Experimental work on the alkali metals seems to show that, even when the spectral peak for the volume photoelectric effect is appreciable, it lies so much farther in the ultraviolet than the peak for the surface effect that the two do not overlap. For this reason, the two effects can be discussed separately in these simple metals.   In addition, we know from the work of preceding chapters that the properties of alkali metals usually conform closely to those of the

[1] I. TAMM and S. SCHUBIN, *Z. Physik*, **68**, 97 (1931).

[2] K. MITCHELL, *Proc. Roy. Soc.*, **146**, 442 (1934); **153**, 513 (1936); *Proc. Cambridge Phil. Soc.*, **31**, 416 (1935).

[3] R. D. MYERS, *Phys. Rev.*, **49**, 938 (1936); A. G. HILL, *Phys. Rev.*, **53**, 184 (1938).

simple free-electron model so that it should apply to them. Recent experimental work on the photoelectric effect in barium by Cashman and Bassoe[1] shows that the surface and volume peaks of this metal overlap. Thus, the two effects would have to be discussed simultaneously in this case; moreover, a more complicated model would have to be employed, for the electrons in divalent metals are not nearly free.

To begin with, Hill computed the energy distribution function of electrons that are emitted by light of a given frequency and compared the computed function with observed ones for the case of sodium. Although the two types of curve agree well at the high-energy end, the agreement at low energies is poor, for the theoretical curves start out linearly whereas the observed curves start out nearly quadratically. The most reasonable explanation of these discrepancies is that the surface on which the measurements were made was sufficiently contaminated so that either the work function varied from point to point or the electronic transmission coefficient was different from the computed value. The way in which these quantities can affect emission was discussed in connection with thermionic emission at the end of Sec. 30.

In addition, Hill compared the observed and calculated spectral distribution functions, that is, the functions giving the dependence of the total current per unit light intensity on the frequency of the radiation. The observed curve possesses a peak that is much sharper than the peak of the theoretical curve, as may be seen from Fig. 18. A possible explanation of this discrepancy lies in the fact that the detailed optical properties of the metal were neglected in Hill's treatment. This possibility was first realized by Mitchell, but Schiff and Thomas[2] have furnished more direct evidence for its importance in a computation that is based on a semiclassical treatment of radiation. This topic has also been discussed, more recently, by Makinson.[3]

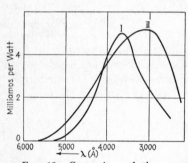

Fig. 18.—Comparison of the computed and observed spectral distribution curves for the photoelectric yield in sodium. I is the experimental curve and II is the theoretical curve. (*After Hill.*)

b. Breadth of Optical Absorption and Emission Bands.—If the atoms of an insulating crystal were held rigidly during a change in electronic

[1] R. J. Cashman and E. Bassoe, *Phys. Rev.*, **55**, 63 (1939).
[2] L. I. Schiff and L. H. Thomas, *Phys. Rev.*, **47**, 860 (1935).
[3] R. E. B. Makinson, *Proc. Roy. Soc.*, **162**, 367 (1937).

state involving absorption and emission of radiation, the frequency distribution of absorbed or emitted light would have only the natural width (see Sec. 148, part *b*). The actual emission and absorption spectra of solids exhibit a broadening that increases with increasing temperature. The primary source of this breadth is the fact that the vibrational modes of the crystal also may be stimulated during an electronic transition. Since the amount of vibrational energy that may be involved has a finite range, the allowed optical emission or absorption frequencies also extend over a finite range.

A rudimentary treatment of the theory of broadening has been given by Peierls;[1] the salient points of his work are as follows. The elastic constants and the equilibrium atomic positions are usually different for the normal and excited electronic states of an insulator. For this reason, the systems of vibrational wave functions for the normal and excited states are not identical. If the difference between the atomic potential energies for the normal and excited states is designated by $\Delta E(\xi_1, \cdots, \zeta_f)$ where $\xi_1, \ldots, \zeta_f$ are the configurational coordinates of the atoms, the vibrational wave functions χ'_n for the excited electronic state may be expressed in terms of the vibrational wave functions χ_m for the normal state by means of the perturbation equation

$$\chi'_n = \chi_n + \sum_m{}' \chi_m \frac{\int \chi_m{}^* \Delta E \chi_n d\tau}{E_n - E_m}. \tag{1}$$

The indices n and m correspond to sets of vibrational quantum numbers. Now, if χ_m is the vibrational wave function of the system before the transition, the final state may be any state χ'_n for which the integral

$$\int \chi_m{}^* \chi'_n d\tau \tag{2}$$

does not vanish, if we assume that the electronic transition is allowed. The integration in (2) takes place over the configurational coordinates. According to (1), the integral (2) is equal to

$$\frac{\int \chi_m{}^* \Delta E \chi_n d\tau}{E_n - E_m}.$$

An analysis of this integral that is based on a power series expansion of ΔE shows that at low temperatures the absorption or emission bands should consist of a sharp strong line which has a companion band on its long wave-length side whose shape is simply related to the vibrational frequency spectrum. At high temperatures, the structure is more complicated.[2]

[1] R. Peierls, *Ann. Physik*, **13**, 905 (1932).
[2] A treatment of this problem for the case of metals has been given by T. Muto, *Sci. Papers Inst. Phys. Chem. Res.*, **27**, 179 (1935).

c. The Fluorescence of Crystals.—Many simple crystals fluoresce when illuminated with ultraviolet light or bombarded with electrons. Although a large number of these *phosphors* have been prepared for commercial purposes, only a very few have been investigated with sufficient thoroughness to make a discussion of the mechanism of luminescence feasible.[1] Three substances in the second class are the zinc sulfide phosphors, willemite, which is a form of zinc silicate, and the alkali halide thallium phosphors, which are alkali halide crystals containing a small amount of thallium halide. We shall discuss briefly the properties of the first of these, which is typical of the set.[2]

The zinc sulfide phosphors are prepared most simply by heating zinc sulfide alone or with a small amount of another heavy metal sulfide, such as the sulfides of copper, silver, and manganese. The pure phosphor fluoresces with a light-blue color under near-ultraviolet light, whereas the other materials have different colors that depend upon the impurity atoms. The quantum efficiency of this luminescence usually is nearly unity at room temperature. The materials usually are strongly phosphorescent; that is, some of the light is emitted after excitation ceases. The length of time required for emission of this stored light increases as the temperature is lowered.

It is found that these fluorescent zinc sulfide materials are photoconducting and that the spectral sensitivity curve for stimulating photoconductivity extends over essentially the same region as the corresponding curve for stimulation of luminescence. On the basis of facts of this kind and a knowledge of the behavior of impurity atoms in semi-conductors (*cf.* Secs. 110 to 112), it has been concluded that the stimulating ultraviolet light liberates electrons from neutral interstitial atoms of the impurity metal, or of zinc in the pure phosphor, and that light is emitted when the electron and interstitial ion recombine, the color of the emitted light depending upon the kind of interstitial atom that does the emitting.

Since the freed electron may be trapped before returning, the crystal is phosphorescent. The decay of phosphorescence is temperature-dependent, since the trapped electrons must be freed thermally. On the basis of a detailed examination of this decay, Johnson[3] has concluded that there are at least two types of trapping center.

The zinc sulfide phosphors may be stimulated to a lesser extent by ultraviolet light that lies in the fundamental absorption band of zinc

[1] See the reviews by F. Seitz, *Trans. Faraday Soc.*, **35**, 74 (1939); H. W. Leverenz and F. Seitz, *Jour. Applied Phys.*, **10**, 479 (1939).

[2] This picture was presented independently by A. Schleede, *Angew. Chem.*, **50**, 908 (1937), and by the writer, *Jour. Chem. Phys.*, **6**, 454 (1938).

[3] R. P. Johnson, *Jour. Opt. Soc. Am.*, **29**, 387 (1939).

sulfide or by bombardment with cathode rays. Although the energy efficiency of this type of excitation is of the order of one-tenth the efficiency of near-ultraviolet excitation, it is about one thousand times higher than it would be if only the centers that are ionized by direct absorption were responsible for the light. If the absorption of energy in the fundamental absorption band produces excitons, as in the case of the alkali halides, we may conclude that a fraction of the excitons eventually give their energy to the interstitial atoms by a process analogous to a collision of the second kind. It is possible in zinc sulfide, however, that absorption in the fundamental region produces free electrons instead of excitons and that these excite the interstitial atoms by a collision of the first kind.

The wave length of the emitted radiation is always longer than that of the exciting radiation; the reason for this is probably that given in Sec. 108. In addition, the emission band consists of a broad band at room temperature. This band becomes narrower as the temperature is lowered[1] and usually consists of a single sharp line and several weak satellites at very low temperatures. The explanation of the sharpening undoubtedly is that given in part *b* of this section; however, the fine structure has not yet been completely interpreted, although it probably is also connected with the stimulation of the vibrational modes of the crystal.[2]

Willemite resembles the sulfide phosphors closely, for photoconductivity[3] accompanies luminescence in this case as well. The alkali halide phosphors, however, belong in a different category, for they are not photoconductors.[4] A fuller discussion of these materials may be found in the articles listed in footnote 2, page 672.

d. The Photolysis of Crystals.—Many crystals become colored or decompose when they are irradiated with light of suitable wave length. In this connection, we have already discussed the discoloration produced in alkali halides by X rays (*cf.* Sec. 111), which is due to the transfer of electrons from inner shells to vacant negative-ion sites.

The most important and useful photolytic process occurring in a simple crystal is that responsible for the latent photographic image in silver halide crystals. If silver chloride or silver bromide crystals are exposed for a short time to light lying in the visible or near ultraviolet region of the spectrum, a visually imperceptible change is produced in them; however, when the crystals are placed in certain reducing agents,

---

[1] J. T. RANDALL, *Nature* **142**, 113 (1938); *Trans. Faraday Soc.*, **35**, 2 (1939).

[2] F. SEITZ, *Trans. Faraday Soc.*, **35**, 1 (1939).

[3] R. HOFSTADTER, *Phys. Rev.*, **54**, 864 (1938).

[4] See *Jour. Chem. Phys.*, **6**, 150 (1938).

known as "developers," the irradiated parts of the crystal proceed to decompose with the production of free silver and the corresponding halogen. The same decomposition may be produced by continuous irradiation without development—a process known as the "print-out effect."

The credit for unraveling the fundamental processes in the darkening of the silver halides belongs to a large number of workers whose contributions extend over many years of intensive work.[1] After it had been definitely established that the decomposition products of the print-out effect are free silver and halogen gas, Fajans suggested that the fundamental action of the light is to transfer an electron from a halogen ion to a silver ion, producing free silver in accordance with the reaction

$$Ag^+ + Br^- \rightarrow Ag^+ + Br + electron \rightarrow Ag + Br. \qquad (3)$$

This hypothesis was supported by the observation of Toy and Harrison (*cf.* Sec. 135) that photoconductivity accompanies the absorption of light in the region of wave lengths in which the latent image is produced. After the development of the zone theory of solids, Webb employed this scheme to describe the freeing of electrons and their subsequent trapping in the lattice. Although this work went a long way toward explaining the initial steps in the darkening process, it left unexplained the manner in which the silver ions migrate to a given point in order to form a clump of free silver. The final steps were developed by Gurney and Mott[2] who were able to give a fairly complete description of the darkening process. Briefly, the picture is as follows:

1. After being freed, the photoelectron wanders about through the crystal and ultimately becomes trapped at a point near the surface. It is believed that the most likely trapping center is a speck of silver sulfide, the reason for this being that extensive chemical work has shown that the gelatin of photographic emulsions must contain a small amount of a sulfur compound if the latent image is to be produced. Presumably, a small amount of this substance is used in the production of silver sulfide. Gurney and Mott suggest that the work function of silver sulfide is enough larger than the work functions of the silver halides so that a small speck of the former substance should be a good trapping center.

2. The trapped electron attracts the silver ions in its vicinity, and these ions migrate toward it by the ordinary process of ionic conductivity. One of the silver ions reaches the trapped electron and is neutralized, producing an atom of silver. This is the essential point in Gurney and

---

[1] See the survey by J. H. Webb, *Jour. App. Phys.*, **11**, 18 (1940).
[2] R. W. Gurney and N. F. Mott, *Proc. Roy. Soc.*, **164**, 151 (1938).

Mott's picture and is supported by the fact that the silver halides have an appreciable ionic conductivity at room temperature (*cf.* Fig. 66, Chap. I).   It is assumed that there are only one or two trapping positions in the small crystals that occur in ordinary photographic emulsions, so that practically all the free electrons produced in a given crystal go to the same point.   Thus, one atom of free silver is formed at the trapping center for each photoelectron released.   It is believed that small specks of silver formed in this way represent the latent image.

3. It should be added that the free halogen atoms produced during the formation of the latent image presumably diffuse out of the crystal. The probability that they will run into the latent image and interact with it is small.

4. Gurney and Mott suggest that in the early stages of the formation of the latent image the trapped electrons have an appreciable chance of evaporating and returning to the halogen atoms from which they were originally released, thereby reversing the process.   As the amount of free silver grows, however, the work function of the trapping centers should approach the value of about 4 ev for metallic silver, making it more and more difficult for reversal to occur.   According to the most reliable measurements it requires between five and ten quanta per grain to form a stable latent image under the most favorable conditions.   This fact indicates that the work function of a clump of five or ten silver atoms is sufficiently large to prevent reversal at ordinary temperatures.

5. Extensive experimental investigation by Webb and others has shown that the efficiency for producing the latent image decreases with decreasing light intensity at very weak intensities.[1]   This fact is an immediate consequence of the possibility of the reversal discussed under 4, for if the light intensity is sufficiently weak each silver atom may dissociate thermally before another is formed.   The efficiency for producing the latent image does not continually increase with increasing light intensity, however, for it is found to fall at high intensities.   Limited ionic conductivity presumably is responsible for this effect.   Unless the charges of the trapped electrons are neutralized as fast as they are trapped, some of the electrons will be repelled from the trapping center and will recombine with the holes.   In this connection, Webb[2] has shown that the efficiency for production of the latent image attains a low value that is independent of light intensity at liquid-air temperature.   Presumably, both the probability of thermal dissociation of the silver atoms and the ionic conductivity are vanishingly small at this temperature so that

[1] This type of change of efficiency with intensity is related to reciprocity-law failure of ordinary photographic plates for exposures with light.

[2] J. H. Webb and C. H. Evans, *Jour. Optical Soc. Am.*, **28**, 249 (1938).

the light simply charges the centers to a point where all other electrons are repelled.   When the crystal is warmed, enough silver ions migrate to the electrons to neutralize the charge, which happens to be sufficient to form a latent image.   If it were not sufficient, no latent image would be formed as a result of low-temperature illumination.

6. If illumination is continued after the latent image has formed, the amount of free silver continues to grow and eventually an appreciable fraction of the crystal is decomposed corresponding to the results of the print-out effect.

7. Since the latent image is near the surface of the crystal, it comes in contact with the developer when the crystal is immersed.   Apparently, the silver atoms of the latent image oxidize the developer molecules and thus obtain a negative charge which attracts silver ions and causes the amount of free silver to grow just as if illumination had been continued.

Another type of darkening process which has been studied extensively[1] is that occurring in zinc sulfide which has been suitably heated.   Since zinc sulfide is most commonly used either as a paint pigment or as a luminescent material, this darkening usually is a disadvantage.

---

[1] This work is reviewed in a paper by N. T. Gordon, F. Seitz, and F. Quinlan, *Jour. Chem. Phys.*, **7**, 4 (1938).

APPENDIX

DERIVATION OF HARTREE'S AND FOCK'S EQUATIONS

a. Hartree's Equations.—Hartree's equations are based upon an eigenfunction of the type

$$\Psi = \psi_1(\mathbf{r}_1)\psi_2(\mathbf{r}_2) \cdots \psi_n(\mathbf{r}_n), \tag{1}$$

in which it is assumed that the ψ_i satisfy the normalization condition

$$\int |\psi_i|^2 d\tau_i = 1 \tag{2}$$

but are not necessarily orthogonal.

From the variational theorem, we should expect the "best" ψ_i to be those for which

$$\delta \int \Psi^* H \Psi d\tau(x_1, \cdots, z_n, \zeta_1, \cdots, \zeta_n) = 0 \tag{3}$$

with the auxiliary condition (2). We shall write H in the form

$$H = \sum_i H_i + \frac{1}{2}{\sum_{i,j}}' \frac{e^2}{r_{ij}} \qquad (i, j = 1, \cdots, n) \tag{4}$$

where H_i depends only upon the variables $\mathbf{r}_i$ and is the same function of these as H_j is of $\mathbf{r}_j$. Equation (3) may then be written in the form

$$\sum_i \int \Big[\psi_1^*(\mathbf{r}_1) \cdots \psi_n^*(\mathbf{r}_n)\Big(\sum_j H_j + \frac{1}{2}{\sum_{j,k}}' \frac{e^2}{r_{jk}}\Big) \cdot$$

$$\psi_1(\mathbf{r}_1) \cdots \psi_{i-1}(\mathbf{r}_{i-1})\psi_{i+1}(\mathbf{r}_{i+1}) \cdots \psi_n(\mathbf{r}_n)\delta\psi_i(\mathbf{r}_i) \Big] d\tau' +$$

$$\sum_i \int \Big[\psi_1^*(\mathbf{r}_1) \cdots \psi_{i-1}^*(\mathbf{r}_{i-1})\psi_{i+1}^*(\mathbf{r}_{i+1}) \cdots \psi_n^*(\mathbf{r}_n)\delta\psi_i^*(\mathbf{r}_i) \cdot$$

$$\Big(\sum_j H_j + \frac{1}{2}{\sum_{j,k}}' \frac{e^2}{r_{jk}}\Big)\psi_1(\mathbf{r}_1) \cdots \psi_n(\mathbf{r}_n) \Big] d\tau' = 0. \tag{5}$$

When integrated, this reduces to

$$\sum_i \int \psi_i^* \Big[\sum_{k \neq i} \int \psi_k^* H_k \psi_k d\tau_k + \frac{1}{2}{\sum_{j,k \neq i}}' e^2 \int \frac{|\psi_j(\mathbf{r}_j)|^2 |\psi_k(\mathbf{r}_k)|^2}{r_{jk}} d\tau_{jk} +$$

$$H_i + {\sum_j}' e^2 \int \frac{|\psi_j|^2}{r_{ij}} d\tau_j \Big]\delta\psi_i + \text{(a symmetrical expression in } \delta\psi_i^*) = 0. \tag{6}$$

The variational equivalent of (2) is

$$\lambda_i(\int \psi_i{}^* \delta\psi_i d\tau + \int \delta\psi_i{}^* \psi_i d\tau) = 0. \tag{7}$$

The result of adding (7) to (6) with Lagrangian multipliers, λ_i, is

$$\sum_i \int \psi_i{}^* \left[\sum_{j \neq i} \int \psi_j{}^* H_j \psi_j d\tau_j + \frac{1}{2} \sum_{j,k \neq i}{}' e^2 \int \frac{|\psi_j(\mathbf{r}_j)|^2 |\psi_k(\mathbf{r}_k)|^2}{r_{jk}} d\tau_j + \right.$$

$$\left. H_i + \sum_j{}' e^2 \int \frac{|\psi_j|^2}{r_{ij}} d\tau_j + \lambda_i \right] \delta\psi_i d\tau_i +$$

$$\text{(a symmetrical expression in } \delta\psi_i{}^*) = 0. \tag{8}$$

If the condition that H is Hermitian is used, the position of $\delta\psi_i$ and $\psi_i{}^*$ may be reversed in the written term of (8). Since $\delta\psi_i$ and $\delta\psi_i{}^*$ are independent variations and are independent of the variations of $\delta\psi_k$ and $\delta\psi_k{}^*(k \neq i)$, the necessary and sufficient condition that (8) be satisfied is that the coefficient of each $\delta\psi_i$ and each $\delta\psi_i{}^*$ be zero. These conditions are

$$H_i\psi_i + \left(\sum_j{}' \int \frac{|\psi_j|^2}{r_{ij}} d\tau_j \right)\psi_i + \left(\sum_{j \neq i} \int \psi_j{}^* H \psi_j d\tau_j + \right.$$

$$\left. \sum_{j,k \neq i}{}' e^2 \int \frac{|\psi_j|^2|\psi_k|^2}{r_{jk}} d\tau_{jk} + \lambda_i \right)\psi_i = 0 \tag{9}$$

and

$$H_i\psi_i{}^* + \left(\sum_j{}' \int \frac{|\psi_j|^2}{r_{ij}} d\tau_j \right)\psi_i{}^* + \left(\sum_{j \neq i} \int \psi_j{}^* H \psi_j d\tau_j + \right.$$

$$\left. \sum_{j,k \neq i}{}' e^2 \int \frac{|\psi_j|^2|\psi_k|^2}{r_{jk}} d\tau_{jk} + \lambda_i \right)\psi_i{}^* = 0.$$

Obviously, only one of these need be considered since the two equations are complex conjugates. We obtain Hartree's equations

$$H_i\psi_i + \left(\sum_{j \neq i} \int \frac{|\psi_j|^2}{r_{ij}} d\tau_j \right)\psi_i + \epsilon_i\psi_i = 0 \tag{10}$$

by setting

$$\left(\sum_{i \neq j} \int \psi_j{}^* H \psi_j d\tau_j + \sum_{j,k \neq i}{}' e^2 \int \frac{|\psi_j|^2|\psi_k|^2}{r_{jk}} d\tau_{jk} + \lambda_i \right) = \epsilon_i. \tag{11}$$

If we require, in addition to (2), that

$$\int \psi_i{}^* \psi_j d\tau = \delta_{ij} \tag{12}$$

and if we add the variational equivalent of this to (8), we find that a term $\sum_{j \neq i} \lambda_{ij} \psi_j$ is added to (11).

b. Fock's Equations.—The derivation of Fock's equations may proceed along exactly the same lines as those employed in part *a* with the exception that the basic wave function is

$$\Psi = \frac{1}{\sqrt{n!}} \sum_{P_v} (-1)^{p_v} P_v [\psi_1(\mathbf{r}_1) \cdots \psi_n(\mathbf{r}_n) \eta_1(\zeta_1') \cdots \eta_n(\zeta_n')] \quad (13)$$

where

$$\psi_1 = \psi_2, \qquad \psi_3 = \psi_4, \cdots, \qquad \psi_{n-1} = \psi_n$$

and the spin functions η are opposite for the members of each of these pairs of equal functions. In this case, we have

$$\delta\Psi = \frac{1}{\sqrt{n!}} \sum_{P_v} (-1)^{p_v} P_v \Big[\sum_i \psi_1(\mathbf{r}_1) \cdots$$

$$\psi_{i-1}(\mathbf{r}_{i-1}) \psi_{i+1}(\mathbf{r}_{i+1}) \cdots \psi_n(\mathbf{r}_n) \delta\psi_i(\mathbf{r}_i) \eta_i(\zeta_i') \cdots \eta_n(\zeta_n') \Big]. \quad (14)$$

We shall substitute this in the equation

$$\delta \int \Psi^* H \Psi d\tau = \int \Psi H \delta\Psi d\tau + \int \delta\Psi^* H \Psi d\tau = 0 \quad (15)$$

where H is given by (4). After integrating, summing over spin, and using the condition $\int \psi_i^*(\mathbf{r}) \psi_j(\mathbf{r}) d\tau = 0$, for $\psi_i \neq \psi_j$, and $\int |\psi_i|^2 d\tau = 1$, we obtain

$$\sum_i \int \Big(\psi_i^*(\mathbf{r}_1) \Big[\sum_j \int \psi_j^*(\mathbf{r}_2) H_2 \psi_j(\mathbf{r}_2) d\tau_2 +$$

$$\frac{1}{2} \sum_{j,k}' e^2 \int \frac{|\psi_j(\mathbf{r}_2)|^2 |\psi_k(\mathbf{r}_3)|^2}{r_{23}} d\tau_{23} -$$

$$\frac{1}{2} \sum_{\substack{j,k \\ \| \text{ spins}}}' e^2 \int \frac{\psi_j^*(\mathbf{r}_2) \psi_k^*(\mathbf{r}_3) \psi_j(\mathbf{r}_3) \psi_k(\mathbf{r}_2)}{r_{23}} d\tau_{23} + H_i +$$

$$e^2 \sum_j \frac{|\psi_j(\mathbf{r}_2)|^2}{r_{12}} d\tau_2 + \lambda_{12}' \Big] - \Big\{ \sum_{\substack{j \\ \| \text{ spins}}}' \psi_i^*(\mathbf{r}_1) \Big[e^2 \int \frac{\psi_i^*(\mathbf{r}_2) \psi_j(\mathbf{r}_2)}{r_{12}} d\tau_2 +$$

$$\int \psi_i^*(\mathbf{r}_2) H_2 \psi_j(\mathbf{r}_2) + \lambda_{ji}' \Big] \Big\} \Big) \delta\psi_i(\mathbf{r}_1) d\tau_1 +$$

[a symmetrical expression in $\delta\psi_i^*(\mathbf{r}_1)$] = 0. (16)

The λ'_{ji} are the Lagrangian multipliers for the orthogonality and normalization conditions. We shall set

$$\lambda_{ii} = \lambda'_{ii} + \sum_j \int \psi_j{}^*(\mathbf{r}_2)H_2\psi_j(\mathbf{r}_2)d\tau_2 + \frac{1}{2}\sum_{j,k}{}' e^2 \int \frac{|\psi_j(\mathbf{r}_2)|^2|\psi_k(\mathbf{r}_3)|^2}{r_{23}}d\tau_{23} -$$

$$\frac{1}{2}\sum_{\substack{j,k \\ \parallel \text{ spins}}} e^2 \int \frac{\psi_j{}^*(\mathbf{r}_2)\psi_k{}^*(\mathbf{r}_3)\psi_j(\mathbf{r}_3)\psi_k(\mathbf{r}_2)}{r_{23}}d\tau_{23} \quad (17)$$

and

$$\lambda_{ij} = \lambda'_{ij} + \sum_j \int \psi_i{}^*(\mathbf{r}_2)H_2\psi_j(\mathbf{r}_2)d\tau_2. \quad (18)$$

We find, upon equating the coefficient of $\delta\psi_i(\mathbf{r}_1)$ to zero, as before, that

$$\left[H_1 + \sum_j e^2 \int \frac{|\psi_j(\mathbf{r}_2)|^2}{r_{12}}d\tau_2 + \lambda_{ii}\right]\psi_i(\mathbf{r}_1) -$$

$$\sum_{\substack{j \\ \parallel \text{ spins}}}{}' \left[e^2 \int \frac{\psi_j{}^*(\mathbf{r}_2)\psi_i(\mathbf{r}_2)}{r_{12}}d\tau_2 + \lambda_{ij}\right]\psi_j = 0. \quad (19)$$

Equation (19) is valid only for those states of zero multiplicity which correspond to a complete set of paired ψ_i. In other cases, these equations will be modified in a way that depends upon the type of wave function. We shall not discuss these cases since the one leading to (19) is sufficiently general for our needs.

NAME INDEX

A

Abraham, M., 211, 213, 637
Ahlberg, J. E., 114*ff.*
Akulov, N. S., 22*ff.*
Allen, J. F., 485
Allison, S. K., 288, 475
Anderson, C. T., 13, 58
Andrews, D. H., 114*ff.*
Austin, J. B., 15

B

Baber, W. G., 540
Baedeker, K., 69, 70
Balamuth, L., 95
Banks, F., 495
Bardeen, J., 340, 348, 352, 354, 374, 381*ff.*, 395, 397*ff.*, 400, 421*ff.*, 520*ff.*, 530*ff.*
Barkhausen, H., 21
Barnes, R. B., 95, 125, 666, 667
Barnett, S. J., 426
Barrer, R. M., 494
Bartlett, J. H., 251
Bassoe, E., 670
Baumbach, H. H. v., 71
Bearden, J. H., 440
Becker, G., 45*ff.*
Becker, J. A., 146, 162, 168, 404
Becker, R., 20, 211, 213, 627, 637
Beeman, W. W., 440
Bethe, H., 92, 140, 141, 147, 166, 185, 233, 314, 505, 507, 511, 520, 613*ff.*
Beyer, J., 556
Bichowsky, F. R., 3, 46, 72
Biltz, W., 38
Birtwistle, G., 137
Bitter, F., 18, 617
Black, M. M., 250, 251
Blackman, M., 100, 116, 117, 120, 133*ff.*, 597*ff.*, 667
Bleick, W. E., 89, 265, 269, 383, 393, 493

Bloch, F., 140, 251, 301*ff.*, 303*ff.*, 314*ff.*, 319, 516, 518, 520, 530, 531*ff.*, 602, 617*ff.*, 642, 649*ff.*
Boas, W., 98
Bohr, N., 235, 237
Boltzmann, L., 139, 143, 169, 319, 479, 516*ff.*, 585*ff.*
Borelius, G., 41
Born, M., 76, 79, 82*ff.*, 85*ff.*, 91, 97, 99*ff.*, 111, 112*ff.*, 117, 118, 124*ff.*, 138, 265, 271, 272, 470, 493, 553, 667
Bose, 209
Bottema, J. A., 39
Bouckaert, L. P., 275
Bozorth, R., 21, 23, 625
Bragg, W. L., 505, 507, 509, 513, 544, 604, 612
Brattain, R. R., 95, 125, 666, 667
Brattain, W. H., 168, 404
Breit, G., 210, 643
Bridgman, P. W., 5, 137, 180, 374, 381*ff.*
Briggs, H. B., 652, 653*ff.*
Brillouin, L., 100, 141, 209, 234, 272, 283*ff.*, 384, 520, 581
Brody, E., 138
Brönsted, J. N., 483, 484
Brooks, H., 426, 628
Brown, F. W., 246, 251, 265
Brown, W. F., 20
Bruch-Willstäter, M., 393*ff.*
Buehl, R. C., 562, 563
Burrau, O., 255*ff.*
Burton, E. F., 485

C

Campbell, L. L., 68
Carrard, A., 58
Cashman, R. J., 401, 670
Channel-Evans, K. M., 28, 32
Chodorow, M. I., 275, 430*ff.*
Clark, A. R., 485
Clark, C. W., 15*ff.*, 57, 114, 117, 136, 157
Clusius, K., 13, 74, 485

SUBJECT INDEX

A

Absorption coefficient, 631*ff.*, 635, 646*ff.*
  alkali metals, 652
  semi-conductors, 662
  transition metals, 656
Absorption spectra, alkali halides, 410, 446, 657*ff.*
  F centers, 662
  infrared, 665
  ionic crystals, 408*ff.*
  metals, 651*ff.*
  semi-conductors, 661*ff.*
Acceleration in band scheme, 315*ff.*
Accidental degeneracy, 290
Activation energy, reactions in solids, 474, 550
  semi-conductors, 459*ff.*
Additivity, atomic heats, 38
  ionic radii, 51
  ionic susceptibilities, 59
  refractivities, 660
Adiabatic approximation, 470*ff.*
Alkali halides, absorption spectra, 408*ff.*, 446
  bands, 441
  charge distribution, 444
  cohesion, 80*ff.*
  conduction levels, 446
  discoloration, 460
  F centers, 457
  Hall effect, 563
  ionic conductivity, 55, 385*ff.*
  lattice defects, 548*ff.*
  Madelung constants, 78
  optical properties, 657*ff.*
  photoconductivity, 413, 446, 459, 563
  semi-conductors, 457
  vacancies. 458*ff.*
Alkali metals, Bloch functions, 350*ff.*
  cohesion, 348*ff.*, 366
  correlation energy, 366
  coulomb field, 349
  effective mass. 353

Alkali metals, elastic constants, 116
  electronic structure, 420*ff.*
  exchange energy, 359, 421
  ion-core field, 348
  level density, 366
  optical properties, 423, 652
  paramagnetism, 599
  simple treatment, 382
  specific heat, 116, 421
  total wave function, 308*ff.*
  work function, 399
Alkaline earth metals, bands, 424*ff.*
  level density, 424
Alkaline earth salts, absorption spectra, 408*ff.*
  cohesion, 81*ff.*
  excitation states, 413*ff.*
  Hall coefficient, 467
  photoconductivity, 413
Allotropy, carbon, 484
  cobalt, 8, 487
  helium, 485
  ionic crystals, 89*ff.*
  iron, 8, 487
  metals, 2
  sulfur, 484
  theory, 473*ff.*, 478*ff.*
  tin, 8, 483
Alloys, 25*ff.*, conductivity, 541*ff.*
  Curie point, 45, 624
  diamagnetism, 595
  diffusion, 495
  equilibrium conditions, 500
  exchange integral, 624
  ferromagnetism, 45, 623*ff.*
  filling of levels, 434, 501
  heat of formation, 38
  Hume-Rothery rules, 28, 30*ff.*
  interstitial, 25*ff.*
  magnetic susceptibilities, 42*ff.*, 595
  ordered, 37, 502
  phase boundaries, 499
  phase changes, 500
  quenching of magnetization, 44*ff.*

687

A CATALOG OF SELECTED
DOVER BOOKS
IN ALL FIELDS OF INTEREST

A CATALOG OF SELECTED DOVER
BOOKS IN ALL FIELDS OF INTEREST

DRAWINGS OF REMBRANDT, edited by Seymour Slive. Updated Lippmann, Hofstede de Groot edition, with definitive scholarly apparatus. All portraits, biblical sketches, landscapes, nudes. Oriental figures, classical studies, together with selection of work by followers. 550 illustrations. Total of 630pp. 9⅛ × 12¼.
21485-0, 21486-9 Pa., Two-vol. set $25.00

GHOST AND HORROR STORIES OF AMBROSE BIERCE, Ambrose Bierce. 24 tales vividly imagined, strangely prophetic, and decades ahead of their time in technical skill: "The Damned Thing," "An Inhabitant of Carcosa," "The Eyes of the Panther," "Moxon's Master," and 20 more. 199pp. 5⅜ × 8½. 20767-6 Pa. $3.95

ETHICAL WRITINGS OF MAIMONIDES, Maimonides. Most significant ethical works of great medieval sage, newly translated for utmost precision, readability. Laws Concerning Character Traits, Eight Chapters, more. 192pp. 5⅜ × 8½.
24522-5 Pa. $4.50

THE EXPLORATION OF THE COLORADO RIVER AND ITS CANYONS, J. W. Powell. Full text of Powell's 1,000-mile expedition down the fabled Colorado in 1869. Superb account of terrain, geology, vegetation, Indians, famine, mutiny, treacherous rapids, mighty canyons, during exploration of last unknown part of continental U.S. 400pp. 5⅜ × 8½. 20094-9 Pa. $6.95

HISTORY OF PHILOSOPHY, Julián Marías. Clearest one-volume history on the market. Every major philosopher and dozens of others, to Existentialism and later. 505pp. 5⅜ × 8½. 21739-6 Pa. $8.50

ALL ABOUT LIGHTNING, Martin A. Uman. Highly readable non-technical survey of nature and causes of lightning, thunderstorms, ball lightning, St. Elmo's Fire, much more. Illustrated. 192pp. 5⅜ × 8½. 25237-X Pa. $5.95

SAILING ALONE AROUND THE WORLD, Captain Joshua Slocum. First man to sail around the world, alone, in small boat. One of great feats of seamanship told in delightful manner. 67 illustrations. 294pp. 5⅜ × 8½. 20326-3 Pa. $4.50

LETTERS AND NOTES ON THE MANNERS, CUSTOMS AND CONDITIONS OF THE NORTH AMERICAN INDIANS, George Catlin. Classic account of life among Plains Indians: ceremonies, hunt, warfare, etc. 312 plates. 572pp. of text. 6⅛ × 9¼. 22118-0, 22119-9 Pa. Two-vol. set $15.90

ALASKA: The Harriman Expedition, 1899, John Burroughs, John Muir, et al. Informative, engrossing accounts of two-month, 9,000-mile expedition. Native peoples, wildlife, forests, geography, salmon industry, glaciers, more. Profusely illustrated. 240 black-and-white line drawings. 124 black-and-white photographs. 3 maps. Index. 576pp. 5⅜ × 8½. 25109-8 Pa. $11.95

CATALOG OF DOVER BOOKS

THE BOOK OF BEASTS: Being a Translation from a Latin Bestiary of the Twelfth Century, T. H. White. Wonderful catalog real and fanciful beasts: manticore, griffin, phoenix, amphivius, jaculus, many more. White's witty erudite commentary on scientific, historical aspects. Fascinating glimpse of medieval mind. Illustrated. 296pp. 5⅜ × 8¼. (Available in U.S. only)　　　24609-4 Pa. $5.95

FRANK LLOYD WRIGHT: ARCHITECTURE AND NATURE With 160 Illustrations, Donald Hoffmann. Profusely illustrated study of influence of nature—especially prairie—on Wright's designs for Fallingwater, Robie House, Guggenheim Museum, other masterpieces. 96pp. 9¼ × 10¾.　　25098-9 Pa. $7.95

FRANK LLOYD WRIGHT'S FALLINGWATER, Donald Hoffmann. Wright's famous waterfall house: planning and construction of organic idea. History of site, owners, Wright's personal involvement. Photographs of various stages of building. Preface by Edgar Kaufmann, Jr. 100 illustrations. 112pp. 9¼ × 10.
23671-4 Pa. $7.95

YEARS WITH FRANK LLOYD WRIGHT: Apprentice to Genius, Edgar Tafel. Insightful memoir by a former apprentice presents a revealing portrait of Wright the man, the inspired teacher, the greatest American architect. 372 black-and-white illustrations. Preface. Index. vi + 228pp. 8¼ × 11.　　　24801-1 Pa. $9.95

THE STORY OF KING ARTHUR AND HIS KNIGHTS, Howard Pyle. Enchanting version of King Arthur fable has delighted generations with imaginative narratives of exciting adventures and unforgettable illustrations by the author. 41 illustrations. xviii + 313pp. 6⅛ × 9¼.　　　21445-1 Pa. $5.95

THE GODS OF THE EGYPTIANS, E. A. Wallis Budge. Thorough coverage of numerous gods of ancient Egypt by foremost Egyptologist. Information on evolution of cults, rites and gods; the cult of Osiris; the Book of the Dead and its rites; the sacred animals and birds; Heaven and Hell; and more. 956pp. 6⅛ × 9¼.
22055-9, 22056-7 Pa., Two-vol. set $20.00

A THEOLOGICO-POLITICAL TREATISE, Benedict Spinoza. Also contains unfinished Political Treatise. Great classic on religious liberty, theory of government on common consent. R. Elwes translation. Total of 421pp. 5⅜ × 8½.
20249-6 Pa. $6.95

INCIDENTS OF TRAVEL IN CENTRAL AMERICA, CHIAPAS, AND YUCATAN, John L. Stephens. Almost single-handed discovery of Maya culture; exploration of ruined cities, monuments, temples; customs of Indians. 115 drawings. 892pp. 5⅜ × 8½.　　22404-X, 22405-8 Pa., Two-vol. set $15.90

LOS CAPRICHOS, Francisco Goya. 80 plates of wild, grotesque monsters and caricatures. Prado manuscript included. 183pp. 6⅞ × 9⅜.　　22384-1 Pa. $4.95

AUTOBIOGRAPHY: The Story of My Experiments with Truth, Mohandas K. Gandhi. Not hagiography, but Gandhi in his own words. Boyhood, legal studies, purification, the growth of the Satyagraha (nonviolent protest) movement. Critical, inspiring work of the man who freed India. 480pp. 5⅜ × 8½. (Available in U.S. only)
24593-4 Pa. $6.95

ILLUSTRATED DICTIONARY OF HISTORIC ARCHITECTURE, edited by Cyril M. Harris. Extraordinary compendium of clear, concise definitions for over 5,000 important architectural terms complemented by over 2,000 line drawings. Covers full spectrum of architecture from ancient ruins to 20th-century Modernism. Preface. 592pp. 7½ × 9⅝. 24444-X Pa. $14.95

THE NIGHT BEFORE CHRISTMAS, Clement Moore. Full text, and woodcuts from original 1848 book. Also critical, historical material. 19 illustrations. 40pp. 4⅝ × 6. 22797-9 Pa. $2.25

THE LESSON OF JAPANESE ARCHITECTURE: 165 Photographs, Jiro Harada. Memorable gallery of 165 photographs taken in the 1930's of exquisite Japanese homes of the well-to-do and historic buildings. 13 line diagrams. 192pp. 8⅜ × 11¼. 24778-3 Pa. $8.95

THE AUTOBIOGRAPHY OF CHARLES DARWIN AND SELECTED LETTERS, edited by Francis Darwin. The fascinating life of eccentric genius composed of an intimate memoir by Darwin (intended for his children); commentary by his son, Francis; hundreds of fragments from notebooks, journals, papers; and letters to and from Lyell, Hooker, Huxley, Wallace and Henslow. xi + 365pp. 5⅜ × 8. 20479-0 Pa. $5.95

WONDERS OF THE SKY: Observing Rainbows, Comets, Eclipses, the Stars and Other Phenomena, Fred Schaaf. Charming, easy-to-read poetic guide to all manner of celestial events visible to the naked eye. Mock suns, glories, Belt of Venus, more. Illustrated. 299pp. 5¼ × 8¼. 24402-4 Pa. $7.95

BURNHAM'S CELESTIAL HANDBOOK, Robert Burnham, Jr. Thorough guide to the stars beyond our solar system. Exhaustive treatment. Alphabetical by constellation: Andromeda to Cetus in Vol. 1; Chamaeleon to Orion in Vol. 2; and Pavo to Vulpecula in Vol. 3. Hundreds of illustrations. Index in Vol. 3. 2,000pp. 6⅛ × 9¼. 23567-X, 23568-8, 23673-0 Pa., Three-vol. set $36.85

STAR NAMES: Their Lore and Meaning, Richard Hinckley Allen. Fascinating history of names various cultures have given to constellations and literary and folkloristic uses that have been made of stars. Indexes to subjects. Arabic and Greek names. Biblical references. Bibliography. 563pp. 5⅜ × 8½. 21079-0 Pa. $7.95

THIRTY YEARS THAT SHOOK PHYSICS: The Story of Quantum Theory, George Gamow. Lucid, accessible introduction to influential theory of energy and matter. Careful explanations of Dirac's anti-particles, Bohr's model of the atom, much more. 12 plates. Numerous drawings. 240pp. 5⅜ × 8½. 24895-X Pa. $4.95

CHINESE DOMESTIC FURNITURE IN PHOTOGRAPHS AND MEASURED DRAWINGS, Gustav Ecke. A rare volume, now affordably priced for antique collectors, furniture buffs and art historians. Detailed review of styles ranging from early Shang to late Ming. Unabridged republication. 161 black-and-white drawings, photos. Total of 224pp. 8⅜ × 11¼. (Available in U.S. only) 25171-3 Pa. $12.95

VINCENT VAN GOGH: A Biography, Julius Meier-Graefe. Dynamic, penetrating study of artist's life, relationship with brother, Theo, painting techniques, travels, more. Readable, engrossing. 160pp. 5⅜ × 8½. (Available in U.S. only) 25253-1 Pa. $3.95

HOW TO WRITE, Gertrude Stein. Gertrude Stein claimed anyone could understand her unconventional writing—here are clues to help. Fascinating improvisations, language experiments, explanations illuminate Stein's craft and the art of writing. Total of 414pp. 4⅝ × 6⅝. 23144-5 Pa. $5.95

ADVENTURES AT SEA IN THE GREAT AGE OF SAIL: Five Firsthand Narratives, edited by Elliot Snow. Rare true accounts of exploration, whaling, shipwreck, fierce natives, trade, shipboard life, more. 33 illustrations. Introduction. 353pp. 5⅜ × 8½. 25177-2 Pa. $7.95

THE HERBAL OR GENERAL HISTORY OF PLANTS, John Gerard. Classic descriptions of about 2,850 plants—with over 2,700 illustrations—includes Latin and English names, physical descriptions, varieties, time and place of growth, more. 2,706 illustrations. xlv + 1,678pp. 8½ × 12¼. 23147-X Cloth. $75.00

DOROTHY AND THE WIZARD IN OZ, L. Frank Baum. Dorothy and the Wizard visit the center of the Earth, where people are vegetables, glass houses grow and Oz characters reappear. Classic sequel to *Wizard of Oz*. 256pp. 5⅜ × 8. 24714-7 Pa. $4.95

SONGS OF EXPERIENCE: Facsimile Reproduction with 26 Plates in Full Color, William Blake. This facsimile of Blake's original "Illuminated Book" reproduces 26 full-color plates from a rare 1826 edition. Includes "The Tyger," "London," "Holy Thursday," and other immortal poems. 26 color plates. Printed text of poems. 48pp. 5¼ × 7. 24636-1 Pa. $3.50

SONGS OF INNOCENCE, William Blake. The first and most popular of Blake's famous "Illuminated Books," in a facsimile edition reproducing all 31 brightly colored plates. Additional printed text of each poem. 64pp. 5¼ × 7. 22764-2 Pa. $3.50

PRECIOUS STONES, Max Bauer. Classic, thorough study of diamonds, rubies, emeralds, garnets, etc.: physical character, occurrence, properties, use, similar topics. 20 plates, 8 in color. 94 figures. 659pp. 6⅛ × 9¼. 21910-0, 21911-9 Pa., Two-vol. set $14.90

ENCYCLOPEDIA OF VICTORIAN NEEDLEWORK, S. F. A. Caulfeild and Blanche Saward. Full, precise descriptions of stitches, techniques for dozens of needlecrafts—most exhaustive reference of its kind. Over 800 figures. Total of 679pp. 8⅛ × 11. Two volumes. Vol. 1 22800-2 Pa. $10.95
Vol. 2 22801-0 Pa. $10.95

THE MARVELOUS LAND OF OZ, L. Frank Baum. Second Oz book, the Scarecrow and Tin Woodman are back with hero named Tip, Oz magic. 136 illustrations. 287pp. 5⅜ × 8½. 20692-0 Pa. $5.95

WILD FOWL DECOYS, Joel Barber. Basic book on the subject, by foremost authority and collector. Reveals history of decoy making and rigging, place in American culture, different kinds of decoys, how to make them, and how to use them. 140 plates. 156pp. 7⅞ × 10¾. 20011-6 Pa. $7.95

HISTORY OF LACE, Mrs. Bury Palliser. Definitive, profusely illustrated chronicle of lace from earliest times to late 19th century. Laces of Italy, Greece, England, France, Belgium, etc. Landmark of needlework scholarship. 266 illustrations. 672pp. 6⅛ × 9¼. 24742-2 Pa. $14.95

ILLUSTRATED GUIDE TO SHAKER FURNITURE, Robert Meader. All furniture and appurtenances, with much on unknown local styles. 235 photos. 146pp. 9 × 12. 22819-3 Pa. $7.95

WHALE SHIPS AND WHALING: A Pictorial Survey, George Francis Dow. Over 200 vintage engravings, drawings, photographs of barks, brigs, cutters, other vessels. Also harpoons, lances, whaling guns, many other artifacts. Comprehensive text by foremost authority. 207 black-and-white illustrations. 288pp. 6 × 9. 24808-9 Pa. $8.95

THE BERTRAMS, Anthony Trollope. Powerful portrayal of blind self-will and thwarted ambition includes one of Trollope's most heartrending love stories. 497pp. 5⅜ × 8½. 25119-5 Pa. $8.95

ADVENTURES WITH A HAND LENS, Richard Headstrom. Clearly written guide to observing and studying flowers and grasses, fish scales, moth and insect wings, egg cases, buds, feathers, seeds, leaf scars, moss, molds, ferns, common crystals, etc.—all with an ordinary, inexpensive magnifying glass. 209 exact line drawings aid in your discoveries. 220pp. 5⅜ × 8½. 23330-8 Pa. $3.95

RODIN ON ART AND ARTISTS, Auguste Rodin. Great sculptor's candid, wide-ranging comments on meaning of art; great artists; relation of sculpture to poetry, painting, music; philosophy of life, more. 76 superb black-and-white illustrations of Rodin's sculpture, drawings and prints. 119pp. 8⅜ × 11¼. 24487-3 Pa. $6.95

FIFTY CLASSIC FRENCH FILMS, 1912–1982: A Pictorial Record, Anthony Slide. Memorable stills from Grand Illusion, Beauty and the Beast, Hiroshima, Mon Amour, many more. Credits, plot synopses, reviews, etc. 160pp. 8¼ × 11. 25256-6 Pa. $11.95

THE PRINCIPLES OF PSYCHOLOGY, William James. Famous long course complete, unabridged. Stream of thought, time perception, memory, experimental methods; great work decades ahead of its time. 94 figures. 1,391pp. 5⅜ × 8½. 20381-6, 20382-4 Pa., Two-vol. set $19.90

BODIES IN A BOOKSHOP, R. T. Campbell. Challenging mystery of blackmail and murder with ingenious plot and superbly drawn characters. In the best tradition of British suspense fiction. 192pp. 5⅜ × 8½. 24720-1 Pa. $3.95

CALLAS: PORTRAIT OF A PRIMA DONNA, George Jellinek. Renowned commentator on the musical scene chronicles incredible career and life of the most controversial, fascinating, influential operatic personality of our time. 64 black-and-white photographs. 416pp. 5⅜ × 8¼. 25047-4 Pa. $7.95

GEOMETRY, RELATIVITY AND THE FOURTH DIMENSION, Rudolph Rucker. Exposition of fourth dimension, concepts of relativity as Flatland characters continue adventures. Popular, easily followed yet accurate, profound. 141 illustrations. 133pp. 5⅜ × 8½. 23400-2 Pa. $3.50

HOUSEHOLD STORIES BY THE BROTHERS GRIMM, with pictures by Walter Crane. 53 classic stories—Rumpelstiltskin, Rapunzel, Hansel and Gretel, the Fisherman and his Wife, Snow White, Tom Thumb, Sleeping Beauty, Cinderella, and so much more—lavishly illustrated with original 19th century drawings. 114 illustrations. x + 269pp. 5⅜ × 8½. 21080-4 Pa. $4.50

SUNDIALS, Albert Waugh. Far and away the best, most thorough coverage of ideas, mathematics concerned, types, construction, adjusting anywhere. Over 100 illustrations. 230pp. 5⅜ × 8½. 22947-5 Pa. $4.00

PICTURE HISTORY OF THE NORMANDIE: With 190 Illustrations, Frank O. Braynard. Full story of legendary French ocean liner: Art Deco interiors, design innovations, furnishings, celebrities, maiden voyage, tragic fire, much more. Extensive text. 144pp. 8⅜ × 11¾. 25257-4 Pa. $9.95

THE FIRST AMERICAN COOKBOOK: A Facsimile of "American Cookery," 1796, Amelia Simmons. Facsimile of the first American-written cookbook published in the United States contains authentic recipes for colonial favorites—pumpkin pudding, winter squash pudding, spruce beer, Indian slapjacks, and more. Introductory Essay and Glossary of colonial cooking terms. 80pp. 5⅜ × 8½. 24710-4 Pa. $3.50

101 PUZZLES IN THOUGHT AND LOGIC, C. R. Wylie, Jr. Solve murders and robberies, find out which fishermen are liars, how a blind man could possibly identify a color—purely by your own reasoning! 107pp. 5⅜ × 8½. 20367-0 Pa. $2.00

THE BOOK OF WORLD-FAMOUS MUSIC—CLASSICAL, POPULAR AND FOLK, James J. Fuld. Revised and enlarged republication of landmark work in musico-bibliography. Full information about nearly 1,000 songs and compositions including first lines of music and lyrics. New supplement. Index. 800pp. 5⅜ × 8¼. 24857-7 Pa. $14.95

ANTHROPOLOGY AND MODERN LIFE, Franz Boas. Great anthropologist's classic treatise on race and culture. Introduction by Ruth Bunzel. Only inexpensive paperback edition. 255pp. 5⅜ × 8½. 25245-0 Pa. $5.95

THE TALE OF PETER RABBIT, Beatrix Potter. The inimitable Peter's terrifying adventure in Mr. McGregor's garden, with all 27 wonderful, full-color Potter illustrations. 55pp. 4¼ × 5½. (Available in U.S. only) 22827-4 Pa. $1.75

THREE PROPHETIC SCIENCE FICTION NOVELS, H. G. Wells. *When the Sleeper Wakes, A Story of the Days to Come* and *The Time Machine* (full version). 335pp. 5⅜ × 8½. (Available in U.S. only) 20605-X Pa. $5.95

APICIUS COOKERY AND DINING IN IMPERIAL ROME, edited and translated by Joseph Dommers Vehling. Oldest known cookbook in existence offers readers a clear picture of what foods Romans ate, how they prepared them, etc. 49 illustrations. 301pp. 6⅛ × 9¼. 23563-7 Pa. $6.00

SHAKESPEARE LEXICON AND QUOTATION DICTIONARY, Alexander Schmidt. Full definitions, locations, shades of meaning of every word in plays and poems. More than 50,000 exact quotations. 1,485pp. 6½ × 9¼. 22726-X, 22727-8 Pa., Two-vol. set $27.90

THE WORLD'S GREAT SPEECHES, edited by Lewis Copeland and Lawrence W. Lamm. Vast collection of 278 speeches from Greeks to 1970. Powerful and effective models; unique look at history. 842pp. 5⅜ × 8½. 20468-5 Pa. $10.95

THE BLUE FAIRY BOOK, Andrew Lang. The first, most famous collection, with many familiar tales: Little Red Riding Hood, Aladdin and the Wonderful Lamp, Puss in Boots, Sleeping Beauty, Hansel and Gretel, Rumpelstiltskin; 37 in all. 138 illustrations. 390pp. 5⅜ × 8½. 21437-0 Pa. $5.95

THE STORY OF THE CHAMPIONS OF THE ROUND TABLE, Howard Pyle. Sir Launcelot, Sir Tristram and Sir Percival in spirited adventures of love and triumph retold in Pyle's inimitable style. 50 drawings, 31 full-page. xviii + 329pp. 6½ × 9¼. 21883-X Pa. $6.95

AUDUBON AND HIS JOURNALS, Maria Audubon. Unmatched two-volume portrait of the great artist, naturalist and author contains his journals, an excellent biography by his granddaughter, expert annotations by the noted ornithologist, Dr. Elliott Coues, and 37 superb illustrations. Total of 1,200pp. 5⅜ × 8.
Vol. I 25143-8 Pa. $8.95
Vol. II 25144-6 Pa. $8.95

GREAT DINOSAUR HUNTERS AND THEIR DISCOVERIES, Edwin H. Colbert. Fascinating, lavishly illustrated chronicle of dinosaur research, 1820's to 1960. Achievements of Cope, Marsh, Brown, Buckland, Mantell, Huxley, many others. 384pp. 5¼ × 8¼. 24701-5 Pa. $6.95

THE TASTEMAKERS, Russell Lynes. Informal, illustrated social history of American taste 1850's–1950's. First popularized categories Highbrow, Lowbrow, Middlebrow. 129 illustrations. New (1979) afterword. 384pp. 6 × 9.
23993-4 Pa. $6.95

DOUBLE CROSS PURPOSES, Ronald A. Knox. A treasure hunt in the Scottish Highlands, an old map, unidentified corpse, surprise discoveries keep reader guessing in this cleverly intricate tale of financial skullduggery. 2 black-and-white maps. 320pp. 5⅜ × 8½. (Available in U.S. only) 25032-6 Pa. $5.95

AUTHENTIC VICTORIAN DECORATION AND ORNAMENTATION IN FULL COLOR: 46 Plates from "Studies in Design," Christopher Dresser. Superb full-color lithographs reproduced from rare original portfolio of a major Victorian designer. 48pp. 9¼ × 12¼. 25083-0 Pa. $7.95

PRIMITIVE ART, Franz Boas. Remains the best text ever prepared on subject, thoroughly discussing Indian, African, Asian, Australian, and, especially, Northern American primitive art. Over 950 illustrations show ceramics, masks, totem poles, weapons, textiles, paintings, much more. 376pp. 5⅜ × 8. 20025-6 Pa. $6.95

SIDELIGHTS ON RELATIVITY, Albert Einstein. Unabridged republication of two lectures delivered by the great physicist in 1920–21. *Ether and Relativity* and *Geometry and Experience*. Elegant ideas in non-mathematical form, accessible to intelligent layman. vi + 56pp. 5⅜ × 8½. 24511-X Pa. $2.95

THE WIT AND HUMOR OF OSCAR WILDE, edited by Alvin Redman. More than 1,000 ripostes, paradoxes, wisecracks: Work is the curse of the drinking classes, I can resist everything except temptation, etc. 258pp. 5⅜ × 8½. 20602-5 Pa. $3.95

ADVENTURES WITH A MICROSCOPE, Richard Headstrom. 59 adventures with clothing fibers, protozoa, ferns and lichens, roots and leaves, much more. 142 illustrations. 232pp. 5⅜ × 8½. 23471-1 Pa. $3.95

PLANTS OF THE BIBLE, Harold N. Moldenke and Alma L. Moldenke. Standard reference to all 230 plants mentioned in Scriptures. Latin name, biblical reference, uses, modern identity, much more. Unsurpassed encyclopedic resource for scholars, botanists, nature lovers, students of Bible. Bibliography. Indexes. 123 black-and-white illustrations. 384pp. 6 × 9. 25069-5 Pa. $8.95

FAMOUS AMERICAN WOMEN: A Biographical Dictionary from Colonial Times to the Present, Robert McHenry, ed. From Pocahontas to Rosa Parks, 1,035 distinguished American women documented in separate biographical entries. Accurate, up-to-date data, numerous categories, spans 400 years. Indices. 493pp. 6½ × 9¼. 24523-3 Pa. $9.95

THE FABULOUS INTERIORS OF THE GREAT OCEAN LINERS IN HISTORIC PHOTOGRAPHS, William H. Miller, Jr. Some 200 superb photographs capture exquisite interiors of world's great "floating palaces"—1890's to 1980's: *Titanic, Ile de France, Queen Elizabeth, United States, Europa,* more. Approx. 200 black-and-white photographs. Captions. Text. Introduction. 160pp. 8⅞ × 11¾. 24756-2 Pa. $9.95

THE GREAT LUXURY LINERS, 1927–1954: A Photographic Record, William H. Miller, Jr. Nostalgic tribute to heyday of ocean liners. 186 photos of Ile de France, Normandie, Leviathan, Queen Elizabeth, United States, many others. Interior and exterior views. Introduction. Captions. 160pp. 9 × 12. 24056-8 Pa. $9.95

A NATURAL HISTORY OF THE DUCKS, John Charles Phillips. Great landmark of ornithology offers complete detailed coverage of nearly 200 species and subspecies of ducks: gadwall, sheldrake, merganser, pintail, many more. 74 full-color plates, 102 black-and-white. Bibliography. Total of 1,920pp. 8⅝ × 11¼. 25141-1, 25142-X Cloth. Two-vol. set $100.00

THE SEAWEED HANDBOOK: An Illustrated Guide to Seaweeds from North Carolina to Canada, Thomas F. Lee. Concise reference covers 78 species. Scientific and common names, habitat, distribution, more. Finding keys for easy identification. 224pp. 5⅜ × 8½. 25215-9 Pa. $5.95

THE TEN BOOKS OF ARCHITECTURE: The 1755 Leoni Edition, Leon Battista Alberti. Rare classic helped introduce the glories of ancient architecture to the Renaissance. 68 black-and-white plates. 336pp. 8⅜ × 11¼. 25239-6 Pa. $14.95

MISS MACKENZIE, Anthony Trollope. Minor masterpieces by Victorian master unmasks many truths about life in 19th-century England. First inexpensive edition in years. 392pp. 5⅜ × 8½. 25201-9 Pa. $7.95

THE RIME OF THE ANCIENT MARINER, Gustave Doré, Samuel Taylor Coleridge. Dramatic engravings considered by many to be his greatest work. The terrifying space of the open sea, the storms and whirlpools of an unknown ocean, the ice of Antarctica, more—all rendered in a powerful, chilling manner. Full text. 38 plates. 77pp. 9¼ × 12. 22305-1 Pa. $4.95

THE EXPEDITIONS OF ZEBULON MONTGOMERY PIKE, Zebulon Montgomery Pike. Fascinating first-hand accounts (1805–6) of exploration of Mississippi River, Indian wars, capture by Spanish dragoons, much more. 1,088pp. 5⅜ × 8½. 25254-X, 25255-8 Pa. Two-vol. set $23.90

A CONCISE HISTORY OF PHOTOGRAPHY: Third Revised Edition, Helmut Gernsheim. Best one-volume history—camera obscura, photochemistry, daguerreotypes, evolution of cameras, film, more. Also artistic aspects—landscape, portraits, fine art, etc. 281 black-and-white photographs. 26 in color. 176pp. 8⅜ × 11¼. 25128-4 Pa. $12.95

THE DORÉ BIBLE ILLUSTRATIONS, Gustave Doré. 241 detailed plates from the Bible: the Creation scenes, Adam and Eve, Flood, Babylon, battle sequences, life of Jesus, etc. Each plate is accompanied by the verses from the King James version of the Bible. 241pp. 9 × 12. 23004-X Pa. $8.95

HUGGER-MUGGER IN THE LOUVRE, Elliot Paul. Second Homer Evans mystery-comedy. Theft at the Louvre involves sleuth in hilarious, madcap caper. "A knockout."—Books. 336pp. 5⅜ × 8½. 25185-3 Pa. $5.95

FLATLAND, E. A. Abbott. Intriguing and enormously popular science-fiction classic explores the complexities of trying to survive as a two-dimensional being in a three-dimensional world. Amusingly illustrated by the author. 16 illustrations. 103pp. 5⅜ × 8½. 20001-9 Pa. $2.00

THE HISTORY OF THE LEWIS AND CLARK EXPEDITION, Meriwether Lewis and William Clark, edited by Elliott Coues. Classic edition of Lewis and Clark's day-by-day journals that later became the basis for U.S. claims to Oregon and the West. Accurate and invaluable geographical, botanical, biological, meteorological and anthropological material. Total of 1,508pp. 5⅜ × 8½.
21268-8, 21269-6, 21270-X Pa. Three-vol. set $25.50

LANGUAGE, TRUTH AND LOGIC, Alfred J. Ayer. Famous, clear introduction to Vienna, Cambridge schools of Logical Positivism. Role of philosophy, elimination of metaphysics, nature of analysis, etc. 160pp. 5⅜ × 8½. (Available in U.S. and Canada only) 20010-8 Pa. $2.95

MATHEMATICS FOR THE NONMATHEMATICIAN, Morris Kline. Detailed, college-level treatment of mathematics in cultural and historical context, with numerous exercises. For liberal arts students. Preface. Recommended Reading Lists. Tables. Index. Numerous black-and-white figures. xvi + 641pp. 5⅜ × 8½.
24823-2 Pa. $11.95

28 SCIENCE FICTION STORIES, H. G. Wells. Novels, *Star Begotten* and *Men Like Gods*, plus 26 short stories: "Empire of the Ants," "A Story of the Stone Age," "The Stolen Bacillus," "In the Abyss," etc. 915pp. 5⅜ × 8½. (Available in U.S. only)
20265-8 Cloth. $10.95

HANDBOOK OF PICTORIAL SYMBOLS, Rudolph Modley. 3,250 signs and symbols, many systems in full; official or heavy commercial use. Arranged by subject. Most in Pictorial Archive series. 143pp. 8⅛ × 11. 23357-X Pa. $5.95

INCIDENTS OF TRAVEL IN YUCATAN, John L. Stephens. Classic (1843) exploration of jungles of Yucatan, looking for evidences of Maya civilization. Travel adventures, Mexican and Indian culture, etc. Total of 669pp. 5⅜ × 8½.
20926-1, 20927-X Pa., Two-vol. set $9.90

DEGAS: An Intimate Portrait, Ambroise Vollard. Charming, anecdotal memoir by famous art dealer of one of the greatest 19th-century French painters. 14 black-and-white illustrations. Introduction by Harold L. Van Doren. 96pp. 5⅜ × 8½.
25131-4 Pa. $3.95

PERSONAL NARRATIVE OF A PILGRIMAGE TO ALMANDINAH AND MECCAH, Richard Burton. Great travel classic by remarkably colorful personality. Burton, disguised as a Moroccan, visited sacred shrines of Islam, narrowly escaping death. 47 illustrations. 959pp. 5⅜ × 8½. 21217-3, 21218-1 Pa., Two-vol. set $17.90

PHRASE AND WORD ORIGINS, A. H. Holt. Entertaining, reliable, modern study of more than 1,200 colorful words, phrases, origins and histories. Much unexpected information. 254pp. 5⅜ × 8½. 20758-7 Pa. $4.95

THE RED THUMB MARK, R. Austin Freeman. In this first Dr. Thorndyke case, the great scientific detective draws fascinating conclusions from the nature of a single fingerprint. Exciting story, authentic science. 320pp. 5⅜ × 8½. (Available in U.S. only) 25210-8 Pa. $5.95

AN EGYPTIAN HIEROGLYPHIC DICTIONARY, E. A. Wallis Budge. Monumental work containing about 25,000 words or terms that occur in texts ranging from 3000 B.C. to 600 A.D. Each entry consists of a transliteration of the word, the word in hieroglyphs, and the meaning in English. 1,314pp. 6⅜ × 10.
23615-3, 23616-1 Pa., Two-vol. set $27.90

THE COMPLEAT STRATEGYST: Being a Primer on the Theory of Games of Strategy, J. D. Williams. Highly entertaining classic describes, with many illustrated examples, how to select best strategies in conflict situations. Prefaces. Appendices. xvi + 268pp. 5⅜ × 8½. 25101-2 Pa. $5.95

THE ROAD TO OZ, L. Frank Baum. Dorothy meets the Shaggy Man, little Button-Bright and the Rainbow's beautiful daughter in this delightful trip to the magical Land of Oz. 272pp. 5⅜ × 8. 25208-6 Pa. $4.95

POINT AND LINE TO PLANE, Wassily Kandinsky. Seminal exposition of role of point, line, other elements in non-objective painting. Essential to understanding 20th-century art. 127 illustrations. 192pp. 6½ × 9¼. 23808-3 Pa. $4.50

LADY ANNA, Anthony Trollope. Moving chronicle of Countess Lovel's bitter struggle to win for herself and daughter Anna their rightful rank and fortune—perhaps at cost of sanity itself. 384pp. 5⅜ × 8½. 24669-8 Pa. $6.95

EGYPTIAN MAGIC, E. A. Wallis Budge. Sums up all that is known about magic in Ancient Egypt: the role of magic in controlling the gods, powerful amulets that warded off evil spirits, scarabs of immortality, use of wax images, formulas and spells, the secret name, much more. 253pp. 5⅜ × 8½. 22681-6 Pa. $4.00

THE DANCE OF SIVA, Ananda Coomaraswamy. Preeminent authority unfolds the vast metaphysic of India: the revelation of her art, conception of the universe, social organization, etc. 27 reproductions of art masterpieces. 192pp. 5⅜ × 8½.
24817-8 Pa. $5.95

CHRISTMAS CUSTOMS AND TRADITIONS, Clement A. Miles. Origin, evolution, significance of religious, secular practices. Caroling, gifts, yule logs, much more. Full, scholarly yet fascinating; non-sectarian. 400pp. 5⅜ × 8½.
23354-5 Pa. $6.50

THE HUMAN FIGURE IN MOTION, Eadweard Muybridge. More than 4,500 stopped-action photos, in action series, showing undraped men, women, children jumping, lying down, throwing, sitting, wrestling, carrying, etc. 390pp. 7⅞ × 10⅝.
20204-6 Cloth. $19.95

THE MAN WHO WAS THURSDAY, Gilbert Keith Chesterton. Witty, fast-paced novel about a club of anarchists in turn-of-the-century London. Brilliant social, religious, philosophical speculations. 128pp. 5⅜ × 8½. 25121-7 Pa. $3.95

A CEZANNE SKETCHBOOK: Figures, Portraits, Landscapes and Still Lifes, Paul Cezanne. Great artist experiments with tonal effects, light, mass, other qualities in over 100 drawings. A revealing view of developing master painter, precursor of Cubism. 102 black-and-white illustrations. 144pp. 8¾ × 6⅜. 24790-2 Pa. $5.95

AN ENCYCLOPEDIA OF BATTLES: Accounts of Over 1,560 Battles from 1479 B.C. to the Present, David Eggenberger. Presents essential details of every major battle in recorded history, from the first battle of Megiddo in 1479 B.C. to Grenada in 1984. List of Battle Maps. New Appendix covering the years 1967–1984. Index. 99 illustrations. 544pp. 6½ × 9¼. 24913-1 Pa. $14.95

AN ETYMOLOGICAL DICTIONARY OF MODERN ENGLISH, Ernest Weekley. Richest, fullest work, by foremost British lexicographer. Detailed word histories. Inexhaustible. Total of 856pp. 6½ × 9¼.
21873-2, 21874-0 Pa., Two-vol. set $17.00

WEBSTER'S AMERICAN MILITARY BIOGRAPHIES, edited by Robert McHenry. Over 1,000 figures who shaped 3 centuries of American military history. Detailed biographies of Nathan Hale, Douglas MacArthur, Mary Hallaren, others. Chronologies of engagements, more. Introduction. Addenda. 1,033 entries in alphabetical order. xi + 548pp. 6½ × 9¼. (Available in U.S. only)
24758-9 Pa. $11.95

LIFE IN ANCIENT EGYPT, Adolf Erman. Detailed older account, with much not in more recent books: domestic life, religion, magic, medicine, commerce, and whatever else needed for complete picture. Many illustrations. 597pp. 5⅜ × 8½.
22632-8 Pa. $8.50

HISTORIC COSTUME IN PICTURES, Braun & Schneider. Over 1,450 costumed figures shown, covering a wide variety of peoples: kings, emperors, nobles, priests, servants, soldiers, scholars, townsfolk, peasants, merchants, courtiers, cavaliers, and more. 256pp. 8⅜ × 11¼. 23150-X Pa. $7.95

THE NOTEBOOKS OF LEONARDO DA VINCI, edited by J. P. Richter. Extracts from manuscripts reveal great genius; on painting, sculpture, anatomy, sciences, geography, etc. Both Italian and English. 186 ms. pages reproduced, plus 500 additional drawings, including studies for *Last Supper, Sforza* monument, etc. 860pp. 7⅞ × 10¾. (Available in U.S. only) 22572-0, 22573-9 Pa., Two-vol. set $25.90

THE ART NOUVEAU STYLE BOOK OF ALPHONSE MUCHA: All 72 Plates from "Documents Decoratifs" in Original Color, Alphonse Mucha. Rare copyright-free design portfolio by high priest of Art Nouveau. Jewelry, wallpaper, stained glass, furniture, figure studies, plant and animal motifs, etc. Only complete one-volume edition. 80pp. 9⅜ × 12¼. 24044-4 Pa. $8.95

ANIMALS: 1,419 COPYRIGHT-FREE ILLUSTRATIONS OF MAMMALS, BIRDS, FISH, INSECTS, ETC., edited by Jim Harter. Clear wood engravings present, in extremely lifelike poses, over 1,000 species of animals. One of the most extensive pictorial sourcebooks of its kind. Captions. Index. 284pp. 9 × 12.
23766-4 Pa. $9.95

OBELISTS FLY HIGH, C. Daly King. Masterpiece of American detective fiction, long out of print, involves murder on a 1935 transcontinental flight—"a very thrilling story"—NY Times. Unabridged and unaltered republication of the edition published by William Collins Sons & Co. Ltd., London, 1935. 288pp. 5⅜ × 8½. (Available in U.S. only) 25036-9 Pa. $4.95

VICTORIAN AND EDWARDIAN FASHION: A Photographic Survey, Alison Gernsheim. First fashion history completely illustrated by contemporary photographs. Full text plus 235 photos, 1840–1914, in which many celebrities appear. 240pp. 6½ × 9¼. 24205-6 Pa. $6.00

THE ART OF THE FRENCH ILLUSTRATED BOOK, 1700–1914, Gordon N. Ray. Over 630 superb book illustrations by Fragonard, Delacroix, Daumier, Doré, Grandville, Manet, Mucha, Steinlen, Toulouse-Lautrec and many others. Preface. Introduction. 633 halftones. Indices of artists, authors & titles, binders and provenances. Appendices. Bibliography. 608pp. 8⅜ × 11¼. 25086-5 Pa. $24.95

THE WONDERFUL WIZARD OF OZ, L. Frank Baum. Facsimile in full color of America's finest children's classic. 143 illustrations by W. W. Denslow. 267pp. 5⅜ × 8½. 20691-2 Pa. $5.95

FRONTIERS OF MODERN PHYSICS: New Perspectives on Cosmology, Relativity, Black Holes and Extraterrestrial Intelligence, Tony Rothman, et al. For the intelligent layman. Subjects include: cosmological models of the universe; black holes; the neutrino; the search for extraterrestrial intelligence. Introduction. 46 black-and-white illustrations. 192pp. 5⅜ × 8½. 24587-X Pa. $6.95

THE FRIENDLY STARS, Martha Evans Martin & Donald Howard Menzel. Classic text marshalls the stars together in an engaging, non-technical survey, presenting them as sources of beauty in night sky. 23 illustrations. Foreword. 2 star charts. Index. 147pp. 5⅜ × 8½. 21099-5 Pa. $3.50

FADS AND FALLACIES IN THE NAME OF SCIENCE, Martin Gardner. Fair, witty appraisal of cranks, quacks, and quackeries of science and pseudoscience: hollow earth, Velikovsky, orgone energy, Dianetics, flying saucers, Bridey Murphy, food and medical fads, etc. Revised, expanded In the Name of Science. "A very able and even-tempered presentation."—The New Yorker. 363pp. 5⅜ × 8.
20394-8 Pa. $5.95

ANCIENT EGYPT: ITS CULTURE AND HISTORY, J. E Manchip White. From pre-dynastics through Ptolemies: society, history, political structure, religion, daily life, literature, cultural heritage. 48 plates. 217pp. 5⅜ × 8½. 22548-8 Pa. $4.95

SIR HARRY HOTSPUR OF HUMBLETHWAITE, Anthony Trollope. Incisive, unconventional psychological study of a conflict between a wealthy baronet, his idealistic daughter, and their scapegrace cousin. The 1870 novel in its first inexpensive edition in years. 250pp. 5⅜ × 8½.   24953-0 Pa. $4.95

LASERS AND HOLOGRAPHY, Winston E. Kock. Sound introduction to burgeoning field, expanded (1981) for second edition. Wave patterns, coherence, lasers, diffraction, zone plates, properties of holograms, recent advances. 84 illustrations. 160pp. 5⅜ × 8¼. (Except in United Kingdom)   24041-X Pa. $3.50

INTRODUCTION TO ARTIFICIAL INTELLIGENCE: SECOND, EN-LARGED EDITION, Philip C. Jackson, Jr. Comprehensive survey of artificial intelligence—the study of how machines (computers) can be made to act intelligently. Includes introductory and advanced material. Extensive notes updating the main text. 132 black-and-white illustrations. 512pp. 5⅜ × 8½.   24864-X Pa. $8.95

HISTORY OF INDIAN AND INDONESIAN ART, Ananda K. Coomaraswamy. Over 400 illustrations illuminate classic study of Indian art from earliest Harappa finds to early 20th century. Provides philosophical, religious and social insights. 304pp. 6⅜ × 9⅜.   25005-9 Pa. $8.95

THE GOLEM, Gustav Meyrink. Most famous supernatural novel in modern European literature, set in Ghetto of Old Prague around 1890. Compelling story of mystical experiences, strange transformations, profound terror. 13 black-and-white illustrations. 224pp. 5⅜ × 8½. (Available in U.S. only)   25025-3 Pa. $5.95

ARMADALE, Wilkie Collins. Third great mystery novel by the author of *The Woman in White* and *The Moonstone.* Original magazine version with 40 illustrations. 597pp. 5⅜ × 8½.   23429-0 Pa. $7.95

PICTORIAL ENCYCLOPEDIA OF HISTORIC ARCHITECTURAL PLANS, DETAILS AND ELEMENTS: With 1,880 Line Drawings of Arches, Domes, Doorways, Facades, Gables, Windows, etc., John Theodore Haneman. Sourcebook of inspiration for architects, designers, others. Bibliography. Captions. 141pp. 9 × 12.   24605-1 Pa. $6.95

BENCHLEY LOST AND FOUND, Robert Benchley. Finest humor from early 30's, about pet peeves, child psychologists, post office and others. Mostly unavailable elsewhere. 73 illustrations by Peter Arno and others. 183pp. 5⅜ × 8½.
                                                   22410-4 Pa. $3.95

ERTÉ GRAPHICS, Erté. Collection of striking color graphics: *Seasons, Alphabet, Numerals, Aces* and *Precious Stones.* 50 plates, including 4 on covers. 48pp. 9⅜ × 12¼.   23580-7 Pa. $6.95

THE JOURNAL OF HENRY D. THOREAU, edited by Bradford Torrey, F. H. Allen. Complete reprinting of 14 volumes, 1837–61, over two million words; the sourcebooks for *Walden,* etc. Definitive. All original sketches, plus 75 photographs. 1,804pp. 8½ × 12¼.   20312-3, 20313-1 Cloth., Two-vol. set $80.00

CASTLES: THEIR CONSTRUCTION AND HISTORY, Sidney Toy. Traces castle development from ancient roots. Nearly 200 photographs and drawings illustrate moats, keeps, baileys, many other features. Caernarvon, Dover Castles, Hadrian's Wall, Tower of London, dozens more. 256pp. 5⅜ × 8¼.
                                                   24898-4 Pa. $5.95

AMERICAN CLIPPER SHIPS: 1833–1858, Octavius T. Howe & Frederick C. Matthews. Fully-illustrated, encyclopedic review of 352 clipper ships from the period of America's greatest maritime supremacy. Introduction. 109 halftones. 5 black-and-white line illustrations. Index. Total of 928pp. 5⅜ × 8½.
25115-2, 25116-0 Pa., Two-vol. set $17.90

TOWARDS A NEW ARCHITECTURE, Le Corbusier. Pioneering manifesto by great architect, near legendary founder of "International School." Technical and aesthetic theories, views on industry, economics, relation of form to function, "mass-production spirit," much more. Profusely illustrated. Unabridged translation of 13th French edition. Introduction by Frederick Etchells. 320pp. 6⅛ × 9¼. (Available in U.S. only) 25023-7 Pa. $8.95

THE BOOK OF KELLS, edited by Blanche Cirker. Inexpensive collection of 32 full-color, full-page plates from the greatest illuminated manuscript of the Middle Ages, painstakingly reproduced from rare facsimile edition. Publisher's Note. Captions. 32pp. 9⅜ × 12¼. 24345-1 Pa. $4.50

BEST SCIENCE FICTION STORIES OF H. G. WELLS, H. G. Wells. Full novel *The Invisible Man*, plus 17 short stories: "The Crystal Egg," "Aepyornis Island," "The Strange Orchid," etc. 303pp. 5⅜ × 8½. (Available in U.S. only)
21531-8 Pa. $4.95

AMERICAN SAILING SHIPS: Their Plans and History, Charles G. Davis. Photos, construction details of schooners, frigates, clippers, other sailcraft of 18th to early 20th centuries—plus entertaining discourse on design, rigging, nautical lore, much more. 137 black-and-white illustrations. 240pp. 6⅛ × 9¼.
24658-2 Pa. $5.95

ENTERTAINING MATHEMATICAL PUZZLES, Martin Gardner. Selection of author's favorite conundrums involving arithmetic, money, speed, etc., with lively commentary. Complete solutions. 112pp. 5⅜ × 8½. 25211-6 Pa. $2.95
THE WILL TO BELIEVE, HUMAN IMMORTALITY, William James. Two books bound together. Effect of irrational on logical, and arguments for human immortality. 402pp. 5⅜ × 8½. 20291-7 Pa. $7.50

THE HAUNTED MONASTERY and THE CHINESE MAZE MURDERS, Robert Van Gulik. 2 full novels by Van Gulik continue adventures of Judge Dee and his companions. An evil Taoist monastery, seemingly supernatural events; overgrown topiary maze that hides strange crimes. Set in 7th-century China. 27 illustrations. 328pp. 5⅜ × 8½. 23502-5 Pa. $5.00

CELEBRATED CASES OF JUDGE DEE (DEE GOONG AN), translated by Robert Van Gulik. Authentic 18th-century Chinese detective novel; Dee and associates solve three interlocked cases. Led to Van Gulik's own stories with same characters. Extensive introduction. 9 illustrations. 237pp. 5⅜ × 8½.
23337-5 Pa. $4.95

Prices subject to change without notice.
Available at your book dealer or write for free catalog to Dept. GI, Dover Publications, Inc., 31 East 2nd St., Mineola, N.Y. 11501. Dover publishes more than 175 books each year on science, elementary and advanced mathematics, biology, music, art, literary history, social sciences and other areas.